Chemical Reactor Design and Technology

NATO ASI Series

Advanced Science Institutes Series

A Series presenting the results of activities sponsored by the NATO Science Committee, which aims at the dissemination of advanced scientific and technological knowledge, with a view to strengthening links between scientific communities.

The Series is published by an international board of publishers in conjunction with the NATO Scientific Affairs Division

A	**Life Sciences**	Plenum Publishing Corporation
B	**Physics**	London and New York
C	**Mathematical and Physical Sciences**	D. Reidel Publishing Company Dordrecht and Boston
D	**Behavioural and Social Sciences**	Martinus Nijhoff Publishers Dordrecht/Boston/Lancaster
E	**Applied Sciences**	
F	**Computer and Systems Sciences**	Springer-Verlag Berlin/Heidelberg/New York
G	**Ecological Sciences**	

Series E: Applied Sciences – No. 110

Chemical Reactor Design and Technology

Overview of the New Developments of Energy and Petrochemical Reactor Technologies. Projections for the 90's

edited by

Hugo I. de Lasa

Faculty of Engineering Science
The University of Western Ontario
London, Ontario
Canada N6A 5B9

1986 **Martinus Nijhoff Publishers**
Dordrecht / Boston / Lancaster
Published in cooperation with NATO Scientific Affairs Division

Proceedings of the NATO Advanced Study Institute on "Chemical Reactor Design and Technology", London, Ontario, Canada, June 2-12, 1985

Library of Congress Cataloging in Publication Data

NATO Advanced Study Institute on "Chemical Reactor
Design and Technology (1985 : London, Ont.)

(NATO ASI series. Series E, Applied sciences ; 110)
"Proceedings of the NATO Advanced Study Institute on
"Chemical Reactor Design and Technology", London,
Ontario, Canada, June 2-12, 1985"--T.p. verso.
"Published in cooperation with NATO Scientific
Affairs Division."
1. Chemical reactors--Design and construction--
Congresses. I. De Lasa, Hugo I. II. North Atlantic
Treaty Organization. Scientific Affairs Division.
III. Series: NATO ASI series. Series E, Applied
sciences ; no. 110.
TP157.N296 1985 660.2'81 86-8345

ISBN-13: 978-94-010-8457-4 **e-ISBN-13: 978-94-009-4400-8**
DOI: 10.1007/978-94-009-4400-8

Distributors for the United States and Canada: Kluwer Academic Publishers, 190 Old Derby Street, Hingham, MA 02043, USA

Distributors for the UK and Ireland: Kluwer Academic Publishers, MTP Press Ltd, Falcon House, Queen Square, Lancaster LA1 1RN, UK

Distributors for all other countries: Kluwer Academic Publishers Group, Distribution Center, P.O. Box 322, 3300 AH Dordrecht, The Netherlands

NATO ADVANCED STUDY INSTITUTE

ON

CHEMICAL REACTOR DESIGN AND TECHNOLOGY

Overview of the New Developments of Energy and Petrochemical Reactor Technologies. Projections for the 90's

DIRECTOR

Prof. Hugo de Lasa, The University of Western Ontario, Canada

CO-DIRECTOR

Prof. Alirio Rodrigues, Universidade do Porto, Portugal

ORGANIZING ADVISORY COMMITTEE

Prof. M.A. Bergougnou, The University of Western Ontario, Canada
Prof. J.R. Grace, The University of British Columbia, Canada
Dr. R.M. Koros, Exxon Research and Engineering, U.S.A.
Dr. M. Ternan, Energy, Mines and Resources, Canada

PREFACE

Today's frustrations and anxieties resulting from two energy crises in only one decade, show us the problems and fragility of a world built on high energy consumption, accustomed to the use of cheap non-renewable energy and to the acceptance of existing imbalances between the resources and demands of countries. Despite all these stressing factors, our world is still hesitating about the urgency of undertaking new and decisive research that could stabilize our future. Could this trend change in the near future? In our view, two different scenarios are possible. A renewed energy tension could take place with an unpredictable timing mostly related to political and economic factors. This could bring again scientists and technologists to a new state of shock and awaken our talents. A second interesting and beneficial scenario could result from the positive influence of a new generation of researchers that with or without immediate crisis, acting both in industry and academia, will face the challenge of developing technologies and processes to pave the way to a less vulnerable society.

Because Chemical Reactor Design and Technology activities are at the heart of these required new technologies the timeliness of the NATO-Advanced Study Institute at the University of Western Ontario, London, was very appropriate. For instance, processes such as the transformation of renewable resources (wood and biomass), the conversion of less critical ones (coal, natural gas, heavy crudes) and the chemical changes following these transformations (synthesis and conversion of methanol, synthesis of hydrocarbons, synthesis of ammonia) could be objectives for new chemical reactors. We perceive this hour as a time of action for a generation of researchers who could innovate in Chemical Reactor Engineering. We believe as well that these novel approaches will result in a more effective interaction and communication between industry and academia, between various countries and between researchers in organizations. It will possibly be necessary in the future to reformulate the research action placing more emphasis upon an interdisciplinary approach. With these goals in mind we planned the NATO Advanced Study Institute meetings involving lecturers and participants from several countries, from both university and industry, and from various organizations.

It should be stressed that Chemical Reactor Technology for processes involving catalytic and non-catalytic reactions evolved in recent years in numerous directions and options. Because of the diversity of topics, the selection of subjects and lecturers proved to be a challenging and complex task. We decided to settle on a compromise, focusing the scope of the ASI on the more representative types of reactors and the more significant associated technological problems. The main areas of concentration for the ASI were then: fixed bed reactors, fluidized beds (two and three phase), - slurry reactors, - trickle beds, - transport in catalysts, - kinetic data analysis, - mixing and tracer techniques, - bubble column reactors, - specific applications (hydrocracking of petroleum residua, ammonia and methanol synthesis reactors).

The final program for the NATO-ASI covered in summary the present status of Chemical Reactor Design and Technology, novel ideas, new strategies and their projections for the coming decade. These matters were considered through an intense program of activities which included 15 formal lectures, two poster sessions with 23 presentations, 3 laboratory demonstrations with 6 experiments, 3 special panel discussions and 2 talks delivered by special invited speakers.

The participants had the opportunity to contribute to the NATO-ASI during the Poster Session. Posters covering a wide range of subjects were presented: non-isothermal trickle beds, catalyst deactivation, mixing in fluidized beds, control of chemical reactors, maldistribution in chemical reactors, cyclic operation in trickle beds, polymerization reactors, etc. The most relevant contributions of the Poster Session were selected to be included in this NATO-ASI Proceedings.

Another important activity of the NATO-ASI was the Panel Discussion. Three panels were organized on tracer techniques, trickle beds and slurry reactors and fluidized bed reactors. Each panelist delivered a short presentation indicating points to be clarified, crucial issues, novel ideas. The short talks were followed by discussion periods with active participation of the audience and panelists. Six laboratory demonstrations were also presented at the University of Western Ontario on the following subjects: - entrainment and grid leakage in fluidized beds, - ultrapyrolysis process for the conversion of biomass, - fast catalytic cracking and regeneration using the pulse technique, - mass transfer and bubble phenomena in three-phase fluidized beds, - novel configuration for fixed bed reactors. It was possible to show with these laboratory demonstrations basic principles, advanced instrumentation, novel reactors, applications, strategies for scaling-up.

The dinners of the NATO-ASI Conference provided a forum for the review of future trends and strategies. Two Special Invited Speakers, Dr. J. Grace and Dr. J. Wright, delivered talks about Research Needs in Chemical Reactor Engineering and Role of Universities and Research in High Technology Development in Canada.

In summary, the NATO-ASI held at Spencer Hall, London, Ontario was a massive learning experience where both participants and lecturers found the appropriate atmosphere for fruitful technical exchanges and for visualizing the technological changes in Reactor Engineering for the 90's.

Here I would like to express my gratitude to the members of the Advisory Committee, Prof. A. Rodrigues, Prof. J.R. Grace, Prof. M.A. Bergougnou, Dr. R. Koros and Dr. M.A. Ternan who co-operated in many ways in the difficult task of selecting the lecturers and organizing the NATO-ASI meeting.

My special thanks should be addressed to the Local Organizing Committee, graduate students of the University of Western Ontario who helped so effectively with the numerous organizational tasks of the NATO-ASI Conference.

A special acknowledgement to the participants from industry and academia who contributed with remarkable enthusiasm and interest to make our meeting a successful forum for the exchange of technical views in Chemical Reactor Engineering.

I would like to express my sincere thanks to the NATO Scientific Affairs Division who supported financially the NATO-ASI Conference. I would like to gratefully acknowledge as well the financial contribution of the Natural Sciences and Engineering Research Council of Canada.

My appreciation to Prof. K. Shelstad, Faculty of Engineering Science, The University of Western Ontario, for his most valuable contribution in the process of reviewing the papers included in this book.

Finally my deepest gratitude to my wife, Graciela, who provided all the cooperation and inspiration needed for the success of this event.

London, Ontario, Canada Hugo de Lasa

TABLE OF CONTENTS

TRANSPORT PROCESSES IN CATALYST PELLETS

Alírio RODRIGUES, Carlos COSTA and Rosa FERREIRA

Department of Chemical Engineering
University of Porto, Porto, Portugal

INTRODUCTION

The subject of reaction and diffusion in porous catalysts is now a well established branch of knowledge discussed in several books such as by Aris [1] and Jackson [2]. Its practical importance has long been recognized since the pioneer work by Thiele [3].

In this paper we will first review some basic concepts and apply them to the design of isothermal reactors working in the diffusional regime. Then we will concentrate our attention on the problem of intraparticle convection in large pore catalysts. Several aspects of this question will be dealt with - effectiveness factors for iso - thermal and nonisothermal catalysts, measurement of effective diffusivities and the implication of intraparticle convection effects on the design and operation of fixed bed catalytic reactors.

From a qualitative analysis of the competition between reaction and diffusion in an isothermal pellet one easily recognizes that the parameter governing the steady state behavior of the pellet is the ratio between time constants for diffusion and reaction, i.e., τ_d/τ_r. If $\tau_d \ll \tau_r$ the reaction rate is much slower than the diffusion rate; the concentration profile inside the pellet is then almost flat and equal to the external surface concentration. The effectiveness factor is around unity. However, when $\tau_d \gg \tau_r$ the concentration inside the pellet will be lower than the external surface concentration and the effectiveness factor will be lower than one provided the reaction order is $n > 0$. In the first situation the catalyst is working in the kinetic controlled regime; in the second case the catalyst is operating in the diffusion controlled regime. We can easily recognize that the Thiele modulus ϕ is such that $\phi^2=\tau_d/\tau_r$ with $\phi=\ell(kc_s^{n-1}/D_e)^{1/2}$ for irreversible nth order reactions.

The mathematical model for diffusion and reaction in a homogeneous isothermal catalyst where a first order, irreversible reaction occurs is:

$$\frac{d^2f}{dx^2} + \frac{s-1}{x}\frac{df}{dx} + \phi^2 f = 0 \qquad (1a)$$

$$x=1 \quad , \quad f=1 \qquad (1b)$$

$$x=0 \quad , \quad df/dx=0 \qquad (1c)$$

where $f=c/c_s$ is the reduced concentration inside the pellet, $x=z/\ell$ is the reduced position in the pellet (ℓ is the half thickness of a slab or the radius of a sphere) and s is a shape factor (s=1,2 and 3 for slab, infinite cylinder and sphere, respectively).

For a slab catalyst the concentration profile inside the pellet is:

$$f = \cosh(\phi x) \; / \; \cosh \phi \qquad (2)$$

and the effectiveness factor is:

$$\eta = r_{obs}/r_s(c_s) = \mathrm{th}\phi/\phi \qquad (3)$$

For strong diffusional regime $\eta \simeq 1/\phi$. Whatever the shape is we get:

$$\eta \simeq s/\phi \qquad (4)$$

For zero order reactions one should take into account that r=k only if c> 0; otherwise r=0. In a situation where the concentration reaches a zero value inside the particle at a point x=x* we should replace the boundary condition (1c) by f=0 and df/dx=0 at x=x*. The effectiveness factor is now simply the ratio of the "utilized" particle volume and the total particle volume, i.e., $\eta=1-x^*$. In the kinetic controlled regime $\eta \simeq 1$ and in the pure diffusional regime $\eta \simeq \sqrt{2}/\phi$. Again for any shape we get:

$$\eta \simeq s\sqrt{2}/\phi \qquad (5)$$

In general for irreversible n^{th} order reactions the catalyst effectiveness factor in the diffusional regime is:

$$\eta \simeq s\sqrt{2/(n+1)}\;/\phi \qquad (5a)$$

Before going on to discuss the importance of the effectiveness factor for reactor design let us briefly emphasize the point that we need in the previous treatment to know the Thiele modulus (and hence the kinetic constant k) in order to calculate η. In practice the true kinetic constant can be obtained from various experiments carried out with different particle sizes. However this is a time consuming procedure. What we need is a method to calculate the effectiveness factor by doing just one kinetic experiment with a real catalyst and thus measuring r_{obs}. This can be done because as pointed out by Weisz [4] $\eta\phi^2=\ell^2 r_{obs}/D_e c_s$ involves only measurable quantities and the function $\eta=f(\eta\phi^2)$ is known.

Let us discuss now the implication of the catalyst effectiveness factor on reactor design. The simplest situation is obviously that in which there are no diffusional limitations. Then the amount of catalyst needed to get a given conversion is:

- for a CSTR

$$W_A = F_{Ao}\, X_e \;/\; r'(X_e) \tag{6a}$$

- for a plug flow reactor

$$W_P = F_{Ao}\int_0^{X_e} \frac{dX}{r'(X)} \tag{6b}$$

with the reaction rate r' is expressed in mole/gcat.s .

If mass transfer resistances are important then the actual mass of catalyst needed to get the same conversion is $W'=W/\eta$. In summary, with a given amount of catalyst the conversion at the outlet can be calculated through the so-called "design equations" extended below to the case where we have strong diffusion effects:

o <u>Zero order reaction $r'=k$;slab catalyst</u> (Rodrigues et al.[5])

<u>Perfectly mixed reactor or CSTR</u>

- kinetic controlled regime $X_{Ak} = N_r$ (7a)
- diffusion controlled regime $X_{Ad} = \sqrt{\alpha^2 + 2\alpha} - \alpha$ (7b)

In these formulae $N_r = Da_I = \tau/\tau_r = k\rho_b\tau/c_{in}$ (number of reaction units or Damkholer number), $\alpha = N_r N_d = (Da_I/\phi)^2$, $N_d = \tau/\tau_d = D_e\tau/\ell^2$ (number of diffusional mass transfer units). Equation (7b) is only valid for $N_r \geqslant 2N_d/(1+2N_d)$.

<u>Plug flow reactor</u>

- kinetic controlled regime $X_{Pk} = N_r$ (8a)
- diffusion controlled regime $X_{Pd} = \sqrt{2\alpha} - \alpha/2$ (8b)

Equation (8b) is valid for situations in which $N_r \geq 2N_d/(1+N_d/2)^2$. Figure 1 shows the conversion obtained in a CSTR and a plug flow reactor when the catalyst is working in the pure diffusional regime.

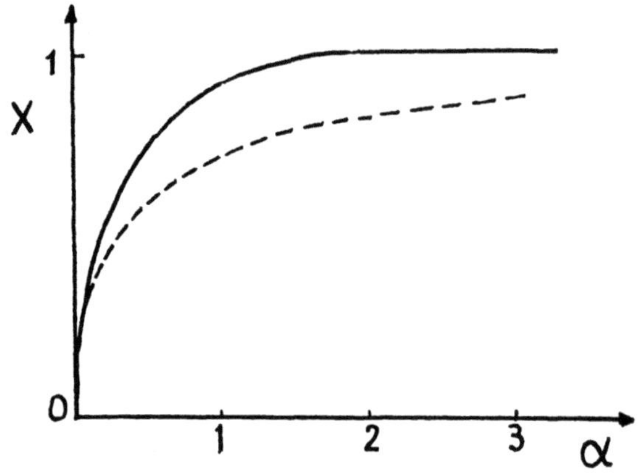

Figure 1 - Conversion X as a function of α (catalyst working in the diffusional regime) [5]

____ plug flow reactor

---- CSTR

We observe that only one parameter α governs the steady state behavior of ideal reactors provided the catalyst is working in pure diffusion regime. Moreover, in a plug flow reactor, there is a critical value for α (or in practical terms a critical height for the reactor) at which complete conversion is obtained, i.e., $\alpha=2$. This is because for the diffusional regime a zero order reaction is equivalent to a 1/2 order reaction in the kinetic regime.

o First order reaction $r'=kc$; slab catalyst

CSTR

The conversion is now $X_A = \eta Da_I/(1+\eta Da_I)$ where $Da_I=N_r=k\rho_b\tau$ and $\eta = th(\phi)/\phi$. The limiting cases are:

-kinetic controlled regime $$X_{Ak} = \frac{Da_I}{1+Da_I} \tag{9a}$$

- diffusion controlled regime $X_{Ad} = Da_I/(\phi + Da_I) = \sqrt{\alpha}/(1+\sqrt{\alpha})$ (9b)

Plug flow reactor

The conversion is now $X_P = 1 - \exp(-\eta\, Da_I)$ .The limiting situations are:

- kinetic controlled regime $X_{Pk} = 1 - \exp(-Da_I)$ (10a)

- diffusion controlled regime $X_{Pd} = 1 - \exp(-Da_I/\phi) = 1 - \exp(-\sqrt{\alpha})$ (10b)

If we know the conversion for an ideal plug flow reactor without diffusion limitations X_{Pk} and the actual Thiele modulus ϕ we can easily calculate X_{Pd} by taking into account that $1 - X_{Pd} = (1 - X_{Pk})^{1/\phi}$.

In summary:the steady state behavior of these reactors is then governed by three time constants:

τ - space time or time constant for the reactor

τ_r - time constant for the reaction

τ_d - time constant for diffusion

INTRAPARTICLE AND EXTERNAL CONCENTRATION AND TEMPERATURE GRADIENTS

At a point in the reactor one should consider the competition between reaction,mass and heat transfer inside the catalyst and also in the film around the catalyst particles.An important step in reactor design is the a priori estimate of external and intraparticle gradients of concentration and temperature.These values can be a guide for the choice of a reactor model.

Let us recall how to calculate such gradients:

Concentration gradients

o Intraparticle

$$\frac{c_{Ab} - c_{As}}{c_{Ab}} = \frac{r_{obs}\,\ell}{k_f\, c_{Ab}} = \bar{\eta}\,\overline{Da}_{II} \qquad (11a)$$

o External (film)

$$\frac{c_{As}}{c_{Ab}} = 1 - \bar{\eta}\,\overline{Da}_{II} \qquad (11b)$$

where $\bar{\eta} = r_{obs}/r_b(c_{Ab}, T_b)$ and $\overline{Da}_{II} = k(T_b)\ell/k_f$.

Temperature gradients

o Intraparticle

$$\frac{T_{max}-T_s}{T_s} = \beta \quad \text{or} \quad \frac{T_{max}-T_s}{T_b} = \bar{\beta}(1-\bar{\eta}\ \overline{Da}_{II}) \tag{11c}$$

o External(film)

$$\frac{c_{As}}{c_{Ab}} = \bar{\beta}\ b \tag{11d}$$

where $\beta=(-\Delta H)D_e c_{As}/\lambda_e T_s$ (Prater thermicity factor) ,$\bar{\beta}$ is referred to bulk conditions and $b=Bi_m/Bi_h=(k_f\ell/D_e)/(h\ell/\lambda_e)$.

The formulae presented above are based on ideas put forward by Damkholer.From analysis of the competition between reaction,heat and mass transfer (diffusion) inside catalysts he obtained an equation which relates temperature and concentration at a point within the pellet,i.e.,

$$T-T_s = \frac{(-\Delta H)D_e}{\lambda_e}(c_{As}-c) \tag{12}$$

Obviously the maximum temperature inside the pellet occurs at c=0.

Similarly, the temperature difference in the film is related to the concentration difference by:

$$T_s-T_b = \frac{(-\Delta H)\ k_f}{h}(c_{Ab}-c_{As}) \tag{13}$$

where k_f is the film mass transfer coefficient and h is the film heat transfer coefficient.

Practical values of the parameters governing the behavior of a catalyst pellet are:

Prater thermicity factor ($\bar{\beta}$)	$-1 < \bar{\beta} < 1$
Arrhenius number ($\bar{\gamma}$)	$5 < \bar{\gamma} < 40$
Thiele modulus (ϕ)	$10^{-2} < \phi < 30$
Lewis number for the catalyst, $Le_c=(\lambda_e/\rho_c c_{pc})/D_e$	$10 < Le_c < 10^4$
Adiabatic temperature rise (dimensionless) $B=(-\Delta H)c_o/\rho_f c_{pf} T_o$	$10^{-2} < B < 10$
$b=Bi_m/Bi_h$	$10 < b < 10^4$ gas/solid $10^{-4} < b < 10^{-1}$ liquid/solid

INTRAPARTICLE FORCED CONVECTION IN ISOTHERMAL CATALYSTS

Most studies on transport processes and reaction in porous catalysts do not include forced convection inside the pores. However, this effect should be considered in large pore catalysts. In recent years a number of theoretical and experimental studies in this area have been reported (Nir and Pismen [6], Nir [7], Cogan et al. [8], Rodrigues et al. [9,10]).

Large pore catalysts or supports are used in selective oxidations and enzyme immobilization. The effect of intraparticle forced convection should be addressed from two aspects:

- Experimental measurements of effective diffusivities
- Implication in reactor design of neglecting intraparticle convection.

Let us first look at the measurement of effective diffusivities in catalysts. Several systems can be considered which have been reviewed by Cresswell and Orr [11].

Starting with a CSTR system containing porous catalysts and assuming that experiments are carried out with a nonadsorbable tracer at constant temperature we get:

- Overall transient mass balance for the CSTR

$$c_{f,in}=c_f+\varepsilon \frac{dc_f}{d\theta} + \varepsilon_p(1-\varepsilon) \frac{d\langle c\rangle}{d\theta} \tag{14a}$$

- Mass balance in a volume element of the catalyst

$$\frac{\partial^2 c}{\partial x^2} - \lambda \frac{\partial c}{\partial x} = \frac{1}{N_d} \frac{\partial c}{\partial \theta} \tag{14b}$$

In the above equations c_f is the outlet tracer concentration in the fluid phase, ε_p is the particle porosity, ε is the reactor porosity, $\theta=t/\tau$ (with τ defined as the ratio between the reactor volume and the flowrate) and $\lambda=v_o \ell/D_e$ is the intraparticle Peclet number (v_o is the intraparticle convective velocity).

The boundary conditions for this problem are:

$$\begin{aligned} x&=0 \quad , \quad c=c_s \\ x&=2 \quad , \quad c=c_s \\ \theta&=0 \quad , \quad c_f=\bar{c}=0 \end{aligned} \tag{14c}$$

Several transfer functions can be calculated:

i) Transfer function relating the catalyst average concentration and the surface concentration

$$g_p(s) = \overline{\langle c\rangle} / \bar{c}_s = M(s)/(s/N_d) \quad (15)$$

where

$$M(s) = \frac{(e^{2r_2}-1)(e^{2r_1}-1)}{e^{2r_2} - e^{2r_1}} \sqrt{(\lambda/2)^2 + (s/N_d)}$$

$$r_1, r_2 = \frac{\lambda}{2} \pm \sqrt{(\lambda^2/4) + s/N_d}$$

ii) Transfer function relating the tracer concentration at the reactor outlet to the concentration at the inlet.

$$G(s) = \bar{c}_f / \bar{c}_{f,in} = \{1 + \varepsilon s + (1-\varepsilon)\varepsilon_p s \ \ g_p(s)\}^{-1} \quad (16)$$

The moments of the impulse response of the CSTR are:

$$\mu_o = 1 \quad (17a)$$

$$\mu_1 = \varepsilon + (1-\varepsilon)\ \varepsilon_p = \varepsilon_t \quad (17b)$$

$$\mu_2 = 2\mu_1^2 + \frac{2}{3}(1-\varepsilon)\varepsilon_p \frac{f(\lambda)}{N_d} \quad (17c)$$

with

$$f(\lambda) = \frac{3}{\lambda} \left(\frac{1}{\text{th}\ \lambda} - \frac{1}{\lambda}\right) \quad (18)$$

The variance is then:

$$\sigma^2 = \varepsilon_t^2 + \frac{2}{3}(1-\varepsilon) \frac{f(\lambda)}{N_d} \varepsilon_p \quad (19)$$

In the absence of intraparticle convection, $\lambda=0$ and we have $f(\lambda)=1$. It follows that if convection were neglected we really get:

$$\tilde{g}_p(s) = \text{th}(\sqrt{(s/\tilde{N}_d)})/\sqrt{(s/\tilde{N}_d)} \quad (20a)$$

$$\tilde{G}(s) = \{1 + \varepsilon s + (1-\varepsilon)\varepsilon_p s\ \tilde{g}_p(s)\}^{-1} \quad (20b)$$

$$\tilde{\sigma}^2 = \varepsilon_t^2 + \frac{2}{3}(1-\varepsilon)\varepsilon_p \frac{1}{\tilde{N}_d} \quad (20c)$$

where $\tilde{N}_d$ is based on an <u>apparent</u> effective diffusivity $\tilde{D}_e$.

If we compare Equations (19) and (20c) we get:

$$1/\tilde{N}_d = f(\lambda)/N_d \quad (21)$$

and finally:

$$\tilde{D}_e = D_e / f(\lambda) \qquad (22)$$

This result shows that increasing the flowrate (or the Reynolds number Re) the intraparticle velocity v_o increases also (and so does λ) and the apparent effective diffusivity also increases.

Experiments for measuring effective diffusivities in catalysts can also be carried out in fixed bed systems such as wide body or single pellet string reactors (SPSR).

Assuming the axial dispersion model for fluid flow and film mass transfer resistance we can now write:

-Mass balance in a volume element of the reactor

$$\varepsilon D_{ax} \frac{\partial^2 c_f}{\partial z^2} = u_o \frac{\partial c_f}{\partial z} + \varepsilon \frac{\partial c_f}{\partial t} + (1-\varepsilon)\varepsilon_p \frac{\partial \langle c \rangle}{\partial t} \qquad (23)$$

-Kinetic law for film mass transfer

$$k_f a_p (c_f - c_s) = \varepsilon_p \frac{\partial \langle c \rangle}{\partial t} \qquad (24)$$

Introducing dimensionless quantities $Pe = u_o L/\varepsilon D_{ax}$ (Peclet number), $z^* = z/L$ and $Bi_m = k_f \ell / \varepsilon_p D_e$ we can get the transfer function for the fixed bed:

$$G(s) = \frac{\bar{c}_{f,out}}{\bar{c}_{f,in}} = \exp \left\{ \frac{Pe}{2} \left(1 - \sqrt{1 + \frac{4N(s)}{Pe}}\right) \right\} \qquad (25)$$

with

$$N(s) = \varepsilon s + (1-\varepsilon)\varepsilon_p s \frac{M(s)}{\frac{s}{N_d}\left(1 + \frac{M(s)}{Bi_m}\right)}$$

The moments of the impulse response of the reactor are then:

$$\mu_1 = \varepsilon_t \qquad (25a)$$

$$\mu_2 = \varepsilon_t^2 \left(1 + \frac{2}{Pe}\right) + \frac{2}{3}(1-\varepsilon)\varepsilon_p \frac{f(\lambda) + (3/Bi_m)}{N_d} \qquad (25b)$$

Again the same conclusion can be drawn, that is, the apparent effective diffusivity (obtained from a model which did not include intraparticle convection) increases when the flowrate increases, as shown from Equation (22).

However, experiments can be carried out in such a way that intraparticle convective transport is not important. This is the case in the dynamic Wicke-Kallenbach apparatus used by different authors (Dogu and Smith [12], Furusawa et al. [13]) and shown in Figure 2.

In this method the pressure in both chambers is kept constant so convective transport across the pellet does not occur. At high flowrates the first moment of the response of the bottom chamber is:

$$\mu_1 = L^2 \varepsilon_p / 6D_e \tag{26}$$

Coming back to the effect of intraparticle convection on the measured value of the effective diffusivity let us mention two experimental examples. The first one deals with a dynamic chromatographic method in a SPSR. The characteristics of the catalyst and reactor are listed below [9].

o <u>Example 1 - Catalyst BM 329 (Rhone-Poulenc)</u>

This is a catalyst based on vanadium and phosphorus oxides developed for partial oxidation of hydrocarbons.

particle diameter	d_p=0.45 cm
particle porosity	ε_p=0.537 cm^3 voids/cm^3 pellet
solid density	ρ_c=2.84 g/cm^3

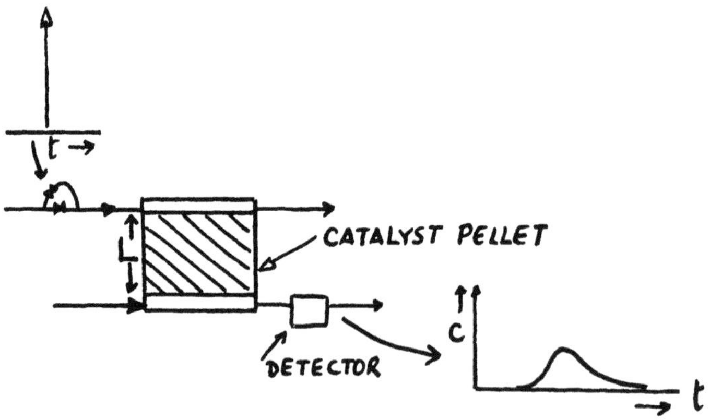

Figure 2 - Dynamic Wicke-Kallenbach cell for measuring effective diffusivities in catalysts.

apparent density ρ_a=1.31 g/cm^3 pellet

S_{BET} =4.3 m^2/g

mean pore diameter $\bar{d}$ =10^4 Å

The reactor used in the experiments was 30 cm long with a diameter of 2.12 cm and porosity of 0.676.

The apparent effective diffusivity reported as a function of the Reynolds number is shown in Figure 3.

These results were obtained from two signals of tracer concentration at two different positions in the reactor.The model did not consider intraparticle convective transport so what is measured is not the true effective diffusivity but only an apparent effective diffusivity.

In order to reconcile experiments and theory we developed a model in which intraparticle convection was included.First of all it will be interesting to be able to predict the intraparticle convective velocity v_o as a function of the superficial velocity u_o. We start with an Ergun type equation for the pressure drop across the bed,i.e.,

$$\Delta P/L = a\, u_o + b\, u_o^2 \qquad (27)$$

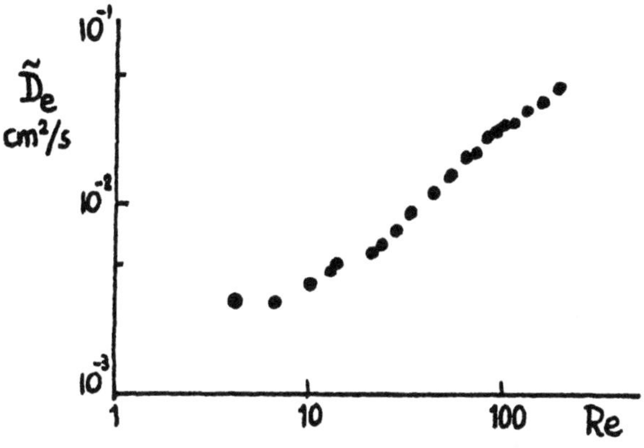

Figure 3 - Apparent effective diffusivity De as a function of the Reynolds number (tracer gas:hydrogen in nitrogen as carrier gas; ambient temperature) [9]. Reproduced by permission of the American Institute of Chemical Engineers.

The pressure drop across a pellet can be written as:

$$\Delta p/\ell = \mu\, v_o/B \qquad (28)$$

From Equations (27) and (28) we get:

$$v_o = a_1 B\, u_o + a_2 B\, u_o^2$$

where B is the permeability of the pellet.

Now the model tells us that the apparent time constant for diffusion $\tilde{\tau}_d$ is $\tilde{\tau}_d = \tau_d\, f(\lambda)$ .Then

$$\tilde{\tau}_d\ (Re) = f(Re;B,\tau_d) \qquad (29)$$

From experimental data ($\tilde{\tau}_d$,Re) it is now possible to calculate the permeability B and the true time constant for diffusion τ_d and hence the true effective diffusivity.

In Figure 4 the intraparticle velocity v_o is shown as a function of the superficial velocity u_o.The agreement between experimental results and model calculations is quite satisfactory.

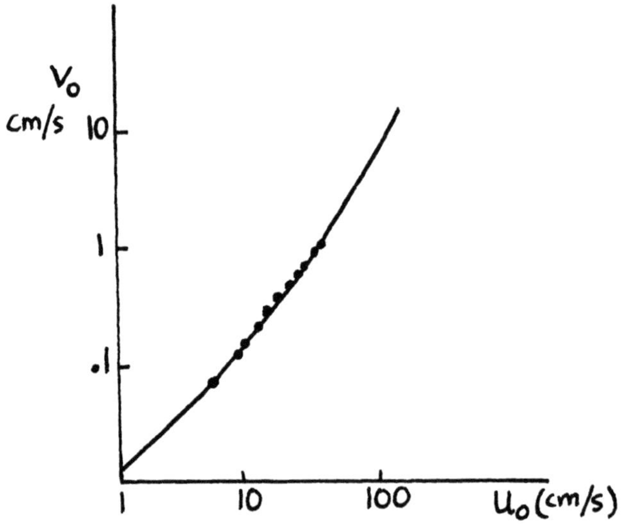

Figure 4- Intraparticle velocity v_o versus superficial velocity u_o(tracer gas:H_2;catalyst BM 329)

o Example 2 - Support for ethylene oxidation: α-alumina

The second example deals with measurements made by Cheng et al. [14] using different tracer gases (helium,argon,hydrogen) in a carrier gas (nitrogen).The catalyst has the following characteristics:

$d_p = 0.766$ cm

$\rho_a = 1230$ kg/m^3

$\varepsilon_p = 0.58$

$\bar{d}_p = 5\times10^{-6}$m

The reactor had a length L=1.5 m and a diameter $d_T = 8.9\times10^{-3}$m. The study zone was 1m long.

In Figure 5 we show the apparent effective diffusivity as a function of the Reynolds number for several tracer gases.Again $\tilde{D}_e$ increases with flowrate.

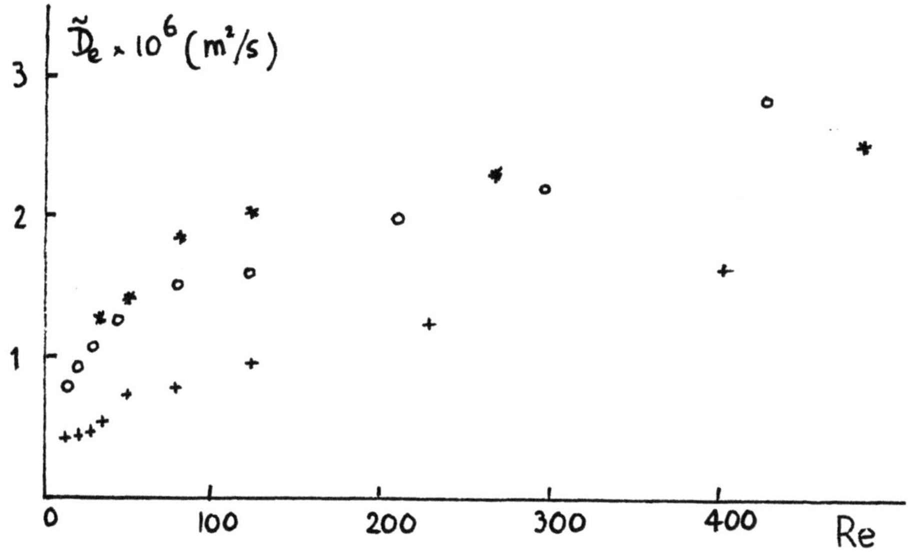

Figure 5- Apparent effective diffusivities of various tracers in α-alumina versus Reynolds number

ooo Helium
+++ Argon
*** Hydrogen

At T=20 C we have the following estimates for gas-phase diffusion coefficients:

$D(He-N_2) = 70.5 \times 10^{-6}\ m^2/s$

$D(Ar-N_2) = 18.75 \times 10^{-6}\ m^2/s$

$D(H_2-N_2) = 74.1 \times 10^{-6}\ m^2/s$

We see on Fig.5 that the apparent effective diffusivity for hydrogen is higher than for argon and helium. However, the ratio of apparent diffusivities for two tracer gases is lower than the ratio of the corresponding diffusion coefficients. This may be due to the influence of intraparticle convective velocity. In fact, let us take a time constant for diffusion τ_d=1.2 s for $He-N_2$ in alumina. The corresponding value for $Ar-N_2$ in alumina should be:

$$\tau_d(Ar-N_e) = \tau_d(He-N_2)\ \frac{D_{He-N_2}}{D_{Ar-N_2}} = 4.512\ s$$

or

$$\tau_d(Ar-N_2)/\ \tau_d(He-N_2) = 3.76$$

However from experimental results at Re=200 we get a ratio of the apparent time constants for diffusion

$$\tilde{\tau}_d(Ar-N_2)/\tilde{\tau}_d(He-N_2) = 1.60$$

If we take into account intraparticle convection we get a ratio of 1.40 in reasonable agreement with the experimental value.

One should be careful when using a measured value of the effective diffusivity by a dynamic chromatographic method for reactor design. As said before we get at a given Reynolds number an apparent effective diffusivity $\tilde{D}_e$ (as a result of a model which does not include convective transport inside the pellet). How will the reactor design be affected by using this value for $\tilde{D}_e$ instead of that of the true effective diffusivity and the intraparticle Peclet number for convection? In order to answer this question let us define a quantity

$$\tilde{E} = \eta_{dc}/\tilde{\eta}_d \tag{30}$$

where

$$\eta_{dc} = \frac{\frac{1}{r_1} - \frac{1}{r_2}}{\coth r_1 - \coth r_2} \tag{31}$$

is the catalyst effectiveness factor when pore diffusion and convection in the pores are both taken into account and $\tilde{\eta}$ is the effectiveness factor based on an apparent effective diffusivity $\tilde{D}_e$ for a slab catalyst and irreversible first order reaction, i.e.,

$$\tilde{\eta} = \frac{\text{th } \tilde{\phi}}{\tilde{\phi}} \tag{32}$$

It follows that the quantity $\tilde{E}$ is then:

$$\tilde{E} = \frac{\sqrt{f(\lambda)}}{\text{th}(\phi \sqrt{f(\lambda)})} \sqrt{1 + \frac{\lambda^2}{4\phi^2}} \tag{33}$$

In Figure 6a we show $\tilde{E}$ as a function of λ with the Thiele modulus as a parameter, for first order irreversible reactions. Similar plots can be calculated for zero order reactions as shown in Figure 6b.

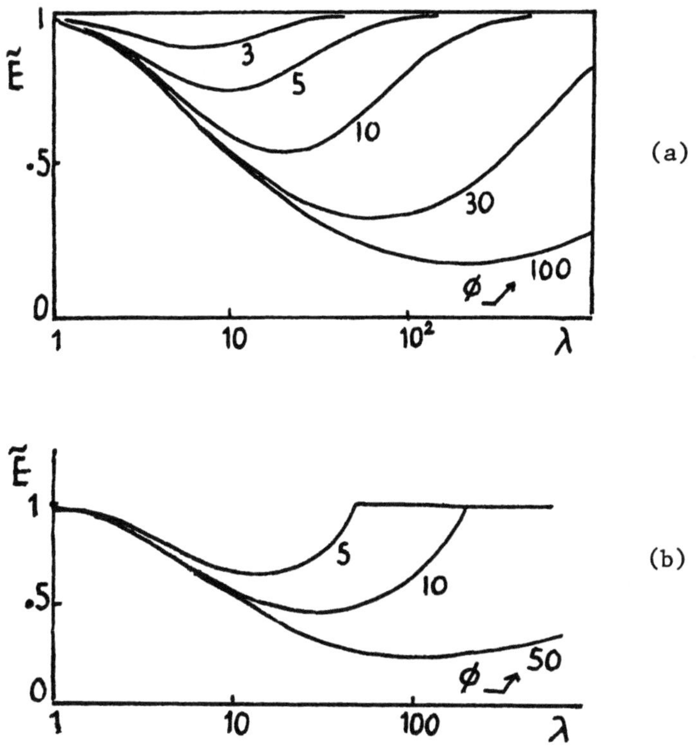

Figure 6 - $\tilde{E}$ versus λ with ϕ as a parameter [10]
(a) first order irreversible reaction
(b) zero order reaction

Let us consider a situation in which the intraparticle Peclet number λ=10 and the Thiele modulus is ϕ=10.For a first order irreversible reaction we get $\tilde{E}$ around 0.5.This means that in order to get the desired product specification the reactor should have a length two times longer than that calculated on the basis of the apparent diffusivity $\tilde{D}_e$.

The theoretical analysis of diffusion,convection and reaction in porous catalysts can also be carried out by comparing the effectiveness factor η_{dc} with the effectiveness factor η_d based on the true effective diffusivity.Thus by defining a quantity E as:

$$E= \eta_{dc}/\eta_d \tag{34}$$

and plotting E as a function of the Thiele modulus ϕ with λ as a parameter we observe that in kinetic and pure diffusional regimes there is no effect of convective transport in the pores.However, in the intermediate regime there is a better catalyst utilization due to intraparticle convection. This is shown in Figures 7a and 7b for first and zero order reactions respectively.We should emphasize that in the intermediate regime the concentration profiles within the pellet are not symmetrical. Typical concentration profiles are shown in Figure 8 for zero order reactions.

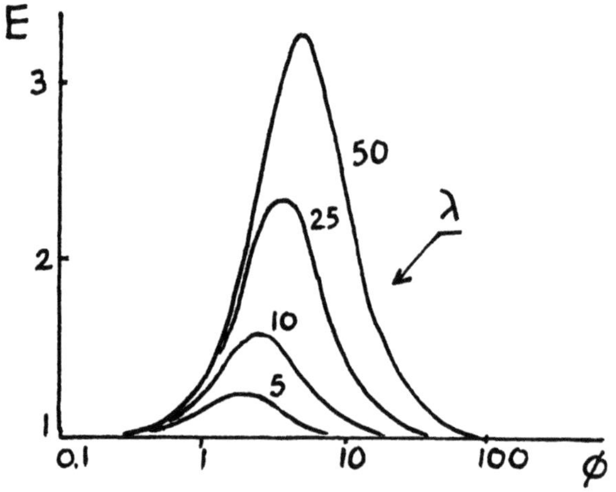

Figure 7a - E versus ϕ for first order irreversible reactions in isothermal slab catalysts.

For first order reactions (irreversible) in isothermal slab catalysts it is possible to get analytical solutions for the concentration profile inside the particle,i.e.,

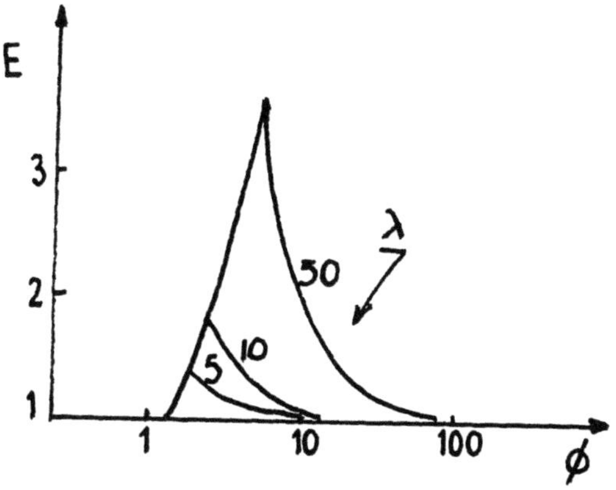

Figure 7b - E versus ϕ for zero order reactions in isothermal slab catalysts. [10]

$$f= \frac{1}{e^{2r_2} - e^{2r_1}} \{(e^{2r_2} -1)e^{r_1 x} +(1-e^{2r_1})e^{r_2 x} \} \qquad (35)$$

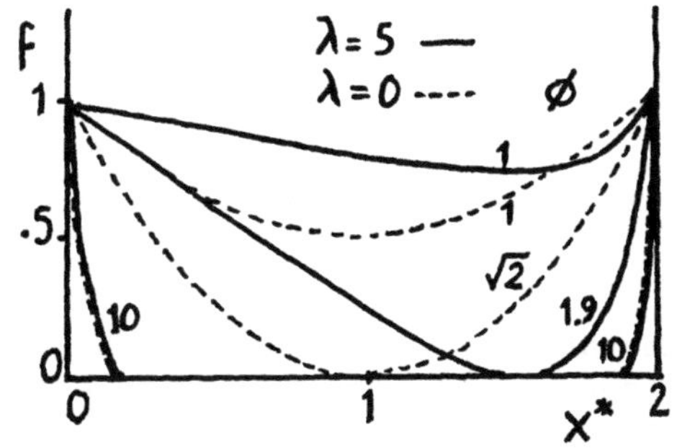

Figure 8- Concentration profiles inside the pellet for zero order reactions. [10]

INTRAPARTICLE CONVECTION, DIFFUSION AND REACTION IN NONISOTHERMAL CATALYSTS.

Let us now extend the analysis to the case of a nonisothermal catalyst following our previous work (Ferreira et al. |15|). The model equations for a volume element of a slab catalyst in the case of a first order irreversible reaction are:

o Mass balance

$$\frac{d^2 f}{dx^{*2}} - 2\lambda_m \frac{df}{dx^*} - 4\phi_s^2 \, e^{-\gamma(\frac{1}{T^*} - 1)} f = 0 \qquad (36a)$$

o Thermal balance

$$\frac{d^2 T^*}{dx^{*2}} - 2\lambda_h \frac{dT^*}{dx^*} + 4\phi_s^2 \beta \, e^{-\gamma(\frac{1}{T^*} - 1)} f = 0 \qquad (36b)$$

o Boundary conditions

$$x^*=0 \ , \ f=1 \text{ and } T^*=1 \qquad (36c)$$

$$x^*=1 \ , \ f=1 \text{ and } T^*=1 \qquad (36d)$$

In the above equations $x^*=z/2\ell$ (we reduced the space coordinate for the pellet by its thickness just because it is convenient for the numerical procedure developed below), $T^*=T/T_s$, $f=c/c_s$, $\phi_s = \ell\sqrt{k_s(T_s)/D_e}$, β is the Prater thermicity factor evaluated at the particle surface conditions, $\lambda_m=v_o\ell/D_e$ is the mass intraparticle Peclet number and $\lambda_h=v_o\ell/(\lambda_e/\rho_f c_{pf})$ is the thermal intraparticle Peclet number.

The catalyst effectiveness factor is calculated by:

$$\eta = \int_0^1 e^{-\gamma(\frac{1}{T^*} - 1)} f \, dx^* \qquad (36e)$$

Concentration and temperature profiles inside the catalyst pellet were obtained by using orthogonal collocation for the numerical solution of the model equations. Let us briefly recall how the technique works. The variables f and T* are approximated by polynomials of degree N+1, i.e.,

$$f_{N+1} = \sum_{i=1}^{N+2} \ell_i(x^*) \, f_i \qquad (37a)$$

$$T^*_{N+1} = \sum_{i=1}^{N+2} \ell_i(x^*) \, T^*_i \qquad (37b)$$

where

$$\ell_i(x^*) = P_{N+2}(x^*)/(x^*-x_i^*)\ P_{N+2}^{(1)}\ (x_i^*)$$

are Lagrange interpolation polynomials, x_i^* are N+2 interpolation points and $P_{N+2}(x^*)=(x^*-x_1^*)(x^*-x_2^*)\ldots(x^*-x_{N+2}^*)$.

Then f and T* are replaced by Equations (37a) and (37b), their derivatives are calculated and introduced in the model equations. The residuals at the interpolation points are:

$$R_N(f_j,T_j^*,x_j^*)=\sum_{i=1}^{N+2}(B_{ji}-2\lambda_m A_{ji})f_i-4\phi_s^2 e^{-\gamma(\frac{1}{T_j^*}-1)}f_j \quad (38a)$$

$$R_N'(f_j,T_j^*,x_j^*)=\sum_{i=1}^{N+2}(B_{ji}-2\lambda_h A_{ji})T_i^*+4\phi_s^2\beta\ e^{-\gamma(\frac{1}{T_j^*}-1)}f_j \quad (38b)$$

The collocation method makes the residuals R_N and R_N' equal to zero at each collocation point x_j^*. The system of nonlinear ODEs is thus transformed into a system of 2N nonlinear algebraic equations. This can be solved by some iterative procedure. The effectiveness factor is calculated by:

$$\eta=\sum_{i=1}^{N+2} w_i\ f_i\ e^{-\gamma(\frac{1}{T_i^*}-1)} \quad (39)$$

where

$$w_i=\int_0^1 W(x^*)\ \ell_i(x^*)\ dx^*$$

are the quadrature weights and the weight function $W(x^*)=1$ in our case. The integral involved in the definition of the effectiveness factor was calculated by a Lobatto quadrature with high accuracy by choosing the zeros of the Nth order Jacobi polynomial $P_N^{(1,1)}(x^*)$ as interior quadrature points.

Figure 9 shows the effectiveness factor η as a function of the Thiele modulus ϕ_s for $\lambda_h=0$ and various λ_m.

Concentration and temperature profiles in nonisothermal pellets are shown in Figures 10a and 10b for the corresponding situations.

The maximum temperature inside the catalyst can be predicted as a function of λ_m and λ_h for a given reaction/catalyst system (γ and β are fixed). Results are presented in Figure 11.

Such plots enable us to make a priori estimates of the maximum temperature rise in a complex situation where convection inside the pores takes place.

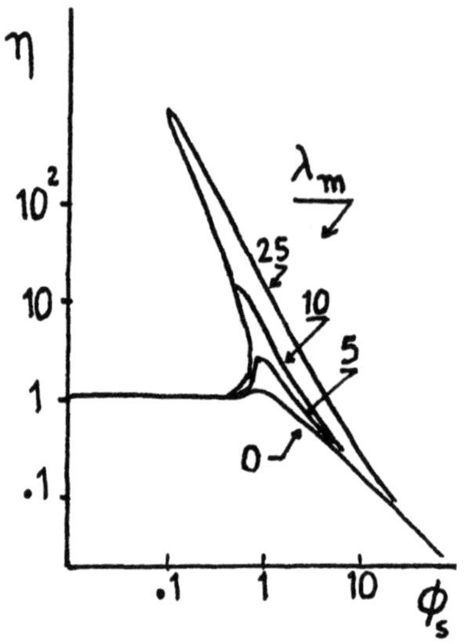

Figure 9 -Effectiveness factor for first order irreversible reaction, convection and diffusion in nonisothermal slab catalysts as a function of the Thiele modulus (γ=20 ; β=0.1 ; λ_h=0)

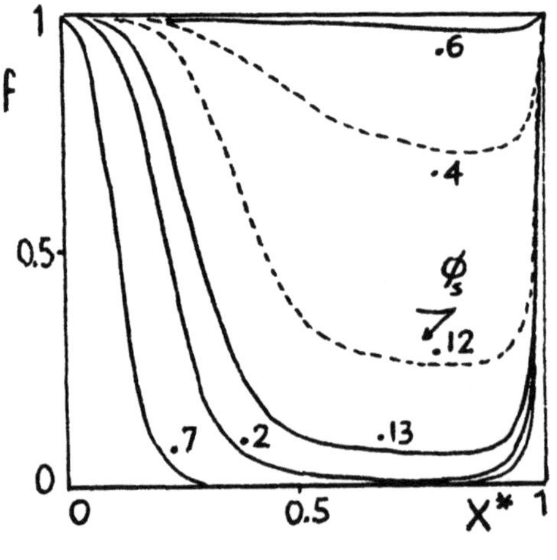

Figure 10a - Concentration profiles inside nonisothermal catalysts (γ=20, β=0.1, λ_h=0, λ_m=25)

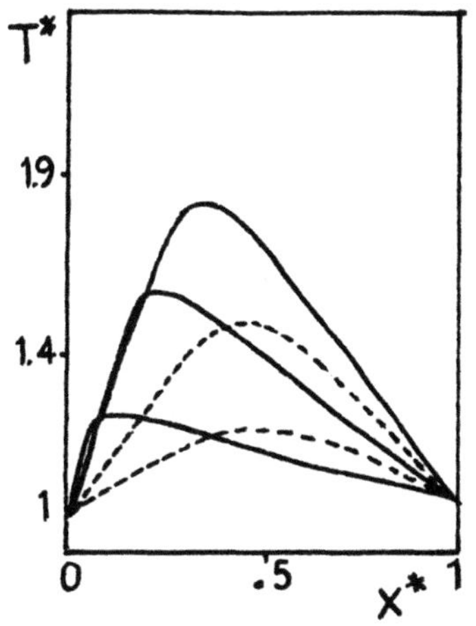

Figure 10b - Temperature profiles inside nonisothermal catalyts (γ=20, β=0.1, λ_h=0, λ_m=25)

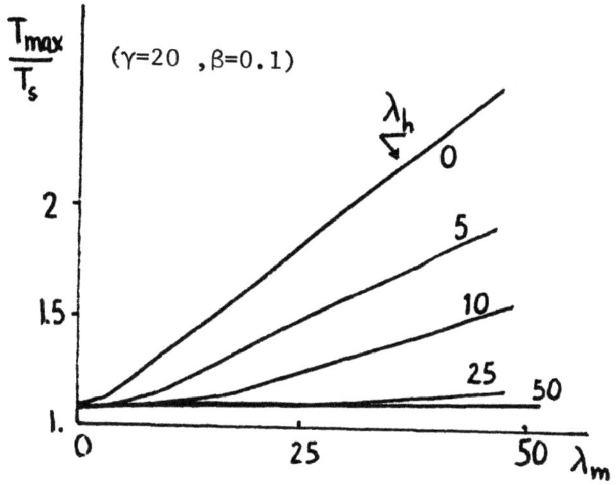

Figure 11- Maximum temperature rise inside catalysts as a function of λ_m and λ_h .

IMPLICATION OF INTRAPARTICLE CONVECTION ON REACTOR DESIGN

Let us discuss this question through a practical example involving simulation of a tubular fixed bed catalytic reactor for the synthesis of phthalic anhydride(PA).

The main route for the production of PA is the selective oxidation of o-xylene. In the BASF process (gas phase process) a V_2O_5 based catalyst is used. The operating temperature is around 375 C-410 C and excess air is used. The reaction is carried out in multitubular reactors with 10000 tubes and each unit has a capacity of 40000 to 50000 t/year.

In the following section we will simulate a reactor tube by using three different models:

- o Model I - Pseudohomogeneous unidimensional model(PH)
- o Model II- Heterogeneous (hybrid) unidimensional model with intraparticle concentration gradients due to diffusion and isothermal particles(HT_d)
- o Model III-Heterogeneous(hybrid) unidimensional model including intraparticle diffusion and convection but assuming isothermal particles (HT_{dc}).

We will look at the influence of model sophistication on the axial concentration and temperature profiles in the reactor and also see how the runaway diagram is modified.

The unidimensional pseudohomogeneous model (PH)

The model assumes that there are no gradients of temperature and concentration in the film and inside the particle. Moreover we assume that the fluid phase is in plug flow.

It can be shown that under typical operating conditions the catalyst particle is almost isothermal. In fact the maximum temperature rise inside the pellet is around 1K. Also the plug flow hypothesis is valid since the axial Peclet number ($Pe=u_o L/\varepsilon D_{ax}$) is around 750.

The model equations in dimensionless form are:

o Mass balance

$$\frac{df_b}{dz^*} = - Da_I \ f_b \tag{40a}$$

o Thermal balance

$$\frac{dT_b^*}{dz^*} = Da_I \ B \ f_b \ - Bi_w \ (T_b^* - T_w^*) \tag{40b}$$

o Boundary conditions

$$z^*=0, \ f_b=1 \ , T_b^*=1 \tag{40c}$$

In the above equations $f_b = c_b/c_o$ is the reduced concentration for the bulk fluid phase, $T_b^* = T_b/T_o$ is the reduced temperature, c_o and T_o are the feed conditions, $z^* = z/L$ and the dimensionless parameters are:

- $Da_I = k\,L/u_o$
- $B = (-\Delta H)\,c_o/\rho_f c_{pf} T_o$
- $Bi_w = 2U(L/u_o)/R_o \rho_f c_{pf}$

The initial value problem was numerically solved by using GEAR routine [16].

Typical axial concentration and temperature profiles are shown in Figures 12a and 12b, respectively illustrating the important aspec of parametric sensivity.

The analysis also shows that the wall temperature T_w is an important parameter. Increasing T_w results in an increase in temperature and conversion and therefore we get a lower runaway limit. The effect of the superficial velocity u_o is similar to that of T_w.

It should be said that in the computer simulation we took into account change of physical properties with temperature. Model equations were written for constant fluid properties for reasons of simplicity.

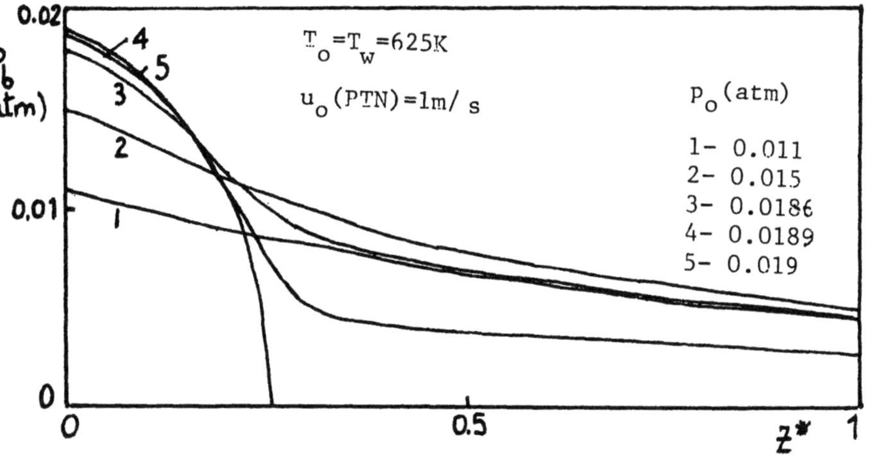

Figure 12 a- Axial partial pressure profiles in the reactor -pseudohomogeneous model

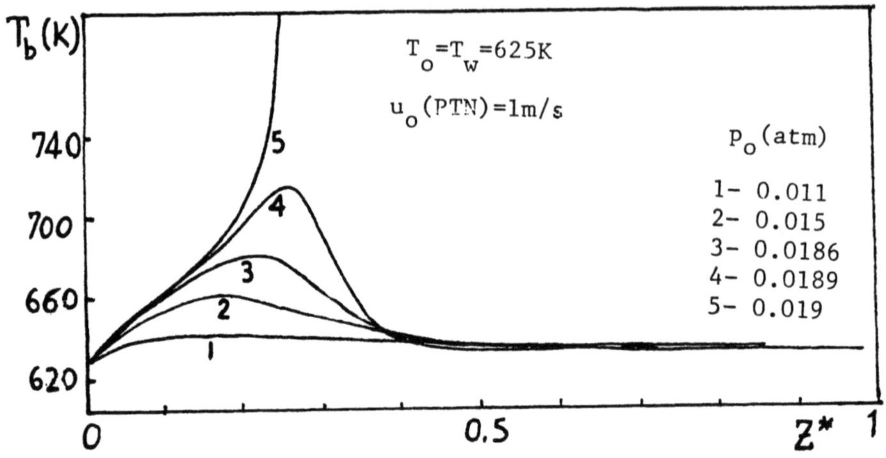

Figure 12b - Axial temperature profiles in the reactor: pseudohomogeneous model

For simulation purposes we used typical operating conditions, fluid properties and correlations listed below.

o kinetic data for the oxidation of o-xylene to PA [17]

Pseudo first order irreversible reaction

$$r' = 0.28\ p^o_{O_2}\ p\ \exp(19.837 - 13636/T)\ [\text{mole/Kg cat.s}] \quad \text{or}$$

$$r = 0.28\ p^o_{O_2} \rho_b RT \exp(19.837 - 13636/T) c_b\ [\text{mole/m}^3\ \text{s}]$$

where

$$k = 0.28\ p^o_{O_2}\ \rho_b\ RT \exp(19.837 - 13636/T)\ [s^{-1}]$$

o Inlet partial pressure of oxygen

$p^o_{O_2} = 0.208$ atm

o Heat of reaction and activation energy

$(-\Delta H) = 307000$ cal/mole ; $E = 27094$ cal/mole

o Reactor characteristics

length $L = 3$ m
radius $R_o = 0.0127$ m
porosity $\varepsilon = 0.42$

- Catalyst particles

  equivalent particle diameter d_p=0.008 m
  bulk density ρ_b=1300 kg/m^3
  particle porosity ε_p=0.725
  tortuosity factor τ_p=7
  specific area S_{BET}=1000 m^2/kg
  average pore radius $\bar{r}$ =6470 Å

- Fluid properties

  ρ_f=353.183/T kg/m^3 (density)

  c_{pf}=238.3 + 3.18x10^{-2} T cal/kg. K (heat capacity)

  u_o(PTN)=1 m/s

  $u_o = \rho_f(PTN)\ u_o(PTN)/\rho_f$

  μ_f =1.674x10^{-6} $\sqrt{T}$ - 1.17x10^{-5} kg/m.s (viscosity)

  λ_f(400 C)=2.27x10^{-2} cal/m.s.K (thermal conductivity)

  D_{12} =5.756 x10^{-10} T$^{3/2}$ m^2/s (molecular diffusivity)

  D_K =9.414 $\bar{r}\sqrt{T}$ m^2/s (Knudsen diffusivity)

- Transport properties of the catalyst

  λ_e(400 C)= 0.22 cal/m.s.K

  $1/D_e = 1/D_{Ke} + 1/D_{12e}$

  with $D_{Ke}=\varepsilon_p D_K/\tau_p$ and $D_{12e} = \varepsilon_p D_{12}/\tau_p$

- Correlations for film mass (heat) transfer coefficients

  $\varepsilon\ j_m$=0.357 Re$^{-0.359}$ 3 < Re < 2000

  with $j_m u_o/k_f = Sc^{2/3}$

  For the calculation of the heat transfer coeeficient in the film h we use the Chilton-Colburn analogy $j_m=j_h$ where

  $j_h = (h/u_o\rho_f c_{pf})Pr^{2/3}$

  with $Re=u_o\rho_f d_p/\mu_f$, $Sc=\mu_f/\rho_f D_{12}$ and $Pr=c_{pf}\ \mu_f/\lambda_f$

Calculation of the overall heat transfer coefficient at the wall was made by :

$$U=\frac{33.921 + 4.678\times10^{-2}\ c_{pf}\ u_o(PTN)}{1.686 + 9.463\times10^{-4}\ c_{pf}\ u_o(PTN)} \quad cal/m^2\ s\ K$$

o Model parameters (at reference conditions T_o=625K)

$Da_I = 0.697$ (Damkholer number)

$\gamma = 21.8$ (Arrhenius number)

$\beta = 0.0053$ (Prater thermicity factor)

$\lambda_m = 10$ (mass intraparticle Peclet number)

The heterogeneous(hybrid)unidimensional model HT_d

The model assumptions are:

- mass transport inside the pellet occurs by diffusion only
- the catalyst particle is isothermal
- concentration gradients inside the particle are important
- concentration and temperature gradients in the film can not be neglected
- the fluid phase is in plug flow

The model equations are now:

o Mass balance for the fluid phase

$$\frac{df_b}{dz^*} = -\bar{\eta}\ Da_I\ f_b \tag{41a}$$

o Thermal balance for the fluid phase

$$\frac{dT_b^*}{dz^*} = \bar{\eta}\ Da_I\ f_b - Bi_w(T_b^* - T_w^*) \tag{41b}$$

o Mass balance at the catalyst boundary

$$k_f a_p (1-\varepsilon)(f_b - f_s) = \eta\ k_s(T_s)\ f_s \tag{41c}$$

o Thermal balance at the catalyst boundary

$$h(T_s^* - T_b^*) = (-\Delta H)\ (c_o/T_o)\ (f_b - f_s) \tag{41d}$$

o Bed boundary conditions

$$z^*=0\ ,\ f_b=1 \text{ and } T_b^*=1 \tag{41e}$$

Since the catalyst particle was assumed to be isothermal the analytical solution of the mass balance for a volume element of the particle is well known leading to an effectiveness factor:

$$\eta = \frac{\tanh \phi_s}{\phi_s} \tag{42}$$

The effectiveness factor referred to the bulk conditions is then:

$$\bar{\eta} = \eta \, \frac{f_s}{f_b} \, \frac{k_s(T_s)}{k_b(T_b)} \tag{43}$$

At each integration step along the axial coordinate of the bed Δz^* we calculate f_b and T_b^* by numerically solving Equations (41a) and (41b).Then at that point the catalyst surface conditions f_s and T_s^* are obtained through Equations (41c),(41d) and (42).

In Figures 13a and 13b we show axial concentration (partial pressure of o-xylene) and temperature profiles.For conditions similar to those used in the pseudohomogeneous model it can be seen that the runaway limit is higher.Now we are observing a runaway condition at an initial o-xylene partial pressure of 0.0458 atm instead of 0.019 atm (Figure 12a).

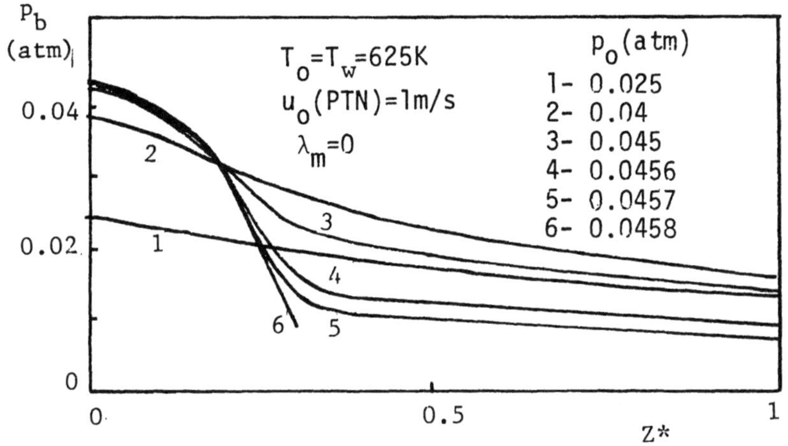

Figure 13 a - Axial partial pressure profiles in the reactor: heterogeneous(hybrid) unidimensional model HT_d

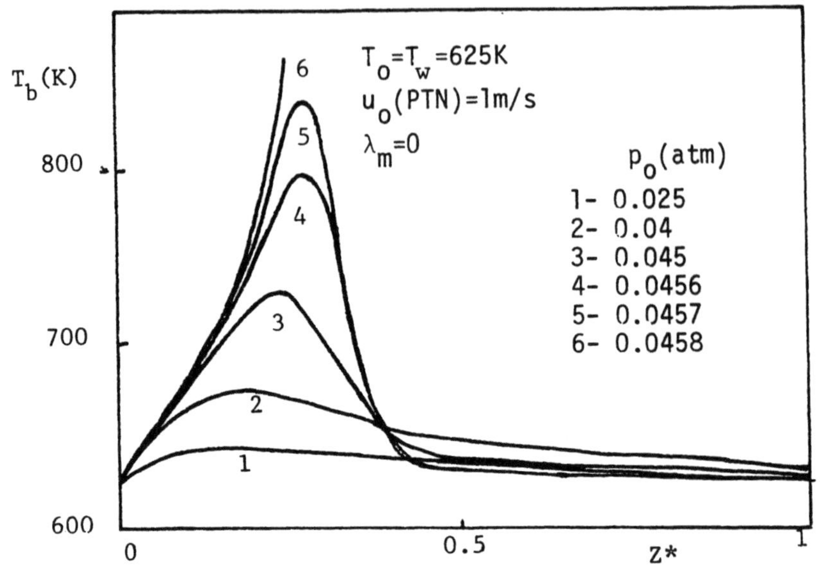

Figure 13b -Axial temperature profiles in the reactor: heterogeneous (hybrid) model:HT_d

The heterogeneous unidimensional model HT_{dc}

The model assumptions are those mentioned in the previous model but now we consider that transport processes inside the pellet are diffusion and convection. Model equations (41a) to (41e) are still valid.The only difference is that the mass balance of a volume element of the pellet is now:

$$\frac{d^2f}{dx^{*2}} - 2\lambda_m \frac{df}{dx^*} - 4\,\phi_s^2\, f = 0 \qquad (44)$$

The analytical solution of Equation (44) with boundary conditions $x^*=0$ and $x^*=1$, $f=1$ is:

$$f=c/c_s=\frac{\sinh r_2\, e^{r_1(2x^*-1)} - \sinh r_1\, e^{r_2(2x^*-1)}}{\sinh(r_2 - r_1)} \qquad (45)$$

The effectiveness factor η referred to the catalyst surface contions is given by Equation (31),i.e.,

$$\eta = \frac{1/r_1 \; - 1/r_2}{\coth r_1 \; - \coth r_2} \tag{31}$$

The effectiveness factor $\bar{\eta}$ referred to the bulk conditions is:

$$\bar{\eta} = \eta \; \frac{f_s}{f_b} \frac{k_s(T_s)}{k_b(T_b)} \tag{43}$$

The numerical procedure is similar to that described before but now the catalyst surface conditions are obtained from Equations (41c), (41d) and (45).

Axial concentration and temperature profiles in the reactor are shown in Figures 14a and 14b.We can see that the runaway limit that corresponds to an initial partial pressure of 0.040 atm is now between those calculated from the pseudohomogeneous and heterogeneous (diffusion) models.

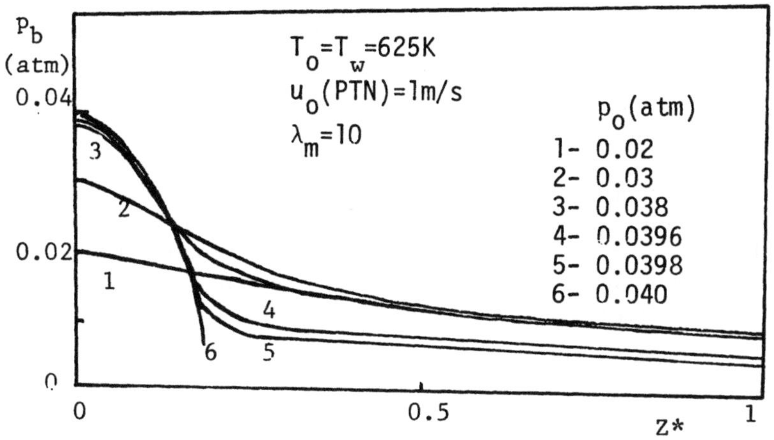

Figure 14a - Axial partial pressure (o-xylene) profiles in the reactor:heterogeneous (diffusion+convection) model HT_{dc}

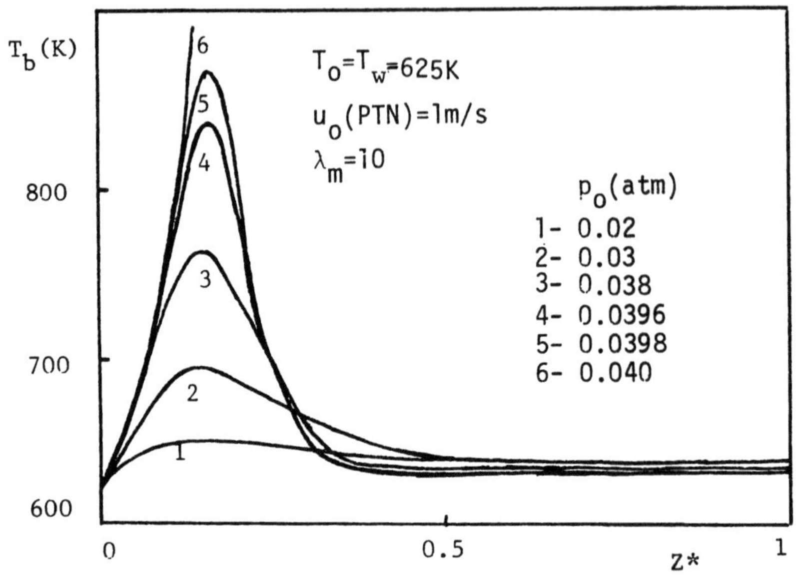

Figure 14b - Axial temperature profiles in the reactor: heterogeneous (diffusion + convection) model HT_{dc}.

From the analysis of reactor operation (influence of feed temperature,wall temperature,inlet o-xylene partial pressure) a runaway diagram can be presented showing the safe region of operation as well as the runaway region.

This diagram has been constructed from all the three models described above.The diagram is shown in Figure 15. For the situation under study the pseudohomogeneous model appears to be the most conservative model.Model sophistication can justify work in more severe conditions without danger of runaway.Moreover runaway limits predicted by the model with convection (HT_{dc}) are intermediate between pseudohomogeneous and heterogeneous (diffusion) models.

The coordinates in this diagram are N/S and S where:

$$N= 2U/R_o \rho_f c_{pf} k \tag{46}$$

$$S= \frac{(-\Delta H)\ p_o}{\rho_f c_{pf}\ RT_o} \quad \frac{E}{RT_o} \tag{47}$$

The diagram was calculated for cases in which feed and wall temratures are the same.

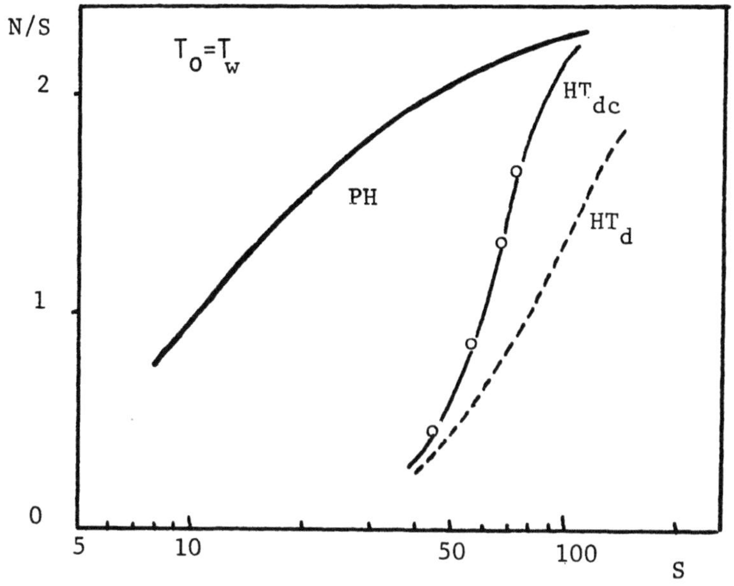

Figure 15 - Runaway diagram : N/S versus S

—— Pseudohomogeneous model
---- Heterogeneous (diffusion) model
oooo Heterogeneous (diffusion + convection)model

CONCLUSIONS

An attempt has been made to briefly review basic concepts involved in the analysis of transport processes and reaction in catalyst pellets and use them for the design of isothermal reactors containing catalysts working in the diffusional regime.

Moreover a detailed account of the importance of intraparticle forced convection when measuring effective diffusivities and designing reactors has been presented.This includes the analysis of non-isothermal catalysts and simulation and operation of a fixed bed catalytic reactor.

ACKNOWLEDGEMENTS

This work was supported by NATO Scientific Affairs Division and Gulbenkian Foundation.

NOTE ADDED IN PROOF

In a recent paper Cresswell |18| discussed the effect of intraparticle convection on the yield of consecutive-parallel reactions. He also measured diffusion and convection rates in unimodal and bimodal catalysts.

NOTATION

a_p - specific area of the catalyst particle
Bi_m - mass Biot number
Bi_h - heat Biot number
Bi_w - wall Biot number
B - adiabatic temperature rise; catalyst permeability
c - concentration inside the catalyst particle
$\langle c \rangle$ - average concentration in the catalyst particle
c_f - fluid phase concentration
c_{Ab} - concentration of reactant A in the bulk of fluid phase
c_{As} - concentration of reactant A at the external catalyst surface
Da_I - Damkholer number $(=k\rho_b\tau)$
Da_{II} - Damkholer number $(=k(T_b)\ell/k_f)$
D_{ax} - axial dispersion
D_e - effective diffusivity
$\tilde{D}_e$ - apparent effective diffusivity
d_p - particle diameter
$\bar{d}$ - average pore diameter
E - enhancement factor defined by Eq.(34)
$\tilde{E}$ - ratio befined by Eq.(30)
f - reduced fluid phase concentration
G - transfer function
h - film heat transfer coefficient
$(-\Delta H)$ - heat of reaction
k - kinetic constant
k_f - film mass transfer coefficient
ℓ - characteristic dimension of the catalyst particle
N - parameter defined by Eq.(46)
N_r - number of reaction units
N_d - number of diffusional mass transfer units
n - reaction order
Pe - Peclet number
Δp - pressure drop across a catalyst pellet
ΔP - pressure drop across the catalytic bed
r - reaction rate expressed in mole/m^{3}s
r' - reaction rate expressed in mole/kgcat.s
r_{obs} - observed rate of reaction
Re - Reynolds number
s - shape factor;Laplace parameter
S - quantity defined by Eq.(47)
T - temperature
T* - reduced temperature
u_o - superficial velocity
v_o - intraparticle convective velocity

X - conversion
x - reduced spatial coordinate for the pellet $(=z/\ell)$
x^* - reduced spatial coordinate for the pellet $(=z/2\ell)$; position inside the catalyst at which $c=0$
z - spatial coordinate for the reactor; spatial coordinate for the catalyst
z^* - reduced spatial coordinate for the reactor
W_A - amount of catalyst in a CSTR
W_P - amount of catalyst in a plug flow reactor

Greek symbols

α - parameter in Eq.(7b) $(=N_r N_d)$
β - Prater thermicity factor (refers to surface conditions)
$\bar{\beta}$ - Prater thermicity factor (refers to bulk conditions)
γ - Arrhenius number $(=E/RT_s)$
$\bar{\gamma}$ - Arrhenius number (E/RT_b)
ε - bed porosity
ε_p - particle porosity
λ - intraparticle Peclet number
λ_m - mass intraparticle Peclet number
λ_h - heat intraparticle Peclet number
λ_e - catalyst thermal conductivity
ϕ - Thiele modulus
τ - space time for the reactor
τ_r - time constant for the reaction
τ_d - time constant for the diffusion process
θ - reduced time $(=t/\tau)$
η - catalyst effectiveness factor in terms of surface conditions
$\bar{\eta}$ - catalyst effectiveness factor in terms of bulk conditions
η_d - effectiveness factor based on the true effective diffusivity
$\tilde{\eta}_d$ - effectiveness factor based on the apparent effective diffusivity
η
ρ_c - density of the catalyst solid
ρ_a - apparent density of the catalyst
ρ_b - bulk density of the catalyst
σ^2 - variance of the impulse response
μ_n - moments of nth order of the impulse response

Subscripts

s - surface conditions
b - bulk conditions
w - wall
d - diffusion
dc - diffusion and convection

REFERENCES

1. Aris,R. The Mathematical Theory of Diffusion and Reaction in Permeable Catalysts,vol 1(Oxford University Press,1975).
2. Jackson,R. Transport in Porous Catalysts(Elsevier,1977).
3. Thiele,E. Relation Between Catalytic Activity and Size of Particle.Ind.Eng.Chem. 31(1939)916
4. Weisz,P. and D.Prater. Interpretation of Measurements in Experimental Catalysis.Advances in Catalysis 6(1954)143
5. Rodrigues,A.,Grasmick,A. and S.Elmaleh. Modelling of Biofilm Reactors.The Chem.Eng.J. 27(1983)B39-B48
6. Nir,A. and L.Pismen.Simultaneous Intraparticle Forced Convection,Diffusion and Reaction in a Porous Catalyst.Chem.Eng. Sci.32(1977)35
7. Nir,A. Simultaneous Intraparticle Forced Convection ,Diffusion and Reaction in a Porous Catalyst II-Selectivity of Sequential Reactions .Chem.Eng.Sci.32(1977)925
8. Cogan,R.,Pipko,G. and A.Nir. Simultaneous Forced Convection, Diffusion and Reaction in a Porous Catalyst III- Depolymerization of paraldehyde. Chem.Eng.Sci. 37(1982)147-151
9. Rodrigues,A.,Ahn,B. and A.Zoulalian.Intraparticle Forced Convection Effect in Catalyst Diffusivity Measurements and Reactor Design. AIChEJ 28(1982)541-546
10. Rodrigues,A.,Orfão,J. and A.Zoulalian.Intraparticle Convection, Diffusion and Zero Order Reaction in Porous Catalysts.Chem. Eng.Communications.27(1984)327-337
11. Cresswell,D. and N.Orr. Measurement of Binary Gaseous Diffusion Coefficients Within Porous Catalysts in Residence Time Distri - bution Theory in Chemical Engineering,ed.A.Petho and R.Noble (Verlag Chemie,1982)
12. Dogu,G. and J.M.Smith. A Dynamic Method for Catalyst Diffusivities. AIChEJ 21(1975)58-61
13. Furusawa,T.,Suzuki,M. and J.M.Smith. Rate Parameters in Heterogeneous Catalysis by Pulse Technique.Cat.Rev.Sci.Eng.13(1976)43
14. Cheng,S.,Rodrigues,A. and A.Zoulalian. Diffusivity Measurements in Large Pore Catalysts by a Dynamic Chromatographic Technique. 9th Ibero-American Symp.on Catalysis(Lisbon,1984)
15. Ferreira,R.,Loureiro,J.,Costa,C. Zoulalian and A.Rodrigues.Convection in Large Pore Catalysts:its Effect on the Transient Behavior of Catalytic Reactors. 9th Canadian Symp. Catalysis (Quebec,1984)
16. Hindmarsh,A. Report UCID-30001,Lawrence Livermore Laboratory (1974).
17. Froment,G. and K.Bischoff.Chemical Reactor Analysis and Design (John Wiley,1979).
18. Cresswell,D. Intraparticle Convection:Its Measurement and Effect on Catalyst Activity and Selectivity. Applied Catalysis 15(1985) 103-116.

TRANSPORT RESTRICTIONS IN CATALYST PARTICLES WITH SEVERAL CHEMICAL REACTIONS

LARS J. CHRISTIANSEN
JØRGEN E. JARVAN

HALDOR TOPSØE A/S

ABSTRACT

This paper describes a mathematical model for a single catalyst particle in which several chemical reactions take place. The model includes transport restrictions against mass and heat transfer in the interior and in the gas film surrounding the particle, and it accepts a general type reaction rate expression such as a power law expression or a Langmuir-Hinshelwood expression. The model is reduced to a number of coupled second order differential equations - one for each reaction - by use of the stoichiometric coefficients.

The differential equations are solved numerically for the catalyst effectiveness factor by means of the orthogonal collocation method.

1. INTRODUCTION

In a heterogeneous catalytic process, the over-all performance of the converter is very dependent on the physical properties of a single catalyst particle.

The mathematical model which allows for differences in concentrations and temperature between the bulk gas stream and the particles is called a heterogeneous reactor model by Froment {1}. This is contrary to a socalled pseudo-homogeneous model for a catalytic reactor, where the concentrations and temperature in the bulk gas phase and in the particles are considered identical.

A general purpose program that performs axial integration of temperature and conversion in a fixed-bed reactor has been described by Jarvan {2} and Kjær {3}. In this program, the converter is divided into a number of zones and in each zone a number of physical transport and thermodynamic properties are considered constant. During the integration of the catalyst bed, the mathematical model for mass and heat transport restrictions in the catalyst particle is solved in the beginning of each integration step, and the result is expressed as the socalled catalyst effectiveness factor.

In the following, it is shown how the mathematical model was formulated for the transport restrictions in a single catalyst particle where several chemical reactions take place. It was required that the model does not impose limitations with respect to the rate controlling transport mechanisms and to the actual form of the rate expressions.

1.1 Transport restrictions in a single catalyst particle

The rate expression for a chemical reaction is in general a function of the concentrations of reactants and products, temperature and pressure. The transport restrictions against mass and heat transport in a single catalyst particle cause a variation in these properties, and hence a variation in the reaction rate. The pressure variation in the catalyst particle is not taken into account, however, because practical experience has shown this effect to be negligible for reactions in ammonia, methanol and hydrogen plants.

As an illustration, consider the simple reversible reaction,

$$A \rightleftarrows B \tag{1}$$

If the reaction is exothermal, the temperature and concentration variation in and around the particle might look as shown in Fig. 1:

FIG. 1

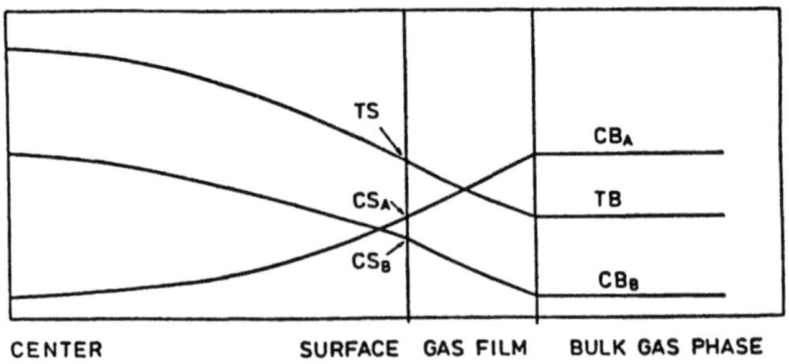

Because of resistance to mass transport through the gas film surrounding the particle, a concentration gradient is required so that the concentration of A is lower on the surface than in the bulk phase. A must diffuse into the particle through the catalyst pores, and this also causes a concentration gradient through the particle. The product B must pass through the same transport restrictions in the opposite direction, and this requires a concentration gradient for B so that the concentration of B is higher in the interior than in the bulk gas phase. The combined effect of these diffusion restrictions is that the reaction rate in the interior of the particle will be lower than for bulk phase conditions, since the reaction rate normally decreases as the reaction proceeds.

Besides the mass transfer restrictions, the heat transfer restrictions should also be taken into account. If the reaction is exothermal, the heat produced must be transported away and this requires a temperature gradient. Since the reaction rate normally increases with increasing temperature, the reaction rate would be higher in the interior of the particle than in the bulk phase, if the concentrations were identical. The effectiveness of the catalyst will then be greater than it would have been, if only mass transfer restrictions were taken into account. The reverse situation occurs, if the reaction is endothermal.

To summarize, the mathematical model considers the following four transport processes: 1) pore diffusion, 2) heat conduction in the catalyst particle, 3) diffusion of mass and 4) diffusion of heat in the gas film surrounding the particle.

The catalyst effectiveness factor is now introduced as a quantitative measure of the combined effects of the transport restrictions. It is defined as the ratio of the average rate of reaction, which would have been obtained, if there had been no transport restriction, i.e. as if the temperature and gas composition had been the same inside the particle as in the bulk gas phase, Satterfield {4}. The effectiveness factor for reaction No. j is called ETA_j, and at a given axial distance in the converter, the effective actual rate $REFF_j$ of reaction j is then calculated from

$$REFF_j = ETA_j \quad RATBLK_j \tag{2}$$

where $RATBLK_j$ is the intrinsic rate for reaction j, i.e. the reaction rate without transport restrictions. The intrinsic rate equation is usually obtained from laboratory experiments with catalyst powder. The mathematical model should be independent of the actual form of the rate equation and should accept rate expressions such as power law expressions or Langmuir-Hinshelwood expressions.

2. MATHEMATICAL MODEL

2.1 Mass and heat balances

The mass and heat balances around a catalyst particle are established by use of the basic transport equations.

There are four equations, all of the same kind, where the flux of either heat or mass is calculated as the product of a driving force and a transport coefficient.

The mass transport of component i in the catalyst pores is described by the basic equation for diffusion flow, Fick's law:

$$N_i = -DEFF_i \frac{dC_i}{dr} \qquad (3)$$

where N_i is the flux of component i, $\frac{dC_i}{dr}$ the concentration gradient, and $DEFF_i$ the effective diffusion coefficient of i in the catalyst pores. $DEFF_i$ is calculated from the bulk gas diffusion coefficient, the pore size distribution and the labyrinth factor by the methods given in Kjær {5}.

The heat transport in the particle is described by the basic equation for heat transport by conduction, Fourier's law:

$$Q = -KPART \frac{dT}{dr} \qquad (4)$$

where Q is the heat flux, $\frac{dT}{dr}$ the temperature gradient, and KPART the thermal conductivity of the catalyst particle.

The basic equation for mass transfer of component i through the gas film is:

$$NS_i = KG_i \, (MS_i - MB_i) \qquad (5)$$

where NS_i is the surface flux of component i, MS_i and MB_i are the mole fractions of i at the particle surface and in the bulk phase, respectively, and KG_i the mass transfer coefficient of i, calculated by the methods given in Kjær {5}.

The last equation is the equation for heat transport through the gas film:

$$QS = HSURF \, (TS \quad TB) \qquad (6)$$

where QS is the surface heat flux, TS and TB the temperature at the surface and in the bulk phase, respectively, and HSURF the heat transfer coefficient through the gas film calculated by the methods given in Kjær {5}.

In establishing the mathematical model for the transport processes in the interior and in the gas film surrounding the catalyst particle, it is assumed that the catalyst particle is a sphere with radius RAD or may be approximated by such. The reacting gas mixture is composed of COMP components and REAC chemical reactions take place in the catalyst.

The actual rate of reaction in the particle $RATE_j$ is the rate of reaction for a component with stoichiometric coefficient +1 and originating from reaction j only, but when the material balance for a single component i is established, it is the rate of production of i that should be used. It can be found from the stoichiometric coeffients as

$$R_i = \sum_{j=1}^{REAC} S_{ij}\, RATE_j \qquad (7)$$

where R_i is the rate of production of i and S_{ij} is the stoichiometric coefficient for component i in reaction j. Note that the rate of production is negative if the component is consumed.

A material balance for component i and a heat balance in a spherical shell of thickness dr and at the distance r from the particle center are now set up. After insertion of (3) and (4) and rearrangements, the following coupled second order differential equations are found:

$$\frac{d^2C_i}{dr^2} + \frac{2}{r}\frac{dC_i}{dr} + \frac{R_i\ DENS}{DEFF_i} = 0 \qquad (8)$$

$$\frac{d^2T}{dr^2} + \frac{2}{r}\frac{dT}{dr} + \frac{\sum_{j=1}^{REAC} RATE_j\ (-\Delta H_j)DENS}{KPART} = 0 \qquad (9)$$

where DENS is the catalyst particle density and $(-\Delta H_j)$ the heat of reaction for reaction j. It should be noted that the reaction rate is a function of radial distance because of the concentration and temperature variation in the catalyst.

The boundary conditions for the differential equations in the particle center and at the particle surface are given by:

$$\frac{dC_i}{dr} = 0, \quad \frac{dT}{dr} = 0 \qquad \text{at } r = 0 \qquad (10)$$

$$C_i = CS_i, \quad T = TS \qquad \text{at } r = RAD \qquad (11)$$

The surface concentrations CS_i and the surface temperature TS are, however, unknown, but they are related to the bulk phase conditions by equations (5) and (6).

One problem arises here since the driving force in equation (5) is expressed in mole fractions and not in concentration, which is the dependent variable in the differential equation. Equation (5) is therefore converted to concentration by multiplication with the total concentration CTOT in the bulk phase:

$$CTOT = \frac{P}{Z\ RG\ TB} \qquad (12)$$

where P is the pressure, Z the compressibility factor and RG the gas constant. Equation (5) can thus be rewritten as:

$$NS_i = \frac{KG_i}{CTOT} (CS_i - CB_i) \qquad (13)$$

The small error introduced by using the bulk phase temperature TB to convert the surface mole fraction MS_i is found to be negligible for reactions in ammonia, methanol and hydrogen plants.

The last equation required is the definition of the catalyst effectiveness factor ETA_j for reaction j. The formula is derived from the definition given in the Introduction. Since the particle is assumed to be spherical, the equation is:

$$ETA_j = \frac{\int_0^{RAD} 4\pi\ r^2\ RATE_j\ DENS\ dr}{\frac{4}{3}\pi\ RAD^3\ RATBLK_j\ DENS} \qquad (14)$$

where $RATBLK_j$ is the value of the reaction rate calculated at bulk phase conditions.

2.2 Simplification of the mathematical model

A mixture with COMP components is fully described by a total of COMP + 1 coupled second order differential equations, one differential equation (8) for each component and one differential equation (9) for the temperature. It is therefore natural to introduce a simplification in order to reduce the number of differential equations, and hereby save computer time. The simplification is based upon the key component concept. Among the components present in the mixture, a key component for each reaction is selected. It will be shown that when the concentrations of the key components are known, it is possible to calculate the concentrations of the other components and the temperature in the catalyst particle.

Normally, the limiting reactants are chosen as key components. The subscript for a key component is k and equation (8) can then be written:

$$\frac{d^2C_k}{dr^2} + \frac{2}{r}\frac{dC_k}{dr} + \frac{R_k\ DENS}{DEFF_k} = 0 \qquad (15)$$

and a total of REAC equations of this type are now considered. The rate of production of k is given by an equation similar to (7):

$$R_k = \sum_{j=1}^{REAC} S_{kj}\ RATE_j \qquad (16)$$

This equation can be solved with respect to $RATE_j$:

$$RATE_j = \sum_{k=1}^{REAC} SINV_{jk}\ R_k \qquad (17)$$

where $SINV_{jk}$ is the inverted stoichiometric matrix for key component k in reaction j.

As an example, consider the methanol synthesis reaction from carbon dioxide coupled with the water gas shift reaction:

$$CO_2 + 3H_2 = CH_3OH + H_2O$$

$$CO + H_2O = CO_2 + H_2$$

with the two reaction rates $RATE_1$ and $RATE_2$, respectively.

The stoichiometric coefficients are:

	REACTION 1	REACTION 2
CH_3OH	+1	0
H_2O	+1	-1
CO	0	-1
H_2	-3	1
CO_2	-1	1

The key components are selected as carbon dioxide for the methanol synthesis reaction and carbon monoxide for the water gas shift reaction. The production of CO_2 and CO can be calculated from equation (16) as:

$$R_{CO_2} = -RATE_1 + RATE_2$$

$$R_{CO} = -RATE_2$$

Solution of these two equations with respect to $RATE_1$ and $RATE_2$ gives:

$$RATE_1 = -R_{CO_2} - R_{CO}$$

$$RATE_2 = -R_{CO}$$

and the inverted stoichiometric matrix for the key components in the example is then:

	CO_2	CO
REACTION 1:	-1	-1
REACTION 2:	0	-1

The rate of production of the other components R_i can now be found by insertion of (17) in (7):

$$R_i = \sum_{j=1}^{REAC} S_{ij} \sum_{k=1}^{REAC} SINV_{jk} R_k \tag{18}$$

Now equations (8), (15), and (18) are all combined giving the following second order differential equation:

$$DEFF_i \left\{ \frac{d^2C_i}{dr^2} + \frac{2}{r} \frac{dC_i}{dr} \right\} =$$

$$\sum_{j=1}^{REAC} S_{ij} \sum_{k=1}^{REAC} SINV_{jk} \quad DEFF_k \left\{ \frac{d^2C_k}{dr^2} + \frac{2}{r} \frac{dC_k}{dr} \right\} \qquad (19)$$

This equation can be integrated analytically by use of the boundary conditions in equations (10) and (11), and the result is:

$$DEFF_i \ (C_i - CS_i) =$$

$$\sum_{j=1}^{REAC} S_{ij} \sum_{k=1}^{REAC} SINV_{jk} \ DEFF_k \ (C_k - CS_k) \qquad (20)$$

It is seen that if the surface concentrations CS_i and CS_k are known, the concentrations of the other components in the interior of the catalyst particle can be calculated from the concentrations of the key component C_k.

A similar equation can be used to calculate the temperature from the key component concentrations. It is derived by combining equations (15), (17), and (9). The result is:

$$KPART \left\{ \frac{d^2T}{dr^2} + \frac{2}{r} \frac{dT}{dr} \right\} =$$

$$\sum_{j=1}^{REAC} (-\Delta H_j) \sum_{k=1}^{REAC} SINV_{jk} \quad DEFF_k \left\{ \frac{d^2C_k}{dr^2} + \frac{2}{r} \frac{dC_k}{dr} \right\} \qquad (21)$$

This equation can also be integrated twice using the boundary conditions in equations (10) and (11), and the integration result is:

$$KPART \ (T - TS) =$$

$$\sum_{k=1}^{REAC} DEFF_k \ (C_k - CS_k) \sum_{j=1}^{REAC} (-\Delta H_j) \ SINV_{jk} \qquad (22)$$

It is hence possible to calculate the temperature in the interior of the catalyst particle, when TS, CS_k and the key component concentrations in the interior of the particle are known.

It has been demonstrated that the number of second order differential equations has been reduced from COMP + 1 to REAC equations by use of the key component concept.

2.3 Calculation of the catalyst effectiveness factor

Assuming that the surface concentrations of the key components CS_k are known, it will now be shown how CS_i and TS can be calculated. The surface flux of key component k is found from:

$$NS_k = \frac{KG_k}{CTOT} (CS_k - CB_k) \tag{23}$$

and the flux of a hypothetical component with stoichiometric coefficient +1 originating from reaction j only can be calculated from an equation similar to equation (17):

$$FLUX_j = \sum_{k=1}^{REAC} SINV_{jk} \, NS_k \tag{24}$$

The fluxes of the individual components can then be calculated from:

$$NS_i = \sum_{j=1}^{REAC} S_{ij} \, FLUX_j \tag{25}$$

In other words, it is now possible to calculate the surface mole fractions MS_i from equation (5).

The heat flux at the particle surface is given by:

$$QS = \sum_{j=1}^{REAC} FLUX_j \, (-\Delta H_j) \tag{26}$$

and the surface temperature TS can now be calculated from (6).

The catalyst effectiveness factor for reaction j can also be calculated from $FLUX_j$. The average rate of formation of j in the particle with transport restrictions is given by:

$$\frac{4}{3} \pi \, RAD^3 \, RATBLK_j \, DENS \, ETA_j \tag{27}$$

and the amount of moles flowing out of the particle originating from reaction j is:

$$4\ \pi\ RAD^2\ FLUX_j \tag{28}$$

By a material balance consideration, these two expressions must be identical and combination gives:

$$ETA_j = \frac{3\ FLUX_j}{RAD\ RATBLK_j\ DENS} \tag{29}$$

The effectiveness factor can thus be calculated when $FLUX_j$ is known.

It has been demonstrated that the surface concentrations, the surface temperature, the effectiveness factor and the surface fluxes can be calculated from the surface concentrations of the key components, CS_k.

If the surface fluxes of the key components NS_k are known, CS_k can be calculated from equation (23). NS_k are, however, also unknown, but an expression can be found by combining the definition of the catalyst effectiveness factor, equation (14) with equation (29). The resulting equation for the surface flux of a hypothetical component with stoichiometric coefficient +1 in reaction j is:

$$FLUX_j = \frac{1}{RAD^2} \int_0^{RAD} r^2\ RATE_j\ DENS\ dr \tag{30}$$

Multiplication of this equation by S_{kj} and summation over the j reactions give the required relationship when equations (25) and (16) are used.

$$NS_k = \frac{1}{RAD^2} \int_0^{RAD} r^2 R_k\ DENS\ dr \tag{31}$$

Combination of this equation with equation (23) eliminates NS_k and the resulting expression contains only CS_k as unknown variable, although it is given implicitly.

$$\frac{KG_k}{CTOT}\ (CS_k - CB_k) = \frac{1}{RAD^2} \int_0^{RAD} r^2\ R_k\ DENS\ dr \tag{32}$$

In evaluation of the integral, the concentrations of the key components at the particle surface, CS_k are used to calculate the reaction rates R_k at r = RAD.

The boundary conditions at the particle surface and the calculation of the catalyst effectiveness factors have then been merged into REAC integral equations with the surface concentrations of the key components as unknown variables.

3. NUMERICAL SOLUTION OF THE MODEL

In the analysis and design of fixed-bed catalytic reactors, catalyst effectiveness factors must be evaluated several times in a single calculation, and it is therefore necessary that the numerical method for solution of the mathematical model is reliable and fast.

It has been shown that the model is a set of REAC coupled second order differential equations with boundary conditions at the interval end-points: particle center and particle surface. This type of ordinary differential equation is conveniently solved by a global method where the dependent variables are approximated in the whole interval by polynominals. The method used here is the orthogonal collocation method, where orthogonal polynominals are used in the approximation.

The basis for this method is given in Villadsen and Stewart {6}, Villadsen {7}, and in Villadsen and Michelsen {8}, where the method is placed in the general subject of solution of differential equations by polynominal approximation. Properties of the orthogonal polynominals, such as the zeros are calculated by the methods given in Michelsen and Villadsen {9}.

In the orthogonal collocation method, the dependent variables are approximated by Lagrange interpolation polynominals, which in turn are constructed from orthogonal polynominals, and the node points of those interpolation polynominals are the zeros of an orthogonal polynominal. The node points are also called collocation points, and in the orthogonal collocation method, the differential quotients are represented as a weighted sum of the values of the dependent variables in the collocation points. The differential quotients are said to be discretized, and the discretization matrices are found from properties of the interpolation polynominals. Insertion of the discretized approximations to the differential quotients in the differential equations reduces these to a set of coupled algebraic equations, and the values of the dependent variables in the collocation points can then be determined by solution of these algebraic equations.

But before the solution can be carried out, the REAC coupled differential equations and the boundary conditions should be transformed to the orthogonal collocation method. The orthogonal polynominals are defined in the interval from zero to one, and the independent variable should therefore be transformed to a dimensionless radius:

$$x = \frac{r}{RAD} \tag{33}$$

Equation (15) is then transformed to:

$$\frac{d^2C_k}{dx^2} + \frac{2}{x}\,\frac{dC_k}{dx} + \frac{R_k\ DENS\ RAD^2}{DEFF_k} = 0 \tag{34}$$

where k = 1,2, ..., REAC.

The boundary condition at x = 0 is then transformed to

$$\frac{dC_k}{dx} = 0 \qquad \text{at } x = 0 \tag{35}$$

and the boundary condition at x = 1 (equation 32) is then

$$\frac{KG_k}{CTOT}\,(CS_k - CB_k) = RAD \int_0^1 x^2 R_k\ DENS\ dx \tag{36}$$

In equation (34), it is seen that it will be an advantage to work with an even function of the independent variable, since the boundary condition at x = 0, equation (35), then automatically is satisfied. Consequently, a new independent variable is introduced

$$u = x^2 \tag{37}$$

and when this is inserted in equation (34), the following equation is obtained:

$$4u\,\frac{d^2C_k}{du^2} + 6\,\frac{dC_k}{du} + \frac{R_k\ DENS\ RAD^2}{DEFF_k} = 0 \tag{38}$$

and the boundary condition at u = 1 becomes

$$\frac{KG_k}{CTOT}\,(CS_k - CB_k) = \frac{RAD\ DENS}{2}\int_0^1 u^{\frac{1}{2}} R_k\ du \tag{39}$$

Let the number of interior collocation points be NCOL, and the number of the surface point be NTCOL where

$$NTCOL = NCOL + 1 \tag{40}$$

If the concentration of key component k in collocation point n is called CK_{kn}, the first and second order derivatives in collocation point m, called $ROOT_m$, are then expressed as:

$$\left(\frac{dC_k}{du}\right)_{u=ROOT_m} = \sum_{n=1}^{NTCOL} A_{mn} \; CK_{kn} \tag{41}$$

$$\left(\frac{d^2C_k}{du^2}\right)_{u=ROOT_m} = \sum_{n=1}^{NTCOL} B_{mn} \; CK_{kn} \tag{42}$$

where A and B are the socalled discretization matrices evaluated from the properties of the orthogonal polynominals.

If the production of key component k in collocation point m is called RC_{km} equation (38) can now be written as:

$$4 \; ROOT_m \sum_{n=1}^{NTCOL} B_{mn} \; CK_{kn} + 6 \sum_{n=1}^{NTCOL} A_{mn} \; CK_{kn}$$

$$+ \frac{DENS \; RAD^2}{DEFF_k} \; RC_{km} = 0 \tag{43}$$

NCOL equations of this type can be written, and it is seen that the solution of the second order differential equation has been reduced to the solution of NCOL coupled algebraic equations with CK_{kn} as unknown variables.

The integral in the boundary condition at u = 1 can in general not be solved analytically and it is therefore solved by a quadrature formula, where the quadrature weights in the collocation points, W_n, are also determined from the properties of the orthogonal polynominals. Equation (39) is then written as:

$$\frac{KG_k}{CTOT} \left(CS_{kNTCOL} - CB_k\right) -$$

$$\frac{RAD\ DENS}{2} \sum_{n=1}^{NTCOL} W_n\, RC_{kn} = 0 \qquad (44)$$

It should be noted that the function $u^{\frac{1}{2}}$ in the integral in equation (39) has been included in the quadrature weights, W_n.

This algebraic equation is solved together with the NCOL algebraic equations in (43) giving a total of NTCOL algebraic equations. In the general case with REAC reactions, a total of REAC NTCOL algebraic equations are set up.

The reaction rate is normally a non-linear function of the temperature and the mole fractions of the active components. An analytical solution of the algebraic equations is therefore not possible. They must be solved numerically and for this purpose the Newton-Raphson method is used.

An initial guess of the concentrations of the key components in the collocation points is given, and the guessed values are varied systematically until all equations of type (43) and (44) become equal to zero within a certain limit. In the Newton-Raphson method, the differential quotient of each equation with respect to tbe values of the dependent variables must be known, and it is seen from the equations that it is necessary to calculate the differential quotients of the reaction rates with respect to the key component concentrations. These differential quotients are calculated numerically by giving the key component concentrations a small increment and then the differential quotients are calculated as difference quotients.

4. DISCUSSION

A general mathematical model for calculation of catalyst effectiveness factors for chemical reactions taking place in a spherical catalyst particle has been developed.

The model comprises a number of coupled second order differential equations which are solved numerically by the orthogonal collocation method.

The solution method has proven its reliability in the daily work at HALDOR TOPSØE A/S, where it is used in the design and analysis of fixed-bed reactors in ammonia and hydrogen plants. Applications of the method are shown in Ib Dybkjær {10}.

Nothing has been said about the possibility of multiple stationary states in catalyst particles, which is the subject of a number of papers in the recent literature. The reason is that the occurence of multiple stationary states in fixedbed reactors working in steady-state under industrial conditions is not very likely.

ACKNOWLEDGEMENT

We wish to express our gratitude to Mr. M.L. Michelsen at the Department of Chemical Engineering, the Technical University of Denmark, for much valuable assistance in the application of the orthogonal collocation method.

NOTATION

A	Discretization matrix for first order derivative
B	Discretization matrix for second order derivative
A, B	Component identifications in Introduction
C	Concentration, $kmol/m^3$ gas
CB	Bulk phase concentration, $kmol/m^3$ gas
CS	Surface concentration, $kmol/m^3$ gas
CK	Concentration in a collocation point, $kmol/m^3$ gas
CA	Concentration of component A in a collocation point. Used in example in Part 3
CTOT	Total concentration in bulk phase, $kmol/m^3$ gas
COMP	Number of components
DEFF	Effective diffusion coefficient, m^3 gas/hr/m solid
DENS	Catalyst particle density, kg/cu.m solid
ETA	Catalyst effectiveness factor
FLUX	Surface flux of hypothetical component with stoichiometric coefficient +1, $kmol/hr/m^2$ solid
HSURF	Surface heat transfer coefficient, $kcal/hr/^oC/m^2$ solid
$-\Delta H$	Heat of reaction, kcal/kmol of a component with stoichiometric coefficient +1
KG	Surface mass transfer coefficient, $kmol/hr/m^2$ solid
KPART	Particle thermal condustivity, $kcal/m/hr/^oC$
MB	Bulk Phase mole fraction
MS	Surface mole fraction
N	Mass flux, $kmol/hr/m^2$ solid
NS	Surface mass flux, $kmol/hr/m^2$ solid
NCOL	Number of interior collocation points
NTCOL	Number of collocation points including surface point
Q	Heat flux, $kcal/hr/m^2$ solid
QS	Surface heat flux, $kcal/hr/m^2$ solid
R	Production rate, kmol/hr/kgcat

REFF	Effective reaction rate, kmol/hr/kgcat
RATE	Rate of reaction for hypothetical component with stoichiometric coefficient +1, kmol/hr/kgcat
RATBLK	Reaction rate at bulk phase conditions, kmol/hr/kgcat
RC	Reaction rate in collocation point, kmol/hr/kgcat
r	Radial distance, m
RAD	Particle radius, m
REAC	Number of chemical reactions
RG	Gas constant, atm m^3 gas/kmol/K
ROOT	Collocation point
S	Stoichiometric coefficient
SINV	Inverted stoichiometric matrix for key components
T	Temperature, K
TB	Bulk temperature, K
TS	Surface temperature, K
u	Dimensionless radial distance, $u=x^2$
w	Radau quadrature weights
x	Dimensionless radial distance
z	Compressibility factor

SUBSCRIPTS

A	Component A
B	Component B
i	Component No. i
j	Reaction No. J
k	Key component No. k
n	Collocation point No. n
m	Collocation point No. m

REFERENCES

1. Froment, G.F. Fixed-Bed Catalytic Reactors. Technological Fundamental Design Aspects. Chem.-Eng.-Techn. 46, 374 (1974).
2. Jarvan, J.E. A General Purpose Program for Calculation and Analysis of Fixed-Bed Catalytic Reactors, Ber. Bunsen Ges. Phys. Chem. 74, 142 (1970).
3. Kjær, J. Computers Methods in Catalytic Reactor Calculations, Haldor Topsøe, Vedbæk/Denmark (1972).
4. Satterfield, C.N. Mass Transfer in Heterogeneous Catalysis, MIT Press (1970).
5. Kjær, J. Computer Methods in Gas Phase Thermodynamics, Haldor Topsøe, Vedbæk/Denmark (1972).
6. Villadsen, J. and Stewart. Solution of Boundary-Value Problems by Orthogonal Collocation. Chem. Eng. Sci. 22, 1483 (1967).
7. Villadsen, J. Selected Approximation Methods for Chemical Engineering Problems. Inst. for Kemiteknik, Danmarks

Tekniske Højskole, Lyngby (1970).
8. Villadsen, J. and Michelsen, M.L. Solution of Differential Equation Models by Polynominal Approximation. Prentice-Hall, N.Y. (1978).
9. Michelsen, M.L. and Villadsen, J. A Convenient Computational Procedure for Collocation Constants. Chem. Eng. J. 4, 64 (1972).
10. Dybkjær, I. Design of Ammonia and Methanol Synthesis Reactors. This course.

CATALYST TECHNOLOGY FOR REACTORS USED TO HYDROCRACK PETROLEUM RESIDUA

Marten Ternan and R. H. Packwood

Energy Research Laboratories, CANMET
Energy Mines and Resources
Ottawa, Ontario, K1A 0G1, Canada

ABSTRACT

In this paper, the phenomena occuring in catalysts used to hydrocrack petroleum residua are discussed. Reaction sites are provided by the catalyst sulphide phase (Mo is the majority cation) and by the catalyst oxide phase (Al is the majority cation). The influence of the promoter cations (typically Co or Ni) is also described. The catalyst is deactivated by coke and by metals. Furthermore, the reaction rate is often controlled by the rate of diffusion of the large carbonaceous molecules in the residua. All of these factors have been considered in mathematical simulations of the phenomena occuring in the catalyst.

INTRODUCTION

Catalysts are an important technological factor in many chemical reaction systems. Although catalysis is not yet a precise science, it has progressed far beyond its former image of being a black art. The catalyst commonly used to hydrocrack heavy oils and petroleum residua, promoted molybdenum-alumina, will be discussed here, in order to illustrate this point.

The technological history of the hydrocarbon processing industry contains several instances of catalytic processes replacing thermal ones. Fluid catalytic cracking, reforming and lube oil processing are examples. This transition may now be occuring in Canada, for heavy oil conversion. Previously the primary processes used to convert the bitumen from the Canadian oil sands were thermal (delayed coking and fluid coking). At the present time two catalytic hydrocrackers which will process residual oil are under construction, and a third one is at the design stage.

Four distinct topics related to the catalytic hydrocracking of residual oil will be discussed, the catalytic mechanism (or how the catalyst works), catalyst deactivation, diffusion in the catalyst pores, and catalyst modelling.

THE CATALYTIC MECHANISM

Several types of reactions occur on residual oil hydrocracking catalysts[1], molecular weight reduction (hydrogenation and bond cleavage), hydrodemetallization (Ni and V), and heteroatom removal (S, N, and O). These DESIRABLE REACTIONS can all be generalized as a reaction between a carbonaceous molecule, CM, and hydrogen, as follows:

$$CM + H_2 \longrightarrow \text{Reaction Products} \qquad (1)$$

Polymerization and dehydrogenation, also occur on residual oil hydrocracking catalysts. These UNDESIRABLE REACTIONS do not involve hydrogen as a reactant and can be generalized as follows:

$$CM \longrightarrow \text{Reaction Products} \qquad (2)$$

In Equation 1, hydrogen can be considered to be chemisorbed by electron holes (represented as $\oplus$) in the catalyst. Wise and co-workers [2-6] have reported experimental evidence for the relationship between desirable reactions and p-conductivity in sulphided cobalt-molybdeum-alumina catalysts.

The carbonaceous molecule in Equation 1 is chemisorbed by electron acceptor states (represented as ☐) in the catalyst. Examples of electron acceptor states include anion vacancies in molybdenum sulphides and Lewis acid sites in alumina. Heckelsberg and Banks [7] have explained catalytic cracking in terms of electron transfer which occurs via electron acceptor states.

In alumina, the Al^{3+} cations are the adsorption sites, or electron acceptor states.

$$O^{2-} \quad \underset{O^{2-}}{\overset{☐}{Al^{3+}}} \quad O^{2-} \quad \xrightarrow{CM} \quad O^{2-} \quad \underset{O^{2-}}{\overset{CM}{Al^{3+}}} \quad O^{2-}$$

Thermodynamic equilibrium calculations for the reaction

$$MoS_2 + H_2 \rightleftharpoons MoS_3 + H_2 \qquad (3)$$

indicate [1] that at reaction conditions, molybdenum sulphides should be predominently present as MoS_2 with a small amount of MoS_3.

The carbonaceous molecule will be adsorbed at anion vacancies [8-10] which in this case are sulphur vacancies in MoS_3. As an example, a carbonaceous molecule in a residual oil would probably be a polycyclic aromatic containing a thiophenic ring, CM-S

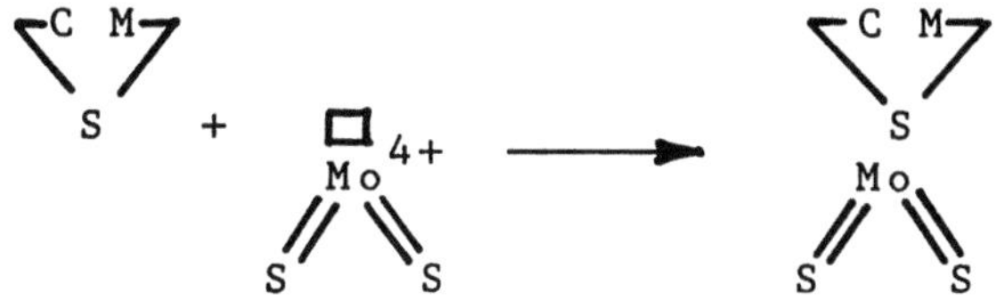

$$\underset{SS}{\overset{C-M}{\underset{Mo}{\diagdown S \diagup}}} + ☐_{Mo}^{4+} \longrightarrow \underset{SS}{\overset{C-M}{\underset{Mo}{\diagdown S \diagup}}} \qquad (4)$$

The sulphur containing carbonaceous molecule could be chemisorbed by an electron acceptor state, which is a sulphur vacancy in MoS_3, as indicated in Equation 4. The chemisorbed species can react with hydrogen as shown in Equation 5.

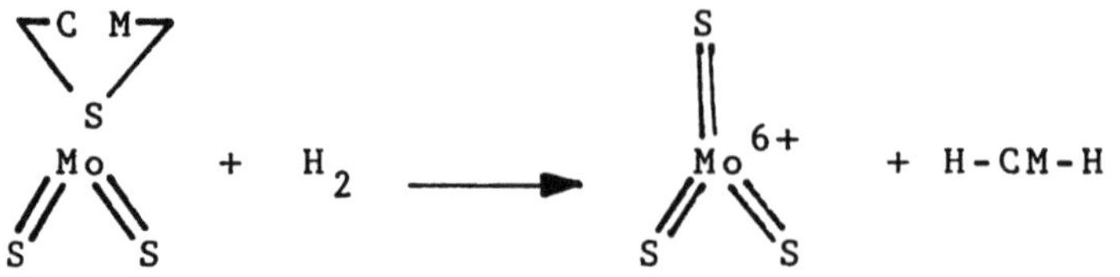

(5)

The hydrogen which reacts is also chemisorbed on the catalyst in a manner to be described below. The reaction products are MoS_3 and a hydrogenated carbonaceous molecule. The MoS_3 can also be considered to be MoS_2 containing non-stoichiometric sulphur. The non-stoichiometric MoS_2, containing excess sulphur anions, must also contain electron holes in order to preserve electron neutrality, as shown in Equation 6

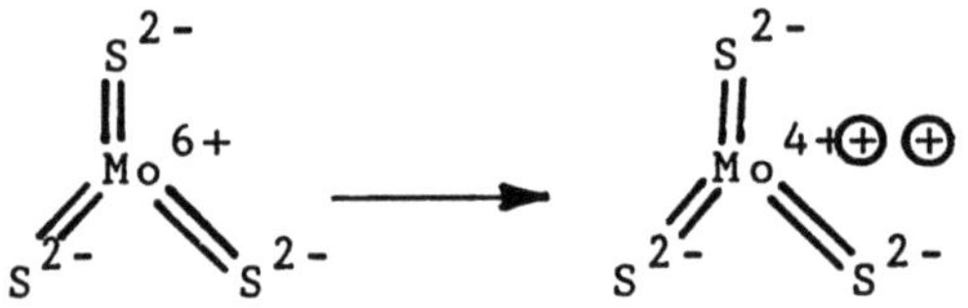

(6)

Hydrogen chemisorption has been associated with electron holes [3] as shown in Equation 7

$$0.5\ H_2 + \oplus \longrightarrow H^+ \qquad (7)$$

Wright et al [11] have reported experimental evidence for hydrogen chemisorption on sulphur anions. These concepts are combined, in Equation 8.

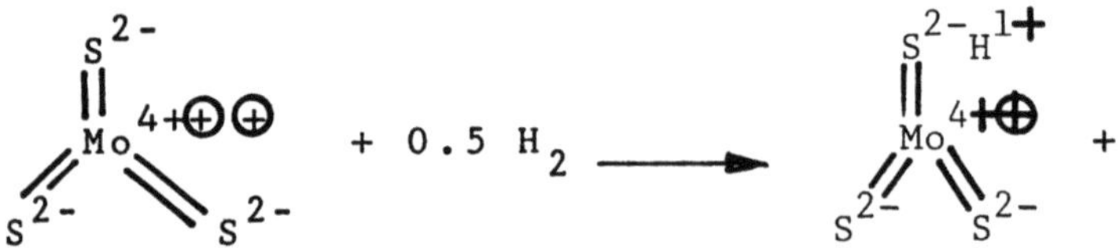

(8)

Continued reaction with chemisorbed hydrogen leads to the formation of hydrogen sulphide and the regeneration of the original electron acceptor state (sulphur anion vacancy), as shown in Equation 9.

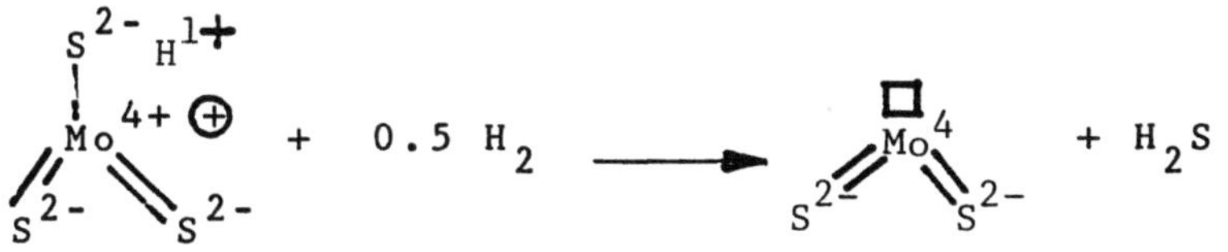

$$\underset{S^{2-}}{\overset{S^{2-}\ H^{1+}}{\underset{\parallel}{Mo^{4+}}\ \oplus}}\ +\ 0.5\ H_2\ \longrightarrow\ \underset{S^{2-}}{\overset{\square_4}{\underset{\parallel}{Mo^{4}}}}\ +\ H_2S \qquad (9)$$

If the desirable reaction, shown in Equation 1, was an elementary reaction, then its rate could be expressed as

$$\text{Rate} \propto [CM]\ [H_2] \qquad (10)$$

In terms of the reaction sites this would be

$$\text{Rate} \propto [\ \square\ ]\ [\ \oplus\ ] \qquad (11)$$

Two site mechanisms of this type have been proposed previously [12-14]

Similarly, the rate of the undesirable reaction can be expressed as

$$\text{Rate} \propto [CM] \qquad (12)$$

Expressed in terms of reaction sites on the catalyst, it becomes

$$\text{Rate} \propto [\ \square\ ] \qquad (13)$$

Equations 10-13 are based on the carbonaceous molecules being chemisorbed by electron acceptor states and on hydrogen being chemisorbed by electron holes.

The role of the promoter ions in the catalyst is to provide additional electron holes [2,6], for the chemisorption of hydrogen, as illustrated in Equation 14.

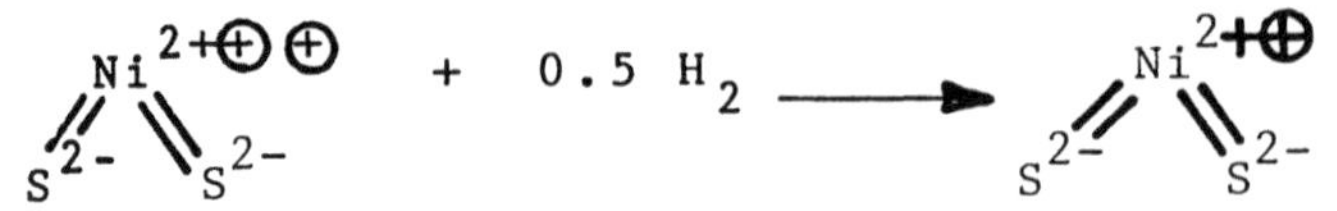

$$\text{Ni}^{2+} \oplus \oplus \underset{S^{2-}}{\overset{}{\diagdown}} \underset{S^{2-}}{\overset{}{\diagup}} + 0.5\, H_2 \longrightarrow \text{Ni}^{2+} \oplus \underset{S^{2-}}{\overset{}{\diagdown}} \underset{S^{2-}}{\overset{}{\diagup}} \qquad (14)$$

As described above both the oxide phase (alumina) and the sulphide phase (MoS_2 containing excess non-stoichiometric sulphur) of the catalyst contain electron acceptor states ($\square$). The data in Figure 1 are for catalysts having different amounts of sulphide and oxide phases. As the proportion of molybdenum increases, the number of molybdenum electron acceptor states will increase and the number of alumina electron acceptor states will decrease.

Figure 1 shows that the sulphur conversion of Athabasca bitumen (see Table 1) increases as the molybdenum content of the catalyst increases. By increasing the molybdenum content of the catalyst, both the number of molybdenum electron acceptor states and the number of electron holes ($\oplus$) increase, as indicated in Equations 4 to 9. Equation 10 shows that both types of reaction sites are required for desirable reactions. Therefore the increase in sulphur conversion with with molybdenum content of the catalyst is consistent with Equation 10.

Figure 1 shows that greater sulphur conversions were attained by adding nickel promoter ions to the catalyst. This experimental result is also consistent with Equation 10 which shows that conversion should increase when the number of electron holes ($\oplus$) increase. Promoter ions introduce electron holes into the catalyst, as indicated in Equation 14.

The above reaction mechanism (Equation 10) has also been used to describe the influence of hydrogen sulphide [1], alkali metal ions [15], and halide ions [16,17].

CATALYST DEACTIVATION

Deactivation during catalytic hydrocracking of residual oils is caused by the formation of coke, by the deposition of metals (Ni and V), and sometimes by the deposition of clay or sand particles that may be present in the feedstock [18].

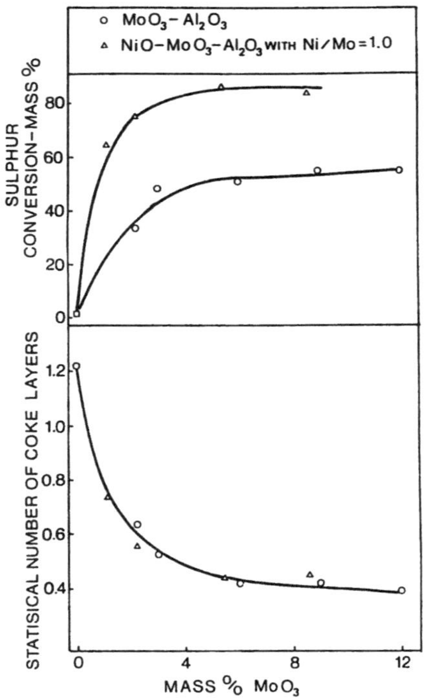

Figure 1 Conversion of sulphur containing carbonaceous species at 400°C (mass %) and statistical number of monolayers of coke on the catalyst versus MoO_3 concentration in the oxide form of the catalyst. Triangles represent $NiO-MoO_3-Al_2O_3$ catalysts having a Ni/Mo ratio (atomic) of 1, circles represent MoO_3 catalysts. Used with permission (ref 1).

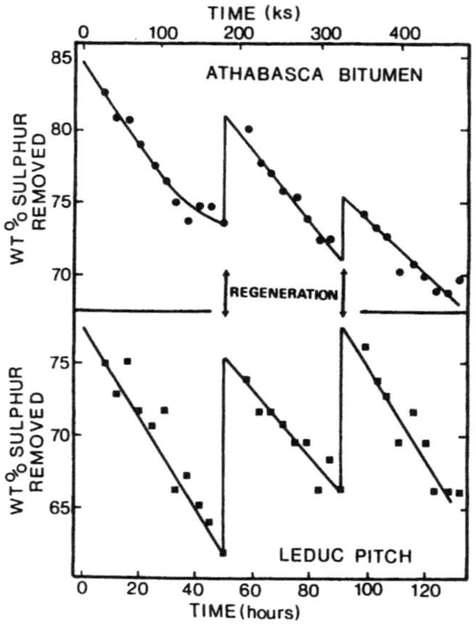

Figure 2 Mass % sulphur removed versus time, for Athabasca bitumen and Leduc pitch. Regenerations were performed after 180 ks (50 h), 328 ks (91 h) and 475 ks (132 h). The reaction temperatures were 450°C and 460°C for Athabasca bitumen and Leduc pitch respectively. Reprinted by permission of the copyright owner (ref 18).

The lower part of Figure 1 shows the amount of coke on the catalyst as a function of the catalyst molybdenum content. Equation 13 indicates that undesirable reactions are only dependent upon the number of electron acceptor states (□) in the catalyst. In Figure 1, as the number of sulphide phase electron states increase, the number of oxide phase (alumina) electron acceptor states decrease. It has been shown [1] that the Fermi level of the oxide phase is much lower than the Fermi level of the sulphide phase. This means that coke will be bonded much more strongly to the oxide phase. Therefore, as the number of oxide phase electron acceptor states decrease, the quantity of coke on the catalyst also decreases, as shown in Figure 1.

The lower part of Figure 1 also shows that the presence of nickel promoter ions in the catalyst had no influence on the catalyst coke content. This is consistent with Equation 13 which shows that undesirable reactions are not influenced by the number of electron holes (⊕) in the catalyst.

The separate effects of coke formation and metals deposition have been described in a study with two different residual oil feedstocks [18]. Their properties are listed in Table 1. They have a similar tendency to form coke (Conradson Carbon Residue) but much different contents of organometallic compounds (nickel and vanadium).

TABLE 1 FEEDSTOCK PROPERTIES

	Athabasca Bitumen	Leduc Pitch
Specific Gravity 15/15°C	1.000	0.991
Ash (m %) 700°C	0.70	0.013
Nickel (ppm)	68	13
Vanadium (ppm)	189	10
Iron (ppm)	358	28
Condradson Carbon Residue (m %)	12.6	14.7
Pentane Insolubles (m %)	15.8	6.1
Benzene Insolubles (m %)	0.90	0.09
CS_2 Insolubles (m %)	0.88	0.05
Sulphur (m %)	4.72	0.92
Nitrogen (m %)	0.42	0.55
Residuum (+525°C)	51	63

Results from reaction experiments are shown in Figure 2. For both feedstocks, the sulphur conversions decrease with increasing time on stream. After 50 hours on stream, both catalysts were regenerated by burning the coke. When the catalysts were put back on stream, deactivation continued. Regeneration was performed again, after 90 hours on stream. This was followed by another on stream period, during which deactivation continued.

After each deactivation, the catalyst processing the Leduc pitch returned to its original sulphur conversion. In contrast after each regeneration, the catalyst processing the Athabasca bitumen produced progressively lower and lower sulphur conversions. The Leduc pitch contained no metals. Therefore after the coke was removed by regeneration, the catalyst returned to its original conversion level. However the Athabasca bitumen contained large quantities of metals. Even after the coke was removed by regeneration, the metals would remain on the catalyst. The metals which were not removed were responsible for the lower sulphur conversions obtained after each regeneration.

Deactivation by metals deposition is also reflected in the measured metals contents of the catalysts. Figures 3 and 4 show electron microprobe measurements of metals as a function of radial position in the catalyst. It is apparent that large amounts of metals are deposited on the outer edge of the catalyst pellets, while the interior contains very little metal. At sufficiently long times on stream, the pores at the outer edge of the catalyst pellets become completely blocked. This phenomena is referred to as pore mouth plugging. When the pores become plugged there is a catastrophic decrease in conversion.

In Figures 3 and 4 the nickel and vanadium were originally present in the feedstock as organometallic compounds. The iron was probably associated with the clay or sand. It is reflected in the ash content of the bitumen in Table 1 being greater than that of the pitch.

The carbon content of the catalyst, determined by a special electron microprobe technique, is shown

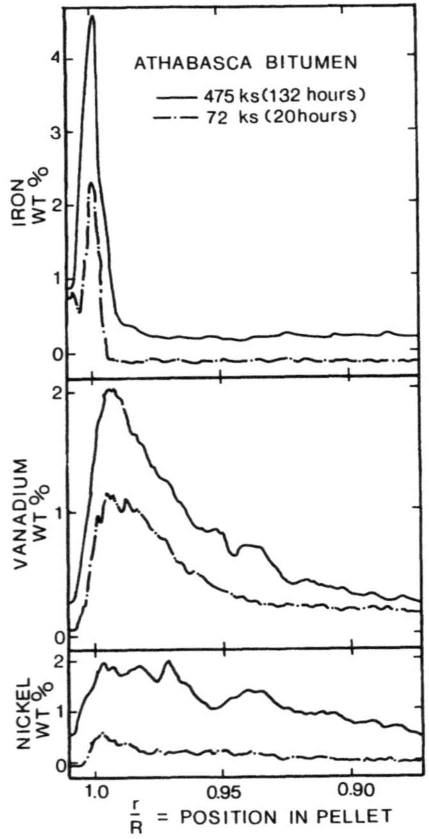

Figure 3 Fe, V and Ni concentrations versus radial position in catalyst pellets used to process Athabasca bitumen

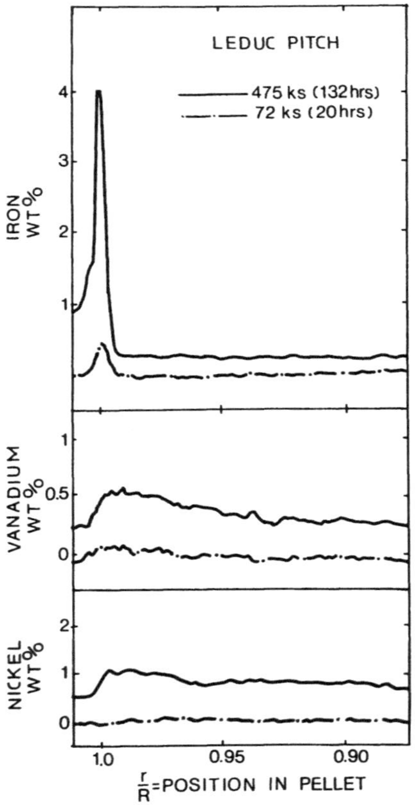

Figure 4 Fe, V and Ni concentrations versus radial position in catalyst pellets used to process Leduc pitch.

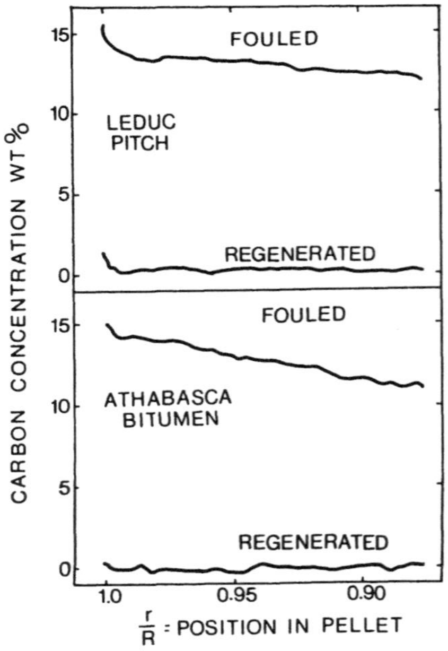

Figure 5 Carbon concentration versus radial position in the catalyst pellet

in Figure 5. It is apparent that carbon deposition is similar with both feedsocks. Furthermore there is very little change in carbon content from the exterior of the catalyst pellet to the interior. This is consistent with Auger results reported by Pollack and co-workers[19]. However these workers also used transmission electron microscopy which is capable of greater resolution. In this way, they were able to determine that on a microscopic scale, the carbon was deposited in patches on the catalyst surface.

DIFFUSION IN CATALYST PORES

Many investigators have reported diffusional limitations in catalysts used for hydroprocessing residual feedstocks [20-22]. Effectiveness factors varying from 0.7 to 0.2 have been reported [23-24].

The Ni and V metals profiles in Figures 3 and 4 can also be interpreted in terms of diffusion. Metals are deposited at the reaction sites at which the large organometallic compounds are converted into smaller molecules. The large quantity of metal deposits at the external edge of the catalyst pellets indicate that the organometallic compounds are reacting near the exterior of the catalyst pellet before they have been able to diffuse to the center.

Most of the measurements on diffusivities of compounds in the liquid phase, in catalyst pores, have been made with small molecules. Recently Chantong and Massoth [26] measured diffusivities of non-metallic porphyrin molecules. Baltus and Anderson [27] made diffusivity measurements using narrow molecular weight fractions of a petroleum residuum. In their work tetrahydrofuran asphaltenes were separated into five fractions by gel permeation chromatography. The polystyrene equivalent average molecular weights they reported were 3000, 6000, 12000, 24000, and 48000.

All of these measurements have been correlated by similar empirical equations having empirical constants. These equations are generally of the form indicated here in Equation 15,

$$D_{eff} = D_o \exp(-4.0\ d_m/d_p) \qquad (15)$$

where, D_{eff} is the effective diffusivity in the catalyst pore, D_o is the diffusivity in bulk solution, d_m is the molecular diameter, and d_p is the catalyst pore diameter. The molecular diameter used was the diameter of the model compounds in studies which used pure compounds. For asphaltene fractions, the molecular diameter was calculated using the Stokes Einstein equation with measured values of bulk diffusivities[27].

Almost all of the experimental diffusion measurements have been made at room temperature, in extremely dilute solutions (less than 3 m %). Information at reaction temperatures, in real solutions containing dissolved hydrogen, is sparse [28,29]. One known complication is that the molecules tend to agglomerate to form micelles when the solutions are not extremely dilute. For example, values from 885 to 10910 have been reported for average molecular weights of Athabasca bitumen asphaltenes, depending on the concentration of the solution used [30]. Experiments performed by Moschopedis et al [31] have shown that as the solution concentration increased the molecular weight increased, presumably as a result of micelle formation. Because the diameter of the diffusing species will increase when micelles are formed (as shown by an increase in molecular weight of the diffusing species), slower diffusion rates will be attained, in accordance with equation 15. However, the higher thermal energies in the molecules at reaction temperature may tend to break up micelles. The effects of these factors on diffusion rates is not known quantatively.

CATALYST MODELLING

A number of papers modelling conversion and deactivation during catalytic residuum hydrocracking have appeared in the literature [32-44]. The conversion has been predicted as a function of pore radius [34,40]. The amount of the deposit has also been predicted as a function of pore radius [33,44]. Profiles have been predicted as a function of radial distance through the catalyst pellet [36]. Deactivation rates have been modelled as a function of catalyst pore size [44] up to the point of pore plugging.

Recent investigations, both experimental measurements [45,46] and simulation studies (42) have indicated that bimodal catalysts have considerable potential. The large pores provide greater rates of diffusion than the smaller pores. However the small pores provide greater surface area on which the conversion reactions can occur. Ideally, if both large and small pores are combined, an optimum can be attained.

REFERENCES

1. Ternan, M., Catalysis, Molecular Weight Change and Fossil Fuels Can. J. Chem. Eng. 61 (1983) 133-147.
2. Aoshima, A. and H. Wise., Hydrodesulphurization Activity and Electronic Properties of Molybdenum Sulphide Catalyst, J. Catal. 34 (1975) 145-151.
3. Wentrcek, P. R. and H. Wise., Defect Control of Hydrogenation Activity of Molybdenum Sulphide Catalyst, J. Catal. 45 (1976) 349-355.
4. Wentrcek, P. R. and H. Wise., Hydrodesulphurization Activity and Defect Structure of Co-Mo Sulphide Catalyst, Preprints, Am. Chem. Soc. Div. Petrol. Chem. 22 (1975) 525-532.
5. Wise, H., Defect Structure of Co-Mo Sulphide Catalyst and Hydrodesulphurization Activity, Proc. 2nd Intern. Conf. Chemistry and Uses of Molybdenum (eds. P. C. H. Mitchell and A. Seaman), p. 160-163, (London, Climax Molybdenum, 1977).
6. Wentrcek, P. R. and H. Wise, Hydrodesulphurization Activity and Defect Structure of Co-Mo Sulfide Catlayst, J. Catal. 51 (1978) 80-85.
7. Heckelsberg, L. F. and R. L. Banks, Electron Transfer Cracking, Preprints, Petrol. Chem. Div., Am. Chem. Soc. 22 (1977) 19-25.
8. Lipsch, J. M. J. G. and G. C. A. Schuit, The $CoO-MoO_3-Al_2O_3$ Catalyst Properties, J. Catal. 15 (1969) 179-189.
9. Schuit, G. C. A. and B. C. Gates, Chemistry and Engineering of Catalytic Hydrodesulphurization, AIChE J 19 (1973) 417-438.
10. Voorhoeve, R. J. H. and J. C. M. Stuiver, Kinetics of Hydrogenation on Supported and Bulk Nickel Tungsten Sulphide Catalysts, J. Catal 23 (1971)

228-235.
11. Wright, C. J., C. Sampson, D. Fraser, R. B. Moyes, P. B. Wells, and C. J. Rickel, Hydrogen Sorption by Molybdenum Sulphide Catalysts,J.Chem. Soc. Faraday I, 76 (1980) 1585-1598.
12. Desikan, P. and C. H. Amberg, Catalytic Hydrodesulphurization of Thiophene, Selective Poisoning and Acidity of the Catalyst Surface, Can. J. Chem. 42 (1964) 843-850.
13. Tanaka, K., T. Okuhara, S. Sato, and K. Miyahara, Evidence for Two Kinds of Active Sites on a Molybdenum Sulphide, J. Catal. 43 (1976) 360-362
14. Massoth, F. E., Studies of Molybdena Alumina Catalysts, Kinetics of Thiophene Hydrogenolysis, J. Catal. 47 (1977) 316-327.
15. Kelly, J. F. and M. Ternan, Hydrocracking Athabasca Bitumen with Alkali Metal Promoted $CoO-MoO_3-Al_2O_3$ Catalysts, Can. J. Chem. Eng. 57 (1979) 726-733.
16. Boorman, P. M., J. F. Kriz, J. R. Brown, and M. Ternan, $Co-Mo/Al_2O_3$ Catalysts in the Hydrocracking of Athabasca Bitumen: Effects of the Presence of Na and F on their Composition and Performance, Proc. 4th Intern. Conf. Chem. Use Molybdenum, (eds. H. F. Barry and P. C. H. Mitchell), p. 192-196 (Ann Arbor, Climax Molybdenum, 1982).
17. Boorman, P. M., J. F. Kriz, J. R. Brown, and M. Ternan, Hydrocracking Bitumen Derived from Oil Sands with Sulphided MoO_3-CoO Catalysts having Supports of Varying Composition, Proc. 8th Intern. Congr. Catal. vol. II, p.281-291, (Frankfurt, Verlag Chemie 1984).
18. McColgan, E. C. and B. I. Parsons, "The Hydrocracking of Residual Oils and Tars, Part 6 Catalyst De-activation by Coke and Metals Deposition," Mines Branch Research Report R-273, Department of Energy Mines and Resources Canada, Ottawa, 1974.
19. Sanders, J. V., J. A. Spink, and S. S. Pollack, The Structure of Coke Deposits on HDS Catalysts, Appl. Catal. 5 (1983) 65-84.
20. Ternan, M., Catalytic Hydrogenation and Asphaltene Conversion of Athabasca Bitumen, Can. J. Chem. Eng. 61 (1983) 689-696.
21. Green, D. C. and D. H. Broderick, Residuum Hydroprocessing, Chem. Eng. Progr. 77(12) (1981) 33-39.
22. Dautzenberg, F. M., J. Van Klinken, K. M. Pronk,

S.T. Sie, and J.B. Wiffels, Catalyst Deactivation through Pore Mouth Plugging during Residue Desulphurization, in "Chemical Reaction Engineering - Houston" (eds. V. W. Weekman and P. Luss), p. 254-267, ACS Symposium Series 55, 1978.
23. Shah, Y. T. and J. A. Paraskos, Intraparticle Diffusion Effects in Residue Hydrodesulphurization, Ind. Eng. Chem. Proc. Des. Dev. 14 (1975) 368-372.
24. Newson, E., Catalyst Deactivation Due to Pore Plugging by Reaction Products, Ind. Eng. Chem. Proc. Des. Dev. 14 (1975) 27-33.
25. Hardin, A. H. and M. Ternan, The Effects of Pore Size in MoO_3-CoO-Al_2O_3 Hydrocracking Catalysts for Bitumen Conversion, Preprints, 2nd World Congr. Chem. Eng., 3 (1981) 134-138.
26. Chantong, A. and F. E. Massoth, Restrictive Diffusion in Aluminas, AIChE J. 29 (1983) 725-731.
27. Baltus, R. E. and J. L. Anderson, Hindered Diffusion of Asphaltenes through Microporous Membranes, Chem, Eng. Sci. 38 (1983) 1959-1969.
28. Thrash, R. J. and R. H. Pildes, The Diffusion of Petroleum Asphaltenes through Well Characterized Porous Membranes, Preprints, Div. Petrol. Chem. Am. Chem. Soc. 26(2) (1981) 515-525.
29. Seo, G. and F. E. Massoth, Effect of Pressure and Temperature on Restrictive Diffusion of Solutes in Aluminas, AIChE J. 31 (1985) 494-496.
30. Champagne, P. J., E. Manolakis, and M. Ternan, Molecular Weight Distribution of Athabasca Bitumen Fuel, 64 (1985) 423-425.
31. Mochopedis, S. E., J. F. Fryer, and J. G. Speight, Investigation of Asphaltene Molecular Weights, Fuel, 55 (1976) 227-232.
32. Nam I. S. and J. R. Kittrell, Use of Catalyst Coke Content in Deactivation Modeling, Ind. Eng. Chem. Proc. Des. Dev., 23 (1984) 237-242.
33. Oyekunle, L. O. and R. Hughes, Metal Deposition in Residuum Hydrodesulphurization Catalysts, Chem. Eng. Res. Dev., 62 (1984) 339-343.
34. Rajagopaian, K. and D. Luss, Influence of Catalyst Pore Size on Demetallation Rate, Ind. Eng. Chem. Proc. Des. Dev., 18 (1979) 459-465.
35. Beeckman, J. W. and G. F. Froment, Catalyst Deactivation by Site Coverage and Pore Blockage, Chem. Eng. Sci., 35 (1980) 805-815.
36. Chiou M. J. and J. H. Olson, Effect of Catalyst Pore Structure on the Hydroprocessing of Coal Derived Liquids, Am. Chem. Soc. Preprints Div

Petrol. Chem., 23(4) (1978) 1421-1436.
37. Do D. D., Catalytic Conversion of Large Molecules: Effect of Pore Size and Concentration, AIChE J., 30 (1984) 849-853.
38. Ahn, B. J. and J. M. Smith, Deactivation of Hydrodesulphurization Catalysts by Metals Deposition, AIChE J., 30 (1984) 739-746.
39. Nalitham, R. V., A. R. Tarrer, J. A. Guin and C. W. Curtis, Application of a Catalyst Deactivation Model for Hydrotreating Solvent Refined Coal Feedstocks, Ind. Eng. Chem. Proc. Des. Dev., 22 (1983) 645-653.
40. Haynes, H. W. and K. Leung, Catalyst Deactivation by Pore Plugging and Active Site Poisoning Mechanisms, Chem. Eng. Commun., 23 , (1983) 161-179.
41. Chrostowski, J. W. and C. Georgakis, Pore Plugging Model for Gas Solid Reactions, Am. Chem. Soc. Symp. Ser., 65 (1978) 225-237.
42. Hughes, C. C. and R. Mann, Interpretation of Catalyst Deactivation by Fouling from Interactions of Pore Structure and Foulant Deposit Geometrics, Am. Chem. Soc. Symp. Ser., 65 , (1978) 201-213.
43. Papayannakos, N. and J. Marangozis, Kinetics of Catalytic Hydrodesulphurization of a Petroeum Residue in a Batch Recycle Trickle Bed Reactor, Chem. Eng. Sci. 39 (1984) 1051-1061.
44. Leung, K. and H. W. Haynes, Catalyst Deactivation by Pore Plugging and Active Site Poisoning Mechanisms: Parallel Poisoning in Bidisperse Structured Catalyst, Chem. Eng. Commun. 31 (1984) 1-20.
45. Plumail, J. C., Y. Jacquin and H. Toulhoat, Influence of the Carrier Pore Structure on the Initial Activity of Co-Mo/Al_2O_3 Catalysts in the Hydrotreatment of a BOSCAN Heavy Crude: Methodology and Results, Proc. 4th Intern. Conf. Chem. Use of Molybdenum (eds, H. F. Barry and P. C. H. Mitchell), p.389-393, (Ann Arbor, Climax Molybdenum, 1982).
46. Suzuki, M. and K. Onuma, Hydrotreating Catalysts for Heavy Oils: Reaction Performance and Aging Behavior of Bimodal Alumina Based Catalysts for Desulfurization and Demetallation, J. Japan Petrol. Inst. 27(5) (1984) 420-428.

KINETIC DATA ANALYSIS AND PARAMETER ESTIMATION

Prof. Dr. Hanns Hofmann

University of Erlangen-Nürnberg, Germany

1. INTRODUCTION

The rational design, simulation and optimization of an industrial reactor has to be based on accurate, reliable data for mass, energy and momentum transport as well as chemical kinetics. For this purpose specially designed laboratory or pilot experiments are needed, as only in very rare cases is a theoretical evaluation possible. The aim of any kinetic data analysis in reaction engineering is to find adequate reaction model equations and to arrive at reliable parameter estimates dependent on the operating variables of the reactor, like pressure, temperature, flow conditions etc. These parameters then will be used in a reactor model, i.e., a set of equations that describe the behavior of a reactor with sufficient accuracy.

The scope of this paper is to discuss the present status of the art and to pinpoint problems in kinetic data analysis and parameter estimation. It can be regarded as a continuation of an earlier paper concerning the same subject /1/ and is concentrated on heterogeneous catalytic vapor phase reactions. Since most industrial reacting systems are of sufficient complexity, the experimental program can be very expensive and special attention should be given to the correct experimental set-up. Furthermore, as experimental results always contain errors, the experiments should be designed for error minimization. Finally, the data analysis has to arrive at parameters allowing a safe but straight forward design, simulation and optimization of the technical unit.

Fluid phase

$$u\frac{\partial y_i}{\partial z} = \varepsilon D_{er}\left[\frac{1}{r}\frac{\partial y_i}{\partial r}+\frac{\partial^2 y_i}{\partial r^2}\right] + k_s a_s (y_i - f_{is})$$

$$u\rho c_p\frac{\partial \vartheta}{\partial z} = \lambda_{er}\left[\frac{1}{r}\frac{\partial \vartheta}{\partial r}+\frac{\partial^2 \vartheta}{\partial r^2}\right] + h_s a_s (\Theta_s - \vartheta)$$

Catalyst pellet

$$\frac{d^2 f_i}{dy^2}+\frac{2}{y}\frac{df_i}{dy} - \left[\Phi^2 f_i^n \exp\left\{\gamma\left(1-\frac{1}{\theta}\right)\right\}\right] = 0$$

$$\frac{d^2 \theta}{dy^2}+\frac{2}{y}\frac{d\theta}{dy} + \left[\beta\Phi^2 f_i^n \exp\left\{\gamma\left(1-\frac{1}{\theta}\right)\right\}\right] = 0$$

Φ = Thiele-modulus β = Prater-number γ = Arrhenius-number

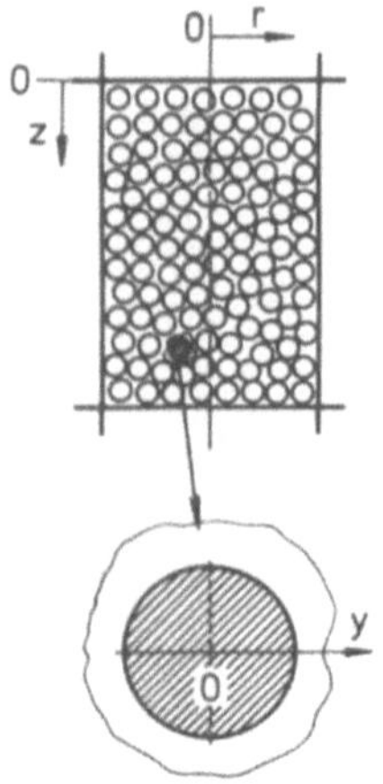

Fig. 1: Two-dimensional heterogeneous reactor model for a catalytic fixed-bed reactor

2. A FUNDAMENTAL DILEMMA IN KINETIC DATA ANALYSIS

A fundamental dilemma in kinetic data analysis can be demonstrated by the well-known static two-dimensional heterogeneous reactor model for the industrially important catalytic fixed-bed reactor (Fig. 1) /2/.

The model consists of a set of partial differential equations for the fluid as well as for the solid catalyst phase with special parameters for the different mechanisms acting in the reactor, such as forced convection, dispersion, mass and heat transfer and the chemical conversion. In case of a complex reaction like the butene oxidation to maleic anhydride /3/ depicted in Figure 2 (with 18 identified reaction steps and even more reactants), the reaction rate term in the reactor model becomes very complex and the number of equations is increased since for any stoichiometrically independent reactant a separate mass balance in both phases is needed. Also an equation describing the pressure drop across the catalyst bed has to be added.

A detailed analysis of the radial heat transfer in the catalyst bed /4,5/ shows (see Fig. 3) that e.g. the dispersion parameters in the balance equations for the fluid phase in Fig. 1 are dependent on other parameters characteristic of more fundamental transport mechanisms and, moreover, the heat conduction through the porous particles itself consists again of four basic mechanisms (Fig. 4), each having its own characteristic parameter /6/.

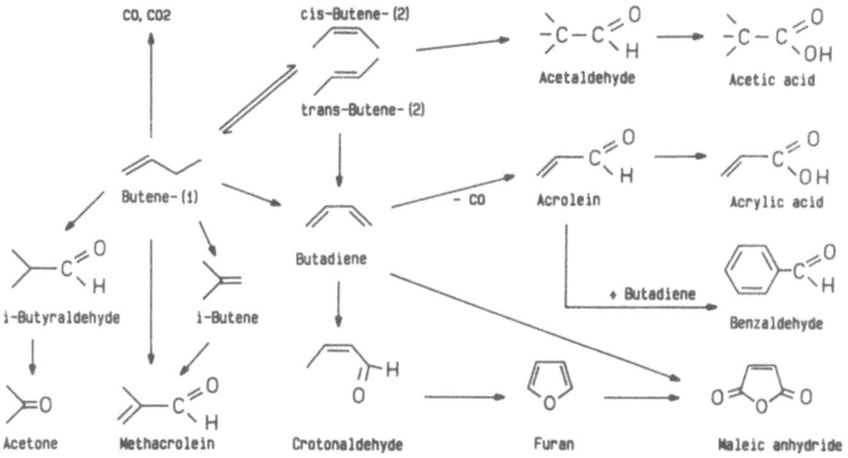

Fig. 2: Complex reaction scheme for the catalytic oxidation of butene

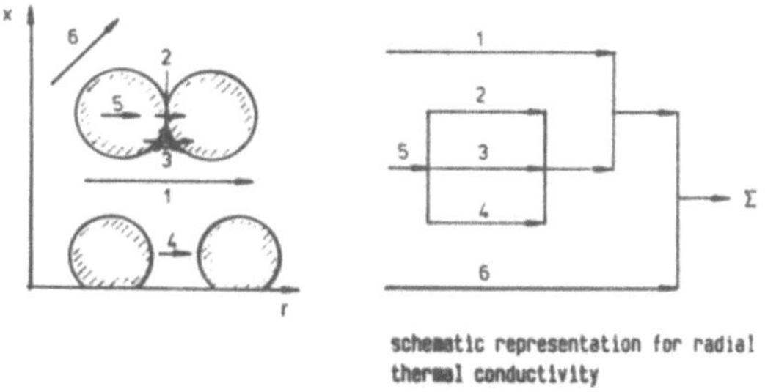

1 radiation and heat conduction between neighbour voids of the packing (λ, α_{SM})
2 conduction through the contact surface between particles (α)
3 conduction in the stagnant film in the vicinity of the contact surface (λ_M)
4 radiation from particle to particle (α_{SFM})
5 conduction through the particles (λ_p)
6 heat transport by forced convection

$$\frac{\lambda_{er}}{\lambda_{fluid}} = \varepsilon\left[1 + \beta\frac{\alpha_{SM}\cdot dp}{\lambda_M}\right] + \frac{\beta(1-\varepsilon)}{\frac{1}{\Phi} + \frac{dp}{\lambda_M}(\alpha + \alpha_{SP})\frac{2\lambda_M}{3\lambda_F}} + \delta\cdot\gamma\cdot Re_0\cdot Pr$$

$\beta = 1.0$ and $\Phi = \Phi_1(\lambda_p/\lambda_M)$ for loose packing

$\beta = .895$ and $\Phi = \Phi_2(\lambda_p/\lambda_M)$ for dense packing

$\delta\,\gamma = 0.10 - 0.16$

Fig. 3: Mechanisms contributing to the effective radial thermal conductivity (for nomenclature see list of symbols)

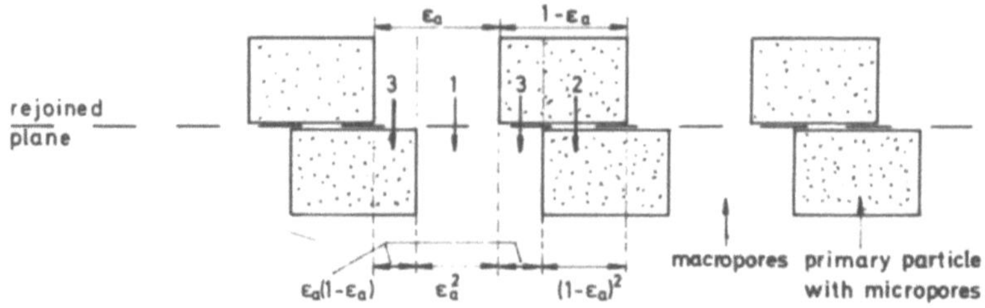

4 Mechanisms for heat conduction in bidisperse porous particles

$$\lambda_p = \underbrace{\varepsilon_a^2 \lambda_a}_{\text{mech. 1}} + \underbrace{(1-\varepsilon_a)^2 \lambda_i}_{\text{mech. 2}} + \underbrace{2\varepsilon_a(1-\varepsilon_a)^2 \lambda_{series}}_{\text{mech. 3}} + \underbrace{(1-\varepsilon_a)^2 \lambda_s}_{\text{solid conduction}}$$

$$\lambda_a = \frac{1}{\frac{1}{\lambda_m} + \frac{2\beta L}{d_a \cdot \lambda_m}} \qquad \lambda_i = \frac{\varepsilon_i^2/(1-\varepsilon_a)^2}{\frac{1}{\lambda_m} + \frac{2\beta L}{d_i \cdot \lambda_m}} \qquad \lambda_{series} = \frac{2}{\frac{1}{\lambda_i + \lambda_s} + \frac{1}{\lambda_a}} \qquad \lambda_s = \frac{\varepsilon_s^2}{(1-\varepsilon_a)^2} \cdot \lambda_s'$$

$$\log(\lambda_s' \cdot 10^5) = 0.859 + 3.12\,\frac{\lambda_{SW}}{\sigma}$$

Fig. 4: Effective thermal conductivity of porous solids according to the random pore model (for nomenclature see list of symbols)

This shows that from a practical point of view the data analysis in reaction engineering never can hope to estimate the characteristic parameters of all basic transport and reaction steps. Neither would it be possible to study experimentally each basic step isolated, nor does it seem possible to perform enough experiments to get reliable values for these basic parameters.

Therefore, it is good engineering practice to use in reactor models so-called effective parameters by lumping several basic mechanisms or reaction steps into an overall phenomenon described by one single term in the respective reactor or reaction model. In this way, the parameters are always bound to the model equations by which they have been defined. The parameter values determined by regression analysis cannot be used independently in other model equations like physical properties, flow velocity etc., measurable independently without any model definition. This fundamental dilemma has some consequences in kinetic data analysis and parameter estimation that should always be kept in mind:

- First, a priori criteria are needed for a rational lumping of basic mechanism and/or model simplification (see section 3).

- Secondly, the experimental set-up should be selected in such a way that kinetic parameters can be determined under conditions as close as possible to the technical conditions (see section 4).

- Thirdly, transport parameters for heat and mass determined under non-reacting conditions should be used with care in reactor design /7/.

3. RATIONAL MODEL REDUCTION

On the other hand, the dilemma described above is mitigated by the fact that according to experience in experimental set-ups used today in research and development, kinetic parameters can never be determined more precisely than $\pm$ 5 to 10 % relative. Therefore, very often because of economic reasons sophisticated details cannot be considered in practical reactor modeling. It is even questionable whether in practical cases a reactor model with such detail as the two-dimensional heterogeneous model given in Fig. 1 is necessary, in view of the experimental and computational effort needed for its parameter estimation and reactor simulation. On the contrary, a priori criteria are welcome, according to which the importance of a specific term in the reactor model can be estimated to decide whether the term can be dropped or not. Some typical examples of such criteria based on the assumption of less than 5 % deviation are depicted in Fig. 5.

$$\left|\frac{(-\Delta H_R)\, r_{eff}\, d_p}{2\, h_s T_s}\right| < \frac{0{,}15}{Ar} \quad \text{and} \quad \frac{r_{eff}\, d_p}{2 k_s c_s} < \frac{0{,}15}{n}$$

$$Da\text{II}\, \left| n - Ar\, Pt \right| < 1$$

$$\left| \frac{n\, Da\text{I}}{Bo_a} - \frac{Ar\, Da\text{III}}{Pe_a} \right| < 0{,}05$$

$$Pe_a\, Pe_r\, (d_t/d_p)^2 < \frac{16}{1 + 8/Bi_w}$$

Fig. 5: Typical examples of a priori criteria developed for simplifications of heterogeneous catalytic fixed-bed reactor models (for nomenclature see list of symbols)

If the first two inequalities hold /8,9/, it can be expected that heat and mass transport outside the catalyst particle is not limiting and therefore there is no need for a heterogeneous model; a pseudo-homogeneous model may be sufficient, in which the whole catalyst particle is regarded as a sink and/or source for heat and mass in the fluid phase. This results in a drastic reduction of the number of model equations, and the transfer terms in the fluid phase equations are lumped into effective reaction rate terms. If the second inequality holds /9/, pore diffusion is not limiting. The third inequality is valid if axial dispersion can be neclected /10/ and the fourth if radial dispersion is of no importance /10/.

In the case of complex reactions, usually also a simplification of the reaction model is needed to arrive at workable rate terms with a limited number of regression parameters. This is all the more, necessary, as kinetic parameters like the preexponential factor and activation energy of a reaction step are strongly correlated to each other /11/. Wei /12/ and many other authors /13,14,15/ have developed lumping rules for complex reactions and shown the consequences on the overall kinetics, resulting from the so-called exact, approximate or proper lumping /16/. Unfortunately, these rules are limited mainly to first order steps, whereas in many cases of practical importance, rate laws of different orders prevail.

The rational basis of any reaction lumping is the concentration-reactiontime-diagram of the respective reaction as it is derived from experimental results /17/ as e.g. in Fig. 6 for the butene oxidation.

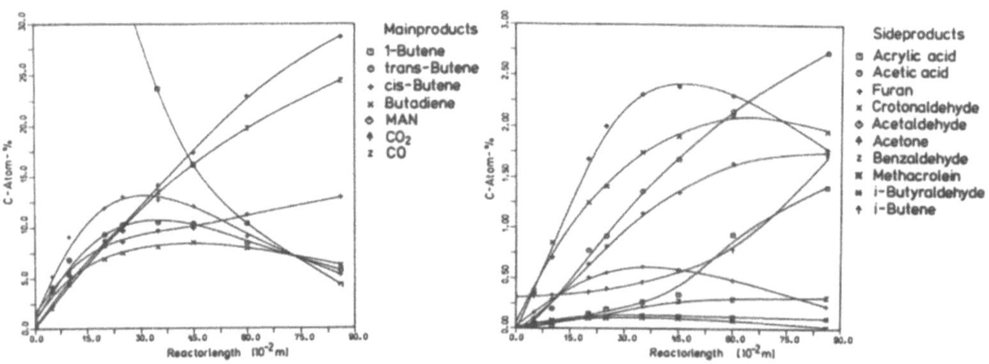

Fig. 6: Concentration-reactiontime-diagrams for the butene oxidation (Inlettemperature 663 K, P_2O_5/V_2O_5 - catalyst, V=420Nl/h)

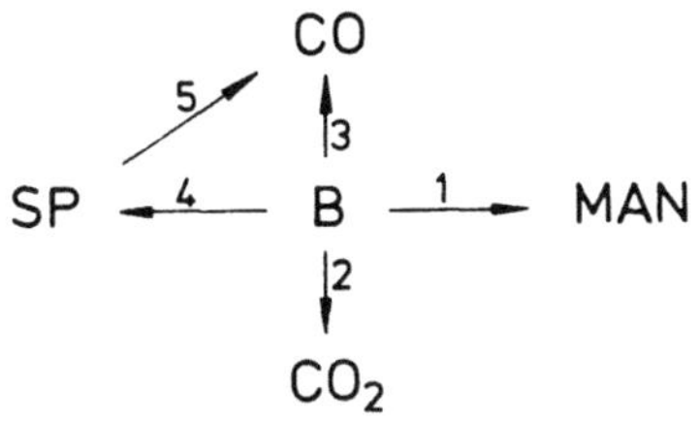

B = butene-(1) + butene-(2) + butadiene + furan + crotonaldehyde
SP = sideproducts (aldehydes, ketones, carbonic acids)

1.	B + 3 O_2 -----> MAN + 3 H_2O	1315 kJ/mole
2.	B + 6 O_2 -----> 4 CO_2 + 4 H_2O	2540 kJ/mole
3.	B + 4 O_2 -----> 4 CO + 4 H_2O	1404 kJ/mole
4.	B + O_2 -----> 2 SP	316 kJ/mole
5.	SP + 1.5 O_2 -----> 2 CO + 2 H_2O	544 kJ/mole

Fig. 7: Simplified reaction scheme for the catalytic oxidation of butene on a catalyst

Usually all intermediates having low concentrations are lumped together as they are consumed rather rapidly and are not limiting. But on the other hand, as a general rule, at least three independent lumps are defined, one for the initial material (educt), one for the desired main product and one consisting of the sum of all side products. If special side products are involved, like CO, CO_2 and H_2O, it may be of interest to keep them separate and hence have additional reactions. This has been done in the catalytic oxidation of butene (Fig. 7) /2/, where all intermediates on the butene-maleic anhydride route have been lumped together with butene in the B-lump, all side products into the SP-lump, but CO and CO_2 have been kept separate, because once they have been formed from a C_4-molecule they are of no value whereas these steps contribute significantly to the energy balance of the reaction.

4. EXPERIMENTAL TOOLS

Concerning the experimental determination of effective transport and/or chemical parameters, one has to distinguish between four different categories, i.e.

- packing related parameters, like effective radial thermal conductivity λ_{er} or wall heat transfer coefficient h_w
- particle related parameters, like pellet thermal conductivity λ_p or effective pore diffusivity D_{eff}
- exchange parameters between phases, like volumetric heat transfer coefficient $h_s a_s$ or volumetric mass transfer coefficient $k_s a_s$

- reaction related parameters, like rate constants k, activation energy E or adsorption constants K.

For each type of parameter special laboratory equipment is in use, aiming to determine the different parameters under the most realistic reaction conditions.

Packing related parameters certainly should be determined in highly instrumented modular integral reactors /6,18/, equipped with special probes for radial as well as axial temperature and concentration measurements as shown in Fig. 8.

In order to minimize the experimental error, special probes were used: 25 miniature thermocouples located at 9 different radial positions introduced in an asbesteous honeycomp in a single crosssection of a 50 mm diameter column or 6 capillaries for concentration sampling at 6 radial positions. The large amount of simultaneously obtained data was recorded automatically on a data storage device and subsequently processed with the help of the computer program FIBSAS /19/ for a two-dimensional pseudo-homogeneous dispersion model of a fixed-bed reactor.

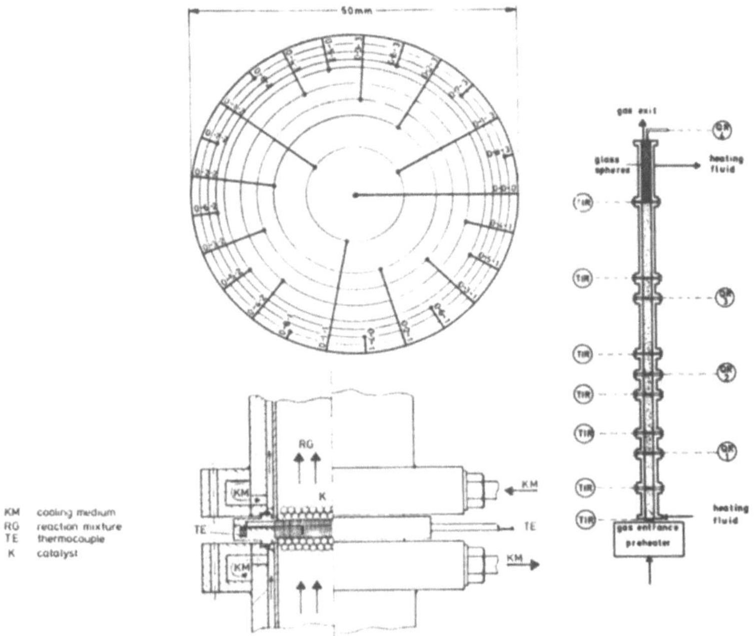

Fig. 8: Laboratory integral reactor for the experimental determination of packing parameters (top left: distribution of thermocouples, bottom left: detail of probe crossection, right: whole reactor

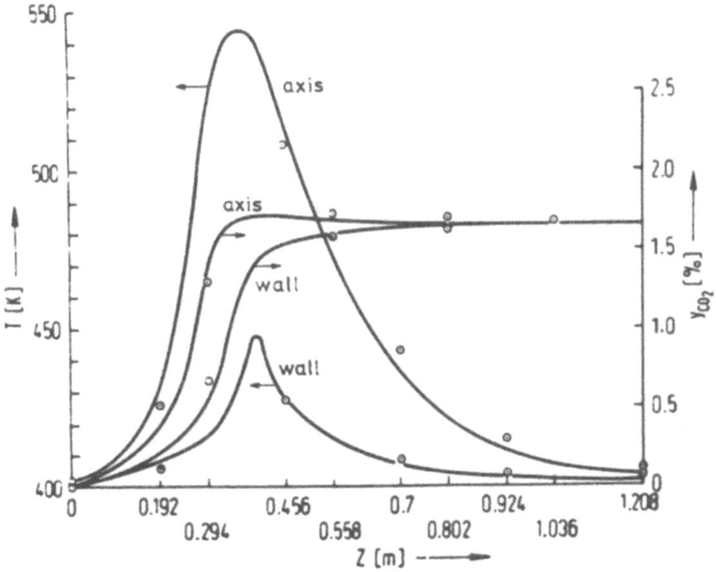

Fig. 9: Temperature and concentration profiles along the axis and near the wall of a fixed-bed reactor

To measure e.g. heat transfer parameters under reacting and non-reacting conditions, the operating conditions of the reactor were selected so that in the first zone of the 1200 mm long packed bed, chemical reaction was dominant, whereas in the second zone pure heat transfer took place. The experimental data shown in Fig. 9 relate to the oxidation of CO with air using 5 mm particles of silica supported CuO-catalyst. The radial temperature and concentration profiles in the different measurement planes are depicted in Figure 10.

These curves clearly reflect the stronger reaction intensity along the reactor axis and the subsequent equalization of concentration with a slight fall in the product concentration (CO_2) towards the end of the reactor.

From the measured temperature profiles in the zone of pure heat transfer, λ_{er} and h_w were determined first. Next, the optimum parameter combination (k_o^w, E, λ_{er} and h_w) for the reaction zone was estimated to determine to what extent the heat transfer parameters would be adversely affected by possible weaknesses of the model /18/.

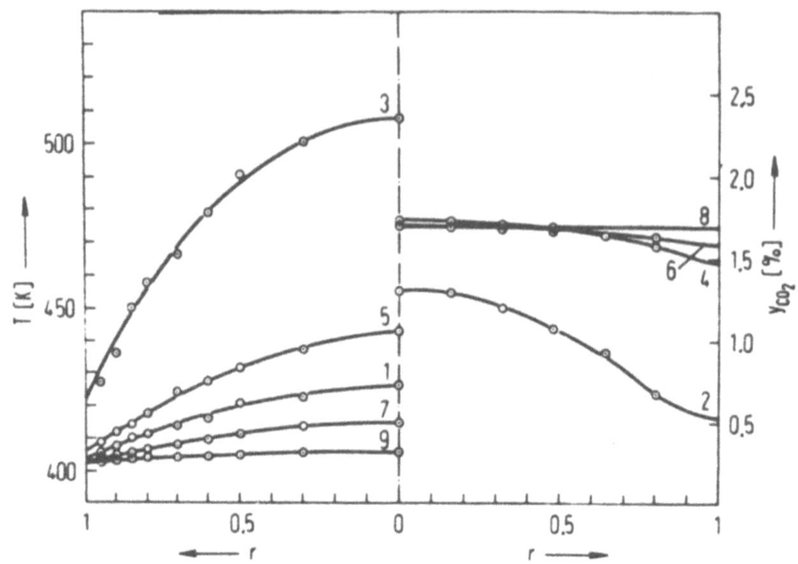

Fig. 10: Radial temperature and concentration distributions in a 50 mm diameter packed bed (numbers belong to the different cross sections shown in Fig. 8).

Fig. 11 shows as one result of this procedure the dimensionless effective radial thermal conductivity obtained by regression with FIBSAS for reacting conditions compared to values calculated for non-reacting conditions. Due to the interaction between the heat transfer and the chemical kinetic parameters, both λ_{er} and h_w changed by about 20 to 30 %. The deviation increased with decreasing flux of reactants, possibly due to lumping of further (heterogeneous) effects. As these interactions vary from reaction to reaction, their origins may also be different and practically impossible to identify.

Particle related parameters should be determined in a so-called single pellet reactor /20,21/ as shown schematically in Fig. 12 for the determination of effective pellet diffusivity D_{eff}. Similar equipment is in use for the determination of the effective pellet thermal conductivity /22/.

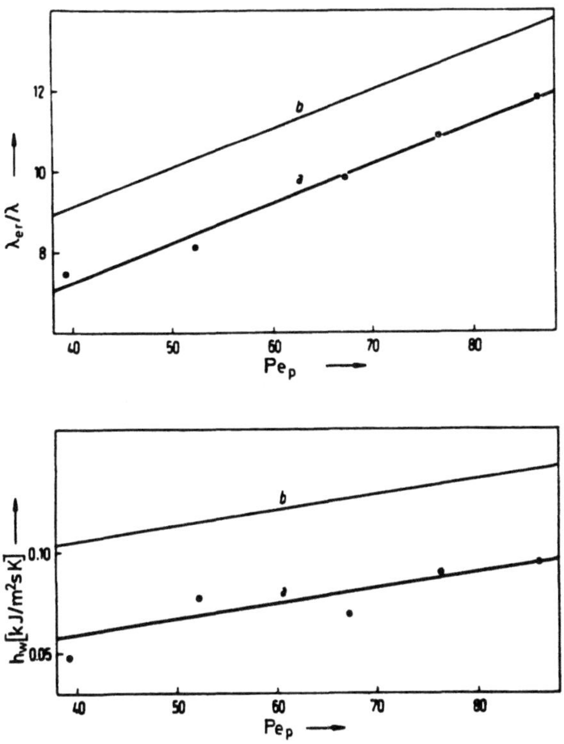

Fig. 11: Dimensionless effective radial thermal conductivity λ_{er} and effective wall heat transfer coefficient h_w as function of Peclet-number (a: measured under reacting conditions, b: calculated according to /42/for non-reacting conditions)

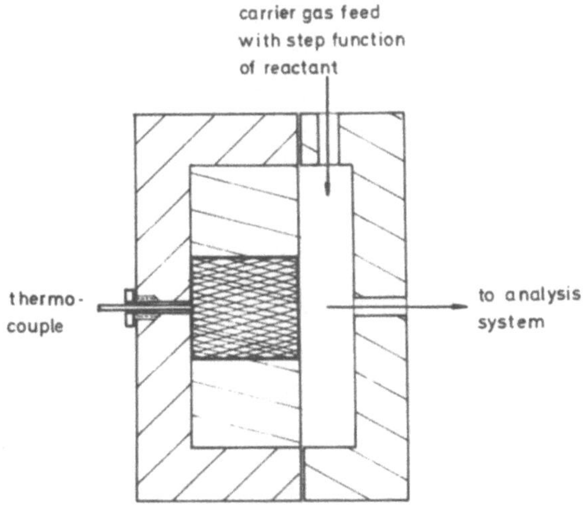

Fig. 12: Single pellet reactor for determination of effective pellet diffusivity under reaction conditions

		integral reactor	gradientless reactor	differential reactor	special types
	stationary operated	standard industrial laboratory reactor, catalyst desactivation, single tube of technical multitube reactor	model discrimination, parameter estimation	model discrimination, parameter estimation	identification of intermediates by infrared cells
instationary operated	concentration pulse	catalyst screening, catalyst activity, product spectrum	catalyst screening, catalyst activity, prod.spectr.	determination of transport and kinetic parameters	---
instationary operated	concentration step	evaluation of the reaction scheme microkinetics	sorption kinetics	---	---
instationary operated	temperature ramp	---	---	evaluation of elementary steps	---

Table I: Classification of laboratory reactors for evaluation of reaction related parameters

For the evaluation of reaction related parameters, a great variaty of laboratory reactors has been developed which can be classified according to the matrix given in Table I.

Dependent on the particular purpose, integral, gradientless, differential or other reactor types are preferred. Also the type of operation - stationary or non-stationary - may be different. In industry the standard laboratory reactor for heterogeneous catalytic gasphase reactions is the stationary operated integral reactor. E.g., continuously operated infrared flow cells filled with catalyst and heated up to reaction temperature are used in research laboratories to identify reaction intermediates chemisorbed on the catalyst surface. Also, non-stationary operated gradientless reactors are used to study sorption kinetics. As a general rule the more simplified the data analysis,the more sophisticated the type of reactor and its operation has to be. E.g., a gradientless reactor (also called recycle reactor) mainly used in kinetic model screening and parameter estimation needs a temperature resistant recirculation pump and different feed compositions (see Fig. 13), but can be modeled by simple algebraic equations, whereas an integral reactor is most easily operated but can be modeled only by a set of differential balance equations like those given in Fig. 1.

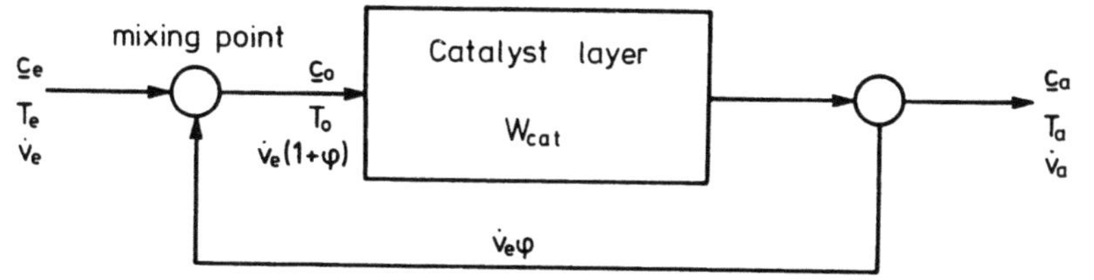

effective rate of reaction for component i (for reactions without volume change)

$$r_{i,eff}(\underline{c}_a, T_a) = \frac{c_{i,a} - c_{i,e}}{v_i\,(1+\varphi)\,\tau'} \qquad \left[\frac{\text{kmole}}{\text{m}^3\,\text{s}}\right]$$

for sufficiently hight reflux ratio φ:

$$r'_{i,eff}(\underline{c}_a, T_a) = \frac{\dot{v}\,(c_{i,a} - c_{i,e})}{v_i\,W_{cat}} \qquad \left[\frac{\text{kmole}}{\text{kg s}}\right]$$

Fig. 13: Gradientless reactor and its basic equations (for nomenclature see list of symbols)

Today a trend can be seen to computer aided experimentation. Not only the operation and control of the reactor but also automatic data sampling, analysis and statistically based sequential design of the experiments are performed by micro-computers /23,24,25/. In general, the experimental effort is appreciable and any method is welcome which can reduce the number of necessary experiments without loss of required accuracy.

5. STATISTICALLY BASED METHODS FOR EXPERIMENTAL DESIGN AND DATA ANALYSIS

A generally accepted method to reduce the number of experiments required for a given accuracy of parameters to be estimated or models to be screened is the statistically based design and analysis of experiments /26/. This technique also has the advantage of providing simultaneously a measure of reliability of the estimates. This is an important point, because in the literature very often (still today) parameter values can be found without any indication of confidence limits. Fig. 14 shows two different confidence limits in use and two criteria for the sequential design of experiments for parameter estimation to minimize the joint linearized confidence region or to maximize the smallest eigenvalue of the squared Jakobian ($\underline{J}^T\underline{J}$), respectively. The basic philosophy is to get the highest increase in parameter precision with one additional experiment. The Jakobian is formed by the n x p matrix of the model derivatives with respect to the p parameters /27,28/.

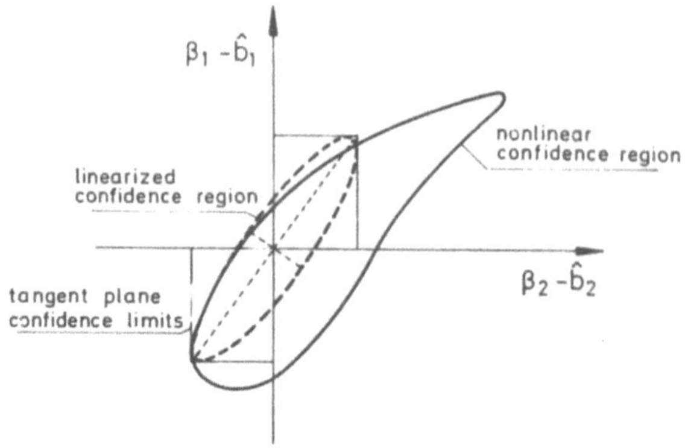

Criteria:

a) Minimization of the joint linearized confidence region (Box - Lukas)

$$\det(\underline{J}^T\underline{J}) \overset{!}{=} \underset{\underline{x}}{Max}$$

b) Maximization of the smallest eigenvalue of the linearized confidence region (sphericity) (Hosten)

$$\lambda_{min(\underline{J}^T\underline{J})} \overset{!}{=} \underset{\underline{x}}{Max}$$

Fig. 14: Confidence limits and criteria for the sequential design of experiments in parameter estimation.

Both criteria can be used to gain the highest information with the smallest number of experiments.

The regression analysis of experimental data then is performed according to one of the objective functions given in Fig. 15 for non-linear regression in different multiresponse situations. The selection of a specific objective function depends on the information available a priori concerning the different variances as given in the right column of Fig. 15 /29/. Naturally, from the statistical point of view, the different criteria are more efficient if more information about the variances is available.

In the course of non-linear regression itself and the statistical design of experiments, one is very often confronted with severe numerical problems in optimization (and integration) caused by the shape of the response surface of the objective function /30/.

General assumptions: $\varepsilon_{iu} \sim N(0, \sigma_{iu}^2)$

$E(\varepsilon_{iu}\,\varepsilon_{jv}) = \sigma_{uv}\,\delta_{ij}$

Criteria	special assumptions
$\psi_1 = \sum_{u=1}^{r} \sum_{v=1}^{r} \sigma^{uv}\left[\sum_{i=1}^{n} (y_{iu} - \eta_{iu})(y_{iv} - \eta_{iv})\right]$	all σ_{uv} known; not necessarily zero
$\psi_2 = \sum_{u=1}^{r} \frac{1}{\sigma_{uu}} \sum_{i=1}^{n} (y_{iu} - \eta_{iu})^2$	σ_{uu} known; $\sigma_{uv} = 0;\ u \neq v$
$\psi_3 = \sum_{u=1}^{r} \sum_{i=1}^{n} (y_{iu} - \eta_{iu})^2$	$\sigma_{uu} = \sigma^2;\ \sigma_{uv} = 0;\ u \neq v$
$\psi_4 = \sum_{u=1}^{r} \ln \sum_{i=1}^{n} (y_{iu} - \eta_{iu})^2$	σ_{uu} unknown, $\sigma_{uv} = 0;\ u \neq v$
$\psi_5 = \det(\underline{D})$	all σ_{uv} unknown, not necessarily zero
$\psi_6 = \det(\underline{D}) \overset{!}{=} \underset{\underline{\beta}, \underline{y}_m}{\mathrm{Min}}$	all σ_{uv} unknown; not necessarily zero few data points missing
$\psi_7 = \prod_{u=1}^{r} (c_{uu})^{n_u/2}$	σ_{uu} unknown; $\sigma_{uv} = 0;\ u \neq v$; many data points missing

Fig. 15: Objective functions for non-linear parameter estimations in multiresponse situations

According to our experience these can be overcome using a combination of different methods; e.g. either

- a so-called "optical optimization" /31/ where the engineer interacting with a graphic display in a first step can change the numerical values of the parameters at will and only in a second step mathematical optimization routines like the Marquardt algorithm /32/ are used for the final fit, or

- a linearization of the underlying model in the first step to provide initial guesses for the parameters followed by an optimization algorithm applied to the non-linear model, or

- the use of a relatively robust routine like the "Complex" algorithm /33/ or even an evolutionary method /34,35/ followed by Marquardt's routine

Concerning the numerical routines available for simultaneous parameter estimation in differential equations with non-linear boundary values, like those given in Figure 1, an interesting review about practical experiences can be found in /36/.

The situation is even more complex in the case of reaction related parameters. Independent of the appropriate reactor model, for complex reactions not only the reaction scheme but also the adequate type of rate equation for each reaction step has to be chosen before the parameters can be estimated. As a rule, in reaction engineering only the analytically measureable reactants (and reactions) should form the basis of the reaction network in contrast to physico-chemical research where the "true" reaction mechanism (involving radical intermediates or active complexes) is sought. Certainly, the stoichiometry of the experimental product spectrum is important, but also the concentration/reactiontime dependencies like those given in Figure 6 are helpful. In contrast to parameter estimation and model discrimination, there exists no unique and straight forward analytic procedure for the built-up of even a simplified reaction scheme. The intuition of the chemical reaction engineer is therefore heavily relied upon.

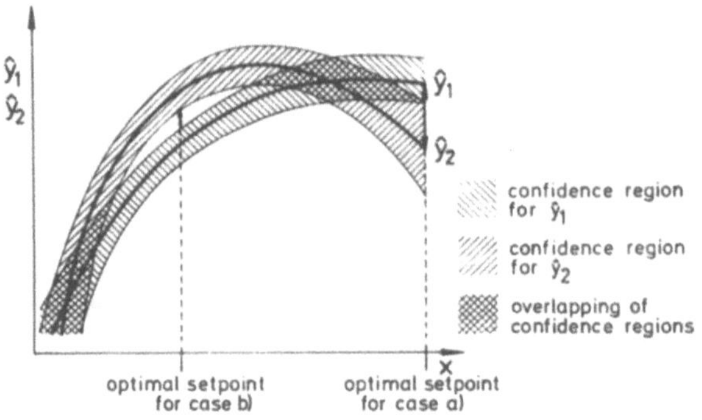

Criteria:

a) maximum difference between calculated responses

$$(\hat{y}_1 - \hat{y}_2)^2 \overset{!}{=} \underset{x}{\mathrm{Max}}$$

(Hunter-Reiner)

b) maximum difference between the confidence region of the responses

$$(\hat{y}_1 - \hat{y}_2)^2 \cdot \left(\frac{1}{\sigma^2+\hat{\sigma}_1^2} + \frac{1}{\sigma^2+\hat{\sigma}_2^2}\right) + \frac{(\hat{\sigma}_2^2 - \hat{\sigma}_1^2)^2}{(\sigma^2+\hat{\sigma}_1^2)(\sigma^2+\hat{\sigma}_2^2)} \overset{!}{=} \underset{x}{\mathrm{Max}}$$

(Box - Hill)

or

$$(\hat{y}_1 - \hat{y}_2)^2 \cdot \frac{1}{2\sigma^2 + \hat{\sigma}_1^2 + \hat{\sigma}_2^2} \overset{!}{=} \underset{x}{\mathrm{Max}}$$

(Buzzi-Ferraris)

Fig. 16: Design criteria for discrimination between different models (y_1, y_2: response of model 1 and 2 respectively)

Once the reaction scheme is available, an appropriate rate equation for each step of the network must be found. For this purpose, a very effective sequential model discrimination technique is available, also based on a statistical approach. The basic idea is to design the next experiment at a level of the independent variables such that the expected difference between the responses y_i of rival models /37/ or their confidence regions /38,39/ is a maximum (see Fig. 16).

The most simple and easy to use criterion is that of Hunter and Reiner /37/. The criteria of Box and Hill /38/ or of Buzzi-Ferraris /38/ require additional information on error variances ^2_i of the ith model and the experimental error ^{2}, respecively and also take into consideration errors caused by imprecise parameter estimates. Furthermore, the last two criteria suppress inadequate models in the design of the next experiment.

The rational application of the above-mentioned criteria both for model discrimination as well as for parameter estimation requires the aid of a computer in connection with a suitable reactor, most preferably a gradientless reactor.

6. EXAMPLES FOR ILLUSTRATIONS

Having indicated tools and methods for kinetic data analysis and parameter estimation, practical examples from our own laboratory will be presented in this section to elucidate further aspects of the problem.

6.1. Catalytic dehydrogenation of cyclohexanol

The catalytic dehydrogenation of cyclohexanol to cyclohexanone is performed on a zinc oxide catalyst at temperatures of 570-650 K /40/. Fig. 17 shows the intrinsic rate equation together with the corresponding rate parameters determined using a laboratory micro-integral reactor, operated isothermally and filled with catalyst of 0.3 mm particle diameter.

The rate equation was selected from over 50 rival equations by model discrimination based on 36 experimental runs at 6 different temperature levels.

To determine wether this rate equation is able also to describe the behavior of a tube bundle reactor of industrial dimensions, a second set of experiments was performed with catalyst cylinders 4 x 6.2 mm in in the modular integral reactor of Figure 8 (length 1640 mm, diameter 50 mm).

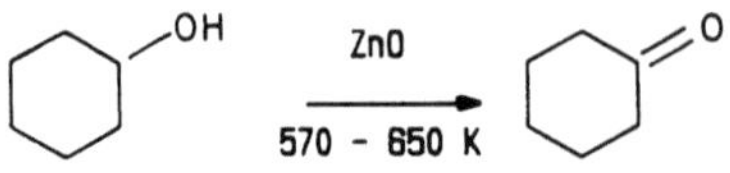

$+ H_2 \quad (-\Delta H_R) = -6.47\ E{+}7\ J/kmole$

Intrinsic rate equation

$$r_{OL} = \frac{k_1\, p_{OL}^2}{[\,1 + K_2\, p_{OL}\,]^2}$$

Reaction rate parameters on 95% confidence level

	estimated value	lower limit	upper limit
k_{01}	8.14 E+27	7.32 E+27	8.45 E+27
E_1 /R	38.834	38.827	38.861
K_{02}	3.41 E+15	3.37 E+15	3.52 E+15
E_2 /R	21.046	21.016	21.060

Effective rate parameters from a pilot unit

k_{01eff}	=	7.92 E+17 [gmoles/g cat, h, bar]
$(E_1)_{eff}/R$	=	28.458 [K]
K_2		remains the same

Fig. 17: Intrinsic and effective rate parameters for the catalytic cyclohexanol dehydrogenation

Under these conditions it was expected that pore diffusion would be a limiting step and therefore effective rate parameters (see also Fig. 17) had to be used in the simulation. The simulation was based on a pseudo-homogeneous two-dimensional reactor model using the standard routine program FIBSAS /19/.

The calculated and experimental temperature distribution is shown in Fig. 18. The fit of the axial concentration profile was even better. A perfect fit cannot be expected because of the simplified reactor model and reaction model and the use of constant average parameter values throughout the whole reactor. On the other hand, the agreement between simulation and experiment seems to be sufficient for a study of the behavior of industrial multitube reactors with larger tube diameters. It turned out that for a maximum allowable entrance temperature of 640 K and a conversion larger than 95 %, as demanded by economic considerations, the tube diameter has to be limited to 10 cm, a result that is in excellent agreement with reports on technical units.

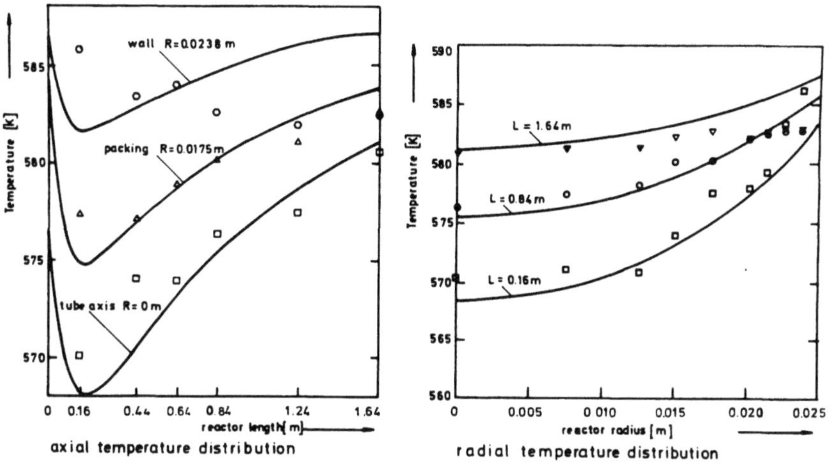

Fig. 18: Experimental and simulated radial and axial temperature profiles of a pilot scale reactor for cyclohexanol dehydration

6.2. Butene oxidation to maleic anhydride

As a second example, the kinetic analysis of butene oxidation to maleic anhydride is presented /3/. This is a rather complex reaction (see Fig. 2) and it will be demonstrated how kinetic parameters can be evaluated even from polytropic integral reactor data, provided the experimental information is detailed enough. For this an integral reactor with side stream analysis (length 900 mm, inner diameter 20 mm) filled with 0,5 mm particles of different V_2O_5/P_2O_5 catalysts has been used. Fig. 19 shows the schematic diagram of the experimental equipment.

Ten cooled gas sampling probes were distributed along the catalyst packing and a movable thermocouple was installed in the axis as well as in the wall of the tube. A typical axial product distribution (analysed by gaschromatography) has been already shown in Fig. 6. These products have been lumped according to the simplified reaction scheme given in Fig. 7. All reaction steps could be sufficiently described by the rate equation

$$r = \frac{k_0 \cdot \exp(-E/RT) \cdot x^n}{1 + k\, x} \qquad \text{with } n = 1 \text{ or } 2 .$$

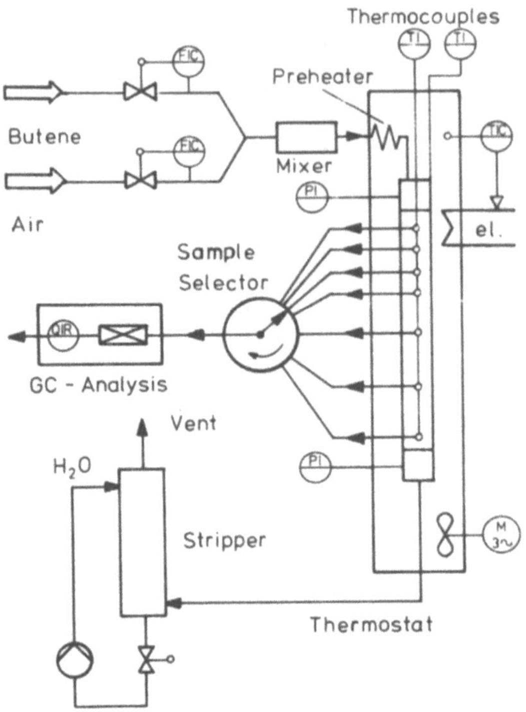

Fig. 19: Schematic diagram of experimental equipment for butene oxidation to maleic anhydride

where x represents the mole fraction of the lumped species B for reactions numbered 1 to 4 and SP for reaction number 5. Parameter estimation was performed in a first stage using a modified evolutionary strategy /41/ with contraction/expansion procedures for the allowed search region in the parameter space according to

$$\underline{D} = \underline{D}^{*} \left(0.5 + \frac{(\underline{b}^{*} - \underline{b})}{\underline{D}^{*}}\right)^{2}$$

where $\underline{D}$ is the allowed search region, $\underline{b}$ is the parameter vector and the asterix indicates the previous values. For the second step the Simplex algorithm /42/ was used to arrive finally at the parameters given in Table II for three different catalysts.

Reaction number	Catalyst A				Catalyst B				Catalyst C			
i	k_o	E	n	K	k_o	E	n	K	k_o	E	n	K
1	1.1E3	54	1	950	7.1E7	87	2	416	1.1E4	66	1	477
2	1.4E5	68	2	0	2.1E2	56	1	0	1.1E2	52	1	0
3	2.7E4	85	1	0	1.4E4	80	1	0	1.5E4	81	1	0
4	85.E9	149	1	8.8E3	3.6E5	81	1	8.8E3	3.5E5	81	1	7.9E3
5	2.5E3	34	2	0	3.9E4	34	2	0	2.1E4	32	2	0

Table II: Kinetic parameters in the rate equations for butene oxidation.

The fit between simulation (using a pseudohomogeneous reactor model) and experiment is shown in Fig. 20. The fit seems to be rather good; however it should be mentioned that $_{er}$ and h_w had to be optimized simultaneously with the kinetic parameters, which is not surprising regarding the results shown in Fig. 11. Values for these

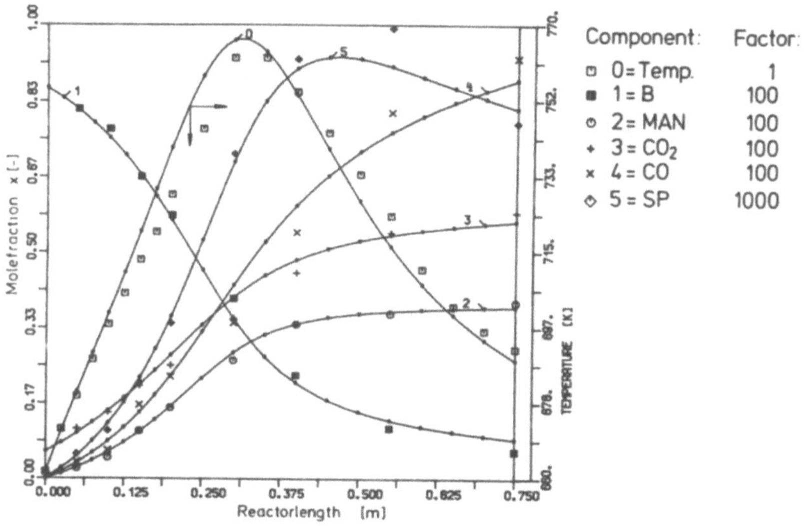

Fig. 20: Experimental and simulated axial concentration and temperature distributions for butene oxidation

parameters for non-reacting conditions calculated according to /43/ were too low by a factor up to 6. The validity of this kinetic analysis has been checked successfully by using the same lumped reaction scheme and rate expressions in the simulation of a pilot reactor (length 2200 mm, inner diameter 25 mm). The final aim of this study was to predict the selectivity behavior of a technical unit for the three different catalysts. For this, a relatively tall sidestream reactor has been selected, as only an overall (effective) kinetic study was required. Its advantage is the simple construction and operation as well as the large amount of data gained in a single experiment.

6.3. Computer aided kinetic data analysis

The most appropriate laboratory reactor for detailed kinetic investigations is the continuously operated, gradientless recycle reactor. A large number of different constructions is described in the literature /44/. We have developed and successfully used for many years for heterogeneously catalysed vapor phase reactions such a reactor with internal recirculation (Fig. 21). It can be operated up to 800 K and 50 bars (catalyst volume 10 cm^3).

For model discrimination and parameter estimation this reactor is part of a closed-loop-online microprocessor based computer system whose software structure is shown in Fig. 22. The three main program routines communicate via data files and a common part. This system can be operated interactively via a dialog routine and has implemented the above-mentioned criteria and numerical routines for the sequential design of experiments on model discrimination and

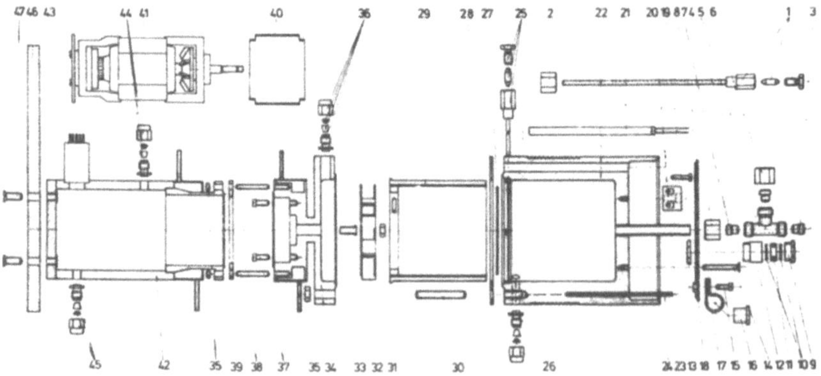

Fig. 21: Construction details of a gradientless laboratory reactor with internal recirculation (29 catalyst basket, 32 turbine, 41 motor)

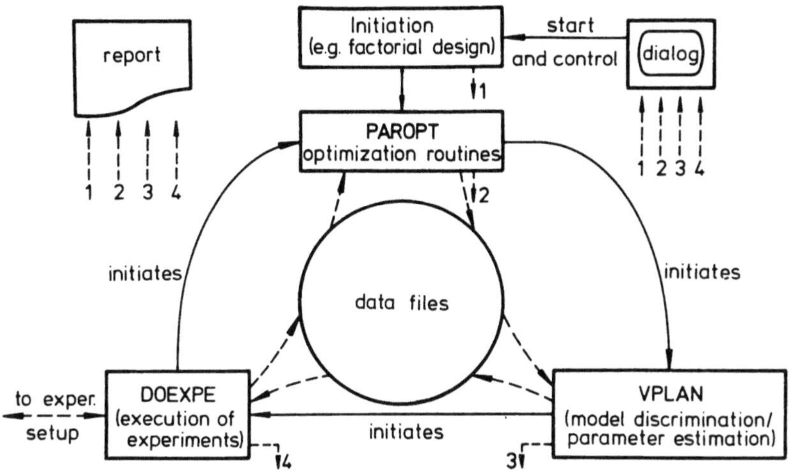

Fig. 22: Software structure for a computer controlled recycle reactor

parameter estimation. As an example the automatic performance of a factorial design for the estimation of starting values in parameter estimation is depicted in Fig. 23.

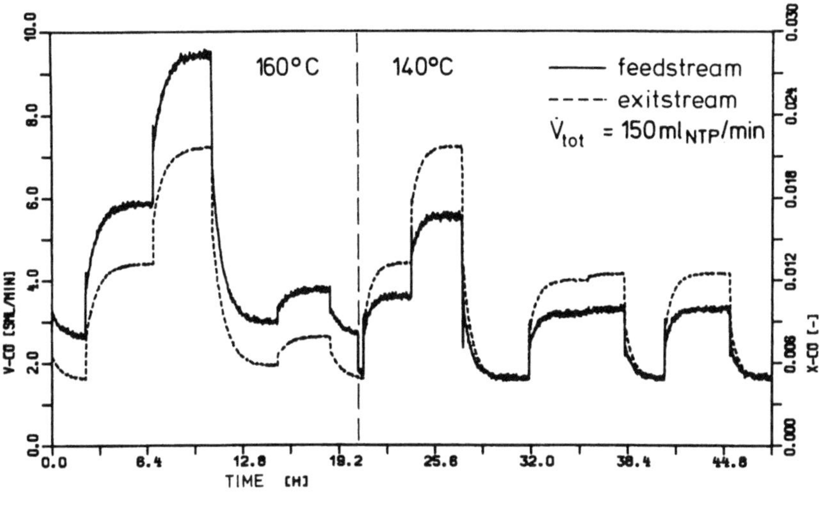

Fig. 23: Computer controlled factorial design for preliminary parameter estimates

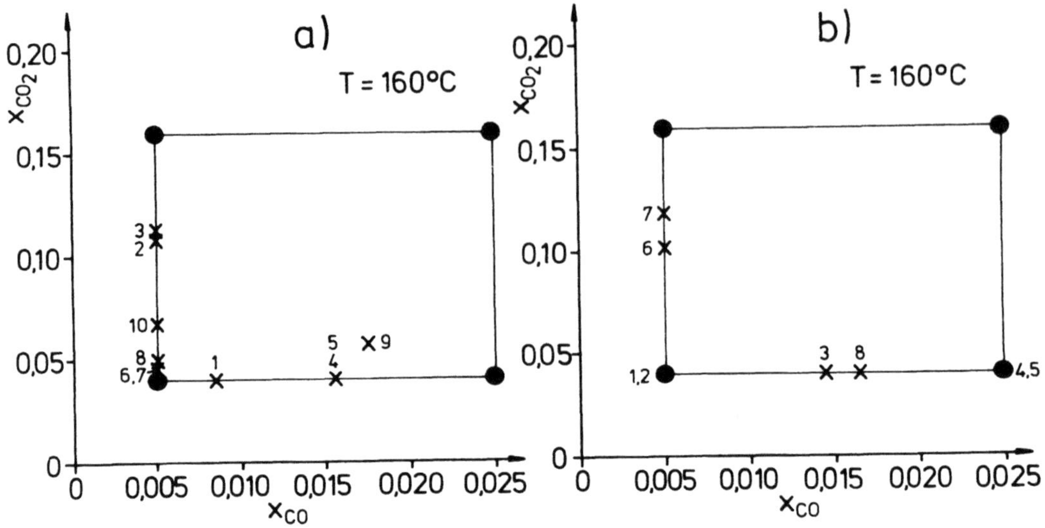

Fig. 24: Location of sequentially designed experiments for model discrimination in the variable space. a) according to Box-Hill, b) according to Buzzi-Ferraris

The reaction studied was the highly exothermic CO-oxidation ($-\Delta H_R$ = 283 kJ/mol). The aim was to discriminate between three different reactor models via the above-mentioned criteria of (a) Box-Hill and (b) Buzzi Ferraris, respectively. The experiments designed according to both criteria alternatively used are located exclusively on the boundaries of the variable space (see Fig. 24 a and b).

Fig. 25 shows that 4 additionally designed experiments were sufficient to identify model 3 according to both criteria as superior on the basis of the variance ratio $\hat{\sigma}_{min}/\hat{\sigma}_1$ (bold lines) where $\hat{\sigma}_{min}$ is the smallest variance of all rival models, or the a posteriori probability p_1 (thin lines).

According to our experience, this microprocessor controlled experimental set-up is a very efficient tool for kinetic investigations. The investment in hardware and software may be regarded as comparatively expensive, but it will pay off. Properly designed experiments and absence of subjective errors are the main reasons for efficiency. A certain constraint can arise in complex reaction systems with the chemical analysis (e.g. via mass-spectrometry) as well as in the proper design of the concentration control in the case the of deactivating catalysts. /45,36/.

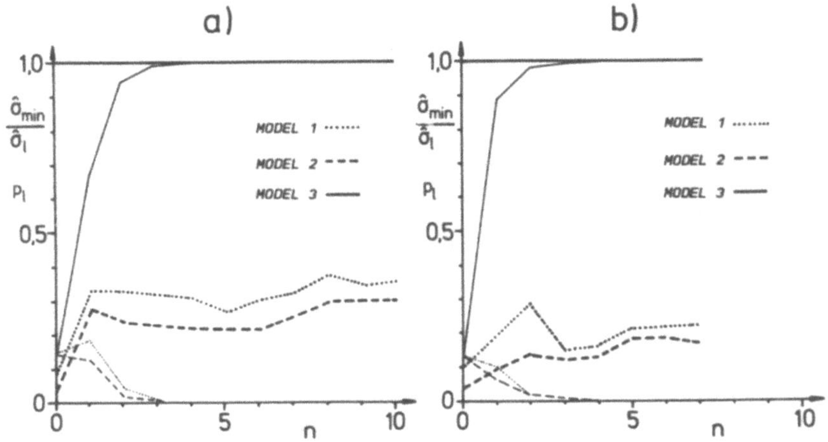

Fig. 25: Variance ratio $_{min}/_{l}$ and a posteriori probability p_l for model discrimination a) according to Box-Hill, b) according to Buzzi-Ferraris

mod.1: $r=k_o\ X_{CO}/(1+K_1\ X_{CO}+K_2\ X_{CO})^2$

mod.2: $r=k_o\ X_{CO}/(1+K_1\ X_{CO}+K_2\ X_{CO};$

mod.3: $r=k_o\ X_{CO}^{0,75}/(1+K_1\ X_{XO}\ )^2$

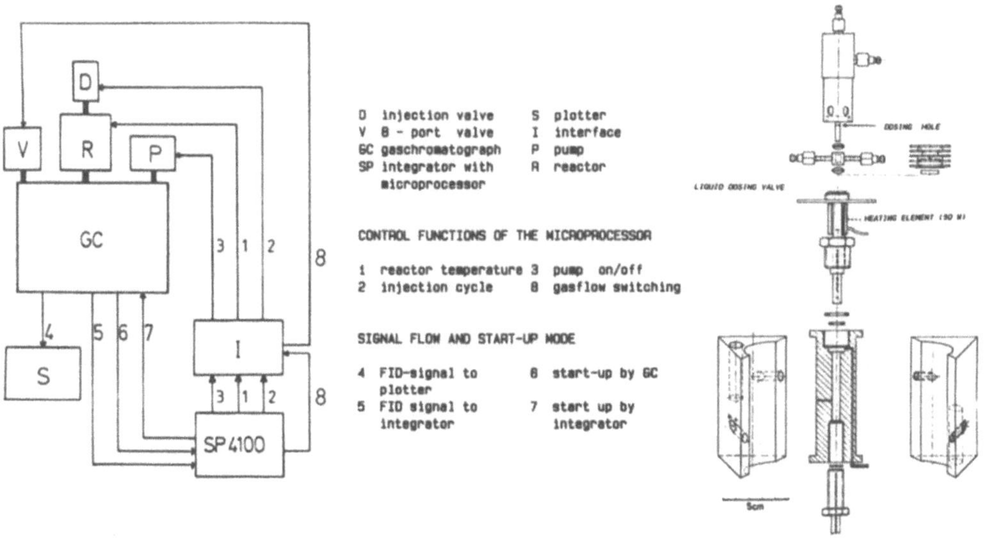

Fig. 26: Flow diagram of the experimental set-up and construction details of a microprocessor controlled pulse reactor /47/

6.4. Pulse reactor for catalyst screening and characterization

To study details of the activity and selectivity behavior of a catalyst, or to get a reaction engineering fingerprint of different catalysts a microprocessor controlled pulse reactor combined with a gaschromatograph or mass-spectrometer is a very powerful tool, especially if it is operated under temperature programmed conditions. Fig. 26 shows the flow diagram of such a unit developed in our laboratory /46/.

A typical time sequence for one experiment and the corresponding control functions of the microprocessor is shown in Fig. 27. The reaction studied in this case was the oxydehydrogenation of isobutyraldehyde and isobutyric acid to methacrylic acid, using a heteropolyacid as catalyst. A typical result (fingerprint of one catalyst) is the conversion-selectivity/temperature-diagram for isobutyric acid oxydehydrogenation shown in Fig. 28.

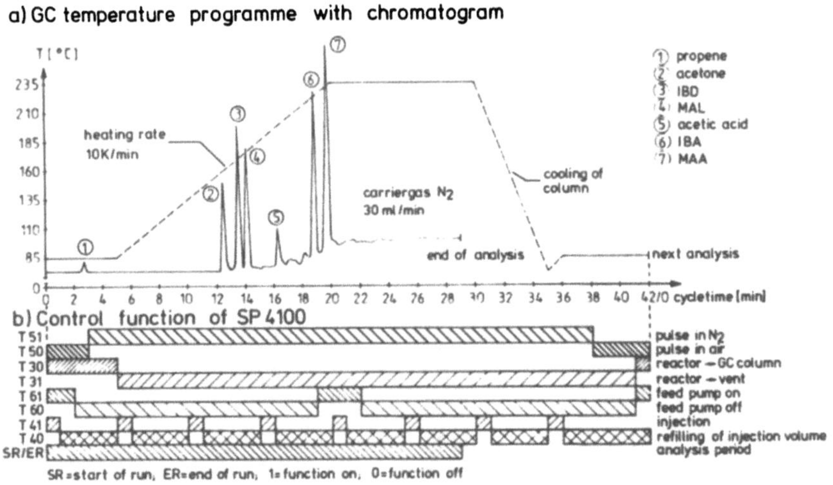

Fig. 27: Time sequence for one pulse experiment and the corresponding control functions of the microprocessor /47/

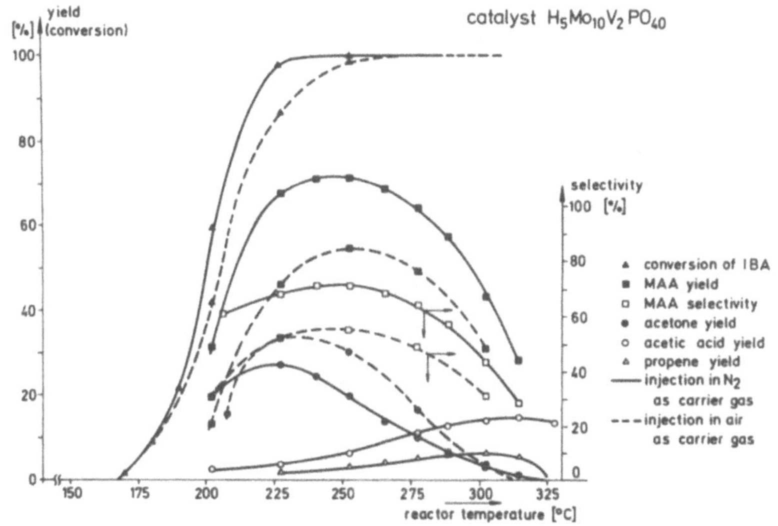

Fig. 28: Conversion-selectivity-diagram for oxydehydrogenation of isobutyric acid (carrier gas flow rate 30 ml/min, pulse volume 0.6 l NTP isobutyric acid) /47/

It is shown that complete conversion at an acceptable selectivity level (60 %) can be reached at about 250°C. It was proposed that the reaction proceeds according to a redox mechanism. If this is really so, it should be possible to separate the two reaction steps - the reduction of the catalyst by reactant in the absence of free oxygen and its reoxidation in air - experimentally. Such experimental results, also obtained in the above-mentioned pulse reactor, are depicted in Figure 29.

As a conclusion, pulse reactor experiments give valuable information about reaction mechanisms as well as catalyst behavior at the very beginning of its life. Both lead to better understanding of the kinetics and/or better catalyst design. The principle disadvantage of a pulse reactor is that kinetic information is not obtained. If the reactor is stationarily operated under temperature programmed conditions (TPD, TPR or TPO) further valuable information can be received for reaction modeling, like adsorption and desorption of reactants, including heat and energy of catalyst-substrate-bonding etc. /47,48/.

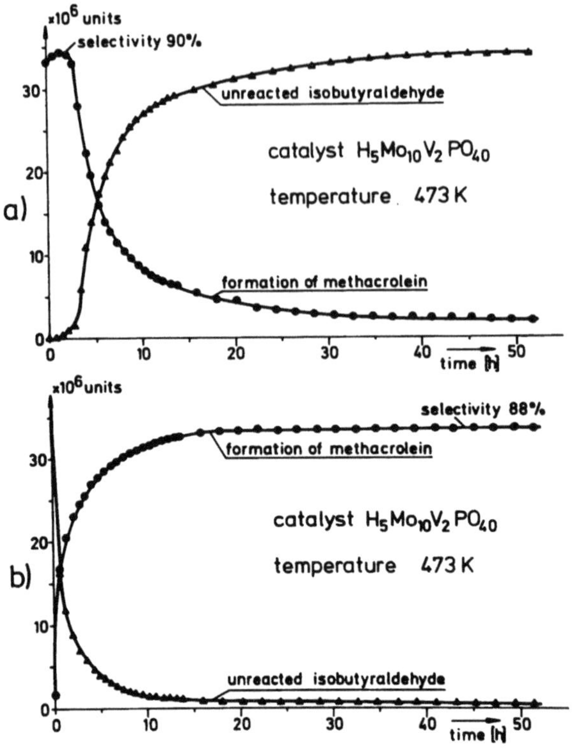

Fig. 29: Separate reduction (a) and reoxidation (b) of a heteropoly acid catalyst in isobutyraldehyde oxydehydrogenation /47/

6.5. Wavefront analysis, a dynamic method for kinetic investigations

A similar non-stationary method used for the elucidation of reaction schemes and kinetic analysis of individual reaction steps in complex systems is the so-called wavefront analysis /49/. The experimental set-up consists of an adiabatically operated packed bed reactor with a dosing section able to produce a step disturbance at the entrance and an analysis section, e.g. a mass spectrometer, to follow the front of the transition function at the exit (Fig. 30).

The transient responses of different reactants are shown in Fig. 31 for the watergas shift reaction performed on a CuO/ZnO-catalyst. In the special case depicted in Fig. 31, the catalyst was pretreated with a H_2O/H_2-mixture having a given p_{H_2O}/p_{H_2}-ratio =10, leading to a high and uniform oxidation level of the whole catalyst packing. After a step change of the inlet concentration

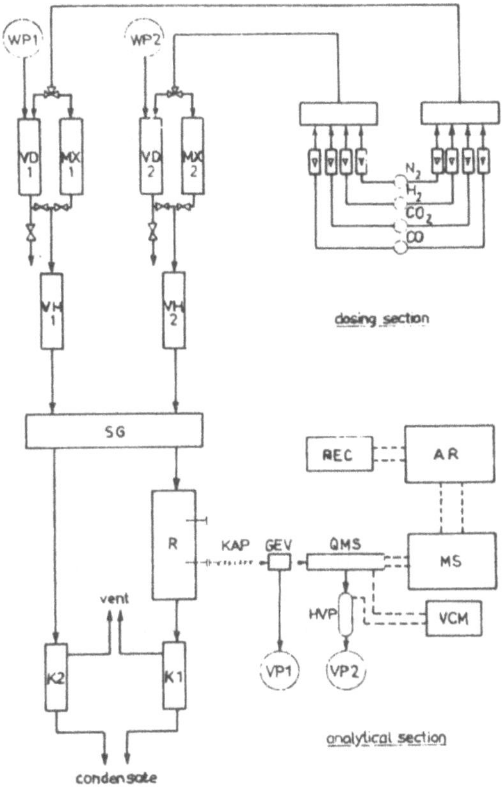

Fig. 30: Experimental set-up for the wavefront analysis (R: reactor, SG: signal generator, WP: water pump, VD: evaporator, MX: mixer, VH: preheater, K: condensor, VP: vacuum pump, GEV: gas inlet valve, QMS: mass spectrometer, MS: mass selector, REC: recorder)

into the reaction mixture at t=0 the transient was recorded. From the characteristic elements of the wavefront analysis (amplitude damping of the disturbance, time delay between entrance and exit and slope of the transient at the wavefront) as well as additional qualitative information gained from the whole transient, a certain reaction scheme could be established consisting of two redundant Eley-Rideal-mechanisms with different adsorbed species. Furthermore, the oxidation level of the solid catalyst depends on a redox-mechanism as shown in Fig. 32.

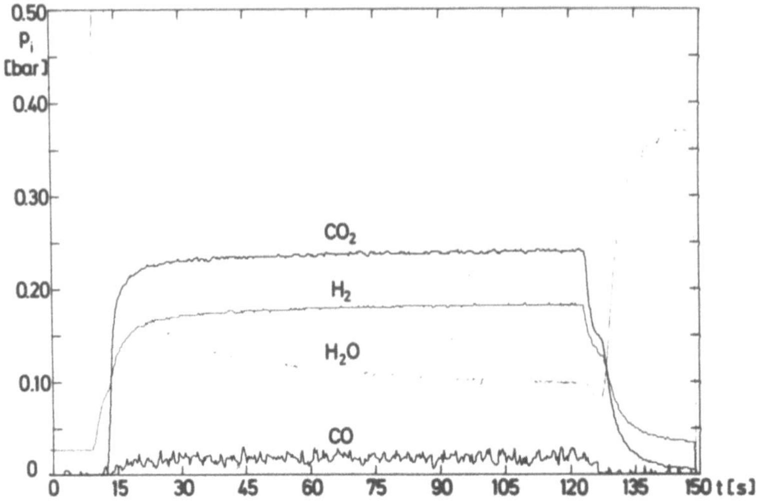

Fig. 31: Transient responses for the watergas shift reaction (catalyst at high oxidation level, T=280°C, P_{H_2O}/P_{H_2}=10)

(1) Eley-Rideal-Mechanism

$$H_2O + (CO)^* \xrightarrow{k_1} H_2 + CO_2$$

(2) Eley-Rideal-Mechanism

$$CO + (H_2O)^* \xrightarrow{k_2} CO_2 + H_2$$

(3) Redox-Mechanism

$$\text{(A)}\quad CO + (O) \xrightarrow{k_3} CO_2 + (\ )$$

$$\text{(B)}\quad H_2O + (\ ) \xrightarrow{k_4} H_2 + (O)$$

Fig. 32: Reaction scheme of the watergas shift reaction according to the wavefront analysis /49, 50/.

The degree to which the different partial mechanisms contribute to the overall conversion depends strongly on the temperature but also on the oxidation level of the catalyst as shown in Fig. 33 /50/.

The advantages of this technique are

- The wavefront "sees" only a uniformly pretreated and isothermal catalyst independent of a temperature change if the front has passed, since the thermal wave travels much slower than the concentration wave because of the higher heat capacity of the packing compared to its mass capacity.

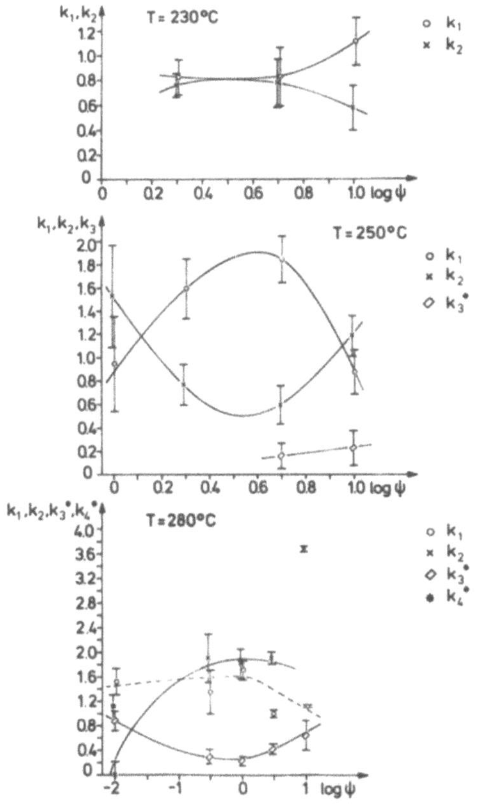

Fig. 33: Kinetic parameters for the different reaction steps in the watergas shift reaction (see Fig. 32).

- it is possible to disturb anyone of the reactants at the entrance to the reactor. Therefore, different steps can be stimulated and kinetically analysed separately for their kinetic parameters.

- In non-linear situations the transient may also be dependent on the step height, introducing in this way a new experimental variable.

- Depending upon whether a positive or a negative step was applied to the system different transients may be received (shock wave and ordinary wave respectively).

In fact, wavefront analysis yields much more detailed qualitative and quantitative information than the analysis of stationary operated reactors. Naturally, the technique also has its limits. E.g., if complex reaction systems including molecules of higher molecular weight are involved, the mass-spectrometer data are probably difficult to analyse quantitatively for all reactants and products.

7. CONCLUSIONS

This review, which certainly cannot claim to have covered the whole spectrum of problems, shows that considerable progress has been made in the last decade in kinetic data analysis and parameter estimation. Specially designed reactors, dynamic analysis techniques, as well as computer aided experimentation are some of the highlights. But reaction analysis and reactor simulation for heterogeneously catalysed reactions are still a challenging task that needs the intelligent action of the chemical reaction engineer.

Acknowledgment

The valuable discussion of the subject with Prof. G. Emig and other members of our laboratory is highly appreciated.

LIST OF SYMBOLS

a	area m^2
Ar	Arrhenius number (dimensionless)
$\underline{b}$	regression parameters
Bi	Biot number (dimensionless)
Bo	Bodenstein number (dimensionless)
c	concentration $kmole/m^3$
c_p	specific heat kJ/kg, K
d	diameter m
D	dispersion coefficient m/s^2

$\underline{D}$	matrix; search region
E	activation energy kJ/mole ; expected value
f	fraction (dimensionless)
h	heat transfer coefficient $kJ/m^2,s,K$
ΔH_R	heat of reaction kJ/kmole
$\underline{J}$	Jacobian matrix
k_o	preexponential factor (dimension same as reaction rate constant)
k	reaction rate constant; mass transfer coefficient
K	adsorption equilibrium constant bar^{-1}
L	length m
N	normally distributed
n	order of reaction (dimensionless); number of experiments
p	partial pressure bar
p_1	a posteriori probability
Pe	Peclet number (dimensionless)
Pr	Prandtl number (dimensionless)
Pt	Prater number (dimensionless)
r	radius m , reaction rate $kmole/m^3,s$
R	universal gas constant kJ/mole, K
Re	Reynolds number (dimensionless)
T	temperature K
u	flow velocity m/s
v,V	flow rate Nml/min
w	weight kg
x	mole fraction (dimensionless)
y	pellet radius (dimensionless), mesured response value
z	reactor length (dimensionless)

Greek Symbols

α	heat transfer coefficient $kJ/m^2,s,K$
β	Prater number (dimensionless); factor in fig. 3; true parameter
γ	Arrhenius number (dimensionless); factor in fig. 3
ϵ	void fraction (dimensionless); error
η	calculated response value
θ	relative fluid temperature (dimensionless)
λ	thermal conductivity kJ/m,s,K; eigenvalue
ν	stoichiometric coefficient
σ^2	variance
τ	mean residence time s
[illegible]	recycle ratio (dimensionless)
Ψ	criterion in experimental design for parameter estimation

Subscript

a	exit, axial
c	catalyst
e	entrance
eff	effective
er	effective radial
i	i-th reactant, i-th experiment
l	l-th model
min	smallest
p	pellet
r	radial
t	tower
u	u-th response
v	v-th response
w	wall

Superscript

T	transformed
$\wedge$	calculated
uv	between response u and v
*	previous value

REFERENCES

/1/ H. Hofmann: Industrial process kinetics and parameter estimation, Adv.Chem.Ser. 109(1972)519-534

/2/ P. Trambouze, H. van Landeghem and J.P. Wauquier: Les reacteurs chimiques, Edition Technip, Paris 1984

/3/ H. Hofmann, G. Emig and W. Röder: The use of an integral reactor with sidestream-analysis for the investigation of complex reactions, EFCE Publ.Ser. 37(1984)419-426

/4/ S. Yagi and D. Kunii: Studies on the effective thermal conductivities in packed beds, AIChE J. 3(1957)373-381

/5/ D. Kunii and J.M. Smith: Heat transfer characteristics of porous rocks, AIChE J. 6(1960)71-78

/6/ N. Wakao and J.M. Smith: Diffusion in catalyst pellets, Chem.Eng.Sci. 17(1962)825-834

/7/ H. Hofmann: Progress in modeling of catalytic fixed-bed reactors, Germ.Chem.Eng.2(1979)258-267

/8/ D.E. Mears: Diagnostic Criteria for heat transport limitation in fixed-bed reactors, J.Catal.20(1971) 127-131

/9/ D.E. Mears: Tests for transport limitations in experimental catalytic reactors, Ind.Eng.Chem.Proc.Des.Dev.10(1971)541-547

/10/ D.E. Mears: On criteria for axial dispersion in nonisothermal packed bed catalytic reactors, Ind.Eng.Chem.Fund. 15(1976)20-23
/11/ M. El-Sawi, G. Emig and H. Hofmann: A study of the kinetics of vinyl acetate synthesis, Chem.Eng.J. 13(1979)201-211
/12/ J. Wei and J.C.W. Kuo: A lumping analysis of monomolecular reaction systems I,II, Ind.Eng.Chem.Fund. 8(1969)114-123,124-133
/13/ D. Luss and P. Hutchinson: Lumping of mixtures with many parallel first order (n-th order) reactions, Chem.Eng.J. 1(1970) 129-136, 2(1971)172-178
/14/ H.H.Lee: Synthesis of kinetic structure of reaction mixtures of irreversible first-order reaction, AIChEJ 24(1978)116-123
/15/ S.M. Jacob, B. Gross, St.E. Voltz and V.W. Weekman, jr.: A lumping and reaction scheme for catalytic cracking, AIChEJ. 22(1976)701-713
/16/ U. Hoffmann and H. Hofmann: Kinetik von Reaktionssystemen mit vielen Komponenten und vielen Reaktionen, sowie deren vereinfachende Behandlung, Chem.Ing.Techn. 48 (1976) 465-469
/17/ W. Röder: Diss. Univ. of Erlangen-Nürnberg 1986
/18/ U. Fiand: Diss. Univ. of Erlangen-Nürnberg 1978
/19/ U. Hoffmann, G. Panthel and G. Emig: FIBSAS standard program for the Design and Analysis of Fixed-Bed Reactors, Erlangen 1977
/20/ E. Aust: Diss. Univ. of Erlangen-Nürnberg 1986
/21/ J.R. Balder and E.E. Petersen: Application of the single pellet reactor for direct mass transfer studies I and II, J.Catal. 11(1968)195-201, 202-210
/22/ U. Hoffmann, G. Emig and H. Hofmann: Comparison of different determination methods for effective thermal conductivity of porous catalysts, ACS Symp.Ser. 65(1978)189-200
/23/ R. Broucek, G. Emig and H. Hofmann: Rechnergesteuerter Kreislaufreaktor für kinetische Untersuchungen,Chem.Ing.Techn. 56 (1984) 236-237
/24/ J. Mandler, R. Lavie and M. Sheintuch: An automated catalytic system for the sequential optimal discrimination between rival models, Chem.Eng.Sci. 38(1983)979-990
/25/ L.M. Rose, C. Schifferli and D.W.T. Rippin: Computer application in the analysis of chemical data and plants, Proc.Chemdata Helsinki (1977)132-136
/26/ D.M. Himmelblau: Process analysis by statistical methods, John Wiley and Sons, New York 1970
/27/ G.E.P. Box and H.L. Lucas: Design of experiments in non-linear situations, Biometrika 46(1959)77-90
/28/ L.H. Hosten: A sequential experimental design procedure for precise parameter estimation based upon the shape of the joint confidence region, Chem.Eng.Sci. 29(1974)2247-2252

/29/ G. Emig: Planung und Auswertung reaktionskinetischer Versuche zur Parameterermittlung aus integralen Labordaten, Fortschritts-Berichte der VDI-Zeitschriften, Reihe 3, Nr. 43, VDI-Verlag 1976

/30/ G. Emig, U. Hoffmann and H. Hofmann: Handbuch zum DECHEMA-Kurs "Planung und Auswertung von Versuchen zur Erstellung mathematischer Modelle", DECHEMA, Frankfurt/Main 1975

/31/ P. Gans: Visually interactive parameters' estimation and refinement - A technique for data-fitting, Comp.&Chem. 1 (1977) 291-293

/32/ D.W. Marquardt: An algorithm for least square estimation of non-linear parameters, Soc.Ind.Appl.Math.J. 11(1963)431-441

/33/ U. Hoffmann and H. Hofmann: Einführung in die Optimierung; Verlag Chemie, Weinheim 1971

/34/ M.W. Heuckroth, J.L. Gaddy and L.D. Gaines: An examination of the adaptive random search technique, AIChEJ. 22(1976)744-750

/35/ D.L. Martin and J.L. Gaddy: Process optimization with the adaptive randomly directed search, AIChE Symp.Ser. 78 No. 214(1982)99-107

/36/ A. Löwe: Laborreaktoren für heterogene Gasreaktionen - Neue Entwicklungen; Chem.Ing.Techn. 51 (1979)779-788

/37/ W.G. Hunter and A.M. Reiner: Design for discriminating between two rival models, Technometrics 7(1965)307-323

/38/ G.E.P. Box and W.J. Hill: Discrimination among mechanistic models, Technometrics 9(1967)57-71

/39/ G. Buzzi-Ferraris, P. Forzatti, G. Emig and H. Hofmann: Sequential experimental design for model discrimination in the case of multiple response, Chem.Eng.Sci. 39(1984)81-85

/40/ F. Garcia-Ochoa, M. Lenz, G. Emig and H. Hofmann: Reaktionstechnische Untersuchungen der katalytischen Dehydrierung von Cyclohexanol zu Cyclohexanon, Chem.Ztg. 105(1981)349-354

/41/ P. Koeppe and C. Hamann: Ein Basic-Programm zur nichtlinearen Regressionsanalyse, Markt und Technik 10 (1981)60-64

/42/ J.A. Nelder and R. Mead: A simplex method for function minimization, Comp.J. 7(1965)308-313

/43/ VDI-Wärmeatlas, 3. Auflage 1976, Gg 1-15

/44/ H. Jankowski, J. Nelles, R. Adler, B. Kubis and Ch. Salzer: Experimentelle und Auswertungsmethoden zur reaktionstechnischen Untersuchung heterogen-gaskatalytischer Prozesse; Chem.Techn. 30 (1978) 441-446

/45/ R. Broucek, Diss. Univ. Erlangen-Nürnberg 1985

/46/ K. Kürzinger, Diss. Univ. Erlangen-Nürnberg 1983

/47/ K. Kürzinger, G. Emig and H. Hofmann: Oxydehydrogenation of isobutyraldehyde and isobutyric acid as a new route to methacrylic acid, Proc. ICC8, Vol. V, p. 499-507, Berlin 1984

/48/ J.L. Falconer and J.A. Schwarz: Temperature programmed desorption and reaction application to supported catalysts; Catal.Rev.-Sci.Eng.25(1983)141-227

/49/ E. Fiolitakis and H. Hofmann: Wavefront analysis, a specific analysis of transient responses for the investigation of continuously operated distributed heterogeneous reaction systems, Catal.Rev.Sci.Eng. 24 (1982) 113-157 and ACS Symp.Ser. 178 (1982) 277-301

/50/ R.J. Kalenczuk, E. Fiolitakis and H. Hofmann: Untersuchungen zur Kinetik der Tieftemperatur-Konvertierung, Wellenfrontanalyse der Ortsrelaxation, Chem.Ing.Techn.52 (1980) 966-968

TRACER METHODS IN CHEMICAL REACTORS. TECHNIQUES AND APPLICATIONS.

M. P. Duduković

Chemical Reaction Engineering Laboratory
Washington University, St. Louis, MO 63130

1. INTRODUCTION

Tracer methods are encountered in many areas of science and engineering. The diversity of their uses is illustrated by measurement of blood flow and capillary permeability of the microcirculation in medicine and by flow visualization in channels and around airplane wings in mechanical and aerospace engineering. Other applications are flow and transport measurements in rivers in hydrology, transport measurements of pollutants in soils in civil engineering, and measurements of spreading of plumes in the atmosphere in environmental engineering. Additional uses involve identification of reaction mechanisms of chemical and catalytic reactions, measurement of diffusion rates, etc. All these methods rely on perturbing the system under investigation and monitoring the system's response to such perturbations. This response is then interpreted. Some conclusions can be obtained on a model-free basis, others are model dependent.

Reynolds' dye experiment on transition to turbulence in pipe flow and G. I. Taylor's experiments on axial dispersion in laminar flow represent the early use of tracers in flow visualization and transport parameters evaluation in chemical engineering. A more widespread use in chemical reactors dates to the work of Danckwerts (1). He realized that the performance of process equipment depends on the residence time distribution of process fluid, and that this information can be obtained by tracer methods. Residence time distributions are now discussed in standard chemical reaction engineering texts (2,3) and are well summarized in a recent excellent monograph by Nauman and Buffham (4). Tracer methods, residence time distributions and the related subjects of mixing in chemical reactors, reactions in turbulent flow, micromixing, etc., have also been treated thoroughly in a number of recent review papers (5-14). This leaves little new ground for another review. Therefore, no attempt is made here to cover the literature exhaustively. Instead, a personal view of the utility of various theoretical concepts and practical applications of the tracer methods in reaction engineering is given. It is hoped that this will be of use both to students and practicing engineers. The emphasis is placed on the approaches that have proven to be useful in the past and on the areas where additional work is needed in order to bring about new breakthroughs. We start by reviewing the theory and use of tracer methods in homogeneous reaction systems and then present selected studies of heterogeneous systems. Distinction is made between results that can be interpreted on a model-free basis as opposed to those that are model dependent.

2. REACTOR MODELING FOR HOMOGENEOUS REACTIONS

Two key tasks typically arise in chemical reaction engineering:

1. Reactor design is required for a given reaction system and "a priori" prediction of reactor performance is called for from a reactor model or scale-up effort.

2. Assessment of performance of existing reactors is asked for, and suggestions for improvement are solicited.

We will call the first one the design problem, and the second one the trouble shooting problem.

How can tracer methods help us in solving these two problems? We know that reactor performance, as measured by conversion of the limiting reactant or by product selectivity, is a function of kinetics, flow pattern and mixing pattern in the reactor. The flow and mixing phenomena in various reactor geometries are complex, and we are currently unable to characterize them completely (at an economical cost). The only reactors that we know how to design, predict their performance and scale up with confidence, are those that behave as the two ideal reactor types, i.e. the plug flow (PFR) and the continuous flow stirred tank reactor (CSTR). The former is based on the assumption of instantaneous and perfect mixing on the molecular scale in directions perpendicular to the main direction of flow, with no mixing in the direction of flow. This results in one-dimensional concentration gradients. The latter assumes instantaneous and perfect mixing on the molecular scale throughout the reactor volume. No spatial concentration gradients exist in the ideal CSTR. Tracer methods can certainly help us in troubleshooting on real reactors in identifying whether they deviate from the assumed ideal reactor behavior.

To illustrate briefly why it is important to assess such deviations, let us consider a simple single reaction, at isothermal conditions and at constant density, of the following stoichiometry:

$$A + B = P \tag{1}$$

For an irreversible reaction of overall order n, with stoichiometric feed reactant ratio, the design equations for the two ideal reactors are well known. They are:

$$Da_n = k\, C_{Ao}^{\,n-1}\, \bar{t} = \frac{x_A}{(1-x_A)^n} \quad ; \quad \text{for a CSTR} \tag{2}$$

$$Da_n = k\, C_{Ao}^{\,n-1}\, \bar{t} = \frac{1}{1-n}\left[1 - (1-x_A)^{1-n}\right] \quad ; \quad \text{for a PFR} \tag{3}$$

Given one of the two ideal flow patterns, reactor conversion is determined by the Damkohler number, Da_n, which is the ratio of the characteristic process time $t_p = \bar{t} = V/Q$ (mean residence time) and the characteristic reaction (kinetic) time $t_k = 1/(k\, C_{Ao}^{\,n-1})$.

Unfortunately, real reactors do not behave as ideal ones. The first type of problem that arises is related to flow maldistribution resulting in stagnant flow zones or flow bypassing. If a tubular or mixed flow reactor has a stagnant zone, i.e. a region through which flow is extremely slow, then its effective characteristic process time is reduced, i.e. $\bar{t}_{eff} = V_{active}/Q$. It is intuitively obvious, and mathematically evident from eqs. (2) and (3), that a reduction in Damkohler number results in a decrease of reactant conversion. On the other hand, if some fraction $(1-\alpha)$ of the feed bypasses the reaction zone completely, the Damkohler number increases for the reacting fraction of the flow and its conversion increases. However, conversion at reactor exit for the overall combined reactor and bypass flow is reduced for reaction orders larger than zero. For a zeroth order reaction, conversion remains unaffected by bypassing as long as $Da_o/\alpha \leq 1$, and is reduced for $Da_o/\alpha \geq 1$ when it becomes proportional to the reacting flow fraction. Pathological negative order reactions can even benefit from bypassing and exhibit optimal conversion at certain α! This last result, while mathematically interesting, is of little practical utility because if the reactor was designed properly to operate at complete conversion any degree of bypassing will adversely affect its performance. Bypassing and stagnancy are just some extreme deviations from ideal flow pattern behavior. Tracer methods, as will be shown in a later section, can be used to identify these problems and characterize the reactor flow pattern. One should also note that, contrary to what one usually learns in undergraduate courses, the ideal reactor formulas do not bracket (bound) performance of the real reactor if stagnancy and bypassing are present.

The other problem that arises in real reactors is related to the assumption of premixed feed. In deriving eqs. (2) and (3) we have tacitly assumed that the reactor feed can be intimately premixed, down to the molecular scale, without reaction. This implies that the mixing time in generating a homogeneous mixture is infinitesimally short compared to the characteristic reaction time. When this is not the case, we have to consider the characteristic mixing time (or mixing intensity which is proportional to its reciprocal) and the scale of mixing in addition to reaction and process time. Unfortunately, mixing is difficult to describe by a single parameter since it occurs by several mechanisms in parallel such as convection, turbulent mixing and molecular diffusion. If we consider an extremely simplistic approach, we can visualize in the Lagrangian sense that the scale of segregation of the fluid, i.e. the typical size of fluid eddies, is reduced from a macroscale, L_s, to a scale approaching the microscale, λ, in a characteristic mixing time, T_m. Based on an assumed homogeneous turbulence field this characteristic time can be related to a characteristic distance in the direction of mean flow. Beyond that time (distance) only molecular diffusion is at work in reducing the intensity of segregation with a characteristic mixing time $\tau_m \propto \lambda^2/D_m$. Again, tracer methods are very useful in determining characteristic mixing times and in obtaining various correlation functions that describe the turbulence field (7,8,10, 15,16). Inert and reactive tracers can be used effectively in assessing macro-and-micro-scale mixing patterns in reactors (17).

This means that tracers can be used to provide the information both about the global flow pattern in the vessel and about intimate mixing details on a microscale. Clearly, when $t_k >> T_m$, $t_k >> \tau_m$ the information about the global mixing pattern (macromixing) is sufficient to describe the reactor, while when the opposite is true, a detailed description of all scales of mixing is essential. However, when characteristic reaction times are so fast compared to mixing that the two reactants cannot coexist together, then reaction occurs on surfaces, i.e. in very thin layers of fluid. Under such circumstances it is hardly appropriate calling such a system homogeneous.

3. RESIDENCE TIME DISTRIBUTIONS: DEFINITIONS AND EXPERIMENTAL DETERMINATION FROM TRACER RESPONSES

As already mentioned, Danckwerts (1) realized that reactor performance depends on the distribution of residence times of the fluid and introduced this concept to the chemical engineering literature. By now this material is covered in standard undergraduate courses but for completeness a brief review is provided here.

For an isothermal, homogeneous system of constant density with well-defined boundaries and a single inlet and outlet stream, a residence time distribution, F(t), is defined as the probability that a fluid element has a residence time less than t. At steady state F(t) is the fraction of the fluid in the outflow that has resided in the vessel less than time t. Here we assume that each fluid element has a definite point of entry into the vessel and that age (residence time) is acquired by the fluid element only while it resides within the system's boundaries.

F(t) is a probability distribution which can be obtained directly from measurements of the system's response in the outflow to a step-up tracer input in the inflow. Consider that at time $t = 0$ we start introducing a red dye at the entrance of the vessel into a steady flow rate Q of white carrier fluid. The concentration of the red dye in the inlet flow is C_o. At the outlet we monitor the concentration of the red dye, C(t). If our system is closed, i.e. if every molecule of dye can have only one entry and exit from the system (which is equivalent to asserting that input and output occur by convection only), then $QC(t)/QC_o$ is the residence time distribution of the dye. This is evident since all molecules of the dye appearing at the exit at time t must have entered into the system between time 0 and time t and hence have residence times less than t. Only if our red dye is a perfect tracer, i.e.. if it behaves identically to the white carrier fluid, then we have also obtained the residence time distribution for the carrier fluid and $F(t) = C(t)/C_o$. To prove that the tracer behaves ideally and that the F curve is obtained, the experiment should be repeated at different levels of C_o. The ratio $C(t)/C_o$ at a given time should be invariant to C_o, i.e. the tracer response and tracer input must be linearly related. If this is not the case, then $C(t)/C_o$ is only the step response for the tracer, which includes some nonlinear effects of tracer interactions in the system, and which does not represent the true residence time distribution for the system.

Another simple test of tracer response linearity is to follow the previously described step-up experiment with a step-down one, i.e. when the whole system has reached concentration C_o one should reduce the dye concentration in the inflow to zero. Call that $t = 0$ and monitor the outflow concentration $C_d(t)$. Now, if the system is at steady state and the tracer is perfect, $C_d(t)/C_o$ will yield the washout function, W(t), which is the fraction of the fluid outflow with residence times larger than t.

From the definition of probability distributions, it is evident that:

$$F(t) + W(t) = 1 \tag{4}$$

and

$$\lim_{t \to \infty} F(t) = 1 \tag{5a}$$

$$\lim_{t \to \infty} W(t) = 0 \tag{5b}$$

If. eq. (4) is not satisfied for step-up and step-down tracer tests, then the proper distribution functions have not been obtained by the tracer test. The cause might be either that the system was not at steady state, or that the tracer did not behave perfectly and underwent a nonlinear process within the system. Another requirement for establishing the F(t) curve correctly from tracer step-up or step-down experiments, which is particularly important for systems with laminar flow, is that tracer must be introduced proportionally to flow (flow tagging) and that its mixing cup concentration must be monitored at the outflow (18-21).

Since F(t) is a probability distribution on $t \varepsilon [o,\infty)$, it must have an associated probability density function so that $F(t) = \int_o^t E(\tau)d\tau$. This residence time density function or exit age density function, E(t), was unfortunately originally introduced in the chemical engineering literature under the misleading name of residence time distribution while it is actually a derivative of the distribution. We will continue to designate this density function by E(t) as traditionally recommended in chemical engineering literature (22) as opposed to a more generally acceptable mathematical notation of lower case f(t). E(t)dt is the fraction of the outflow that has residence times between t and t + dt. Under the previously quoted assumptions E(t) can be obtained directly by injecting an instantaneous impulse of tracer of mass m_T, and by monitoring the tracer concentration at the outflow, $C_p(t)$. It can also be obtained indirectly by differentiation of the F(t) curve:

$$E(t) = \frac{Q\, C_p(t)}{m_T} = \frac{dF}{dt} \tag{6}$$

Integration of eq. (6) over all times and the property of probability density functions provides a check on the tracer experiment:

$$\frac{m_T}{Q} = \int_0^\infty C_p(t)dt \tag{7}$$

Equation (7) has been used traditionally in medicine as the Stewart-Hamilton indicator-dilution technique (23-25) to measure the unknown flow rate through isolated organs. In this method a known bolus, m_T, of tracer is injected into an artery and the tracer dilution curve, $C_p(t)$, is measured at the outflow in a major vein. In chemical engineering systems eq. (7) should be used as a basic mass balance consistency check on the accuracy of the tracer experiment. Failure to satisfy eq. (7) indicates errors and nonidealities, and requires repeats of the experiment at different m_T s to establish linearity. On occasion, only a response R(t), which is linearly proportional to tracer concentration, is measured and the mass injected m_T is unknown. Then the experiment should be repeated to establish linearity of R(t) with respect to m_T. The obtained age density function $R(t)/\int_0^\infty R(t)dt$ should be invariant to the mass injected. Although the above simple rules are rather evident, they are all too often neglected by practitioners. Unfortunately, not all tracers at concentrations used behave ideally, and erroneous conclusions result when the nonideal tracer residence time distribution is accepted as the F curve of the system. The tracer tests needed to determine E(t), F(t) and W(t) functions are schematically represented in Figure 1.

In addition to the F, W and E curves, all of which are based on the sample space (population) of the exiting fluid, one can introduce an internal age density function based on the fluid within the system. I(t)dt is defined as the fraction of fluid elements in the system of age between t and t + dt. The relationships between I, E and other functions are readily obtained and are reported in all standard texts (2-4). I(t) can be obtained directly by injecting a radioactive tracer (of long half life) at time t = 0 and by monitoring the radioactivity of the whole system. Similarly, the internal age distribution (the integral of I(t)) can be obtained by measuring the tracer concentration throughout the system upon step-up tracer injection.

Recently, Buffham (26) completed the family of the various density functions by introducing $\chi(t)dt$ as the fraction of the fluid in the system of residence time around t and snowed that:

$$\chi(t) = (t/\bar{t})\, E(t) \tag{8}$$

We should recall that each fluid element is characterized by its age α and life expectancy λ which sum to yield its residence time. The population of elements in the system characterized by $\chi(t)$ has therefore a whole spectrum of ages.

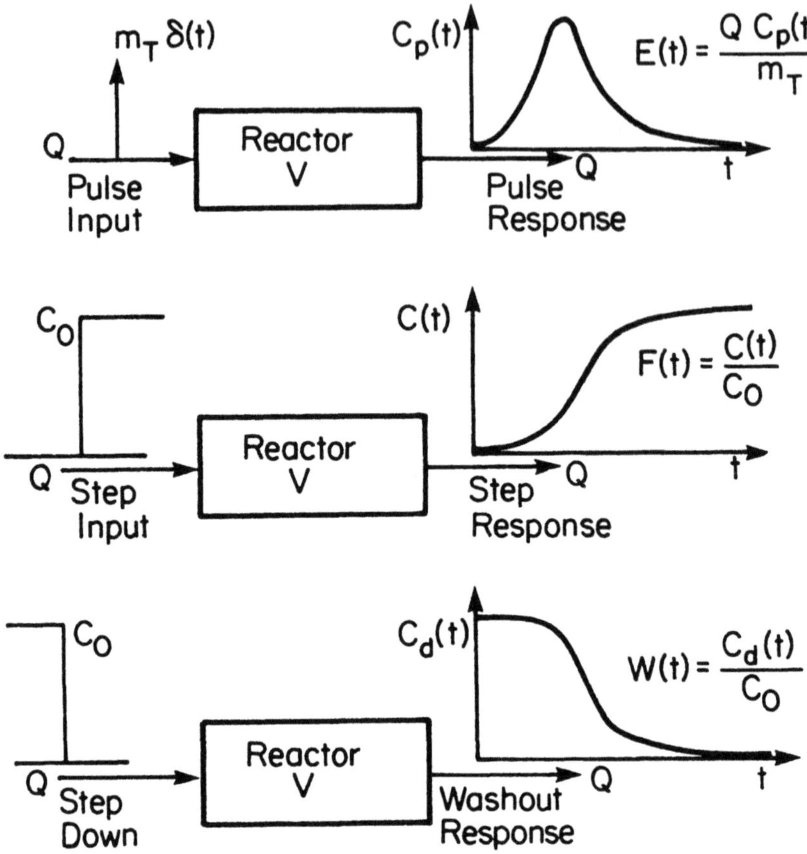

Figure 1. Tracer Tests and Residence Time Distributions (Systems with One Inlet and Outlet).

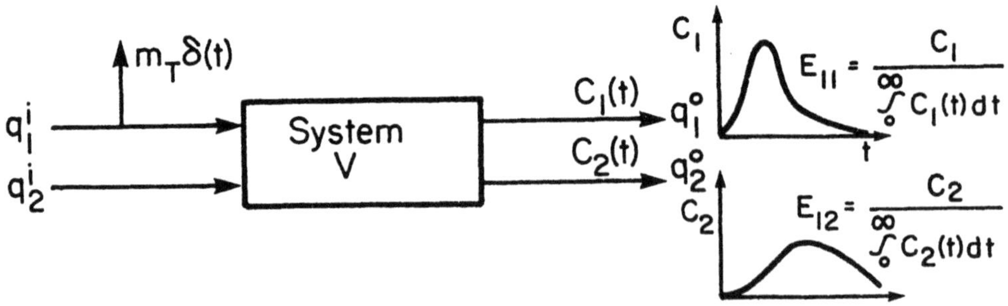

Figure 2. Tracer Tests and Residence Time Distributions (Systems with Multiple Inlets and/or Outlets).

Of all the experimental means for determining the above density or distribution functions direct measurement of I(t) or of W(t) is preferred as inherently the most accurate. The former is not practically feasible in chemical engineering but is practiced in biomedicine. Direct determination of both F(t) and W(t) is an attractive technique in laboratory work but
industrial practice because long term perturbations (tagging) of the process cannot be tolerated. Hence, most frequently E(t) is determined from pulse experiments, often without appropriate checks on tracer and response linearity. Frequency responses are not usually used but are reviewed by Shinnar (12).

Extensions of residence time distributions to systems with multiple inlets and outlets have been described (27-29). If the system contains M inlets and N outlets one can define a conditional density function $E_{ij}(t)$ as the normalized tracer impulse response in outlet j to input in inlet i as shown schematically in Figure 2.

$$E_{ij}(t) = \frac{C_j(t)}{\int_o^{\infty} C_j(t)dt} \tag{9}$$

The exit age density function for the system is given by:

$$E(t) = \sum_{i=1}^{M} \sum_{j=1}^{N} q_{ij} E_{ij}(t)/Q \tag{10}$$

where $Q = \sum_{i=1}^{M} q_i^i = \sum_{j=1}^{N} q_j^o$ is the total flow rate through the system and q_{ij} is the flow rate from inlet i to outlet j. Since q_{ij} s are unknown, they have to be determined from steady state tracer measurements by monitoring the tracer concentration in various outlets while tracer is injected at a steady rate in a single inlet port.

Extensions of the theory to systems with oscillating flows have also been reported (30,31) as well as a stochastic approach to modeling of general compartmental systems (32).

All types of substances can be used as tracers in obtaining residence time distributions as long as the assumption of the linearity of the tracer response is satisfied. In liquids: dyes, organic species with double bonds, electrolytes and radioactive isotopes have been used as tracers. Spectrophotometry, calorimetry, differential refractometry, electroconductivity and radioactive measurements have been used for detection. As long as the system is flow dominated, most of these tracers will behave like the carrier fluid, and the greatest concern is to establish the linearity of the detection system and of the tracer response. In systems where diffusion is dominant even radioactive isotopes may not behave like the carrier fluid due to the difference in the diffusion coefficient for the tracer and for the traced species (33). In gases, radioisotopes and different gas species have been used as tracers, while thermal conductivity, mass spectroscopy and flame ionization have been employed for detection. Again, in flow dominated systems many tracers are satisfactory. In situations where diffusion dominates extra caution should be taken.

There is no reference that summarizes all tracer techniques. Some information is available for inert tracers (34), for chemically reactive tracers (17), for tracers suitable for flow visualization (35) and for radioisotopes as tracers (36-38). In Chapter 2 of Wen and Fan (58) most of the reported tracer studies in chemical reaction engineering are summarized. This includes numerous studies in various reactors using the above described tracers and techniques.

4. MODEL INDEPENDENT APPLICATIONS OF RTDS

4.1 Mean Residence Time, Stagnancy and Bypassing

We have introduced a set of probability density functions for characterization of flow patterns in process vessels and have established that they can be evaluated by properly executed tracer tests. An important question is: What model-free information do they provide about the reactor (process vessel) which is being investigated? Even a quick visual observation of E(t) curves (or F curves) can establish departure from ideal reactors and give guidance as to the degree of mixing present in the system, as illustrated in Figure 3a. Presence of stagnancy, bypassing or internal recirculation is also sometimes evident (Figure 3b). This is already powerful information in troubleshooting. It can be used as a diagnostic test to monitor what effect changes in reactor geometry or baffles positioning or mixing rate, etc. might have on the tracer response curve. That is the approach used in industrial practice. Frequently the problem is solved by trial and error when conditions are found that yield an RTD close to the one of the desired ideal reactor. What quantitative conclusions can we derive on a model-free basis from these tracer experiments?

It can readily be shown that the mean (centroid) of the exit-age density function is the mean residence time, V/Q.

$$\bar{t} = \mu_{1E} = \int_0^\infty tE(t)dt = V/Q \tag{11}$$

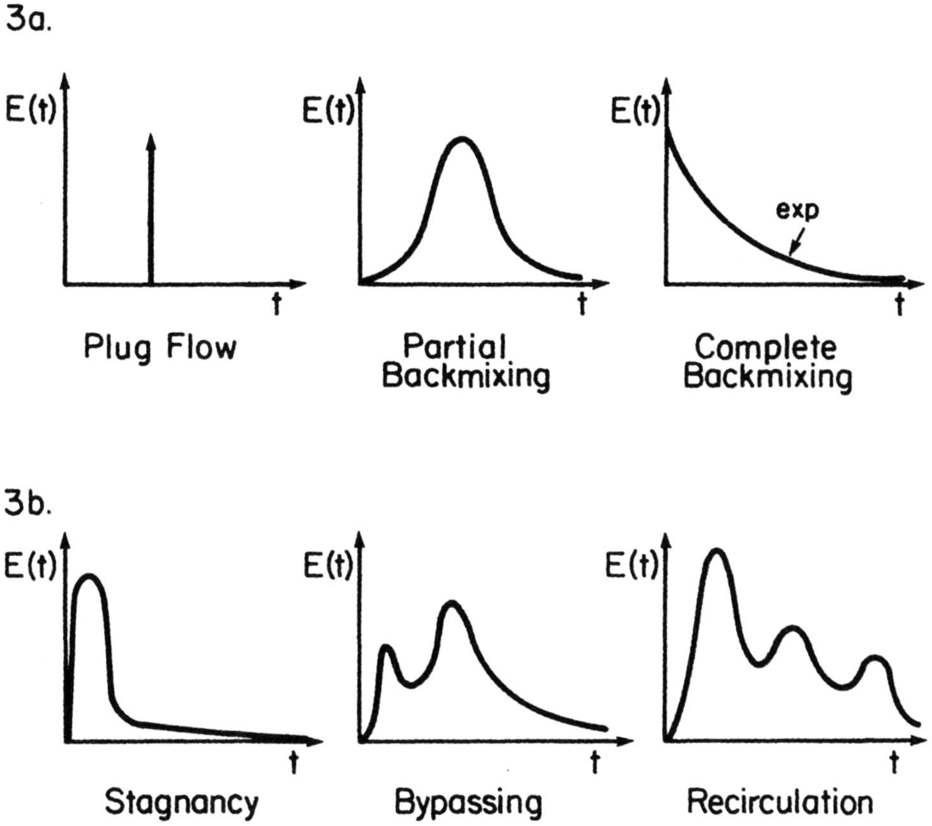

Figure 3. Detection of Typical Mixing Patterns by Tracer Tests.

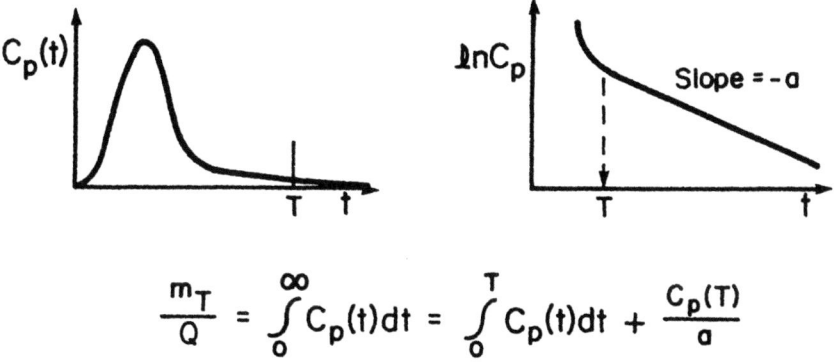

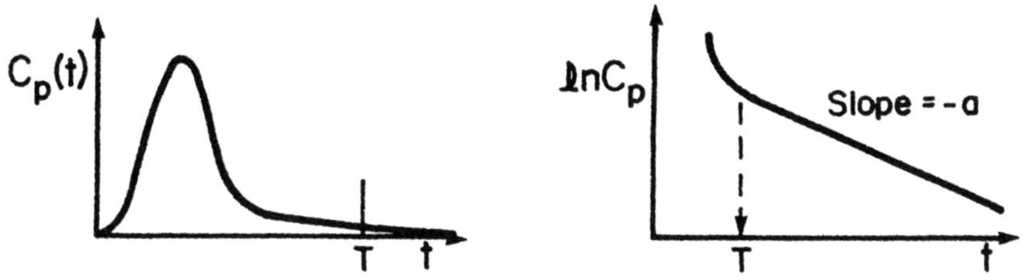

$$\frac{m_T}{Q} = \int_0^{\infty} C_p(t)dt = \int_0^{T} C_p(t)dt + \frac{C_p(T)}{a}$$

Figure 4. Mass Balance for Pulse Tests and Exponential Fit of the Tail.

This relationship has been proven for closed systems independently by several investigators in the chemical engineering literature (1,39,40); yet, it was known at least several decades earlier to medical investigators as the central volume principle (41,42). The necessary assumptions of system closedness (no diffusion across flow boundaries) and an elegant and concise proof of the central volume principle have been recently summarized by Aris (5).

Equation (11) is model independent and provides the means for determining the unknown volume of the system. The question arises in practice whether $\bar{t}$ can be evaluated accurately from eq. (11). Small errors in the tracer mass balance, eq. (7), can lead to large errors in $\bar{t}$ and even larger errors in the estimation of the variance of the E curve (43). Due to the relationship of E(t) and F(t) the mean residence time can also be obtained directly from a step-up or a step-down test:

$$\frac{V}{Q} = \bar{t} = \int_0^\infty [1 - F(t)]\, dt = \int_0^\infty W(t)\, dt \tag{12}$$

Determination from the step-down test is the most accurate (43).

The central volume principle can be extended to a multiphase system consisting of P various immiscible subregions but with a single flowing phase (5,44). Then, if the processes for tracer exchange among various phases are linear and the equilibrium distribution of tracer is linear, the mean of an impulse tracer response is:

$$\frac{V_f + \sum_{i=1}^{n} K_{fi} V_i}{Q_f} = \frac{\int_0^\infty t\, R(t)dt}{\int_0^\infty R(t)dt} \tag{13}$$

Here K_{fi} is the equilibrium constant, i.e. the ratio of tracer concentration in phase i and the flowing phase f when the two are in equilibrium, V_f and Q_f are the volume and flow rate of the flowing phase f, and V_i is the volume of phase i in the system. Equation (13) then describes the mean residence time of a particular tracer in the system. Its mean residence time consists of the sum of residence times in the various regions. These may be different from tracer to tracer, since K_{fi} will vary depending on the tracer properties. Equation (13) has been used extensively in medicine to measure the volumes of distribution of various substances in different organs such as the liver, brain, heart muscle, etc. (23-25). Its use in chemical reactors will be illustrated later when dealing with heterogeneous systems. This equation is used in practice to determine a system's inventory as shown for chlor-alkali plants by Hines (45) who used it to find the mercury content in the cells.

Stagnancies are major problems in real systems. However, theoretically, as indicated by eqs. (11) and (12), no stagnancy can ever exist in the system, in as much that all the system's volume is accessible to tracer at least by diffusional processes. In practice, however, the tail of the exit age density function, or of the washout curve, is difficult to measure accurately. Often when the measurement noise masks the response the tail is truncated. The truncated tail represents the contribution of tracer from regions that are accessible by diffusion only. The first moment of the truncated response $\bar{t}_{app} < V/Q$ and the difference is attributed to stagnant volume $V_s = (\bar{t} - \bar{t}_{app})\,Q$. Clearly, this is not a precise definition of stagnancy since the stagnant volume would depend on who did the experiment, what detector was used, what was the level of noise and when was the response truncated. If the tail is truncated, then the tracer impulse response should be normalized based on the area under the curve to obtain a proper density function that approximately describes the distribution of residence times in regions through which there is active flow. Such an experimental density function should not be matched to a model predicted one to evaluate diffusional parameters of the model since most of the diffusional information is lost in the truncated tail.

It is preferred to extrapolate the tail of the exit age density curve or washout curve by an exponential decay (4,46) and to satisfy the mass balance, eq. (7), exactly, as shown in Figure 4. The theoretical justification for this relies on the fact that the exit age density function always must have an exponential tail (4,24,47). This is obviously true if we represent the system by a set of interchanging mixed compartments in series or parallel (24), or if we model it as a fully developed convective flow profile with diffusional exchange with stagnant zones (47). Only convective patterns which neglect molecular diffusion can result in non-exponential tails for E or W curves (4). The exponential tail will contain the smallest eigenvalue which is either inversely proportional to the largest local mean residence time of a particular compartment, or inversely proportional to the characteristic diffusional time constant for the largest diffusional distance perpendicular to flow. The question still arises whether we can experimentally observe the time behavior when the smallest eigenvalue becomes dominant and correctly extrapolate the tail. If that is not the case, and the extrapolation was done at earlier times, the mass balance eq. (7) will not be 100% satisfied and eq. (11) will determine the mean residence time to be smaller than V/Q. When tracer experiments are done carefully and precisely, not only the mass balance, eq. (7), but also the central volume principle, eq. (19), is always satisfied.

For example, in Figure 5 the first moment of a tracer impulse

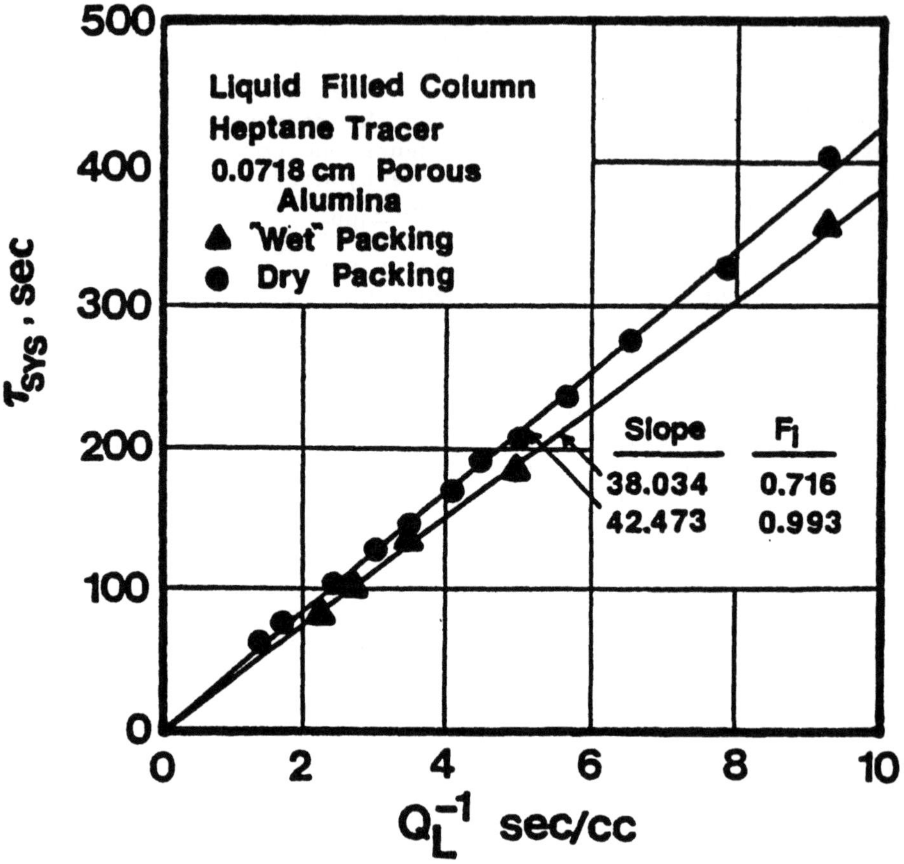

$\tau_{sys} = \mu_{1E}$ = mean residence time

Q_L = liquid flow rate

F_i = fractional pore fill-up

Figure 5. The Effect of Accessible Bed Porosity on Tracer Mean Residence Time.

response for a liquid flow through a packed bed of alumina particles is plotted versus reciprocal flow rate. The slope is the volume of the system accessible to tracer which is $\varepsilon_T V$ or $[\varepsilon_B + (1-\varepsilon_B)\,\varepsilon_p]\,V$ where ε_T is the total porosity, ε_B is the external bed porosity, ε_p is the particle porosity and V is the reactor volume. One curve was obtained with deactivated alumina which contained adsorbed water and hence reduced ε_p, the other curve was obtained with activated alumina from which all water was desorbed resulting in larger ε_p (48). Due to the accuracy of the tracer method employed (49) not only the tracer tests agreed within 2% of the independently measured bed total porosity but the difference between two tracer curves agreed within 5% with the volume occupied by water which was independently determined by gravimetric means (50). This demonstrates that tracer experiments, when properly executed, are very powerful but that many variables can affect their results (in this case the state of activation of the packing). Some doubts about obtaining the same mean residence time from pulse and step tracer experiments were expressed by Schiesser and Lapidus (51,52) but Rothfeld and Ralph (53) demonstrated that when experiments are properly conducted eqs. (11) and (12) yield identical results. Curl and McMillen (43) showed that Shiesser and Lapidus' anomalous results were caused by their inability to measure the tail of the impulse response accurately.

The only violation of the central volume principle was reported by Awasthi and Vasudeva (54) who unfortunately did not demonstrate the linearity of their tracer responses and who dealt with two moving phases and two inlets and outlets. Their anomalous results could have been caused by poorly executed experiments, by nonlinearity of tracer responses (since the symmetry of W(t) and F(t) and the integrability of the impulse response E(t) to obtain F(t) were not checked), or by the movement of tracer in and out of the system by two flowing phases.

Given the proper exit age density function which satisfies eq. (7) and eq. (11), the mean age of the fluid in the system can also be evaluated. By definition it is given by the first moment of the internal age density function, I(t):

$$\mu_{1I} = \int_0^\infty t\, I(t)dt = \frac{\bar{t}}{2}\,[1 + \bar{\sigma}_E^2] \tag{14}$$

where $\bar{t}$ is the mean residence time for the system and $\bar{\sigma}_E^2$ is the dimensionless variance of the exit age density function:

$$\bar{\sigma}_E^2 = \int_0^\infty (t - \bar{t})^2\, E(t)dt/\bar{t}^2 \tag{15}$$

The mean age in a CSTR ($\bar{\sigma}_E^2 = 1$) is the same as the mean residence time since, due to the perfect mixing assumption, there is no difference in age distributions in the vessel and in the outflow. The mean age in a PFR ($\bar{\sigma}_E^2 = 0$) is half the mean residence time. Systems that have mean ages in excess of the mean residence time ($\bar{\sigma}_E^2 > 1$) may be defined as exhibiting stagnancy. However, large values of the variance are difficult to obtain accurately from pulse experiments due to the inherent inaccuracies in the tail of the E(t) curve discussed previously. A somewhat more accurate expression for the variance may be obtained from:

$$\bar{\sigma}_E^2 = 2 \int_0^\infty tW(t)dt/\bar{t}^2 - 1 \tag{16}$$

Clearly the variance of residence times in the outflow is an important measure of the flow pattern in the system. A small dimensionless variance indicates little deviation from plug flow. Large variance, but less than unity, indicates imperfect mixing or distributed flow pathways with intermixing, while a variance larger than one points to stagnant zones.

So far we have reviewed a powerful model-independent formula for estimation of the volume of distribution of various substances in hetero- or homogeneous systems with a single flowing phase, eq. (13). We have also introduced two semi-quantitative ways of assessing stagnancy in the system. One involves truncating the tail of the pulse response or washout curve and attributing the difference between the actual mean residence time, V/Q, and the calculated mean residence time, $\bar{t}_{app}$, to stagnancy. The second involves extrapolation of the tail of the exit age density function so that both the mass balance, eq. (7), and the central volume principle, eq. (11), are satisfied. Examination of the variance of the E curve can then provide information on stagnancy ($\bar{\sigma}_E^2 > 1$). Neither method is truly satisfactory in a quantitative sense. The first is rather arbitrary, the second, while theoretically more satisfying, is plagued by the problem that small errors in satisfying eq. (7) and eq. (11) may still lead to large errors in eq. (15) for the variance (43). Wherever possible, step tests, eq. (12) and eq. (16), are recommended. Both methods are useful and informative in a semiquantitative sense.

As a semiquantitative, model-independent diagnostic of stagnancy one can use the approach of Naor and Shinnar (55) based on the intensity function. The approach relies on the realization that a well behaved flow pattern will have a monotonically increasing probability for fluid elements to exit in the next time interval as their age in the system increases. The intensity function is the following conditional probability density function:

$$\Lambda(t) = \frac{E(t)}{W(t)} \tag{17}$$

The intensity function for an ideal CSTR is $1/\bar{t}$ since the probability of exiting is independent of fluid age. In case of a PFR $\Lambda(t)$ is a delta function at the mean residence time since all elements of that age must exit. For any well behaved flow with no bypassing or stagnancy $\frac{d\Lambda}{dt} > 0$ and $\frac{d^2\Lambda}{dt^2} > 0$. When bypassing is present a local maximum in Λ may develop while stagnancy or bypassing will also cause a decaying portion of Λ over a certain range of t. Sketches of a few models and corresponding intensity functions are shown in Figure 6.

The intensity function is a powerful tool in identifying flow branching into dramatically unequal residence time paths (bypassing) or in detecting slow exchange between mean flow and regions perpendicular to the main flow direction (stagnancy). However, it does not provide a rigid rule for quantification of stagnancy or bypassing. This is all that can be expected from the tracer method since the meaning of stagnancy and bypassing depends on the reaction time scale at hand. The fraction of flow that spends ten seconds in a reactor of 1 minute mean residence time may be considered as bypassing if the reaction characteristic time is 1 minute. However, if the reaction time is ten seconds, it is the other fraction of flow that spends a long time in the reactor that is stagnant. A volume is stagnant if the flow that sweeps it has residence times orders of magnitude larger than the characteristic reaction time but may not be considered stagnant if the reaction time is of a comparable order of magnitude.

For further interpretation of stagnancy or bypassing we need a flow model. At that point, however, all information generated loses its generality and becomes model dependent. We will consider nonideal reactor models later after we review some other model independent features of the RTDs.

4.2 Reactor Performance for First-Order Reactions

Experimentally determined residence time distribution do not describe the flow pattern uniquely. They only give the information on the distribution of residence times for all fluid elements. This distribution can be achieved by an infinite variation of mixing patterns. For example, all three systems in Figure 7 have the RTD of a perfect CSTR but different mixing patterns.

The information on RTD alone is sufficient to completely determine reactor performance for first-order reactions. In a first-order process the remaining reactant concentration in any fluid element depends only on the residence time of the element in the reactor and is independent of the environment and intermixing that the element experienced while in the reactor. This is so

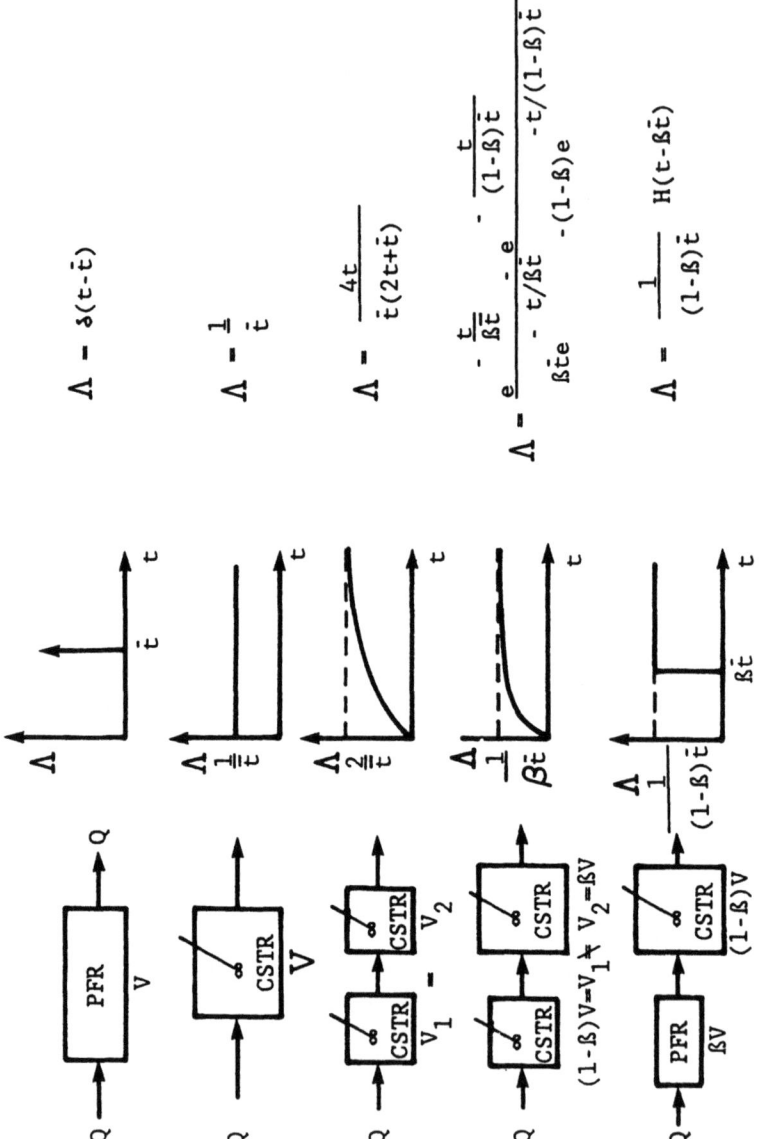

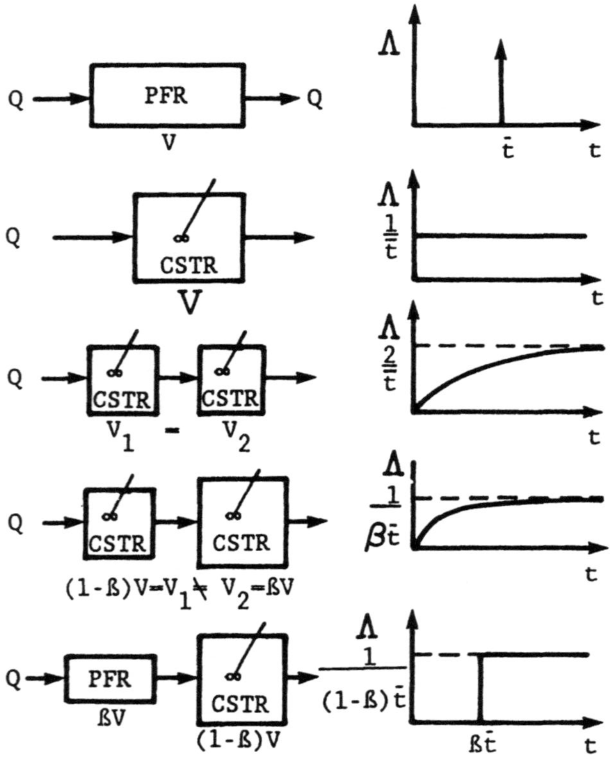

$$\Lambda = \delta(t-\bar{t})$$

$$\Lambda = \frac{1}{\bar{t}}$$

$$\Lambda = \frac{4t}{\bar{t}(2t+\bar{t})}$$

$$\Lambda = \frac{e^{-\frac{t}{\beta\bar{t}}} - e^{-\frac{t}{(1-\beta)\bar{t}}}}{\beta\bar{t}e^{-t/\beta\bar{t}} - (1-\beta)e^{-t/(1-\beta)\bar{t}}}$$

$$\Lambda = \frac{1}{(1-\beta)\bar{t}} H(t-\beta\bar{t})$$

Figure 6. The Intensity Function for Some Simple Flow Models.

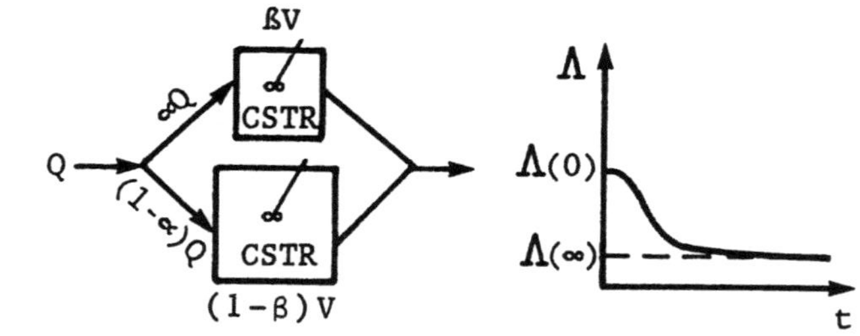

$$\Lambda = \frac{\frac{\alpha^2}{\beta} e^{-\alpha t/\beta\bar{t}} + \frac{(1-\alpha)^2}{1-\beta} e^{-(1-\alpha)t/(1-\beta)\bar{t}}}{\alpha e^{-\alpha t/\beta\bar{t}} + (1-\alpha)\, e^{-(1-\alpha)t/(1-\beta)\bar{t}}}$$

$$\Lambda(0) = \frac{\alpha^2}{\beta\bar{t}} + \frac{(1-\alpha)^2}{(1-\beta)\bar{t}}$$

$$\Lambda(\infty) = \frac{\alpha}{\beta\bar{t}} \qquad \frac{\beta}{\alpha}\bar{t} \le \frac{\bar{t}}{2}$$

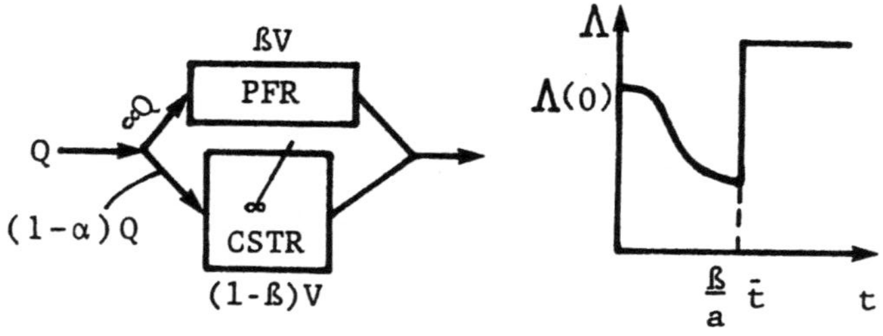

$$\Lambda = \frac{(1-\alpha)/(1-\beta)\bar{t}}{1 + \frac{\alpha}{1-\alpha}\, e^{(1-\alpha)t/(1-\beta)\bar{t}}}$$

For $\quad t < \frac{\beta}{\alpha}\bar{t}$

$$\Lambda(0) = (1-\alpha)^2/(1-\beta)\bar{t}$$

$$\Lambda = (1-\alpha)/(1-\beta)\bar{t}\ ;\ t \ge \frac{\beta}{\alpha}\bar{t}$$

Figure 6. (Continued).

because each molecule behaves independently of its neighbor and undergoes changes completely independent of any other molecule. It is a well known property of the linear rate process that the average rate for N unmixed individual fluid elements is the same as the rate of the completely mixed elements, as shown by eq. (18):

$$\bar{r} = \frac{\sum_{i=1}^{N} kC_i V_i}{\sum_{i=1}^{N} V_i} = k \frac{\sum_{i=1}^{N} C_i V_i}{\sum_{i=1}^{N} V_i} = k\bar{C} \tag{18}$$

In order to calculate the exit reactant concentration for a first-order irreversible reaction in a reactor of known RTD, one must find the reactant concentration in a fluid element of residence time t, i.e. in a batch reactor after time t, multiply it by the fraction of fluid elements of that residence time and sum (integrate) over all permissible residence times. This gives:

$$\frac{C_A}{C_{Ao}} = 1 - x_A = \int_o^\infty e^{-kt}\ E(t)dt = \mathcal{L}\{E(t)\}_{s=k} = \bar{E}(s=k) \tag{19}$$

or

$$\frac{C_A}{C_{Ao}} = \int_o^\infty e^{-Da_1\theta}\ E_\theta(\theta)d\theta = \bar{E}_\theta\ (s=Da_1) \tag{19b}$$

where $\theta = t/\bar{t}$ and $E_\theta(\theta)$ is the normalized density function.

Reactor performance is given by the Laplace transform of the exit age density function evaluated at $s = k$, or at $s = Da_1$ for a normalized E_θ curve. Repeating a first-order reaction at different temperatures, i.e. changing k, one can generate $\bar{E}(s)$ which upon inversion can give E(t). This reactive tracer method is discussed by Shinnar (12). Its practical utility is limited.

Equation (19) is readily extended to a system of n reversible first-order reactions:

$$\underline{C} = \int_o^\infty \underline{C}_b\ E(t)dt \tag{20}$$

Since an element of the column vector of batch concentrations $\underline{C}_b$ is:

$$\underline{C}_{jb} = C_{je} + \sum_{k=1}^{n-1} A_{jk}\ e^{-\lambda_k t} \tag{21}$$

where A_{jk}, C_{je} are constants (which depend on initial and equilibrium composition) and the λ_k's are the (n-1) nonzero eigenvalues of the system. The exit concentration of component j is then given by:

$$C_j = C_{je} + \sum_{k=1}^{n-1} A_{jk}\, \bar{E}\,(s = \lambda_k) \tag{22}$$

For the set of first-order reactions knowledge of linear algebra to obtain A_{jk}, C_{je} and λ_k and knowledge of $\bar{E}(s)$ is all that is necessary to predict reactor performance. Conversely, if exit concentrations, C_j, are monitored and C_{je}, A_{jk} and the λ_k's are obtained from batch experiments or analytically, then the (n-1) values $\bar{E}(s = \lambda_k)$ can be calculated from eq. (22) and E(t) constructed by inversion. A large set of linear reactions could provide complete information about the residence time density function in a single experiment. In practice, such a set is difficult to find and the experiment is hard to execute.

Equations (19) and (22) are theoretically pleasing but their practical utility is limited. In a troubleshooting problem eq. (22) would allow us to recover $\bar{E}(s)$ of the system but we rarely deal with a homogeneous system of linear reactions! In a design problem we often do not know the actual RTD and are trying to design an ideal reactor. If the RTD can be predicted based on a nonideal reactor model, then species concentrations can also be calculated based on that model. Then eq. (22) represents at best only a mathematical short-cut!

An important ramification of eq. (19) and eq. (20) is that a plug flow reactor is the optimal reactor for first-order reactions. For a single reaction that is easy to see. If we have a PFR of mean residence time $\bar{t}$ and a reactor of an arbitrary RTD of the same mean residence time, then the PFR will yield the lowest exit reactant concentration. The reactor with the arbitrary RTD can be viewed as a set of PFRs of different residence times in parallel. Some of these with residence times lower than $\bar{t}$ have higher C_{Ab} than the desired value, those with residence times higher than $\bar{t}$ have lower C_{Ab}. However, the concentration of the mixture lies on the tie-lines which connect elements of various residence times on the C_{Ab} vs t curve. Since C_{Ab} is a convex function of time, i.e. $C'_{Ab} < 0$ and $C''_{Ab} > 0$ for all t, then tie-lines always lie above the curve, and C_A of the mixture for arbitrary E(t) is always larger than $C_{Ab}(\bar{t})$. A plug flow reactor is clearly the best RTD selection in maximizing conversion for a given $\bar{t}$ from eq. (19). Similarly, if we desire to achieve the maximum concentration of a product C_j in a set of first-order reactions, there is always a residence time in a PFR at which this is accomplished. All other residence times yield C_j less than the maximal value. Therefore, a single PFR operating at the optimal residence time is again the best RTD selection in maximizing the yield from eq. (20) or eq. (22).

The main limitation of eqs. (19) and (22) are that they are applicable only in truly isothermal and homogeneous systems. Conceptually, the first limitation can be overcome for a single first-order reaction by introducing, after Nauman (56), a thermal age density function $E_T(t_T)$. The thermal time (age) is taken to be (assuming the Arrhenius dependence of the rate constant on temperature):

$$\frac{dt_T}{d\alpha} = e^{-E/RT} \tag{23}$$

This reduces the batch concentration expression to:

$$C_{Ab} = C_{Ao}\, e^{-k_o t_T} \tag{24}$$

and eq. (19) to:

$$\frac{C_A}{C_{Ao}} = \int_0^\infty e^{-k_o t_T}\, E_T(t_T)dt_T = \bar{E}_T(s = k_o) \tag{25}$$

where k_o is the frequency factor of the Arrhenius expressions. In order to obtain $E_T(t_T)$ from tracer information, one needs the information on $dT/d\alpha$, i.e. the variation of temperature with age of the fluid element. This depends on heat transfer properties within the system and can only be approximately developed for ideal flow patterns when external heat transfer dominates the temperature distribution. Practical ways of evaluating the thermal age distribution have not been discovered to date. However, its evaluations based on reactor models in order to assess the reactant temperature time history are a valuable learning tool in laminar polymerization reactors (56).

The constraint that the system must be homogeneous can best be understood by a simple example. Consider a first-order catalyzed reaction occurring in a reactor with an RTD of a single CSTR. Now imagine that the actual system takes one of the forms presented in Figure 7. As long as the catalyst of the same activity is uniformly distributed, the system is truly homogeneous and eq. (19) predicts its performance irrespective of its internal details. For a first-order reaction with Damkohler number of $Da_1 = k\bar{t} = 9$, the reactant conversion is $x_A = 0.9$. Let us assume now that the same uniform catalyst activity of Figure 7a is distributed in a p:1 ratio in tanks 1 and 2, respectively, of models of Figure 7b and 7c. This requires that $k_1\beta\bar{t} + k_2(1-\beta)\bar{t} = k\bar{t}$ and $k_1\bar{t} = pk_2\bar{t}$. The performance of the three models of Figure 7 is now given by:

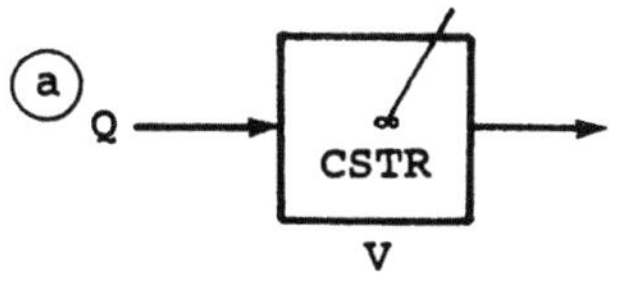

$$\bar{E}(s) = \frac{1}{1 + \bar{t}s}$$

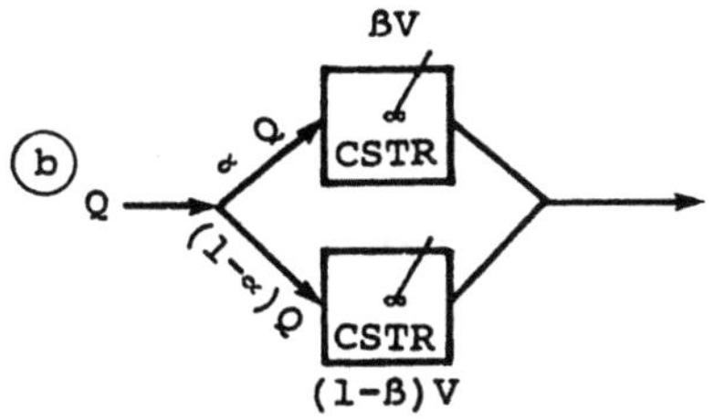

$$\bar{E}(s) = \frac{\alpha}{1 + \frac{\beta}{\alpha}\bar{t}s} + \frac{(1-\alpha)}{1 + \frac{1-\beta}{1-\alpha}\bar{t}s}$$

$$\alpha \equiv \beta \equiv \frac{1}{2}$$

$$\bar{E}(s) = \frac{1}{1 + \bar{t}s}$$

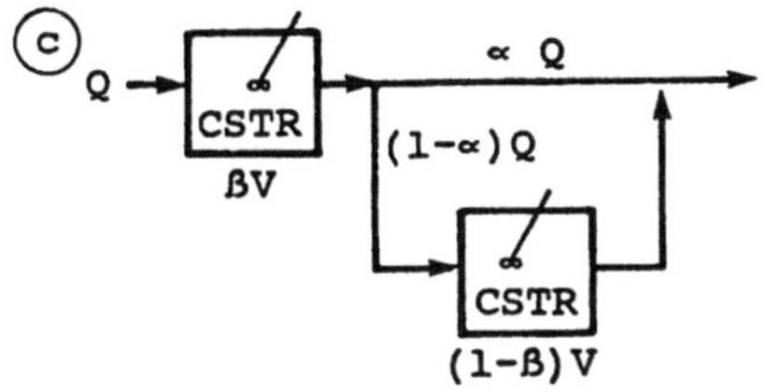

$$\bar{E}(s) = \frac{\alpha}{1 + \beta\bar{t}s} + \frac{(1-\alpha)}{(1+\beta\bar{t}s)\left(1 + \frac{1-\beta}{1-\alpha}\bar{t}s\right)}$$

$$\alpha \equiv \beta \equiv \frac{1}{2}$$

$$\bar{E}(s) = \frac{1}{1 + \bar{t}s}$$

In all cases: $\bar{t} = \frac{V}{Q}$

Figure 7. Some Flow Patterns Leading to an RTD of a Single CSTR.

$$x_{Aa} = 1 - \frac{1}{1 + Da_1} = 0.9 \tag{26a}$$

$$x_{Ab} = 1 - \frac{\alpha}{1 + \frac{\beta}{\alpha} \frac{pDa_1}{1 + (p-1)\beta}} - \frac{1-\alpha}{1 + \frac{Da_1}{1 + (p-1)\beta} \cdot \frac{1-\beta}{1-\alpha}} \tag{26b}$$

$$x_{Ac} = 1 - \frac{\alpha}{1 + \frac{\beta\ pDa_1}{1 + (p-1)\beta}} - \frac{1-\alpha}{(1 + \frac{\beta\ pDa_1}{1 + (p-1)\beta})(1 + \frac{1-\beta}{1-\alpha} \cdot \frac{Da_1}{1+(p-1)\beta})} \tag{26c}$$

An overall RTD of a single CSTR requires that $\alpha = \beta = 0.5$. At an activity ratio of $p = 9$, while still keeping $Da_1 = 9$, we get $x_{Ab} = 0.792$ and $x_{Ac} = 0.925$. Clearly, for a nonhomogeneous system the knowledge of the RTD is not sufficient in predicting reactor performance even for a first-order process. This has important ramifications in dealing with heterogeneous systems.

For heterogeneous systems and nonlinear homogeneous systems more detailed models of the reacting environment during the fluid's passage through the reactor are necessary.

We have shown here that if the system is truly homogeneous its performance can be uniquely determined from its RTD in case of first-order reactions. Thus, tracer methods are of great help in troubleshooting and in determining why the reactor performance is what it is! They are not as helpful in the design problems. If we design a reactor type for which we have no previous experience we have no idea what its RTD may be. We then have several options. We can: (i) solve the detailed hydrodynamic model numerically for the actual reactor geometry, (ii) build a hydrodynamically similar cold model and perform tracer studies to get the RTD or (iii) try to bracket the reactor performance based on some extreme RTDs. The first approach is costly and very difficult to implement for turbulent flows, the third is not accurate enough which leaves only the second approach as the viable one. Prudent designers of new reactor types will follow it.

At first it might seem surprising that the reactor performance for a reactor of a given mean residence time cannot be readily bounded. This seems contrary to what some of our undergraduate students are taught in textbooks. Let us for example treat a single irreversible first-order reaction. The conversion based on the PFR model is:

$$x_{Ap} = 1 - e^{-Da_1} \tag{27}$$

While that based on a CSTR model is:

$$x_{Ac} = \frac{Da_1}{1 + Da_1} \tag{28}$$

Nauman and Buffham (4) report that the difference between eq. (27) and eq. (28) is maximal at $Da_1 = 2.51$ when it reaches the value of 0.204. The relative error between the two conversions, assuming the plug flow value as the design value, peaks at $Da_1 = 1.79$ when it is -23%. Of course, if exactly the same conversion is desired in a CSTR as in the PFR the ratio of the required Damkohler numbers, i.e. the ratio of required reactor volumes, is unbounded as conversion approaches unity:

$$\frac{Da_{1CSTR}}{Da_{1PFR}} = \frac{x_A}{(1-x_A)\ \ell n\ [1/(1-x_A)]} \tag{29}$$

This still does not seem too troublesome since if the desired conversion is x_{Ad} and the maximum acceptable absolute value of the relative error in actual conversion is $|\varepsilon|$, then it seems that the designer could simply conservatively design his reactor based on the CSTR model and select the mean residence time so that:

$$Da_1 \geq \frac{x_{Ad}(1 - |\varepsilon|)}{1 - x_{Ad}\ (1 - |\varepsilon|)} \tag{30}$$

Unfortunately, the RTD of a CSTR does not provide the lower limit on reactor performance. Equation (30) is worthless unless one can guarantee that there is no bypassing and stagnancy in the reactor. To prove that, one needs an experimental RTD and the design is not truly predictive any more. We consider here the RTD based on the generalized tanks in series model, i.e. based on the gamma probability density function, which allows the following representation of the normalized exit age density functions:

$$E_\theta(\theta) = \frac{\left(\frac{1}{\bar{\sigma}_E^2}\right)^{\frac{1}{\bar{\sigma}_E^2}} \theta^{\left(\frac{1}{\bar{\sigma}_E^2} - 1\right)} \exp\ (-\theta/\bar{\sigma}_E^2)}{\Gamma\left(\frac{1}{\bar{\sigma}_E^2}\right)} \tag{31}$$

where $\bar{\sigma}_E^2$ is the dimensionless variance of $E_\theta(\theta)$. When $\bar{\sigma}_E^2 < 1$ this model represents a generalization of the tanks in series model but at $\bar{\sigma}_E^2 > 1$ it becomes an attractive description of

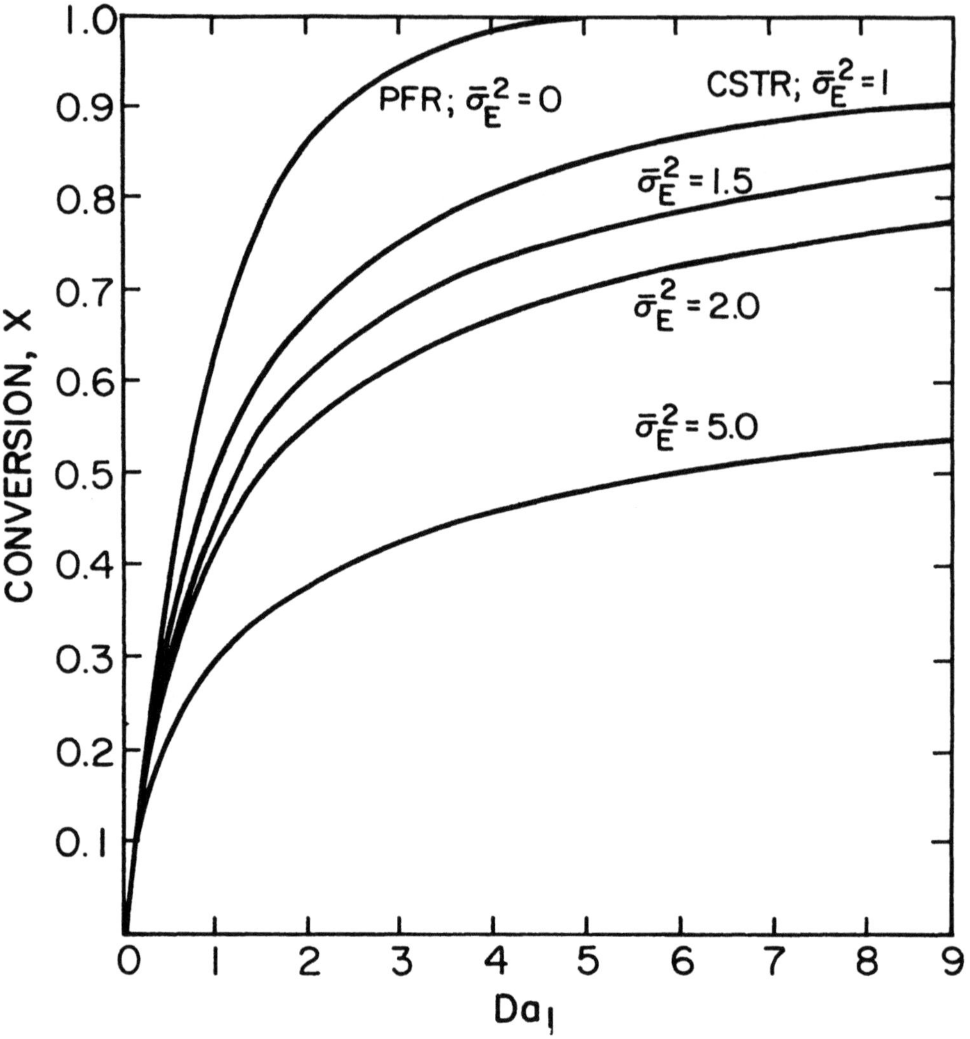

Figure 8. Conversion for a First-Order Reaction.

stagnancy or bypassing. The conversion for a first-order reaction is now given by eq. (19) resulting in:

$$x_A = 1 - \frac{1}{(Da_1 \bar{\sigma}_E^2 + 1)^{1/\bar{\sigma}_E^2}} \tag{32}$$

Equation (32) predicts conversions lower than that in a CSTR when $\bar{\sigma}_E^2 > 1$. At $\bar{\sigma}_E^2 = 1$, CSTR performance is recovered. Figure 8 gives the conversion for a first-order process as a function of Damkohler number as predicted by eqs. (27), (28) and (32). Clearly, only the upper bound on performance can be established in absence of information on the RTD or at least on its variance. This is not of great help to the designer. Thus, we conclude that knowledge of RTDs is essential for predicting reactor performance for first-order reaction systems.

In nonlinear reactions the rate of reaction depends on the rate of encounter of reactant molecules. Therefore, the information on residence time distribution (macromixing), as determined by inert tracer studies, is not sufficient to predict reactor performance uniquely. One must also know what environment was experienced by the reactant molecules during their passage through the reactor, i.e. micromixing information is necessary. Zwietering (39) showed that for a known RTD two extreme models of micromixing (segregated flow and maximum mixedness) bound the reactor conversion for an n-th order reaction. However, more detailed micromixing models are necessary to properly predict selectivity in multiple nonlinear reactions. Various aspects of the micromixing models are considered in this volume by Villermaux. It suffices to add that the broader the RTD, i.e. the larger $\bar{\sigma}_E^2$, the more pronounced micromixing effects can be.

5. FLOW MODELS FOR NONIDEAL REACTORS

Since the RTD of a particular reactor type does not describe its performance uniquely for nonlinear reactions, it is customary to develop flow models for the reactor. The parameters of the model are then determined from tracer studies and the structure of the model is used to calculate reactor performance.

A flow model implies certain knowledge of the basic flow pattern in the reactor. It assigns some detailed mixing characteristics to the particular RTD. For example, sketches in Figure 7 are three different flow models describing an exponential RTD in more detail. The parameters of the models are either obtained based on physical considerations or are estimated by matching model predictions with experimentally observed tracer responses. Tracer may be injected in different parts of the system and monitored at different points as well in order to obtain the desired information. Models frequently encountered in chemical reaction engineering consist of ideal reactor elements arranged in series, parallel or with recycle. Laminar flow with diffusion into stagnant pockets and the axial dispersion model are also popular in describing tubular or packed bed reactors.

For networks consisting of perfect mixers (CSTRs) and plug flow elements (time delays) the overall transfer function $\bar{E}(s)$, i.e. the Laplace transform of the unit impulse response, can be obtained by using the elementary rules for linear systems. Split points, mixing points and connecting lines are assumed to have no increase in residence time associated with them, i.e. their volume to flow rate ratio is negligibly small. Figure 9 describes the situations that are most frequently encountered. For M parallel subsystems, the response is given by:

$$\bar{E}(s) = \sum_{j=1}^{M} \alpha_j \bar{E}_j \left(\frac{\beta_j}{\alpha_j} s\right) \tag{33}$$

For N subsystems in series the response is:

$$\bar{E}(s) = \prod_{j=1}^{N} \bar{E}_j (\beta_j s) \tag{34}$$

Here α_j is the fraction of flow associated with branch j, β_j is the fraction of total volume of the system associated with subsystem j, while $\bar{E}_j(s)$ is the transfer function of subsystem j (assuming all volume V and flow rate Q goes through it). Finally, a subsystem $\bar{E}_f(s)$ containing a fraction β of the total system's volume with recycle through a subsystem $\bar{E}_b(s)$ at recycle ratio R has the transfer function:

$$\bar{E}(s) = \frac{\bar{E}_f\left(\frac{\beta s}{R+1}\right)}{R + 1 - R\bar{E}_f\left(\frac{\beta s}{R+1}\right) \bar{E}_b\left(\frac{(1-\beta) s}{R}\right)} \tag{35}$$

When there is no residence time associated with the recycle loop, i.e. $\beta = 1$ and $\bar{E}_b = 1$, eq. (35) can also take the following form:

$$\bar{E}(s) = \frac{1}{R} \sum_{n=0}^{\infty} \left(\frac{R}{R+1}\right)^n \bar{E}_f^{\,n} \left(\frac{s}{R+1}\right) \tag{36}$$

This represents repeated convolutions of the first passage time density function, $E_f(t)$, upon itself.

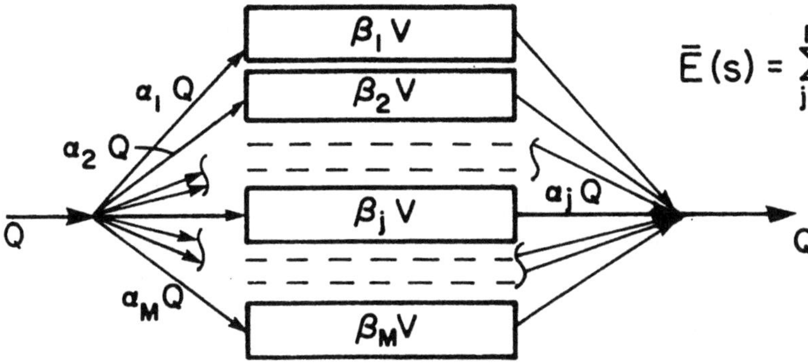

$$\bar{E}(s) = \sum_{j=1}^{M} \alpha_j \bar{E}_j\left(\frac{\beta_j}{\alpha_j} s\right)$$

$$\sum_{j=1}^{M} \alpha_j = 1 \qquad \sum_{j=1}^{M} \beta_j = 1$$

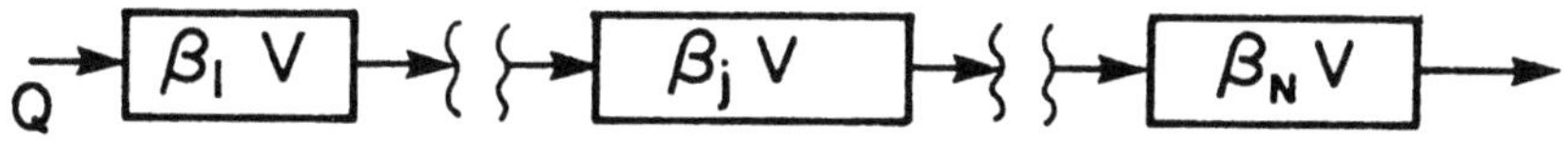

$$\sum_{j=1}^{N} \beta_j = 1 \qquad \bar{E}(s) = \prod_{j=1}^{N} \bar{E}_j(\beta_j s)$$

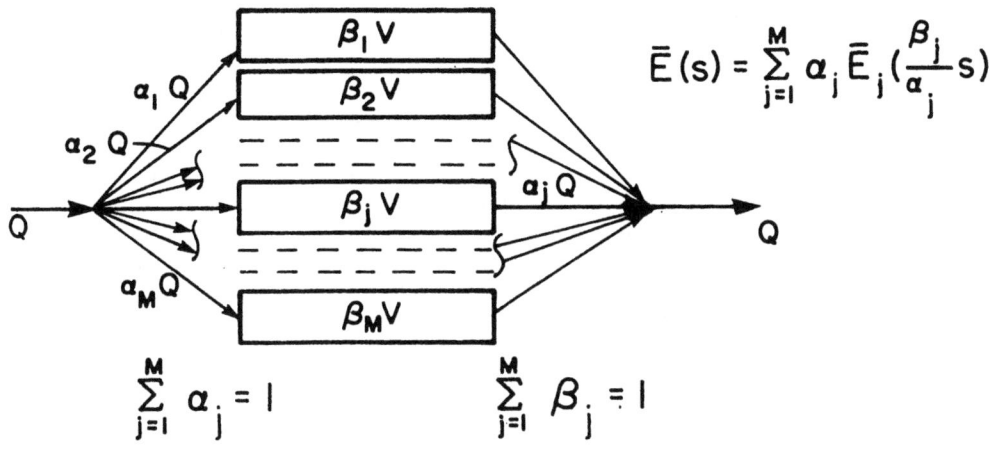

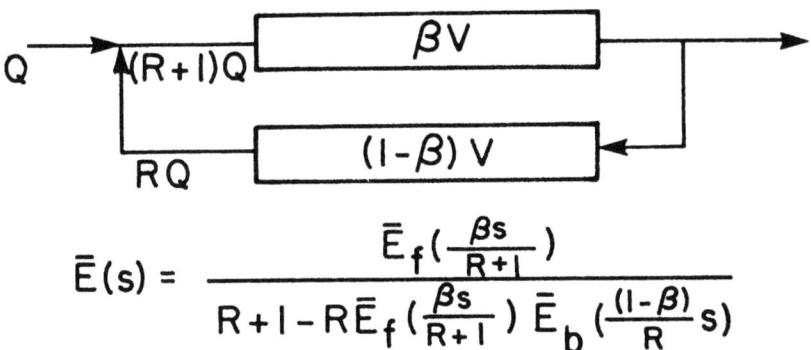

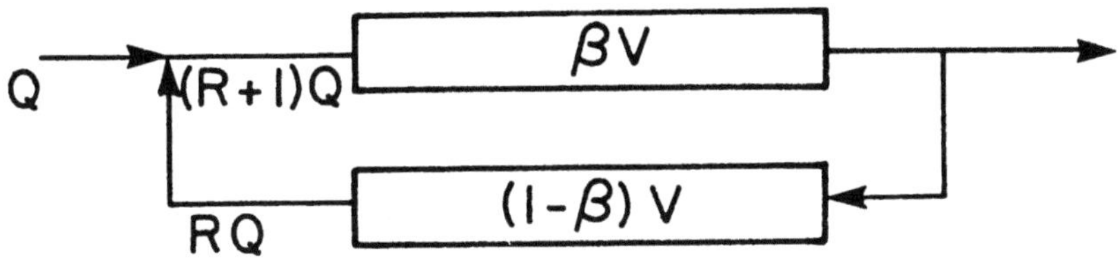

$$\bar{E}(s) = \frac{\bar{E}_f\left(\frac{\beta s}{R+1}\right)}{R+1-R\bar{E}_f\left(\frac{\beta s}{R+1}\right)\bar{E}_b\left(\frac{(1-\beta)}{R} s\right)}$$

Figure 9. Transfer Functions for Systems in Parallel, Series and with Recycle.

For models that involve crossmixing or backmixing against the direction of flow, proper differential equations must be developed and the Laplace transform must be taken and solved for the impulse response at the exit. The same holds true for flow models with diffusion or dispersion when one deals with partial differential equations. Many models which are believed to describe physical reality in some reactor types under certain conditions have been reported in the literature. The standard texts of Levenspiel (2,57) and specialized monographs by Nauman and Buffham (4) and Wen and Fan (58) give a comprehensive treatment of various flow models. Compartmental models are also treated extensively by Sheppard (23) and Lassen and Perl (25). Aris (5) recently listed the available references for various models that involve sequences of stirred tanks in parallel, with bypass, with backmixing, with plug flow, with stagnancy and with recycle. He also reported the references for general networks, arrays of stirred tanks, recycle systems, plug flow with diffusion, flow in helical tubes and stochastic flows. With all this information readily available there is no need for repetition here. Clearly, none of these models is general, but each of them could be used to fulfill a particular goal.

In formulating and using nonideal reactor models one should keep in mind our overall objective which is to build and operate an ideal reactor. Only ideal reactors are scaleable and their performance predictable. The nonideal flow models and experimental RTD curves are needed to assess deviations from ideality. When these deviations are small, a successful one-or-two parameter model can be constructed to interpret them. When deviations are large, one should concentrate on finding ways to diminish them rather than to interpret them. Multiparameter models are difficult to use and have very little value in scaleup. The exceptions are situations when the variations of some model parameters can be predicted independently based on first principles or based on accumulated experimental evidence.

5.1 Stirred Tank Reactor

When designing or operating a stirred tank reactor we would like it to behave as an ideal CSTR. Tracer studies can quickly tell us whether there are deviations from such expected behavior. Semilogarthmic plots of tracer impulse or step-down responses are a particularly useful diagnostic tool in these situations. Any deviations from a single straight line indicates nonideality. If deviations are present the most likely culprits are bypassing or stagnancy. Bypassing may be caused by inlet and outlet ports being in too close proximity. Stagnancy may result from slow exchange between the recirculation flow in the propeller (turbine) region and the flow in the rest of the tank. We can use a model now to quantify these deviations. For example, a two-parameter model of Cholette and Cloutier (59) consists of a well-mixed region, stagnancy and bypass and seems adequate for the task at hand. It can be used in conjunction with tracer studies to investigate the effect of the propeller and turbine type, baffle positioning and rotational speed on model parameters, i.e. on the fraction of stagnant volume and fraction of flow that is bypassing. Such systematic studies have been reported by Nagata (60). The results are certainly useful in a qualitative sense. They help eliminate bad mixer designs and poor baffling combinations. However, results are model-dependent and little or no information is obtained for scaleup. Other dead zone models (61), the fractional tubularity model (60), recycle models (62,63), or complex multiparameter models (64,65) could also be used to interpret the tracer results. One should remember that little is to be gained by increasing the number of model parameters unless there is good physical evidence for necessity of such an approach.

Model selection also depends on the task at hand. While the Cholette-Cloutier model (59) is adequate for input-output testing of a CSTR, a recycle model (N-tanks + PFR with recycle) might be more appropriate when the tracer response is measured internally at points in the tank or when the goal is to determine recirculation times (66-68). A recycle model seems an attractive and almost natural interpretation for a flow pattern in a stirred tank. It was frequently surmised that in the limit, as the recycle ratio goes to infinity, every recycle system exhibits an exponential RTD (69,70). Proving that this is not necessarily the case was the subject of a series of papers (71-74). The recent paper by Buffham and Nauman (75) summarizes the findings which are of importance to practicing engineers. Such a practitioner would normally increase the recycle ratio, R, at constant system flow rate by increasing r.p.m. in hopes that ideal CSTR behavior will be obtained. Indeed, if increased r.p.m. guaranteed that flows everywhere are substantially higher than Q, the resulting RTD would be exponential. However, if there are inaccessible regions (behind baffles, etc.) through which material continues to pass at rates comparable to Q, the RTD will deviate from the expected exponential (a two tank model may be required) but the actual $\bar{t} = V/Q$ will be recovered. If there are regions accessible by diffusion only, the RTD will be a defective exponential with $\bar{t}_{app}$ less than V/Q. This is caused by the fact that the tail of the E curve associated with regions reached only by diffusion will not be detectable and will contain a vanishingly small number of tracer molecules. We have discussed this problem earlier.

Inert tracers are mainly used to characterize nonideal model parameters for stirred tanks. However, attempts were made to use the fluctuation of the tracer concentration response, measured optically or by electroconductivity, to evaluate a micromixing parameter of the model such as the degree of segregation. Leitman and Ziegler (76) developed an empirical RTD curve to describe the tracer response in their laboratory stirred tank. This RTD contained information on segregation and its parameters were correlated with rotational speed. Attempts to use the model in predicting conversion of a second-order reaction (saponification of ethyl acetate) were inconclusive (77). The problem is that second-order reactions are not sensitive enough to segregation. Hanley and Mischke (78) used tracer fluctuations in the outlet to infer the degree of segregation in the stirred tank. Studies of reaction conversion did not test the model critically.

Spencer et al. (79,80) used reactive tracers to identify the so-called segregation function that describes transition from segregation by age to that by life expectancy. The predictive ability of the model remains uncertain. Lintz and Weber (81) used an autocatalytic reaction to evaluate parameters of a micromixing model. Flow visualization studies using the clock reactions of Denbigh and Danckwerts (82,83), whereby color change occurs after a specified time upon mixing of reactants, and regional tracer dilution studies by gamma camera (84) have also been employed to characterize mixing in stirred tank reactors. This does not include tracer studies aimed at solely identifying micromixing and turbulence parameters which will be briefly mentioned later.

For interpretation of RTDs in systems with multiple impellers or in systems agitated by gas, such as a bubble column, one should consult Nagata (60), Wen and Fan (58) and Meclenburgh and Hartland (85). For description of liquid backmixing in gas-stirred reactors the axial dispersion model dominates the literature (86). Nevertheless, depending on the tasks at hand, a model for mixed tanks with backmixing of Klinkenberg (87), a recycle model with cross-mixing (88) and the plume model (89,90) are worth considering.

5.2 Tubular and Packed Bed Reactors

These reactors are built with the purpose of approaching plug flow. A simple one parameter model should be used in assessing deviations from expected plug flow behavior for tubular reactors with or without inert nonporous packing. The only exceptions are short beds with large deviations from plug flow or beds packed with porous particles. In the former case a two-dimensional model is needed while in the latter case diffusional contributions have to be accounted for by additional parameters.

The transfer functions and normalized exit age density functions for three plausible one-parameter models are listed in Table 1. They are: the axial dispersion model, the N-stirred tanks in series model and the gamma probability density model. The dimensionless variance of the experimentally measured $E_\theta(\theta)$ curve yields the parameter for the chosen model:

Table 1. One-Parameter Models for Tubular Reactors

1. The Axial Dispersion Model

$$\bar{E}(s) = \frac{4\sqrt{1+\frac{4s}{Pe_{ax}}}\ \exp\{\frac{Pe_{ax}}{2}[1-\sqrt{1+\frac{4s}{Pe_{ax}}}]\}}{(1+\sqrt{1+\frac{4s}{Pe_{ax}}})^2 - (1-\sqrt{1+\frac{4s}{Pe_{ax}}})^2 \exp\{-Pe_{ax}\sqrt{1+\frac{4s}{Pe_{ax}}}\}}$$

$$E(\theta) = 2e^{Pe_{ax}/2} \sum_{n=1}^{\infty} \frac{\omega_n \sin\omega_n\ [Pe_{ax}^2 + 4\omega_n^2]\ \exp\{-\frac{Pe_{ax}^2 + 4\omega_n^2}{4Pe_{ax}}\theta\}}{Pe_{ax}\ [Pe_{ax}^2 + 4Pe_{ax} + 4\omega_n^2]}$$

Approximate Solution for Long Columns ($Pe_{ax} > 10$)

$$E(\theta) = \sqrt{\frac{Pe_{ax}}{4\pi\theta}}\ \exp[-\frac{Pe_{ax}(1-\theta)^2}{4\theta}]$$

2. The N-Stirred Tanks in Series Model

$$\bar{E}(s) = \frac{1}{(1+\frac{s}{N})^N} \quad ; \quad E(\theta) = \frac{N^N \theta^{N-1} e^{-N\theta}}{(N-1)!}$$

3. The Gamma Density Model

$$\bar{E}(s) = \frac{1}{(1+\frac{s}{M})^M} \quad ; \quad E(\theta) = \frac{M^M \theta^{M-1} e^{-M\theta}}{\Gamma(M)}$$

$$M = 1/\bar{\sigma}_E^2$$

$$\bar{\sigma}_E^2 = \frac{1}{p} = \frac{2D_{ax}}{\bar{u}L} = \frac{2}{Pe_{ax}} \approx \frac{1}{N} \tag{37}$$

Here $\bar{\sigma}_E^2$ is the experimentally measured variance of the $E_\theta(\theta)$ curve, N is the number of tanks in series, p is the parameter of the gamma density model, D_{ax} is the axial dispersion coefficient based on total cross-sectional area, $\bar{u}$ is the superficial velocity and L is the total reactor length. It is unfortunate that $D_{ax}/\bar{u}L$, which is the intensity of dispersion, is identified with the reciprocal of the axial Peclet number, Pe_{ax}. To make matters worse the axial Peclet number is often defined based on tube diameter or particle diameter generating additional confusion among students of the subject. The Peclet number should be reserved to represent the product of the Schmidt and Reynolds numbers.

The axial dispersion model has been discussed exhaustively in the literature. The reader is referred to Levenspiel (57), Nauman and Buffham (4), Wen and Fan (58), and Levenspiel and Bischoff (91) for numerous available references. The appropriateness of various boundary conditions has been debated for decades (92-95) and arguments about their effect on reactor performance continue to the present day (96). We now know that the Danckwerts boundary conditions make the model closed so that a proper residence time distribution can be obtained from the model equations given below (when the reaction rate term is set to zero):

$$\frac{1}{Pe_{ax}} \frac{\partial^2 C}{\partial z^2} - \frac{\partial C}{\partial z} + R = \frac{\partial C}{\partial \theta} \tag{38}$$

$$z = 0 \quad ; \quad \frac{1}{Pe_{ax}} \frac{\partial C}{\partial z} - C = C_o \tag{39}$$

$$z = 1 \quad ; \quad \frac{\partial C}{\partial z} = 0 \tag{40}$$

What boundary conditions are satisfied in practice and how to obtain the RTD experimentally is less clear. The uses of two-place measurements and the extension of impulse response theory to open systems are discussed in (4). For small intensity of dispersion (large Pe_{ax}) the boundary conditions effect diminishes and approximate expressions can be used for the E-curve (see Table 1).

The theoretical justification and experimental verification for the interpretation of tracer responses in laminar and turbulent flow by the axial dispersion model are provided by G. I. Taylor (97-99), Aris (100) and others. In fully-developed laminar flow in a circular tube the axial dispersion coefficient is (97,100):

$$D_{ax} = D_m + \frac{\bar{u}^2\, d_t^2}{192 D_m} = D_m \left(1 + \frac{1}{192} Pe^2\right) \tag{41}$$

where Pe = Re·Sc.

For fully-developed turbulent flow in a smooth cylindrical pipe theory (98,99) suggests:

$$D_{ax} = k\, d_t \bar{u} \sqrt{f} = k' d_t \bar{u} \cdot Re^{-1/8} \tag{42}$$

where

$$k = 7.14 \text{ to } 10.1,\ k' = 2 \text{ to } 2.8$$

In the above d_t is the tube diameter and D_m is the molecular diffusivity. For fully developed laminar flow over stagnant pockets of fluid the following expression is suggested for the axial dispersion coefficient in the flowing fluid (100):

$$D_{ax} = D_m \left[1 + \frac{1 + 3\beta + 11\beta^2}{192\,(1 + \beta)} Pe^2\right] \tag{43}$$

Here β is the fraction of fluid volume that is stagnant and Pe = Re·Sc. Perturbation theory has been applied to the fundamental equations for fluid flow in packed beds in order to generate the axial and radial dispersion coefficients. A recent study (101) suggests the following value for the dispersion coefficients at large Peclet numbers:

$$D_{ax} = D_m \left[1 + \frac{3}{4} Pe + \frac{1}{6} \pi^2 (1-\varepsilon)\, Pe \ell n Pe + \frac{1}{15} \left[\frac{K D_m / D_e\, (1-\varepsilon)}{(1-(1-\varepsilon)(K-1))^2} Pe^2\right] \right] \tag{44}$$

$$D_r = D_m \left[1 + 0.278 \sqrt{(1-\varepsilon)\, Pe}\right] \tag{45}$$

Here ε is bed porosity, K is the partition coefficient for tracer concentration between particles and fluid, D_m is molecular diffusivity and D_e is the effective diffusivity in the particle. Contributions to the axial dispersion coefficient arise (in order of appearance in Eq. (44) from molecular diffusion, mechanical mixing due to random velocity profiles, diffusion contributions due to delays in stagnant zones and due to diffusion into porous particles. The radial dispersion coefficient contains only molecular and mechanical contributions. One should note that the asymptotic value of the radial dispersion coefficient is reached rapidly on the order of the mechanical dispersion characteristic time $d_p/\sqrt{1-\varepsilon}\,\bar{u}$. The axial dispersion coefficient, on the other hand, reaches its long-time behavior after a longer time on the order of the characteristic particle diffusion time. In long beds we could combine radial diffusion effects into a single effective axial dispersion coefficient following Bischoff (102).

The above formulas are provided as theoretical guidance for the use of the dispersion model. For evaluation of actual coefficients the reader can consult the numerous experimental studies and correlations for tubes, packed and fluidized beds presented by Wen and Fan (58). One should remember that theory only justifies the use of the axial dispersion model at large Peclet numbers ($Pe \gg 1$) and at small intensities of dispersion, i.e. $D_{ax}/\bar{u}L < 0.15$. Therefore, attempts in the literature to apply the dispersion model to small deviations from stirred tank behavior, i.e. for large intensities of dispersion, $D_{ax}/\bar{u}L > 1$, such as in describing liquid backmixing in bubble columns, should be considered with caution. Better physical models of the flow patterns are necessary for such situations and the dispersion model should be avoided.

Because the parameter of the axial dispersion model, as observed from numerous experimental studies (58), has been so extensively correlated with Peclet number, designers consider the model useful for scaleup and use it for reactor calculations. The model gives a nice analytical expression for prediction of conversion of a single, irreversible first-order reaction ($\bar{E}(s)$ in Table 1 with Da_1 replacing s). The expressions for exit concentrations for a system of reversible first-order reactions with the same axial dispersion coefficient (turbulent flow) are much more complex and their evaluation is computationally demanding. In cases of different dispersion coefficients (laminar flow), the analytical problem is close to intractable. For nonlinear reactions the axial dispersion model leads to a set of two-point boundary value problems which must be solved by an appropriate iterative numerical scheme. This is a great disadvantage of the model. We conclude that the axial dispersion model is cumbersome in reactor type calculations and should be abandoned. The reasons for this can be stated as follows.

The dispersion model is at best only an approximate representation of physical reality, often in contradiction with selected physical evidence. For example, the dispersion model predicts the appearance of tracer upstream of the injection point, yet all the experimental evidence in packed beds points to downstream spreading of the dye only. The boundary conditions of the dispersion model are often difficult to meet in practice. Considering the above discrepancies between the model and physical reality it seems hardly justified to do the elaborate boundary value problem calculations. This is especially true since for RTDs close to PFR micromixing effects cannot be pronounced. It seems more appropriate to relate the intensity of dispersion as given by eq. (37) to an equivalent number of tanks in series. One can select the first larger, N_ℓ, and first smaller, N_s, integers that bracket $\bar{u}L/2\,D_{ax}$ and perform the initial value problem calculations for reactor performance. These integrations of ODEs are well-suited for modern computers. Results obtained with N_ℓ and N_s should be within a few percent of each other. If that is not the case, the intensity of dispersion is too large, N is too small and a more realistic model of the reactor flow pattern should be sought. Similarly one can calculate $p = \bar{u}L/2D_{ax}$ and use the segregated flow model to calculate by integration the reactor performance while approximating $E_\theta(\theta)$ by the gamma density function. When the dispersion intensity is small this should work. When the intensity is large even the dispersion model itself may not represent physical reality well and a multi-dimensional model is needed.

One might object to sacrificing the accuracy in describing physical reality for computational convenience. That is not the case. There is nothing that makes the axial dispersion model a superior representation of mixing in packed beds than approaches advanced by other models. It is the large amount of information compiled on intensity of dispersion and investigators' inertia that propagate the use of this model.

If one is ill at ease with the N-CSTRs in series or the gamma density model as a physical picture of the packed bed (although both have a physical basis) there are other good models to choose from. The Deans' cell model (103) is a proven representation of packed beds which can also be arrived at by probability arguments (47). It leads to an initial value problem of the tanks in series type for reactor performance calculations. From the physical point of view Levich et. al. (47) described well the basic features of flow in a packed bed of nonporous particles. Some of the fluid passes through the bed by random flowpaths of various lengths. The RTD of this portion of the fluid, if the bed is of sufficient length, is approximately Gaussian. The other portion of the fluid gets delayed by multiple excursions into nonflowing or stagnant zones. This contributes a prolonged tail to the actual RTD of the system. Levich showed how to decompose the two effects. His model consisted of a series of N mixed cells each containing a fraction β of stagnant fluid. By equating the variance of ages for this model and the axial dispersion model he established the relationship among model parameters. In our notation this becomes:

$$D_{ax} = \frac{\bar{u}L}{2N} + \frac{\bar{u}L\ \beta^2 Q}{Nq} \tag{46}$$

where $\bar{u}$ is the superficial velocity, N is the number of cells in series, L is the reactor length, β is the fluid volume fraction of stagnancy, Q is the total flow rate and q is the volumetric exchange rate between flowing and stagnant zones. The model is valid for long beds only when $\bar{u}L/D_{ax} > 10$, and it contains three parameters: N, q and β. Usually N is of the order of bed length divided by particle diameter which leaves β and q to be determined. This perhaps can be done by performing a set of tracer experiments on short beds ($10 < N < 10\ Q/q$) at different flow rates Q. Then the position of the maximum of the tracer pulse response occurs at $(1-\beta)\ L/\bar{u}$. The variance of the symmetric part of the response is $1/[N\ (1-\beta)]$. The long tail of the response (not accounted for in the above variance) decays as $\exp(-q\ \bar{u}\ Nt/Q\beta L)$. This provides the relationships for estimation of model parameters q and β which do depend on flow rate Q, particle size, method of packing and physical properties of the fluid but which should not be affected much by the scale of the bed (provided the beds are larger than some minimum length and the fluid is uniformly distributed).

Equation (46) is a cornerstone which establishes a key relationship between the axial dispersion model and other fluid exchange type models. Among them the cross flow models of Hoogendorn and Lips (104) and of Hochman and Effron (105), the probabilistic time delay model of Buffham et. al. (106), or the hopping model of Rathor et. al. (107) are also good representations of flow in packed beds. All of them contain at least two parameters, but lead to initial value problems in reactor calculations. Schwartz and Roberts (108) selected the parameters of various models in such a way to get the same variance of the exit age density function. Under such conditions, when the variance was small enough, all models predicted the same conversion (within negligible differences) for first- and second-order reactions. Their paper points out an intuitively evident result that the nature of the backmixing model for small deviations from plug flow is unimportant. For short beds and/or for nonlinear rates that are affected profoundly by local mixing patterns, such as the reactant inhibited Langmuir-Hinshelwood rate forms, the nature of the mixing model is very important and a one-dimensional model may be inadequate.

One of the recent versions of the cross flow model deserves attention. That is the model of Hinduja et. al. (109) which is based on a set of reasonable physical assumptions. The bed is visualized as consisting of layers perpendicular to the direction of flow. Layer thickness, λ, is a function of bed packing and is of the order of the diameter of the packing. The tracer (reactant) is transported from the flowing fluid element by: convective transport within the flowing stream and exchange of fluid with stagnant regions in the same layer and in the adjacent layers. This results in a model with three dimensionless parameters:

$S = \varepsilon_s/\varepsilon$ = holdup of stagnant fluid/bed porosity

$F = d_p q_f/\bar{u}$ = $\dfrac{\text{(particle diameter) x (exchange flow rate)}}{\text{mean superficial velocity}}$

$L = (\lambda/d_p)^2$ = characteristic dimensionless distance

By equating the rate of change in fluid momentum caused by the exchange flow with the inertial pressure drop term in the Ergun equation a relationship between F and S is established:

$$F = 1.75\ (1-S)^2\ \frac{1-\varepsilon}{\varepsilon} \tag{47}$$

By equating the variance of the model to that for the axial dispersion model one obtains

$$F = S^2\ \left(\frac{\bar{u}L}{D_{ax}}\right) = S^2\ Pe_{ax} \tag{48}$$

The intensity of dispersion Pe_{ax}^{-1} is available as a function of Reynolds number. The evaluation of the final parameter λ requires information on radial spreading of the tracer and could be obtained from reported data on radial Peclet numbers to give

$$Pe_r = \frac{\bar{u}\ d_p}{D_r} = \frac{2}{FL} \tag{49}$$

This allows "a priori" use of the model and results in an initial value problem for reactor performance calculations. This model is recommended when deviations from plug flow are not overly large.

There are no universally accepted models for short beds, fast reactions and unpremixed feeds when deviations from plug flow are appreciable. These situations should be avoided whenever possible, or they must be modeled on a case by case basis, while fully accounting for entrance effects, manner of fluids introduction to the bed, etc. At low Reynolds number numerical schemes based on finite element analysis are the recommended approach. At high Reynolds number one should consult the turbulence literature.

Again, tracer tests can now be invaluable in determining what is going on. However, input-output tests are not sufficient because too many parameters have to be evaluated. Monitoring of localized and regional spreading of tracer is now required in conjunction with plausible models.

6. RESIDENCE TIME DISTRIBUTION THEORY AND TRACER METHODS IN HETEROGENEOUS SYSTEMS

Most of the industrially important processes are conducted in heterogeneous reactors. Gas-solid catalytic reactions are handled in packed-beds, fluidized beds and entrained solids reactors. Gas-liquid and gas-liquid-solid reactions are performed in bubble columns, mechanically stirred reactors, packed-beds with two-phase flow, ebullated beds, etc. The same two key problems of reaction engineering, introduced in Section 2, arise, i.e. the problem of "a priori" design and scale-up and that of troubleshooting. Heterogeneous systems are much more complex than homogeneous ones. This is caused by numerous transport processes, i.e. diffusion, adsorption-desorption, mass and heat transfer across boundary layers, etc., which in addition to kinetics can affect and determine overall reaction rates. Only plug flow or perfectly mixed stirred tank reactors can be designed and scaled up with confidence. Tracer methods are indirectly of great help in "a priori" design and scaleup through their use in cold flow modeling and in the quantification of flow patterns in the proposed reactor geometry. They are again invaluable for troubleshooting in existing reactors. It is important then to assess what information can be obtained based on the use of tracers on a model-free basis and which quantities will be model dependent. We need to know which of the conclusions and methods that were valid in homogeneous systems can be directly extended to heterogeneous ones.

6.1 Residence Time Distributions in Systems With Several Environments

We can think of a heterogeneous reactor as a system with several environments. Each environment is a space within the reactor that is occupied by a particular phase (e.g. gas, solid, liquid). There may be a single inlet and outlet with only one phase flowing through the system, as would be the case for packed-beds, fluidized-beds with no solids carryover, semi-batch bubble columns or stirred slurry reactors. In other cases, e.g. packed-beds with two-phase flow, continuous bubble columns, etc., there may be more inlets and outlets and two or more phases might be entering and leaving the system. The reaction of interest will often only occur in one phase, i.e. in the reaction environment, but requires the transport of reactants from other environments into the reaction environment. Thus, in order to calculate the reactor performance, we should know the residence time distribution of reactants in the reaction environment and the micro-mixing properties of the reaction environment. This leads us to the realization that the residence time distribution for the system cannot be defined since all tagged components may exhibit different residence time distributions.

In a homogeneous system we were able to define the RTD for the system and to determine it experimentally by the use of perfect tracers. Each species in a homogeneous system has the same RTD equal to the RTD of the system. The above argument is based on the fact that RTDs in homogeneous systems of interest, such as reactors, are flow dominated. Convection and turbulence determine the RTD and do not allow differences in residence times to develop due to differences in species diffusivities. In heterogeneous systems various components in the feed phase (e.g. gas) may have different adsorption affinities for the solid or different solubilities in the liquid. Therefore, they will have different residence times in the reactor. Measuring all these RTDs would definitely be impractical.

Let us consider then what can be done to characterize the heterogeneous systems in a practical way. From a theoretical standpoint it seems that we should determine the RTD of reactants in the reacting environment. Then, by analogy to homogeneous systems, we should be able to predict reactor performance in case of first-order reactions and possibly bound it for n-th order reactions. From the practical standpoint it is clear that we can perform tracer experiments using tracers that distribute themselves differently between various environments. In the limit we can use some that cannot enter at all a particular environment and hence trace the flow of a single phase only. We will now examine whether such tracer studies could yield the information of interest.

The notion of calling time by different names in different environments seems to date back to a seminar that John Beek gave at several universities (110,111). Aris (6, 110) introduced the proper terminology from probability theory and defined (while using different symbols than those used below) for a system with a single inlet and outlet and n-environments:

E_n $(t_1,t_2,t_3 \ldots t_n)$ dt_1 $dt_2 \ldots dt_n$ = (fraction of the fluid in the outflow that has spent time between t_i and $t_i + dt_i$ in each i-th environment for all $i = 1,2 \ldots n$) (50)

This is a multivariable joint probability density function and therefore:

$$\underbrace{\int_o^\infty \int_o^\infty}_{n} E_n\,(t_1,t_2 \ldots t_n)\; dt_1 dt_2 \ldots dt_n = 1 \tag{51}$$

The marginal probability density $E_{1i}(t_i)dt_i$ is the fraction of the outflow that has spent time t_i to $t_i + dt_i$ in the i-th environment irrespective of the times spent in other environments. It is obtained by integrating E_n over all the t_js except for t_i.

$$E_{1i}(t_i) = \underbrace{\int_o^\infty \int_o^\infty}_{n-1} E_n(t_1,t_2 \ldots t_n)\; dt_1\; dt_2\; dt_{i-1} dt_{i+1} \ldots dt_n \tag{52}$$

The density function for the distribution of total time $t = \sum_{j=1}^{n} t_j$ is given by:

$$E(t) = \underbrace{\int \int}_{n-1} E_n\,(t-t_2-\ldots-t_n,t_2,t_n)\; dt_2 \ldots dt_n \tag{53}$$

where the integral is performed over the region in the positive orthant enclosed by the planes $t_i \geq 0$, $i = 1,2,3$ n and the hyperplane $\sum_{j=1}^{n} t_j = t$.

The expected values of t_i give the mean residence times spent in each i-th environment ($i = 1,2 \ldots n$), the variances σ_i^2 measure the spread of residence times in each environment while the correlation coefficient for the covariance measures the dependence of residence times in one environment on those in another.

$$\mu_{1i} = \int_o^\infty t_i E_{1i}(t_i)\; dt_i \tag{54}$$

$$\sigma_i^2 = \int_o^\infty (t_i-\mu_{1i})^2\; E_{1i}(t_i)\; dt_i \tag{55}$$

$$\sigma_i \sigma_j \rho_{ij} = \underbrace{\int_o^\infty \int_o^\infty}_{n} (t_i-\mu_{1i})(t_j-\mu_{1j}) \; E_n \; (t_1,t_2,..t_n) \; dt_1 dt_2 ... dt_n \tag{56}$$

The overall mean residence time is:

$$\mu_1 = \sum_{j=1}^{n} \mu_{1j} \tag{57}$$

and the variance of the total residence times is related to the variances of residence times in various environments by:

$$\sigma^2 = \sum_{i=1}^{n} \sum_{j=1}^{n} \rho_{ij} \sigma_i \sigma_j = \sum_{i=1}^{n} \sigma_i^2 + \sum_{i=1}^{n} \sum_{\substack{j=1 \\ j \neq i}}^{n} \sigma_{ij} \sigma_i \sigma_j \tag{58}$$

Since one is often interested in deriving the above multivariable and marginal probability density functions (p.d.f.) from models, it is useful to introduce the multivariable transform function:

$$\bar{E}_n \; (s_1,s_2,...s_n) = \underbrace{\int_o^\infty ... \int_o^\infty}_{n} e^{-(s_1 t_1 + s_2 t_2 + ... + s_n t_n)} E_n(t_1,t_2,..t_n) \; dt_1 ... dt_n \tag{59}$$

It is readily seen that the transform of the marginal density function is given by:

$$\bar{E}_{1i} \; (s_i) = \bar{E}_n \; (0,..., \; s_i,...0) \tag{60}$$

when all the $s_j = 0$; $j = 1,2,...n$ except for $j = i$ for the environment of interest. The Laplace transform of the exit age density function is now given by:

$$\bar{E}(s) = \bar{E}_n \; (s, \; s, \; ...s) \tag{61}$$

when all the s_is are set equal to s.

The moments can be derived from the appropriate derivatives of the transforms:

$$\mu_{1i} = - \frac{\partial \bar{E}_n}{\partial s_i} (0,0,...0) \tag{62}$$

$$\sigma_i^2 = \frac{\partial^2 \bar{E}_n}{\partial s_i^2} (0,0,...0) - \mu_{1i}^2 \tag{63}$$

$$\sigma_i\sigma_j\rho_{ij} = \frac{\partial^2\bar{E}_n}{\partial s_i\,\partial s_j}(0,0,\ldots 0) - \mu_i\mu_j \tag{64}$$

One can also use conditional probability functions. A joint conditional probability density function of $t_2, t_3, \ldots t_n$ given t_1 is defined by:

$$E_n\,(t_2,t_3,\ldots t_n \mid t_1) = \frac{E_n\,(t_1,t_2\ldots t_n)}{E_{1n}\,(t_1)} \tag{65}$$

Its means and variances can be defined in the usual way and are called the conditional means and conditional variances. Shinnar et. al. (12,112) utilized the conditional probabilities in their description of multienvironment systems.

If we consider again our multienvironment system consisting of n regions Ω_i (i=1,2,3...n) and boundaries between regions $\partial\Omega_{ij}$ with i,j = 0,1...n,n+1, where 0 indicates the inlet boundary and (n+1) the exit boundary, it is reasonable to assume that transport within each region can be described by the usual convective diffusion equations. If to each region we assign a different time, then these transport equations take the following form:

$$\frac{\partial c_i}{\partial t_i} = \nabla\cdot(D_i\nabla c_i) - (\underline{u}_i\cdot\nabla)\,c_i, \text{ in } \Omega_i \tag{66}$$

Here c_i is the local tracer concentration in environmental i, $\underline{u}_i$, and D_i are the velocity vector and diffusivity in environment i, and t_i is the residence time in environment i. Proper boundary conditions require no accumulation of tracer at the boundaries and continuity of tracer flux. Closedness of the system on the boundaries with the inlet and exit environment is required also, i.e. the net input and output of tracer occurs by flow only. To obtain directly from the model the joint p.d.f. a normalized unit impulse input is required. First of all it is readily apparent that if we take the Laplace transform of the above equation we get:

$$\nabla\cdot(D_i\nabla\bar{c}_i) - (\underline{u}_i\cdot\nabla)\,\bar{c}_i - s_i\bar{c}_i = 0 \tag{67}$$

This is a description of a system at steady state in which first order reaction takes place, the rate constant s_i being different in each environment. The transform of the multivariable joint p.d.f can then be readily obtained since the system of equations is linear. Moreover, obtaining the transform of a marginal p.d.f. simply requires assigning no reaction terms (s_i = 0) in all environments except in the region of interest and solving the problem for the properly dimensionalized tracer concentration at the exit boundary. This provides a powerful tool for evaluation of various transforms of marginal, joint and overall p.d.f.s from models of the system (5,110). We should also note that if $s_i = s$ ($t_i = t$) we look for the overall RTD although we still deal with the heterogeneous system since tracer can be partitioned non-uniformly in various regions, i.e. $c_{i_{eq}} = K_{ij}c_{j_{eq}}$, $K_{ij} \neq 1$. From this formulation the multiphase central volume principle, eq. (13), can readily be proven. This has been done for the case when a single phase flows through a multiphase system, other phases being confined to the system (5,44). By using the above formulation and the extension of the RTD theory to multiport systems (27-29), it seems intuitively apparent that the proof of the central volume principle can be extended to an arbitrary number of inlet and exit ports or incoming and outgoing phases. This, to the best of our knowledge, has not been formally done. Again, it is clear that in heterogeneous systems the central volume principle, eq. (13), is not a property of the system at a given flow rate but is the property of the species traced since K_{ij}s are likely to depend greatly on the choice of tracer.

6.1.1 Two-environment systems with one flowing phase. It is instructive to consider some examples of models for two-phase systems, examine the multivariable and marginal p.d.f.s and see how they relate to reactor performance and how they can be measured. We start with the system schematically shown below in Figure 10.

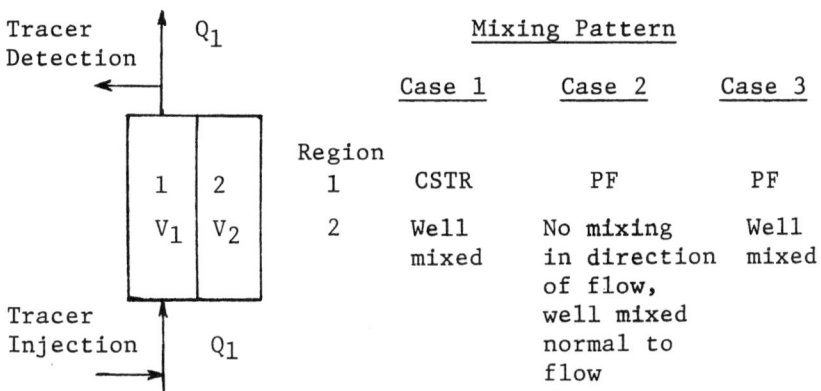

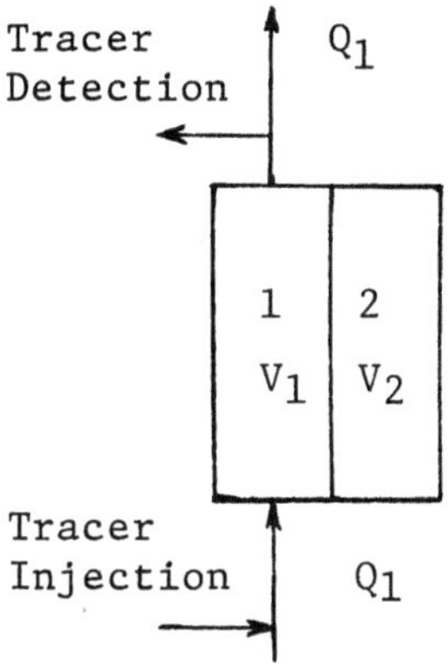

Mixing Pattern

Region	Case 1	Case 2	Case 3
1	CSTR	PF	PF
2	Well mixed	No mixing in direction of flow, well mixed normal to flow	Well mixed

Figure 10. Schematic of a Two-Phase System with a Single Flowing Phase 1 and Possible Mixing Patterns

Phase 1 flows at the constant flow rate Q_1 through the reactor; phase 2 is contained within the reactor. Total volume in the reactor of the flow-through phase is V_1 and of the contained phase is V_2. This can represent a number of reactors such as: catalyst slurry (phase 1 is liquid and phase 2 is solid), semibatch bubble columns (phase 1 is gas, phase 2 is liquid), packed-bed or fluidized bed (phase 1 is gas, phase 2 is solid), etc. We consider the three limiting ideal flow patterns after Aris (5). In Case 1 both phase 1 and 2 are considered well mixed. This could represent a small L/D sparged or mechanically agitated bubble column. In Case 2 the flowing phase is in plug flow and there is no mixing (or transport) in phase 2 in the direction of flow while there is instantaneous equilibration perpendicular to flow. Packed beds with negligible diffusional or adsorption resistances could be approximated in this way. Case 3 envisions the flowing phase in plug flow and the other phase as well mixed. Fluidized beds and bubble columns under certain conditions may approach this limit. In all three cases it is assumed that the transport flux from one phase to the other can be expressed by a product of a mass transfer coefficient, k_m, and the appropriate local concentration difference $(c_1 - c_2/K)$. For brevity and illustrative purposes we will consider only Case 2 here.

Case 2. Here we deal with plug flow of the flow-through phase and the model in the transform domain is given by:

$$\frac{d\bar{c}_1}{d\xi} = -\lambda\,(\bar{c}_1-\bar{c}_2/K) - s_1\theta_1\bar{c}_1 \tag{68}$$

$$\lambda\,(\bar{c}_1-\bar{c}_2/K) = s_2\theta_2\bar{c}_2 \tag{69}$$

$$\xi = 0 \quad ; \quad \bar{c}_1 = 1 \tag{70}$$

The solution for the various transforms are:

$$\bar{E}_2\,(s_1,s_2) = e^{-\left(s_1\theta_1 + \frac{\lambda s_2\theta_2 K}{\lambda + s_2\theta_2 K}\right)} \tag{71}$$

$$\bar{E}_{11}\,(s_1) = e^{-s_1\theta_1} \tag{72a}$$

$$\bar{E}_{12}\,(s_2) = e^{-\lambda + \lambda^2/(\lambda + s_2\theta_2 K)} \tag{72b}$$

$$\bar{E}\,(s) = e^{-\theta_1\left(s + \frac{\lambda s\,\theta_2 K/\theta_1}{\lambda + s\theta_2 K}\right)} \tag{73}$$

Note that $\lambda = k_m A/Q_1$ and $\theta_i = V_i/Q_1$.

The moments are:

$$\mu_{11} = \theta_1; \ \mu_{12} = K\theta_2; \ \sigma_1^2 = 0; \ \sigma_2^2 = 2K^2\theta_2^2/\lambda, \ \rho_{12} = 0 \tag{74}$$

The overall mean and variance are:

$$\mu_1 = \theta_1 + K\theta_2 \tag{75a}$$

$$\sigma^2 = 2K^2\theta_2^2/\lambda \tag{75b}$$

The inversion of the overall p.d.f. yields:

$$E(t) = e^{-\lambda\left(1+\frac{t-\theta_1}{K\theta_2}\right)} \{\delta(t-\theta_1) + \frac{\lambda}{\sqrt{(t-\theta_1)K\theta_2}}$$

$$\cdot I_1\left(2\lambda\sqrt{\frac{t-\theta_1}{K\theta_2}}\right) H(t-\theta_1)\} \tag{76}$$

while the marginal p.d.f.s are:

$$E_{11}(t_1) = \delta(t_1-\theta_1) \tag{77a}$$

$$E_{12}(t_2) = e^{-\lambda\left(1+\frac{t_2}{K\theta_2}\right)} I_1\left(2\lambda\sqrt{\frac{t_2}{K\theta_2}}\right) \tag{77b}$$

We recognize here the cross-flow model for packed-beds discussed earlier in Section 5.2 (104-107). This represents its generalization for $K \neq 1$. Let us consider for the moment our system to be a tube packed with nonporous catalyst particles with plug flow of gas. Then λ represents the ratio of the combined maximum volumetric mass transfer rate to the catalyst surface and adsorption rate at the surface to the flow through the tube. If we know the true rate constant, k_t, for surface reaction and the partition coefficient K for adsorption, then conversion for a first-order process is given by $x_A = 1 - \bar{E}_{12}(k_t)$. However $E_{12}(t_2)$ cannot be obtained directly by a tracer experiment. It can be approximately evaluated by using highly adsorbing tracers. Since we know how to scaleup only ideal reactors we would like to represent the behavior of this system as a plug-flow reactor. Thus, we would normally state for the performance of our pilot unit that $1-x_A = e^{-k_s\theta_2}$ where k_s is the apparent rate constant and $\theta_2 = S/Q_1$

with S being the catalyst area. Using the relationship between conversion and $\bar{E}_{12}(k_t)$ based on the segregated flow model, we get:

$$k_s = \frac{k_t K}{1 + \frac{k_t \theta_2 K}{\lambda}} \tag{78}$$

Only if $\lambda >> k_t \theta_2 K$ is the use of our ideal reactor model for the pilot unit justified and good scaleup assured.

We did not need to restrict ourselves to nonporous catalyst. We can consider plug flow through a packed bed of porous catalyst particles where external mass transfer, diffusion into the pores of the particles, and adsorption-desorption in the walls of the pores occur. Models of this type have been extensively exploited in the literature for evaluation of various parameters such as diffusivity, adsorption constants, etc. (113). Only at fast adsorption-desorption rates one obtains the usual relationship between the apparent rate constant, the true rate constant and the overall catalyst effectiveness factor. A recent paper discussed the conditions when adsorption-desorption effects need not be accounted for explicitly (114). In a related field of biomedical engineering this model was studied in relation to the Krogh cylinder models of the capillary exchange in the microcirculation (115). It was shown (115) that the above model leads to equivalent RTDs when either all the resistance to transport is lumped into a permeability coefficient at the capillary wall or when the transport resistance in the region surrounding the capillary is distributed and ascribed to diffusion. Examples of the $E(t/\theta_1)$ functions for the two models are shown in Figure 11a and 11b. Whenever λ or its equivalent becomes very large, overall plug flow behavior is approached with the mean residence time of $\theta_1 + K\theta_2$.

We note that in this model when we have strictly plug flow and transport resistances only perpendicular to flow that then $E_{11}(t_1)$ and $E_{12}(t_2)$ are stochastically independent, i.e. $\rho_{12} = 0$. Shinnar et. al. (112) arrived at that conclusion by probabilistic arguments. This means that we can find the residence time density on the catalyst, i.e. in the reacting environment, by performing tracer experiments with either nonadsorbing or adsorbing tracers. As long as the nonadsorbing tracer experiences plug flow only, we know from the above that:

$$\bar{E}(s) = \bar{E}_{11}(s) \cdot \bar{E}_{12}(s) = e^{-\theta_1 s} \cdot \bar{E}_{12}(s) \tag{79}$$

which can be interpreted as:

Legend for Figure 11

The Krogh Cylinder Model

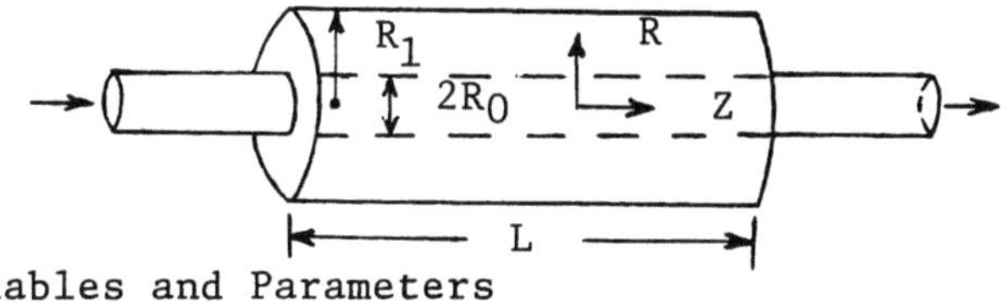

Variables and Parameters

$$t = \frac{t_{actual}}{\bar{t}_c} \;;\; z = \frac{Z}{L} \;;\; r = \frac{R}{R_1} \;;\; \bar{t}_c = \frac{V_c}{Q} = \frac{\pi R_o^2 L}{Q}$$

$$\rho = \frac{R_o}{R_1} \;;\; b = \frac{2L}{R_o} \;;\; \ell = \frac{L}{R_1}$$

u,v = dimensionless concentration in capillary and surrounding cylinder, respectively

$$\gamma = \frac{\ell}{Pe_r} = \frac{V_c}{Q}\frac{D_r}{R_1^2} = \frac{\text{mean transit time in capillary}}{\text{characteristic diffusion time}}$$

$\beta = \frac{PS}{Q}$ = St = Stanton number for transfer across capillary wall

Case A

$$\frac{\partial u}{\partial t} + \frac{\partial u}{\partial z} = \frac{b}{Pe_r}\left.\frac{\partial v}{\partial r}\right|_{\rho} \qquad u(o,z) = v(o,z,r) = 0$$

$$u(t,o) = \delta(t)$$

$$\frac{\partial v}{\partial t} = \frac{\ell}{Pe_r r}\frac{\partial}{\partial r}\left(r\frac{\partial v}{\partial r}\right) \qquad u(t,z) = v(t,z,\rho)$$

$$\left.\frac{\partial v}{\partial r}\right|_{r=1} = 0$$

Case B

$$\frac{\partial u}{\partial t} + \frac{\partial u}{\partial z} = -\beta(u-v) \qquad u(o,z) = v(o,z) = 0$$

$$\frac{\partial v}{\partial t} = \frac{\beta\rho^2}{1-\rho^2}(u-v) \qquad u(t,o) = \delta(t)$$

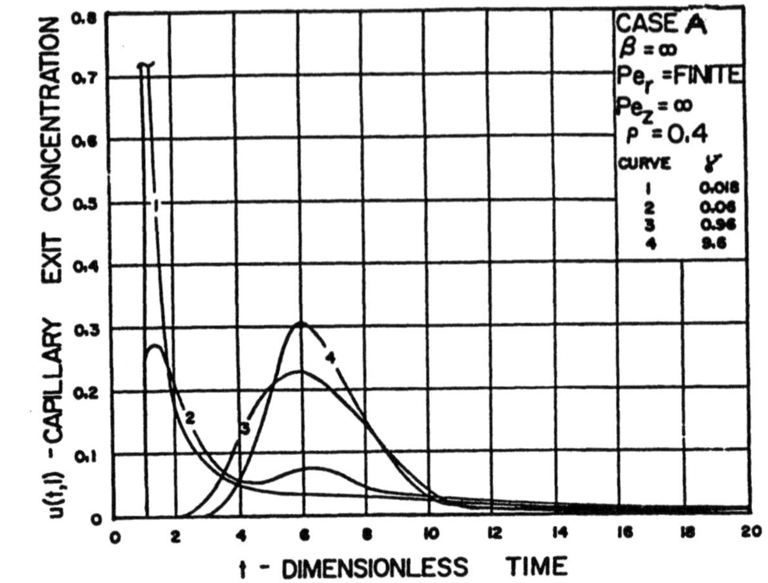

Figure 11a. Impulse Response for the Krogh Cylinder Model with Plug Flow in the Capillary (Phase 1), Finite Diffusional Resistance Perpendicular to Flow Direction in the Tissue (Phase 2) and No Resistance at the Capillary Wall (Phase Boundary). Equivalent to Case 2.

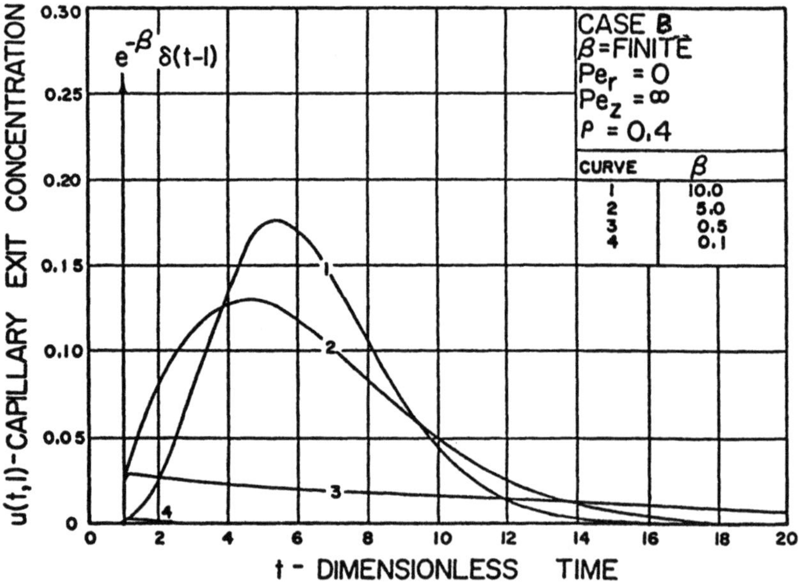

Figure 11b Impulse Response for the Krogh Cylinder Model with Plug Flow in the Capillary (Phase 1), Finite Permeability at the Capillary Wall (Phase Boundary) and no Diffusional Resistance Perpendicular to Flow Direction in the Tissue (Phase 2). Equivalent to Case 2.

$$\bar{E}_a(s) = \bar{E}_{na}(s)\cdot\bar{E}_{12}(s) = e^{-\theta_1 s}\cdot\bar{E}_{12}(s) \tag{80}$$

where $\bar{E}_a(s)$ and $\bar{E}_{na}(s)$ are transfer functions for the adsorbing and nonadsorbing tracers. In order to predict reactor performance we only need to know $\bar{E}_{12}(s)$ which can be obtained from eq. (80). In principle $E_{12}(t)$ can be obtained by deconvolution of eq. (80). However, since $\bar{E}_{na}(s)$ is just a time delay it is clear that $E_{12}(t) = E_a(t + \theta_1)$. Thus, the adsorbing tracer curve, for the tracer with K equal to that of the reactant, can be used directly, by substituting $t + \theta_1$ for time t, to obtain the RTD for the reacting environment. However, in practice it may be difficult to find a tracer with the adsorption equilibrium constant K equal to that of the reactant, and the following approach is preferred.

From Eq. (72), eq. (73) and eq. (80) we note that $\bar{E}_a(s)$ can be represented as follows:

$$\bar{E}_a(s) = e^{-(\theta_1 s + H(s))} \tag{81}$$

where H(s) is defined by:

$$H(s)\cdot\bar{c}_1 = \lambda\,(\bar{c}_1 - \bar{c}_2/K) \tag{82}$$

which is the term for the dimensionless volumetric exchange flux between the two environments. The form of $\bar{E}_{na}(s) = e^{-\theta_1 s}$ is the result of the first-order differential operator $\mathcal{L}_{1\xi}(\bar{c}_1) = \frac{d\bar{c}_1}{d\xi}$ on the left-hand side of eq. (68). If we replaced that by a second-order operator $\mathcal{L}_{2\xi}$ (for example the one that describes the axial dispersion model), the form of $\bar{E}_{na}(s)$ changes (e.g. becomes a sum of exponentials for the axial dispersion model). The response for a nonadsorbing tracer is then described by:

$$\mathcal{L}_{2\xi}(\bar{c}_1) = -\,s\theta_1\bar{c}_1 \tag{83}$$

while the response of the adsorbing tracer can be obtained by solving:

$$\mathcal{L}_{2\xi}(\bar{c}_1) = -\,(s\,\theta_1 + H(s))\,\bar{c}_1 \tag{84}$$

The boundary conditions are the same in both cases. This shows that the transform of the response for the adsorbing tracer, $\bar{E}_a(s)$, can be obtained from the transfer function for the nonadsorbing one, $\bar{E}_{na}(s)$, by simply replacing s with $s + H(s)/\theta_1$, i.e.:

$$\bar{E}_a(s) = \bar{E}_{na}\ (s + \frac{H(s)}{\theta_1}) \tag{85}$$

To obtain the transform of the p.d.f. for residence times on the catalyst, i.e. in the reacting environment, we would have to solve:

$$\mathcal{L}_{2\xi}\ (\bar{c}_1) = -\ H(s)\bar{c}_1 \tag{86}$$

which results in:

$$\bar{E}_r(s) = \bar{E}_{12}(s) = \bar{E}_{na}\ (\frac{H(s)}{\theta_1}) \tag{87}$$

This is a general result that allows us to find the p.d.f. of sojourn times in the reacting environment from p.d.f.s for tracers that do not penetrate into this environment at all. The expression given by eq. (87) is valid only when a tracer or reactant leaves and re-enters the moving phase at the same point, i.e. there are no fluxes in the reacting environment parallel to the flow direction. This, for example, precludes surface diffusion in the direction of flow. This constraint keeps the marginal p.d.f.s stochastically independent, i.e. $\rho_{12} = 0$. We also note that our global (design) rate constants are introduced by:

$$k_w\,\frac{W}{Q} = k_s\,\frac{S}{Q_1} = H\,(s = k_t) = \frac{k_t K\theta_2}{1 + k_t K\theta_2/\lambda} \tag{88}$$

This last equation is only valid for plug flow of gas.

This demonstrates that successful scaleup must duplicate conditions on the microscopic (single catalyst particle) level which results in the same $H(k_t)$ upon scaleup, and on the macroscale (flow pattern) which results in the same $E_{na}(t)$. Therefore, nonadsorbing tracers provide very useful information with respect to the macroscopic aspect of scaleup. The limiting case of the above development is our simple model of Case 2 with plug flow of the nonadsorbing species, i.e. $\bar{E}_{na}(s) = e^{-\theta_1 s}$. The transform of the density function of sojourn times in the reaction environment then is:

$$\bar{E}_r(s) = \bar{E}_{12}(s) = e^{-\ H(s)/\theta_1} \tag{89}$$

In addition, for any form of $\bar{E}_{na}(s)$, if the adsorption-desorption rate is very fast, $\lambda \to \infty$, then $H(s) \to K\theta_2 s$. The transform for the p.d.f. of sojourn times on the catalyst is:

$$\bar{E}_r(s) = \bar{E}_{na}\left(\frac{K\theta_2}{\theta_1} s\right) \tag{90}$$

By the elementary rule of Laplace transform inversion we have:

$$E_{12}(t) = E_r(t) = \frac{\theta_1}{K\theta_2} E_{na}\left(\frac{\theta_1}{K\theta_2} t\right) \tag{91}$$

Thus the p.d.f for nonadsorbing tracers can be used directly to estimate the p.d.f. of sojourn times on the catalyst surface if the partition coefficient, K, for the reactant is known and adsorption-desorption is fast.

When we deal with beds of porous catalyst particles we often assume that we have plug flow for a long pilot scale reactor. We cannot find a tracer that does not diffuse into the particles and hence even the $E_{na}(t)$ curve for a nonadsorbing tracer is not a delta function. However, if the reactor is long enough, $\bar{t} >> d_p^2/D_e$, the response is close to plug flow and we interpret reactor performance by the PFR model. For a first-order catalytic reaction this means that we are implicitly asserting:

$$\bar{E}_{na_p}\left(\frac{H(k_t)}{\theta_1}\right) = e^{-k_w (W/Q_1)_p} \tag{92a}$$

where subscript p indicates the values in the pilot plant reactor and W is the mass of catalyst. Conversion in the large reactor (subscript L) at the same temperature and feed conditions and using the same catalyst is now predicted by:

$$1 - x_A = e^{-k_w (W/Q_1)_L} \tag{92b}$$

This really means that we are expecting the following equality to be true:

$$\bar{E}_{na_L}\left(\frac{H(k_t)}{\theta_1}\right)_L = e^{-\left(\frac{Q_1}{W}\right)_p \left(\frac{W}{Q_1}\right)_L} \bar{E}_{na_p}\left(\frac{H(k_t)}{\theta_1}\right)_p \tag{93}$$

Clearly, many things can go wrong which destroy the equality given by eq. (93). Temperature and catalyst activity may be different resulting in different $H(k_t)$, improper catalyst amount may have been used plus not keeping W/Q_1 constant, or the flow pattern may have changed resulting in a different $E_{na}(t)$. Equation (93) does not provide us with new tools in dealing with heterogeneous systems but allows us to approach the problem in a systematic manner. After we have eliminated the possibilities of the differences on microscopic (particle) scale we can explore the possibility of flow or contacting maldistribution by tracer methods using non-adsorbing and adsorbing tracers.

The above approach could possibly be generalized further to include two-dimensional space operators $\mathcal{L}_{2\xi\eta}$ in the description of the reactor provided that the tracer that exits from the flowing phase into other phases at a particular point re-enters at the same point in the flowing phase space. This has not been attempted. The first introduction of the H(s) function in evaluating the moments of various dispersion models with diffusion into the catalyst seems to be due to Miller and Bailey (116). Shinnar et. al. (112) and Rumschitzki and Shinnar (117) have also pointed out the possibilities offered by eq. (87). The approach presented is well suited for analysis and scaleup of packed-beds and trickle-beds with nonvolatile liquid rate limiting reactants. It cannot be used in fluidized beds, unless by extensive baffling plug flow is approached, because in fluid-beds the tracer can also move with the solids and $\rho_{12} \neq 0$. The function H(s) can be generalized to include external mass transfer, internal diffusional and adsorption-desorption resistance terms. Although not presented in such a form its elements are available in the literature (113).

In the above discussion the behavior of nonlinear reaction schemes was not considered. Conventional approaches to reactor modeling based on ideal reactor models are suggested and tracer studies with nonadsorbing tracers should be conducted to determine whether the assumptions of ideal flow patterns are satisfied. Adsorbing tracers can be used to assess catalyst contacting through the use of the central volume principle.

<u>6.1.2 Multi-environment systems with two flowing phases</u>. These systems are perhaps of most interest in reaction engineering applications since they include the most frequently used multi-phase reactors. Gas-liquid bubble columns, ebullated beds, three-phase fluidized beds, gas-lift slurry reactors, trickle-bed reactors, pneumatic transport reactors, etc. fall into this category. Some of the developments presented in Section 6.1.1 can be extended to treat these systems. The multivariable joint p.d.f. has to be defined taking into the account that the system has multiple inlets and outlets, <u>i.e</u>. by following the rules established in Section 3 by the appropriate extension of eqs. (9) and (10). However, this approach has not been presented or used to date. The main reason is that the transforms do not have a readily useable analytical form and are functions of many system

parameters. Their analytical inversion does not seem possible even for the simplest flow patterns of both phases such as co-current plug flow of two phases, countercurrent plug flow of two phases, plug flow of one phase and well-mixedness of the other phase.

Let us illustrate some of the difficulties involved by considering a simple example of a two-phase system where both phases can be assumed in the first approximation to be in cocurrent plug flow. The governing equations in the multivariable transform plane are:

$$\frac{d\bar{c}_1}{d\xi} + \lambda\ (\bar{c}_1 - \frac{\bar{c}_2}{K}) + s_1\theta_1\bar{c}_1 = 0 \qquad (94)$$

$$\frac{d\bar{c}_2}{d\xi} - \lambda\beta\ (\bar{c}_1 - \frac{\bar{c}_2}{K}) + s_2\theta_2\bar{c}_2 = 0 \qquad (95)$$

$$\xi = 0 \quad ; \quad \bar{c}_1 = p_1\frac{M_i}{\theta_1} \quad ; \bar{c}_2 = (1-p_1)\ \beta\frac{M_i}{\theta_i} \qquad (96)$$

where $\bar{c}_1$ and $\bar{c}_2$ are the multivariable transforms of the tracer concentration in phases 1 and 2, respectively; $\theta_1 = \varepsilon_1 V/Q_1$ and $\theta_2 = \varepsilon_2 V/Q_2$ with V being the total volume of the system; and ε_1 and ε_2 are the holdup of phases 1 and 2 respectively. The flow rates in phases 1 and 2 are Q_1 and Q_2 respectively. The flow rate ratio is $\beta = Q_1/Q_2$. The exchange coefficient is $\lambda = k_m aV/Q_1$ where $k_m a$ is the volumetric mass transfer coefficient, M_i is the amount of tracer that is instantaneously injected at t = 0 ($t_1 = t_2 = 0$) with fraction p_1 being injected into stream (phase) 1. Transform variables s_1, s_2 correspond to residence time t_1 and t_2 in the two respective environments. The transforms for the exit concentration $\bar{c}_1$ (ξ = 1, s_1, s_2) and $\bar{c}_2$ (ξ = 1, s_1, s_2) can be obtained from the above system of linear equations and are functions of the system parameters θ_1, θ_2, λ, β and K. They are lengthy expressions which are not presented here for the sake of brevity. One should note, however, that $\bar{c}_1$ $(1,s_1,s_2)$ and $\bar{c}$ $(1,s_1,s_2)$ are not transforms of p.d.f.s. The transform of a joint p.d.f. can be defined by:

$$\bar{E}_2(s_1,s_2) = [Q_1\bar{c}_1(1,s_1,s_2) + Q_2\bar{c}_2\ (1,s_1,s_2)]/M_i \qquad (97)$$

The transform of the overall residence time density function $\bar{E}(s)$ is obtained from eq. (97) when $s=s_1=s_2$. $\bar{E}(s)$ depends on the injection of tracer, i.e. on the fraction p_i injected into each stream. This becomes apparent by considering the mean of the $\bar{E}(s)$ which is:

$$\mu_1 = \frac{(\varepsilon_1 + K\varepsilon_2)\,V}{Q_1 + KQ_2} + [\frac{p_1 KQ_2\theta_1 + (1-p_1)\,KQ_2\theta_2}{\lambda\,(Q_1 + KQ_2)} -$$

$$\frac{(\varepsilon_1 + K\varepsilon_2)\,VKQ_2}{\lambda\,(Q_1 + K\theta_2)^2}]\mathrm{x}\;[1 - e^{-\lambda\,(Q_1 + KQ_2)/VQ_2}] \tag{98}$$

Only when the tracer is introduced proportionally to the capacity of the flow for the tracer, i.e. $p_1 = Q_1/(Q_1 + KQ_2)$, is the mean reduced to a simple expression:

$$\mu_1 = \frac{(\varepsilon_1 + K\varepsilon_2)\,V}{Q_1 + KQ_2} \tag{98a}$$

This represents the generalization of the central volume principle and of flow tagging for single-phase systems. When the tracer is introduced into each phase in proportion to the capacity of the phase for the tracer then the mean of the weighted response is the weighted volume of tracer distribution divided by the weighted flow rate through the system. In practice eq. (98a) can be achieved only if a rapidly equilibrating tracer is introduced so that $\lambda \to \infty$ or at least $\lambda > KQ_2/(Q_1 + KQ_2)$.

More frequently the tracer is injected into one phase only and its response is measured at the exit of the same phase. Both the zeroth and first moment are then functions of ε_1, ε_2, K, and λ. Detailed equations are presented elsewhere (118-120). A nonadsorbing tracer can then be used to determine the holdup of the liquid, ε_1, while gas holdup is obtained by difference $\varepsilon_2 = 1 - \varepsilon_1$. An absorbing tracer can then be used to evaluate K and λ from the zeroth and first moment, or λ directly from the zeroth moment if K is known. Examples of the use of various moments in determination of system parameters are presented by Mills et. al. (118) and Ramachandran and Smith (119,120).

We should note that here, as before, reactor performance at steady state for a first-order reaction with rate constants k_1 and k_2 in each respective phase is given by $x_A = 1 - \bar{E}_2\,(k_1,k_2)$. If the reaction takes place only in phase 2 then $x_A = 1 - \bar{E}_2\,(0,k) = 1 - \bar{E}_{12}(k)$. Thus, obtaining the density function of sojourn times in the reacting environment, $E_{12}(t)$, would be useful but is difficult in practice. In principle $\bar{E}_{12}(k)$ could be obtained by varying the temperature (assuming this does not affect the holdups ε_1 and ε_2 and the transport parameter λ) by measuring conversion for a first-order process in the liquid and by inverting the result. This is rarely attempted in practice.

We note that due to the fact that the tracer does not leave and re-enter a phase at the same point along the flow direction that it is no longer possible to infer the performance of the system from the nonadsorbing tracer response. No attempts at generalizations in these systems with two moving phases have been reported. Tracers that cross phase boundaries are mainly used to evaluate the various transport parameters. Tracers confined to a single phase are used to check the flow pattern of that phase. The equations for the moments of tracer responses can be derived for counter-current flow (118), as well as mixedness of one phase and axial dispersion in one or both phases (119,120). The interested reader can consult the literature. A good summary is provided by Ramachandran and Chaudhari (121). We will now briefly review the reported uses of tracers in heterogeneous systems.

6.2 Use of Tracers in Multiphase Reactors

We have illustrated above that a general theory that ties reactor performance to the RTD of a tracer is lacking in multi-phase systems. Nevertheless tracer responses can be and are used effectively to determine various parameters in multiphase systems. The interpretation of some results relies on single-phase theory and is model independent. Most results are interpreted based on matching reactor model response and tracer response. The most commonly used tracer experiments involve determination of holdup, backmixing in a phse, transport coefficients among phases and contacting of one phase by the other. Holdup of a phase is readily determined by using a tracer confined to that particular phase and applying the central volume principle to the response. Evaluation of backmixing is done best by again utilizing a tracer confined to the phase of interest. The resulting RTD is then matched to an appropriate backmixing model. Transport coefficients can be evaluated (once backmixing in each phase has been determined) by matching model predictions to the response of a tracer that was injected in one phase but distributes itself into both. Phase contacting can also be inferred from the response of a tracer that distributes itself through the phases of interest. The mathematics involved depends on the model chosen, is simple in principle due to the linearity of the system of equations but might be algebraically tedious. The papers by Ramachandran and Smith (120) and Mills et. al. (118) illustrate that the method of moments can quickly become algebraically tedious when one deals with two moving phases. Analytical solutions in the time domain are also difficult to obtain due to algebraic complexity and might not be readily convergent at small values of time if a series representation is used. Any further development of the method of moments or time domain solutions for more complex reactors with multiple moving phases would profit tremendously by application of the symbolic computer mathematical manipulation languages such as REDUCE,

MACSYMA, etc. We expect that this will occur in the near future.

Let us now briefly consider some of the tracer applications in particular reactor types. No attempt was made to do a comprehensive literature survey and only illustrative examples are cited.

6.2.1 Trickle-bed reactors. Nonvolatile, nonadsorbing tracers have been traditionally used to evaluate liquid holdup and backmixing. These studies have been summarized by Schwartz et. al. (122), by Duduković and Mills (123) and by Ramachandran and Chaudhari (121). Usually the axial dispersion model has been used to interpret liquid backmixing. The dispersion number is then obtained either from the variance of the tracer response curve or by time domain matching of the model predicted and the measured tracer response. Other models for the liquid phase in trickle-beds have been used and tracer studies are employed to determine model parameters such as the exchange coefficient between the flowing and the stagnant liquid, etc. (123-126). Tracers are also used to evaluate liquid distribution and spreading in trickle-beds (123,127).

Schwartz et. al. (128), Colombo et. al. (129) and Mills and Duduković (48) have shown that nonvolatile nonadsorbing and linearly adsorbing tracers can be used to determine the liquid-solid contacting efficiency. By the central volume principle the first moment of the normalized tracer impulse response for a nonvolatile tracer is:

$$\mu_1 = \frac{VH_L}{Q_L} + \eta_c \frac{S_g V \rho_p \ (1-\varepsilon_B)\ K}{Q_L} \tag{99}$$

where H_L is liquid holdup, Q_L is liquid flow rate, V is total volume of the reactor, S_g is catalyst total surface area (internal + external) per gram, ρ_p is catalyst density, ε_B is bed porosity, an $K = c_S/c_L$, with units of length, is the equilibrium constant for the tracer between the catalyst surface and the liquid. Contacting efficiency η_c is the fraction of the total catalyst area contacted by liquid. When a nonadsorbing tracer is used, K = 0, and the second term in eq. (99) vanishes. Holdup data are then obtained from the measured mean. The use of an adsorbing tracer at the same flow rate gives full eq. (99) for the mean. From the difference of the two means η_c can be obtained provided S_g, V, ε_B, ρ_p, and K are known. In practice it is often difficult to know K precisely because it depends on the packing temperature-time and concentration-exposure history. In laboratory reactors accurate values of η_c can be obtained by running tracer tests with liquid flow only in liquid-filled columns at the same values of Q_L as in the two-phase tests. Then (123):

$$\eta_c = \frac{(\mu_{1_a} - \mu_{1_{na}})_{TP}}{(\mu_{1_a} - \mu_{1_{na}})_{LF}} \tag{100}$$

where subscripts a and na denote an adsorbing and nonadsorbing tracer, respectively, while subscripts TP and LF refer to the two-phase and liquid-full cases, respectively. In industrial practice it is more difficult to execute eq. (100) since liquid-full runs can only be done at start-up or shut-down time. Unfortunately, for direct determination of contacting from eq. (99) K must be known for the packing. This is not an easy task considering that K for the tracer is often a strong function of temperature and of poisons to which the packing was exposed. There is sufficient experimental evidence, however, to indicate that internal pore fillup of catalyst particles is complete ($\eta_c = 1$) except in the presence of highly exothermic reactions (123). Tracer volatility can significantly alter equation (99) as shown by Schwartz et. al. (128) and indicated in Section 6.1.2.

While total catalyst contacting, η_c, can be obtained on a model-free basis by eq. (99), evaluation of the fraction of external catalyst area in contact with the flowing liquid is model dependent. It was shown that an apparent effective diffusivity can be extracted from the variance of the tracer impulse response (48,123,129). This diffusivity is always smaller in trickle flow than for the same liquid-full bed at the same liquid flow rate. The dispersion model with particle diffusion always assumes complete external contacting of particles by liquid which may not be the case in trickle flow. This means that the effective diffusional time constant is increased in trickle flow resulting in a reduced apparent effective diffusivity which is based on the total external surface area. Using this diffusivity in the expression for the Thiele modulus, and equating it to the modulus defined for trickle-bed operation by Duduković (130) results in the following estimate of the external catalyst contacting efficiency, η_{CE}.

$$\eta_{CE} = \sqrt{\frac{De_{TP}}{De_{LF}}} \tag{101}$$

The results obtained by the use of eq. (101) agreed well with estimates of external contacting obtained from reaction studies (48,123). This allows tracer determination of external contacting in laboratory reactors. A complete transient model for tracer responses in partially externally wetted catalyst beds is not available to date.

In industrial practice catalyst contacting is often identified with catalyst utilization. All deviations of the commercial scale reactor from the performance predicted based on the pilot plant plug-flow reactor data are ascribed to incomplete contacting (incomplete catalyst utilization). Typically, both global liquid maldistribution (channeling, bypassing or stagnancy) and local incomplete irrigation of catalyst particles leads to inferior catalyst utilization (123). Whenever reactor malfunctioning is suspected tracer studies are performed to identify the flow pattern of the liquid phase. Murphee et. al. (131) showed that kinetic effects and contacting effects can be decoupled. They arbitrarily defined contacting efficiency as the ratio of θ_p to $\bar{\theta}$. Here $\bar{\theta}$ is the mean of the tracer-determined RTD of the liquid (clearly the tracer must be nonvolatile and nonadsorbing such as C^{14} tagged octadecane used by Murphee et. al. (131) to trace diesel fuel and gas oil in a desulfurization unit). θ_p is the residence time required in a plug flow reactor to reach a conversion of 90% for a first-order reaction using the same rate constant as needed to reach 90% conversion in a reactor of the actual RTD with mean residence time of $\bar{\theta}$. This rate constant is obtained by trial and error from:

$$0.1 = \int_0^\infty E(t)e^{-kt}dt = \bar{E}(k) \tag{102}$$

where E(t) is experimentally determined. Then:

$$\theta_p = \frac{1}{k} \ln 10 \tag{103a}$$

and:

$$\bar{\theta} = \int_0^\infty tE(t)dt \tag{103b}$$

This method was used successfully to interpret data from a malfunctioning commercial unit. Repacking of the catalyst remedied the problem.

Other studies also showed that tracing of the liquid phase can provide clear indications of stagnancy or bypassing by use of the intensity function. More quantitative analysis of the data requires a reactor model (34).

The fact, mentioned earlier, that in multiphase systems one obtains the RTD for the tracer, not the system, is illustrated by Figure 12. In this figure the tracer tests from a hydrodesulfurization unit are represented. Nonvolatile and volatile tracers are used and their response is clearly different. Only the nonvolatile tracer curve can be interpreted in terms of an axial dispersion model to assess the liquid flow pattern. Even then diffusion into the particles and adsorption on the catalyst must be included in the model. These two effects can contribute more

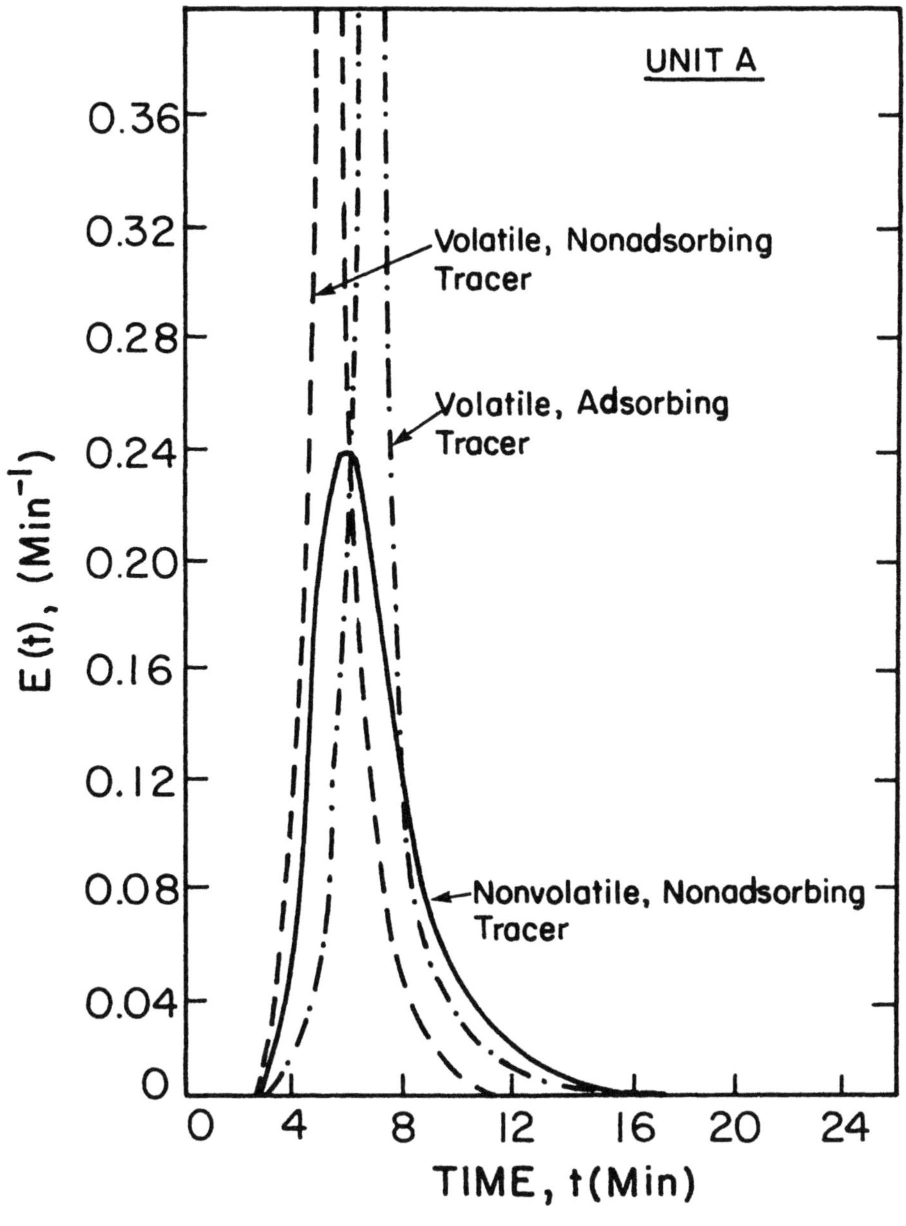

Figure 12. Nonvolatile and Volatile Tracer Impulse Responses in a Hydrodesulfurization Unit.

to the spread of the curve than axial dispersion itself. Figure 13 shows the nonvolatile tracer response from two units processing the same number of barrels a day. Unit B, however, has a three times lower superficial velocity and flow maldistribution is suspected. The corresponding intensity functions for the two units at operating conditions of Figure 13 and the nonvolatile tracer are shown in Figure 14. Indeed the curve for Unit B indicates flow maldistribution, i.e. stagnancy or bypassing. However, if the intensity function is plotted for the volatile tracer of Unit A, as shown in Figure 15, bypassing is indicated. This is to be expected due to the ability of the volatile tracer to "bypass" the liquid through the vapor phase. Figure 15 also illustrates that data truncation, as opposed to exponential extrapolation of the tail, affects the magnitude of Λ but not its basic shape. Only nonvolatile, nonadsorbing tracers really "trace" the liquid phase. Insoluble tracers yield information on the gas flow pattern. For accurate evaluation of liquid or gas holdup we should:

- use a tracer completely confined to the phase of interest (i.e. nonvolatile or insoluble)
- accurately measure the flow rate of the phase of interest
- accurately determine the mass of tracer injected and check the mass balance on tracer

This applies to other multiphase reactors, not just trickle-beds. When the reactions in trickle-beds involve nonvolatile liquid reactants which limit the reaction rate, then the interpretation of liquid-phase tracer data in a TBR can be approached based on Case 2 discussed in Section 6.1.1. Clearly, scale-up is possible under conditions discussed in that section. In case of volatile liquid reactants or rates governed by both gas and liquid reactants the trickle-bed must be considered as a system with two flowing phases discussed in Section 6.1.2 and scale-up is difficult.

6.2.2 Bubble columns. Tracers are used in bubble columns and gas-sparged slurry reactors mainly to determine the backmixing parameters of the liquid phase and/or gas-liquid or liquid-solid mass transfer parameters. They can be used for evaluation of holdup along the lines reviewed in the previous Section 6.2.1. However, there are simpler means of evaluating holdup in bubble columns, e.g. monitoring the difference in liquid level with gas and without gas flow. Numerous liquid phase tracer studies of backmixing have been conducted (132-149). Steady-state or continuous tracer inputs (132,134,140,142) as well as transient studies with pulse inputs (136,141,142,146) were used. Salts such as $KC\ell$ or $NaC\ell$, sulfuric acid and dyes were employed as tracers. Electroconductivity detectors and spectrophotometers were used for tracer detection. The interpretation of results relied on the axial dispersion model. Various correlations for the dispersion

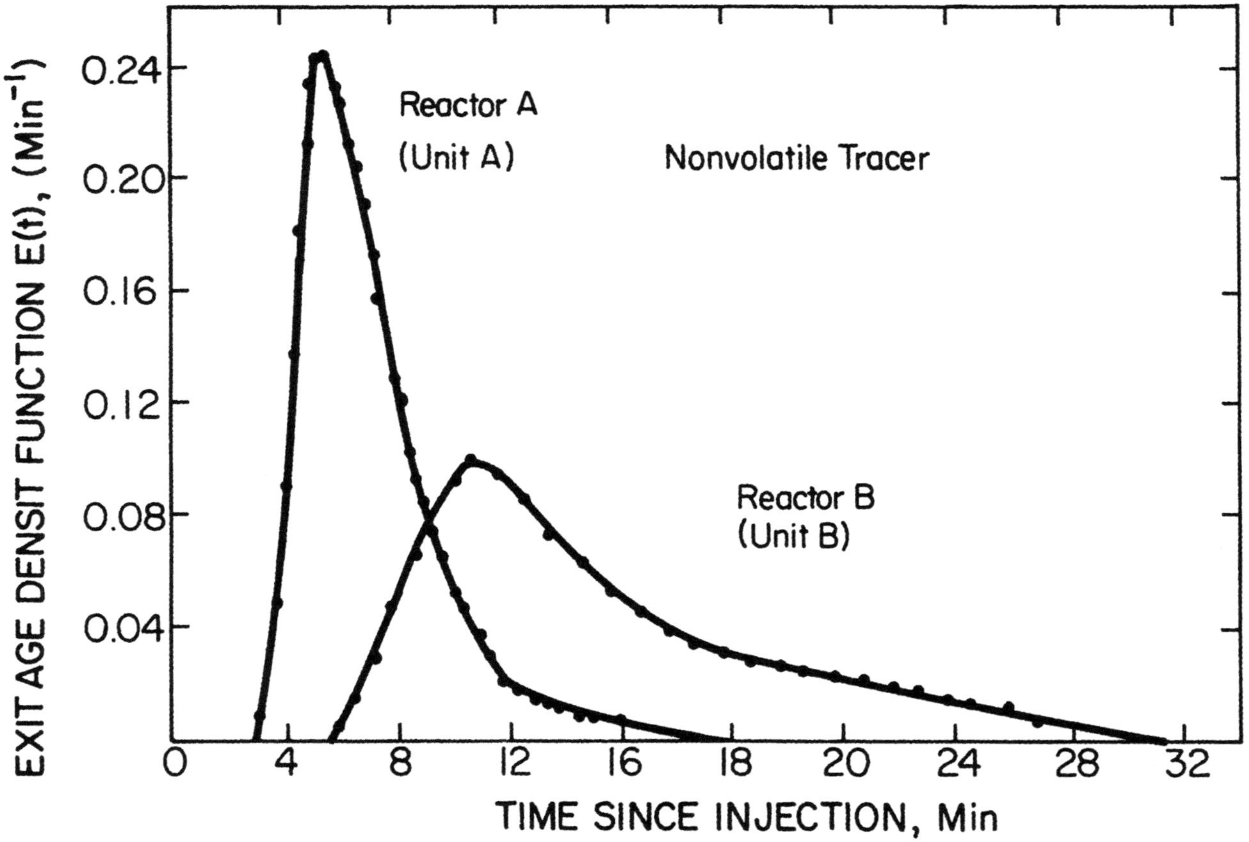

Figure 13. Impulse Response of a Nonvolatile Tracer in Two Hydrodesulfurization Units.

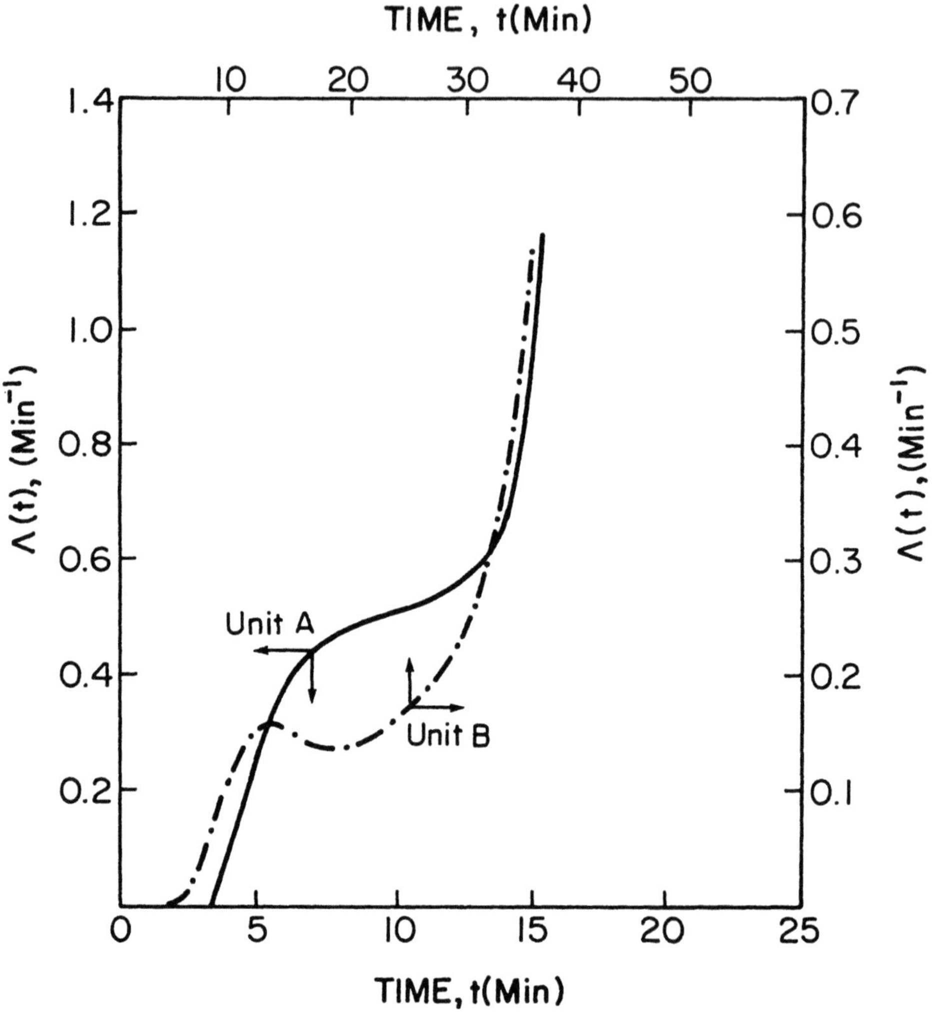

Figure 14. The Intensity Function for the Nonvolatile Tracer in Hydrodesulfurization Reactors A and B.

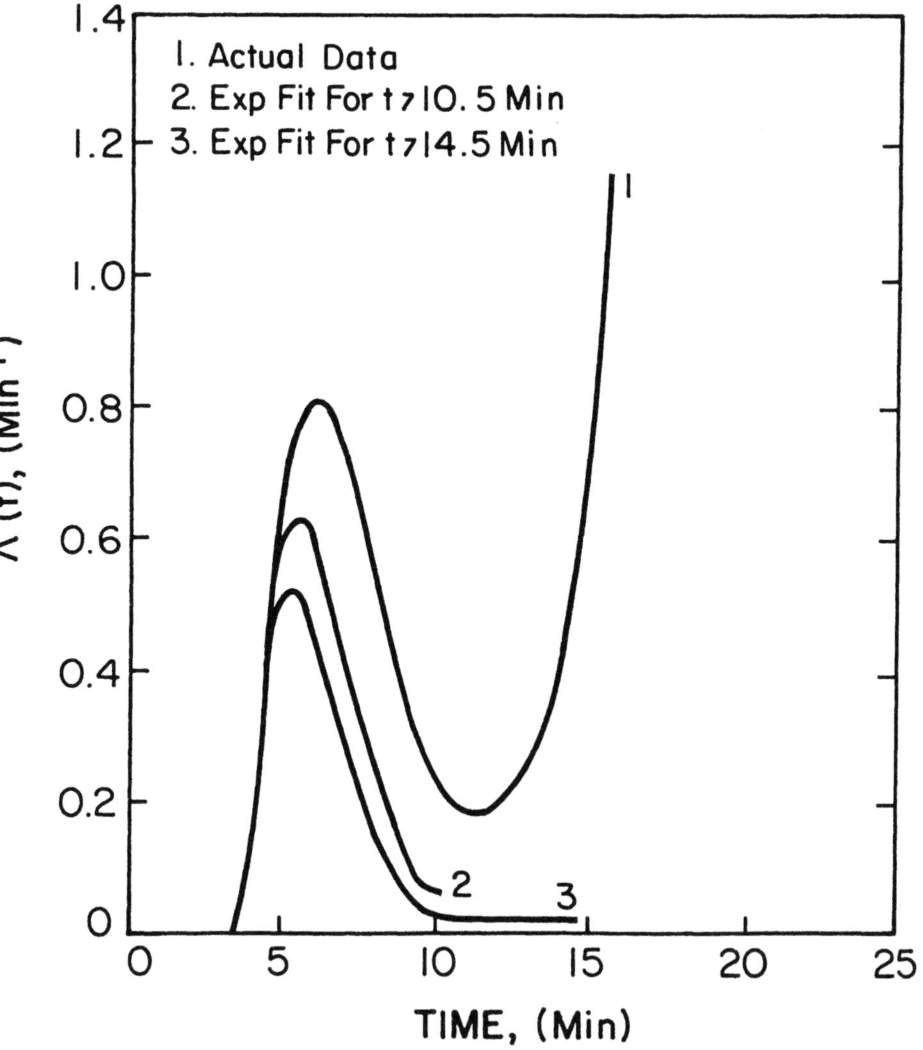

Figure 15. The Intensity Function for the Volatile Tracer in Hydrodesulfurization Unit A.

coefficient were developed as a function of gas velocity and physical properties. The effect of various internals such as packing (150), screens (151), and vertically suspended tubes (152) on the axial dispersion coefficient was also studied by tracer methods. For a summary of various correlations of liquid backmixing the reader is referred to the review by Shah et. al. (86) or to the monograph by Ramachandran and Chaudhari (121). However, due to the fact that the liquid phase is close to complete backmixing in bubble columns the axial dispersion model seems ill suited to describe it. Nevertheless, it is predominantly used!

Gas can exhibit various flow patterns depending on operating conditions, and tracer studies have considered the problem of gas backmixing using insoluble tracers (or slightly soluble ones like argon, helium, oxygen, etc.) (142,147,153). The use of volatile or absorbing tracers can yield mass transfer coefficients. Traditionally, these studies are conducted at steady state and one refers to them as mass transfer studies rather than tracer studies. Mass transfer coefficients depend on the model selected for interpretation of results. However, Deckwer et. al. (154) pointed out recently that the value for the Stanton number, containing the mass transfer coefficient, determined from tracer studies is not affected much by the amount of axial dispersion assumed for the liquid when a pure stream of absorbing gas is used. Mills et. al. (118) have indicated that the zeroth moment of dynamic response for the absorbing tracer provides all the steady-state information. Higher moments provide additional information. Since the flow patterns in a semi-batch bubble column approach those of Case 3 of Section 6.1.1, scale-up is difficult. The same is true of continuous bubble columns. Recently more sophisticated experimental techniques such as the gamma camera have been used to scan bubble columns and determine holdup profiles and backmixing (155).

6.2.3 Fluidized-bed reactors. Tracers have been used to evaluate backmixing of the gas, gas-solid contacting, solids backmixing, and the contact time distribution, i.e. the probability density function of sojourn times in the reaction environment. The literature in this area is rather abundant and unfortunately to some extent confused. No attempt is made here at a comprehensive review. Rather, we want to point out some highlights and indicate that the interpretation of all reported results is straightforward in view of the theory reviewed in Section 6.1.1.

In an often quoted study, Overcashier et. al. (156) used inert nonadsorbing tracers and showed that the use of horizontal baffles in the bed allows one to approach the plug flow RTD for the gas. In an unbaffled bed the gas RTD was close to that of a perfectly mixed tank but did not fit completely the CSTR behavior. The physical interpretation of these findings is clear. In an unbaffled bed at some multiples of the minimum fluidization velocity large bubbles are formed which cause extensive mixing of the emulsion phase. Some gas may bypass the bed, the rest is close to a well mixed state. In a baffled bed bubble growth is hindered and limited and the rate of solids circulation suppressed. This approximates gas flow through a set of well mixed stages and hence ultimately approaches plug flow. The implications for scaleup are important. Baffled beds, with plug flow of the flowing phase and partial containment of the solids within baffles, are scaleable by methods discussed in Section 6.1.1 Case 2. High conversions can be successfully achieved with confidence in such reactors as illustrated recently by Mobil's methanol to gasoline (MTG) process. The unbaffled beds with gas bypassing and high solids circulation are difficult to scale up, as decades of research in this area clearly demonstrate, and should be avoided when possible.

Orcutt et. al. (157) seem to have been first to notice that fluidized bed performance for first-order solids catalyzed reactions can be given by the analogy to the segregated flow model for homogeneous reactions provided an appropriate contact time density function, i.e. probability density of sojourn times in the reaction environment is defined. They showed (using somewhat different symbols) that:

$$1-x_A = \bar{E}_r \left(k \frac{W}{Q}\right) = \int_0^\infty e^{-k\,(W/Q)\theta} E_r(\theta)d\theta \qquad (104)$$

where $E_r(\theta)$ is the p.d.f. of sojourn times for the reaction environment. If gas bypassing is suspected, with f being the fraction of flow bypassing, the above eq. (104) can be represented by:

$$\bar{E}_r \left(k \frac{W}{Q}\right) = f + (1-f) \int_0^\infty e^{-k\,(W/Q)\theta}\, G(\theta)d\theta \qquad (105)$$

where $G(\theta)$ is the continuous portion of $E_r(\theta)$. Orcutt et. al. (157) used ozone decomposition on iron oxide particles at different temperatures to obtain values of $\bar{E}_r\left(k \frac{W}{Q}\right)$ at various values of the rate constant k. Inversion of the Laplace transform is in principle possible but was not attempted. Instead the parameters of $E_r(\theta)$ in terms of the assumed models were evaluated from the data. It was also shown that the fraction of gas bypassing can be obtained from eq. (105) at high values of k, i.e. for a fast reaction. Clearly, the method of Orcutt et. al. (157) represents a particular application of the marginal p.d.f. discussed in Section 6.1.1. Yates and Constans (158) experimentally confirmed that the moments for RTD of nonadsorbing and adsorbing tracers differ widely in a fluidized bed and that the use of strongly adsorbing tracers can give good indications about the degree of solids backmixing.

Nauman and Collinge (159,160) expanded upon the concept of contact time distributions introduced by Orcutt et. al. (157) and showed how to obtain it by use of an adsorbing and a nonadsorbing tracer or of a pair of adsorbing tracers with different adsorption rate constants. The tracers must adsorb linearly and reversibly. The theory as presented does not state clearly the most important limitation of the method, i.e. the assumptions made with regard to the flow pattern of the two phases. It also represents an example of the utilization of methods reviewed in Section 6.1.1. The theory holds for two situations: i) when the gas is backmixed to any extent (from complete backmixing to plug flow) but solids mixing is limited and tracer exits and re-enters the gas at the same point along the bed, or ii) the gas phase and the solids are perfectly mixed. This means that the contact time distribution theory can be used for baffled beds. Since perfect mixing is never quite achieved in unbaffled beds, the use of theory here is doubtful. Further use of the contact time distributions has been reviewed by Nauman and Buffham (4) and by Dohein and Collinge (161,162). Adsorbing tracers can also be used to evaluate transport coefficients based on a plausible model for the reactor.

Some recent applications of tracers in fluidized beds include: measurement of solids flow patterns (163), evaluation of the RTD in an entrained flow gasifier (164), study of lateral solids mixing in a packed fluidized bed (165), investigation of gas distribution (166) and gas and solids mixing (167).

6.2.4 Other reactors and uses of tracers. Tracers are used extensively in all other two-phase (gas-liquid, gas-solid, liquid-solid) and three-phase reactor types (gas-liquid-solid, liquid-liquid-solid). Tracers confined to a single phase are used to determine the RTD of that phase and evaluate its flow pattern. Tracers that can be transported from one phase to another are frequently used for evaluation of various rate parameters and transport coefficients such as: mass transfer coefficients, particle effective diffusivity, adsorption rate constants, kinetic rate constants, etc. The interpretation of tracer studies in evaluation of the above parameters is always dependent on the selected model for the system. We do not attempt to review this vast literature but will just cite a few examples as good starting points for the interested reader.

Effective diffusivity can be evaluated in a single pellet (113,168,169) or in a stirred vessel (115,170,171) or in a packed bed (113). The method of moments is usually used in parameter estimation. The method of moments and the theory of chromatography were also used extensively by Smith and coworkers in packed bed systems (113) for estimation of adsorption rate constants, particle diffusivity (172) and kinetic constants (173). Use of tracers in slurries and the dynamic response of three-phase slurry systems was recently updated by Datta and Rinker (174).

The whole subject of linear chromatography was systematically reviewed by Villermaux (175). This work is strongly recommended to those who intend to use tracers in packed-beds. Villermaux's chapter contains a concise review of the classical results for a sequence of equilibrated and nonequilibrated stages and the continuum approach as well. Most importantly, Villermaux clearly indicates the important results of Section 6.1.1 and shows that when the exchange processes occur only perpendicularly to flow direction (a good assumption for a packed column) the column response can be constructed from the knowledge of the nonadsorbing tracer response and the local exchange transfer function. Villermaux also deals with parallel and series distribution of adsorption sites (i.e. capacitances for the tracer) and shows how to use appropriate approximate solutions and when such solutions are possible.

6.2.5 Use of tracers in mechanistic investigations. In this review we have concentrated on describing the use of tracers in identification of flow patterns and other global reactor parameters. Another important area of tracer use is in discrimination of kinetic mechanisms for surface-catalyzed reactions. Radioactive tracers are always employed. For a good survey of this application of tracers the reader is referred to the reviews by Happel (176) and Bennett (177). Happel et. al. (178,179) have been recently employing these techniques in CO/H_2 methanation over a nickel catalyst.

7. STOCHASTIC MODELS

This review did not cover extensively the stochastic approach in reactor modeling and in tracer studies interpretation. Some references were given with respect to the work of Shinnar and coworkers (12,31,55,112). Recently this stochastic approach has been pursued extensively by Fan and coworkers (180-183) and the interested reader is referred to it. A promising model towards a generalized RTD for a reactor was also reported by Glasser and Jackson (184).

8. CONCLUSIONS

An attempt was made to survey the state-of-the-art in the use of tracers in chemical reactors and the associated areas of reactor modeling. Due to the vastness of the subject a personal view was given of the methods that seem to have found acceptance in laboratory and industrial practice.

In summary, we can reiterate the most important points of this review. The theory for RTDs in homogeneous systems is well developed. It allows us, knowing the RTD of the system, to calculate reactor behavior exactly for first-order reactions and to bound it for most nonlinear simple reactions. Tracers prove to be a unique experimental tool in obtaining RTDs and in analyzing the flow pattern departure from ideality. They are also useful in identifying parameters for the assumed reactor models.

In heterogeneous systems the theory is not complete but the introduction of multivariable density functions and their transforms and of the associated marginal probability densities seems promising in generalization of various concepts. The powerful central volume principle results from the theory and allows us to evaluate holdups of various phases. For systems with a single flowing phase and transport perpendicular to main flow direction, i.e. for stochastically independent residence time distributions in the two environments, the RTD of a tracer confined to the flowing phase and the knowledge of transport or kinetic parameters is sufficient to characterize the system completely for linear reactions. This concept has been propagated under the name of contact time distribution. In other situations tracer studies have to be coupled with plausible reactor models in order to evaluate model parameters. Tracers are again invaluable in detecting flow maldistribution and in identifying transport rates.

Overall, tracers have been and continue to be used in all homogeneous and heterogeneous reactor types. They remain an irreplacable tool in scaleup and cold flow modeling on which all prudent design engineers rely. The broad spectrum of their use is illustrated by this review.

9. ACKNOWLEDGEMENT

My gratitude goes to Mr. Kevin Myers who has dedicated unselfishly many hours to proofreading and correcting this manuscript.

10. REFERENCES

1. Danckwerts, P. V., Continuous Flow Systems: Distribution of Residence Times. Chem. Eng. Science 2 (1953) 1.
2. Levenspiel, O., Chemical Reaction Engineering (New York, Wiley, 2nd ed., 1972).
3. Froment, G. F. and K. B. Bischoff. Chemical Reactor Analysis and Design (New York, Wiley, 1979).
4. Nauman, E. B. and B. A. Buffham. Mixing in Continuous Flow Systems (New York, Wiley, 1983).

5. Aris, R. The Scope of R.T.D. Theory, Residence Time Distribution Theory in Chemical Engineering. (Petho, A. and R. D. Noble, eds., Weinheim, Verlag-Chemie, 1982).
6. Aris, R. Residence Time Distribution with Many Reactions and in Several Environments, Residence Time Distribution Theory in Chemical Engineering (Petho, A. and R. D. Noble, eds., Weinheim, Verlag-Chemie, 1982).
7. Brodkey, R. S. Fundamentals of Turbulent Motion, Mixing and Kinetics. Chem. Eng. Communications 8 (1981) 1.
8. Bourne, J. R. Micromixing Revisited, ISCRE 8 (Edinburgh, September, 1984).
9. Nauman, E. B. Residence Time Distributions and Micromixing. Chem. Eng. Communications 8 (1981) 53.
10. Patterson, G. K. Application of Turbulence Fundamentals to Reactor Modelling and Scaleup. Chem. Eng. Communications 8 (1981) 25.
11. Shinnar, R. Tracer Experiments in Chemical Reactor Design, Levich Birthday Conference on Physical Chemistry and Hydrodynamics (Oxford, July, 1977).
12. Shinnar, R. Use of Residence Time and Contact Time Distributions in Reactor Design, Chemical Reaction Engineering Handbook (Carberry, J. J. and A. Varma, eds., New York, Marcel Dekker, 1985).
13. Villermaux, J. Mixing in Chemical Reactors. Am. Chem. Soc. Symp. Series 226 (1983) 135.
14. Waldram, S. P. Nonideal Flow in Chemical Reactors, Comprehensive Chemical Kinetics, Vol. 23, Ch. 6 (G. H. Bamford, et. al. eds., Elsevier, Amsterdam, 1985).
15. Brodkey, R. S., ed. Turbulence in Mixing Operations: Theory and Applications to Mixing and Reaction (New York, Academic Press, 1975).
16. Hinze, J. O. Turbulence (New York, McGraw Hill, 1975).
17. Danckwerts, P. V. Tracers, Residence Times, Mixing and Dispersion. Lecture Notes (Pittsburgh, University of Pittsburgh, 1976).
18. Gonzales-Fernandez, J. M. Theory of the Measurement of the Dispersion of an Indicator in Indicator-Dilution Studies. Circulation Research 10 (1962) 409.
19. Levenspiel, O. and J. C. R. Turner. The Interpretation of Residence Time Experiments. Chem. Eng. Science 25 (1970) 1605.
20. Turner, J. C. R. The Interpretation of Residence-Time Measurements in Systems with and without Mixing. Chem. Eng. Science 26 (1971) 549.
21. Buffham, B. A. On the Residence Time Distribution for a System with Velocity Profiles in its Connection with the Environment. Chem. Eng. Science 27 (1972) 987.

22. Villermaux, J. Nomenclature and Symbols Recommended by the Working Party on Chemical Reaction Engineering of the E. F. Ch. E., Chem. Eng. Science 35 (1980) 2065.
23. Sheppard, C. W. Basic Principles of the Tracer Method (New York, Wiley, 1962).
24. Jacquez, J. A. Compartmental Analysis in Biology and Medicine (Amsterdam, Elsevier, 1972).
25. Lassen, N. and W. Perl. Tracer Kinetic Methods in Medical Physiology (New York, Raven, 1979).
26. Buffham, B. A. Internal and External Residence Time Distributions. Chem. Eng. Communications 22 (1983) 105.
27. Ritchie, B. W. and A. H. Tobgy. Residence Time Analysis in Systems Having Many Connections with their Environment. Ind. Eng. Chem. Fundamentals 17 (1978) 287.
28. Buffham, B. A. and H. W. Kropholler. The Washout Curve, Residence Time Distribution and F Curve in Tracer Kinetics. Math. Biosciences 6 (1970) 179.
29. Zhuang, Z. Model for Flow System Having Multiple Inlet and Outlet Streams. Scientia Simica 24 (1981) 626.
30. Sherman, H. On the Theory of Indicator-Dilution Methods Under Varying Blood-Flow Conditions. Bull. Math. Biphysics 22 (1960) 417.
31. Krambeck, F. J., R. Shinnar and S. Katz. Interpretation of Tracer Experiments in Systems with Fluctuating Throughput. Ind. Eng. Chem. Fundamentals 8 (1969) 431.
32. Krambeck, F. J., R. Shinnar and S. Katz. Stochastic Models for Chemical Reactors. Ind. Eng. Chem. Fundamentals 6 (1967) 276.
33. Cussler, E. L. Diffusion-Mass Transfer in Fluid Systems (Cambridge, Cambridge University Press, 1984).
34. Bischoff, K. B. and E. A. McCracken. Tracer Tests in Flow Systems. Ind. Eng. Chemistry 58 (1966) 18.
35. Merzkirch, W. Flow Visualization (New York, Academic Press, 1974).
36. Wang, C. H. and D. L. Willis. Radiotracer Methodology in Biological Science (Englewood Cliffs, Prentice Hall, 1965).
37. Burton, B. S. Take a Look at Nuclear Gauges. Instr. Control Systems. December (1976) 41.
38. Hines, D. B. Some Applications of Radioisotopes in Chemical and Engineering Research, AIChE One Day Symposium, St. Louis, April 1978.
39. Zwietering, T. N. The Degree of Mixing in Continuous Flow Systems. Chem. Eng. Science 11 (1959) 1.
40. Spalding, D. B. A Note on Mean Residence Times in Steady Flows of Arbitrary Complexity. Chem. Eng. Science 9 (1958) 74.
41. Stewart, G. N. The Pulmonary Circulation Time, the Quantity of Blood in the Lungs and the Output of the Heart. Am. J. Physiology 58 (1921) 20.

42. Hamilton, W. F., J. W. Moore, J. M. Kinsman and R. G. Spurling. Simultaneous Determination of the Pulmonary and Systematic Circulation Times in Man and of a Figure Related to the Cardiac Output. Am. J. Physiology 84 (1928) 84.
43. Curl, R. and M. L. McMillan. Accuracy in Residence Time Measurements. AIChE J. 12 (1966) 819.
44. Duduković, M. P. Tracer Analysis of the Microcirculation (Ph.D. Thesis, Chicago, IIT, 1972).
45. Hines, D. B. Chlor-alkali Plant Mercury Inventory with Mercury-203 Radioisotope. Proc. 15th Meeting Chlorine Plant Managers (The Chlorine Institute, Inc., February, 1972), p. 12.
46. Duduković, M. P. and R. M. Felder. Mixing Effects in Chemical Reactors. AIChE Modular Instruction, Series E. Kinetics, Vol. 4 (New York, AIChE, 1983).
47. Levich, V. G., V. S. Markin and Y. A. Chrismadzkev. On Hydrodynamic Mixing in a Model of a Porous Medium with Stagnant Zones. Chem. Eng. Science 22 (1967) 1357.
48. Mills, P. L. and M. P. Duduković. Evaluation of Liquid-Solid Contacting in Trickle-Bed Reactors by Tracer Methods. AIChE J. 27 (198k) 893; AIChE J 28 (1982) 526.
49. Mills, P. L. and M. P. Duduković. Modified Differential Refractometer for Continuous Liquid-Phase Residence Time Distribution Studies. Ind. Eng. Chem. Fundamentals 18 (1979) 292.
50. Mills, P. L. Catalyst Effectiveness and Solid-liquid Contacting in Trickle-Bed Reactors (D.Sc. Thesis, St. Louis, Washington University, May 1980).
51. Lapidus, L. Flow Distribution and Diffusion in Fixed-Bed Two-Phase Reactors. Ind. Eng. Chemistry 49 (1957) 1000.
52. Schiesser, W. E. and L. Lapidus. Further Studies of Fluid Flow and Mass Transfer in Trickle-Beds. AIChE J. 7 (1961) 163.
53. Rothfeld, L. B. and J. L. Ralph. Equivalence of Pulse and Step Residence Time Measurements in a Trickle-Phase Bed. AIChE J. 9 (1963) 852.
54. Awasthi, R. C. and K. Vasudeva. On Mean Residence Times in Flow Systems. Chem. Eng. Science 38 (1983) 313.
55. Naor, P. and R. Shinnar. Representation and Evaluation of Residence Time Distributions. Ind. Eng. Chem. Fundamentals 2 (1963) 278.
56. Nauman, E. B. Nonisothermal Reactors: Theory and Applications of Thermal Time Distribution. Chem. Eng. Science 32 (1977) 359.
57. Levenspiel, O. The Chemical Reactor Omnibook (Corvallis, Oregon State University Bookstores, 1979).
58. Wen, C. Y. and L. T. Fan. Models for Flow Systems and Chemical Reactors (New York, Marcel Dekker, 1975).

59. Cholette, A. and L. Cloutier. Mixing Efficiency Determination for Continuous Flow Systems. Can. J. Chem. Engineering, 37 (1953) 105.
60. Nagata, S. Mixing: Principles and Applications (New York, Halsted Press, 1975).
61. Corrigan, T. E. and W. O. Beavers. Dead Space Interaction in Continuous Stirred Tank Reactors. Chem. Eng. Science 23 (1968) 1003.
62. Norwood, K. W. and A. B. Metzner. Flow Patterns and Mixing Rates in Agitated Vessels. AIChE J. 6 (1960) 432.
63. Gibilaro, L. G. The Recycle Flow Mixing Model. Chem. Eng. Science 26 (1971) 299.
64. Van de Vusse, J. G. A New Model for the Stirred Tank. Chem. Eng. Science 17 (1962) 507.
65. Moo-Young, M. and K. W. Chan. Non-Ideal Flow Parameters for Viscous Fluids Flowing Through Stirred Tanks. Can. J. Chem. Engineering 49 (1971) 187.
66. Holmes, D. B., R. M. Voncken and J. A. Dekker. Fluid Flow in Turbine-Stirred Baffled Tanks. I. Circulation Time. Chem. Eng. Science 19 (1964) 201.
67. Voncken, R. M., D. B. Holmes and H. W. den Hartog. Fluid Flow in Turbine-Stirred Baffled Tanks. II. Dispersion During Circulation. Chem. Eng. Science 19 (1964) 209.
68. Khang, S. J. and O. Levenspiel. New Scaleup and Design Method for Stirrer Agitated Batch Mixing Vessels. Chem. Eng. Science 31 (1976) 569.
69. Rippin, D. W. T. The Recycle Reactor as a Model of Incomplete Mixing. Ind. Eng. Chem. Fundamentals 6 (1967) 488.
70. Fu, B., H. Weinstein, B. Bernstein and A. B. Shaffer. Residence Time Distributions of Recycle Systems - Integral Equation Formulation. Ind. Eng. Chem. Process Design Development 10 (1971) 501.
71. Buffham, B. A. and E. B. Nauman. On the Limiting Form of the Residence Time Distribution for a Constant Volume Recycle System. Chem. Eng. Science 30 (1975) 1519.
72. Nauman, E. B. and B. A. Buffham. Limiting Forms of the Residence Time Distribution for Recycle Systems. Chem. Eng. Science 32 (1977) 1233.
73. Nauman, E. B. and B. A. Buffham. A Note on Residence Time Distributions in Recycle Systems. Chem. Eng. Science 34 (1979) 1057.
74. Rubinovitch, M. and U. Mann. The Limiting Residence Time Distribution of Continuous Recycle Systems. Chem. Eng. Science. 34 (1979) 1309.
75. Buffham, B. A. and E. B. Nauman. Residence-Time Distribution at High Recycle Ratios. Chem. Eng. Science 39 (1984) 841.
76. Leitman, R. H. and E. N. Ziegler. Stirred Tank Reactor Studies: Part I: Mixing Parameters. Chem. Eng. Journal 2 (1971) 252.

77. Leitman, R. H. and E. N. Ziegler. Stirred Tank Reactor Studies: Part II. Conversion Models. Chem. Eng. Journal 3 (1972) 245.
78. Hanley, T. R. and R. A. Mischke. A Mixing Model for a Continuous Flow Stirred Tank Reactor. Ind. Eng. Chem. Fundamentals 17 (1978) 51.
79. Spencer, J. L., R. R. Lunt and S. A. Leshaw. Identification of Micromixing Mechanisms in Flow Reactors: Transient Inputs of Reactive Tracers. Ind. Eng. Chem. Fundamentals 19 (1980) 135.
80. Spencer, J. L. and R. R. Lunt. Experimental Characterization of Mixing Mechanisms in Flow Reactors Using Reactive Tracers. Ind. Eng. Chem. Fundamentals 19 (1980) 142.
81. Lintz, H. G. and W. Weber. The Study of Mixing in a Continuous Stirred Tank Reactor Using an Autocatalytic Reaction. Chem. Eng. Science 35 (1980) 203.
82. Denbigh, K. G., N. Dombrowski, A. J. Kisiel and E. R. Place. The Use of the "Time Reaction" in Residence Time Studies. Chem. Eng. Science 17 (1962) 573.
83. Danckwerts, P. V. and R. A. M. Wilson. Flow Visualization by Means of a Time Reaction. J. Fluid Mechanics 16 (1963) 412.
84. Castellana, F. S., M. I. Friedman and J. L. Spencer. Characterization of Mixing in Reactor Systems Through Analysis of Regional Tracer Dilution Data Obtained with a Gamma Camera. AIChE Journal 30 (1984) 207.
85. Mecklenburgh, J. C. and S. Hartland. The Theory of Backmixing. (London, Wiley, 1975).
86. Shah, Y. T., G. J. Stiegel and M. M. Sharma. Backmixing in Gas-Liquid Reactors. AIChE Journal 24 (1978) 369.
87. Klinkenberg, A. Distribution of Residence Times in a Cascade of Mixed Vessels with Backmixing. Ind. Eng. Chem. Fundamentals 5 (1966) 283.
88. Hochman, J. M. and J. R. McCord. Residence Time Distribution in Recycle Systems with Crossmixing. Chem. Eng. Science 25 (1970) 97.
89. Bhavaraju, S. M., T. W. F. Russell and H. W. Blanch. The Design of Gas Sparged Devices for Viscous Liquid Systems. AIChE Journal 24 (1978) 454.
90. Maruyama, T., N. Kamishima and T. Mizuchina. Investigation of Bubble Plume Mixing by Comparison with Liquid Jet Mixing. J. Chem. Eng. Japan 17 (1984) 120.
91. Levenspiel, O. and K. B. Bischoff. Patterns of Flow in Chemical Process Vessels. Advances in Chemical Engineering, Vol. 4 (Drew, T. B., ed., New York, Academic Press, 1963).
92. Van der Laan, E. T. Notes on the Diffusion-Type Model for the Longitudinal Mixing of Fluids in Flow. Chem. Eng. Science 7 (1958) 187.

93. Wehner, J. F. and R. H. Wilhelm. Boundary Conditions of Flow Reactor. Chem. Eng. Science 6 (1956) 89.
94. Bischoff, K. B. A Note on Boundary Conditions for Flow Reactors. Chem. Eng. Science 16 (1961) 131.
95. Choi, C. Y. and D. D. Perlmutter. A Unified Treatment of the Inlet Boundary Condition for Dispersive Flow Models. Chem. Eng. Science 31 (1976) 250.
96. Parulekar, S. J. and D. Ramkrishna. Analysis of Axially Dispersed Systems with General Boundary Conditions I, II and III. Chem. Eng. Science 39 (1984) 1571, 1581, 1599.
97. Taylor, G. I. Dispersion of Soluble Matter in Solvent Flowing Slowly through a Tube. Proc. Roy. Society (London) A219 (1953) 186.
98. Taylor, G. I. The Dispersion of Matter in Turbulent Flow through a Pipe. Proc. Roy. Society (London) A223 (1954) 446.
99. Taylor, G. I. Diffusion and Mass Transport in Tubes. Proc. Physical Society (London) B67 (1954) 857.
100. Aris, R. On the Dispersion of a Solute in a Fluid Flowing through a Tube. Proc. Roy. Society A235 (1956) 67.
101. Koch, D. L. and J. F. Brady. Dispersion in Fixed Beds. J. Fluid Mechanics (1985). To appear.
102. Bischoff, K. B. A Note on Gas Dispersion in Packed Beds. Chem. Eng. Science 24 (1969) 607.
103. Deans, H. A. A Mathematical Model for Dispersion in the Direction of Flow in Porous Media. Soc. Petr. Eng. Journal 3 (1963) 49.
104. Hoogendorn, C. J. and J. Lips. Axial Mixing of Liquid in Gas-Liquid Flow through Packed Beds. Can. J. Chem. Engineering 43 (1965) 125.
105. Hochman, J. M. and E. Effron. Two-Phase Countercurrent Downflow in Packed Beds. Ind. Eng. Chemistry Fundamentals, 8 (1969) 63.
106. Buffham, B. A., L. G. Gibilaro and M. N. Rathor. A Probabilistic Time Delay Description of Flow in Packed Beds. AIChE Journal 16 (1970) 218.
107. Rathor, M. N., L. G. Gibilaro and B. A. Buffham. The Hopping Model for Residence Time Distributions of Systems with Splitting and Merging Streams. AIChE Journal 31 (1985) 330.
108. Schwartz, J. G. and G. W. Roberts. An Evaluation of Models for Liquid Backmixing in Trickle-Bed Reactors. Ind. Eng. Chem. Process Design Development 12 (1973) 262.
109. Hinduja, M. J., S. Sundaresan and R. Jackson. A Cross-flow Model of Dispersion in Packed Bed Reactors. AIChE Journal 26 (1980) 274.
110. Aris, R. Residence Times in Several Environments. Recent Advances in Engineering Analysis of Chemically Reacting Systems. (Doraiswamy, L. V., ed., New Delhi, Wiley Eastern, 1984).
111. Levenspiel, O. Private Communications. March, 1983.

112. Shinnar, R., P. Naor and S. Katz. Evaluation of Multiple Tracer Experiments. Chem. Eng. Science 27 (1972) 1627.
113. Furusawa, T., M. Suzuki and J. M. Smith. Rate Parameters in Heterogeneous Catalysis by Pulse Techniques. Cat. Rev. Sci. Eng. 13 (1976) 43.
114. McCoy, B. J. Approximation of a Heterogeneous Chemical Reaction with a Fluid Phase Reaction. Chem. Eng. Science 39 (1984) 1524.
115. Weinstein, H. and M. P. Duduković. Tracer Methods in the Circulation, Topics in Transport Phenomena, Ch. 4 (Gutfinger, C., ed., New York, Hemisphere Pub. Corp. 1975).
116. Miller, G. A. and J. E. Bailey. Some New Results for Chromatographic Kinetics Studies. AIChE Journal 19 (1973) 876.
117. Shinnar, R. and D. Rumschitzki. The Use of Residence Time Distributions in Heterogeneous Reactor Modeling, Design and Scaleup. 76th AIChE Annual Meeting, San Francisco, November 1984. paper 139b.
118. Mills, P. L., W. P. Wu and M. P. Duduković. Tracer Analysis in Systems with Two-Phase Flow. AIChE Journal 25 (1979) 885.
119. Ramachandran, P. A. and J. M. Smith. Dynamics of Three-Phase Slurry Reactors. Chem. Eng. Science 32 (1977) 873.
120. Ramachandran, P. A. and J. M. Smith. Dynamic Behavior of Trickle-Bed Reactors. Chem. Eng. Science 34 (1979) 75.
121. Ramachandran, P. A. and R. V. Chaudhari. Three-Phase Catalytic Reactors. (New York, Gordon & Breach, Pub., 1983).
122. Schwartz, J. G., E. Weger and M. P. Duduković. Liquid Holdup and Dispersion in Trickle-Bed Reactors. AIChE Journal 22 (1976) 953.
123. Duduković, M. P. and P. L. Mills. Contacting and Hydrodynamics in Trickle-Bed Reactors, Encyclopedia of Fluid Mechanics (Cheremisinoff, N. P., ed., New York, Gulf Publ. Corp., 1985).
124. Sicardi, S., G. Baldi and V. Specchia. Hydrodynamic Models for the Interpretation of the Liquid Flow in Trickle-Bed Reactors. Chem. Eng. Science 35 (1980) 1775.
125. Eroglu, I. and T. Dogu. Dynamic Analysis of a Trickle-Bed Reactor by Moment Technique. Chem. Eng. Science 38 (1983) 801.
126. Kan, K. M. and P. F. Greenfield. Residence-Time Model for Trickle-Flow Reactors Incorporating Incomplete Mixing in Stagnant Regions. AIChE Journal 29 (1983) 123.
127. Herskowitz, M. and J. M. Smith. Liquid Distribution in Trickle-Bed Reactors - 2. Tracer Studies. AIChE Journal 24 (1978) 450.
128. Schwartz, J. G., E. Weger and M. P. Duduković. A New Tracer Method for Determination of Liquid-Solid Contacting Efficiency in Trickle-Bed Reactors. AIChE Journal 22 (1976) 894.

129. Colombo, A. J., G. Baldi and S. Sicardi. Solid-Liquid Contacting Effectiveness in Trickle-Bed Reactors. Chem. Eng. Science 31 (1976) 1101.
130. Duduković, M. P. Catalyst Effectiveness Factor and Contacting Efficiency in Trickle-Bed Reactors. AIChE Journal 23 (1977) 940.
131. Murphee, E. V., A. Voorhies, Jr. and F. X. Mayer. Application of Contacting Studies to the Analysis of Reactor Performance. Ind. Eng. Chem. Proc. Des. Develop. 3 (1964) 381.
132. Argo, W. B. and D. R. Cova. Longitudinal Mixing in Gas-Sparged Tubular Vessels. Ind. Eng. Chem. Process Des. Develop. 4 (1965) 352.
133. Bischoff, K. B. and J. B. Phillips. Longitudinal Mixing in Orifice Plate Gas-Liquid Reactors. Ind. Eng. Chem. Process Des. Develop. 5 (1966) 416.
134. Reith, T., S. Renken and B. A. Israel. Gas Holdup and Axial Mixing in Fluid Phase of Bubble Columns. Chem. Eng. Science 23 (1968) 619.
135. Kunigita, E., M. Ikura and T. Otake. Liquid Behavior in Bubble Column. J. Chem. Eng. Japan 3 (1970) 24.
136. Ohki, Y. and H. Inoue. Longitudinal Mixing of the Liquid Phase in Bubble Columns. Chem. Eng. Science 25 (1970) 1.
137. Eissa, S. H., M. M. El-Halwagi and M. Saleh. Axial and Radial Mixing in a Cocurrent Bubble Column. Ind. Eng. Chem. Process Des. Develop. 10 (1971) 31.
138. Deckwer, W. D., U. Graeser, H. Langemann and Y. Serpemen. Zones of Different Mixing in Liquid Phase Bubble Columns. Chem. Eng. Science 28 (1972) 1223.
139. Chen, B. H. Effects of Liquid Flow on Axial Mixing of Liquid in a Bubble Column. Can. J. Chem. Engineering 50 (1972) 436.
140. Deckwer, W. D., R. Burckhart and G. Zoll. Mixing and Mass Transfer in Tall Bubble Columns. Chem. Eng. Science 29 (1974) 2177.
141. Hikita, H. and H. Kikukawa. Liquid-Phase Mixing in Bubble Columns: Effect of Liquid Properties. Chem. Eng. Journal 8 (1974) 191.
142. Towell, G. D. and G. H. Ackerman. Axial Mixing of Liquid and Gas in Large Bubble Reactors. Proc. Symp. Chem. Reaction Engineering (Amsterdam, 1972, pp B3-1-B3-13).
143. Alexander, B. F. and Y. T. Shah. Axial Dispersion Coefficients in Bubble Columns. Chem. Eng. Journal 11 (1976) 153.
144. Gondo, S., S. Tanaka, K. Kazikuri and K. Kusunoki. Liquid Mixing by Large Gas Bubbles in Bubble Columns. Chem. Eng. Science 28 (1973) 1437.
145. Eissa, S. H. and K. Schügerl. Holdup and Backmixing. Investigations in Cocurrent and Countercurrent Bubble Columns. Chem. Eng. Science 30 (1975) 1251.

146. Rice, R. G., J. M. I. Tupperainen and R. M. Hedge. Dispersion and Holdup in Bubble Columns - Comparison of Rigid and Flexible Spargers. Can. J. Chem. Engineering 59 (1981) 677.
147. Field, R. W. and J. F. Davidson. Axial Dispersion in Bubble Columns. Trans. Instn. Chem. Engineers 58 (1980) 228.
148. Kelkar, B. G., S. R. Phulgaonkar and Y. T. Shah. The Effect of Electrolyte Solutions on Hydrodynamic and Backmixing Characteristics in Bubble Columns. Chem. Eng. Journal 27 (1983) 125.
149. Kelkar, B. G., Y. T. Shah and N. L. Carr. Hydrodynamics and Axial Mixing in a Three Phase Bubble Column - Effects of Slurry Properties. Ind. Eng. Chem. Process Des. Develop. 23 (1984) 308.
150. Stiegel, G. J. and Y. T. Shah. Backmixing and Liquid Holdup in Gas-Liquid Cocurrent Upflow Packed Column. Ind. Eng. Chem. Process Des. Develop. 16 (1977) 37.
151. Chen, B. H. Holdup and Axial Mixing in Bubble Columns Containing Screen Cylinder. Ind. Eng. Chem. Process Des. Devel. 15 (1976) 20.
152. Shah, Y. T., C. A. Ratway and H. G. Mcilvried. Backmixing Characteristics of a Bubble Column with Vertically Suspended Tubes. Trans. Instn. Chem. Engineers 56 (1978) 107.
153. Pilhofer, Th., H. F. Bach. and K. H. Mamgartz. Determination of Fluid Dynamic Parameters in Bubble Column Design. ACS Symp. Series 65 (1978) 372.
154. Deckwer, W. D., K. Nguyen-Tien, B. G. Kelkar and Y. T. Shah. Applicability of Axial Dispersion Model to Analyze Mass Transfer Measurements in Bubble Columns. AIChE Journal 29 (1983) 915.
155. Vasalos, I. A., E. M. Bild, D. N. Rundell and D. F. Tatterson. Experimental Techniques for Studying the Fluid Dynamics of the H-Coal Reactor. Coal Processing Technology V6 (New York, AIChE J., 1980, pp. 226).
156. Overcashier, R. H., D. E. Todd and R. B. Olney. Some Effects of Baffles on a Fluidized System. AIChE Journal 5 (1959) 54.
157. Orcutt, J. C., J. F. Davidson and R. L. Pigford. Reaction Time Distributions in Fluidized Catalytic Reactors. Chem. Engr. Progress Symp. Series No. 38, Vol. 58 (1962) 1.
158. Yates, J. G. and J. A. P. Constans. Residence Time Distributions in a Fluidized Bed in which Gas Adsorption Occurs: Stimulus-Response Experiments. Chem. Eng. Science 28 (1973) 1341.
159. Nauman, E. B. and C. N. Collinge. The Theory of Contact Time Distributions in Gas Fluidized Beds. Chem. Eng. Science 23 (1968) 1309.
160. Nauman, E. B. and C. N. Collinge. Measurement of Contact Time Distribution in Gas Fluidized Beds. Chem. Eng. Science 23 (1968) 1317.

161. Dohein, M. A. and C. N. Collinge. Contact Time Distribution in Fluidized-Bed Reactors, Part I. Measurement at Room and Higher Temperatures. Chem. Eng. Journal 19 (1980) 39.
162. Dohein, M. A. and C. N. Collinge. Contact Time Distribution in Fluidized-Bed Reactors. Part II. Application of Mathematical Models and Parameter Estimation. Chem. Eng. Journal 19 (1980) 47.
163. Baba, T., M. Nakajima, S. Morooka and H. Matsuyama. New Measuring System for Flow Patterns of Solid Particles in Gas-Solid Fluidized Bed. J. Chem. Eng. Japan 17 (1984) 275.
164. Rabbits, M. C., G. J. Van Den Houten, D. Glasser and A. W. Bryson. Modeling of Residence Time Distribution in an Entrained Flow Coal Gasification Reactor. Chemsa 9 (1983) 220.
165. Kato, K., D. Taneda, Y. Sato and M. Maa. Lateral Solid Mixing in a Packed Fluidized Bed. J. Chem. Eng. Japan 17 (1984) 78.
166. Bauer, W. and J. Werther. Role of Gas Distribution in Fluidized Bed Chemical Reactor Design. Chem. Eng. Communications 18 (1982) 137.
167. Wippern, D., K. Wittman, J. Kuehne, H. Helmrich and K. Schügerl. Characterization of Fluidized Bed Reactors with Gas Tracer Measurements. Chem. Eng. Communications 10 (1981) 307.
168. Burghardt, A. and J. M. Smith. Dynamic Response of a Single Catalyst Pellet. Chem. Eng. Science 34 (1979) 267.
169. Duduković, M. P. An Analytical Solution for the Transient Response in a Diffusion Cell of the Wicke-Kallenbach Type. Chem. Eng. Science 37 (1982) 153.
170. Dutta, R., B. Croes and R. G. Rinker. Transient Response of Continuous Flow Stirred Reactors Containing Heterogeneous Systems for Catalysis or Sorption. Chem. Eng. Science 38 (1983) 885.
171. Midoux, N. and J. C. Charpentier. Apparent Diffusivity and Tortuosity in a Liquid Filled Porous Catalyst used for Hydrodesulfurization of Petroleum Products. Chem. Eng. Science 28 (1973) 2108.
172. Schneider, P. and J. M. Smith. Adsorption Rate Constants from Chromatography. AIChE Journal 14 (1968) 762.
173. Suzuki, M. and J. M. Smith. Kinetic Studies by Chromatography. Chem. Eng. Science 26 (1971) 221.
174. Dutta, R. and R. G. Rinker. Transient Response of Three Phase Slurry Reactors. Chem. Eng. Science 39 (1984) 893.
175. Villermaux, J. Theory of Linear Chromatography, Percolation Processes: Theory and Applications (Rodriguez, A. E. and Toudeur, D., eds., Alphen eau den (Rijn, the Netherlands, Sijthoff & Noordhoff, 1981, p. 83).
176. Happel, J. Study of Kinetic Structure Using Marked Atoms. Catalysis Reviews 6 (1972) 221.

177. Bennett, C. O. The Transient Method and Elementary Steps in Heterogeneous Catalysis. Catal. Rev. - Sci. Eng. 13 (1976) 121.

178. Happel, J., I. Suzuki, P. Kokayeff and V. Fthemakis. Multiple Isotope Tracing of Methanation over Nickel Catalyst. J. Catalysis 65 (1980) 59.

179. Happel, J., H. Y. Cheh, M. Otarad, S. Ozawa, A. J. Severdia, T. Yoshido and V. Fthemakis. Multiple Isotope Tracing of Methanation over Nickel Catalyst II. Deuteromethanes Tracing. J. Catalysis 75 (1982) 314.

180. Nassar, R., L. T. Fan and J. R. Too. A Stochastic Treatment of Unimolecular Reactions in an Unsteady State Continuous Flow System. Chem. Eng. Science 36 (1981) 1307.

181. Nassar, R., J. R. Too and L. T. Fan. Stochastic Modeling of Polymerization in a Continuous Flow Reactor. J. Applied Polymer Science 26 (1981) 3745.

182. Too, J. R., L. T. Fan and R. Nassar. Markov Chain Models of Complex Chemical Reactions in Continuous Flow Reactors.

183. Fox, R. O. and L. T. Fan. A Master Equation Formulation for Stochastic Modeling of Mixing and Chemical Reactions in Inter-Connected Continuous Stirred Tank Reactors. Instn. Chem. Engrs. Symp. Series 87 (1984) 561.

184. Glasser, D. and R. Jackson. A Generalized Residence Time Distribution Model for a Chemical Reactor. Instn. Chem. Engrs. Symp. Series 87 (1984) 535.

11. NOMENCLATURE

A - area for mass transfer
$\underline{C}$ - concentration vector
$\underline{C}_b$ - concentration vector in a batch
C_i - concentration in region i
C_A - concentration of reactant A
C_{A_b} - concentration of reactant A in a batch
C_{A_o} - feed or initial concentration of A
C_o - normalizing tracer concentration
$C(t)$ - tracer concentration at the exit in response to set-up in the inlet
$C_d(t)$ - tracer concentration at the exit in response to step-down in the inlet
$C_p(t)$ - tracer concentration at the exit in response to an impulse injection
D_m - molecular diffusivity
De_{TP} - effective diffusivity under two-phase flow condition
De_{LF} - effective diffusivity under liquid only flow
D_e - effective diffusivity in particles
D_{ax} - axial dispersion coefficient
Da_n - Damkohler number for n-th order reaction, eq. (2)

d_t - tube diameter
d_p - particle diameter
$E(t)$ - exit age (residence times) density function
$E_{na}(t)$ - exit age density function for nonadsorbing tracer
$E_a(t)$ - exit age density function for adsorbing tracer
$E_r(t)$ - probability density of sojourn times in reacting environment
$\overline{E}(s)$ - Laplace transform of the exit age density function
$E_{ij}(t)$ - conditional density function, eq. (9)
F_i - fractional pore fillup
$F(t)$ - residence time distribution
$H(x-a)$ - Heaviside's unit step function
$H(s)$ - transfer function defined by eq. (82)
$I(t)$ - internal age density, eq. (14)
K - partition coefficient between phases 1 and 2
K_{fi} - partition coefficient for tracer between flowing phase f and phase i
k_o - frequency factor in the Arrhenius rate constant
k - rate constant for n-th order reaction
k_m - mass transfer coefficient
k_t - true first order rate constant
k_s - apparent first order rate constant
k_w - first order rate constant per unit catalyst mass
L - reactor length
L_s - macroscale of mixing
m_T - amount of tracer injected
N - number of mixed cells
n - reaction order
Pe - Peclet number (Re.Sc)
Pe_{ax} - Peclet number for axial dispersion
Q - volumetric flow rate
Q_L - liquid flow rate
Q_f, Q_i - volumetric flow rate of flowing phase or of phase i, respectively
q - volumetric exchange rate between flowing and stagnant phase
q_{ij} - flow rate from inlet i to outlet j
$q_{i_o}^{i}$ - flow rate from inlet i to all outlets
q_i - flow rate to outlet i from all inlets
$\underline{R}$ - recycle ratio
$\overline{r}$ - average first order reaction rate, eq. (18)
$R(t)$ - linear response to tracer concentration
s - Laplace transform variable
T_m - characteristic macromixing time
t - time
t_k - characteristic reaction time

$\bar{t}$ - mean residence time, V/Q
t_{app} - apparent mean residence time
$\underline{t}_T$ - thermal age, eq. (23)
u - mean velocity
V - reactor volume
V_f, V_i - volume of flowing phase and phase i, respectively
V_s - stagnant volume
W - mass of catalyst in the bed
W(t) - washout curve, eq. (4)
x_A - fractional conversion of A
z - dimensionless axial distance

Greek Symbols

α - internal age, also fraction of flow that does not bypass
α_j - fraction of flow rate in stream j
β_j - fraction of volume in subsystem j
ε_i - holdup of phase i
ε_B - bed (external) porosity
ε_p - particle porosity
ε_T - total porosity
η_{CE} - fraction of externally wetted catalyst
θ - dimensionless time, $t/\bar{t}$
θ_i - regional mean residence time, V_i/Q_1
λ - life expectancy, also dimensionless mass transfer mass transfer coefficient $k_m A/Q_1$, and also mixing microscale
$\Lambda(t)$ - intensity function, eq. (17)
μ_{1E}, μ_{1I} - first moment of the E and I function, respectively
$\bar{\bar{\sigma}}_E^2$ - dimensionless variance of the exit age density function, eq. (15)
$\chi(t)$ - probability density of residence times for fluid in the reactor, eq. (8)

MACRO AND MICROMIXING PHENOMENA IN CHEMICAL REACTORS

Jacques VILLERMAUX

Laboratoire des Sciences du Génie Chimique - Centre National de la Recherche Scientifique - Ecole Nationale Supérieure des Industries Chimiques, NANCY - FRANCE

I. INTRODUCTION

Attempts to describe mixing in chemical reactors is one of the most striking examples which show that Chemical Reaction Engineering is not a mere combination of chemistry and applied physics, but an original scientific discipline which has developed its own concepts.

Up until 1950 mixing was considered to be a problem in fluid mechanics. In single fluids, for instance, mixing and chemical reaction could be accounted for by solving the Navier-Stokes equations with appropriate boundary conditions and source terms for chemical production. Unfortunately, although this approach is conceivable for the simplest situations, the method becomes very difficult to apply to the complex flow patterns encountered in industrial reactors and almost impracticable in multiphase flow. And even if these equations could be solved by using a large capacity computer they would provide chemical engineers with a host of useless analytical data which would mask the overall behaviour of the reactor and have to be averaged to yield practical input/output relationships.

Major breakthroughs came in the fifties from the work of Danckwerts, Zwietering, Levenspiel and others, who considered the reactor as a system and introduced novel concepts such as residence time distribution and macro and micromixing, ideas which are now familiar to most chemical engineers. This new "Systems Approach" was immediately recognized as very fruitful, not only to solve industrial problems by simple and elegant methods, but also to better understand the fundamental behaviour of chemical reactors. Followwing these pioneering papers, basic concepts were progressively enriched and extended to more complex systems by systematic use of population balance methods, and by a Lagrangian description of fluid mixing on a microscopic level. For the last thirty years mixing in chemical reactors has been and still is a very active area for research and industrial applications.

The aim of the present paper is to give a general outline of the subject, recalling classical concepts, presenting recent results and trying to recognize frontiers where further progress is still required.

2. MACROMIXING

Macromixing is the process whereby parts of the fluid having different histories come into contact on the macroscopic scale. Danckwerts [1] proposed to characterize macromixing in continuous flow systems by the Residence Time Distribution (RTD) of the fluid, a concept which became very famous and is now the basis of most reactor models. RTD theory can be found in many classical textbooks [2] [3] [4]. Usually, simplifying assumptions are made such as

- steady state and deterministic flow
- incompressible fluid and constant flow-rate
- one single inlet and one outlet
- plug flow in inlet and outlet ports
- isothermal conditions

With these assumptions, which may appear somewhat restrictive, RTD theory makes it possible to solve many industrial problems. The theory will be only briefly outlined here, mainly to introduce recent extensions which allow more complex situations to be dealt with.

2.1. Classical distribution functions

The age α of a fluid (or solid) particle is defined as the time elapsed since it entered the reactor. Its life expectancy or residual lifetime λ is the time it has still to spend before leaving the reactor. The residence time $t_s = \alpha + \lambda$ is the age of the leaving particle in the outlet section. As the reactor contains particles with different histories, all these quantities are distributed. $E(t_s)$ is the RTD such that the fraction of leaving flow-rate Q containing particles with a residence time between t_s and $t_s + dt_s$ is $dQ = Q\, E(t_s)dt_s$. A related function is the cumulative

RTD $F(t_s) = \int_0^{t_s} E(t)dt$. $I(\alpha)$ is the Internal Age Distribution (IAD) such that $VI(\alpha)d\alpha$ represents the volume fraction of the reactor occupied by particles with an age between α and $\alpha + d\alpha$.

A less known distribution is the Intensity Function $\Lambda(\alpha)$ or $\Lambda(t_s)$ such that $\Lambda(\alpha)d\alpha$ is the fraction of the population of age $\alpha = t_s$ which is to leave in the next $d\alpha$ time interval. These functions are not independent from each other and it can be easily shown that

$$E(t) = -\tau \frac{dI(t)}{dt} = \frac{dF(t)}{dt} \tag{2-1}$$

$$\Lambda(t) = \frac{E(t)}{\tau I(t)} = \frac{E(t)}{1 - F(t)} \tag{2-2}$$

where $\tau = V/Q$ is the space time in the reactor.

What made Danckwert's theory very successful is that $E(t)$ and the related distributions can be experimentally measured by means of tracers. In particular, $E(t)$ appears as the response of the reactor to a unit-pulse injection of tracer. The flow-system may thus be considered as a linear dynamic system (with respect to tracer concentration) and all the resources of Systems Dynamics can be used for identifying $E(t)$ from tracer experiments. In particular, the response $y(t)$ to an imperfect pulse injection is given by the convolution product

$$y(t) = \int_0^t x(t')\, E(t - t')\, dt' \tag{2-3}$$

The practical implementation of such methods is discussed in another chapter of this book [5] to which the reader is referred.

Other familiar quantities are the moments of the RTD, especially the mean residence time

$$\overline{t_s} = \int_0^\infty t_s\, E(t_s)\, dt_s \tag{2-4}$$

and the second order central moment, or variance

$$\sigma^2 = \int_0^\infty (t_s - \overline{t_s})^2\, E(t_s)\, dt_s \tag{2-5}$$

Higher order moments are difficult to interpret owing to large uncertainty in their experimental determination.

It results from (2-3) that $\overline{t_s} = \overline{t_y} - \overline{t_x}$ and $\sigma_s^2 = \sigma_y^2 - \sigma_x^2$. Under the simplifying assumptions above, and in the absence of by-pass or dead volume, $\overline{t_s} = \tau$.

2.2. The Bundle of Parallel Tubes (BPT) model

A very useful representation of the flow pattern within the reactor can be obtained as follows: the fluid is assumed to flow at constant velocity into small tubes whose length are thus proportional to the residence time. The tubes are piled up according to increasing residence times. The internal volume is thus reorganized in the form of a bundle of tubes. In the elementary tube of residence time (or "length") between t_s and $t_s + dt_s$, the fractional flowrate is $dQ/Q = E(t_s)\ dt_s$. Finally, one obtains the sketch shown in Fig. 1 which has interesting properties. If all the tube inlets are lined up as shown in the figure, all particles having the same age remain on a vertical line and the locus of tube extremities is an exact picture of the F-distribution. The meaning of $\Lambda(t_s)dt_s$, fraction of fluid of age $(t_s, t_s + dt_s)$ which leaves in the next dt_s, also appears in the figure. There are two versions of the model. In the first one, the length of a tube is t_s and its width is dQ/Q. The height of the pile is then 1 and the surface of the bundle in the sketch is equal to $\overline{t_s} = \tau = V/Q$. In the second one, which will be used later, the width of the tube is dQ so that the height of the pile is equal to the volumetric flowrate Q. The surface of the bundle then represents the reactor volume V.

2.3. Macromixing in non-ideal flow

Two ideal and extreme situations may occur - plug flow where all particles have the same residence time ($E(t_s) = \delta(t_s - \tau)$), and perfectly mixed flow where the leaving probability is the same for each particle ($E(t_s) = (1/\tau)\exp(-t_s/\tau)$).

In real situations, an intermediate behaviour is generally observed with a distribution of residence times about the mean. Tracer experiments first yield information about hydrodynamic parameters. Let y(t) be the response to a pulse injection of tracer. Several cases may be considered:

(a) Neither the exact amount of tracer nor the absolute calibration of the detector are known (y is only assumed to be proportional to the concentration of tracer). The only attainable information is the mean residence time

$$\overline{t_s} = \int_o^T t\ y(t)\ dt \ \Big/ \int_o^T y(t)\ dt \qquad (2\text{-}6)$$

The response curve is truncated at time T when the signal is undistinguishable from the base-line. Of course, this experiment also yields the shape of $E(t) = y(t)/ \int_o^T y(t)\, dt$.

(b) The amount of injected tracer (n moles) and the absolute calibration of the detector are known (C(t) is the true tracer concentration). By a straightforward tracer balance both the flowrate Q and the volume V can be determined.

$$Q = n/ \int_o^T C(t)\, dt \qquad (2\text{-}7)$$

$$V = Q\, \overline{t_s} \qquad (2\text{-}8)$$

This method may be very helpful when no other measurement of Q is possible.

(c) Independent values for the flowrate, say Q', and the reactor volume, say V', are known from direct measurements. If Q and V differ significantly from Q' and V', this may reveal operating errors such as by-passing or dead-volume. In the latter case, the response curve generally exhibits tailing and the dead volume V' - V may depend on the choice of T.

2.4. Macromixing models

After obtaining the RTD from tracer experiments, it is interesting to represent it by a flow model, either for further calculations of chemical conversion or as a tool for scale-up.

In the case of small deviations from plug flow or simple dispersion within the reactor volume, one-parameter models may suffice to account for the RTD. Two of them are widely used:

- The dispersion model (axially dispersed plug flow)

This simply assumes that axial dispersion (D_A $m^2.s^{-1}$) is superimposed onto plug flow. Axial dispersion may be caused by a velocity profile in the radial direction or statistical dispersion in a packing or turbulent diffusion or by any physicochemical process which delayes some particles with respect to others. The model parameter is the axial PECLET number, $Pe = uL/D_A$, or its reciprocal, the dispersion number, D_A/uL. Depending on the boundary conditions assumed at the reactor inlet and outlet (which are different from those of the simple assumptions above), a lot of mathematical formulae can be found in the literature for the RTD [3]. This is often academic as in the range of usefulness of the model (small deviation from plug flow, say Pe > 20) all conditions lead to response curves which cannot be distinguished from one another on account of the accuracy of the experimental data (6). The RTD is then close to

$$E(t_s) = \frac{1}{2}\left(\frac{Pe}{\pi\ \tau\ t_s}\right)^{1/2} \exp\left(-\frac{Pe(\tau - t_s)^2}{4\ \tau\ t_s}\right) \qquad (2\text{-}9)$$

the mean residence time is $\overline{t_s} = \tau$ and the variance is approximately

$$\sigma^2 = 2\ \sigma^2/Pe \qquad (2\text{-}10)$$

- The mixing tanks in series model

The RTD is taken to be equal to that of a cascade of J equal stirred tanks in series, which is a Gamma distribution:

$$E(t_s) = \left(\frac{J}{\tau}\right)^J \frac{t_s^{J-1}\ \exp(-Jt_s/\tau)}{(J-1)!} \qquad (2\text{-}11)$$

The mean residence time is $\overline{t_s} = \tau$ and the variance is

$$\sigma^2 = \tau^2/J \qquad (2\text{-}12)$$

This model can be extended to non-integer values of J upon replacing in eq. (2-11) (J-1)! by the Gamma function

$$\Gamma(J) = \int_0^\infty \exp(-x)\ x^{J-1}\ dx \qquad (2\text{-}13)$$

J is then a parameter which has no direct physical meaning. Both axial dispersion and mixing tanks in series are approximately equivalent, especially at high Pe (or J) values. By setting Pe = 2J, or better Pe = 2(J - 1), as Pe → 0 then J → 1. The mixing tanks in series model is very popular for representing "regularly" dispersed RTD curves.

However, it may happen that even by equating variances, it is impossible to find a curve from these simple models which matches the shape of the experimental RTD. This is generally the case when complex flow patterns with by-passing, recycle and exchange with stagnant zones are involved. More sophisticated models can be built by associating elementary flow patterns such as plug flow zones (with or without axial dispersion), well mixed zones, stagnant zones, etc. according to various modes of connection - such as, in series, in parallel, with exchange, by-pass, recycle streams, etc. A host of these models can be found in the recent literature on reactor modeling. Fig. 2 shows, for instance, the classical model that Cholette and Cloutier proposed for non-ideal stirred tanks. In addition to the space time τ, there are 3 parameters characterizing, respectively, by-pass $(1 - \alpha)$, dead zone $(1 - \beta)$ and mass exchange between zones (γ).

Another example will be developed in Sec. 2-6.

2.5. Determination of macromixing parameters from tracer experiments

After assuming a flow model based both on the physical structure of the reactor and the characteristic features of the experimental RTD curve, mass balance equations are written for the dispersion of an inert tracer among the various zones involved in the model. These equations are linear differential equations (ODE or PDE) owing to the linear character of mixing processes. By solving the equations in the Laplace domain, the theoretical transfer function $G^*(s,p_i)$ is obtained, which is nothing but the Laplace transform of the theoretical RTD $E^*(t,p_k)$ where p_k are the parameters of the model

$$\int_0^\infty E^* (t,p_k)\, e^{-st}\, dt = G^*(s,p_k) = \frac{\bar{\bar{C}}_{out}}{\bar{\bar{C}}_{in}} \tag{2-14}$$

In the simplest cases, $G^*(s,p_k)$ can be inverted to yield an analytical expression for $E^*(t,p_k)$ in the time domain. With most models, this can only be done numerically. In order to determine the set of p_k, various methods are possible and used in the literature. They generally rely on least square fitting between theoretical expressions and experimental data. Some of these methods are the following:

a) In the Laplace domain, but there is no guarantee that matching $G^*(s,p_k)$ and $G(s)$ will provide a good fit between $E^*(t,p_k)$ and $E(t)$. This is equivalent to saying that model and experimental reactor yield the same conversion for a first order reaction (see Sec. 4).

b) In the Fourier domain, by matching both real and imaginary parts of $G^*(i\omega,p_k)$ and $G(i\omega)$. Parseval's theorem then makes sure that $E^*(t,p_k)$ and $E(t)$ are also at the closest in the least square sense.

c) In the real domain, when $E^*(t,p_k)$ is available. Least square fitting of parameters may also be carried out by comparing $y(t)$ to $y^*(t,p_k)$ calculated via the convolution integral [2-3].

d) From the moments of the RTD by solving equations

$$\mu_n = \mu_n^{*}(p_k) \tag{2-15}$$

$$\text{where } \mu_n = \int^{\infty} t^n E(t)\, dt \tag{2-16}$$

The advantage of this method is that $\mu_n^{*}(p_k)$ can be directly deduced from the transfer function by Van der Laan's theorem even if $E^{*}(t,p_k)$ is unknown:

$$\mu_n^{*}(p_k) = (-1)^n \left(\frac{\partial^n G^{*}(s,p_k)}{\partial s^n}\right)_{s=0} \tag{2-17}$$

As pointed out above, it is difficult to go farther than the third order moment owing to experimental inaccuracies. Exploitation of the variance $\sigma^2 = \mu_2 - \mu_1^2$ is especially popular. For "well shaped" dispersion peaks, this can be determined without any integration (Fig. 3). In practice all these methods make it possible to determine 2 or 3 unknown parameters. If the flow model involves more parameters, they cannot be determined with confidence from the RTD alone and additional observations are required.

2.6. Example: Van Deemter's model for fluidized beds

Modelling of gas solid fluidized bed catalytic reactors has been a challenge to chemical reaction engineers for a long time. Numerous papers have been published on this subject owing to the difficulty of the scale-up problem. At low fluidization velocity the main gas upflow takes place in the form of large bubbles which are renewed and finally burst at the upper surface of the bed. Two families of models have been proposed - the hydrodynamic and the phenomenologic models. The prototype of the latter is the so-called MAY-VAN DEEMTER model describing the flow pattern of the gas phase (Fig. 4). The gas in the bubble phase is assumed to be in plug flow and to exchange with the gas in the dense phase. The gas in the dense phase is mixed by axial dispersion. This is a 3-parameter model (in addition to space time): the parameter are the relative volume of both phases (β), the mass transfer coefficient between phases (K), and the axial dispersion in the dense phase (D). Chemical reaction is assumed to take place only in the dense phase. In the presence of a first order chemical reaction, mass balances are written (see notations in Figure 4):

$$\begin{cases} u\Omega_1 \dfrac{\partial C_1}{\partial z} + K\,(C_1 - C_2) + \Omega_1 \dfrac{\partial C_1}{\partial t} = 0 & \text{(2-18)} \\[2ex] \Omega_2\, D \dfrac{\partial^2 C_2}{\partial z^2} = K\,(C_2 - C_1) + k\,\Omega_2\, C_2 + \Omega_2 \dfrac{\partial C_2}{\partial t} & \text{(2-19)} \end{cases}$$

Defining reduced variables

$x = z/H$, $\theta = ut/H$, $\beta = \Omega_2/\Omega_1$

$N = KH/u\Omega_1 = KH/u_o\Omega$ = number of transfer units

$Pe = u\Omega_1 H/D\Omega_2 = u_o H(1 + \beta)/D$ = axial Peclet number

$Da = kH\Omega_2/u\Omega_1 = kH/(1 + \beta)u_o\Omega = k_m m/Q$ = Damköhler number

The mass balance equations are written in reduced from

$$\begin{cases} \dfrac{\partial C_1}{\partial x} + N(C_1 - C_2) + \dfrac{\partial C_1}{\partial \theta} = 0 & \text{(2-20)} \\[2ex] \dfrac{1}{Pe} \dfrac{\partial^2 C_2}{\partial x^2} = N(C_2 - C_1) + Da.C_2 + \beta \dfrac{\partial C_2}{\partial \theta} & \text{(2-21)} \end{cases}$$

and in the Laplace domain

$$\begin{cases} \dfrac{d\bar{\bar{C}}_1}{dx} + N(\bar{\bar{C}}_1 - \bar{\bar{C}}_2) + s\,\bar{\bar{C}}_2 = 0 & \text{(2-22)} \\[2ex] \dfrac{1}{Pe} \dfrac{d_2\bar{\bar{C}}_2}{dx^2} = N(\bar{\bar{C}}_2 - \bar{\bar{C}}_1) + Da.\bar{\bar{C}}_2 + \beta\, s\, \bar{\bar{C}}_2 & \text{(2-23)} \end{cases}$$

This assumes that the reactor is initially free from tracer. If $Da = 0$ these equations represent experiments with inert tracers. $s = 0$ corresponds to chemical reaction at steady state.

Two limiting cases are generally considered:

Model A. Very high mass transfer between bubble phase and dense phase ($N \to \infty$), $C_1 \approx C_2$. The spread in residence time is only due to axial dispersion in the dense phase. From (2-22) and (2-23) with appropriate boundary conditions ($x \to \infty$, $\bar{\bar{C}}_1 \to 0$), the transfer function is derived

$$G_A(s) = \frac{\overline{\overline{C}}_{1\ out}}{\overline{\overline{C}}_{1\ in}} = \exp\left[\frac{Pe}{2} - \left(\frac{Pe^2}{4} + Pe(Da + s(1+\beta))\right)^{1/2}\right] \quad (2\text{-}24)$$

the reduced moments of the RTD are found from Van der Laan's relation

$$\left.\begin{aligned} &\mu_{1A} = 1 + \beta,\ \sigma_A^2 = \frac{2}{Pe}(1+\beta)^2 \\ &\sigma_A^2/\mu_{1A}^2 = 2/Pe \end{aligned}\right\} \quad (2\text{-}25)$$

Conversion at steady state is given by

$$\left(\frac{C}{C_o}\right)_A = \exp\left[\frac{Pe}{2} - \left(\frac{Pe^2}{4} + Pe.Da\right)^{1/2}\right] \quad (2\text{-}26)$$

Model B. Plug flow of bubbles, no mixing in the dense phase ($Pe = \infty$) and limited transfer rate. The spread in residence time is due to reversible mass transfer. The transfer function is then

$$G_B(s) = \frac{\overline{\overline{C}}_{1\ out}}{\overline{\overline{C}}_{1\ in}} = \exp\left[-\frac{\beta s^2 + s(Da + N(1+\beta)) + N.Da}{N + Da + s}\right] \quad (2\text{-}27)$$

with the moments

$$\left.\begin{aligned} &\mu_{1B} = 1 + \beta,\ \sigma_B^2 = 2\ \beta^2/N \\ &\sigma_B^2/\mu_{1B}^2 = \frac{2}{N}\left(\frac{\beta}{1+\beta}\right)^2 \end{aligned}\right\} \quad (2\text{-}28)$$

The conversion at steady state is

$$\left(\frac{C}{C_o}\right)_B = \exp\left[-\frac{N\ .\ Da}{N + Da}\right] \quad (2\text{-}29)$$

An equivalence between models A and B may be obtained by matching variances $\sigma_A^2 = \sigma_B^2$

$$Pe = N(1 + \beta)^2/\beta^2 \qquad (2\text{-}30)$$

Figure 5 shows the shape of the RTD obtained by numerical inversion of G_A and G_B (Fast Fourier Transform algorithm) for typical values of the parameters (β = 5, N = 4, Pe = 5.76). Although both curves have the same mean and variance, their shape is quite different. In particular, the curve for model B exhibits a sharp pulse at the beginning caused by the bubbles which flow directly upwards without exchanging with the dense phase. By studying the dependence of model parameters upon the size of the reactor, upon particle size distribution and upon experimental conditions in cold flow experiments, it is possible to establish correlations for scaling-up the reactor. This method was successfully used in the design of commercial reactors [7] [8]. The way to use this kind of model for predicting chemical conversion is discussed in Sec. 4.

2.7. Reactor with two inlets

This case is interesting in practice when the reactor is fed by two separate streams of reactants. Under the same assumptions as above, two RTDs can be defined, one for each stream, i.e. $E_1(t_s)$ and $E_2(t_s)$. Let Q_1 and Q_2 be the respective flowrates. The BPT model is easily generalized to this situation (Fig. 6). If one imagines that each streamlet keeps its identity, a distinct bundle of tubes may be associated with each feedstream. The overall RTD (integral distribution) is built by associating elementary tubes having the same residence time, from which its results that (Fig. 6)

$$QF = Q_1F_1 + Q_2F_2 \qquad (2\text{-}31)$$

Representations in which the bundles are placed side by side, fluid particles having the same age or the same life expectancy being located on the same vertical are very useful for building micromixing models as will be shown in Sec. 3.

2.8. General case : Population balance formulation

As pointed out in the introduction to this section, the assumptions on which the classical RTD theory are based, are restrictive. However, it is still possible to use concepts like RTD or IAD when these assumptions are no longer valid - i.e. at unsteady state, in a compressible fluid, or in a variable volume reactor having large inlet and outlet ports - by using the "population balance" method.

The basic quantity is the distribution function $\psi(x,y,z,\alpha,t)$ which is such that $\psi d\alpha dV$ is the number of particles in volume $dv = dxdydz$ having an age between α and $\alpha + d\alpha$ at time t. ψ is some kind of hyper-concentration in a space with 4 dimensions x,y,z and α. The usual concentration at this point is $C = \int_0^\infty \psi d\alpha$, and the normalized age distribution is $f = \psi/C$. At steady state, $f(\alpha)$ would be the impulse response to an injection of tracer at the inlet.

The general population balance method considers entities flowing with velocity $\vec{u}$ and having m properties ζ_i which change with time according to $v_i = d\zeta_i/dt$. Introducing the distribution $\psi(x,y,z, \zeta_1, \zeta_2 \dots \zeta_m, t)$ and the rate of generation g of entities within intervals dx, dy, dz, $d\zeta_1 \dots d\zeta_m$, the balance equation of entities is written [6]

$$\frac{\partial \psi}{\partial t} + \mathrm{div}(\psi \vec{u}) + \sum_{i=1}^{m} \frac{\partial}{\partial \zeta_i}(v_i \psi) - g = 0 \tag{2-32}$$

In the case of fluid particles flowing in a reactor, there is a single property, the age $\zeta = \alpha$ and $v = d\alpha/dt = 1$. The particles are neither created nor destroyed, so that $g = 0$. The population balance equation for ψ defined above is then

$$\frac{\partial \psi}{\partial t} + \mathrm{div}\ (\psi u) + \frac{\partial \psi}{\partial \alpha} = 0 \tag{2-33}$$

Average quantities may be defined as follows:

$$\overline{\psi} = \frac{1}{V}\int_V \psi\, dV, \quad \overline{\psi}_{in} = \frac{1}{Q_{in}}\int_{A_{in}} u\psi dA, \quad \overline{\psi}_{out} = \frac{1}{Q_{out}}\int_{A_{out}} u\psi dA \tag{2-34}$$

$$\overline{C} = \int_\alpha \overline{\psi} d\alpha, \quad \overline{C}_{in} = \frac{1}{Q_{in}}\int_{A_{in}} uCdA, \quad \overline{C}_{out} = \frac{1}{Q_{out}}\int_{A_{out}} uCdA \tag{2-35}$$

Upon averaging (2-33) over the reactor volume, one obtains

$$\frac{1}{V}\frac{\partial}{\partial t}(V\overline{\psi}) + \frac{Q_{out}\,\overline{\psi}_{out} - Q_{in}\,\overline{\psi}_{in}}{V} + \frac{\partial \overline{\psi}}{\partial \alpha} = 0 \tag{2-36}$$

$\overline{\psi}/\overline{C} = I(\alpha,t)$ represents the instantaneous internal age distribution,

$\overline{\psi}_{out}/\overline{C}_{out} = E(\alpha,t) = E(t_s,t)$ is the instantaneous RTD,

$\overline{\psi}_{in}/\overline{C}_{in} = \delta(\alpha)$ is a Dirac Delta function as the inlet surface is the origin of ages.

Introducing

$$\tau_{in} = \frac{V}{Q_{in}}, \quad \tau'_{in} = \frac{N}{F_{in}} = \tau_{in} \frac{\overline{C}}{\overline{C}_{in}} \tag{2-37}$$

$$\tau_{out} = \frac{V}{Q_{out}}, \quad \tau'_{out} = \frac{N}{F_{out}} = \tau_{out} \frac{\overline{C}}{\overline{C}_{out}} \tag{2-38}$$

where $N = V\overline{C}$ is the total number of particles in the reactor, (2-36) becomes

$$\frac{\partial I}{\partial t} + I \left(\frac{1}{\tau'_{in}} - \frac{1}{\tau'_{out}}\right) + \frac{E}{\tau'_{out}} - \frac{\delta(\alpha)}{\tau'_{in}} + \frac{\partial I}{\partial \alpha} = 0 \tag{2-39}$$

This equation links I and E for any kind of flow and regime. It is interesting to introduce the moments of these distributions, i.e.

$$\mu'_n(t) = \int_0^{\infty} \alpha^n \, I(\alpha,t) \, d\alpha \tag{2-40}$$

$$\mu_n(t) = \int_0^{\infty} \alpha^n \, E(\alpha,t) \, d\alpha \tag{2-41}$$

The relationship between these moments is obtained from (2-39) by multiplying each term of this equation by α^n and by integrating each term with respect to α

$$\frac{d\mu'_n}{dt} + \mu'_n \left(\frac{1}{\tau'_{in}} - \frac{1}{\tau'_{out}}\right) + \frac{\mu_n}{\tau'_{out}} - n\,\mu'_{n-1} = 0 \tag{2-42}$$

Consider the following special cases:

- Incompressible fluid. $C = C_{in} = C_{out} = N/V$ $\tau'_{in} = \tau_{in} = V/Q_{in}$, $\tau'_{out} = \tau_{out} = V/Q_{out}$. (2-36) is then written

$$\frac{1}{V}\frac{\partial}{\partial t}(VI) + \frac{Q_{out}\,E - Q_{in}\,\delta(\alpha)}{V} + \frac{\partial I}{\partial \alpha} = 0 \tag{2-43}$$

- Steady flow, any kind of fluid. $\tau'_{in} = \tau'_{out} = \tau'$. (2-39) and (2-42) become

$$\frac{E - \delta(\alpha)}{\tau'} + \frac{dI}{d\alpha} = 0 \tag{2-44}$$

$$\underbrace{\mu_n}_{\text{RTD}} = n\,\tau'\,\underbrace{\mu'_{n-1}}_{\text{IAD}} \tag{2-45}$$

For instance, for n = 1 and 2, one obtains

$$\mu_1 = \overline{t}_s = \tau' \text{ and } \mu_2 = 2\tau'\overline{\alpha}$$

- Constant volume reactor. (2-42) is unchanged. In particular, for n = 1

$$\frac{d\overline{\alpha}}{dt} + \overline{\alpha}\left(\frac{1}{\tau'_{in}} - \frac{1}{\tau'_{out}}\right) + \frac{\overline{t}_s}{\tau'_{out}} - 1 = 0 \tag{2-46}$$

which gives $t_s = \tau'$ at steady state as would be expected.

2.9. Example of application of population balance method

A liquid is poured with flowrate Q(t) into a semibatch reactor (initially empty). What is the expression I(α,t) for the IAD of the fluid in the reactor at time t ?

For an incompressible fluid and since $\tau'_{out} = \infty$, the equation for I is obtained from (2-39)

$$\left.\begin{aligned} &\frac{\partial I}{\partial \alpha} + \frac{\partial I}{\partial t} + \frac{Q}{V}\left[I - \delta(\alpha)\right] = 0, \quad Q = \frac{dV}{dt} \\ &\int_{\alpha=0}^{\alpha=t} I(\alpha,t)\, d\alpha = 1 \end{aligned}\right\} \tag{2-47}$$

With change of variable t' = t - α, (2-47) becomes

$$\left(\frac{\partial I}{\partial t}\right)_{t'} + \frac{Q(t)}{V(t)}\left[I(t,t') - \delta(t-t')\right] = 0 \tag{2-48}$$

which can be integrated by the method of separation of variables to give

$$I = \frac{Q(t')}{V(t)}\left[H(t') + H(t - t')\right] \tag{2-49}$$

The solution is, hence

$$I(\alpha,t) = \frac{Q(t-\alpha)}{V(t)} \left[H(t - \alpha) + H(\alpha)\right] \tag{2-50}$$

This result is obvious as the fraction of fluid having an age α at time t is proportional to the amount of fluid (the flowrate) which entered the reactor at time $t - \alpha$, all ages being comprised between 0 and t. In particular, if Q is constant, the IAD is simply a square wave

$$I(\alpha,t) = \frac{1}{t} \left[H(t - \alpha) + H(\alpha)\right] \tag{2-51}$$

3. MICROMIXING

Micromixing is concerned with all the processes whereby parts of a fluid initially unmixed come into contact and mix on a molecular scale. However, this definition has to be specified. Let p(C) be the local concentration distribution of a given component in an imperfect mixture. The average concentration is

$$\bar{C} = \int_0^{C_{max}} C\, p(C)\, dC \tag{3-1}$$

and the variance of distribution is

$$\sigma^2 = \int_0^{C_{max}} (\bar{C} - C)^2\, p(C)\, dC \tag{3-2}$$

If σ_o^2 is the variance at the beginning of the mixing process, $I_s = \sigma^2/\sigma_o^2$, sometimes called "intensity of segregation", is an index characterizing the quality of mixing. The smaller I_s is the better the achievement of mixing. In Lagrangian coordinates (attached to the fluid particles), $\tau_s = -\sigma^2/(d\sigma^2/dt)$ is a time constant for the rate of segregation decay, i.e., for the rate of mixing. Let $c(x) = C - \bar{C}$ be the deviation from the mean at point x and c(x,r) be the same quantity at distance r from this point. The size of segregated regions mav be characterized by the concentration macroscale

$$L_s(x) = \int_0^{\infty} \frac{\overline{c(x)\ c(x+r)}}{\sigma^2}\, dr \tag{3-3}$$

It is clear that the experimentally measured degree of mixing depends on the spatial resolution (compared to L_s) and on the time resolution (compared to τ_s) of the probe used to estimate local concentrations in the mixture. The mixture appears as uniform if the scale of concentration fluctuations is smaller than the resolution of the probe. Generally, it is admitted that micromixing takes place in the range between molecular dimensions up to the scale where non-uniformities can be detected by usual means of observation, i.e., a few hundred micrometers in a liquid. Processes occurring beyond this range may be considered as macromixing.

The fine texture of a fluid is characterized by the state of segregation. A totally segregated fluid is supposed to be made of small groups of particles ("aggregates") which keep their identity and do not mix with each other upon macromixing. Such a fluid is called a macrofluid. At the opposite extreme, a well micromixed fluid, also called a microfluid, consists of individual molecules which are free to move with respect to each other. Partial segregation generally exists in real fluids, which may thus be considered as a mixture of macro- and microfluid, or as a collection of small aggregates exchanging material with each other.

Mixing may also be considered from the viewpoint of ages of fluid particles. Danckwerts [9] and Zwietering [10] have shown that the IAD or the RTD of the fluid, which characterize macromixing, do not give any information on micromixing. There are for instance an infinite number of ways to arrange fluid particles of different ages with respect to each other within the same IAD (or RTD). In continuous reactors, the limiting state of Minimum Mixedness is achieved when particles which are to leave the reactor together come into contact at the latest moment. Conversely, in the limiting state of Maximum Mixedness, this contact occurs at the earliest moment. This concept of "mixing earliness" is distinct from that of segregation, but both are generally described within the frame of micromixing processes.

3.1. Mixing "earliness" in continuous reactors : Entering and Leaving Environments

From the definition given above, in the state of Minimum Mixedness (Min Mix) neighbouring aggregates have the same age α. Conversely, in the state of Maximum Mixedness (Max Mix) they have the same life expectancy or residual lifetime λ. More generally, fluid aggregates, which have different residence times $t_s = \alpha + \lambda$, cannot spend their whole life in the reactor with neighbours having simultaneously the same α and the same λ. They start in an Entering Environment (E.E.) in a state of minimum mixedness (same α) and gradually pass on to a Leaving Environment (L.E.) in a state of maximum mixedness (same λ). This is independent of the state of segregation of the fluid which may be partially segregated in both Environments. However, it is often implicitly assumed that the E.E. is totally segregated (macrofluid) and the L.E. perfectly micromixed (microfluid). The mechanism for micromixing is then confounded with that for passing from the E.E. to the L.E., but this assumption is not necessary. The BPT-model is very helpful for visualizing these phenomena. In the Min Mix state, the tubes are piled up in such a way that aggregates of the same age are on the same vertical line. In the Max Mix state, the arrangement of tubes is symmetrical and aggregates having the same life expectancy are located on the same vertical (Fig. 7).

A great number of models have been proposed in the literature to represent the transfer of fluid aggregates from the E.E. to the L.E. They are all more or less equivalent and are special cases of the general model of Spencer, Leshaw et al. [11] which is briefly described below.

Along the axis of a small tube of the BPT-model, the fluid gradually passes from the E.E. (Min Mix) to the L.E. (Max Mix) (Fig. 8).

The flow transferred in the interval $d\lambda$ is

$$d^2Q = Q\ A(\lambda,t_s)\ d\lambda\ dt_s \tag{3-4}$$

$A(\lambda,t_s)$ is a mixing function also expressed by introducing $h(\lambda,t_s)$ and the RTD.

$$A(\lambda,t_s) = h(\lambda,t_s)\ E(t_s) \tag{3-5}$$

The fraction of fluid of life expectancy λ in the E.E. is

$$s(\lambda,t_s) = s(o,t_s) + \int_o^\lambda h(\lambda,t_s)\ d\lambda \tag{3-6}$$

The complementary fraction in the L.E. is $1 - s$. $s(\lambda,t_s)$, which may also be written $s(\lambda,\alpha)$, is known as the "segregation function". This term is somewhat confusing as the process above is concerned with segregation of ages (mixing earliness). However, $s(\lambda,\alpha)$ also describes the decay of segregation in the physical sense under the additional (and often made) assumption that E.E. = macrofluid and L.E. = microfluid. In the particular tube considered above, the flowrate of the Max Mix fluid is $dQ_{Max} = Q\ g(\lambda)d\lambda$ with

$$g(\lambda) = \int_\lambda^\infty A(\lambda,t_s)\ dt_s$$

In the whole reactor the Max Mix flowrate of fluid of life expectancy λ is, thence $Q_{Max}(\lambda) = Q \int_\lambda^\infty g(\lambda')d\lambda'$.

This model is represented in Fig. 9. The fluid flows from left to right and passes from EE (unshaded tubes) to LE (shaded area). Transfer between the two Environments may also be described as a Lagrangian process where a small band of volume $dV_{Max} = Q_{Max}d\lambda$

moves along the axis in the direction of decreasing λ. Spencer et al. [11] consider this volume as an "accumulator" whose dimensionless volume v is increasing according to

$$\frac{dv}{dt} = \int_{z=-\infty}^{t} A(-z,t)\, dz \tag{3-7}$$

The accumulator leaves the reactor and discharges at time t = 0. The way to use this model for calculating chemical conversion will be described in Sec. 4.2. Unsteady regimes can also be dealt with thanks to this model.

Most models for "intermediate micromixing" can be compared by means of their "segregation" function $s(\lambda,\alpha)$. Here are some examples taken from the literature:

α^*-Model of Spencer et al. [11]
$s = 1$ for $\lambda > \alpha^* t_s$ (i.e. $\alpha < (1 - \alpha^*)t_s$) and
$s = 0$ for $\lambda < \alpha^* t_s$

τ_D-Model of Spencer et al. [11] which is also the Series-Model of Weinstein and Adler [12].

$s = 1$ for $\alpha < \tau_D$ and $s = 0$ for $\alpha > \tau_D$

Parallel model of Weinstein and Adler [12]

$s = 1$ for $t_s < \tau_P$ and $s = 0$ for $t_s > \tau_P$

Model of Villermaux and Zoulalian [13]
s is only a function of t_s, e.g., $s = \exp(-Kt_s)$

Model of Ng and Rippin [14]
s only depends on the age α, e.g. $s = \exp(-R_s\alpha)$. Plasari et al. [15] have used a model of this kind (Shrinking Aggregate) to represent the reduction of size of entering particles in an unmixed feedstream, $s = (1 - \alpha/t_e)^3$ (see Sec. 3.2 below).

Model of Valderrama and Gordon [16] [17]
The model depends on two parameters β and w

$$s = 1 \text{ for } \alpha < \alpha^* = -\ln(1 - \beta)$$

$$s = \frac{1 - \beta - w}{1 - \beta} \text{ for } \alpha > \alpha^*$$

Most of these models (except the last one) depend on a single parameter. As it is generally assumed that the E.E. consists of a macrofluid and the L.E. of a microfluid, the model also represents the decay of segregation in the reactor (see next Section). A graphical representation of the segregation function s(λ,α) and the corresponding pattern in the BPT model can be found in ref. [18]. Numerical simulations and theoretical arguments based on concentration variance and chemical conversion lead to the conclusion that all single parameter models yield approximately equivalent results provided the parameters obey equivalent relationships. Some examples are given below. In addition, the models for mixing earliness that can be found in the literature often resemble mathematical games involving parameters which have no real physical meaning. Therefore, it is useless to develop new descriptions of this kind. The choice among those that already exist is usually a simple matter of computation convenience.

The case of reactors with two unmixed feedstreams is interesting for practical applications. The BPT-model offers a convenient representation of the reactor where both streams have their own RTD (see above). The pattern of Figure 1 can easily be extended to the case of separate feed streams by joining the corresponding Entering and Leaving Environments (Fig. 10). There are now two E.E. and two L.E. Interaction between aggregates is allowed to occur according to mechanisms discussed in the next Section (e.g. Random-Coalescence-Dispersion or Exchange with the Mean Environment). In the E.E. aggregates are generally allowed to interact if they have both the same age and the same life expectancy. In the L.E. interaction takes place between aggregates having only the same life expectancy. Care must be taken to respect these constraints when developing a new model. Otherwise, the second law of thermodynamics is violated (aggregates cannot unmix spontaneously) and the model is obviously irrelevant. Many models based on the above principles are available in the literature [19] [20] [21] [49].

Mixing earliness may be experimentally determined by injecting two tracers A and B which can react with each other. Close to Minimum Mixedness, the best discrimination is obtained by injecting two sharp pulses of A and B separated by a short interval. Close to Maximum Mixedness, the recommended injection is a square pulse containing both A and B. The reaction extent at the reactor outlet is then strongly dependent on micromixing. The determination is simple if the reaction products are coloured (e.g., hydrolysis of p-nitrophenyl acetate) [22] [23].

3.2. Mechanism and modeling of Segregation decay

The last section describes processes whereby aggregates having the same life expectancy come into contact; these processes are concerned with mixing in the age space. Mixing in the physical space has now to be achieved by actual micromixing of neighbouring fluid aggregates and reduction of concentration gradients.

When segregation decay occurs via a single mechanism, micromixing may be represented by one-parameter phenomenological models which are briefly presented below.

1) Random Coalescence-Dispersion (C-D model) [24]. The basic assumptions of the model are: (i) aggregates of equal size, (ii) same probability of coalescence for all aggregates independent of time and chemical composition, (iii) immediate redispersion after coalescing and formation of a uniform concentration, producing again two aggregates of the same volume, (iv) the coalescence-dispersion frequency is ω, which means that each aggregate undergoes $\omega\Delta t$ C-D processes during time interval Δt.

This random-coalescence process can be represented by deterministic (discrete or continuous) equations. For instance, in a well macromixed stirred tank, the concentration distribution $p(C,t)$ at time t in t tank and $p_{in}(C,t)$ in the inlet stream are connected by Curl's equation:

$$\frac{\partial p}{\partial t} + \frac{\partial (p \mathcal{R})}{\partial C} + \frac{p - p_{in}}{\tau} = \omega \; J \qquad (3\text{-}8)$$

where J is an interaction term given by

$$J = 4 \int_{0}^{C} p(C' \;\; p(2C - C') \, dC' - p(C) \qquad (3\text{-}9)$$

and $\mathcal{R}(C)$ is the rate of chemical reaction producing the species of concentration C in the aggregates.

In a batch reactor ($\tau = \infty$) without chemical reaction, the intensity of segregation (see above) decreases upon coalescence-dispersion according to

$$I_s = \exp \left[-\omega t/2\right] \qquad (3\text{-}10)$$

where $I = \omega\tau$ is a dimensionless micromixing parameter.

The solution of (3-8), which is an integro-differential equation, is difficult. It is often simpler to rely on the stochastic character of C-D processes and to simulate interaction by the method of MONTE-CARLO. For instance, in a set of n aggregates, each containing a concentration C_i ($i = 1,2,3 \ldots n$), a pair of aggregates (concentrations C_i and C_j) are chosen at random and the concentrations are replaced by $(C_i + C_j)/2$ in both particles. As there are $\omega n/2$ C-D per unit time in the reactor this process is repeated $\omega n\Delta t/2$ times (supposed an integer) to obtain the concentration distribution at time $t + \Delta t$. In a CSTR, $\omega n\Delta t/2$ aggregates are drawn at random during the same time Δt and replaced by entering aggregates having the appropriate inlet distribution. In addition, if a chemical reaction is taking place within the aggregates, the chemical species are allowed to react batchwise for the period of time between two successive coalescences. This method was first proposed by Spielman and Levenspiel [25] and extensively used later by many authors [26] [27].

The method can be extended to reactors with two unmixed feedstreams. Figure 11 shows a model of the kind proposed by Kattan and Ader [26] to simulate micromixing in a reactor fed with separate streams of flowrates Q_1 and Q_2. In the BPT representation all fluid elements on a given vertical have the same life expectancy and are allowed to coalesce. Micromixing is simulated by choosing two fluid elements at random (shaded blocs) and causing them to coalesce and redisperse with frequency ω. Simultaneously, the fluid flows from left to right and chemical reaction occurs within the stream between coalescences. Fluid elements move along the tubes in the direction of decreasing λ. The composition at the reactor outlet is obtained by averaging the concentrations at $\lambda = 0$. $\omega = 0$ corresponds to a macrofluid and $\omega = \infty$ to a microfluid. This method may be used for simulating interaction between aggregates in E.E. and L.E. of reactors as explained in Section 3.1.

2) Interaction with a Mean Environment (IEM-model) [43]. The rigorous treatment of the C-D model requires large computer capacity. A much simpler deterministic model can be imagined where interaction between aggregates is accounted for by an equivalent mass transfer between one aggregate and its mean environment (Interaction by Exchange with the Mean or IEM-model). This mass exchange is characterized by a micromixing time t_m. The basic equation for the change of concentration C_j of a species A_j in an aggregate of age α is written

$$\frac{\partial C_j}{\partial \alpha} = \frac{\langle C_j \rangle - C_j}{t_m} + \mathcal{R}_j \qquad (3\text{-}12)$$

$\langle C_j \rangle$ is the mean concentration of A_j in the neighbouring aggregates with which interaction takes place and $\mathcal{R}_j$ is the rate of production of A_j by chemical reaction. $t_m = 0$ in a macrofluid and $t_m = \infty$ in a microfluid. $\langle C_j \rangle$ is defined by the condition that the net sum of all exchange fluxes over the whole population is equal to zero. For instance in a perfectly macromixed CSTR

$$\langle C_j \rangle = \frac{\int_o^\infty C_j \, t_m^{-1} \exp(-\alpha/\tau)\, d\alpha}{\int^\infty t_m^{-1} \exp(-\alpha/\tau)\, d\alpha} \qquad (3\text{-}13)$$

If t_m is constant and independent of α this reduces to

$$\overline{C}_j = \frac{1}{\tau}\int_0^\infty C_j(\alpha)\ \exp(-\alpha/\tau)d\alpha \tag{3-14}$$

where $\overline{C}_j$ is the average concentration in the tank. The solution thus requires an iterative procedure but the convergence is usually very fast. The IEM model may be used with any kind of RTD. Considering first a reactor with one single inlet and outlet (Fig. 7), and interaction between aggregates of same life expectancy in a maximum mixedness configuration (Leaving Environment), the IEM equation is written for constant t_m

$$\left.\begin{aligned} &\frac{\partial C_j}{\partial \alpha} = -\frac{\partial C_j}{\partial \lambda} = \frac{\overline{C}_j(\lambda) - C_j(\lambda,t_s)}{t_m} + \mathfrak{R}_j \\ &\lambda = t_s\ ,\quad C_j = C_{j,in} \end{aligned}\right\} \tag{3-15}$$

This equation has to be integrated along a tube of residence time $t_s = \alpha + \lambda$, with

$$\left.\begin{aligned} &\overline{C}_j(\lambda) = \frac{1}{1 - F(\lambda)}\int_\lambda^\infty C_j(\lambda,t_s)\ E(t_s)\ dt_s \\ &\text{where } F(\lambda) = \int_0^\lambda E(t_s)\ dt_s \\ &C_{j,out} = \overline{C}_j\ (\lambda = 0) \end{aligned}\right\} \tag{3-16}$$

In the case of a reactor with two inlets, (3-15) still holds in each stream but the mean value $\overline{C}_j(\lambda)$ at a given life expectancy must comprise contributions from both feedstreams 1 and 2.

$$\overline{C}_j(\lambda) = \frac{1}{1 - \mathcal{F}_1(\lambda) - \mathcal{F}_2(\lambda)}\left[\int_\lambda^\infty C_{j1}(\lambda,t_s)\ a_1E_1(t_s)dt_s + \int_\lambda^\infty C_{j2}(\lambda,t_s)\ a_2E_2(t_s)\ dt_s\right] \tag{3-17}$$

with $a_1 = Q_1/(Q_1 + Q_2)$, $\mathcal{F}_1(\lambda) = a_1 \int_0^\lambda E_1(t_s)\, dt_s$

and similar definitions for feedstream 2. Actually equations (3-15) to (3-17) are easily solved numerically in a discrete form where the λ-axis is divided into finite $\Delta\lambda$-increments. Fig. 12 shows an example of such a calculation for a reactor with two inlets having RTDs equivalent to that of 2 and 6 tanks in series, respectively.

By comparing the predictions of C-D and IEM models for the variances of concentration distributions in a batch reactor and in a CSTR fed with two populations of tracer, it can be shown that both models give the same results provided that

$$\omega t_m = 4 \tag{3-18}$$

This equivalence also holds for predicting chemical conversion as can be seen in Fig. 12 where reaction extent was calculated for a second order reaction with unmixed feedstreams. The agreement is excellent. More generally, equivalence relationships can be established between all one-parameter micromixing models. For instance, the various models cited above yield approximately the same results under the equivalence conditions:

$$\omega/2 = 2/t_m = R_s = 7/t_e \tag{3-19}$$

(CD) (IEM) (NG-RIPPIN) (Shrinking Aggregate)

Owing to its simplicity, the IEM model can be recommended for chemical engineering calculations. In a CSTR, the dimensionless micromixing parameter is $b = \tau/t_m$.

However, the models described above are purely phenomenological. What is the underlying mechanism for reduction of segregation and how do micromixing parameters depend on mechanical and physicochemical data ?

It is convenient to distinguish three successive stages in the mixing process [29].

(i) Distribution of one fluid through the other and formation of a uniform average composition without decreasing local concentration gradients.

(ii) Reduction of the size of the regions of uniform composition and increase of contact between regions of different composition (aggregates).

(iii) Mixing by molecular diffusion.

The first stage pertains to macromixing. Micromixing is concerned mainly with the last two processes.

Turbulence theory provides a classical approach to single newtonian fluids. The main quantities from this theory in the case of homogeneous isotropic turbulence are recalled in Table 1. The mixing process starts with large eddies (macroscales) whose size is gradually reduced upon mixing while energy and segregation is transferred to smaller structures. Maximum dissipation occurs close to Taylor (Corrsin) microscales and the process ends with the smallest eddies (Kolmogorov microscale). Below this size energy loss occurs by viscous dissipation and mass transfer by molecular diffusion. Identification of these "eddies" with the "aggregates" invoked in the Lagrangian description is not obvious. However, turbulence theory provides a sound basis for interpreting the dependence of aggregate properties upon physicochemical parameters.

Coming back to the aggregate concept, the following mechanisms have been assumed to account for mixing in the second stage (ii):

(a) Convective mixing [30] [31]. Fluid aggregates are stretched out an folded up and the mixture finally exhibits a lamellar or striated structure. The characteristic time for stretching or for the formation of vortices engulfing surrounding material is in the order of $(\nu/\varepsilon)^{1/2}$, where ν is the kinematic viscosity and ε the local rate of energy dissipation. The ultimate striation thickness is generally accepted to be of the order of the Kolmogorov microscale λ_K.

(b) Erosive or dispersive mixing [15]. Fluid aggregates are gradually peeled off and yield smaller fragments by friction at their external surface. In the Shrinking-Aggregate model the size of particles linearly decreases with age $\ell = \ell_o(1 - \alpha/t_e)$ but the expression for the shrinking time t_e is only tentative. In liquids it would be in the order of $\ell_o\lambda_k/\mathfrak{D}$ where ℓ_o is the initial size and $\mathfrak{D}$ the molecular diffusivity.

(c) Coalescence-Dispersion. Fluid aggregates undergo coalescence-redispersion with frequency ω, as already explained above. It is generally accepted that $1/\omega$ is proportional to Corrsin's time constant τ_s. This may also be represented by the IEM model with $t_m \sim \tau_s$.

In the third and last stage (iii), the only mechanism left for mixing is molecular diffusion with a characteristic time constant $t_D = \mu L^2/\mathfrak{D}$. μ is a shape factor (1/3 for slabs, 1/2 for long cylinders, 3/5 for spheres) and L is the ratio of the volume to the external surface area of the particle (if the particle "thickness"

TABLE I

Homogeneous Isotropic Turbulence

Velocity $U = \bar{U} + u$		Concentration $C = \bar{C} + c$	
Mean square fluctuation	$u' = (\overline{u^2})^{1/2}$		$c' = (\overline{c^2})^{1/2}$
autocorrelation	$f(r) = \frac{\overline{u(x)\ u\ (x+r)}}{u'^2}$		$g(r) = \frac{\overline{c(x)\ c(x+r)}}{c'^2}$
Macroscale	$L_f = \int_0^\infty f\ (r)\ dr$		$L_s = \int_0^\infty g(r)\ dr$
Kinetic energy of turbulent motion	$q = \frac{3}{2} u'^2$	Segregation intensity	$c'^2/c_o'^2$
Turbulent energy dissipation	$\varepsilon = - \frac{dq}{dt} = - \frac{3}{2} \frac{du'^2}{dt}$	Segregation dissipation	$\varepsilon_s = - \frac{dc'^2}{dt}$
Spectra Three dimensional $E(k)$, One dimensional $E_1(k_1)$	$u'^2 = \frac{2}{3} \int_0^\infty E(k)\ dk = 2 \int_0^\infty E_1(k_1)\ dk_1$ $\varepsilon = 2\ \nu \int_0^\infty k^2 E(k) dk = 30\ \nu \int_0^\infty k_1^2\ E_1(k_1)\ dk_1$		$E_s\ (k)$ $c'^2 = \int_0^\infty E_s(k)\ dk$ $\varepsilon_s = 2\mathcal{D} \int_0^\infty k^2\ E_s\ (k)\ dk$
Taylor microscale	$\lambda_f^2 = \frac{30\,\nu\, u'^2}{\varepsilon} = \frac{2 \int_0^\infty E_1(k_1) dk_1}{\int_0^\infty k_1^2\ E_1(k_1) dk_1} = \frac{10 \int_0^\infty E(k) dk}{\int_0^\infty k^2 E(k)\ dk}$	Corrsin microscale	$\lambda_s^2 = \frac{12 \mathcal{D} c'^2}{\varepsilon_s} = \frac{6 \int_0^\infty E_s(k)\ dk}{\int_0^\infty k^2\ E_s(k) dk}$
Taylor time constant	$\tau_f = \left[- \frac{1}{u'^2} \frac{du'^2}{dt} \right]^{-1} = \frac{\lambda_f^2}{10\nu}$	Corrsin time constant	$\tau_s = \left[- \frac{1}{c'^2} \frac{dc'^2}{dt} \right]^{-1} = \frac{\lambda_s^2}{12\mathcal{D}} \approx 2 \left(\frac{L_s^2}{\varepsilon}\right)^{1/3}$
	$L_f = \frac{\pi E(o)}{u'^2} \sim \frac{q^{3/2}}{\varepsilon}$	Small Sc :	$\lambda_s^2 / \lambda_f^2 = \mathcal{D}/\nu$, $\tau_s = \tau_f$ (Corrsin)
Viscous dissipation			
Kolmogorov microscale	$\lambda_K = (\nu^3/\varepsilon)^{1/4}$	Corrsin microscale	$\lambda_C = (\mathcal{D}^3/\varepsilon)^{1/4}$
		Batchelor microscale	$\lambda_B = (\mathcal{D}^2 \nu/\varepsilon)^{1/4}$
Viscous dissipation time constant	$\tau_K = (\nu/\varepsilon)^{1/2}$		$\tau_C = (\mathcal{D}/\varepsilon)^{1/2}$

is $\ell = 2R$, then $L = R$ for a slab, $R/2$ for a long cylinder and $R/3$ for a sphere). In aggregates close to the maximum of dissipation, $\ell \sim \lambda_s$ and consequently

$$t_D \sim \tau_s = \frac{\lambda_s^2}{12\mathcal{D}} \approx 2 \left(\frac{L_s^2}{\varepsilon}\right)^{1/3} + 0.5 \left(\frac{\nu}{\varepsilon}\right) \ell n \frac{\nu}{\mathcal{D}} \quad (3\text{-}20)$$

The first term is predominant in liquids. In aggregates reaching the viscous dissipation zone, ℓ may be identified with one of the viscous dissipation microscales of Table 1. The most frequent assumption [31] is that ultimate aggregates have the dimension in the order of Kolmogorov microscale $\lambda_K = (\nu^3/\varepsilon)^{1/4}$ and undergo subsequent stretching by viscous friction. In liquids, λ_K and λ_s lie between 10 and 100 μm. $\lambda_K^2/\mathcal{D}$ and τ_s are of the same order of magnitude between 0.1 and 1 second.

4. MIXING AND CHEMICAL REACTIONS

As far as first order reactions are concerned, the only quantity determining conversion is the time spent by reacting species in the reactor. Therefore, conversion can be calculated from batch kinetics - e.g. $C_B(t)$ - and from macromixing characteristics only - e.g., RTD or IAD. For instance, in a continuous reactor, by merging all streamlets issuing from the tubes of the BPT-Model, one obtains the well-known relationship

$$C_{out} = \int_0^\infty C_B(t_s)\, E(t_s)\, dt_s \quad (4\text{-}1a)$$

or, more generally, for a reaction extent X_i

$$X_{i,out} = \int_0^\infty X_{i,B}(t_s)\, E(t_s)\, dt_s \quad (4\text{-}1b)$$

In a semi-batch reactor

$$C(t) = \int_0^t C_B(\alpha)\, I(\alpha,t)\, d\alpha \quad (4\text{-}2a)$$

or $$X_i(t) = \int_0^t X_{i,B}(\alpha)\, I(\alpha,t)\, d\alpha \quad (4\text{-}2b)$$

If exponential decay functions appear in the expression for

$C_B(t_s)$, conversion can be calculated directly from the transfer function for an inert tracer. For instance, in the case of a first order isothermal reaction, $C_B = C_{in} \exp(-kt_s)$. G(s) being the Laplace Transform of $E(t_s)$, (4-1-a) leads to

$$f = C_{out}/C_{in} = \int_o^{\infty} \exp(-kt_s)\, E(t_s)\, dt_s = G(k) \qquad (4\text{-}3)$$

An example was developed in Sec. 2-6 (Van Deemter's model) where the transfer function $G_R(s,k)$ was derived for the reactant considered as a reactive tracer. The concentration reduction at steady state for this first order reaction is thus given by

$$f = G_R(0,k) = G(k) \qquad (4\text{-}4)$$

where G(s) is the transfer function for an inert tracer. Figure 14 shows f_A and f_B versus the Damköhler number obtained by this method for the two limiting versions of Van Deemter's model and the parameter values of Fig. 5.

When chemical reactions with orders different from one are involved the residence time alone is no longer sufficient to predict conversion since this also depends on the history of concentrations encountered by the reactants in the reactor. Additional information about mixing earliness and segregation must be known.

4.1. Use of macromixing models

Flow models presented in Sec. 2.4 may be used for predicting chemical conversion. Elementary patterns involved in the model are generally assumed to behave as ideal well micromixed reactors. Mixing earliness is implicitly accounted for by the arrangement of these elementary zones with respect to each other and the internal streams connecting them. This method is very popular and is successful for representing and scaling up chemical reactors provided the model has a sound physical basis.

4.2. Use of RTD or IAD plus an assumption about mixing earliness

As already mentioned above, the usual assumption is that the reactor comprises two Environments. The Entering (or Initial) Environment, which consist of a macrofluid, and the Leaving (or Final) Environment where the fluid behaves as a microfluid. The "segregation function" presented above (Sec. 3-1) describes the transfer between these Environments. The first limiting case is that of Minimum Mixedness where the Entering Environment spreads out into the whole reactor. The tubes of the BPT-model remain insulated and (4-1) is still valid. The conversion for a single reaction is written

$$\left.\begin{aligned} C_j &= \int_o^\infty C_{j,B}(\alpha)\; E(\alpha)\; d\alpha \\ \frac{dC_{j,B}}{d\alpha} &= \mathfrak{R}_j(C_k), \quad \alpha = 0, \quad C_{j,B} = C_{jo} \end{aligned}\right\} \qquad (4\text{-}5)$$

Conversely, in the second limiting case, Maximum Mixedness is achieved and the reactor only contains the Leaving Environment. From the BPT-model, Zwietering's famous equation is easily derived:

$$\left.\begin{aligned} & -\frac{dC_j}{d\lambda} = \mathfrak{R}_j + (C_{jo} - C_j)\,\frac{E(\lambda)}{1 - F(\lambda)} \\ & \text{Integration has to be performed from } \lambda = \infty \text{ to } \lambda = 0 \end{aligned}\right\} \qquad (4\text{-}6)$$

Conversion for intermediate states of micromixing may be calculated by application of the phenomenological models presented in Sec. 3.1. In particular, the "accumulator" model presented in Sec. 3.1 leads to the following equation (see eq. 3-7)

$$\frac{dC_j}{dt} = \frac{1}{v}\int_{z=-\infty}^{t} A(-z,-t)\left[C_{j,B}(C_k(z),\, t-z) - C_j\right] dz + \mathfrak{R}_j \qquad (4\text{-}6\text{-}2)$$

which is a generalization of (4-6-1). $C_{j,B}$ is the familiar batch expression and the equation has to be integrated from $t = -\infty$ $(vC_j \to 0)$ to $t = 0$.

Figure 13 shows an example of results obtained with the IEM-model in a CSTR for zero-order and second-order reactions (unmixed and premixed feed). The following conclusions can be drawn from such simulations:

(i) Segregation favours higher order reactions. Conversely, perfect micromixing leads to greater conversion with low order (e.g., zero order) reactions.

(ii) For a given RTD, the difference between the two limiting conversions increases with the spread in residence times; it is a maximum in the case of a CSTR.

(iii) Under usual conditions, the effect of micromixing on conversion is small when reactants are premixed (not more than 10 % on X = 1-f for a second order reaction). Consequently, equation (4-1) may be used for a first estimation except at high conversion. The effect is much more pronounced on reactions with unmixed feedstreams as mixing earliness controls contacting of the reactants.

(iv) Yield and selectivity in multiple reaction systems may strongly depend on micromixing, especially for fast reactions.

This can be understood if one recalls that segregated aggregates behave as small batch reactors where the yield is different from that in well mixed regions. In particular, the distribution of products of fast consecutive competing reactions may serve as an index for the local state of micromixing, as will be shown below (Sec. 4.5). In the same way, segregation influences molecular weight distribution in polymer reactors. However, no general rule can be stated and each case must be examined in itself.

4.3. Generalization: "Reaction Time Distribution" [4] [6]

The RTD concept in continuous reactors may be extended to non-uniform and non-isothermal media. Consider a chemical reaction with rate $r = k\, f(C_j)$ and assume that the rate constant k is locally variable owing to a spatial distribution of extragranular porosity ε (catalytic reaction), temperature, etc... Let $\bar{k}$ be the average value of k over the whole reactor, and $h = k/\bar{k}$ a local parameter. It is possible to define a generalized reaction time t_i such that $dt_i = h dt_s$ where h is the local value of $k/\bar{k}$ along a tube of the BPT-model (Fig. 1). At the tube outlet, $t_i = \langle h \rangle\, t_s$ where $\langle h \rangle$ is the average value of h along the tube axis. Now considering the whole bundle of tubes, there exists a "reaction time distribution" $H(t_i)$ similar to the RTD $E(t_s)$. The conversion in the tube can be calculated by integrating

$$C_o \frac{dX}{dt_i} = k\ f\ (C_j) \tag{4-7}$$

and the overall extent of reaction (under the assumption of minimal mixedness) is

$$X = \int_o^\infty X(t_i)\ H(t_i)\ dt_i \tag{4-8}$$

which generalizes (4-1). Notice that $\bar{t}_i = \bar{t}_s$. For homegeneous isothermal reactors $h \equiv 1$ and $H(t_i)$ is the RTD. For isothermal heterogeneous reactors $h = (1 - \varepsilon)/\varepsilon$ and $H(t_i)$ is the "Contact-time distribution" (CTD) introduced by Nauman and Collinge [4]. In the case of non-isothermal reactors, $h = k(T)/\overline{k(T)}$ and $H(t_i)$ is a "thermal time distribution". The potential applications of this promising concept is limited by the availability of suitable me-methods for the experimental determination of $H(t_i)$.

4.4. Segregation and chemical reaction

When one single mechanism for segregation decay is involved, micromixing may be characterized by one time constant t_m. Simultaneously, the chemical process is characterized by some reaction time t_R (e.g., $t_R^{-1} = k\, C_o^{n-1}$ for a nth order reaction). The key parameter for discussing which process is controlling is the ratio t_R/t_m. If t_R/t_m is small, the chemical reaction is faster than the mixing and microscopic concentration gradients appear; i.e., the fluid becomes segregated. Conversely, when t_R/t_m is large, the fluid may be considered well micromixed from the viewpoint of chemical reaction.

The competition between reaction and diffusion can be represented by the IEM-model. t_m is identified with a diffusion time $t_D = \mu L^2/\mathcal{D}$ (see Sec. 3.2 above) where different diffusivities $\mathcal{D}_j$ and hence different micromixing times t_{mj} may be used for each species. This simple lumped parameter model gives results comparable to those of more sophisticated distributed models, at least for reaction systems which are not too "stiff". An interesting property is revealed by numerical simulations. The simplest way to represent partial segregation in a fluid is to consider that it consists of a mixture of macrofluid (fraction β) and microfluid (fraction $1 - \beta$). It turns out that the ratio $(1 - \beta)/\beta$ is always close to that of two characteristic times [28] [18]. In the case of erosive mixing of two reactants in a CSTR (erosion controls mixing and the product of erosion is a microfluid), one finds

$$\frac{1 - \beta_e}{\beta_e} \approx 3.5\, \frac{\tau}{t_e} \qquad (4\text{-}9)$$

For stoichiometric amounts of premixed reactants in a CSTR, an empirical correlation is found

$$\frac{1 - \beta_D}{\beta_D} = a(t_R/t_m)^m \qquad (4\text{-}10)$$

On average $a \approx 2$ and $m \approx 0.8$. In the simplest case $a = 1$ and $m = 1$. The ratio of microfluid to macrofluid is thus roughly equal to the ratio of reaction time to micromixing time - an interesting "rule of thumb" for rapid estimation of segregation extent.

More sophisticated mechanisms may exist, where chemical reaction takes place during successive stages of micromixing, or combinations of these.

For instance, when t_e, t_D and t_R are of the same order of magnitude, reaction occurs both during the erosion and diffusion stages. The fluid behaves as a mixture of fully segregated aggregates undergoing erosion (volume fraction β_e) and a population of smaller aggregates produced by erosion (volume fraction $1 - \beta_e$) and interacting in the reaction/diffusion regime with a premixed composition. This population is considered as a mixture of macrofluid (volume fraction β_D) and microfluid (volume fraction $1 - \beta_D$). From the above relationships one finds

$$\beta_e = \frac{t_e/\tau}{3.5 + t_e/\tau} \quad \text{and} \quad \beta_D \approx \frac{1}{1 + t_R/t_D} \tag{4-11}$$

from which the chemical conversion can be calculated [28].

Another example is that of reaction and diffusion taking place in stretching aggregates. $\delta(\alpha)$ being the striation thickness, decreasing with age, some authors [32] have introduced a "warped" time t_w defined by

$$dt_w/d\alpha = \mathscr{D}/\delta^2(\alpha) \tag{4-12}$$

The reaction/diffusion equations are then easier to solve as a function of t_w. The IEM-model may also be used with

$$t_m(\alpha) = t_m(0)\ \delta^2(\alpha)/\delta^2(0) \tag{4-13}$$

Examples of these methods can be found in ref. [28] and [33].

4.5. Experimental determination of the state of segregation

By mixing two solutions containing different amounts of tracer, microscopic concentration gradients arise owing to partial segregation of the fluid. However, to deduce micromixing times from these experiments, p(C) should be measured with microprobes having both space and time resolution smaller than that of segregation microscales (e.g., down to the Kolmogorov microscale, 10 µm in liquids). Optical methods seem to be promising but reliable methods are still lacking for measuring the fluctuating concentration in the required frequency range. In principle, chemical methods should be more powerful as chemical reactions play the role of molecular probes. However, these methods are less direct as interpretative models are required to extract micromixing parameters from observed conversions and yields. Either instantaneous or finite rate reactions may be used, provided they are sensitive to micromixing.

Fast consecutive-competing reactions

$$\left.\begin{array}{l} A + B \xrightarrow{k_1} R \\ R + B \xrightarrow{k_2} S \end{array}\right\} \qquad (4-14)$$

with $k_1 >> k_2$ and k_2 still large are very interesting candidate test-reactions to study the local state of segregation. If a small amount of B is mixed into a large excess of A, R is instantaneously formed. If the fluid is segregated, R stays at the contact of B and is immediately converted to S. If the fluid is well micromixed, R and B are dispersed in the whole volume but as the concentration of B is very small, R may subsist until B is totally consumed. Therefore, the amount of S formed is some kind of segregation index. A major advantage is that the system keeps a memory of the micromixing process close to the point of injection. The amount of S can be determined "off-line" by subsequent analysis of the fluid composition. Let the yield of S be $X_S = 2C_S/(2C_S + C_R)$. The fraction of macrofluid is

$$\beta = \frac{X_S - X_{SM}}{1 - X_{SM}} \qquad (4-15)$$

where X_{SM} is the yield one would observe in a well micromixed fluid. This method was extensively exploited by Bourne and coworkers [35] who used azo-coupling of 1-naphthol (A) with diazotised sulphanilic acid (B). Another reaction of this kind has been proposed [34], namely, the precipitation of barium sulphate (S) from a basic EDTA-complex (A) under the influence of an acid (B), (R) being the exactly neutralized complex. $BaSO_4$ is measured after completion of the reaction by sampling and turbidimetry. The development of this method could lead to a simple and inexpensive test to study the quality of mixing in industrial reactors.

5. MIXING IN STIRRED TANKS

Stirred reactors are basic devices for industrial operations like mixing, blending or chemical reaction. The stirred tank is often considered as perfectly macromixed, concentration gradients are ignored and micromixing is characterized by average parameters without any spatial distribution. Actually, the real behaviour of the fluid in the tank is much more complex. Using local probes, it is possible to determine the internal age distribution at various points and to study the circulation pattern of the fluid within the tank. Circulation times t_c can be measured with radio flow-followers. It is found that t_c is generally log-normally distributed around a mean value which is inversely proportional to the stirring speed N:

$$\overline{t}_c = \frac{V}{N_Q\ N\ d^3} \tag{5-1}$$

with values for N_Q between 0.3 and 1 in most cases. A related quantity is the terminal mixing time θ_m, defined as the time constant for exponential decay of fluctuations after injection of a tracer in the tank. Khang and Levenspiel [36] propose for turbines ($Re > 10^4$)

$$N\theta_m\ (d/d_T)^{2/3} = 0.5 \approx 0.1\ P\ (\rho N^3 d^5)^{-1} \tag{5-2}$$

and for propellers

$$N\theta_m\ (d/d_T)^2 = 0.9 \approx 1.5\ P/(\rho N^3 d^5)^{-1} \tag{5-3}$$

where d is the impeller diameter, d_T the tank diameter and P the power input.

Measurements of velocity and concentration fluctuations have revealed that turbulence parameters were also distributed within the tank volume. In a typical situation (aqueous newtonian liquid), the turbulence intensity with respect to turbine tip velocity, $u'/(\pi Nd)$, is between 5 % and 30 %. The turbulent energy dissipation per unit mass, ε, usually varies between 0.2 and 2.5 (some authors say 0.1 and 10) times the average value $\overline{\varepsilon}$, which represents the main part of the mechanical power input, $P/(\rho V) \approx \varepsilon$. This proves that energy dissipation is essentially utilised to promote turbulence and not to sustain the average velocity field. The local variations of ε induce corresponding variations of micromixing intensity from point to point in the tank, as micromixing times depend on ε (see Table 1). The most popular method to predict distribution of turbulence parameters makes use of the so-called "k - ε" models which link the turbulent kinetic energy per unit mass, $k = 1/2\ u'^2$, to its rate of dissipation ε [37] [38] [39]. These models are still partly empirical.

In order to predict chemical conversion in stirred tanks several authors [37] [40] have proposed cell models in which the tank volume is divided into segments characterized by their turbulence level (e.g. solution of a k-ε model). The segments are connected by deterministic or stochastic flowrates accounting for internal recirculation. Conversion is obtained by integrating continuity equations and mass-balances from segment to segment.

Simpler models may also be used where the tank content is divided into two or three zones exchanging material (macromixing) and where micromixing times are different (small t_m in the stirrer zone, larger t_m in other zones). Micromixing and chemical reaction is accounted for by the IEM model.

This poses the problem of prediction and scale-up of mixing parameters (especially micromixing times). Macromixing is closely related to circulation times, which vary as N^{-1} (5-1).

If micromixing is controlled by turbulence the following assumptions seem reasonable for newtonian incompressible media, and for homothetical tanks.

(i) Homogeneous isotropic turbulence is achieved almost everywhere in the tank.

(ii) The average velocity field is not distorted by scale-up and all velocities are proportional to $\overline{U} = \pi N d$.

(iii) The macroscales L_f and L_s are equal and proportional to d.

(iv) The turbulence intensity $u'/\overline{U}$ is roughly constant at a given position, whatever the reactor size.

(v) $\overline{\varepsilon} = P/(\rho V)$

From these assumptions, it can be shown (41) that $\varepsilon \sim N^3 d^2$. Consequently, $\lambda_K = (\nu^3/\varepsilon)^{1/4} \sim N^{-3/4} d^{-1/2} \nu^{3/4}$ and the diffusion time t_D in Kolmogorov aggregates varies as $N^{-3/2}\ d^{-1/2}\ \nu^{3/2} \mathcal{D}^{-1}$. Corrsin time constant τ_S only depends on N^{-1} (as does the circulation time $\overline{t}_C$). Recalling that the power input is $P = N_p \rho N^3 d^5$, it follows that, keeping ε constant requires $P \sim L^3$ and keeping N constant requires $P \sim L^5$, where $L \sim d$ is a characteristic linear dimension of the tank. Therefore, keeping the parameters constant for the diffusion stage of micromixing is less demanding ($P \sim L^3$) than keeping those for the second stage ($P \sim L^5$). It may be pointed out that the rule for scale-up at constant stirred tip velocity (N_d = constant) leads to $P \sim L^2$. Average velocities are then preserved but micromixing characteristics are not.

6. EXAMPLES OF MIXING EFFECTS ON PERFORMANCE OF CHEMICAL REACTORS

The preceeding sections are mainly concerned with mixing in homogeneous or pseudo-homogeneous systems. This does not mean that mixing effects are unimportant in multiphase reactors. Coalescence and break-up of droplets are essential features of liquid-liquid reactors. Likewise, micromixing plays an essential role in large scale fermentors where the concentration of dissolved oxygen has to be kept constant in spite of a strong rate of consumption. However, little is still known and/or published on these subjects which are open to future research.

The effect of macromixing on the performance of chemical reactors is dealt with in standard text books [2] [4] [6] and will not be discussed here. Practical examples of micromixing effects are less known. From the discussion presented in Sec. 4 and influence of micromixing may be expected when a controlling step of the chemical process (time constant t_R) competes with the micromixing process (time constant t_m). This may happen especially in three domains: fast and complex reaction systems, polymerization reactions, and precipitation (crystallization) reactions. Three examples of such micromixing effects are presented below.

6.1. Effect of micromixing on the yield of fast consecutive-competing reactions

As already mentioned in Sec. 4.5, such reactions are very sensitive to mixing effects. For instance, Bourne & coworkers [42] found that in the bromination of resorcin, the amount of 2-4 dibromoresorcin might vary from 30 % to 60 % when the stirring speed was increased from 0 to 360 R.P.M. Fig. 15 shows results obtained by Baldyga, Bourne et al. [43] in azo coupling of 1-naphthol (A) with diazotised sulphanilic acid (B). Under specified conditions (room temperature, pH = 10, dilute solutions), the reaction system conforms to model (4-14) with $k_1 = 3840\ k_2$. The yield of S, $X_S = 2\ C_S/(2C_S + C_R)$ is an index for the degree of segregation of the fluid, as explained above. With initial concentrations after mixing of 0.5 $mol.m^{-3}$ both reaction are complete after 10 ms and the half life-time of B is 0.3 ms. In Fig. 15 X_S is plotted against $t_D/t_R = k_2\ C_{Bo}\ \delta_o^2/\mathcal{D}_A$ where $\delta_o = 0.5\ \lambda_K$. It is not surprising that segregation (X_S) increases with t_D/t_R, i.e., when reaction 2 becomes more and more rapid with respect to diffusion which causes micromixing. Experimental data were obtained in a semibatch reactor and in a CSTR. They are well accounted for by a model of reaction/diffusion in stretching laminae initially of the Kolmogorov size undergoing periodic mixing with the surrounding material after sudden bursts of vorticity. This model only required that the value of $\varepsilon/\overline{\varepsilon}$ be found at the injection point.

6.2. Partial segregation in a polymer reactor

It is well known that segregation of the polymerizing mixture may strongly affect molecular weight distribution (and thereby polydispersity) of the polymer, especially when branching processes are involved [18]. An example is presented below where imperfect mixing may also influence the behaviour of a high pressure polyethylene reactor where polymerization is induced by a fast decomposing initiator (A). As far as monomer consumption only is concerned a very simple kinetic model may be used as complex processes like chain transfer, β-scission etc... globally conserve the concentration of free radicals. This model is written

$$A \xrightarrow{f,k_d} 2R \qquad \text{chemical initiation} \tag{6-1}$$

$$R + M \xrightarrow{k_p} R \qquad \text{propagation} \tag{6-2}$$

$$R + R \xrightarrow{k_t} \text{polymer} \qquad \text{termination} \tag{6-3}$$

In the long chain approximation and at quasi-steady state for free radicals (R) the rate of monomer (M) consumption is

$$r = k_p (k_t)^{-1/2} (f\, k_d)^{1/2} C_A^{1/2} C_M \tag{6-4}$$

$k_g = k_p(k_t)^{-1/2}$ is a characteristic rate constant which must be known to calculate the production of polymer. Experiments for the determination of k_g were carried out in a bench scale CSTR fed with separate feedstreams of initiator and monomer under controlled temperature and pressure [44] [45]. The reactor could be considered as well macromixed but, owing to viscosity and closeness to critical conditions, micromixing effects might occur at two levels. These are control of the initial mixing of the initiator feedstream into the bulk phase and the subsequent state of the reacting mixture which might behave as a micro- or a macrofluid.

Representation of partial segregation of the initiator feedstream is possible by the IEM-model. The concentration C_A of initiator in fresh aggregates is given by

$$\frac{dC_A}{d\alpha} = \frac{\overline{C}_A - C_A}{t_m} - k_d C_A \;;\; \alpha = 0,\ C_A = C_{Ao} \tag{6-5}$$

where $\overline{C}_A$ is the mean concentration in the bulk phase.

The average concentration in the partially segregated phase is given through the age distribution by

$$\langle C_A \rangle = (1/\tau) \int_o^\infty C_A(\alpha) \exp(-\alpha/\tau) d\alpha \tag{6-6}$$

The initiator feed flowrate is F_{Ao} but the effective flowrate actually transferred to the bulk by the net exchange is only F_A, such that

$$F_{Ao} = q\, C_{Ao} = F_A + q \langle C_A \rangle + k_d \langle C_A \rangle v \tag{6-7}$$

where q and v are the flowrate and volume occupied by the segregated phase, respectively. The mass balance in the bulk phase is written

$$F_A = Q\,\overline{C}_A + k_d\,\overline{C}_A\,V \qquad (6\text{-}8)$$

Owing to perfect macromixing, $v/q = V/Q = \tau$. In addition, it may be assumed that $v << V$ and $q << Q$. After a few mathematical manipulations the available initiator fraction is obtained from (6-7) and (6-8).

$$\phi = F_A/F_{Ao} = \left[1 + t_m(k_d + 1/\tau)\right]^{-1} \qquad (6\text{-}9)$$

Knowing the initiator input standard mass balances for the monomer can be written in the bulk phase of the CSTR.

For a microfluid (M) the initiator and monomer balances are, respectively:

$$\phi F_{Ao} - k_d\,\overline{C}_A\,V = Q\,\overline{C}_A \qquad (6\text{-}10)$$

$$F_{MO} - r_V = Q\,\overline{C}_M = F_M \qquad (6\text{-}11)$$

where r is given by (6-4).

The conversion X is then calculated as

$$1 - X = F_M/F_{MO} = \left[1 + k_g\tau\,(f\,k_d\,C_A)^{1/2}\right]^{-1} \qquad (6\text{-}12)$$

where $\overline{C}_A$ is deduced from (6-10).

Conversely, the bulk phase may be considered as a macrofluid. Partial segregation of the initiator is not taken into account. It is simply assumed that initiator and monomer are premixed at the reactor inlet and that polymerization subsequently occurs in a fully segregated macrofluid. The standard solution for batch polymerization is

$$C_M = C_{Mo}\exp\left[-\,2\,k_g\,(f\,C_{Ao}/k_d)^{1/2}\,(1 - \exp(-\,k_d t/2))\right] \qquad (6\text{-}13)$$

The conversion is obtained upon averaging over the RTD.

$$1 - X = F_M/F_{Mo} =$$

$$(1/\tau)\exp\left[-\,2k_g(f\,C_{Ao}/k_d)^{1/2}\right]\int_o^\infty \exp\left[2\,k_g(f\,C_{Ao}/k_d)^{1/2}\right.$$

$$\left. \times \exp(-\,k_d\,t/2) - t/\tau\right]\,dt \qquad (6\text{-}14)$$

In order to compare these models, a set of 55 runs were considered in the ranges $433 < T < 533$ K, $1100 < p < 1950$ bars. The aim was to determine the "true value" of k_g accounting at best for observed conversions. This rate constant was represented by

$$k_g = k_g^o \exp\left(-\,\frac{E + p\Delta V}{RT}\right) \quad (m^3.mol^{-1}.s^{-1})^{1/2} \qquad (6\text{-}15)$$

Typical results from optimal fitting of parameters are reported in Table 2.

Clearly, model (2) is better, showing that partial segregation of initiator occurs. This model involves 6 parameters - 3 for k_d (Table 2) and 3 for t_m which was found to obey an Arrhenius-like relationship:

$$t_m = 1.1 \times 10^{-9} \exp\left[-\,(-\,10100 + 0.57\,p/T)\right] \qquad (6\text{-}16)$$

TABLE 2	k_g^o	E $kJ.mol^{-1}$	$\Delta v \times 10^6$ $m^3.mol^{-1}$	Mean relative deviation estimate
(1) Well micromixed CSTR eq(6-10)(6-11) $\phi = 1$	432	40.2	-33.5	21.9 %
(2) IEM model for initiator, well micromixed bulk phase eq (6-9) to (6-12)	33.4	27.0	-22.6	17 %
(3) Segregated CSTR eq. (6-13), (6-14)	645	36.5	-32.8	30.8 %

This micromixing time ranges between 10^{-2} s and 10 s in the useful experimental range. Segregation occurs because t_m is of the same order of magnitude as the time of decomposition of the initiator,

$1/k_d$. One may then consider that relationship (6-15) with values found by application of model (2) are "free" from mixing effects. Actually, this value was used to account for the production of polymer. This study was the basis for a long term research program aiming at deriving a predictive model of the quality of Low Density Polyethylene produced in industrial reactors. Micromixing is an essential feature of this model.

6.3. Micromixing and precipitation

Precipitation is a special case of crystallization induced by supersaturation arising from a chemical reaction upon mixing of two reactants. Therefore, the rate of precipitation and the crystal size distribution depend on the intensity of mixing which influences contacting of reactants and subsequent processes of nucleation and growth. Both macro- and micromixing effects must be taken into account as they control the local value of supersaturation, as pointed out for instance by Garside and Tavare [46]. Pohorecki and Baldyga [47] have derived a model accounting for the batch precipitation of barium sulphate. It has been shown in Sec. 4.5 that this reaction, carried out in the presence of EDTA, could serve as a test for estimating the intensity of segregation.

The example of simulation below shows the influence of mixing time on crystal size distribution.

A precipitation reaction A + B → P (solid) is carried out in a stirred batch reactor upon simultaneous mixing of A and B. The rates of nucleation (r_1) and growth (r_o) are assumed to be, respectively:

$$r_1 = k_1 \, C^{a_1} (1 + k_2 \, C_c^{a_2}) \qquad m^{-3} \, s^{-1} \tag{6-17}$$

$$r_o = k_o \, C^{a_o} \qquad m \, s^{-1} \tag{6-18}$$

where C is the supersaturation defined by $C = (C_A C_B)^{1/2} - C_S$, C_S the solubility of P and C_C the concentration of crystals. The incoming streams of A and B first remain segregated and then mix (characteristic time t_m) into the bulk phase which is assumed to be well micromixed. A dimensionless formulation of the problem is obtained by introducing a characteristic crystallisation time $t^* = r_o^{*3/4} \, r_1^{*-1/4}$ and a characteristic size $\ell^* = r_o^{*1/4} \, r^{*-1/4}$. r_1^* and r_o^* are the values of r_1 and r_o for $C = C_S$ and $C_C = 0$. Details of the model, based on a crystal population balance, are given elsewhere [48]. Results of a simulation where equal amounts of A and B are initially added during a short time are presented below for $a_o = 3$, $a_1 = 6$, $k_2 = 0$.

	After completion of precipitation	
Mixing time t_m/t^*	Mean size of crystals $\bar{\ell}/\ell^*$	Reduced variance of crystal size distribution $\sigma^2/\bar{\ell}^2$
0.002	0.19	0.06
0.01	0.14	0.09
0.05	0.11	0.18

This shows a noticeable influence of mixing intensity on crystal size distribution. The smaller the mixing time, the smaller the crystals, but the broader the distribution (for the particular set of parameters chosen here).

There is still little work published on this subject but it may be anticipated that precipitation and control of the quality of crystals is one of the areas where micromixing theories will find industrial applications.

7. CONCLUSION

The main conclusions are drawn from ref. [18].

Methods for deriving good macromixing models based on RTD experiments are now well established. The "Population Balance" approach provides a theoretical basis for extending the RTD theory to more complex situations such as variable volume, density and flowrate, multiple inlet/outlet reactors. In this respect, the IAD concept is helpful and seems to have been somewhat overlooked to this day. However, simple procedures, easy to use by practitioners for identifying characteristic parameters of these distributions would still be welcome. This is especially true for the promising "generalized reaction time distribution". Reactive tracers might bring out interesting information.

Turning towards micromixing, it is clear that models for mixing earliness describing transfer between Entering and Leaving Environments are superabundant and sometimes too academic because of dealing with age segregation. One parameter models are more or less equivalent. Either IEM or CD models should be used for simulations.

Micromixing in the physical space (mechanism for segregation decay) deserves more attention. Several stages, each with their own time constant, may interfere, in series or in parallel. When one single mechanism is involved interaction between fluid particles can be conveniently represented by simple models (again IEM or CD models). The method of "characteristic times" is helpful for determining which processes are controlling. Such time constants are:

(a) For macromixing, the space time τ in a continuous reactor, or the circulation time t_c in a stirred tank.

(b) For micromixing, one or several micromixing times; e.g., $t_\delta \sim (\nu/\varepsilon)^{1/2}$ (stretching and vorticity), $\tau_s \sim (L_s^2/\varepsilon)^{1/3}$ (segregation dissipation), $t_D \sim \delta^2/\mathfrak{D}$ ($\delta \sim \lambda_K$, molecular diffusion).

(c) For chemical reaction, one or several reaction times, e.g., $t_R = 1/kC_o^{n-1}$ for the fastest step.

By comparing all these time constants the controlling mixing regime can be determined, sometimes quantitatively. For instance, it has been shown that the micro/macrofluid volume ratio is roughly equal to t_R/t_D.

In spite of recent progress [31] much has still to be done to come to a universal theory making it possible to make a priori predictions. In the Introduction (Sec. 1) a distinction was made between the approach to mixing problems by the methods of fluid mechanics and those of chemical engineering. A gradual and fortunate merging of these two descriptions is presently in progress in the frame of turbulence theory. In order to achieve this experimental data are urgently needed, especially concerning concentration fluctuations in reactive media.

The remarks above essentially apply to homogeneous or pseudo-homogeneous media. As far as multiphase systems, non newtonian liquids or large scale reactors are concerned, the situation is much less bright. Simple and realistic models based on experimental work are required in order to meet industrial needs in this area.

Macromixing models play an essential role in modeling and scale-up of chemical reactors. Besides, the areas where micromixing is a controlling factor, are now clearly identified: selectivity in complex reactions involving at least a fast step between unmixed reactants, crystallization and precipitation reactions, oxygen supply in fermentors ... this list is not exhaustive.

Chemical Reaction Engineering is more and more involved in the production of fine chemicals and sophisticated products. In this respect, conversion of reactants is indeed an important requirement but the quality of products is still more important; i.e., selectivity, molecular weight distribution, sequencing of co-monomers, crystal size distribution etc. Mixing, which is most of the time a key factor for controlling such qualities will thus remain an open area for research and applications in the future.

NOTATION

$a_1 = Q_1/(Q_1 + Q_2)$	flowrate ratio
$A(\lambda, t_s)$	mixing function
c	concentration fluctuation
C	concentration (tracer)
C_1	concentration in bubble phase (fig. 4)
C_2	concentration in dense phase (fig. 4)
d	impeller diameter
D, D_A	axial dispersion coeffient
$\mathcal{D}$	molecular diffusivity
Da	Damköhler number
E	activation energy
$E_1(k_1)$	velocity fluctuation spectral density (one direction)
$E(k)$	velocity fluctuation spectral density (all directions)
$E_s(k)$	concentration fluctuation spectral density
$E(t_s)$	residence time distribution density function (RTD)
$f = \psi/C$	internal age distribution (local normalized density function)
$f(r)$	autocorrelation function (velocity fluctuation)
$f = C_{out}/C_{in}$	residual concentration ratio
$F(t)$	cumulative residence time distribution
F_A	molecular flowrate
$\mathcal{F}_1(\lambda)$	partial cumulative RTD (Eq. 3-17)
g	rate of generation of entities (population balance)
$g(\lambda)$	fractional flowrate of the Max. Mix. fluid
$G(s)$	transfer function
$h(\lambda, t_s)$	mixing function (model of Spencer and Leshaw)
H	total height of a fluidized bed (fig. 4)
$H(t)$	Heaviside step function
$H(t_i)$	reaction time distribution density function
$I(\alpha)$	internal age distribution (IAD)
I_s	intensity of segregation
J	number of mixing tanks in series
J	interaction term (Eq. 3-9)
$k, k_d, k_g, k_m, k_p, k_t$	kinetic rate constants
K	mass transfer coefficient (May Van Deemter model)
ℓ	size of an aggregate, of a crystal
L_f, L_s	fluctuation macroscales
m	mass of catalyst
n	number of moles
N	number of transfer units (May Van Deemter model)
N	number of particles

N	stirring speed
N_Q	circulation number
$p(C)$	concentration distribution
p_k	parameter of a flow model
P	stirring power
$Pe = uL/D$	Peclet number
$q = 3/2\ u'^2$	kinetic energy of turbulent motion
Q	volumetric flowrate
r	rate of reaction
r_o, r_1	rate of growth, of nucleation
R_s	micromixing parameter (model of NG and RIPPIN)
R	ideal gas constant
$\mathfrak{R}$	rate of chemical production
s	Laplace parameter
$s(\lambda, t_s)$	age segregation function
$Sc = \nu/\mathcal{D}$	Schmidt number
t	time
t_c	circulation time
t_D	diffusion time
t_e	erosion time
t_m	micromixing time (IEM model)
t_R	reaction time
t_s	residence time
t_w	warped time (Eq. 4-12)
T	time period
T	temperature
u	intersticial velocity, velocity fluctuation
u_o	superficial velocity
u'	root mean square velocity fluctuation
U	local fluid velocity
v	accumulator volume (model of Spencer and Leshaw)
$v_i = d\zeta_i/dt$	intrinsic velocity of change of property ζ_i
V	volume
w	parameter (model of Valderrama and Gordon)
$x = z/H$	reduced height (fluidized bed)
$x(t)$	inlet tracer concentration
$y(t)$	outlet tracer concentration
z	height (fluidized bed)

Greek symbols

α	age, parameter
β	parameter
$\beta = \Omega_2/\Omega_1$	parameter (May Van Deemter model)
β_D, β_e	segregated volume fraction
γ	parameter
$\Gamma(J)$	Gamma function
$\delta(t)$	Diract delta function (pulse)
δ	striation thickness

ε	turbulent kinetic energy dissipation per unit mass (Table 1)
ε_s	segregation dissipation (Table 1)
ζ_i	intrinsic property of an entity
θ	reduced time
θ_m	terminal mixing time
λ	life expectancy
$\lambda_K, \lambda_f, \lambda_s, \lambda_B$	Kolmogorov, Taylor, Corrsin, Batchelor turbulent microscales (Table 1)
$\Lambda(t)$	intensity function (Eq. 2-2)
μ_n	n^{th} order moment of the RTD about the origin
μ'_n	n^{th} order moment of the IAD about the origin
ν	kinematic viscosity
ρ	volumetric mass
σ^2	variance of distribution
$\tau = V/Q$	space time } Eq. (2-37) and (2-38)
$\tau' = N/F$	space time } Eq. (2-37) and (2-38)
τ_D, τ_P	parameters (mixing earliness models)
τ_f, τ_s	Taylor, Corrsin time constants (Table 1)
τ_K, τ_C	dissipation time constants (Table 1)
ϕ	initiator availability
ψ	distribution of entities (population balance)
ω	interaction (coalescence-dispersion) frequency
Ω_1, Ω_2	cross section of bubble phase, dense phase (Van Deemter Model, fig. 4)

Superscripts

$\bar{A}$	average value (over ages)
$\bar{\bar{A}}$	Laplace transform
$\langle A \rangle$	average value (in space)

Subscripts

A_j	for component j
A_{in}	at reactor inlet
A_{out}	at reactor outlet

REFERENCES

1. Danckwerts, P.V. Continuous flow systems. Distribution of residence time, Chem. Eng. Sci., 2 (1953), 1
2. Levenspiel, O. Chemical Reaction Engineering, Wiley (1972)
3. Wen, C.Y. and L.T. Fan. Models for flow systems and chemical reactors, Dekker (1975)
4. Nauman, E.B. and B.A. Buffham. Mixing in continuous flow systems, Wiley (1983)
5. Dudukovic, M.P. Tracer methods in chemical reactors (this book).

6. Villermaux, J. Génie de la Réaction Chimique. Conception et fonctionnement des réacteurs, Technique et Documentation, Paris (1982)
7. Van Swaaij, W.P.M. and F.J. Zuiderweg. Investigation of ozone decomposition in fluidized beds on the basis of a two-phase model. Proceed fifth europ. second Int. Symp. Chem. React. Eng. Amsterdam, B-9-25 (1972)
8. De Vries, R.J., W.P.M. Van Swaaij, C. Mantovani and Heijkoop. Design criteria and performance of the commercial reactor for the Shell chlorine process. Ibid. B-9-59
9. Danckwerts, P.V. The definition and measurement of some characteristics of mixtures. Appl. Sci. Research, 3 (1952), 279-298
10. Zwietering, T.N. The degree of mixing in continuous flow systems. Chem. Eng. Sci. 11 (1959), 1-15
11. Spencer, J.L., R. Lunt and S.A. Leshaw. Identification of micromixing mechanisms in flow reactors : transient inputs of reactive tracers. Ind. Eng. Chem. Fundam. 19 (1980), 135-141
12. Weinstein, H. and J. Adler. Micromixing effects in continuous chemical reactors. Chem. Eng. Sci. 22 (1967), 65-75
13. Villermaux, J. and A. Zoulalian. Etat de mélange dans un réacteur continu. A propos d'un modèle de Weinstein et Adler. Chem. Eng. Sci. 24 (1969), 1513-1517
14. Ng, D.Y.C. and D.W.T. Rippin. The effect of incomplete mixing on conversion in homogeneous reactions. Proceed 3rd Europ. Symp. Chem. React. Eng., Amsterdam, 161-165, Pergamon Press (1965)
15. Plasari, E., R. David and J. Villermaux. Micromixing phenomena in continuous stirred reactors using a Michaelis-Menten reaction in the liquid phase. Chem. React. Eng. Houston 1978, ACS Symp. Series, 65, 126-139
16. Valderrama, J.L. and A. Gordon. Mixing effects on homogeneous p-order reactions. A two-parameter model for partial segregation. Chem. Eng. Sci. 34 (1979), 1097-1103
17. Ibid. A two parameter model for partial segregation. Application to flow reactors with pre- and unmixed feed. 36 (1981), 839-844
18. Villermaux J. Mixing in chemical reactors, ACS symposium Series, 226 (1983), 135-186
19. Ritchie, B.W. and A.H. Tobgy. A three-environment micromixing model for chemical reactors with arbitrary separate feedstreams. The Chem. Eng. Journal, 17 (1979), 173-182
20. Ritchie, B.W. Simulating the effects of mixing on the performance of unpremixed flow chemical reactors. The Canad. J. of Chem. Eng. 58 (1980), 626-633
21. Mehta and Tarbel. Four environment model of mixing and chemical reaction. A.I.Ch.E. J. 29(1983), 320
22. Spencer, J.L., R. Lunt and S.A. Leshaw. Identification of micromixing mechanisms in flow reactors: transient inputs of reactive tracers, Ind. Eng. Chem. Fundam. 19 (1980) 135-141
23. Spencer, J.L. and R.R. Lunt. Experimental characterization of mixing mechanisms in flow reactors using reactive tracers. Ind.

Eng. Chem. Fundam. 19 (1980), 142-148

24. Curl, R.L. Dispersed phase mixing theory and effects in simple reactors. A.I.Ch.E. J. 9 (1963), 175
25. Spielman, L.A. and O. Levenspiel. A Monte-Carlo treatment for reacting and coalescing dispersed phase systems. Chem. Eng. Sci. 20 (1965), 247-254
26. Kattan, A. and R.J. Adler. A conceptual framework for mixing in continuous chemical reactors. Chem. Eng. Sci. 27 (1972), 247-254
27. Treleaven, C.R. and A.H. Tobgy. Monte-Carlo methods of simulating micromixing in chemical reactors. Chem. Eng. Sci. 27 (1972) 1497-1513
28. Villermaux, J. and R. David. Recent advances in the understanding of micromixing phenomena in stirred reactors. Chem. Eng. Comm. 21 (1983), 105-122
29. Beek Jr, J. and R.S. Miller. Turbulent transport in chemical reactors. Chem. Eng. Prog. Symp. Series, 55 (1959), 23-28
30. Ottino, J.M. Lamellar mixing models for structured chemical reactions and their relationship to statistical models: macro and micromixing and the problem of averages. Chem. Eng. Sci., 35 (1980), 1377-1391
31. Bourne, J.R. Micromixing revisited. Proceed. ISCRE 8, Edimburgh 1984 I. Chem. Symposium Series 87
32. Ranz, W.E. Applications of a stretch model to Mixing. Diffusion and reaction in laminar and turbulent flows. A.I.Ch.E. J. 25 (1979), 41-47
33. Bourne, J.R. and S. Rohani. Mixing and fast chemical reaction. VII Deforming Reaction zone model for the CSTR. Chem. Eng. Sci. 38 (1983), 911
34. Barthole, J.P., R. David and J. Villermaux. A new chemical method for the sutdy of local micromixing conditions in industrial stirred tanks. ACS Symp. Series, 196 (1982), 545
35. Bourne, J.R., F. Kozicki and P. Rys. Mixing and fast chemical reactions. I. Test reactions to determine segregation. Chem. Eng. Sci. 36 (1981), 1643-1648
36. Khang, S.J. and O. Levenspiel. New scale-up and design method for stirred agitated batch mixing vessels. Chem. Eng. Sci. 31 (1976), 569-577
37. Patterson, G.K. Application of turbulence fundamentals to reactor modelling and scale-up. Chem. Eng. Commun. 8 (1981), 25-52
38. Harvey, P.S. and M. Greaves. Turbulent flow in an agitated vessel. I: Predictive model. Trans. I. Chem. E. 60 (1982), 195
39. Ibid. Turbulent flow in an agitated vessel. II: numerical solution and model predictions. 60 (1982), 201
40. Mann, R. Gas liquid contacting in mixing vessels. I. Chem. E. Industrial research fellowship report, 1983
41. Villermaux, J. Micromixing phenomena in stirred reactors. Encyclopedia of Fluid Mechanics (to be published)
42. Bourne, J.R., P. Rys and K. Suter. Mixing effects in the bromination of resorcin. Chem. Eng. Sci. 32 (1977), 711-716

43. Baldyga, J. and J.R. Bourne. A fluid mechanical approach to turbulent mixing and chemical reaction. Part III: computational and experimental results for the new micromixing model. Chem. Eng. Comm. 28 (1984), 259-281
44. Villermaux, J., M. Pons and L. Blavier. Comparison of partial segregation models for the determination of kinetic constants in a high pressure polyethylene reactor. I. Chem. E. Symposium Series, 87 (1984), 553-560
45. Villermaux, J., L. Blavier and M. Pons. Polymer Reaction Engineering (Reichert K.H. and Geiseler W. editors) Berlin (1983), 1 Carl Hanser Verlag.
46. Garside, J. and N.S. Tavare. Crystallization and Chemical Reaction Engineering. Proceed. ISCRE 8. Edimburgh, 1984. I. Chem. Symposium Series 87
47. Pohorecki, R. and J. Baldyga. The use of a new model of micromixing for determination of crystal size in precipitation. Chem. Eng. Sci. 38 (1983), 79-83
48. Villermaux, J. to be published
49. Zwietering, Th.N. A backmixing model describing micromixing in single phase continuous-flow systems. Chem. Eng. Sci. 39 (1984) 1765

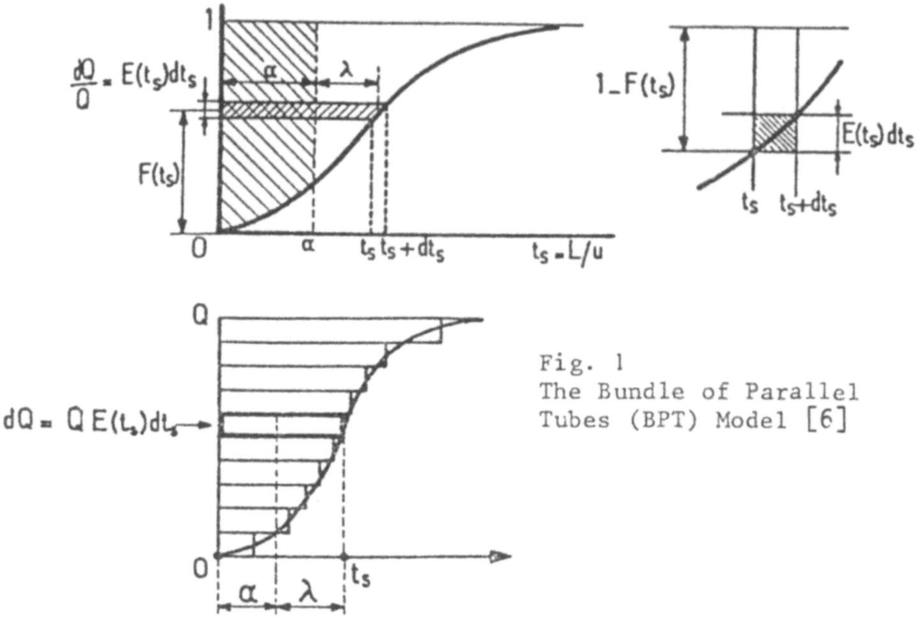

Fig. 1
The Bundle of Parallel Tubes (BPT) Model [6]

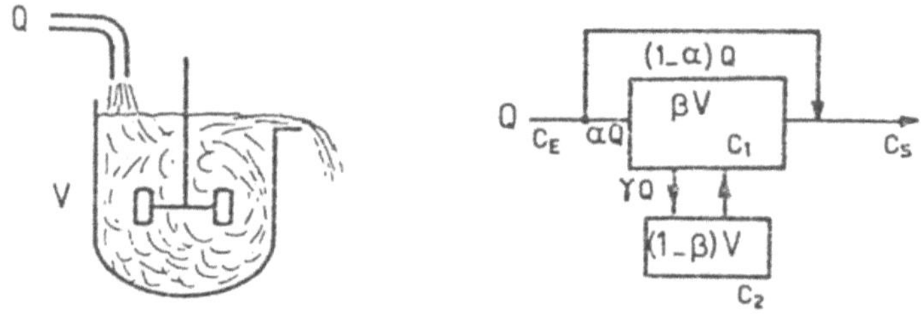

Fig. 2. Model of Cholette and Cloutier [6]

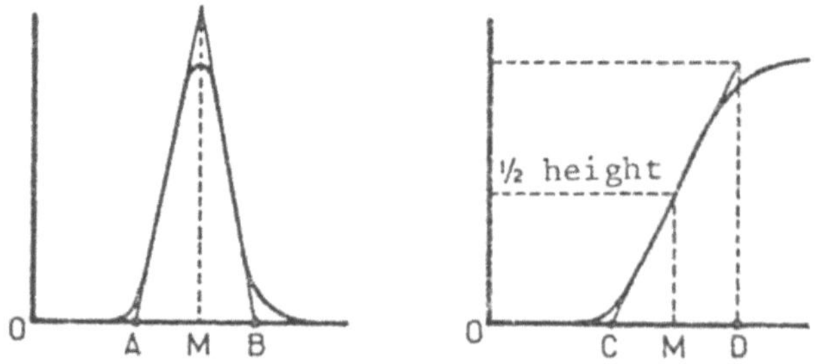

$$4\frac{\sigma}{\bar{t}_s} \approx \frac{AB}{OM} \qquad \frac{\sigma}{\bar{t}_s}\sqrt{2\pi} \approx 2{,}5\frac{\sigma}{\bar{t}_s} \approx \frac{CD}{OM}$$

Fig. 3. Simple determination of RTD variance [6]

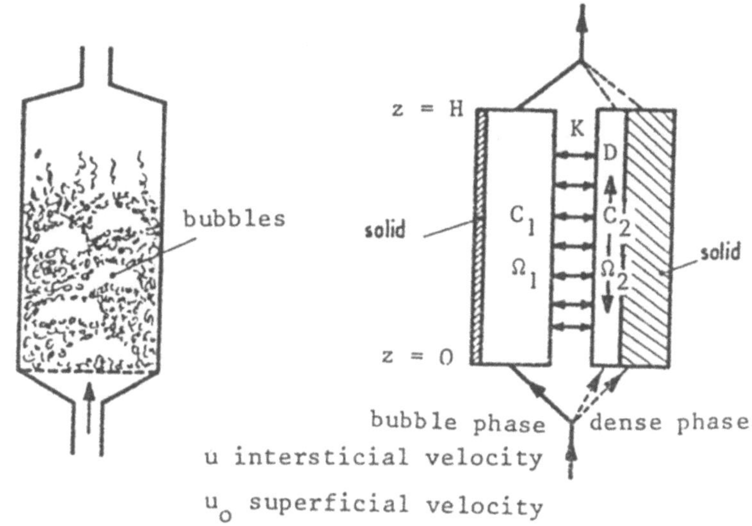

Fig. 4. Model of May-Van Deemter for a fluidized bed. [6]

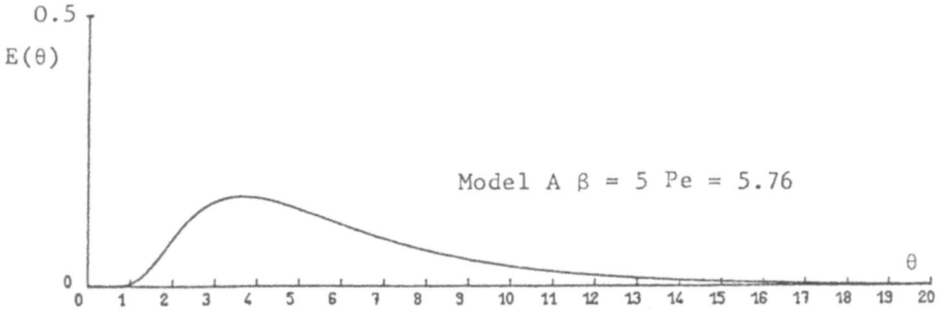

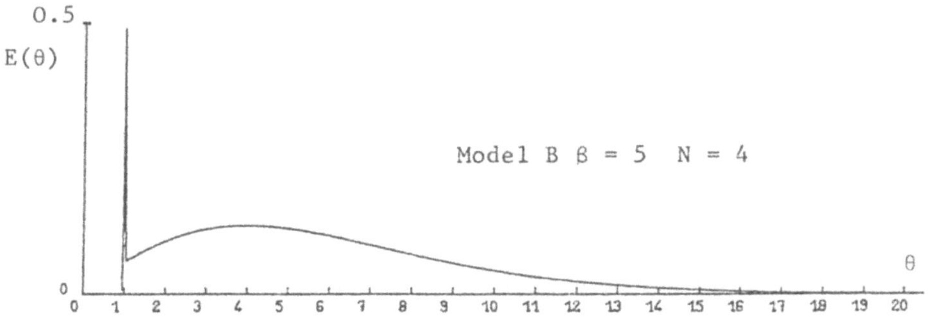

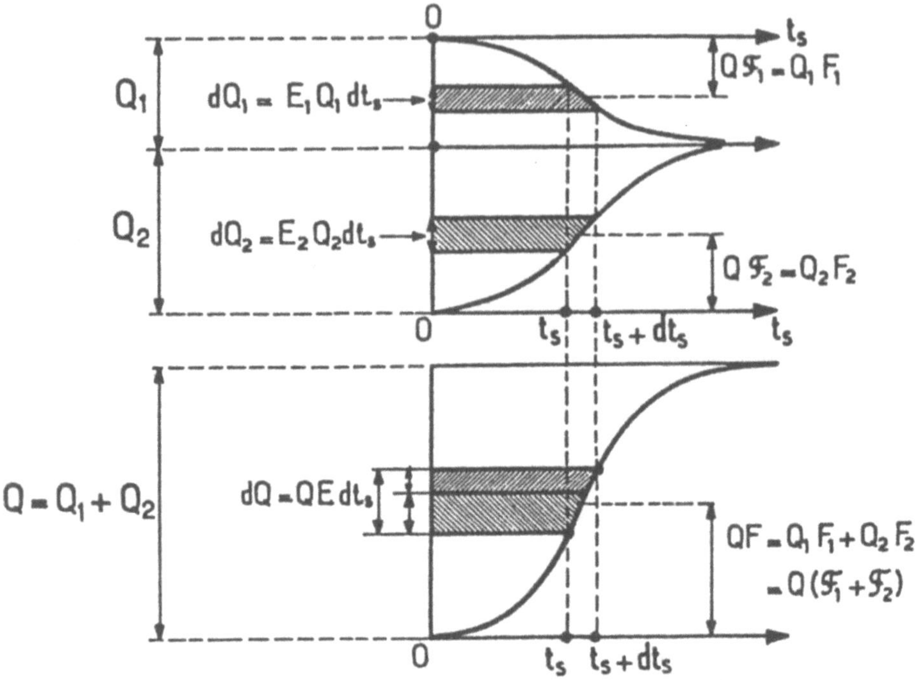

Fig. 6. BPT Model for a reactor with two inlets. [18] Reprinted with the permission from ACS Symposium Series, 226, 135-185 (1983). Copyright (1985) American Chemical Society.

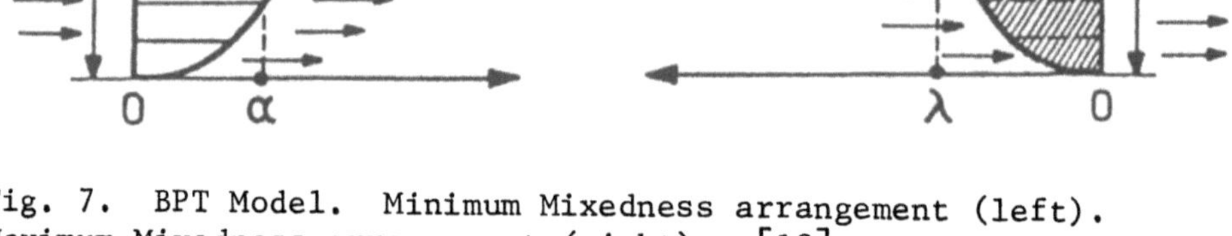

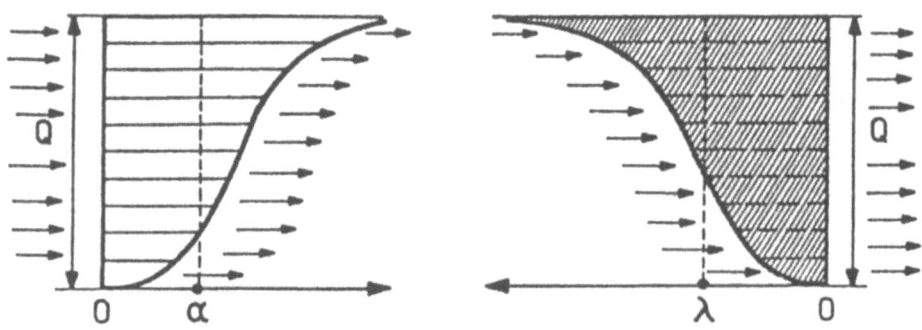

Fig. 7. BPT Model. Minimum Mixedness arrangement (left). Maximum Mixedness arrangement (right). [18] Reprinted with the permission from ACS Symposium Series, 226, 135-185 (1983). Copyright (1985) American Chemical Society.

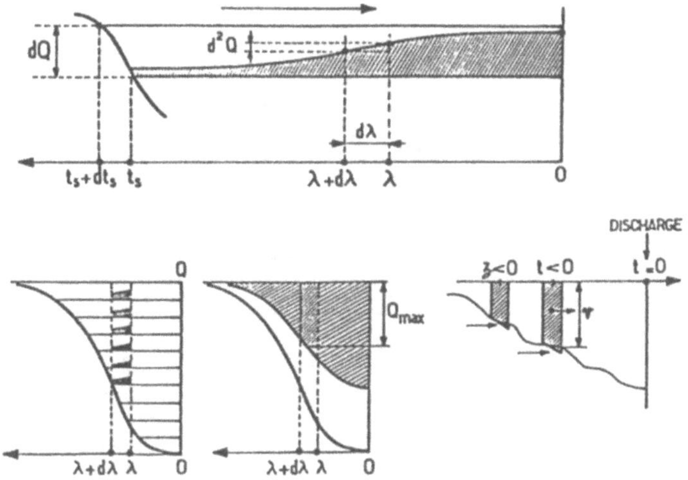

Fig. 8. General model of Spencer and Leshaw. [18] Reprinted with the permission of ACS Symposium Series, 226, 135-185 (1983). Copyright (1985) American Chem. Soc.

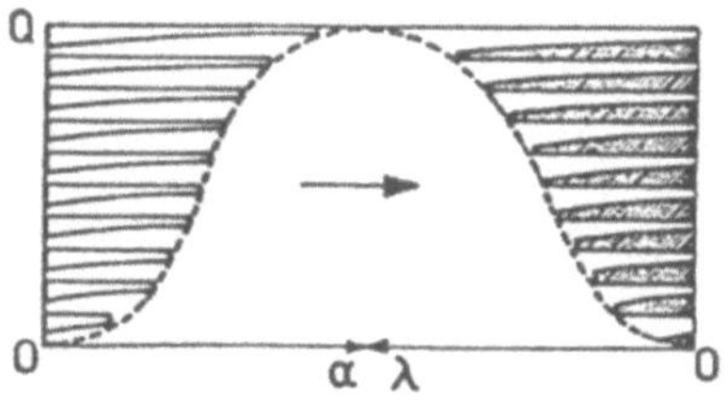

Fig. 9. Passage from Entering to Leaving Environment. [18] Reprinted with the permission of ACS Symposium Series, 226, 135-185 (1983). Copyright (1985) American Chem. Soc.

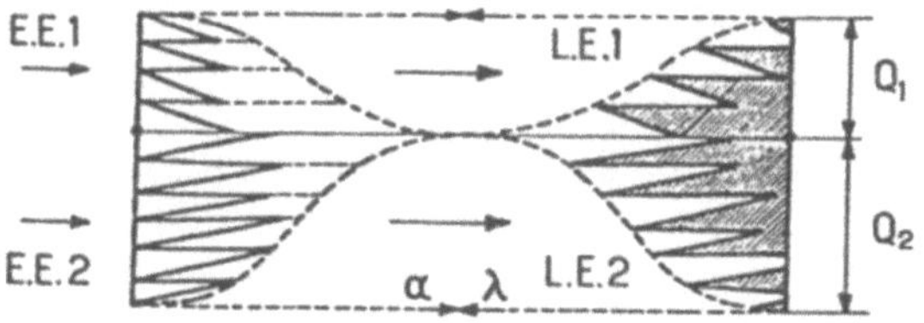

Fig. 10. Reactor with two unmixed feedstreams and four Environments [18] Reprinted with the permission of ACS Symposium Series, 226, 135-185 (1983). Copyright (1985) American Chemical Society.

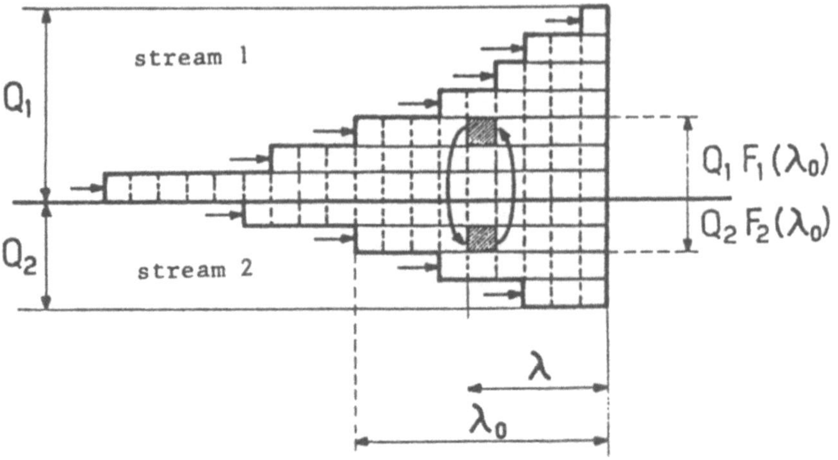

Fig. 11. **Model of Kattan and Adler (26).** Simulation of micromixing by random coalescence. [6]

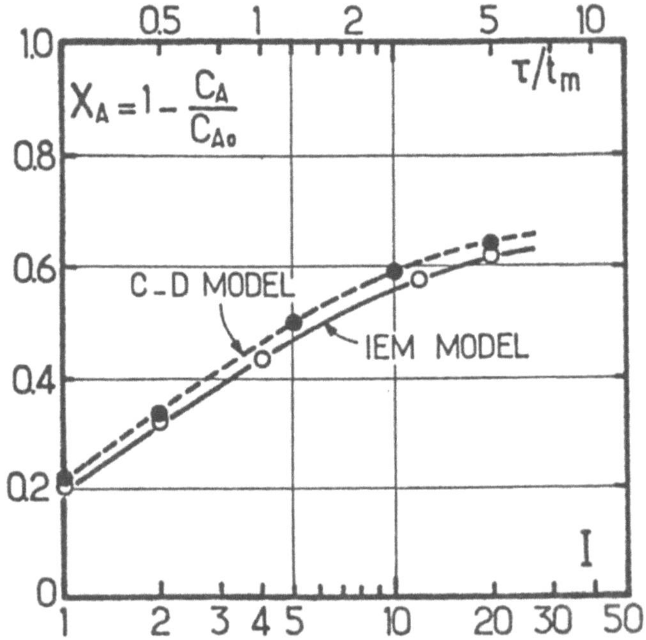

Fig. 12. Equivalence between CD and IEM models.
Reactor with two inlets having different RTDs (2 and 6 tanks in series). Second order reaction $C_{Bo} = 8\ C_{Ao}$, $kC_{Ao}\tau = 0.5$, $I = \omega\tau = 4\ \tau/t_m$ [18] Reprinted with the permission from ACS Symposium Series, 226, 135-185 (1983). Copyright (1985) American Chemical Society.

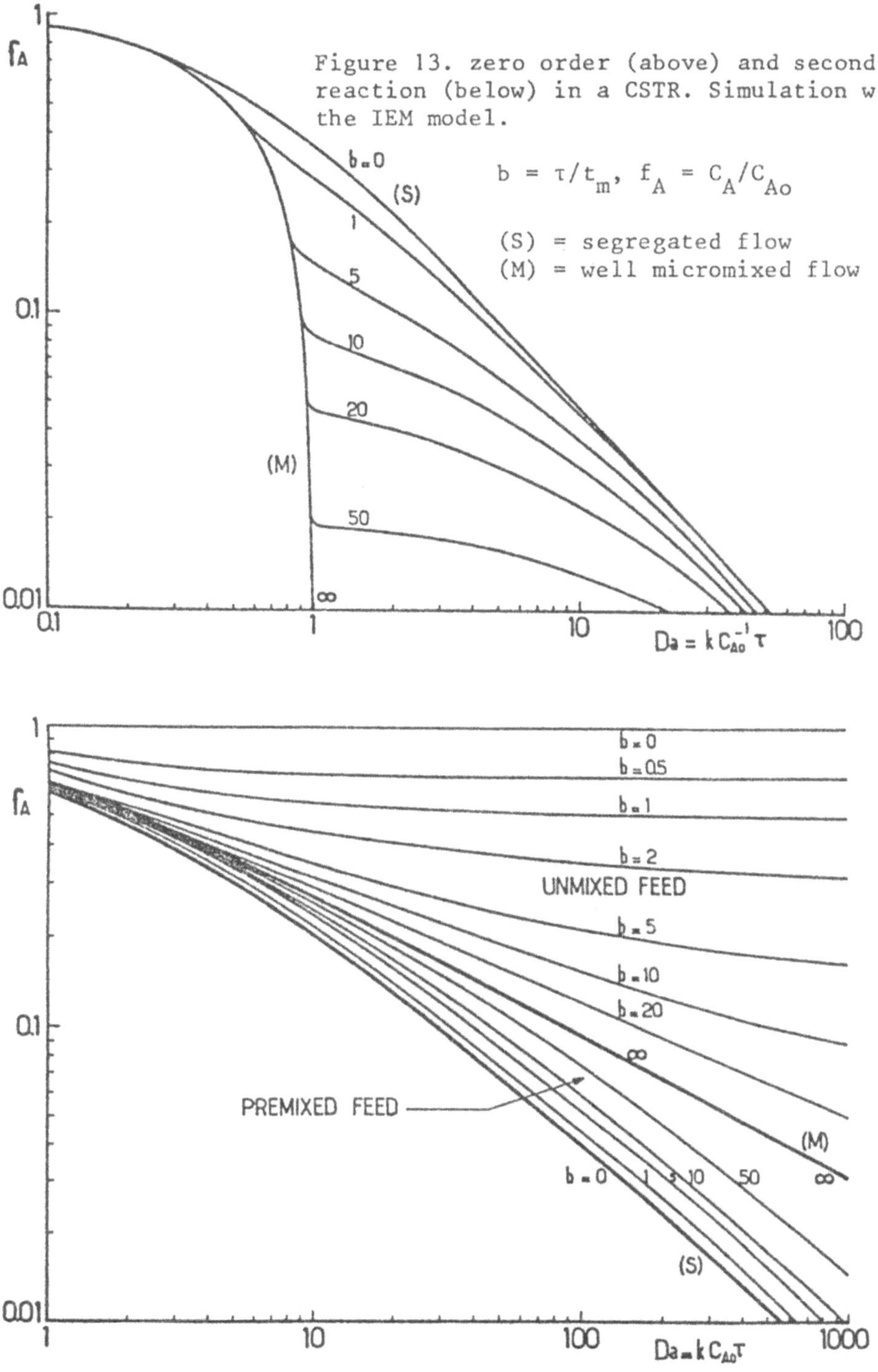

Figure 13. zero order (above) and second order reaction (below) in a CSTR. Simulation with the IEM model.

$b = \tau/t_m$, $f_A = C_A/C_{Ao}$

(S) = segregated flow
(M) = well micromixed flow

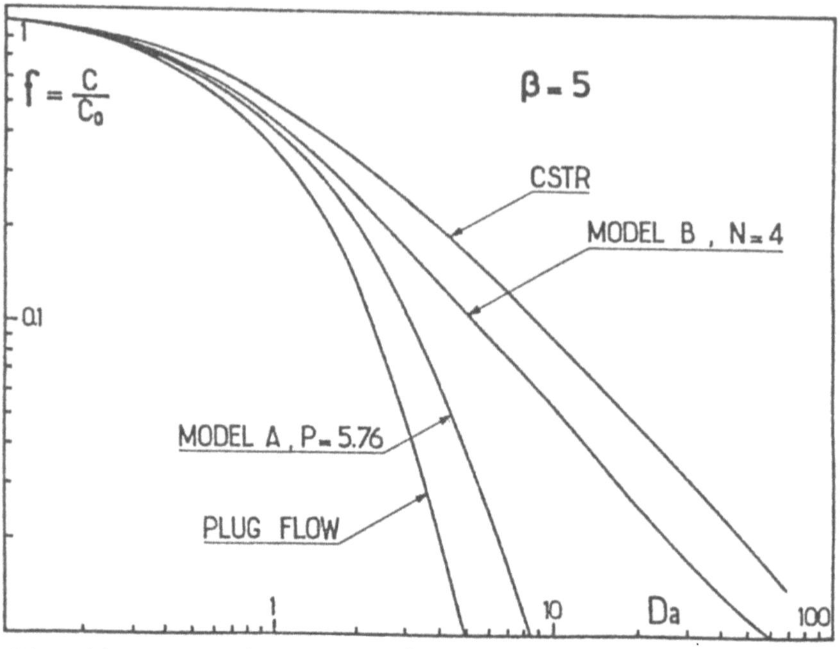

Fig. 14. Conversion for a first order reaction with the May-Van Deemter model (fig. 4).

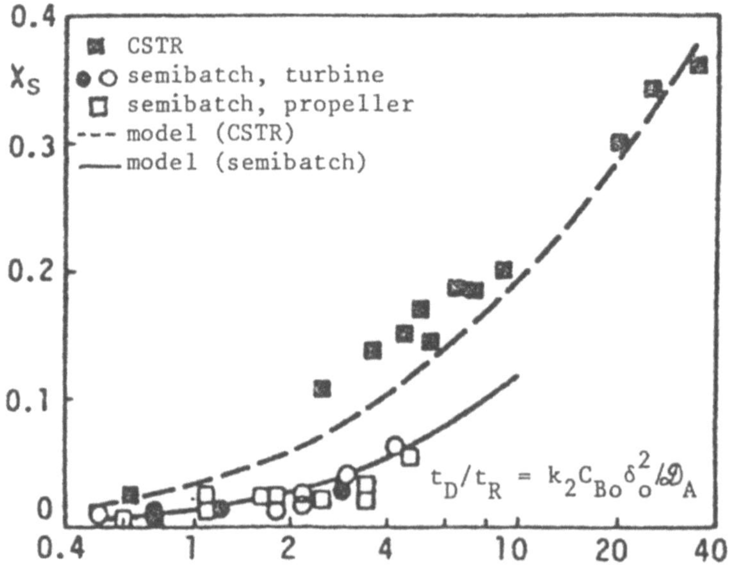

Fig. 15. Yield of S in a consecutive-competing reaction system $A + B \rightarrow R$, $R + B \rightarrow S$
After ref. [31]

MODELLING AND SIMULATION OF TWO-PHASE FLUIDIZED BED REACTORS

John R. Grace

Department of Chemical Engineering
University of British Columbia
Vancouver, Canada V6T 1W5

1 INTRODUCTION

Fluidized bed reactors have unique qualities, notably temperature uniformity, favourable heat transfer and solids mobility, which make them attractive for many catalytic and non-catalytic reactions involving solid particles and a gas or liquid. However, gas fluidized bed reactors present a challenge to the engineer because their patterns of gas-solid contacting and mixing are complex and difficult to characterize. Contacting and mixing are intimately related to the hydrodynamics in the bed itself. Before discussing reactor models and simulation methods, it is essential to provide some basic understanding of some key aspects of the physical behaviour of gas-solid fluidized beds. We begin with a description of the regimes of hydrodynamic behaviour, follow with a general treatment of catalytic reaction in a two-phase, two component reactor, and then consider specific regimes and zones of fluidized beds. Models for gas-solid reactions are treated separately. A brief survey is given of some advanced models which do not fall neatly within the usual range of assumptions or conditions. Emphasis throughout is not only on what is known, but also on challenges and new developments, including the high-velocity fluidization regimes which are of increasing importance in industrial practice.

2 HYDRODYNAMIC REGIMES

2.1 Description

Gas fluidized beds exhibit a number of distinct hydrodynamic regimes which are broadly analogous to those observed with gas liquid systems. The sequence of regimes which may be encountered when increasing the gas flow through a bed of particulate solids is shown schematically in Figure 1. For the purposes of this chapter, we are interested in only four of these regimes (bubbling through fast fluidization). Key features of these regimes, relevant to reactor performance, are as follows:

2.1.1 Bubbling regime. Voids, commonly referred to as bubbles, form at or not far above the gas distributor and rise towards the surface. En route to the surface, they coalesce forming larger bubbles, which may then split and recombine. Bubbles are similar in appearance and behaviour to spherical-cap bubbles in liquids. Their shapes become distorted during coalescence and splitting. The voids are larger as the gas flow rate is increased. The bubbles contain few particles, but their wakes and induced drift cause vigorous vertical mixing and some lateral displacement of the solid particles in the bed.

2.1.2. Slugging regime. If the bubbles are allowed to grow to a sufficient size that they fill most of the column cross-sectional area, the bed exhibits slug flow. The voids are bullet shaped "slugs" which rise in sequence, one after the other. Some coalescence occurs as leading slugs are overtaken from the rear until a relatively stable interslug spacing of about 2 to 5D is achieved. The upper surface of the bed rises steadily, as each slug approaches, then collapses precipitously as the slug breaks through. The slugs are nearly devoid of particles. Wakes are relatively small so that vertical mixing of solids is less effective than in the bubbling regime.

2.1.3. Turbulent fluidization regime. Increasing the superficial gas velocity beyond a certain point, typically of the order of 0.4 to 4 m/s depending on particle size and density, results in a transitional regime which has come to be called "turbulent fluidization". The appearance is chaotic, with small voids appearing, darting obliquely upwards and disappearing.

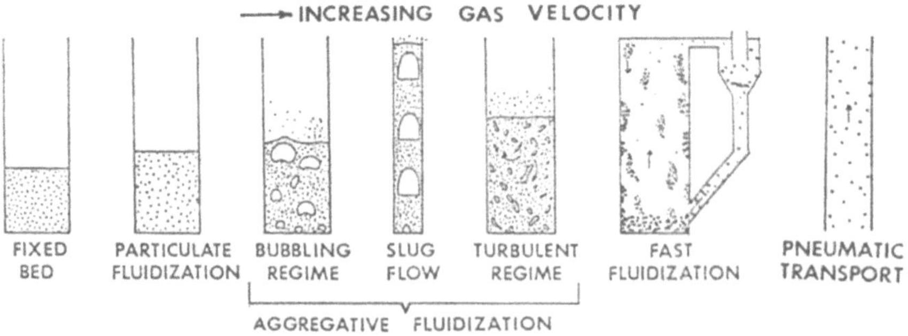

Figure 1: Schematic diagrams showing principal flow regimes when gas flows upwards through a bed of solid particles. (2)

Particle mixing appears to be rapid in the horizontal as well as the vertical direction. The upper surface of the bed is difficult to determine visually, because of the vigorous splashing of particles into the freeboard, but X-ray measurements and pressure profiles indicate that there is a distinct upper surface.

2.1.4 Fast fluidization regime. Further increases in superficial gas velocity to values of the order of 2-5 m/s leads to substantial entrainment of solid particles. However, providing that the solids at the bottom of the vessel are continuously replenished, e.g. by recirculation from a cyclone to form what is termed a "circulating bed" as indicated schematically in Figure 1, a relatively dense suspension can be maintained in the reactor chamber. In this case there is no distinct upper surface, and therefore no bed proper. Instead, particles are entrained upwards by the gas, principally in the interior of the reactor vessel, while many solids move downwards in sheets or strands along the outer wall of the chamber. The overall volume fraction occupied by particles is typically 2 to 20%, decreasing as the gas velocity is increased for a fixed solids feed or recirculation rate. Solids mixing appears to be rapid in both the vertical and horizontal directions.

2.2 Regime Transitions and Criteria

To be able to model fluidized bed reactors, one must first determine the hydrodynamic regime of operation. For an existing unit, this can usually be done experimentally. A rapid response pressure transducer and measurement of time-mean pressure profiles along the height of the reactor are usually sufficient to make this determination (1). Determining features of the signals are as follows:

Particulate fluidization: Very little pressure fluctuation; pressure drop corresponds to bed weight per unit area.

Bubbling regime: Irregular pressure fluctuations of quite large amplitude and frequency of the order of 0.5 to 5 Hz depending on height.

Slug flow: Large amplitude fluctuations of quite regular frequency, typically 0.2 to 1 Hz. Sound emission and column vibration may also indicate slugging.

Turbulent fluidization: Irregular pressure fluctuations of smaller amplitude and larger frequency, usually > 5 Hz. Time-mean pressure profiles show a distinct change in slope, indicating the upper bed surface.

Fast fluidization regime: Rapid fluctuations. Appreciable pressure gradient over entire column height, the gradient indicating an average suspension density of typically 15-200 kg/m^3.

In the absence of operating data, transition velocities and criteria for the various regimes can be estimated as follows:

2.2.1 Bubbling regime. Each of the following three criteria must be satisfied: a) Superficial gas velocity, $U, > U_{mf}$, where U_{mf}, the minimum fluidization velocity, can be predicted (2) from:

$$Re_{mf} = \frac{\rho d_p U_{mf}}{\mu} = \sqrt{(27.2)^2 + 0.0408\ Ar} - 27.2 \qquad (1)$$

with $$Ar = \rho(\rho_p - \rho) g d_p{}^3/\mu^2 \qquad (2)$$

Here d_p is the surface-to-volume mean particle diameter estimated from

$$d_p = 1/\Sigma(x_i/d_{pi}) \qquad (3)$$

b) $U > U_{mb}$ where U_{mb}, the minimum bubbling velocity, is estimated (3) as the greater of U_{mf} or

$$U_{mb} = 2.07\ d_p \rho^{0.06} \mu^{-0.347} \exp\{0.716 m_f\} \quad \text{(SI units)} \qquad (4)$$

where m_f is the fraction of fines, defined as the mass fraction of powder smaller than 45 µm.
c) $U < U_{ms}$ and U_k, defined by Equations (6) and (8) below. If $U_{ms} \leqslant U < U_k$, but the other conditions for slugging enumerated in section 2.2.2 below are not satisfied, the bubbling regime will prevail.

2.2.2 Slug flow regime. All of the following criteria must be satisfied if slug flow is to prevail:
a) $H/D \gtrsim 1.5$ or 2, i.e. the bed must be deep enough for the slugs to form. For a multi-orifice distributor plate having N_{or} holes, a more precise criterion (4) is

$$H \geqslant 3.5D\ (1 - N_{or}{}^{-0.5}) \qquad (5)$$

b) $U \geqslant U_{ms}$, where U_{ms}, the minimum slugging velocity, can be estimated (5) from

$$U_{ms} = U_{mf} + 0.07\ \sqrt{gD} \qquad (6)$$

c) The "maximum stable bubble size",

$$d_{bmax} = 2.0\ (v_T{}^*)^2/g \qquad (7)$$

must be of order D or greater, so that bubbles can grow to fill the column cross-section. In this expression, $v_T{}^*$ is the terminal settling velocity of a spherical particle of diameter $2.7d_p$ in a stagnant gas having the physical properties of the fluidizing gas.
d) $U < U_k$, defined below.

2.2.3 Turbulent regime. The superficial gas velocity must fall in the range $U_k \leqslant U < U_{tr}$. The superficial velocity for the onset of turbulent fluidization has been correlated (2) for the limited range of data available by

$$U_k = 7.0\ \sqrt{\rho_p d_p} - 0.77 \qquad \text{(SI units)} \qquad (8)$$

There is some evidence (6) that U_k decreases as the bed diameter increases. Additional data are required to permit reliable prediction of U_k.

2.2.4 Fast fluidization. To operate in the fast fluidization regime requires both that U exceed the transport velocity, U_{tr}, and that solid particles be fed to the base of the unit with a sufficient flow rate, typically 20-200 kg/m^2-s, that a relatively dense suspension can be maintained in the reactor. Typical reported values (7) of U_{tr} are 1.5 m/s for 49 μm silica alumina cracking catalyst (ρ_p= 1070 kg/m^3) and 3.8 m/s for hydrated alumina particles (ρ_p = 2460 kg/m^3). There are insufficient data to allow U_{tr} to be correlated in a general way.

2.3 Regime Map

For the idealized case of identical spherical particles of diameter d_p in a column of very large D, the bed voidage, ε, is a function of the variables:

$$\varepsilon = f(d_p, \rho, g\Delta\rho, \mu, U) \qquad (9)$$

if interparticle forces can be ignored and $\Delta\rho = \rho_p - \rho$. In dimensionless form, Equation (9) can be rewritten as

$$F(Ar^{1/3}, U^*, \varepsilon) = 0 \qquad (10)$$

where $Ar^{1/3}$, the Archimedes number defined by Equation (2) raised to the power of 1/3, can be regarded as a dimensionless particle diameter, and

$$U^* = U\{\rho^2/\Delta\rho g \mu\}^{1/3} \qquad (11)$$

is a dimensionless superficial gas velocity, independent of d_p. A plot of U* versus $Ar^{1/3}$ appears in Figure 2. The curve for the minimum fluidization condition, derived from Equation (1), separates the fixed bed regime from the fluidization regimes. The curve for the terminal settling velocity of single particles, derived from the standard drag curve of Clift et al.(8), is also plotted. Regions where commercial moving beds, circulating beds, transport reactors and spouted beds operate are also indicated, largely following the boundaries given in a somewhat similar regime diagram by Reh (9). Points are shown indicating minimum

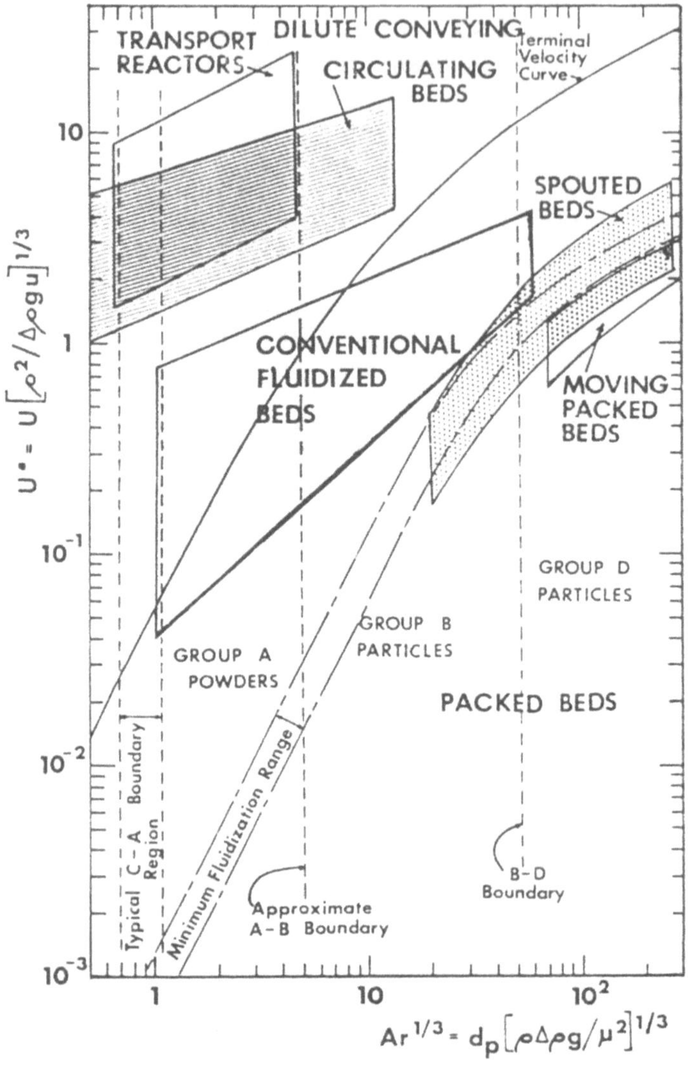

Figure 2: Dimensionless regime diagram for vertical flow of gas through beds of solid particles.

bubbling velocities for Group B powders in the Geldart classification (10). Since interparticle forces and fines content influence U_{mb} in real systems, there is no single curve for minimum bubbling. Approximate boundaries are indicated delineating the Geldart group C (cohesive, fine powders) and Group D (coarse particle) regions. Some data points are also plotted for transitions to turbulent and fast fluidization. Since the column diameter and bed height are not among the variables included in the dimensionless groups, Figure 2 is unable to show the boundary between slugging and non-slugging conditions. However, the diagram gives a convenient method for estimating other regimes of operation.

3 GENERAL TWO-PHASE MODELS

In each of the regimes treated in sections 2.1.1 to 2.1.4, the fluidized bed can be considered to be composed of a pair of distinct "phases" in contact with each other as indicated in Table 1. Each of the phases may contain both gas and solids, but one of these phases, denoted phase 1, has a lower volumetric concentration of solid particles than the other. The grid zone, i.e. the region immediately above the gas distributor, regardless of the hydrodynamic regime prevailing higher in the reactor, may likewise be considered to be composed of two separate phases or regions as indicated in Table 1. Similarly, it is possible to treat the freeboard above fluidized beds operated in the bubbling, slug flow or turbulent fluidization regimes as a combination of two phases.

It is useful to develop, at this stage, a fairly general model which can be applied to each of the two-phase situations listed in Table 1. It is important to recognize that the two "phases" referred to here are not synonymous with components (gas and solid): at any instant phases denote clearly identifiable macroscopic regions of greater or lower solids concentration, not the solid or gas constituents of the mixture. The general model is indicated schematically, in Figure 3 by two parallel chambers

Table 1 - Phases corresponding to regimes of fluidization under consideration

Regime or Zone	1. Dilute Phase	2. Aggregated Phase
Bubbling	Bubbles or voids*	Particles and their interstitial gas surrounding the bubbles.
Slug flow	Slugs (bullet-shaped voids)	Particles and their interstitial gas around and between the slugs.
Turbulent fluidization	Intermittent voids	Surrounding emulsion.
Fast fluidization	Gas containing some individually entrained particles	Sheets, clusters or strands of particles and the gas within them.
Grid region	Grid jets or elongated forming bubbles	Surrounding dense phase.
Freeboard region	Gas containing some individual particles moving upwards or downwards	Packets of ejected solids or clusters and the gas within them.

*Cloud regions are sometimes included with the bubble phase.

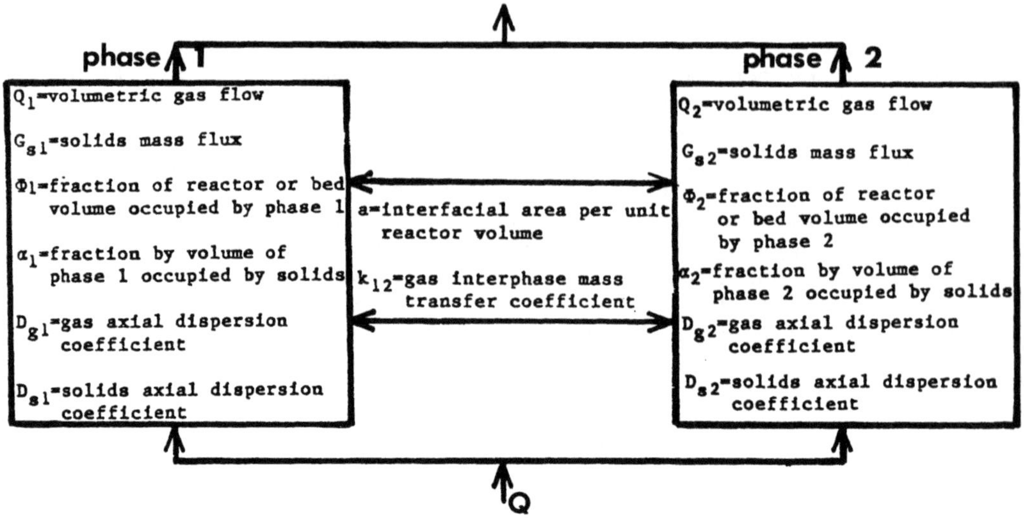

Figure 3: Schematic representation of general two-phase two-component one-dimensional model, with physical variables which must be specified.

representing the two phases or regions, each assumed to have one-dimensional flow of gas and solids through it. There is exchange of gas between the two phases. (Exchange of solids also takes place but may usually be ignored because the differences in solids composition between the phases tend to be small.)

Because of the observed isothermality of gas fluidized beds, we begin by ignoring temperature gradients. This assumption may have to be discarded in some cases, especially for the freeboard and grid regions. For simplicity, we ignore gas volume changes due to reaction at this stage. Physical variables which must be specified for a given model are listed in Figure 3. These variables may be assumed to be invariant with distance in the axial direction, or they may be allowed to vary along the length of the reactor in accordance with hydrodynamic findings. One of the gas flows, Q_1 or Q_2, and either or both of the solids mass fluxes, G_{s1} and G_{s2}, may be negative (downwards). Note that the word "phase" may be replaced by "region" for cases (e.g. grid zone if there are permanent jets) where the dilute and aggregated phases are fixed in space.*

Some of the variables in Figure 3 are constrained. Thus

$$Q_1 + Q_2 = Q = UA \tag{12}$$

*This terminology is, for example, more common and useful for models of spouted bed reactors.

$$G_{s1} + G_{s2} = G_s \tag{13}$$

$$\Phi_1 + \Phi_2 = 1 \tag{14}$$

Hydrodynamic relationships may be used to specify some of the variables. The terminology and methodology of the two-phase flow literature (11) may be used for this purpose. For example, absolute velocities for the gas and solids in phase i can be written

$$u_i = Q_i/[\Phi_i(1-\alpha_i)A]; \quad v_i = G_{si}/[\rho_p \Phi_i \alpha_i] \tag{15}$$

and the relative gas-to-solid velocity in phase i is then simply $u_i - v_i$.

In practice, there are a number of ways in which backmixing of gas and solids may be accounted for in models of the general type represented in Figure 3. These range from plug flow through axial dispersion or stirred tanks in series to perfect mixing. Plug flow with axial dispersion has been adopted here, and it may be extended to the plug flow case by simply setting the relevant axial dispersion coefficient equal to 0. The general model can be derived by writing mole or mass balances. The resulting equation, applicable to each phase, for reaction of a gaseous species is

$$\Phi_i(1-\alpha_i)\frac{\partial c_i}{\partial t} + \frac{Q_i}{A}\frac{\partial c_i}{\partial x} - \Phi_i D_{gi}\frac{\partial^2 c_i}{\partial x^2} + k_{ij}a\,(c_i - c_j) + k_r \Phi_i \alpha_i c_i^{\,n} = 0 \quad (i \neq j) \tag{16}$$

where the five terms account respectively for accumulation, convective or bulk flow, axial dispersion, interphase mass transfer and chemical reaction. The reaction has been assumed to be of order n in the concentration of the reacting gaseous species, and the effective kinetic rate constant, k_r, expressed on a particle volume basis, may depend on the conversion of solid species.

Boundary conditions for the above equation, applied to both phases, depend on the specific details of the model chosen. For the simplest case of plug flow in each phase ($D_{g1} = D_{g2} = 0$) and upflow in each phase, it is sufficient to set $c_1 = c_2 = c_0$ at $x=0$, where c_0 is the initial concentration. When downflow occurs in one of the phases, a split boundary condition problem arises. The Danckwerts boundary condition should be applied where there is appreciable axial dispersion. For transient problems, initial conditions must also be specified. Equation (16) can be applied to gas mixing experiments without reaction by dropping the final term. Equations analogous to Equation (16) can be developed to describe reactions of solid species.

4. BUBBLING REGIME MODELS

4.1 Representative Models

The bubbling regime has been studied more extensively than any other regime. A large number of models have been devised to describe catalytic reactors operating in the bubbling regime. A number of reviews of these models have appeared (e.g. 12-15). Many of the models are of the general form described in the previous section. The dilute phase may describe the bubbles alone, with dispersed solids ignored, or they may include the "cloud" regions, a mantle of solids and interstitial gas through which bubble gas circulates and recirculates when the bubble velocity exceeds the interstitial gas velocity. In almost every case, the models have been developed and tested only under steady state conditions.

For catalytic reactor models, it is not necessary to specify solids mass fluxes and solids axial dispersion coefficients. To illustrate the genesis of models for solid catalysed reactions,

Table 2 - Specification of parameters in general model, as outlined in Section 3 and Table 1, for representative models which apply to the bubbling regime.

Parameter	May model (16)	Orcutt (17,18)	Grace (19)
Q_1	$(U-U_{mf})A$	$(U-U_{mf})A$	UA
Q_2	$U_{mf}A$	$U_{mf}A$	0
Φ_1	ε_b	ε_b	ε_b
Φ_2	$1-\varepsilon_b$	$1-\varepsilon_b$	$1-\varepsilon_b$
α_1	0	0	0.001-0.01
α_2	$1-\varepsilon_{mf}$	$1-\varepsilon_{mf}$	$1-\varepsilon_{mf}$
D_{g1}	0	0	0
D_{g2}	fitted*	0 or ∞	0
a	fitted }*	$6\varepsilon_b/d_b$	$6\varepsilon_b/d_b$
k_{12}	fitted }	$3/4\ U_{mf}+k_D$ **	Equation 18
Boundary conditions:	$c_1=c_o$ @ $x=0$, $c_2 = \frac{D_{g2}\Phi_2 A}{Q_2}\frac{\partial c_2}{\partial x}$ @ $x=0$, $\partial c_2/\partial x = 0$ @ $x=H$	$c_1=c_o$ @ $x=0$, $c_2=c_o$ @ $x=0$ ǂ	$c_1=c_o$ @ $x=0$, $c_2=c_o$ @ $x=0$
Exit Concentration:	$c_H = \frac{Q_1c_1+Q_2c_2}{Q}$	$c_H = \frac{Q_1c_1+Q_2c_2}{Q}$	$c_H = c_1$

* Correlations for these parameters have been proposed by Mireur and Bischoff (22).

ǂ required only for the case where D_{g2} is taken as 0.

** $k_D = 0.975\ (gD_m^2/d_b)^{0.25}$ = diffusional mass transfer.

Table 2 shows how the other parameters and boundary conditions are assigned for four representative models, the May model (16), the two Orcutt models (17,18) (differing only in the assumed dense phase gas mixing pattern), and the two-phase bubbling bed model proposed by Grace (19) as an improved simplification of the three-phase bubbling bed model of Kunii and Levenspiel (20) and as an approximation to the countercurrent backmixing model of Fryer and Potter (21). All of these authors have given solutions for their models for the case of first order irreversible sections. In addition, Grace (15) has given analytic solutions to the Orcutt model with perfect mixing of gas in the dense phase and to his own model for some cases which involve non-first-order, reversible and consecutive reaction.

4.2 Key Parameters and Their Choice

While there is near unanimity on some model parameters, e.g. in setting $D_{g1}=0$, other model parameters have been chosen differently in the various models. Some of the key areas of controversy are as follows:

4.2.1 Choice of phases. While most models specify that the dilute phase contains only bubbles, hence setting $\Phi_1=\varepsilon_b$, a number of models (e.g. 23) include clouds with the bubble phase or constitute them as a third and separate phase (e.g. 20). As discussed in detail elsewhere (15,24), the rate-limiting step for mass transfer between bubbles and remote dense phase resides, under most conditions of practical importance, at the bubble boundaries, not at the cloud surface. Hence the choice of the bubbles (excluding clouds) as the dilute phase is preferred.

4.2.2 Particles in the dilute phase. In practice, some particles rain through bubbles as they rise and coalesce. While the fraction of the bubble volume occupied by particles is small (typically only 0.1 to 1%), even such a small fraction can have an appreciable influence on the reactor performance for fast reactions and slow interphase transfer. For such cases α_1 should be assigned a non-zero value. For slow reactions or fast interphase transfer, it is acceptable to adopt the approximation $\alpha_1=0$.

4.2.3 Division of gas flow. There has been considerable controversy regarding the amount of gas which flows through each phase in a bubbling fluidized bed. The amount of gas carried by translation of bubble voids can be written as

$$Q_b = Y(U-U_{mf})A \tag{17}$$

The so-called "two-phase theory" of Toomey and Johnstone (25), in which Y=1, has been widely adopted, but there is now ample evidence (26) that this overpredicts Q_b. More accurate predictions are obtained if one chooses $Y \approx 0.8$ for Geldart's group A powders, $Y \approx 0.7$ for group B materials, and $Y \approx 0.3$ for group D particles. Note, however, that while Q_b is needed in specifying the bed hydrodynamics, it constitutes only a part of Q_1, the total flow through the dilute phase. The flow is augmented by throughflow of gas in which gas short-circuits through each bubble due to the pressure gradient and the high permeability of the void regions. The extra flow due to throughflow is of order $U_{mf}\varepsilon_b A$ for small $(U-U_{mf})$, but there is evidence that this value may be exceeded considerably at higher flow rates, especially when bubbles become closely packed (e.g. as $\varepsilon_b \rightarrow 0.4$)(27). Hence the approximations shown in Table 2 for Q_1 may be satisfactory for many practical cases.

4.2.4 Interphase mass transfer. Of all the aspects of the model which influence predictions and the accuracy of simulations, interphase mass transfer is the most important. While there have been about a dozen studies of transfer for single bubbles, there are very few in which bubbles have been allowed to interact and coalesce. Interactions have been shown to lead to enhancement in the rate of interphase transfer. Based on their own results and those of other workers, Sit and Grace (28) proposed the following expression for k_{12} for freely bubbling three-dimensional beds:

$$k_{12} = \frac{U_{mf}}{3} + 2\left[\frac{D_m \varepsilon_{mf} u_b}{\pi d_b}\right]^{0.5} \tag{18}$$

where the first term accounts for throughflow and is dominant for large particles while the second term accounts for molecular diffusion effects. Some augmentation in interphase transfer can occur when the gas species being transferred adsorbs strongly on the surface of the particles (29).

4.2.5 Dense phase gas mixing: No fewer than eight different assumptions have been adopted in different models to describe the axial mixing of gas in the dense phase (15). Assumptions range from plug flow in an upwards direction through perfect mixing to downward flow. There is evidence (20) that gas can be dragged downwards by dense phase particles which descend to replace solid material which is transported upwards by bubble wakes and drift, providing that $U/U_{mf} > 6$ to 11. Hence models which assume downflow of gas in the dense phase are appropriate under some conditions. For many cases, the predicted conversions are rather insensitive to the gas mixing pattern assumed (30,31). For this reason, as well as for simplicity and because it represents an intermediate case between upflow and downflow of gas in the dense phase, the adoption of no bulk flow coupled with no axial dispersion in the dense phase is attractive and has been adopted in some models (19,20). The predictions are generally more sensitive to the assumed mixing pattern for high conversions and when predicting selectivity rather than conversions (15,30).

4.2.6 Bubble size predictions. There are many equations available for predicting bubble sizes in fluidized beds. Those which have mechanistic underpinnings (4,32) are preferred to those which are solely empirical. However, the bubble size should not be allowed to exceed the "maximum stable size" predicted from Equation (7). For bundles of horizontal tubes, d_b will also not exceed 1 to 1.5 times the tube-to-tube spacing.

In some reactor models (e.g. 33,34) the increase in bubble diameter as a function of height is simulated in the model. The additional work required to model this variation in the bubble size is most often unwarranted. In any case, variations in bubble sizes with time and radial position at a given height may be as important, and no effort has been made to incorporate these bubble size variations in reactor models. When a single bubble size is adopted to represent the entire bed, calculations (35) indicate that this should be the bubble size at x=0.4H.

4.2.7 Bubble volume fraction. Once the bubble diameter has been predicted, a representative bubble velocity can be estimated from the equation

$$u_b = 0.71\ (gd_b)^{0.5} + (U-U_{mf}) \qquad (19)$$

The bubble volume fraction, ε_b, can then be evaluated as

$$\varepsilon_b = Q_b/(u_b A) \qquad (20)$$

where Q_b/A is estimated from Equation (17) and the expanded bed height is given by

$$H = H_{mf}/(1-\varepsilon_b) \qquad (21)$$

An iterative procedure is required (2) since H is needed so that the representative height (H/2) can be chosen for evaluation of d_b and u_b. This procedure converges quickly. Values of ε_b in excess of about 0.4-0.5 are unrealistic, and probably indicate transition to the turbulent fluidization regime.

4.3 Typical Conversion Results

For an irreversible first order reaction, the two-phase bubbling bed model (15,19) gives the following exit concentration under steady state conditions:

$$\frac{c_H}{c_o} = \exp\left[\frac{-k_1^*[X(\alpha_1\varepsilon_b + (1-\varepsilon_b)(1-\varepsilon_{mf})) + k_1^*\alpha_1\varepsilon_b(1-\varepsilon_b)(1-\varepsilon_{mf})]}{X + k_1^*(1-\varepsilon_b)(1-\varepsilon_{mf})}\right] \quad (22)$$

where $k_1^* = k_r H/U$ (23)

and $X = k_{12} a H/U$ (24)

are a dimensionless kinetic rate constant and a dimensionless interphase mass transfer group (or number of transfer units), respectively. Values of c_H/c_o predicted by the model for $(1-\varepsilon_b)(1-\varepsilon_{mf}) = 0.4$ and $\alpha_1\varepsilon_b = 0$ and 0.005 appear in Figure 4. The curves show clearly the influences of the dimensionless rate constant, mass transfer group and fraction of dilute phase occupied by solids. For slow reactions ($k_1^* \leqslant 0.3$), conversion is insensitive to interphase transfer and to whether or not solids are included with the dilute phase. For intermediate reactions ($0.3 < k_1^* < 10$), the interphase mass transfer group, X, exerts a very significant influence. For fast reactions ($k_1^* > 10$), the group $\alpha_1\varepsilon_b$, giving the fraction of the bed volume occupied by dilute phase solids, also plays a considerable role. In the limit of $X \to \infty$, i.e. for very rapid mass transfer between the phases, the outlet concentration approaches

$$c_H/c_o = \exp\{-N_R\} \quad (25)$$

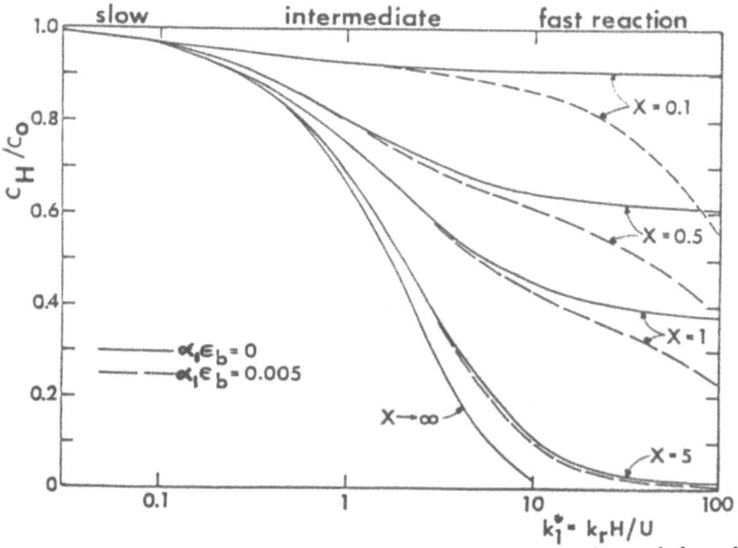

Figure 4: Dimensionless exit concentration predicted by the two-phase bubbling bed model (19) for a first order irreversible gas phase reaction with $(1-\varepsilon_b)(1-\varepsilon_{mf}) = 0.4$ and two different values of $\alpha_1\varepsilon_b$: 0 and 0.005.

where $N_R = k_r H_{mf}(1-\varepsilon_{mf})/U = k_r(\text{Volume of particles})/Q$ (26)

which is the result for a single phase plug flow reactor.

The influence of the order of reaction, as well as the strong role of interphase mass transfer, are shown in Figure 5. Here the conversion is plotted as a function of what it would be in a corresponding plug flow, single phase reactor, where there are no impediments to gas-solid contacting. For simplicity, α_1 has been taken as 0 and $(1-\varepsilon_b)(1-\varepsilon_{mf})$ has been assigned a value of 0.4 as before. Note that the reduction in conversion due to segregation of gas into two separate phases is more significant the higher the order of the reaction.

4.4 Selectivity

Many catalytic reactions which have been attempted in fluidized beds (e.g. manufacture of phthalic and maleic anhydride and the production of acrylonitrile) involve reactions of the type A → B → C, with B being the desired product. Expressions for the concentration of B in the case where both reactions are first order and irreversible have been given by Grace (15,19) for the Orcutt model (17,18) (see Table 2) in which the gas is assumed to be well mixed in the dense phase and for the two-phase bubbling bed model featured in section 4.3. Predictions for the latter model appear in Figure 6 for α_1=0 and for representative values of $(1-\varepsilon_b)(1-\varepsilon_{mf})$ and X. The selectivity for any finite

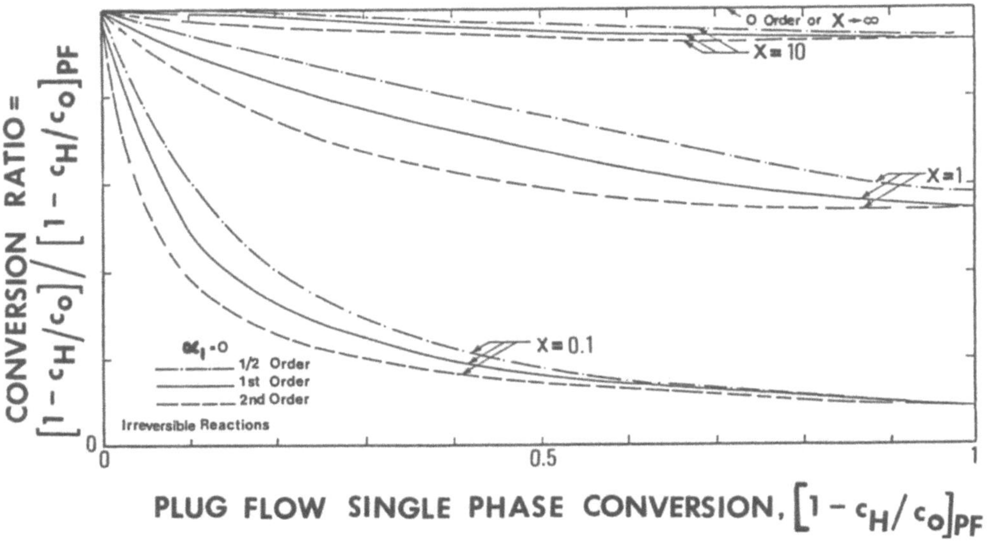

Figure 5: Conversion from the two-phase bubbling bed model (19) relative to corresponding single phase plug flow reactors for irreversible gas-phase reactions of different orders.

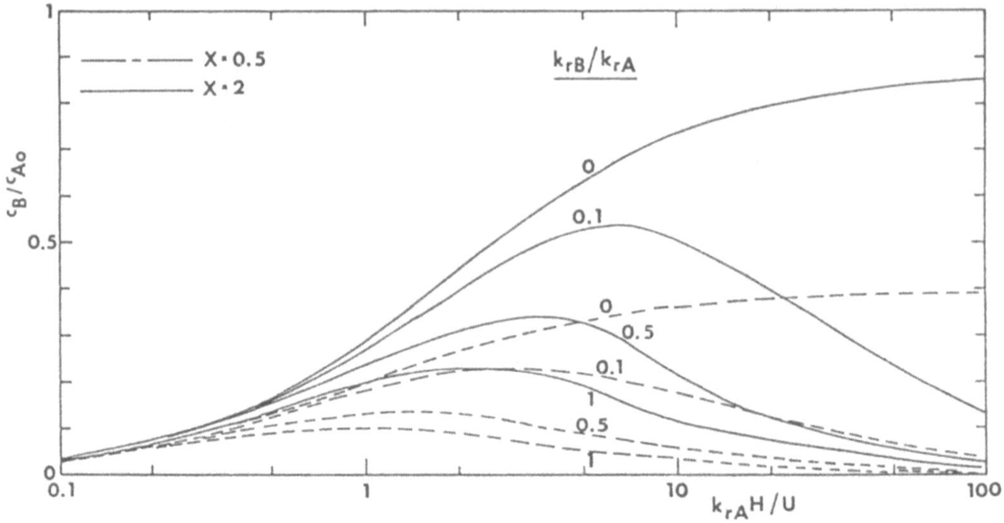

Figure 6: Yield of desired product B calculated from the two-phase bubbling bed model for irreversible consecutive reactions $A \rightarrow B \rightarrow C$; $\alpha_1 = 0$ and $(1-\varepsilon_b)(1-\varepsilon_{mf}) = 0.4$; k_{rA} and k_{rB} are the first order rate constants for disappearance of A and B respectively; k_{12} is assumed to be the same for both A and B. Corresponding C_{AH} values appear in Figure 4.

value of X will always be less favourable than in a corresponding single phase reactor giving the same overall conversion of A. This arises because most of the reaction occurs in the dense phase where the local conversion is high, whereas the overall conversion is determined by the concentration in the bubble phase where the greater part or all of the flow occurs. Selectivities are generally more sensitive to the model adopted and to the interphase transfer than are conversions (30,36,37).

4.5 Experimental Evaluation of Models

In the early days of fluidized bed reactor modelling, experimental testing of models consisted of measuring only overall conversions over severely limited ranges of such variables as particle size, temperature, bed depth, superficial gas velocity and catalyst activity. Since most models had at least one parameter which could be fitted, and since much of the work was for low conversions where predictions are insensitive to the model adopted because of control by kinetic rather than hydrodynamic factors, it was claimed that each model was successful. In the past decade, there has been considerable effort to discriminate between models based on more extensive measurements than conversion alone, such as concentration

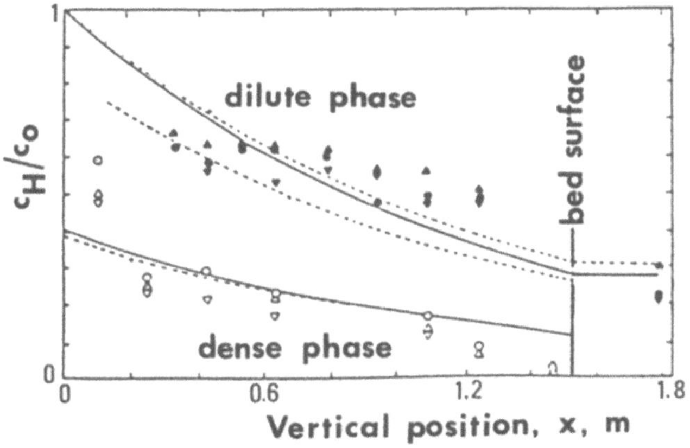

Figure 7: Comparison between experimental and predicted concentration profiles (38) for U = 0.165 m/s, k = 0.2 s^{-1}. Open and blackened symbols are measured concentrations in the dense and bubble phases respectively. Dashed lines are predicted bubble, cloud and emulsion concentrations from the Kunii and Levenspiel (20) model. Solid lines are the predicted bubble and dense phase concentrations from the two-phase bubbling bed model (19).
Reprinted with permission from Ind.Eng. Chem Fundamentals, 14, 75-91(1975). Copyright 1975 American Chemical Society.

profiles in the bed coupled with hydrodynamic measurements (38,39), selectivity determinations coupled with careful kinetic characterization (36,37), or unsteady state tracer measurements in columns of different scale (40). A comparison between experimental concentration profiles determined by Chavarie and Grace (38) and the three-phase (i.e. bubble, cloud and emulsion) (20) and two phase (19) bubbling bed models appears in Figure 7. These two models gave the best agreement among models tested.

Generally speaking, it is fair to say that no single reactor model has proven itself to be applicable in all situations. This arises in part because of the uncertainty in predicting and measuring hydrodynamic variables which influence the reactor performance. Models are generally more successful in simulating existing reactors over a limited range of conditions than in predicting the performance of new fluidized bed reactors. Nevertheless, the simple models discussed above are very helpful in understanding the influence of key variables such as bubble size, particle size and reactor scale.

4.6 Reactor Scale-up

From the earliest days when unsuccessful attempts were made to scale up the Fischer-Tropsch reaction (38), it has been recognized that conversion decreases strongly as many fluidized bed processes are scaled up. The effect of scale on performance according to the two-phase bubbling bed model (19) and the Orcutt model (17,18) with perfect mixing in the dense phase are shown in

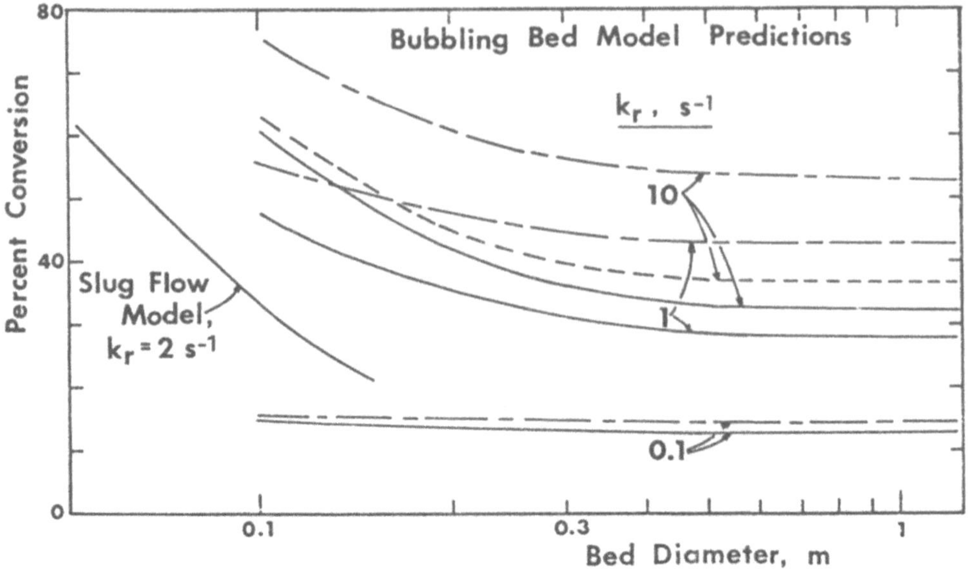

Figure 8: Conversion as a function of reactor diameter. Bubbling bed models: Solid lines: two-phase bubbling bed model (19) with α_1=0; short dashes: same model with α_1=0.005; long-short broken lines: Orcutt model with mixed dense phase. Slug flow model: Hovmand-Davidson (45,46). For conditions see Table 3.

Table 3 - Hydrodynamic variables applicable to Figure 8

	Solid & broken lines	Chain-dotted lines
Model	Bubbling bed, Orcutt	Hovmand slugging bed
U_{mf} (m/s)	0.02	0.009
U (m/s)	0.50	0.20
ε_{mf}	0.48	0.50
D_m (m^2/s)	3.3×10^{-5}	2.0×10^{-5}
Bed depth (m)	H=2.5	H_{mf}=1.0
Y	0.8	1.0
d_{bo} (m)	0.01	N.S.

Figure 8 for operating conditions and variables specified in Table 3. For the smallest diameter reactors, a slugging model is used as discussed in Section 5.1 below. It is seen that conversion is sensitive to both the scale of reactor and the model chosen to represent the reactor for relatively fast reactions and to a lesser extent also for intermediate reactions. (The comparison here is for beds of the same expanded bed height.) For a slow reaction where the reaction is kinetically controlled, the scale of the reactor and the model used to represent it have very little influence on the conversion. Figure 8 suggests that the effect of scale can be profound for relatively small beds, but that there is a levelling off once the bed diameter reaches about 0.5 m, so that this size may be appropriate for pilot plant equipment. This suggestion should, however, be treated with some caution since there is a lack of data available on large diameter beds, since the Mori and Wen correlation which has been used to predict bubble sizes is limited to $D \leqslant 1.22$ m, and since the influence of scale depends on the depth of bed.

The deterioration in reactor performance with increasing scale in the models discussed is associated with the fact that bubbles grow larger as the bed diameter increases. This results in reduced interphase mass transfer, with X approximately proportional to $d_b^{-1.5}$ due to the reduced interfacial area per unit volume and reduced bubble volume fraction as bubbles grow. Note that for small particle systems where bubbles quickly achieve a maximum stable size (see Equation 7), and for low H/D ratios where the walls cannot influence bubble sizes, the influence of bed diameter is predicted to be small. Additional factors, not accounted for in these models, which may also contribute to the reduction in conversion with increasing D are as follows: (i) There is some evidence (e.g. 16,22) that the extent of gas axial dispersion in the dense phase increases as the H/D ratio decreases. (ii) To the extent that clouds are important, they become smaller relative to the bubble sizes as bubbles grow larger. (iii) It is more difficult to ensure uniform introduction of gas at the grid as the reactor is scaled up. (iv) Carryover of solids increases as D increases.

4.7 Effect of Immersed Tubes, Baffles and other Surfaces

One way of countering the adverse influence of scaling up, discussed in the previous section, is to add tubes (horizontal or vertical), baffles or mobile bubble breakers which can control the growth of bubbles. There is some experimental evidence that interphase mass transfer and reactor conversion can be increased in this manner (e.g. 42,43). Some experimental results showing the influence of fixed internals appear in Figure 9. It is seen that considerable improvements in conversion can occur for relatively fast reactions, whereas at low k_1* or low N_R values, where the reaction is kinetically controlled, there is little influence of the internals. There is also little influence on conversion if the solids are fine enough that the bubble size is already limited by splitting (44) (see Equation 7). As already mentioned, it is reasonable to assume an upper limit on d_b of the order of the inter-tube (centre-to-centre) spacing for horizontal tube bundles in applying reactor models. For vertical tubes and internal surfaces other than cylindrical tubes, there are few data available to assist in how much smaller the bubbles will be than in the corresponding bed without internal surfaces.

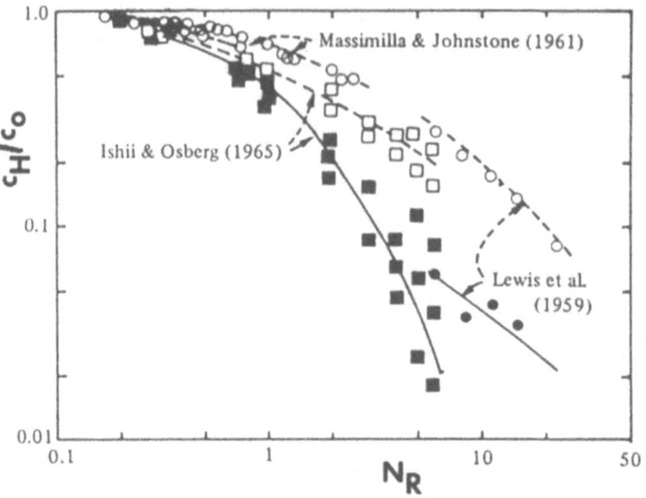

Figure 9: Outlet concentrations for first order reactions with (blacked-in symbols) and without (corresponding open symbols) baffles (42). Reproduced by permission American Institute of Chemical Engineers.

5. MODELS FOR OTHER HYDRODYNAMIC REGIMES

Models similar to those described above for the bubbling regime may also be used to characterize beds operated in the slugging, turbulent and fast fluidization regimes, with appropriate changes in the relationships used to describe the mass transfer and other model parameters. The conditions needed to ensure that the bed is operating in these regimes are presented in Section 2.

5.1 Slug Flow Models

The best known model for slugging fluidized beds is that of Hovmand and Davidson (45,46). This is a variant on the Orcutt model (17,18) (see also Table 2) which assumes plug flow of gas in the dense phase. The interface mass transfer is again composed of two parts, a throughflow component and a diffusional component, with the dimensionless mass transfer coefficient given approximately by

$$X = \frac{H_{mf}A}{u_{s\infty}V_s}\left[U_{mf} + \frac{16\varepsilon_{mf}I}{1+\varepsilon_{mf}}\left(\frac{D_m}{\pi}\right)^{0.5}\left(\frac{g}{D}\right)^{0.25}\right] \tag{27}$$

I is a surface integral with values given in Table 4 as a function of the slug length, h_s, which can be estimated from the quadratic equation:

Table 4 - Values of dimensionless slug surface integral as a function of slug length/bed diameter (46)

h_s/D:	0.30	0.50	1.0	2.0	3.0	4.0	5.0
I :	0.13	0.21	0.39	0.71	0.98	1.24	1.48

$$h_s/D - 0.495\,(1-B)(h_s/D)^{0.5} + 0.061 - 1.939B = 0 \qquad (28)$$

Here $$B = (U-U_{mf})/u_{s\infty} = (U-U_{mf})/[0.35(gD)^{0.5}] \qquad (29)$$

The slug volume V_s can be estimated from the relationship

$$V_s = AD[h_s/D - 0.495\,(h_s/D)^{0.5} + 0.061] \qquad (30)$$

Equation (27) above ignores any interference between the convective and diffusional mass transfer terms. An alternative expression is given (45,46) for the case where there is interaction between the two terms. Equation (28) assumes an interslug spacing of 2D, typical of values observed under most slugging conditions in fluidized beds. The outlet gas concentration for a first order reaction can then be estimated from

$$\frac{c_H}{c_o} = \frac{1}{m_1-m_2}\left[m_1 e^{-m_2 H}\left(1 - \frac{m_2 HU_{mf}}{XU}\right) - m_2 e^{-m_1 H}\left(1 - \frac{m_1 HU_{mf}}{XU}\right)\right] \qquad (31)$$

where m_1 and m_2 are given by

$$m_i = \{X+N_R \pm [(X+N_R)^2 - 4N_R XU_{mf}/U]^{0.5}\}/[2HU_{mf}/U] \qquad (32)$$

and N_R is given by Equation (26). The expanded bed height, H, as slugs break the surface is given approximately by the relationship

$$H = H_{mf}(1 + B) \qquad (33)$$

Comparison between the above model and experimental results in columns of diameter 0.10 m (45) and 0.46 m (47) and with literature data (46) show that there is fair agreement. Agreement is generally improved if a well-mixed entry zone is allowed upstream of the height where slugging begins.

An alternative model for slugging fluidized bed reactors was proposed by Raghuraman and Potter (48). This is an extension of their three-phase (bubbles, cloud-wake phase, emulsion phase) countercurrent backmixing model (21) which accounts for gas downflow in the dense-phase, said to occur for slugging beds when $U > 2.5\ U_{mf}$. Agreement with experimental results is said (48,49) to be somewhat better for this model than for the Hovmand-

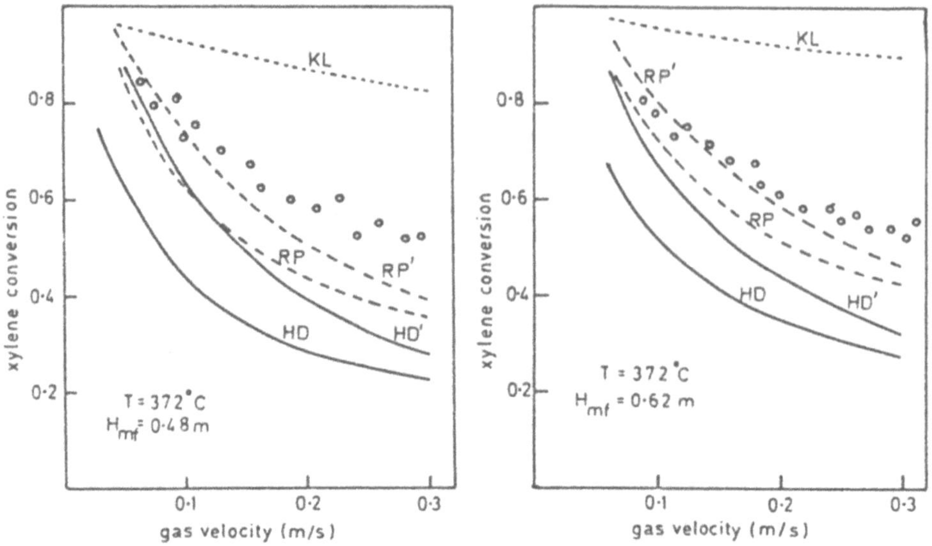

Figure 10: Comparison between experimental data (49) in a 0.10 m reactor under slug flow conditions with reactor models: KL - bubbling bed model (20); HD and HD' - (46) without and with allowance for non-slugging entry region; RP and RP' - (48) without and with allowance for entry zone. U_{mf}=0.02 m/s.

Davidson model. Again, however, it is recommended that separate allowance be made for the bubbling region in the bottom 1.5D height or so of the column before slug flow becomes established. Some comparison between experimental results and models is given in Figure 10. In this comparison, Yates and Grégoire (49) assumed an interslug spacing of 2.8D based on observations with the same particles in a transparent column. The reaction was the oxidation of o-xylene on a vanadium-pentoxide-impregnated catalyst.

Slug flow is most often encountered for laboratory-scale equipment, but it may also occur in vessels 1 m or more in diameter providing that the particles are large enough to allow bubbles to grow to this size and $H \gtrsim 2D$. While slugging is usually thought of as being detrimental to good fluidization, it is important to understand that slug flow may result in fairly high conversions due to retarded void velocities (resulting in increased bed expansions and longer gas residence times in the bed) and to increased interfacial areas due to elongated slug shapes and reduced lateral coalescence. Some predictions from the Hovmand-Davidson (45,46) slug flow model are given in Figure 8 above, demonstrating the enhancement in conversion which occurs as bed diameter decreases in the slugging regime, similar to that for the bubbling bed regime.

5.2 Turbulent Fluidization Regime

The rapid appearance and disappearance of voids in the turbulent fluidization regime (see Section 2.1.3) suggests that there is little or no mass transfer resistance between phases of a turbulent fluidized bed. Hence a single phase model would appear to be appropriate. Fane and Wen (13) propose the use of a single phase plug flow model. For a first order reaction, the outlet concentration is then given by Equation (25). However, experimental data are lacking which would both confirm the assumption of plug flow and confirm the degree of improvement over bubbling beds which this simple model implies. Until such experimental evidence is available, it seems safer to model the turbulent fluidized bed as a single phase CSTR (perfectly mixed stage). For a first order reaction, the outlet concentration is then simply given by

$$c_H/c_o = 1/(1 + N_R) \tag{34}$$

where N_R is given by Equation (26). Alternatively, one can attempt to use gas axial mixing data correlated by Van Deemter (50) or Avidan (51) to estimate effective axial dispersion coefficients which can then be used with a single phase model. Recent experimental measurements (52) have, however, indicated that there are substantial radial gradients in turbulent beds which may necessitate more complex models.

5.3 Fast Fluidized Beds

As outlined in Table 1, a fast fluidized bed can be treated as a dilute upward moving gas-solid suspension in the core combined with a thin layer and strands of denser solids cascading downward along or near the walls. Such "aggregates" are not expected to cause a significant mass transfer resistance between the bulk gas and the solid particle surface (53), especially in view of the stretching and break-up which are observed with high-speed photographs. There are also indications (54) that gas axial dispersion may not be extensive and that near-isothermal conditions prevail. Van Swaaij (55) has presented some calculations in which the fast bed is treated as a single phase plug flow reactor of solids volumetric concentration 18%, with either negligible resistance or a height of mass transfer unit equal to 2 m. Another promising approach would appear to be to use the two-zone (upflow and downflow zones) approach pioneered by Van Deemter (56). Careful experimental data on mixing and reaction are needed to allow models to be tested for this regime.

6 SIMULATION OF ENTRANCE AND FREEBOARD REGIONS

The models considered above make no provision for end effects. There is now ample evidence (e.g. 13,38,57,58) that both the distributor plate zone and freeboard region can play a major role in determining the overall performance of industrial fluid bed reactors.

6.1 Reaction Near the Grid

The first model to consider the grid region as a separate entity (59) adopted the configuration shown schematically in Figure 11a, providing for substantially different interphase mass transfer rates in a lower region corresponding to "grid jets" than in the higher region in which there was a bubbling bed. Other assumptions were that particles were absent from the dilute (jet and bubble) phases and that the entire dense phase acts as a perfect mixer for gas elements. Alternate formulations which remove these latter two constraints have been suggested (13,60) and are indicated schematically in Figure 11b,c and d. For fast reactions, allowance for the grid region in this way can lead to substantial changes in overall conversion, depending largely on the extent of mass transfer between the grid jets and the remaining dense phase.

For grid jets to form at an upward-facing orifice, the ratio d_{or}/d_p must be less than about 25(61). Various expressions for jet penetration have been suggested based on observations in two-

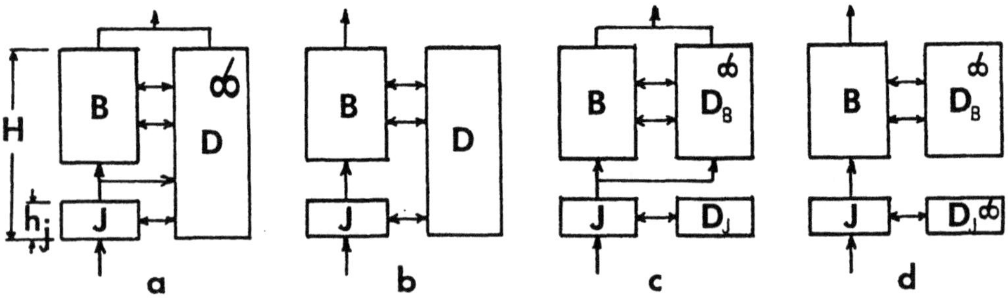

Figure 11: Schematic representation of grid zone models proposed (a) by Behie and Kehoe (59), (b and c) Grace and deLasa (60) and (d) Fane and Wen (13). B,D and J denote bubble phase, dense phase and jets, respectively. Stirrer denotes perfect mixing.

dimensional, three-dimensional and semi-cylindrical columns. The preferred relationship is that given by Merry (62),

$$h_j = 5.2\, d_{or} \left(\frac{\rho d_{or}}{\rho_p d_p}\right)^{0.3} \left[1.3\left(\frac{u_{or}^2}{g\, d_{or}}\right)^{0.2} - 1\right] \quad (35)$$

which implies that jets form only for an orifice velocity, u_{or}, in excess of $0.52\sqrt{gd_{or}}$. If the conditions given above for d_{or}/d_p and u_{or} are not satisfied or if aeration is supplied around the orifice, the gas entering through the orifice will percolate through the particles or form bubbles directly at the gas distributor. When jets do form, they break into voids at the top. The frequency of forming bubbles is generally of the order of 8-16 Hz, and only a part (typically 30-60%) of the gas introduced by the orifice in excess of that required to fluidize the area served by the orifice initially enters the bubble phase.

A simple one-dimensional, steady state model for reaction in the grid region (corresponding to Figure 11b and c) can be derived for permanent jets with the following values in Figure 3: $Q_1=Q=UA$; $Q_2=0$; $\Phi_1=\Phi_j$; $\alpha_1=\alpha_j$; $\alpha_2=1-\varepsilon_{mf}$; $a=a_j$; $D_{g1}=D_{g2}=0$; $k_{12}=k_{jd}$. Here subscript j denotes the jet region. It is assumed that there is plug flow in the jets and stagnant gas in the corresponding dense phase which has a voidage of ε_{mf}. Substitution of these values in Equation (16) for each phase and integration yields the following equation for the concentration at the top of the jets with a first order reaction:

$$\frac{c_{hj}}{c_o} = \exp\left[\frac{-k_j^*[X_j(\Phi_j\alpha_j+(1-\Phi_j)(1-\varepsilon_{mf}))+k_j^*\alpha_j\Phi_j(1-\Phi_j)(1-\varepsilon_{mf})]}{X_j+k_j^*(1-\Phi_j)(1-\varepsilon_{mf})}\right] \quad (36)$$

where $\quad k_j^* = k_r h_j/U;\ X_j = k_{jd}a_j h_j/U \quad (37)$

These equations, first derived by Grace and deLasa (60), are entirely analogous to Equations (22) to (24). For cases where there are appreciable temperature gradients, energy balances can be written in an analogous manner (63). In order to apply the above model, we require values for Φ_j, α_j, a_j and k_{jd} in addition to those of k_r (from chemical kinetic experiments) and h_j (from Equation (35)). Experiments (62) indicate that the fraction α_j of the jet volume occupied by solids is 0.001 to 0.006, too small to influence the overall reaction appreciably except for very fast reactions. In practice, jets always expand after leaving the orifice, with the total jet cross-sectional area typically being twice to four times the fractional free area, ϕ_{or}, of the orifice plate, where

$$\phi_{or} = \pi d_{or}^2 N_{or}/4A = N_{or} d_{or}^2/D^2 \tag{38}$$

If we adopt the intermediate value of three, i.e. $\Phi_j = 3\phi_{or}$, then

$$a_j = 4\sqrt{3}\,\phi_{or}/d_{or} \tag{39}$$

Typically ϕ_{or} is in the range 0.01 to 0.05 for most fluidized bed distributor plates. The above relationships assume that all of the grid holes are operating, which requires (64) that

$$u_{or}^2 \geqslant 0.73\,\rho_p(1-\varepsilon_{mf})gh_j/[\rho(1-U_{mf}^2/U^2)] \tag{40}$$

The key to the influence of the grid region is the value of the jet-to-dense-phase mass transfer coefficient, k_{jd}. Behie and Kehoe suggested values of the order of 1 m/s based on data obtained (65) in a 0.61 m diameter column with cracking catalyst particles. Subsequent work (66,67) has suggested values of k_{jd} an order of magnitude or more lower than this. Profiles of concentration of a reacting gas from Equation (36) are shown in Figure 12 for jet-to-dense-phase mass transfer coefficients of

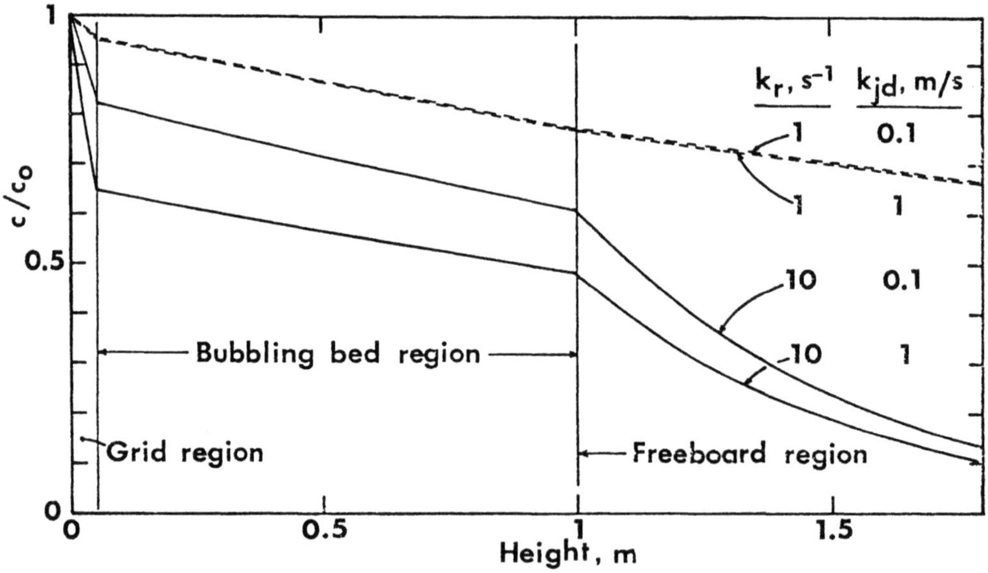

Figure 12: Concentration profiles for gaseous reactant for the conditions: $\phi_{or} = 0.018$, $\alpha_j = 0.005$, $D = 1.0$ m, $d_{or} = 4.0$ mm; $d_p = 160$ μm; $\rho_p = 1100$ kg/m^3, $\rho = 0.80$ kg/m^3, $U = 0.53$ m/s, $\varepsilon_{mf} = 0.48$, $H = 1.0$ m, $\mu = 3.5 \times 10^{-5}$ Ns/m^2, $D_m = 4.3 \times 10^{-5}$ m^2/s.

1 and 0.1 m/s and other typical conditions as specified in the caption. For the higher value of k_{jd} and a fast reaction it is seen that the grid region can give rise to more rapid reaction than in the bubbling region, modelled as described in section 4.3.

The above models assume one-dimensional flow in the dense phase. Models which allow for radial gradients have been proposed by Sit (67) and Piccinini et al (68), the latter for the analogous case of spouted bed reactors. Sit (67) has also measured mass transfer coefficients for forming bubbles and found these to be of the same order of magnitude as for single rising bubbles.

A number of studies (e.g. 47,69-71) have been reported which show that changing the grid can have a significant impact on the overall conversion in fluidized bed reactors. This impact results both from the effect on jetting and hydrodynamics in the grid region and from the influence on the bubble sizes higher in the bed. Work is still needed to characterize and model the grid region, but the models and results summarized above provide a useful starting point for understanding distributor effects.

6.2 Reaction in the Freeboard Region

Although particles tend to be widely dispersed in the freeboard region compared with dense bed zones, gas and particles are in intimate contact there, giving an opportunity for additional reaction. Temperature gradients may also be much larger in the freeboard region than in the bed proper. The effect of the freeboard has been found to be considerable in various processes, for example in determining NO_x emissions from fluidized combustors and afterburning in catalytic cracking regenerators. The additional reactions can be desirable or undesirable.

The earliest published freeboard model due to Miyauchi and Furusaki (72) assumes plug flow of gas and allows for temperature gradients. It was shown that selectivity of an intermediate product could be adversely or favourably influenced by the freeboard region, depending on whether the reactions are exothermic or endothermic. This model requires experimental determination of the solids concentration profiles in the freeboard and makes no provision for recycling of solids carried out of the vessel. Yates and Rowe (73) overcame the first of these limitations by assuming that a fraction ζ of the particles carried in bubble wakes are ejected into the freeboard and that these particles immediately adopt an upward velocity of $U-v_T$. Temperature gradients were ignored and gas was considered either to be in plug flow or perfect mixing within the freeboard region.

With the plug flow assumption, the concentration profile in the freeboard (x>H) is then given for a first order reaction with negligible internal pore diffusion control (15,73) by

$$c/c_H = \exp\left[\frac{-\zeta f_w(Q_b/A)(1-\varepsilon_{mf})(x-H)}{U(U-v_T)[1/k_r+d_p/6k_{gp}]}\right] \qquad (U>v_T) \quad (41)$$

While these simplifications ignore disengagement of small particles, all with terminal settling velocities > U, particle recyle from the cyclones and non-isothermal behaviour, the model allows some estimation of the extent of the freeboard region contribution. Some profiles of gas concentration derived from Equation (41) are given in Figure 12 with typical values of f_w=0.2(2), ζ=0.4(74), Q_b/A from Equation (17) with Y=0.7, and gas-to-particle mass transfer coefficients, k_{gp}, from ref.(8). It is seen that significant additional reaction can occur in the freeboard region, providing that there is an appreciable amount of unreacted gas at x=H.

deLasa and Grace (75) and Chen and Wen (76) have proposed models which attempt to provide an improved simulation of the complex hydrodynamics in the freeboard region. In the former case, particle trajectories were calculated for individual particles ejected into the freeboard. Some allowance was made for dissipation of jets originating with the erupting bubbles. No allowance was made, however, for particle interactions, radial gradients, gas backmixing or non-uniform distributions of particle ejection velocities. The model does make allowance for particles returning to the bed within the freeboard (v_T>U), for temperature gradients within the freeboard region, and for re-introduction of entrained (v_T<U) particles, having been heated or cooled in the freeboard, via the cyclones. Some predictions of coke conversion in an industrial (14m diameter x 10m tall) regenerator appear in Figure 13. The model is seen to predict that the freeboard region exerts a strong influence, especially for shallow beds. Inclusion of the freeboard can be more important than the choice of the model used to simulate in-bed behaviour. For the case shown, coke conversion is remarkably insensitive to the dense bed height. Measurements in the industrial regenerator (58) simulated show that bubbling bed models on their own, with no allowance for end effects, are unable to predict the observed reactor performance.

Chen and Wen (76) tried to improve on this model by allowing for axial dispersion of gas in the freeboard region and by means of assumptions which give more realistic profiles of particle holdup in the dilute phase. Entrainment is calculated from their recent correlation (77) which leads to an exponentially decreasing solids mass flux and makes allowance for solids returning to the bed surface along the vessel walls. Ejected particles were

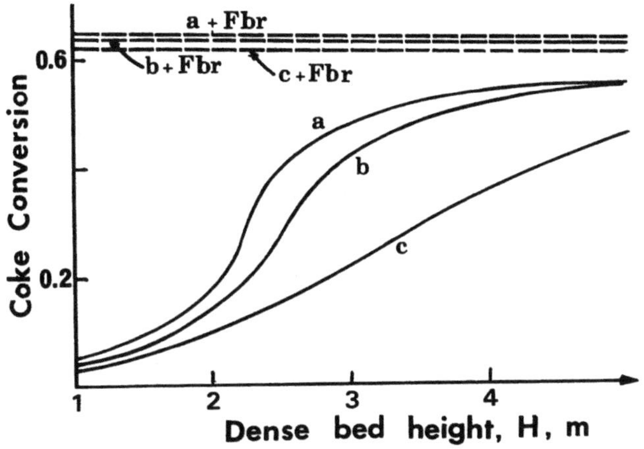

Figure 13: Coke conversions predicted for 14 m dia x 10 m tall catalyst regenerator (75) as a function of dense bed height with and without freeboard region (Fbr) included: a) Dense bed modelled as single-phase CSTR; b) Orcutt model (17,18); c) as in (b) only with lower interphase mass transfer predicted by Kunii and Levenspiel (20). d_p = 60 μm.

Reproduced by permission of the Am. Inst. Chem. Engs.

assumed to have a distribution of initial velocities consistent with the exponential decay of solids flux. Calculations were made of the effect of the freeboard on SO_2 sorption and NO_x reduction in the Babcock and Wilcox 1.8 m square fluidized bed combustor, where in-bed conversions were calculated using the model of Rajan and Wen (78), and axial dispersion coefficients were derived from the Peclet number of gas flow in the freeboard region (79). As shown in Table 5, excellent agreement was obtained with measured outlet concentrations of both SO_2 and NO_x. As much as 40% of the SO_2 capture was found to be achieved within the freeboard region.

Table 5: Comparisons of experimental and predicted SO_2 and NO_x emissions from Babcock and Wilcox 1.8 m square fluid bed combustor (76): T=1121K, H=1.3m, U=2.3 m/s, coal feed rate = 0.25 kg/s, limestone feed rate=0.097 kg/s, total column height=12 m.

S-content of coal	Fines recycle	SO_2 concentration (ppm) Av. expt'l	SO_2 concentration (ppm) Predicted	NO_x concentration(ppm) Av. expt'l	NO_x concentration(ppm) Predicted
3.7%	none	610	603	232	226
3.7%	underbed: 0.153 kg/s	496	501	227	214
4.5%	overbed: 0.277 kg/s	1405	1369	293	301
4.5%	underbed: 0.165 kg/s	867	913	279	280

7 GAS-SOLID REACTIONS

Except for the preceding section, all models treated so far have been for gas-phase reactions catalysed by the fluidized particles. Here we consider reactions of the form

$$\gamma G(\text{gas}) + S(\text{solids}) \rightarrow \text{Products (gas, solids and/or liquids)} \quad (42)$$

where S is a constituent of the bed solids and γ is a stoichiometric constant. Because the particles take part in the reaction, they may undergo physical changes in particle size and/or density. Table 6 lists some gas-solid reactions carried out in fluidized beds and indicates whether the particles grow or shrink.

7.1 Models for Solids Conversion Only

The simplest cases occur where either the gas is inert (i.e. γ=0 in Equation (42) as in limestone calcination) or the gas is present in great excess. In these cases the conversion of gaseous reactants can be ignored, and it is sufficient to consider only the kinetics of the reactions, the residence time distribution of the solids and an overall energy balance. Kunii and Levenspiel (20) and Fane and Wen (13) provide expressions for solids conversions in single or multiple beds for such cases. Each stage is assumed to give perfect mixing of solids. Expressions are tabulated for different governing kinetic rate expressions: homogeneous reaction model and shrinking core model with external mass transfer control, ash diffusion control and chemical reaction control. Population balances are required to

Table 6: Evolution of particle size during gas-solid reactions in fluidized beds. (Largely following Fane and Wen (13)).

Reaction	Particles shrink	Approx. const. size	Particles grow
Calcination		X	
Roasting of ores		X	
Iron ore reduction		X	
Uranium processing	X	X	X
Spent liquor processing			X
Catalyst regeneration		X	
Coal, refuse, biomass combustion	X		
Coal carbonization		X	
Coal, biomass gasification	X	X	
Sulphur capture by limestone, dolomite		X	

account for particles which change their size or density during the process, and may also take into account such factors as elutriation, cyclones of different efficiencies, attrition or agglomeration, inert solid materials, and unsteady feed or withdrawal of solids (20,80-82).

7.2 Models for Calculating Conversion of Both Gas and Solids

In most gas-solid reactions, both the gas and the solids undergo significant changes in composition. Since turnover of solids is usually rapid and gas elements have much shorter residence times than the particles, the solids can generally be considered to be of uniform composition in the vertical direction. The conversion of the gas as it traverses the bed of solids can therefore be predicted using identical methods as for catalytic reactions described in previous sections. The conversion of the solids must also be calculated based on the concentration of gas in contact with the solid particles, kinetic rate expressions and relationships for solids residence time distributions and population balances where required. The method for calculating the solids conversion can be illustrated by considering a polydisperse feed of solids which by sieve analysis is divided into N intervals of representative size d_{pi} and mass fraction x_i. For each d_{pi} we must calculate the conversion of a particle spending time t in the bed from an appropriate kinetic model. For example, for a shrinking core model with chemical reaction control and constant particle size the conversion is

$$\chi_{Si} = 3(t/\theta_i) - 3\,(t/\theta_i)^2 + (t/\theta_i)^3 \qquad (t \leqslant \theta_i) \qquad (43)$$

where θ_i is the time for complete reaction given by

$$\theta_i = \gamma\rho_S d_{pi}/(2k_A\bar{c}_G) \qquad (44)$$

where ρ_S is the molar density of S in the particles, k_A is a rate constant based on the surface area of unreacted core and $\bar{c}_G$ is the mean concentration of G in the gas to which the solids are exposed. For example, $\bar{c}_G$ will simply be the dense phase gas concentration for the Orcutt (17,18) model in which solids are excluded from the bubble phase and gas is perfectly mixed in the dense phase. For the single phase bubbling bed model (15,19) and a reaction which is first order in G,

$$\bar{c}_G = (c_o - c_H)/\{k1^*[\alpha_1\varepsilon_b+(1-\varepsilon_{mf})(1-\varepsilon_b)]\} \qquad (45)$$

where c_H can be obtained from Equation (22).

The mean conversion of particles of size d_{pi} is calculated from

$$\bar{\chi}_{Si} = 1 - \int_0^{\theta_i} (1-\chi_{Si})E(d_{pi},t)dt \qquad (46)$$

where the exit age distribution for most single stage fluidized beds can be assumed to correspond to perfect mixing, so that

$$E(d_{pi},t) = (1/\tau_i)\exp\{-t/\tau_i\} \qquad (47)$$

and $$\chi_{Si} = 3\{\tau_i/\theta_i - 2(\tau_i/\theta_i)^2 + 2(\tau_i/\theta_i)^3[1-\exp(-\theta_i/\tau_i)]\} \qquad (48)$$

The mean residence time for particles of size d_{pi} is given (20) by

$$\tau_i = M/[F_{out} + (1-\eta_i)K_i^{*}A] \qquad (49)$$

where M is the total mass of solids in the bed, F_{out} is the solids outflow from the bed proper, η_i is the separation efficiency for any cyclone or cyclones used for recycle of entrained solids and K_i^* is the elutriation constant which can be calculated from any of a number of correlations. The overall solids conversion is found by summing over all size fractions,

i.e. $$\bar{\chi}_S = \sum x_i\bar{\chi}_{Si} \qquad (50)$$

Iterative solution techniques are commonly required to match the conversion of the gas and solids, which must satisfy the stoichiometry of reactions like that described by Equation (42). While some fairly general models have been proposed for gas-solid reactions in fluidized beds (15,20,83), most models for reactions of this type are specific to a particular reaction or reactor. For example, there have been at least a dozen specific models for fluidized bed combustion of coal and nearly as many for coal gasification. Several references (15,20,84) have provided sample calculations and step-by-step procedures for solving the more general models.

Since general models for gas-solid reactions in fluidized beds involve all the steps required to calculate conversions of gas in catalytic beds, these models involve all of the possible choices and considerations outlined in section 4.2. In addition, assumptions are required regarding the mixing pattern of solids and whether or not population balances will be adopted to account for such factors as elutriation, particle shrinkage (or growth) and different rate-controlling reaction kinetics for particles of differing sizes. Since most gas-solid reactions involve considerable heats of reaction, energy balances are commonly required. In addition, many of the applications involve a series of different reactions, homogeneous as well as heterogeneous, which may need to be modelled separately. For example, in coal combustion it may be necessary to treat coal devolatilization, CO oxidation, SO_2 capture, and NO_x evolution and reduction, in addition to the oxidation of char to give CO and CO_2. Given all of these variables, plus the variety of gas-solid reaction kinetic expressions, it is not surprising that many different reactor models have been proposed. Summaries of the key assumptions underlying various models have been tabulated by Park et al(85), Yates(14) and Grace(15).

Many of the models proposed have been used only to perform parametric studies showing the predicted effect of a host of variables. There are relatively few cases where the models have been compared directly with experimental data. Where such comparisons have been carried out, the models have frequently contained fitting parameters which have allowed good agreement to be obtained over a limited range of experimental conditions. Without such fitted constants, agreement between the models and experimental results has tended to be disappointing. Despite the formidable efforts which have been exerted to draw up reactor models for such reactions as coal and biomass combustion and gasification, it is fair to say that no single model has proved itself as a general purpose tool for gas-solid reactions. That is not to say, however, that models have not been useful in simulating existing reactors, in providing understanding of reactor characteristics and in predicting the influence of changes in operating variables.

Some simulation predictions related to experimental results in equipment of quite large scale have already been presented in Figure 13 and Table 5. As a further example, Figure 14 provides some comparison from a recent paper (86) in which coal char was

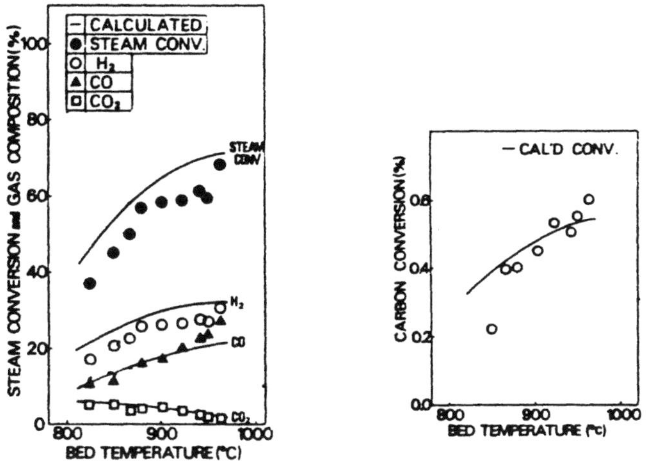

Figure 14: Comparison between bubbling bed model predictions and experimental results obtained in a 76 mm dia x 853 mm long coal char gasifier with steam as the feed gas, U = 0.15 m/s, $U_{mf} \approx 0.03$ m/s (86).

gasified with steam in a small fluid bed reactor, and modelling was carried out using the bubbling bed model of Kunii and Levenspiel (20). Agreement between predictions and experimental results is seen to be good, although the fitted bubble diameter (60 mm) was close enough to the bed diameter (78 mm) that one would have expected slug flow.

7.3 Advanced Models

In addition to the models described above, there are some reactor models which fall outside the usual range of assumptions and methods used to describe gas-solid reactions in fluidized beds. Three of these models are described here.

7.3.1 Shallow bed model. Fan and coworkers (87,88) have treated the case of shallow reactors (bed height low relative to diameter or length for rectangular vessels). While mixing is rapid in the vertical direction, it tends to be limited in the lateral direction so that the residence time distribution for solids can differ considerably from the perfect mixing assumption adopted in most of the models covered in Sections 7.1 and 7.2. Lateral mixing of both solids and gas is described in terms of lateral dispersion coefficients. Unsteady state terms are retained in order to provide dynamic solutions. In other respects the models are "conventional" in their treatment of two phases (bubbles and dense phase). It is shown that lateral

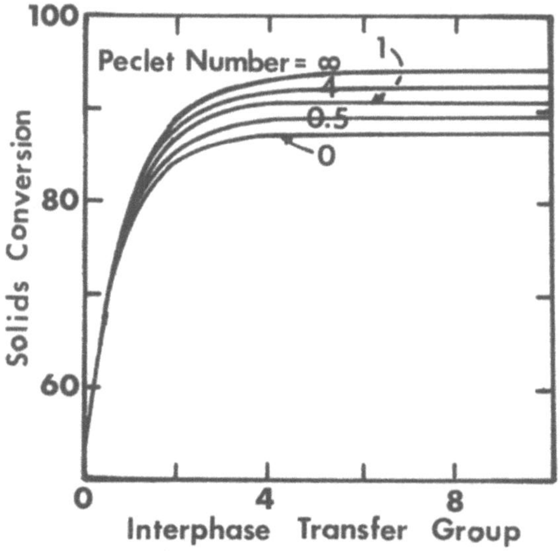

Figure 15: Effect of lateral dispersion of solids on solids conversion for a case considered by Chang et al (88).

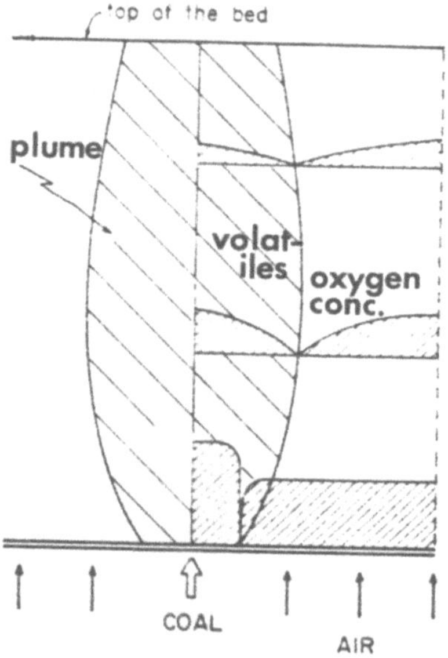

Figure 16: Typical concentration profiles of volatiles (A) and oxygen (B) for plume model (85,89). Reproduced by permiss-

temperature distributions are much more uniform than lateral concentration profiles. The effect of the lateral dispersion Peclet number on solids conversion can be significant as shown in Figure 15 where the calculations are for zinc roasting.

7.3.2 Plume model. Park et al(85,89) argued that devolatilization of coal occurs quickly relative to particle mixing in the vicinity of immersed solid feed inlets in fluidized bed combustion and that one should therefore expect plumes of unburnt volatile gas to originate at each feed point. Oxygen and combustible gases are assumed to diffuse to the boundary of the plume as shown in Figure 16 where they burn, with unburnt gases leaving the bed burning in the freeboard. Char particles are assumed to be well mixed throughout the bed and to burn everywhere except in the plumes. Temperature is also assumed to be uniform throughout the bed. The model can give very different predictions than more conventional models which either have ignored devolatilization or have assumed it to occur uniformly throughout the entire bed.

7.3.3 Numerical modelling approach. Blake and coworkers(90,91) took a more fundamental approach in which they attempted to describe a fluidized bed in terms of continuity, momentum and energy equations for overlapping continua, and then to solve the governing equations numerically by finite difference methods, analogous to numerical solutions of single-phase fluid mechanical problems. Considerable attention was given to obtaining consistent and reasonable constitutive equations and gas-particle interaction coefficients. While promising predictions (e.g. see Figure 17) were obtained showing jet and bubble formation for gas entry through orifices in two-dimensional and axisymmetric geometries, limited progress appears to have been achieved in applying the model to the prediction of chemical reactions. Computer limitations would also appear to rule out application of this approach to fully three-dimensional geometries.

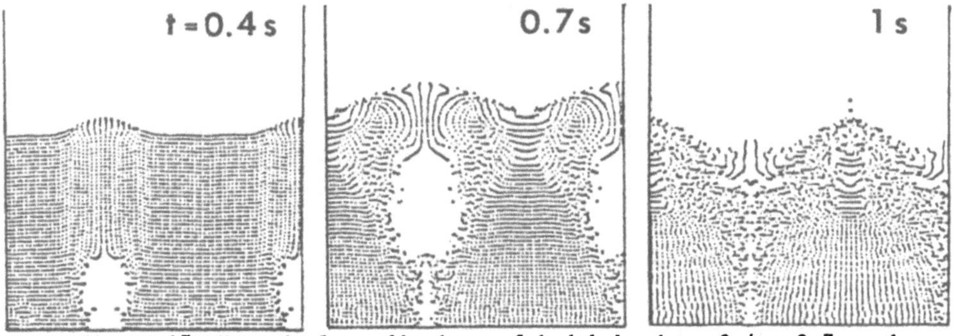

Figure 17: Numerical predictions of bed behaviour 0.4s, 0.7s and 1.0s after establishing a flow of gas (U=3.3 m/s) through orifices in the base of a 0.32 m deep bed of 860 μm glass beads (91). Points represent Lagrangian markers of solids. Reproduced bv permission Am Inst of Chem Engr

7.3.4 Stochastic modelling approach. In addition to the deterministic models reviewed above and employed by the vast majority of those modelling fluidized bed reactors, it is possible to adopt a probabilistic approach which makes allowance for fluctuations in local hydrodynamic properties or interphase exchange (92-95). This approach has some intuitive advantages, since fluidized beds operating in the bubbling, turbulent and fast fluidization regimes certainly exhibit near-random fluctuations of rather large amplitude in most measurable properties. These fluctuations can result in dispersion by processes analogous to Reynolds stresses for turbulent fluids. The stochastic models themselves tend to use many of the elements of deterministic models, with superimposed fluctuations contributing to the mixing and contacting processes. In the absence of data regarding the frequency and amplitude of the fluctuations, different probabilistic models have been adopted in the studies to describe these fluctuations.

8 NEW DEVELOPMENTS AND INNOVATIONS

Clift (96) has argued that models should be no more complex than needed to describe the essentials of the physics which are understood. This view is consistent with the recommendations of this chapter which favour relatively simple models. At this point in time, these often do as well as more sophisticated models and require considerably less computational effort. Nevertheless, there is a constant search for better and more complete models that will provide a more accurate description and predict new variables or results for more complex situations.

Some of the trends in modelling have already been considered in this review:

- There is a need to model higher velocity regimes of fluidization, especially the turbulent (section 5.2) and fast fluidization (section 5.3) regimes: Few industrial beds operate in the bubbling regime, especially for catalytic reactions.

- There is a need to allow for end effects, i.e. the grid region (Section 6.1) and freeboard reactions (Section 6.2). Even when the bed does operate in the bubbling regime, bubbling bed models are often unable to give a good simulation of experimental results.

- There is a need to consider complex reactions and selectivity (section 4.4) as well as simple first order reactions and conversions alone.

- The effects of bed internals (section 4.7) are poorly understood and require further investigation.

Other issues have only been implied or have been omitted, but require further work:

- Almost all data against which reactor models have been tested in the open literature are for relatively small equipment. There is a critical need to test models against data from full-scale industrial reactors.

- Most models have been developed and tested only under steady state conditions. In recent years some attention has been devoted to applying two-phase models for process control and simulation of unsteady conditions (e.g. 37,87,88,97-99). More unsteady state modelling is required, especially since process control models used for fluidized beds are often crude. Further work is also needed to allow the influence of fluctuations in properties and exchange rates to be modelled successfully.

- A number of model studies in recent years (e.g. 30,95-101) have indicated that fluidized bed reactors can exhibit multiple steady states if the reactions are strongly exothermic. The extent of the zone of multiplicity depends on the model. There are no reported cases that I know of where multiplicity has been observed in industrial fluid bed reactors, but this could be a future problem area.

- In many industrial reactions, the number of moles of gas produced differs from the number of moles of reacting gas so that there is a change in volume as the reaction proceeds. Since most reaction takes place in the dense phase while most excess gas flows in the bubble phase, it is not clear how one should model such cases. Some early work on this problem has been reported recently (102-104).

- Fluidized bed reactor models have had some success simulating existing units and very limited success as tools for design of new reactors. As modelling improves, a further objective must be to use the models as tools for process and reactor optimization. We appear to be many years from achievement of this goal.

NOTATION

A	cross-sectional area of fluidized bed (m^2)
Ar	Archimedes number, Equation (2) (-)
a	surface area of interphase boundary per unit bed volume (m^{-1})
B	dimensionless superficial gas velocity, Equation (29) (-)
c	concentration ($kmol/m^3$)

D	bed diameter (m)
D_g	gas axial dispersion coefficient (m^2/s)
D_s	solids axial dispersion coefficient (m^2/s)
D_m	molecular diffusivity (m^2/s)
d_b	bubble diameter (m)
d_{bmax}	maximum stable bubble diameter (m)
d_{bo}	bubble diameter produced at the distributor plate (m)
d_{or}	diameter of orifices in distributor plate (m)
d_p	particle diameter or mean particle diameter (m)
$E(d_p,t)$	exit age distribution function (s^{-1})
F_{out}	mass withdrawal rate of solids from bed (kg/s)
f_w	wake volume/bubble volume (-)
G_s	solids net flux ($kg/m^2/s$)
g	acceleration due to gravity (m/s^2)
H	overall bed height from distributor to bed surface (m)
h_j	jet penetration height (m)
h_s	slug length (m)
I	slug surface integral given by Table 4 (-)
k_A	rate constant based on surface area of unreacted core (m/s)
k_{gp}	gas-to-particle mass transfer coefficient (m/s)
k_{12}	interphase (phase 1 to phase 2) mass transfer coefficient (m/s)
k_r	first order kinetic reaction rate constant (s^{-1})
$k_1{}^*$	dimensionless kinetic rate constant, Equation (23) (-)
$k_i{}^*$	dimensionless kinetic rate constant Equation (37) (-)
K^*	elutriation rate constant (kg/m^2s)
M	total mass of solids in the bed (kg)
m_1,m_2	roots given by Equation (32) (m^{-1})
m_f	mass fraction of powder smaller than 45μm (-)
N_{or}	number of orifices in multi-orifice distributor (-)
N_R	dimensionless kinetic rate constant defined by Equation (26) (-)
n	order of reaction (-)
Q	volumetric gas flow rate (m^3/s)
Q_b	volumetric gas flow due to translation of void units (m^3/s)
Re_{mf}	Reynolds number defined by Equation (1) (-)
t	time (s)
U	superficial gas velocity (m/s)
U^*	dimensionless superficial gas velocity, Equation (11) (-)
U_k	superficial gas velocity at onset of turbulent fluidization (m/s)
U_{mb}	superficial gas velocity at onset of bubbling (m/s)
U_{mf}	minimum fluidization velocity (m/s)
U_{ms}	superficial gas velocity at onset of slug flow (m/s)
U_{tr}	transport velocity (m/s)
u_b	bubble rise velocity (m/s)
u_i,v_i	absolute gas and solids velocities in phase i (m/s)

u_{or}	absolute gas velocity through orifice (m/s)
$u_{s\infty}$	rise velocity of a slug in isolation (m/s)
V_s	slug volume (m^3)
v_T	particle terminal settling velocity (m/s)
v_T^*	terminal settling velocity of particle with diameter $2.7d_p$ (m/s)
X	dimensionless interphase mass transfer group, Equation (24) or (27) (-)
X_j	dimensionless mass transfer group for jet region, Equation (37) (-)
x	vertical distance coordinate, measured upward from distributor (m)
x_i	mass fraction of solid material in size interval i (-)
Y	dimensionless coefficient, Equation (17) (-)
α_i	solids volume fraction in phase i (-)
γ	stoichiometric constant, Equation (42) (k mol G/k mol s)
$\Delta\rho$	density difference, $\rho_p - \rho$ (kg/m^3)
ε	overall bed void fraction (-)
ε_b	fraction of bed volume occupied by bubbles (-)
ε_{mf}	bed voidage at minimum fluidization (-)
η	cyclone separation efficiency (-)
θ	time for 100% conversion of a single particle (s)
ρ	gas density (kg/m^3)
ρ_p	particle density (kg/m^3)
ρ_S	solids molar density of species S ($kmol/m^3$)
μ	gas shear viscosity (Ns/m^2)
τ	mean residence time (s)
Φ_i	fraction of total volume occupied by phase i (-)
ϕ_{or}	fraction of open area in distributor plate (-)
χ_S	conversion of solid species S
ζ	fraction of wake particles ejected into the freeboard

Subscripts and overbar:

o	inlet
1,2	phases 1 and 2
b	bubble
d	dense phase
G	gas reactant species
H	at height H (bed surface)
h_j	at exit from jet
i	particle size range i or phase i
j	jet or jet region; phase j
mf	minimum fluidization
PF	plug flow
p	particle
S	solid reactant species
s	slug
-	average

REFERENCES

1. Grace, J.R. and J. Baeyans. Instrumentation and Experimental Techniques, chapter 13 in Fluid Bed Technology, ed. D. Geldart (New York, Wiley and Sons, 1986).
2. Grace, J.R. Fluidized Bed Hydrodynamics, chapter 8.1 in Handbook of Multiphase Systems, ed. G. Hetsroni (Washington, Hemisphere Publishing, 1982).
3. Geldart, D. and A.R. Abrahamsen. The Effect of Fines on the Behaviour of Gas Fluidized Beds of Small Particles, in Fluidization, ed. J.R. Grace and J.M. Matsen, pp.453-460 (New York, Plenum Publishing, 1980).
4. Darton, R.C., R.D. LaNauze, J.F. Davidson and D. Harrison. Bubble Growth due to Coalescence in Fluidized Beds. Trans. Instn. Chem. Engrs. 55 (1977) 274-280.
5. Stewart, P.S.B. and J.F. Davidson. Slug Flow in Fluidized Beds. Powder Technol. 1 (1967) 61-80.
6. Thiel, W.J. and O.E. Potter. Slugging in Fluidized Beds. Ind. Eng. Chem. Fundam. 16 (1977) 242-247.
7. Yerushalmi, J. and N.T. Cankurt. Further Studies of the Regimes of Fluidization. Powder Technol. 24 (1979) 187-205.
8. Clift, R., J.R. Grace and M.E. Weber. Bubbles, Drops and Particles (New York, Academic Press, 1978).
9. Reh, L. Fluidized Bed Processing. Chem. Engng. Progr. 67, No. 2 (1971) 58-63.
10. Geldart, D. Types of Gas Fluidization. Powder Technol. 7 (1973) 285-293.
11. Zuber, N. and J.A. Findlay. Average Volumetric Concentration in Two-Phase Flow Systems. J. Heat Transf. 87 (1965) 453-468.
12. Grace, J.R. An Evaluation of Models for Fluidized-Bed Reactors. A.I.Ch.E. Symp. Series. 67, No. 116 (1971) 159-167.
13. Fane, A.G. and C.Y. Wen, Fluidized-Bed Reactors, chapter 8.4 in Handbook of Multiphase Systems, ed. G. Hetsroni (Washington, Hemisphere, 1982).
14. Yates, J.G. Fundamentals of Fluidized-Bed Chemical Processes (London, Butterworths, 1983).
15. Grace, J.R. Fluid Beds as Chemical Reactors, chapter 11 in Fluid Bed Technology, ed. D. Geldart (New York, John Wiley,1986).
16. May, W.G. Fluidized-Bed Reactor Studies. Chem. Engng. Progr. 55, No. 12 (1959) 49-56.
17. Orcutt, J.C., J.F. Davidson and R.L. Pigford. Reaction Time Distributions in Fluidized Catalytic Reactors. Chem. Engng. Progr. Symp. Series. 58, No. 38 (1962) 1-15.
18. Davidson J.F. and D. Harrison. Fluidised Particles (Cambridge, Cambridge Univ. Press, 1963).
19. Grace, J.R. Generalized Models for Isothermal Fluidized Bed Reactors, chapter 13 in Recent Advances in the Engineering

Analysis of Chemically Reacting Systems, ed. L.K. Doraiswamy (New Delhi, Wiley Eastern, 1984).
20. Kunii, D. and O. Levenspiel. Fluidization Engineering. (New York, Wiley, 1969).
21. Fryer, C. and O.E. Potter. Countercurrent Backmixing Model for Fluidized Bed Catalytic Reactors. Indust. Engng. Chem. Fundam. 11 (1972) 338-344.
22. Mireur, J.P. and K.B. Bischoff. Mixing and Contacting Models for Fluidized Beds. A.I.Ch.E. Journal 13 (1967) 839-845.
23. Partridge, B.A. and P.N. Rowe. Chemical Reaction in a Bubbling Gas Fluidized Bed. Trans. Instn. Chem. Engrs. 44 (1966) 335-348.
24. Grace, J.R. Fluidized Bed Reactor Modeling: an Overview. A.C.S. Symp. Series. 168 (1981) 3-18.
25. Toomey, R.D. and H.F. Johnstone. Gaseous Fluidization of Solid Particles. Chem. Engng. Progr. 48 (1952) 220-226.
26. Grace, J.R. and R. Clift. On the Two-Phase Theory of Fluidization. Chem. Eng. Sci. 29 (1974) 327-334.
27. Valenzuela, J.A. and L.R. Glicksman. Gas Flow Distribution in a Bubbling Fluidized Bed. A.I.Ch.E. Ann. Mtg. 1981.
28. Sit, S.P. and J.R. Grace, Effect of Bubble Interaction on Interphase Mass Transfer in Gas Fluidized Beds, Chem. Eng. Sci., 36 (1981) 327-335.
29. Bohle W. and W.P.M. Van Swaaij. The Influence of Gas Adsorption on Mass Transfer and Gas Mixing in a Fluidized Bed. In Fluidization, ed. J.F. Davidson and D.L. Keairns, pp.167-172 (Cambridge, University Press, 1978).
30. Bukur, D., H.S. Caram and N.R. Amundson. Some Model Studies of Fluidized Bed Reactors, chapter 11 in Chemical Reactor Theory: a Review, ed. L. Lapidus and N.R. Amundson (Englewood Cliffs, Prentice-Hall, 1977).
31. Jayaraman V.K., B.D. Kulkarni and L.K. Doraiswamy. An Initial Value Approach to the Counter-Current Backmixing Model of the Fluid Bed. A.C.S. Symp. Series, 168 (1981) 19-29.
32. Mori, S. and C.Y. Wen. Estimation of Bubble Diameter in Gaseous Fluidized Beds. A.I.Ch.E. Journal. 21 (1975) 109-115.
33. Kato, K. and C.Y. Wen. Bubble Assemblage Model for Fluidized Catalytic Reactors. Chem. Eng. Sci. 24 (1969) 1351-1359.
34. Peters, M.H., L.S. Fan and T.L. Sweeney. Simulation of Particulate Removal in Gas-Solid Fluidized Beds. A.I.Ch.E. Journal. 28 (1982) 39-49.
35. Fryer, C. and O.E. Potter. Bubble Size Variation in Two-Phase Models of Fluidized Bed Reactors. Powder Technol. 6 (1972) 317-322.
36. Shaw, I., T.W. Hoffman and P.M. Reilly. Experimental Evaluation of Two-Phase Models Describing Fluidized Bed Reactors. A.I.Ch.E. Symp. Series, 70, No. 141 (1974) 41-52.
37. Jaffres, J.L., C. Chavarie, I. Patterson, M. Perrier, L. Casalegno and C. Laguérie. Conversion and Selectivity of the

Oxidation of Benzene to Maleic Anhydride in a Fluidized Bed Reactor. In Fluidization, ed. D. Kunii and R. Toei, pp. 565-573 (New York, Engineering Foundation, 1984).
38. Chavarie, C. and J.R. Grace. Performance Analysis of a Fluidized Bed Reactor. Indust. Engng. Chem. Fundam. 14 (1975) 75-91.
39. Fryer, C. and O.E. Potter. Experimental Investigation of Models for Fluidized Bed Catalytic Reactors. A.I.Ch.E. Journal. 22 (1976) 38-47.
40. Dry, R.J. and M.R. Judd, Fluidized Beds of Fine Dense Powders: Scale-up and Reactor Modelling. Powder Technol. 43, 41-53 (1985).
41. Zenz, F.A. and D.F. Othmer. Fluidization and Fluid-Particle Systems. (New York, Reinhold, 1960).
42. Grace, J.R. Fluidization and its Application to Coal Treatment and Allied Processes. A.I.Ch.E. Symp. Series. 70, No.141 (1974) 21-26.
43. Gbordzoe, E.A.M., H. Littman and M.A. Bergougnou, Gas-Solid Contacting in a Fluidized Bed Containing Mobile Internals. tbp (1986).
44. Botton, R.J. Gas-Solid Contacting in Fluidized Beds. Chem. Engng. Progr. Symp. Series. 66, No. 101 (1970) 8-18.
45. Hovmand, S. and J.F. Davidson, Chemical Conversion in a Slugging Fluidized Bed, Trans. Instn. Chem. Engrs. 46 (1968) 190-203.
46. Hovmand, S. and J.F. Davidson, Pilot Plant and Laboratory Scale Fluidized Reactors at High Gas Velocities: the Relevance of Slug Flow, Chapter 5 in Fluidization, ed. J.F. Davidson and D. Harrison, pp. 193-259 (London, Academic Press, 1971).
47. Hovmand, S., W. Freedman and J.F. Davidson. Chemical Conversion in a Pilot-Scale Fluidized Bed, Trans. Instn. Chem. Engrs. 49 (1971) 149-162.
48. Raghuraman, J. and O.E. Potter. Countercurrent Backmixing Model for Slugging Fluidized-Bed Reactors. A.I.Ch.E. Journal 24 (1978) 698-704.
49. Yates, J.G. and Y.Y. Grégoire, An Experimental Test of Slugging-Bed Reactor Models. in Fluidization, ed. J.R. Grace and J.M. Matsen, pp.581-588 (New York, Plenum, 1980).
50. Van Deemter, J.J. Mixing Patterns in Large-Scale Fluidized Beds. in Fluidization, ed. J.R. Grace and J.M. Matsen, pp.69-89 (New York, Plenum, 1980).
51. Avidan, A. Turbulent Fluid Bed Reactors using Fine Powder Catalysts, A.I.Ch.E.-C.I.E.S.C. Meeting, Beijing (1982).
52. Abed, R. The Characterization of Turbulent Fluid-Bed Hydrodynamics. in Fluidization, ed. D. Kunii and R. Toei, pp.137-144 (New York, Engineering Foundation, 1984).
53. deLasa, H.I. and G. Gau, Influence of Aggregates on the Performance of a Pneumatic Transport Reactor, Chem. Eng. Sci. 28 (1973) 1875-1884.

54. Cankurt, N.T. and J. Yerushalmi, Gas Backmixing in High Velocity Fluidized Beds, in Fluidization, ed. J.F. Davidson and D.L. Keairns, pp. 387-393 (Cambridge, University Press, 1978).
55. Van Swaaij, W.P.M. The Design of Gas-Solids Fluid Bed and Related Reactors. A.C.S. Symp. Series. 72 (1978) 193-222.
56. Van Deemter, J.J. Mixing and Contacting in Gas-Solid Fluidized Beds. Chem. Eng. Sci. 13 (1961) 143-154.
57. Miyauchi, T., S. Furusaki, S. Morooka and Y. Ikeda, Transport Phenomena and Reaction in Fluidized Catalyst Beds, Advances in Chem. Engng. 11 (1981) 275-448.
58. deLasa, H.I., A. Errazu, E. Barreiro and S. Solioz, Analysis of Fluidized Bed Catalytic Cracking Regnerator Models in an Industrial Scale Unit, Can. J. Chem. Eng. 59 (1981) 549-553.
59. Behie, L.A. and P. Kehoe, The Grid Region in a Fluidized Bed Reactor. A.I.Ch.E. Journal. 19 (1973) 1070-1072.
60. Grace, J.R. and H.I. deLasa. Reaction near the Grid in Fluidized Beds. A.I.Ch.E. Journal, 24 (1978) 364-366.
61. Chandnani, P. and N. Epstein, Gas Spouting of Fine Particles, 34th Can. Chem. Engng. Conf., Québec City, October 1984.
62. Merry, J.M.D. Penetration of Vertical Jets into Fluidized Beds, A.I.Ch.E. Journal, 21 (1975) 507-510.
63. Weimar, A.W. and D.E. Clough. The Influence of Jetting-Emulsion Mass and Heat Exchange in a Fluidized Bed Coal Gasifier, A.I.Ch.E. Symp. Ser. 77, No. 205 (1981) 51-65.
64. Davidson, J.F., D. Harrison, R.C. Darton and R.D. LaNauze, The Two-Phase Theory and its Application to Chemical Reactors. in Chemical Reactor Theory, A Review, ed. L. Lapidus and N.R. Amundson, pp.583-685 (Englewood Cliffs, Prentice-Hall, 1977).
65. Behie, L.A., M.A. Bergougnou and C.G.J. Baker. Mass Transfer from a Grid Jet in a Large Gas-Fluidized Bed, in Fluidization Technology, ed. D.L. Keairns. Vol. 1, pp.261-278 (Washington, Hemisphere, 1976).
66. deMichaele, G., A. Elia and L. Massimilla. The Interaction between Jets and Fluidized Beds. Ing. Chim. Ital. 12 (1976) 155-162.
67. Sit, S.P. Grid Region and Coalescence Zone Gas Exchange in Fluidized Beds. Ph.D. dissertation, McGill University, 1981.
68. Piccinini, N., J.R. Grace and K.B. Mathur, Vapour Phase Chemical Reaction in Spouted Beds: Verification of Theory. Chem. Eng. Sci., 34 (1979) 1257-1263.
69. Cooke, M.J., W. Harris, J. Highley and D.F. Williams. Kinetics of Oxygen Consumption in Fluidized Bed Carbonizers. Instn. Chem. Engrs. Symp. Ser. 30 (1968) 21-27.
70. Walker, B.V., Gas-Solid Contacting in Bubbling Fluidized Bed, Ph.D. dissertation, Cambridge Univ., 1970.
71. Bauer, W. and J. Werther. Scale-up of Fluid Bed Reactors with respect to Size and Gas Distributor Design - Measurements and Model Calculations, Proc. 2nd World Congress Chem. Engng. 3 (1981) 69-72.

72. Miyauchi, T. and S. Furusaki. Relative Contribution of Variables Affecting the Reaction in Fluid Bed Contactors. A.I.Ch.E. Journal 20 (1974) 1087-1096.
73. Yates, J.G. and P.N. Rowe. A Model for Chemical Reaction in the Freeboard Region above a Fluidized Bed. Trans. Instn. Chem. Engrs. 55 (1977) 137-142.
74. George, S.E. and J.R. Grace. Entrainment of Particles from Aggregative Fluidized Beds. A.I.Ch.E. Symp. Ser. 74, No. 176 (1978) 67-74.
75. deLasa, H.I. and J.R. Grace. The Influence of the Freeboard Region in a Fluidized Bed Catalytic Cracking Regenerator. A.I.Ch.E. Journal. 25 (1979) 984-991.
76. Chen, L.H. and C.Y. Wen. Model of Solid Gas Reaction Phenomena in the Fluidized Bed Freeboard. A.I.Ch.E. Journal 28 (1982) 1019-1027.
77. Wen, C.Y. and L.H. Chen. Fluidized Bed Freeboard Phenomena-Entrainment and Elutriation. A.I.Ch.E. Journal. 28 (1982) 117-128.
78. Rajan, R.R. and C.Y. Wen. A Comprehensive Model for Fluidized Bed Coal Combustors. A.I.Ch.E. Journal. 26 (1980) 642-655.
79. Wen, C.Y. and L.T. Fan, Models for Flow Systems and Chemical Reactors. (New York, Dekker, 1975).
80. Chen, T.P. and S.C. Saxena. Solids Population Balance for Attrition in Fluidized Beds, Powder Technol. 18 (1977) 279-281.
81. Weimar, A.W. and D.E. Clough. Dynamics of Particle Size/Conversion Distributions in Fluidized Beds. Powder Technol. 26 (1980) 11-16.
82. Overturf, B.W. and G.V. Reklaitis. Fluidized-Bed Reactor Model with Generalized Particle Balances. A.I.Ch.E. Journal. 29 (1983) 813-829.
83. Yoshida, K. and C.Y. Wen. Noncatalytic Solid-Gas Reaction in a Fluidized Bed Reactor. Chem. Eng. Sci. 25 (1970) 1395-1404.
84. Ishida, M. and C.Y. Wen. Effect of Solid Mixing on Noncatalytic Solid-Gas Reactions in a Fluidized Bed. A.I.Ch.E. Symp. Ser. 69, No. 128 (1973) 1-7.
85. Park, D., O. Levenspiel and T.J. Fitzgerald. Plume Model for Large Particle Fluidized Bed Combustors. Fuel, 60 (1981) 295-306.
86. Matsui, I., T. Kojima, T. Furusawa and D. Kunii. Gasification of Coal Char by Steam in a Continuous Fluidized Bed Reactor. in Fluidization, ed. D. Kunii and R. Toei, pp.655-662. (New York, Engineering Foundation, 1984).
87. Fan, L.T. and C.C. Chang. Modeling and Simulation of Dynamic and Steady State Characteristics of Shallow Fluidized Bed Combustors. A.C.S. Symp. Ser. 168 (1981) 95-115.
88. Chang,C.C., L.T. Fan and S.X. Rong. Modelling of Shallow Fluidized Bed Reactors. Can. J. Chem. Eng. 60 (1982) 272-281 and 781-790.

89. Park. D.. O. Levenspiel and T.J. Fitzgerald. A Model for Large Scale Atmospheric Fluidized Bed Combustors. A.I.Ch.E. Symp. Ser. 77, No. 205 (1981) 116-126.
90. Blake, T.R., S.K. Garg, H.B. Levine and J.W. Pritchett. Computer Modelling of Coal Gasification Reactors. Rept. SSS-R-76-2967. (LaJolla, Systems Science & Software, 1976).
91. Pritchett, J.W., T.R. Blake and S.K. Garg. A Numerical Model of Gas Fluidized Beds. A.I.Ch.E. Symp. Ser. 74, No. 176 (1978) 134-148.
92. Krambeck, F.J., S. Katz and R. Shinnar, A Stochastic Model for Fluidized Beds, Chem. Eng. Sci. 24, 2497-1511 (1969).
93. Bywater, R.J., Fluidized Bed Catalytic Reactor according to a Statistical Fluid Mechanics Model. A.I.Ch.E. Symp. Ser. 74, No. 176, 126-133 (1978).
94. Ligon, J.R. and N.R. Amundson, Modelling of Fluidized Bed Reactors: An Isothermal Bed with Stochastic Bubbles. Chem. Eng. Sci. 36, 653-660 (1981).
95. Too, J.R., R.O. Fox, L.T. Fan and R. Nassar, Stochastic Modeling of a Fluidized-Bed Reactor, A.I.Ch.E. Journal 31, 992-998 (1985).
96. Clift, R. An Occumist View of Fluidized-Bed Reactor Modelling. Chem. Engr. (London) 388 (1983) 29-33.
97. deLasa H.I. and A. Errazu. Ignition of a Fluidized Bed Catalytic Cracking Regenerator: Freeboard Region Influence. In Fluidization, ed. J.R. Grace and J.M. Matsen, pp. 563-570 (New York, Plenum, 1980).
98. Fan L-S. and L.T. Fan. Transient and Steady State Characteristics of a Gaseous Reactant in Catalytic Fluidized Bed Reactors. A.I.Ch.E. Journal. 26 (1980) 139-144.
99. Elnashaie, S.S.E.H. and S.H. Elbialy. Multiplicity of Steady States in Fluidized Bed Reactors. Chem. Eng. Sci. 35 (1980) 1357-1368.
100. Kulkarni, B.D., P.A. Ramachandran and L.K. Doraiswamy. Criteria for Temperature Multiplicity in Fluidized Bed Reactors, in Fluidization, ed. J.R. Grace and J.M. Matsen, pp.589-597 (New York, Plenum, 1980).
101. Bukur, D. and N.R. Amundson, Fluidized Bed Char Combustion Kinetic Models. Chem. Eng. Sci. 37 (1982) 17-25.
102. Irani, R.K. Studies in Fluidized Bed Reactors for Complex Systems. Ph.D. dissertation (Bombay, Indian Inst. of Technology, 1980).
103. Corella, J. and R. Bilbao. The Effect of Variation of the Gas Volume and of the Bubble Size on the Conversion in a Fluidized Bed. Intern. Chem. Eng. 24 (1984) 302-310.
104. Kai, T., S. Furusaki and K. Yamamoto. Methanation of Carbon Monoxide by a Fluidized Catalyst Bed. J. Chem. Engng. Japan. 17 (1984) 280-285.

A STOCHASTIC MODEL OF THE BUBBLE POPULATION IN A FLUIDIZED BED

R. O. Fox and L. T. Fan

Department of Chemical Engineering
Kansas State University, Durland Hall
Manhattan, KS 66506

In this paper the generalized master equation is employed to model the bubble population in a bubbling fluidized bed. The random variables are the number of bubbles of different sizes in various compartments. The fluctuations in the total surface area of the bubble phase are also studied by deriving experessions for the mean, variance and correlation function. Details concerning the generalized master equation can be found elsewhere [2].

DERIVATION OF MODEL

We first define a set of random variables as

$$\left\{N_{i,j}: \quad i \in \{1,2,\ldots,M\},\ j \in \{1,2,3,\ldots\}\right\}$$

where $N_{i,j}$ is the number of bubbles in compartment i with volume $j\Delta V$, M is the number of compartments, and ΔV is a small unit of volume which can be taken to be equal to the volume of the smallest bubble. Letting $A_{i,j}$ represent a bubble of size $j\Delta V$ in compartment i, we see that the interactions during an event of coalescence in compartment i can be represented by the following expressions;

$$A_{i,j} + A_{i,k} \rightarrow A_{i,j+k};\ 1 \le i \le M,\ 1 \le j,\ j \le k$$

Note that we have assumed that the final bubble volume is found by simple addition. Other assumptions could also be applied. Indeed, empirical evidence would seem to suggest that the final bubble size will be larger than that found by addition. However, to simplify the present discussion, simple addition is used here. Expressions for the movement of bubbles between compartments can be represented by

$$A_{i,j} \rightarrow A_{i+1,j};\ 1 \leq i < M,\ 1 \leq j$$

In addition to these events, bubbles are added to the bed in the first compartment and leave the bed from the last; expressions for these events are, respectively,

$$X_j \rightarrow A_{1,j} \text{ and } A_{M,j} \rightarrow Y_j;\ 1 \leq j$$

where X_j and Y_j are dummy symbols representing the inlet and outlet environments, respectively. In the following the set of random variables will be denoted by $\{N\}$, and the volume of the system by Ω. However, since the rates of transition are assumed to be dependent on the number density of bubbles in an individual compartment, the reduced volumes $F_i = V_i/\Omega$ need be introduced.

The rate of transition is the specific rate at which the process changes from one state to another state; a state being a specific set of values of the random variables in the random vector $\{N\}$. These changes could be due to coalescence, movement between compartments, or entrance and exit of bubbles from the bed. The rates of transition can thus be written as

$$W_t\left(\begin{Bmatrix}\text{initial state} \\ \text{before transition}\end{Bmatrix}, \begin{Bmatrix}\text{final state} \\ \text{after transition}\end{Bmatrix}\right)$$

Expressions for the rates of transition due to coalescence can now be derived. Letting $\beta_i(j,k)$ be the specific rate constant for coalescence between bubbles of size j and size (k-j) to form bubbles of size k in compartment i, the rates of transition due to coalescence can be written as

$$W_t\left(\{n_{i,k-j}, n_{i,j}, n_{i,k}\}, \{n_{i,k-j}-1, n_{i,j}-1, n_{i,k}+1\}\right)$$

$$= \beta_i(j,k)\ \frac{n_{i,j}n_{i,k-j}}{F_i\Omega} \qquad (1)$$

for all i,j,k; $k \neq 2j$ (coalescing bubbles of different sizes)

$$W_t\left(\{n_{i,j}, n_{i,k}\}, \{n_{i,j}-2, n_{i,k}+1\}\right)$$

$$= \beta_i(j,k)\,\frac{n_{i,j}(n_{i,j}-1)}{F_i\Omega} \tag{2}$$

for all i,j,k; k = 2j (coalescing bubbles of same size)

where all random variables which remain constant during a transition have been omitted from the notation for {n}. Similarly, letting $\gamma_i(j)$ be the rate at which bubbles of size j leave compartment i and enter into compartment (i+1), the rate of transition due to movement between compartment can be written as

$$W_t\left(\{n_{i,j},n_{i+1,j}\},\{n_{i,j}-1,n_{i+1,j}+1\}\right)$$

$$= \gamma_i(j)n_{i,j} \qquad \text{for all } i,j;\ i < M \tag{3}$$

For bubbles exiting from the bed through compartment M, the rates of transition due to exiting are

$$W_t\left(\{n_{M,j}\},\{n_{M,j}-1\}\right) = \gamma_M(j)n_{M,j} \qquad \text{for all } j \tag{4}$$

Finally, for bubbles entering the bed through the first compartment, i.e., compartment 1, immediately above the distributor plate at a rate of $\Omega f(j)$ bubbles of size j per unit time, the rates of transition due to entering bubbles are

$$W_t\left(\{n_{1,j}\},\{n_{1,j}+1\}\right) = \Omega f(j) \qquad \text{for all } j \tag{5}$$

Calculation of Jump Moments

The first and second jump moments, $\tilde{A}_i$ and $\tilde{B}_{ij}$, respectively, are defined as

$$\tilde{A}_i\left(\{\phi(t) + \Omega^{-1/2}z\}\right) = \Omega^{-1}\sum_{\xi_i} \xi_i W_t\left(\{n\};\xi_i\right) \tag{6}$$

and

$$\tilde{B}_{ij}\left(\{\phi(t) + \Omega^{-1/2}z\}\right) = \Omega^{-1}\sum_{\xi_i}\sum_{\xi_j} \xi_i\xi_j W_t\left(\{n\};\xi_i,\xi_j\right) \tag{7}$$

where $\sum_{\xi_i}$ is the summation over all possible values of ξ_i,

$$W_t\left(\{n\};\{\xi\}\right) \equiv W_t\left(\{n\},\{n\}'\right),$$

$\{\xi\} = \{n\}' - \{n\}$, (or $\xi_i = n_i' - n_i$,) and

$$\{n\} = \{\Omega\phi(t) + \Omega^{1/2}z\}$$

To calculate these jump moments for the present example, first the double subscripts are transformed into a single subscript by letting $\ell = j+\varepsilon(i-1)$ where j is the subscript denoting the bubble size, i is the subscript denoting the compartment number, and ε is the volume of the largest compartment divided by ΔV. The variable $n_{i,j}$ now appears in $n_{j+\varepsilon(i-1)}$. Note that each set of values for i and j corresponds to one and only one value of ℓ. Thus using either of these notations in no way changes the meaning of the random variable. Using the single subscript, however, allows for an easier application of Eqs. (6) and (7) which are given for a single subscripted variable.

Equations (6) and (7) imply a summation over all values ξ_i (and ξ_j). Looking at the rates of transition given by Eqs. (1) through (5), we see that the possible values of ξ_ℓ are 1, -1, and -2. To perform the summation implied by Eqs. (6) and (7), we can sum over all values of i, j, and k, and then to use a Kronecker delta to include only those terms which correspond to the proper value of ℓ given in ξ_ℓ. For example, Eq. (5) can be written using $\{\xi\}$ as

$$W_t\left(\{n_{1,j}\};\{1\}\right) = \Omega f(j) \qquad \text{for all } j$$

where $\xi_{1,j} = 1$. In single subscript notation, we then have

$$\xi_{j+\varepsilon(1-1)} = \xi_j = 1.$$

The summation in Eq. (6) for this term can thus be written as

$$\sum_{\xi_\ell} \xi_\ell W_t\left(\{n\};\xi_\ell\right) = \sum_j^{\varepsilon} \delta^k(\ell-j) \sum_{\xi_j} \xi_j\, W_t\left(\{n_j\};\{1\}\right)$$

$$= \sum_j^{\varepsilon} \delta^k(\ell-j)\Omega f(j)$$

since $W_t\left(\{n_j\};\{1\}\right) = \Omega f(j)$ when $\xi_j = 1$ and is equal to zero for all other values of ξ_j.

Expressions for $\tilde{A}_\ell$ and $\tilde{B}_{\ell m}$ are derived from Eqs. (1) and (2) as follows:

$$\tilde{A}_\ell = \Omega^{-1} \sum_i^M \sum_j^{\varepsilon} \left\{\sum_k^{\varepsilon} \frac{1}{2} \left[-\delta^k(\ell-k+j-\varepsilon i+\varepsilon) - \delta^k(\ell-j-\varepsilon i+\varepsilon) + \delta^k(\ell-k-\varepsilon i+\varepsilon)\right]\right.$$

$$[1-\delta^k(k-2j)]\beta_i(j,k)$$

$$\frac{n_{j+\varepsilon(i-1)}n_{k-j+\varepsilon(i-1)}}{F_i\Omega} \quad \begin{pmatrix}\text{bubbles of different}\\ \text{sizes coalescing}\end{pmatrix}$$

$$+\sum_k^{\varepsilon}[-2\delta^k(\ell-j-\varepsilon i+\varepsilon)+\delta^k(\ell-k-\varepsilon i+\varepsilon)]\ \delta^k(k-2j)\beta_i(j,k)$$

$$\frac{n_{j+\varepsilon(i-1)}\left(n_{j+\varepsilon(i-1)}-1\right)}{F_i\Omega} \quad \begin{pmatrix}\text{bubbles of same}\\ \text{size coalesing}\end{pmatrix}$$

$$+[-\delta^k(\ell-j-\varepsilon i+\varepsilon)+\delta^k(\ell-j-\varepsilon i)][1-\delta^k(i-M)]$$

$$\left.\gamma_i(j)n_{j+\varepsilon(i-1)}\right\} \quad \begin{pmatrix}\text{bubbles that transfer}\\ \text{between compartments}\end{pmatrix}$$

$$-\Omega^{-1}\sum_j^{\varepsilon}\delta^k(\ell-j-\varepsilon M+\varepsilon)\gamma_M(j)n_{j+\varepsilon(M-1)} \quad \begin{pmatrix}\text{bubbles that leave}\\ \text{the bed}\end{pmatrix}$$

$$+\sum_j^{\varepsilon}\delta^k(\ell-j)f(j) \qquad \text{(bubbles entering bed)} \qquad (8)$$

$$\tilde{B}_{\ell m}=\Omega^{-1}\sum_i^M\sum_j^{\varepsilon}\left\{\sum_k^{\varepsilon}\frac{1}{2}[-\delta^k(\ell-k+j-\varepsilon i+\varepsilon)-\delta^k(\ell-j-\varepsilon i+\varepsilon)+\delta^k(\ell-k-\varepsilon i+\varepsilon)]\right.$$

$$[-\delta^k(m-k+j-\varepsilon i+\varepsilon)-\delta^k(m-j-\varepsilon i+\varepsilon)+\delta^k(m-k-\varepsilon i+\varepsilon)]$$

$$[1-\delta^k(k-2j)]\beta_i(j,k)$$

$$\frac{n_{j+\varepsilon(i-1)}n_{k-j+\varepsilon(i-1)}}{\Omega F_i} \quad \begin{pmatrix}\text{bubbles of different}\\ \text{sizes coalescing}\end{pmatrix}$$

$$+\sum_k^{\varepsilon}[-2\delta^k(\ell-j-\varepsilon i+\varepsilon)+\delta^k(\ell-k-\varepsilon i+\varepsilon)]$$

$$[-2\delta^k(m-j-\varepsilon i+\varepsilon) + \delta^k(m-k-\varepsilon i+\varepsilon)]\delta^k(k-2j)\beta_i(j,k)$$

$$\frac{n_{j+\varepsilon(i-1)}\left(n_{j+\varepsilon(i-1)} - 1\right)}{F_i\Omega} \quad \begin{Bmatrix}\text{bubbles of same} \\ \text{size coalescing}\end{Bmatrix}$$

$$+ [-\delta^k(\ell-j-\varepsilon i+\varepsilon) + \delta^k(\ell-j-\varepsilon i)]$$

$$[-\delta^k(m-j-\varepsilon i+\varepsilon) + \delta^k(m-j-\varepsilon i)]$$

$$\left.[1-\delta^k(i-M)]\gamma_i(j)n_{j+\varepsilon(i-1)}\right\} \begin{Bmatrix}\text{bubbles that transfer} \\ \text{between compartments}\end{Bmatrix}$$

$$+ \Omega^{-1}\sum_j^{\varepsilon} \delta^k(\ell-j-\varepsilon M+\varepsilon)\delta^k(\ell-m)\gamma_M(j)n_{j+\varepsilon(M-1)} \begin{Bmatrix}\text{bubbles that leave} \\ \text{the bed}\end{Bmatrix}$$

$$+ \sum_j^{\varepsilon} \delta^k(\ell-j)\delta^k(\ell-m)f(j) \qquad \text{(bubbles entering bed)} \tag{9}$$

Employing the change of variables for $\{n\}$ indicated earlier, these expressions must then be expanded in power of Ω to yield

$$\tilde{A}_i\left(\{\phi(t) + \Omega^{-1/2}z\}\right) = \tilde{A}_i\left(\{\phi(t)\}\right) + \Omega^{-1/2}\sum_j z_j\tilde{A}_{ij}\left(\{\phi(t)\}\right) + 0(\Omega^{-1}) \tag{10}$$

and

$$\tilde{B}_{ij}\left(\{\phi(t) + \Omega^{-1/2}z\}\right) = \tilde{B}_{ij}\left(\{\phi(t)\}\right) + 0(\Omega^{-1/2}) \tag{11}$$

which define the functions $\tilde{A}_i\left(\{\phi(t)\}\right)$, $\tilde{A}_{ij}\left(\{\phi(t)\}\right)$, and $\tilde{B}_{ij}\left(\{\phi(t)\}\right)$.

Calculation of Means, Covariances, and Correlation Functions

In the present stochastic model of the bubble population in a fluidized bed, the number of bubbles of any given size is a random variable which can be characterized by its mean value, its covariance with the other random variables, and the time correlation functions. The later are functions which describe the dynamic character of the fluctuations of the random variable. It should be noted that, unlike a deterministic variable, at steady state the random variable will not reach a fixed value but will continue to fluctuate. The mean of the random variable, however, will behave like a deterministic variable and obtain a fixed value. A stochastic model thus can be used to describe systems which exhibit fluctuating behavior during steady-state operation.

For the present model, the means, covariances, and correlation functions can be found from the expanded jump moments (see [2]). In terms of the coefficients given in Eqs. (10) and (11) they are

$$\frac{d\langle N_i\rangle}{dt} = \tilde{A}_i\left(\{\langle N\rangle\}\right) \tag{12}$$

$$\frac{d}{dt}\,\mathrm{Cov}[N_iN_j] = \sum_k \left\{\tilde{A}_{ik}\mathrm{Cov}[N_kN_j] + \tilde{A}_{jk}\mathrm{Cov}[N_kN_i]\right\} + \Omega\tilde{B}_{ij} \tag{13}$$

$$\frac{d}{dt}\,K_{ij}(\tau) = \sum_k A^s_{jk}K_{ik}(\tau) \tag{14}$$

with $K_{ij}(0)$ equal to the steady-state value of $\mathrm{Cov}[N_iN_j]$. For our example we thus find

$$\frac{d}{dt}\langle N_{\ell,m}\rangle$$

$$= -\sum_{j=1}^{\varepsilon-m}\frac{\beta_\ell(m,j+m)}{V_\ell}\langle N_{\ell,m}\rangle\langle N_{\ell,j}\rangle + \sum_{j=1}^{\frac{m}{2}}\frac{\beta_\ell(j,m)}{V_\ell}\langle N_{\ell,j}\rangle\langle N_{\ell,m-j}\rangle$$

$$-\frac{\beta_\ell(m,2m)}{V_\ell}\langle N_{\ell,m}\rangle^2 - \gamma_\ell(m)\langle N_{\ell,m}\rangle + \gamma_{\ell-1}(m)\langle N_{\ell-1,m}\rangle[1-\delta^k(\ell-1)]$$

$$+\,\Omega f(m)\delta^k(\ell-1) \tag{15}$$

$$\frac{d}{dt}\,\mathrm{Cov}[N_{i,j}N_{\ell,m}]$$

$$= -\sum_{k=1}^{\varepsilon-j}\frac{\beta_i(j,j+k)}{V_i}\left\{\langle N_{i,k}\rangle\mathrm{Cov}[N_{i,j}N_{\ell,m}] + \langle N_{i,j}\rangle\mathrm{Cov}[N_{i,j}N_{\ell,k}]\right\}$$

$$- \sum_{k=1}^{\varepsilon-m} \frac{\beta_\ell(m,m+k)}{V_\ell} \left\{ \langle N_{\ell,k} \rangle \mathrm{Cov}[N_{i,j}N_{\ell,m}] + \langle N_{\ell,m} \rangle \mathrm{Cov}[N_{i,j}N_{\ell,k}] \right\}$$

$$+ \sum_{k=1}^{[\frac{j}{2}]} \frac{\beta_i(k,j)}{V_\ell} \left\{ \langle N_{i,k} \rangle \mathrm{Cov}[N_{i,j-k}N_{\ell,m}] + \langle N_{i,j-k} \rangle \mathrm{Cov}[N_{i,k}N_{\ell,m}] \right\}$$

$$+ \sum_{k=1}^{[\frac{m}{2}]} \frac{\beta_\ell(k,m)}{V_\ell} \left\{ \langle N_{\ell,k} \rangle \mathrm{Cov}[N_{\ell,m-k}N_{i,j}] + \langle N_{\ell,m-k} \rangle \mathrm{Cov}[N_{i,j}N_{\ell,k}] \right\}$$

$$- 2\left[\frac{\beta_i(j,2j)}{V_i} \langle N_{i,j} \rangle + \frac{\beta_\ell(m,2m)}{V_\ell}\right] \langle N_{\ell,m} \rangle \mathrm{Cov}[N_{i,j}N_{\ell,m}]$$

$$- [\gamma_i(j) + \gamma_\ell(m)] \mathrm{Cov}[N_{i,j}N_{\ell,m}]$$

$$+ [1-\delta^k(i-1)]\gamma_{i-1}(j) \mathrm{Cov}[N_{i-1,j}N_{\ell,m}]$$

$$+ [1-\delta^k(\ell-1)]\gamma_{\ell-1}(m) \mathrm{Cov}[N_{i,j}N_{\ell-1,m}]$$

$$+ \delta^k(\ell-i) \Big\{ \delta^k(j-m) \Big[\sum_{k=1}^{\varepsilon-j} \frac{\beta_i(m,m+k)}{V_i} \langle N_{i,k} \rangle \langle N_{i,m} \rangle + \frac{\beta_i(j,2j)}{V_i} \langle N_{i,j} \rangle^2$$

$$+ \sum_{k=1}^{[\frac{j}{2}]} \frac{\beta_i(k,j)}{V_i} \langle N_{i,k} \rangle \langle N_{i,j-k} \rangle + \gamma_i(j) \langle N_{i,j} \rangle$$

$$+ [1-\delta^k(i-1)]\gamma_{i-1}(j) \langle N_{i-1,j} \rangle + \delta^k(i-1)\Omega f(j) \Big]$$

$$+ \frac{\beta_i(m,j+m)}{V_i} \langle N_{i,m} \rangle \langle N_{i,j} \rangle - \frac{\beta_i(j,m)}{V_i} \langle N_{i,m-j} \rangle \langle N_{i,j} \rangle$$

$$- \frac{\beta_i(m,j)}{V_i} \langle N_{i,j-m}\rangle\langle N_{i,m}\rangle - \delta^k(j-2m) \frac{\beta_i(m,j)}{V_i} \langle N_{i,m}\rangle^2$$

$$- \delta^k(m-2j) \frac{\beta_i(j,m)}{V_i} \langle N_{i,m}\rangle^2 \Big\}$$

$$- \delta^k(j-m)\{\delta^k(\ell-i+1)\gamma_{i-1}(j)\langle N_{i-1,j}\rangle + \delta^k(\ell-i-1)\gamma_i(j)\langle N_{i,j}\rangle\} \quad (16)$$

$$\frac{d}{d\tau} K_{ij\ell m}(\tau) = -\sum_{k=1}^{\varepsilon-m} \frac{\beta_\ell(m,k+m)}{V_i} [\langle N_{\ell,m}\rangle {}^sK_{ij\ell k} + \langle N_{\ell,k}\rangle {}^sK_{ij\ell m}]$$

$$+ \sum_{k=1}^{\frac{m}{2}} \frac{\beta_\ell(k,m)}{V_\ell} [\langle N_{\ell,k}\rangle {}^sK_{ij\ell m-k} + \langle N_{\ell,m-k}\rangle {}^sK_{ij\ell k}]$$

$$- 2 \frac{\beta_\ell(m,2m)}{V_\ell} \langle N_{\ell,m}\rangle {}^sK_{ij\ell m} - \gamma_\ell(m)K_{ij\ell m}$$

$$+ \gamma_{\ell-1}(m)[1-\delta^k(\ell-1)]K_{ij\ell-1m} \quad (17)$$

The system can now be characterized by the means and the correlation functions of the random variables representing the number of bubbles of different volumes in various compartments.

EXAMPLE CALCULATION FOR THE FIRST COMPARTMENT

The rate-of-transition functions used in this example calculation are

$$\beta_k(i,j) = B_k V_k \Delta V(\frac{i}{j})[\frac{j-i}{V_k - j\Delta V}] \quad (18)$$

$$\gamma_k(j) = \frac{0.7908g^{1/2}(j\Delta V)^{1/6}}{h_k} \quad (19)$$

$$f(j) = \delta^k(j-1) \frac{G}{\Omega \Delta V} \quad (20)$$

where

$$\Delta V = 2.2676\left\{(U - U_{mf}) \frac{S}{N_0 g^{1/2}}\right\}^{6/5} \tag{21}$$

Equation (21) is an expression for the bubble volume at the distributor [1]. Equation (19) follows from considering the superficial velocity of a bubble of volume $j\Delta V$, and Eq. (20) from an assumption that all fluidizing gas above U_{mf} goes into the bubble phase. Further details concerning these expressions can be found elsewhere [2].

The parameters used in this example are

$U - U_{mf} = 2$ cm/s $\Omega f_0 = 120.1\ s^{-1}$

$N_0 = 10$ $B_1 = 330.5\ s^{-1}$

$S = 314\ cm^2$ $h_1 = 9.90$ cm

$g = 980\ cm/s^2$ $\Delta V = 5.23\ cm^3$

$V_1 = 3108.6\ cm^3$

All of these parameters values were chosen, save that for B_1, as being feasible for operating a laboratory scale bed. The value of B_1 was chosen only to illustrate the resulting model. Other values of B_1 could be determined through the fitting of experimental results, or by fundamental studies of bubble dynamics.

Since the rate of coalescence and rate of transfer between compartments are dependent on the compartment height, h_k, this value can be chosen so that the probability of transfer is greater than the probability of coalescence. In this example the value of h_1 has been chosen such that the maximum size bubble in compartment 1 is $4\Delta V$, which results from two coalescence events starting from bubbles of size ΔV. All other sizes would occur with a relatively low probability. Since the maximum number of coalescence events is restricted to two, the following coalescence events are possible:

$$A_{1,1} + A_{1,1} \rightarrow A_{1,2}; \quad A_{1,1} + A_{1,2} \rightarrow A_{1,3}; \quad A_{1,2} + A_{1,2} \rightarrow A_{1,4}.$$

All other coalescence events are considered to occur with very low probability and thus can be neglected.

At steady-state the following values are found from Eqs. (15) and (16),

$\langle N_{1,1}\rangle = 11.333$ $\langle N_{1,2}\rangle = 3.116$

$\langle N_{1,3}\rangle = 3.245$ $\langle N_{1,4}\rangle = 1.277$

$\mathrm{Var}[N_{1,1}] = 9.405$

$\mathrm{Cov}[N_{1,1}N_{1,2}] = -0.643$

$\mathrm{Cov}[N_{1,1}N_{1,3}] = 0.366$

$\mathrm{Cov}[N_{1,1}N_{1,4}] = -0.088$

$\mathrm{Var}[N_{1,2}] = 2.532$

$\mathrm{Cov}[N_{1,2}N_{1,3}] = -1.067$

$\mathrm{Cov}[N_{1,2}N_{1,4}] = -0.082$

$\mathrm{Var}[N_{1,3}] = 2.855$

$\mathrm{Cov}[N_{1,3}N_{1,4}] = -0.143$

$\mathrm{Var}[N_{1,4}] = 1.210$

These values of the covariances can now be considered as the initial conditions for solving Eq. (17) for the correlation functions, two of which are

$$K_{1112}(\tau) = e^{-\theta_1\tau}[-0.6416\cos(\phi\tau) - 10.8546\sin(\phi\tau)]$$

$$K_{1113}(\tau) = e^{-\theta_1\tau}[1.504\cos(\phi\tau) + 7.6055\sin(\phi\tau)] - 1.1385e^{-\theta_2\tau}$$

where

$$\theta_1 = 15.6971\ s^{-1},\ \theta_2 = 3.9565\ s^{-1},\quad \phi = 4.4283\ s^{-1}$$

Calculation of Bubble Phase Surface Area

Often the variable of interest is not the number of bubbles, but some other characteristic, e.g., the total surface area of the bubble phase in a compartment. In this case, the surface area of the bubble phase in compartment 1, S_1, can be related to the number of bubbles of size $j\Delta V$ in compartment 1, $N_{1,j}$. With s_j equal to the surface area of a bubble having a volume of $j\Delta V$,

$$S_1 = \sum_{j=1}^{4} s_j N_{1,j} \tag{22}$$

Assuming that bubbles are spherical, the specified parameters yield the following values for s_j;

s_1=14.570 cm^2, s_2=23.130 cm^2, s_3=30.308 cm^2, s_4=36.716 cm^2.

Then it follows that $\langle S_1\rangle = 382.431$ cm^2

and [see (2)] that

$$K_1(\tau) = e^{-\theta_1\tau}[2584.679\cos(\phi\tau) + 263.483\sin(\phi\tau)]$$

$$+ 1749.778e^{-\theta_2\tau} + 1113.733e^{-\theta_3\tau} \qquad (23)$$

where

$$\theta_1 = 15.6971\ s^{-1},\ \theta_2 = 3.9565\ s^{-1},\ \theta_3 = 1.2775\ s^{-1},\ \phi = 4.4283\ s^{-1}$$

Setting $\tau = 0$ in Eq. (23) yields

$$\text{Var}[S_1] = 548.190\ \text{cm}^4; \quad \text{Sd}[S_1] = 73.812\ \text{cm}^2$$

This indicates that fluctuations in the total surface area of the bubble phase are relatively large since the standard deviation is 19.30 percent of the mean.

A dimensionless correlation function can be found by dividing $K_1(\tau)$ by $\text{Var}[S_1]$, thereby yielding

$$\rho_1(\tau) = e^{-\theta_1\tau}[0.4744\cos(\phi\tau) + 0.0484\sin(\phi\tau)]$$

$$+ 0.3212e^{-\theta_2\tau} + 0.2044e^{-\theta_3\tau} \qquad (24)$$

From this expression, the characteristic time of the fluctuations can be found by integrating from zero to infinity (see [3]). For this example, it is 0.270 sec.

CONCLUSIONS

A stochastic model is presented for the bubble population in a bubbling fluidized bed; it is based on the generalized master equation. Expressions have been found for mean number of bubbles of different sizes in various compartments as well as for the correlation functions of these random variables.

The example calculation has indicated that fluctuations in the total surface area of the bubble phase in the first compartment have a standard deviation of 19.3 percent of the mean and a characteristic time constant of 0.270 seconds. These fluctuations are fairly large and relatively slowly varying compared to white noise.

ACKNOWLEDGEMENTS

This material is based upon work supported under a National Science Foundation Graduate Fellowship awarded to the first author.

NOTATION

$\tilde{A}_\ell$	first jump moment
$\tilde{A}_{\ell m}$	coefficient matrix from the expansion of $\tilde{A}_\ell$
B_k	constant appearing in the expression for the rate of coalescence in compartment k, s^{-1}
$\tilde{B}_{\ell m}$	second jump moment
$Cov[N_{i,j}N_{\ell,m}]$	$\langle N_{i,j}N_{\ell,m}\rangle - \langle N_{i,j}\rangle\langle N_{\ell,m}\rangle$, covariance of $N_{i,j}$ and $N_{\ell,m}$
F_i	V_i/Ω
g	gravitational constant, $cm\cdot s^{-2}$
G	$(U - U_{mf})S$, gas flowrate through bubble phase, $cm^3\cdot s^{-1}$
h_k	height of compartment k, cm
$K_1(\tau)$	correlation function of S_1, cm^4
$K_{ij\ell m}(\tau)$	$\langle N_{i,j}(0)N_{\ell,m}(\tau)\rangle - \langle N_{i,j}(0)\rangle\langle N_{\ell,m}(\tau)\rangle$
M	number of compartments
N_0	number of orifices in the distributor plate
$N_{i,j}$	number of bubbles in compartment i with volume $j\Delta V$
$\langle N_{i,j}\rangle$	expected value of $N_{i,j}$
s_j	surface area of a bubble of volume $j\Delta V$, cm^2
S	cross-sectional area of fluidized-bed, cm^2
S_1	surface area of the bubble phase in compartment 1, cm^2
$\langle S_1\rangle$	expected value of S_1, cm^2
$Sd[S_1]$	$(Var[S_1])^{1/2}$, standard deviation of S_1, cm^2
U	superficial gas velocity, $cm\cdot s^{-1}$
U_{mf}	minimum superficial gas velocity for fluidized, $cm\cdot s^{-1}$
V_i	volume of compartment i, cm^3

$Var[S_1]$	$<S_1^2> - <S_1>^2$, variance of S_1, cm^4
$W_t(\{\cdot\},\{\cdot\}')$	rate of transition, s^{-1}
$Z_{i,j}$	random variable corresponding to the fluctuating component of $N_{i,j}$

Greek Letters

$\beta_i(j,k)$	rate constant for the coalescence between bubbles of size j and size (k-j) in compartment i, $cm^3 \cdot s^{-1}$
$\gamma_i(j)$	rate at which bubbles of size j leave compartment i, s^{-1}
$\delta^k(\cdot)$	Kronecker delta where $\delta^k(0) = 1$ and $\delta^k(x) = 0$ for all $x \neq 0$
ΔV	small unit of volume which can be taken to be equal to the volume of the smallest bubble, cm^3
ε	volume of the largest compartment divided by ΔV
$\theta_1, \theta_2, \theta_3$	time constants appearing in correlation functions, s^{-1}
ξ_i	magnitude of change in random variable N_i
$\rho_1(\tau)$	dimensionless correlation functions
ϕ	time constant appearing in correlation functions, s^{-1}
$\phi_{\ell,m}(t)$	$<N_{\ell,m}>/\Omega$, cm^{-3}
Ω	system size, cm^3
$\Omega f(j)$	rate at which bubbles of size j enter the bed through compartment 1, s^{-1}

REFERENCES

1. Darton, R. C., R. D. Lanuaze, J. F. Davidson, and D. Harrison, 1977, Bubble growth due to coalescence in fluidized beds. Trans. IChemE. 55, 274.

2. Fox, R. O., 1985, Master's Thesis, Kansas State University, Manhattan, Kansas.

3. Horsthemke, W. and R. Lefever, 1984, Noise - Induced Transitions, Springer-Verlag, New York.

DESIGN ASPECTS OF INDUSTRIAL FLUIDIZED BED REACTORS.

RESEARCH NEEDS. SELECTED SUBJECTS

M.A. Bergougnou, C.L. Briens

Chemical Engineering Dept.
University of Western Ontario
Faculty of Engineering Science
London, Ontario, Canada N6A 5B9

D. Kunii

Chemical Engineering Dept.
University of Tokyo
Bunkyo-Ku, Tokyo 113
JAPAN

SUMMARY

Part A gives general guidelines for the design of large commercial fluidized bed reactors with respect to the following aspects: (1) solids' properties and their effect on the quality of fluidization; (2) bubble size control through small solid particle size or baffles; (3) particle recovery by means of cyclones; (4) heat transfer tubes; (5) solids circulation systems; (6) instrumentation, corrosion and erosion, mathematical models, pilot plants and scale-up techniques.

Part B presents a synthesis of the literature available on selected subjects of primary importance to the design of fluidized bed reactors. Correlations and models for the prediction of the minimum fluidization and bubbling conditions, the expansion and voidage of fluidized beds, the size and velocity of gas bubbles are thus presented. Correlations and models to characterize entrainment of solids from fluidized beds and to provide guidelines for the design of gas distributors are also reviewed.

PART A: DESIGN ASPECTS OF INDUSTRIAL FLUIDIZED BED REACTORS

1.0 INTRODUCTION

The reactor is the heart of the chemical process. Although it is only about 20% of the total cost of the plant, it is, by far, the most important piece of equipment. Should it fail to meet conversion or product specifications, no amount of sophistication elsewhere in the process will be of any avail.

Among the very few types of chemical reactors available, the fluidized bed reactor is a major reactor system and one of the workhorses of the petroleum, chemical, mining, nuclear and other industries. Unfortunately, although fluidized solids technology is being applied in an ever increasing number of new applications, the understanding of fluidization phenomena is still and will be for a long time, very rudimentary. Thus, of necessity, the design of industrial fluidized bed reactors is primarily based on past experience and **know-how** . In designing such reactors, the chemical engineer must, judiciously, blend experience with whatever little useful theory is at hand. As in all designs, **attention to details** is most important. It is the aspect which one forgets to deal with which will wreck the process. Indeed it is what one ignores which leads to serious difficulties later on. The greatest problem in this field consists in the fact that fluidized bed reactor development has to be carried out again and again for every new process. Thus, although there are general guidelines for the design of industrial fluidized bed reactors, the interaction between kinetics and fluid mechanics is, most often, of such a complexity that the reaction of interest has to be actually tried in a scaled-down version of the kind of reactor contemplated for the commercial process. This makes process development expensive and time consuming. There is, therefore, a great need to improve our design and scale-up techniques to gain precious lead time and possibly eventually bypass the pilot plant or at least reduce its role and its cost.

The design and scale-up of fluidized bed reactors is all the more difficult that the diversity of applications results in an extremely wide range of operating conditions, e.g. bed depths from a few centimeters to fifteen meters, bed diameters from a few centimeters to about twenty meters, superficial gas velocities from a few centimeters per second to several meters per second (from a few times the minimum fluidization velocity to hundreds of times that value), operating temperatures from the sub-freezing range to 1300°C, particle diameter from a few microns to several millimeters . . .

2.0 OBJECTIVES AND SCOPE OF THE LECTURE

The prime goal here is to outline some of the most important design aspects of industrial fluidized bed reactors. Indeed, it is not feasible or even desirable to review everything as a complete treatise would not even suffice. Thus, background material will be left for the lectures and the workshop. Should some be needed in order to read the paper, well-known textbooks are available [1],[2]. The main idea is to show how to put it all-together in order to have a workable commercial reactor. For those with little or no formal background in fluidization, tutorials will be offered outside lecture hours so that this lecture can be understood. Detailed and numerical recommendations will be given orally during the lecture for two reasons: (1) because they are based mainly on the personal experience of the lecturers; (2) because, not being based on extensive academic experimental work, they have to be heavily qualified and have to be used with judgement. Once given in the written form, people use the numbers and forget the qualifications.

3.0 SUGGESTED APPROACH FOR DESIGNING A LARGE-SCALE INDUSTRIAL FLUIDIZED BED REACTOR.

Because of the particular complexities presented by a new process, industrial fluidized bed reactor design is almost always done as a custom job. <u>A handbook approach is just not possible</u>. If the company does not have extensive experience in the fluidization field (or even if it has), consultants with years of practical design and development experience should be consulted. Management would be foolish, indeed, to embark on a venture which is going to cost possibly hundreds of millions of dollars and require several years of effort without getting its hands on the best information, experience and judgement available.

The next step is to decide the type of reactor needed and which can realistically be afforded. It is always difficult and/or physically or economically impossible to accommodate all the demands one might wish to place on a reactor. A good design is always a good compromise.

Over and above specific process objectives, there is a long list of requirements that a good reactor has to meet, e.g. low investment and operating costs, maximum flexibility and reliability, efficient use of energy and resources, safety, ability to operate virtually unattended, simplicity of start-up and shut-down, maximum turndown possible, maximum shop fabrication of various elements to avoid costly field erection, accessibility for inspection, ability to withstand unskilled operation without serious consequences for the process or from a safety point of view ... A few of the most important design aspects will, now, be reviewed.

4.0 PROPERTIES OF THE SOLIDS

A fluidized reactor has to be designed around specific solids, be they catalytic or not. Their properties are, thus, very important. Therefore, whenever it can be done, e.g. in high performance catalytic reactors where one has some control on the nature of the solids, and unless there are no contra-indications, the **particle size distribution** of the solids should be so tailored as to have a mean around 60-80 microns (fine solids) and a particle size spread such as d_{pmax}/d_{pmin} is between 11 and 25 where d_{pmax} is the diameter of the largest "significant" group of particles [2], dp_{min} is the diameter of the "smallest" significant group of particles. The latter group is very difficult to define because extremely small particles are firmly attached to larger particles through Van der Vaal's or other forces (e.g. electrostatic in certain conditions). The definition of d_{pmin} requires, thus, a lot of judgement. A well-tailored and fine catalyst, all other characteristics being satisfactory, will give good bed fluidity,small bubbles, excellent heat transfer and contacting, easy flow of solids between vessels, good staging (approach to plug flow) on the gas, ...

There is a big need for data on the fluidization characteristics of coarse solids (1 mm particle size and above). For such systems, at low multiples of the minimum fluidization velocity (e.g. around 2 or 3), bubbles are probably of the open kind, i.e. bubbles behave as cavities, the gas flows mainly in-between particles and gas conversions could be high. At high multiples of the minimum fluidization velocity (e.g. above 3), it could well be that gas conversions fall. A systematic study is needed here. Particles as large as 2 cm in diameter ought to be considered. The influence of the particle size distribution on the quality of fluidization ought to be investigated. It is important also to study the heat and mass transfer aspects and the modelling of these systems. Control of the residence time distribution of coarse solids in the reactor is desirable in order to increase the staging (approach to plug flow) on the solids. High stagings are needed in many processes, e.g. iron ore reduction. Equipment is needed which will discharge large particles from high-pressure fluidized beds without taking out too much gas and too many fines.

Particles cannot be too fine, because fine particles are cohesive and will not fluidize well: they will rathole, channel and ball up. As a general rule, cohesive particles, even if not too fine, will fluidize poorly. Sometimes additives can be found which will lead to satisfactory fluidization. Use of vibration or stirring can help fluidize fine, cohesive particles at the price of more equipment complexity and cost.

Mixing characteristics of mixtures of coarse particles ought to be researched. These are of interest in titanium tetrachloride manufacture, in coal combustion and gasification. Generally speaking, there are many applications for large particle systems. For instance, they could be useful for the combustion of coarse coals, the incineration and gasification of solid waste materials, the sintering and direct reduction of ores, the clinkering of cement, the recovery of heat from molten slags and the recovery of waste heat from gases.

When particles grow by deposition of material during reaction, e.g. deposition of coke during fluidized coking of petroleum residual fractions, high velocity jet attriters might have to be used in the bed to keep the particle size from becoming too coarse and provide nuclei for the deposition of the coke. Agglomeration in fluidized beds needs to be studied, especially when the agglomeration material is injected as a liquid or as a slurry. Such systems are not easy to scale-up. There are numerous potential applications for agglomerating fluidization.

Particles should not be too attritable, otherwise excessive amounts of fines will be created and the solids will quickly become too fine and will not fluidize properly. Lots of solids are then lost in the cyclones because of their small particle size. This is especially important for the case of expensive catalysts. It is not economical, in general, to replace the catalytic charge more than each 3 or 4 months. Particles have, thus, every opportunity to attrit excessively if they are too attritable.

If the solids are the desired products of the reaction (e.g. iron oxide being reduced in a fluidized bed reactor by carbon monoxide-hydrogen mixtures), attritability, although still important, is not as critical, because particles do not stay very long in the reactor (hours instead of months for catalysts). However, even in this case, too many fines may play havoc not only with the quality of fluidization but also with the efficiency of the reduction. In a multistage reduction unit, for instance, iron oxide fines will follow the solids down the reactor from stage to stage and will be reduced in the lower stage together with the coarse particles. They will then be reentrained into the upper stage where they will be re-oxidized to iron oxide. This internal cycling of the fines in a multistage reduction unit obviously leads to a loss of reduction efficiency.

Ways ought to be found to deal with attritable, friable solids. In a general way, techniques are needed to treat natural solids with unfavourable characteristics, e.g. friability, stickiness, ... It is desirable to develop prediction methods for such systems, especially when scale-up to large reactor sizes is contemplated.

Particles may become "case hardened" during processing. In many processes, the structure of the particles evolves during their stay in the reactor. For instance, in drying, the outside surface of the particles, when in contact with very hot gases, can become hardened and lose its porosity, thus making further drying of the particle core very difficult. Similarly, during reduction of iron ore, the crystalline structure of the reduced iron shell may collapse under the prolonged action of heat and of the hammering of other particles. This reduced shell becomes, then, quasi-impermeable to reduction gases, trapping a core of unreduced iron oxide at the center of the particle. To prevent this from happening, massive amounts of reduction gases have to be made available to the particles in order to reduce them before they become case hardened. High superficial velocities and pressures are, thus, required and lead to sophisticated and expensive equipment.

Altogether, particle size control is not a trivial problem. More research is needed in this area as fluidization is used to treat solids with increasingly undesirable characteristics.

5.0 BUBBLE SIZE CONTROL

Most commercial fluidized beds have to operate at high velocities, i.e. at high multiples of the minimum fluidization velocity, in order to minimize vessel diameter and reactor cost: For a given production rate, low gas velocities would lead to vessels of excessively large diameter and cost.

At high multiples of the minimum fluidization velocity, most of the gas passes through the bed in the form of bubbles. The feed gas has to be able to get out of the bubbles in order to contact the catalyst and react. Large bubbles move very fast through the bed and make it very difficult for the gas to contact the solids. It is, thus, essential to keep bubble size down if one wants high performance, i.e. high conversion per pass and high selectivity. With coarse solids, such as sand and mineral ores, bubbles can get very big and contacting between gases and solids is poor. This is a case where baffles, among other things, can be useful to break bubbles and keep their size small. Coarse solids are typically above 1 mm. in particle size.

In high conversion reactors, conversion and selectivity correlate pretty much with the percent of fines in the unit. The latter is generally defined as the percent by weight of the catalyst hold-up under 40 microns. For conversions around 98% and above, a reactor would have to have normally from 20 to 35% of 40 micron fines. This is very difficult to do in a commercial unit and requires excellent cyclone design and unit operation. Many fluidized catalytic crackers, for instance, have difficulty in maintaining even 5-10% fines. When the fines get low, more carbon is left on the regenerated catalyst because of large bubbles in the regenerator. This leads to a higher carbon make in the reactor with a serious attendant loss of selectivity. Furthermore, circulation of the catalyst between vessels becomes rough and erratic.

Also, oxygen breakthrough might take place in the regenerator initiating afterburning of the carbon monoxide in the dilute phase above the bed. In combustion units, where liquid or gaseous fuels are injected directly into the bed of coarse particles, large bubbles of cracked fuel vapor move rapidly through the bed, do not have time to mix and react with air bubbles and, thus, inject combustible products into the dilute phase above the bed. The combustion is said to blow out of the bed and high temperatures and possibly a flame are present above the bed. Another possibility with heavy liquid fuels is the formation of soot which leaves with the reactor products and might play havoc with product quality.

Fines might be present inside bubbles. It is very important to know their amount for modelling purposes. Investigators have ignored this aspect which is an important one in the case of highly exothermic and fast reactions. For instance, the combustion of coal particles might well take place only inside and immediately around the bubbles. This is all the more true in the grid zone where the concentration of the reactants is high. In that zone, coal particles, spent cracking catalyst particles (in the regenerator vessel), metal and metal oxide particles in the case of chlorination processes, can become very hot and possibly sinter or at least be damaged. The study of exothermic reactions in the grid jetting region ought to be carried out. Equipment should be developed to study particle concentration inside bubbles and grid jets.

Bubbling should be studied in the difficult conditions encountered in commercial units, e.g. at high temperatures and pressures. Instrumentation should be developed for the easy tracking of bubbles in large units.

6.0 GRID DESIGN

Good grid design is essential for satisfactory operation of large fluidized bed reactors. The effects of bad design in this area cannot, in general, be corrected. The important design criteria for grids should be systematically researched. The case of solids with unfavourable characteristics, e.g. large stones, garbage, wood, sticky materials should be considered with respect to their behaviour in the grid region.

Materials which can fuse may lead to big grid troubles (asphalt, pitch, coal, lignite). In combustion and gasification, ash can fuse, sit on the grid and plug it. The introduction of oxygen is especially critical in these cases. A two-tier gas distribution is preferred. The bed can be fluidized at a lower level at low velocity by steam or producer gas. The oxygen is then injected higher up by special nozzles in a prefluidized bed. Hot spots, sintering, agglomeration are, thus, avoided.

Another example of a two-tier gas distribution system could be in a process of the acrylonitrile synthesis type. Here, two gases have to be injected separately into a bed of fine catalyst, i.e. air on one side, ammonia, propylene and steam on the other side [1]. They cannot be easily pre-mixed beforehand because of the risk of explosion. Air could be injected into the bed by a lower pipe grid ramp; at an appropriate distance above the air injection ramp, another ramp would be located to inject the mixture of ammonia, propylene and steam into a well fluidized mixture of catalyst and air. Here it is essential that air, ammonia and propylene be mixed only in the presence of a well fluidized catalyst; if they were mixed in the absence of catalyst, combustion could take place with attendant loss of selectivity to acrylonitrile. Thus, the catalyst must be already well fluidized at every point where propylene and ammonia are injected. The holes of the pipe grids are generally on the underside of the pipes for greater ease of start-up from the defluidized state. The two grid systems have to be separated enough from each other so as not to be eroded by the grid jets. The two pipe-grid system may be a preferred solution anyway if a large turndown ratio is desired, i.e. some of the pipes may be shut-off to keep the others operating at optimum conditions.

When designing pipe grids, the diameter of the pipe should be large enough so that the velocity head in the pipe and the friction pressure drop along the whole pipe are both under 10% of the value of the pressure drop through the holes. Also when cold gas is fed to a pipe grid in a hot reactor, the effect of the gradual rewarming of the gas along the pipe on gas distribution should be considered and, if necessary, compensated for.

Grid pressure drop should be sufficient to prevent particles from falling under the grid plate or from entering into pipe grids during operation [3]. Probably, all other things being equal, the greater the bed diameter, the worse the dumping and weeping of solids through the grid. Grid weeping and dumping can be very detrimental to grid operation: (a) If the particles are erosive, they can erode grid holes when carried back up through the grid by the gas. Each grid hole could be protected by a special anti-erosion washer but this is expensive; (b) If the solids are sticky, dumped particles will plug grid holes on their way back to the bed; (c) If the dumped particles can react exothermically with the gas, they will do so under the grid and subject the latter to high temperatures for which it was not designed. When this happens, the grid will warp and will be destroyed.

Another important area is the area of grid jet momentum, mass and heat dissipation [4],[5],[6]. The fluidization gas comes with high concentrations of feed components from the grid plenum, jets into the fluidized solids from each grid hole and starts mixing with the already partially reacted gas in the bed. It is necessary to know how fast the concentration, temperature and momentum of this feed dissipates after it exits from the grid hole. For instance, momentum dissipation has to be known when cooling tubes have to be positioned near the grid. If they are placed too close to the grid, they will be eroded by the sand-blasting action of grid jets.If they are placed too far away, the zone of the grid where lots of reaction takes place (due to the high concentration of the feed there), will become too hot. This will lead to undesirable byproducts (catalyst selectivity loss) and to possible deactivation and even fusion of the catalyst.

Grid plugging is an important design item which can lead to extremely short runs when particles have a tendency to be sticky. For instance, in iron ore direct reduction processes, in a multi-stage unit, reduced fines which are not stopped by interstage cyclones tend to climb from bed to bed and plug the interstage grids. Long conical nozzles have been found to practically eliminate the plugging [7].

Constructional aspects of the grid are very important. For instance, one should provide for grid expansion, especially if the bed is very hot and the windbox is cold. A number of designs are available to protect the grid against high temperatures. There are many other important aspects of grid design which cannot be mentioned here for lack of space.

7.0 ENTRAINMENT AND PARTICLE RECOVERY SYSTEM

As already mentioned, fine catalyst is a requirement for high conversion reactors. Also, for reasons of economics and good contacting efficiency, fluidized reactors operate at high velocities (from 0.30 m/sec to as high as 1.5 m/sec for fine catalysts). At the end of the high velocity range, and with fine particles, we no longer have a conventional fluidized bed of the type discussed in the literature. It is an "entrained bed" with the following characteristics: (a) There is no longer a free surface at the top of the bed and bed density decreases continuously from the grid to the cyclones; (b) The particles stay a very short time in the bed at any one time. They go almost right away to the cyclones and are recycled to the grid by cyclone diplegs; this process is repeated over and over again. The whole bed is, thus, in a dynamic state and is sometimes called "dynamic fluidized bed" or "cloud reactor" to distinguish it from low velocity beds; (c) Large particle fluxes are entrained to the cyclones. The hold-up in the dilute phase above the bed is very substantial; as a consequence, a good deal of reaction may take place in the dilute phase.

Control of the entrainment is crucial at all times to maintain adequate amounts of fines in the reactor, say around 20 to 40 percent of particles under 44 microns for high performance, high conversion, high selectivity reactors. This is not easily achieved on a routine basis.

Fines recovered in primary cyclones should, normally, be returned to the grid region (within 30 to 60 cm of the grid), if the bed has a high aspect ratio. For shallow beds, this is less critical, because these beds have a good turnover of solids from top to bottom. It might be enough, in these cases to return fines below the surface of the bed, say, through a trickle valve. The purpose of re-introducing fines as deeply as possible into the bed is to give them a chance to react with fresh feed gas issuing from the grid. Moreover, the presence of fines in the grid region ensures small bubbles, say around 5 cm. in diameter, good bed fluidity and excellent contacting.

Fines collected by cyclones are returned to the region of the grid by means of dipleg tubes. One should make sure that the dipleg diameter is large enough to accommodate the flow of particles coming down from the cyclones; if this is not done, cyclones will flood and will not work. In a commercial unit, as a rule of thumb, dipleg diameter should be at least 10 cm or they might bridge over with particles. Minimum diameters can be computed by assuming that particles go down the dipleg in the unfluidized state. In practice, hopefully, the particles are still in the fluidized state while going down the dipleg or they would bridge. The dipleg cannot be too large, either, or bubbles of entrained gas, instead of following the solids down the dipleg, would climb upwards and interfere with the operation of the cyclone.

When solids do not retain aeration gas for a long time or if they have a tendency to be slightly tacky and cohesive or if the dipleg is very long, one may have to aerate diplegs to prevent solids from bridging there. As one moves down the dipleg towards the grid, pressure goes up due to the hydrostatic pressure of the seal of solids built up in the diplegs; gas entrained in-between the particles sees its volume reduced by the increased pressure to the point where it cannot sustain fluidization anymore. One may have, thus, to inject gas into the dipleg, at appropriate places, to compensate for this contraction of the interstitial gas. Aeration taps may be located every 2 to 3 meters along the dipleg. The dynamics of cyclone diplegs ought to be investigated as cyclones can be a source of great trouble in the field.

When the bed is not fully entrainable at the superficial velocity in the reactor, i.e. when the superficial velocity is lower than the settling velocity of the coarse particles, coarse particles stay in the bed and in the dilute phase up to the transport disengaging height. They, thus, have plenty of time to react. The fines on the contrary spend very little time in the bed and a lot of time in cyclone diplegs. Because of that and paradoxically so, they do not have that much time to react. Their main contribution is probably in maintaining bubble size small in the grid region. As a consequence, in units where the solids take part in the chemical reaction, e.g. in coal combustion and gasification, fines are difficult to react.

There is a need for more phenomenological and theoretical investigations on elutriation and carry-over of particles from large fluidized beds. Among other things one ought to study: (a) The state of dispersion of the solids in the gas stream above the surface of the bed; (b) The behaviour of particle agglomerates in the dilute phase; (c) How to control the rate of particle carry-over from the bed: special internals, floating contactors might be investigated; (d) How to prevent the stripping of fines from the solids going down cyclone diplegs and downcomers without the use of troublesome mechanical devices.

8.0 HEAT TRANSFER

One of the advantages of the fluidized bed reactor is that the heat of reaction is relatively easily removed compared with other reactors, e.g. fixed bed, tubular fixed bed reactors. Heat transfer coefficients are an order of magnitude higher in fluidized bed reactors. Moreover one can, very often, dispense with heat exchanger surfaces, the heat of reaction being removed, then, by injection into the bed of a cold solid or liquid. The situation would be similar for an endothermic reaction.

In general, vertical cooling tubes are used. At high temperatures, say, around 1000°C, they are preferred because any other orientation must support its own weight and material strength is reduced at high temperatures.

When the solids are corrosive, for instance in the oxichlorination of ethylene, horizontal tubes might corrode faster, especially at low superficial velocities when defluidized solids might deposit on top of the tubes. Otherwise, at reasonable superficial velocities, horizontal tubes can be as effective as vertical tubes. In shallow beds (around 1-2 m height), only horizontal tubes will fit. High-pressure systems liberate more heat per cubic meter of bed and so require tall beds to accommodate all the exchanger tubes needed. Furthermore, with tall beds, say, of the order of 10 m in height, the reactor shell diameter is smaller and cheaper. This leads to the preferential use of vertical tubes. For very exothermic reactions, it is the volume of the bed required to house the heat transfer tubes which dictates the height of the bed. In this case, very often, the catalyst has to be diluted to keep the products from reacting further; typical dilutions could be around 50%.

Vertical heat exchanger tubes are generally of the U-tube type and are clustered into bundles. The idea here is to "modularize" the bed in case of a tube leak, for maintenance and for turndown. The corners cannot, generally, be filled by tube bundles; sometimes, dummy tubes are used to keep the geometry of the bed uniform across the reactor cross-section. Each bundle is separately fed with coolant so that it can be taken off operation. The inlet and outlet coolant manifolds to the bundles are staggered in the vertical direction so as not to unduly obstruct the flow of the fluidizing gas. Tube bundles have to be braced and stiffened by crossbars between the tubes so that the natural frequency of the assembly is at least 10 Hz above the vibration induced by bubbles. Since bubble frequency can be at most, say, about 10Hz, this means that the natural frequency of tube bundles should be at least 20 Hz. Heat transfer tubes are often located on a triangular pitch. They cannot be packed too closely because, then, the heat transfer coefficient goes down. Pitch to diameter ratios might be 2 at the lowest end of the range, depending on particle size and other factors. The size of the bundle is dictated by ease of pick-up during maintenance and by the size of the manhole at the top of the reactor through which the bundle has to be pulled out. With vertical tube bundles it is not possible to program more cooling at the bottom of the reactor where heat liberation is highest (this is possible with horizontal tubes). In the case of vertical tubes there may be as much as 10°C temperature difference betwen top and bottom in a high aspect ratio, very exothermic, reactor. One has, thus, the choice between undercooling the bottom or overcooling the top and possibly quenching the reaction there somehow at that point.

If a tube containing high-pressure boiling water breaks, the vessel shell has to be kept from being overpressurized. This is sometimes done through the use of rupture disks on the reactor. In other cases, one assumes that the steam can be safely evacuated through the cyclones and the equipment downstream.

There are considerable uncertainties in the literature about the computation of the heat transfer coefficient. Thus, one has to be generous with heat transfer surface unless one has company data to back up the design. Overdesign may make control and high turndown more difficult to attain.

From an energy conservation point of view one should produce steam or superheated steam inside cooling tubes if the temperature level permits. Eventually, all heat has to be transformed into steam for power production or process heat duties. Because of its high latent heat, steam is an excellent medium for heat transport over long distances in the plant. It is relatively cheap and non-toxic. If any auxiliary coolant is used, more equipment is needed to produce steam and the overall thermodynamic availability is consequently lower. However, DOWTHERM, molten salts might be somewhat safer due to the absence of high-pressures in the tubes.

When water is used as the coolant, one can adopt a pressurized (non-boiling) or a boiling regime. The latter is generally chosen and one tries to be in the nucleate boiling range for high heat transfer coefficients and heat fluxes, thus reducing the size of the heat exchanger. The temperature difference between the wall of the tube and the core of the boiling water may be of the order of 5 to 40°C, depending on the overall operating conditions. According to best heat exchanger practice only a fraction of the water (say vapour is 10% by volume) is allowed to vaporize in the tubes in order to stay in the nucleate boiling range (overall heat transfer coefficients for a nucleate boiling system are about double those for dry steam). In cases where the bed would be very hot, dry steam condition might lead to tube burnout. This cannot happen generally because reactors operate at more moderate temperatures where "dry tubes" are not damaged. Two-phase flow velocities in the tubes have to be high enough, specially in the downleg of the U-tube to prevent liquid-vapour segregation.

For turndown purposes, a few tube bundles can be blocked off. For low aspect ratio beds, lateral temperature gradients might be created because lateral solids mixing is not as good as vertical solids mixing. Horizontal tubes would be better for turndown although they are rarely used because of maintenance and other problems. For maintenance, for instance, the whole top of the reactor has to be taken off to service horizontal tube pancakes.

A bad feature of the boiling water heat exchanger is the possibility of a tube split. Anybody who says that a catastrophic accident cannot happen is not realistic. A scenario should be studied involving the maximum credible accident one can think of. The probability of such an accident should be estimated. These numbers should be used very carefully as the only way to validate these probabilities would be to have a few good accidents. Of course, nobody in his right mind would accept that as a method of validation of accident probabilities. This question is becoming urgent with the generalized use of huge single train fluidized bed processes. Once started, the conflagration could spread to fuel tank farms and to the whole industrial complex. One has to keep this in mind when planning the implantation of a large reactor. A large amount of research work is necessary before one can have truly failsafe and at the same time economical systems, not only with respect to operational safety but also with respect to earthquakes, typhoons or other similar phenomena. As plants gets larger, reliability is an important criterion. Research should be carried out to make it possible to compute, at the design stage, the reliability of a given plant.

Work is needed to get heat transfer coefficients for large plants. So far, one has only correlations obtained in laboratory scale equipment. Here a large national fluidization facility would be invaluable. After this is done, a major effort could be started to develop theory and models for fluidized bed heat transfer. Novel equipment should be invented to measure local heat transfer coefficients in both fixed and fluidized beds. Global heat transfer coefficients for a bed are not enough if one is to obtain satisfactory theory and correlations. The right kind of heat transfer probe might well be able to measure the radiation contribution directly, leading to an estimate of the convection contribution, when operating at high temperatures.

9.0 BAFFLES AND MULTI-STAGE FLUIDIZED BEDS

We have already mentioned that, with coarse solids, say around and above 1 mm in particle size, bubbles are very large, possibly up to 1 m in diameter, and contacting between gases and solids is poor. Very often, e.g. in processing of natural products, mineral ores, one cannot change the characteristics of the solids and contacting has to be improved by other means, e.g. by horizontal baffles. The latter consist generally of perforated plates, with an open area of about 10 to 40% (versus less than 1% for grid support plates), located horizontally across the bed and well sealed at the walls of the vessel. Commercially, there may be a baffle every meter or so in the bed. Each baffle effectively divides the bed into horizontal compartments. A fluidized bed is set up on each baffle with a small dilute phase under the next baffle, (typical dilute phase heights could be of the order of 30 cm). Bubbles coming up through the bed are broken up by the dilute phase zones under the baffles; the latter, thus, prevent bubbles from becoming too large and, in so doing, improve gas-solid contacting very markedly. Well designed horizontal baffles can, in certain cases, bring up the conversion of a large commercial plant operating on coarse solids from 50-70% to what it would be in a laboratory unit, say 95% and more. This is done at the price of more mechanical and operating complexity. Overall, the situation can be summarized by saying that high conversions per pass can be obtained in a commercial unit by using: (a) either a fine (around 60-80 m mean particle diameter) catalyst with a "good" spread of particle sizes; (b) or baffles.

The hydrodynamics of horizontal baffles have been studied recently and the status of the field has been assessed [8]. Horizontal baffles impede axial mixing of the solids and can, thus, create vertical temperature gradients in the reactor unless solids are brought down again, artificially, to the grid level by external transfer tubes or internal draft tubes. With fine catalysts, axial mixing fluxes could be of the order of 50-150kg/m /s. Baffles could cut this flux, typically, by a factor of 5.

Horizontal sieve-tray baffles reduce the requirement for catalyst hold-up in the unit. This can lead to a significant saving if the catalyst is expensive, i.e. around $25/kg. However, the catalyst must be more attrition resistant. Because of the good contacting, coarse catalyst, say around 200 μm in particle size) can be used and catalyst losses in cyclones are decreased.

Disadvantages of sieve tray baffles are increased mechanical complexity and reduced flexibility and turndown of the unit. When starting up from a de-fluidized bed, one has to be extremely careful not to bump the bed through high superficial velocities which might destroy the baffles.

Although horizontal baffles reduce top to bottom mixing of the solids, they do not completely prevent it. There are cases where one needs effective solids segregation between the top and the bottom of the fluidized bed reactor. One has, then, to rely on a brute force approach, i.e. use several superposed beds in the same shell, each one equipped with its own grids, cyclones and solids transfer pipes between beds. Even then, as already mentioned, very fine particles, (in the micron range) escape cyclone collection between stages and are close to being well mixed in the whole reactor (moreover, if fines are sticky, they can plug interstage grids). Top to bottom segregation is almost complete, though, on coarser particles.

10.0 SOLIDS CIRCULATION SYSTEMS

One of the greatest advantages of a fluidized bed system is that solids can be transported easily in and out of vessels and in between vessels for thermal balance and other process requirements. Newby, Katta and Keairns [9] have made a very thorough study of circulation systems for application to fluidized bed coal combustion units. An inter-vessel solid circulation system, e.g. in a catalytic cracker or a fluid coker, must have the following characteristics: (a) The rate of solid flow between the vessels must satisfy the thermal balance and the catalyst requirements of the unit; (b) Solids of a specific size distribution must be transported at specified conditions of temperature and pressure; (c) Must be operable at a range of catalyst flowrates allowing a large turndown ratio for the unit and ease of start-up, shutdown and maintenance; (d) Must respond quickly to a desired change in operating conditions. This is all the more desirable if direct computer control is contemplated; (e) Reliability must be high because many units are large, single-train units at the heart of large industrial complexes. Detrimental phenomena, e.g. agglomeration, plugging, erosion, thermal stress, slide valve malfunction, must be eliminated almost entirely; (f) Catalyst attrition in transport lines must be low compared to attrition elsewhere in the plant; (g) Pressure build-up in standpipes and seals between vessels must be good; (h) Should not involve high-maintenance items or parts, if at all possible, such as high-temperature slide valves; (i) Should minimize length and diameter of high temperature piping; (j) should allow inspection and measurement of the flowrate of solids.

Dense phase transport lines should have gradual curvatures to prevent rough operation. Aeration should be provided to ease the flow of solids in dense phase transport lines and in risers. Blast connections should be available to clear them in case of plugging. Technique should be developed to weld patches on eroded lines during operation.

Maximum flow in standpipes is around 1200 kg/m /s for fine catalyst [10]. Solids velocities could be as high as 1.5 m/s, sufficient to entrain any bubbles downwards. Standpipes up to 1.5 m in diameter and up to 40 m high have been built, able to transfer up to 120 tons/minute of catalyst between vessels. Pressure build-up could be as high as 3 atmospheres. Loss of fines leads to poor circulation, rough operation and poor pressure build-up.

Entry of solids into standpipe takes place, generally, through an overflow well. The splashing and sloshing of large diameter beds throw solids into the well in such a way that the average bed level can be quite a bit below the rim of the overflow well. Appropriate design of the overflow well can prevent this from happening. When solids enter the standpipe at grid level, a slide valve is necessary to control solids flowrate, an undesirable feature in hot units.

Erosion is a serious problem in solids circulation systems. It is minimized by good design and the use of anti-erosion plates at critical points. Without precautions, fluid coke and iron ore, for instance, can erode a transport line in a matter of hours.

Feeding and withdrawing solids from high-pressure units is not a trivial problem. For instance, if one wishes to withdraw hot reduced iron from an iron ore reduction unit operating at high pressure and temperature without withdrawing too much process gas and without plugging the withdrawal line, the operation is a difficult one.

In a catalytic unit, catalyst feed and withdrawal lines have to be provided for start-up, shutdown and emergency operations. The catalyst may have to be cooled before being admitted to the storage silos. The latter are equipped with cyclones to avoid the losses of fines in the transport gas.

Research is needed on solid circulation and slide valve systems. Most of them are used in standpipes and risers in one way or another. Little generalizable knowledge is available on these components, in spite of their wide use in industry. It would be useful to be able to measure simply and reliably the flowrate of solids through a circulating system. The latter should be able to accept coarse lumps or agglomerates without difficulty. It should be fail safe and prevent any flow reversal between two vessels, e.g. a reactor and a burner. Simple and cheap circulation systems should be developed for small plants to be used in developing countries to process, among other things, municipal, agricultural and forestry waste. In a few decades, energy plantations in Brazil, Indonesia, the Phillipines, and Thailand will require skid mounted units on a large scale to make a variety of products, e.g. charcoal for energy and metallurgical use, pyrolytic products. The logistics of transportation are such that a low energy material such as biomass cannot be transported economically very far. It has to be processed on the plantation. Simplicity, safety, versatility and low cost are needed.

11.0 INSTRUMENTATION, CONTROL AND SAFETY

Instrumentation and control are of the utmost importance not only for the smooth operation of the unit but also for safety. Simplicity of control is a must for units to be used in the developing world where highly skilled manpower might not be available.

12.0 CORROSION AND EROSION

Corrosion and erosion go together in fluidized bed units; a corroded patch is soon eroded by the abrasion of the solids. Design has to take this fact into account. Units have been lost in short order because of a combination of corrosion and erosion. Tests should be carried out in the Pilot Plant to see how best to cope with these problems.

If the gases contain water vapor and corrosive compounds, e.g. $HC\ell$, the vessel has to be insulated on the outside to keep hydrochloric acid from condensing on the metal. If this is not done, the vessel will soon be lost.

Vessel and cyclone walls may have to be protected by armored concrete and cement grouting. At the end of a run, the concrete might have disappeared into the catalyst. Bends in pneumatic transfer lines should also be protected by abrasion resistant plates. Any internal should be located away from the grid jet zone or it will be destroyed. Hard facing material should be used whenever erosion can be a problem, e.g. in thermowells or cyclone dustpots.

13.0 ATTRITION

Not only can the unit be lost but also the catalyst. Particle attrition is a very serious problem which can ruin a process. It is a wide open field for research as very little academic work has been published in this area. Attrition can take place at grids, cyclones, in pneumatic transport lines. Thermal shock can lead to particle breakage. Attrition is a very strong function of gas velocity. Direct impingement of particles on a surface must be avoided.

14.0 MATHEMATICAL MODELS AND SCALE-UP

Fluidization is, indeed, a most complex phenomenon. Although it has been and is still being intensively studied, at this point in time and <u>in most</u> cases, nobody would dare build a commercial plant just on the strength of a kinetic study in the laboratory. No mathematical model, at this stage, is powerful enough to allow this kind of scale-up. Almost invariably one has to go through a pilot plant of some kind, a costly and time consuming process. No management would feel confident enough of the final results to agree to the funding and construction of the full-size plant without previous pilot plant work. The reason for this attitude is that commercial fluidized bed plants are very expensive ventures, sometimes in the hundreds of millions of dollars, and their failure can be catastrophic for all but the largest companies.

Present mathematical models deal essentially with the gas exchange and the reactivity of a single and well developed bubble. Most industrial beds are not deep enough to reach that stage and their bubbles almost never become stable and well developed. Moreover, the present bubbling-bed model does not take into account the effect of the grid and of the dilute phase above the bed. Each grid jet can be considered as a microreactor. One has to describe all transfers of heat, mass and momentum in and out of the jet and the conversion taking place in and around grid jets. In a general way, we should recognize that, until direct experimental evidence is available about the detailed flow paterns in a fluidized bed, the parameters in a model, no matter how they are reported, are no more than parameters obtained by a curve-fitting operation, although hopefully, they might be related in some unknown and indirect fashion to the real flow patterns.

If mathematical models do not allow us to go directly from the laboratory bench to the full-scale plant, they can be very helpful though, in the scale-up process. Cold, small-scale mock-ups (e.g. made of plexiglass for better observation of phenomena) of increasing diameters can be built and tested with tracers to see above which diameter the parameters of the mathematical model do not vary any more with size. Presumably, then, the diameter of the pilot plant should be above this critical diameter. Not only the plexiglass mock-ups are tested with tracers, but also eventually the pilot plant. When the latter is run hot, its performance can be compared to the performance predicted by the mathematical model incorporating the basic hydrodynamics of the pilot plant (obtained with tracers) and the kinetics as obtained from the laboratory unit (2 cm or so, in diameter).

Basically, sophisticated tracer experimentation is needed for processes requiring high conversion per pass. If only low conversions are desired, any design (within limits) will do. This was the case for the catalytic cracking process which required about 60% conversion per pass. Scale-up presented no problem and the process was an "instantaneous" success. The latter was deceptive, though, and led to a lot of woe for the developers of the high conversion processes which followed immediately on the footsteps of catalytic cracking (e.g. The Hydrocol hydrocarbon synthesis process at Carthage, Texas, U.S.A. [2].) One of the reasons for the failure of the Hydrocol process was that the bench-scale unit and the pilot plant on one side, the commercial unit on the other side, were operating under two different hydrodynamic regimes: (a) The bench unit and the pilot plant, because of high aspect ratios and of a heavy catalyst (reduced iron mill scale), were operating in the slug-flow regime. Because bubbles were limited in size and were hindered by the walls of the unit, it was easy for the reaction gas to get out of the bubbles and contact the catalyst. Conversion was thus reasonably high; (b) The commercial plant (5 m diameter), on the contrary, was in the free bubble regime. Bubbles could grow to large sizes and move very fast through the bed. In the laboratory, coarse granulometries perform best because of slug flow while fine particles are absolutely necessary for success in the large plant unless baffles are used.

Mathematical models could be checked on a large national or international fluidization facility. As long as physical entities used in the model are not measured on a working unit, they are only "fudge factors" and little can be done to improve the models.

15.0 PILOT PLANTS

We have seen that, for a high conversion process, tracer experimentation on cold mock-ups give the minimum diameter a pilot plant can have. Other considerations dictate its final size. It has to be big enough to allow a fair replication of the grid and of the heat exchanger of the commercial plant. A man has to be able to squeeze himself into the vessel at times to repair various internals, e.g. trickle valves on cyclone diplegs. Sixty centimeters for a minimum diameter is a generally accepted number for large petroleum processes, in spite of the fact that such a diameter makes for an expensive unit when one takes into account feed preparation and product disposal. For small petrochemical processes, when tracer experiments on cold mock-ups show that there is not a big scale-up factor, smaller diameters can be used. For a given process, the size of the pilot plant will depend on the commercialization strategy adopted and on the nature of the process itself.

16.0 OTHER ASPECTS TO BE STUDIED

Many other aspects of fluidized bed operation have, eventually, to be studied. Among these, one could cite the study of the movement of particles inside a fluidized bed, either singly or in bulk (sloshing). Dissipation of the solids and gases inside the bed and grid dumping are strongly affected by these particle movements.

17.0 CONCLUSIONS

The present review of the design and research needs in the field of fluidized bed reactors could not possibly be exhaustive due to space limitations and the vastness of the subject. What has been done was to highlight the very special characteristics of large units compared to small laboratory ones. The small scale laboratory unit is of no use in the commercial field although it is an invaluable tool to find out about fundamentals at the particle level.

PART B: SELECTED SUBJECTS

1. MINIMUM FLUIDIZATION AND BUBBLING CONDITIONS

To fluidize a bed of solid powder, a superficial gas velocity higher than the minimum fluidization velocity must be used. The most obvious use of the minimum fluidization velocity is thus to define the lowest possible gas velocity which can be specified for the design of a fluidized bed. In practice, however, the particle size distribution can vary during operation because of particle attrition, agglomeration, and elutriation. Industrial columns also operate with less than ideal gas distributors which, at low gas velocities, can lead to channelling, formation of defluidized zones and excessive leakage of bed solids into the windbox. Moreover, one usually wants to operate at high gas velocities to maximize the gas flowrate through a given unit. Fluidized beds are therefore usually operated at gas velocities which are significantly higher than the minimum fluidization velocity. An accurate determination of the minimum fluidization velocity would thus seem to be of limited practical value.

However, the minimum fluidization velocity represents a convenient way to characterize the interactions between a given solid powder and a given gas. As such, it is extensively used in semi-empirical models based on the two-phase theory and in empirical correlations which predict bed expansion, heat transfer coefficients, mass transfer coefficients and chemical conversions in fluidized bed reactors (most of these models and correlations also use the bed voidage at minimum fluidization conditions).

Two types of correlations are available for the prediction of the minimum fluidization velocity: purely empirical correlations and correlations based on the Ergun equation for the pressure drop through a fixed bed.

Numerous empirical correlations have been proposed and were reviewed by Thonglimp et al. [15]. The most commonly used correlation was developed by Leva [16] (for SI units):

$$U_{mf} = [7.169 \times 10^{-4}\ g\ d_p^{1.82}\ (\rho_p - \rho_g)^{.94}]/[\rho_g^{.06}\ \mu_g^{.88}] \qquad (1)$$

Many correlations were derived from the correlation developed by Ergun (18) to describe the pressure drop through fixed beds:

$$Ar = 150\ Re_{mf}(1-\varepsilon_{mf})/(\phi_s^2 \varepsilon_{mf}^3) + 1.75\ Re_{mf}^2/(\phi_s\ \varepsilon_{mf}^3) \qquad (2)$$

If the voidage at minimum fluidization conditions and the particle shape factor are known, the minimum fluidization velocity can be directly obtained from the above equations. However, this is usually not the case and empirical formulas are thus derived from the above equations by taking average values for the terms which include the particle shape factor and the voidage at minimum fluidization conditions. Correlations of the following form are thus obtained (see [15] for a review):

$$Re_{mf} = C_1 + [C_1^2 + C_2\ Ar]^{0.5} \qquad (3)$$

Richardson and St. Jeronimo [19] proposed: $C_1 = 25.7$, $C_2 = 0.0365$ and Wen and Yu [20]: $C_1 = 33.7$, $C_2 = 0.0408$.

Recommended correlations are the correlation from Richardson and St. Jeronimo [19] and the correlation from Wen and Yu [20]. For small particles, Leva's correlation [16] can also be used [15],[17]. If the voidage at minimum fluidization and the particle shape factor are known, however, equation (2) is recommended.

Most powders used in industrial fluidized beds consist of multisize particles with a wide size distribution. To calculate the minimum fluidization velocity, one should use the Sauter mean particle diameter [21]. Ideally, one should use the size distribution of the actual bed solids which can be quite different from the solids fed to the bed.

In practice, one can often measure the minimum fluidization velocity of a powder in a bench-scale column with air at room temperature and pressure. The voidage at minimum fluidization conditions can also be measured. Zenz [22] gives a procedure using these data to predict the minimum fluidization velocity under operating conditions. He basically assumes that the voidage at minimum fluidization conditions is not affected by changes in the properties of the fluidizing gas. Although this is an approximation, [21],[23], it should give better results than a correlation. This method should, however, be avoided if the particles become more cohesive or sticky at operating conditions. Practical examples of such particles are coal particles and even silica sand [24].

The minimum fluidization velocity of a given powder/gas system is substantially affected by temperature and pressure. This effect is not only due to the changes in gas viscosity and density but it also results from changes in the bed voidage at minimum fluidization velocity. Chitester et al. [23] conducted some measurements at high pressures and proposed a new correlation which took into account the observed increase with pressure in bed voidage at minimum fluidization conditions.

For irregularly shaped solids, the standard correlations such as the Wen and Yu correlation [20] should only be used to give a rough approximation of the minimum fluidization velocity. Instead the use of measured values of the particle shape factor and of the voidage at minimum fluidization conditions in equation (5) is recommended.

Electrostatic effects [27], particle and gas moisture levels [27] and particle "stickiness" or "adhesiveness" [24] affect the minimum fluidization velocity. Fluctuations in the gas flowrate, which can be caused by some compressors, can have a significant effect on the minimum fluidization velocity.

The apparent particle density is an important parameter for the design of fluidized bed columns. For example, the amount of solids elutriated from a fluidized bed column depends greatly on the apparent particle density of the bed solids. It is quite difficult to measure the apparent particle density of porous particles and the experimental value of the minimum fluidization velocity can be used for its evaluation [28].

Altogether, correlations for the minimum fluidization velocity can lead to erroneous values because they are, of necessity, used for conditions for which they were not tested. It is always desirable to have a file of U_{mf} values for a number of solids. Correlation results should be compared to experimental values for similar solids in the file. Whenever the solids are available, it is always better to measure U_{mf} in the laboratory in a column of large enough diameter, e.g. 0.10 m.

1.1 Bed Voidage at Minimum Fluidization Conditions

Many correlations and models to calculate bed expansion, heat transfer, mass transfer and chemical conversion require knowledge of the voidage at minimum fluidization conditions.

The bed voidage at minimum fluidization conditions is the average value of the fraction of the bed volume occupied by gas at minimum fluidization conditions. Three different techniques are available to measure the bed voidage at minimum fluidization conditions. The first technique uses the experimental value of the minimum fluidization velocity. It also requires knowledge of the fluidizing fluid properties, the particle apparent density, the Sauter mean particle diameter and the particle shape factor. The second technique is based on the measurement of the bed height at minimum fluidization conditions. The minimum fluidization velocity is first determined and the bed height at this velocity is then measured from the visual observation of the bed surface level or obtained from the static pressure profile along the column. V_{mf} is practically always measured with a porous plate or similar grid. This technique also requires knowledge of the apparent particle density and of the total mass of solids present in the bed. It does not require knowledge of either the particle shape factor or the Sauter mean particle diameter. Since the contribution of the gas to the bed weight is negligible (not true for liquid fluidized beds), the total mass of solids in the bed is given by:

$$M_{bed} = \rho_p(1 - \varepsilon_{mf})\ H_{mf} \tag{4}$$

The third technique was proposed by Abrahamsen and Geldart (29) and relates a measured "aerated" voidage to the voidage at minimum fluidization conditions.

The most commonly used correlation was proposed by Wen and Yu (20) and expresses the voidage at minimum fluidization conditions as a function of the particle shape factor:

$$\varepsilon_{mf} = [14\ \phi_s]^{(-1/3)} \tag{5}$$

Although this equation is simple and easy to use, it is not very accurate since it does not take into account the fluid properties, the particle size or the width of the particle size distribution. Another correlation has been proposed by Zenz (22) which takes into account the fluid properties and the particle size. However, it requires an accurate value of the minimum fluidization velocity. If the minimum fluidization velocity has to be measured, then the bed voidage can also be measured at the same time.

The bed voidage at minimum fluidization conditions is affected by the same operating variables which affect the minimum fluidization velocity: temperature and pressure which change the gas properties, adhesiveness or stickiness of the particles, moisture level of the gas and of the particles and electrostatic effects.

1.2 Minimum Bubbling Velocity

The minimum bubbling velocity is defined as the superficial gas velocity at which pockets of almost pure gas or "gas bubbles" will appear in a fluidized bed. As shown earlier (see section 1.1), the height of a fluidized bed can be determined either from visual observations of the bed level or from the static pressure profile along the column. Figure 1 shows how the height of a fluidized bed of cracking catalyst varies with the superficial gas velocity [36]. When the gas velocity is increased above the minimum fluidization velocity, the bed height increases sharply. As the gas velocity is increased further, gas bubbles start appearing and the bed height drops, stabilizes and then increases slowly.

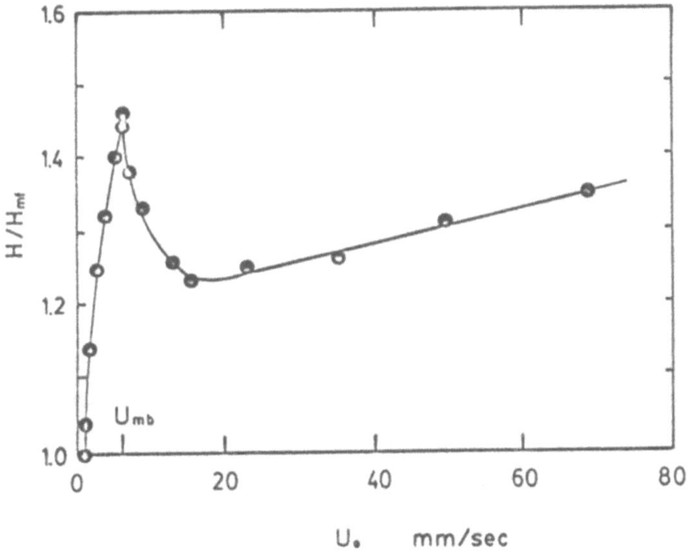

Figure 1 Bed Expansion vs. Superficial Gas Velocity [30]

Two different techniques are used to determine the minimum bubbling velocity. The first technique, which was used by Geldart and Abrahamsen [29] relies on visual detection of the first gas bubble.

A second technique defines the minimum bubbling velocity as the velocity corresponding to the maximum bed height.

The powders which are most commonly used in industrial fluidized beds belong to Geldart's Groups A and B [25]. According to Geldart (25), the best criterion which separates group A powders from group B powders is the difference between the minimum bubbling velocity and the minimum fluidization velocity; this difference is very small for group B powders. This demonstrates the importance of measuring or estimating the minimum bubbing velocity of a powder.

Abrahamsen and Geldart [29] proposed the following correlation for the minimum bubbling velocity:

$$\frac{U_{mb}}{U_{mf}} = \frac{2300\ \rho_g^{0.126}\ \mu_g^{0.523}\ \exp(0.716\ w_f)}{d_p^{0.8}\ g^{0.934}\ (\rho_p - \rho_g)^{0.934}} \tag{6}$$

where w_f is the weight fraction of fines (i.e. of particles with a diameter smaller than 44 μm) in the bed. SI units should be used. Although they do not include the particle shape factor in their correlation, they found that the correlation predictions were verified for angular particles (i.e. particles with shape factors much smaller than 1).

Abrahamsen and Geldart [29] also proposed the following correlation for the bed voidage at minimum bubbling conditions:

$$(1-\varepsilon_{mb})/(1-\varepsilon_{mf}) = [H_{mb}/H_{mf}] = [U_{mb}/U_{mf}]^{0.22} \tag{7}$$

2. FLUIDIZED BED EXPANSION AND VOIDAGE

The average bed voidage can be estimated from the measured bed height. If both the apparent particle density and the total mass of solids present in the bed are known, the bed voidage can be obtained with:

$$\varepsilon = 1 - [M_{bed}/(A\ H\ \rho_p)] \tag{8}$$

or

$$\varepsilon = 1 - [(1 - \varepsilon_{mf})\ H_{mf}/H] \tag{9}$$

2.1 Prediction of the Bed Expansion and the Bed Voidage

Numerous correlations and models have been developed to predict both bed expansion and bed voidage. There are two types of correlations: purely empirical correlations and correlations based on the two-phase theory. Empirical correlations have been reviewed by Thonglimp et al. [15]. One should be aware that most of the correlations which have been published were developed from data collected in small laboratory columns. They thus did not take into account the column diameter which has a very important effect on bed expansion, as shown by Figure 2 [31]. These correlations should thus be used with extreme caution for the design of large industrial columns.

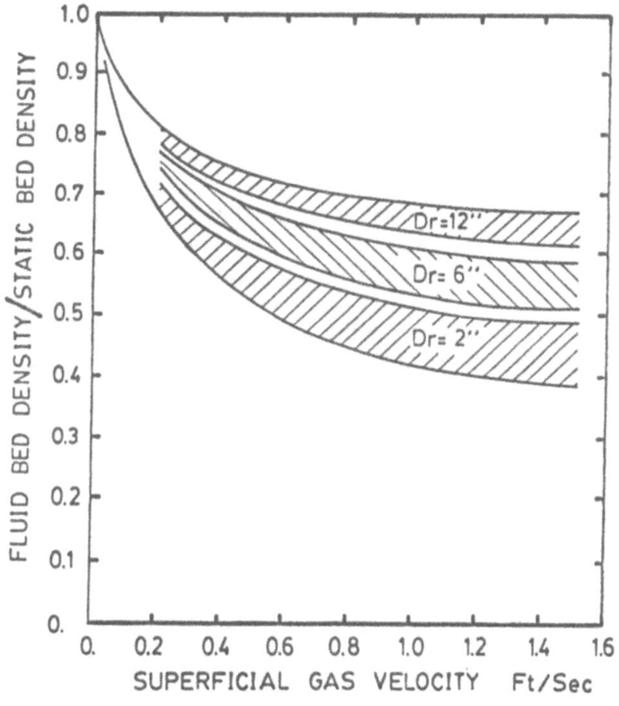

Figure 2 Effect of the column diameter D_c on the variation of the bed density with the superficial gas velocity [31]

From the two-phase theory, the total volume occupied by the gas bubbles in the bed can be obtained:

$$A(H-H_{mf}) = \int_0^H [(V_g-U_{mf})/U_b]\ A\ dz \qquad (10)$$

The bed expansion can thus be expressed as a function of the bed height at minimum fluidization conditions, the minimum fluidization velocity and the bubble velocity.

Because of bubble coalescence, bubble size and, therefore, bubble velocity will vary with the vertical position z. As a first approximation, one can take as average bubble velocity, the bubble velocity at mid-height and the bed expansion can thus be approximated by:

$$[H/H_{mf}] = U_b/(U_b - V_g + U_{mf}) \tag{11}$$

When computing the bubble velocity with a correlation, one should be aware that the bubble size cannot grow without limits. If the calculated bubble size is larger than about 40% of the column diameter, the fluidized bed will operate in the slugging regime. This is specially common in the case of small columns and might explain the effect of the column diameter on bed expansion (see figure 2). In the case of large columns, the bubble diameter cannot grow above the maximum stable bubble size. Many authors have combined the above equation with correlations for bubble velocity and bubble diameter to obtain a correlation for the bed expansion (15). However, these correlations do not take into account the eventual limitation of the bubble diameter by the column diameter in case of slugging and should not be used.

With cohesive powders which belong to Geldart's group C, neither the minimum fluidization velocity nor the minimum bubbling velocity can be accurately measured [32]. Practically no bubbles are formed and the bed expansion is caused by the proliferation and enlargement of horizontal and inclined tracks or channels [32]. Geldart and Wong [32] found that the average bed voidage could then be correlated with a Richardson-Zaki type equation:

$$\varepsilon^n = U_t'/V_g \tag{12}$$

where U_t' is not really the terminal velocity of the particles but is an empirical velocity obtained by extrapolation to $\varepsilon = 1$ of a plot of the logarithm of the experimental values of the gas velocity against the logarithm of the experimental values of the bed voidage. The exponent n is also determined from the experimental data. It is equal to 4.65 for non-cohesive powders and becomes larger as the powder becomes more cohesive [32]. Geldart and Wong [32] also proposed the following correlation:

$$[n/4.65] = 1.26\ [U_t'/U_t]^{0.132} \tag{13}$$

As the gas velocity is increased, the gas bubbles become larger until the slugging regime is reached. If the gas velocity is increased further, the turbulent regime and the fast fluidized bed regimes are successively reached. Avidan and Yerushalmi [33] proposed a technique similar to the technique exposed in the previous section which also relies on an equation of the Richardson-Zaki type. Here again, voidage correlations do, some times, give absurd numbers. It is imperative to have a file of expansion data for typical solids and to make comparisons to see if the numbers obtained by use of the correlations make sense. Even better, if the solids are available, expansion data should be taken in a large laboratory column (Fig. 2) and extrapolated to the conditions of the commercial plant.

2.2 Voidage of the Emulsion Phase

The two-phase theory assumes that the voidage of the emulsion phase, which is also called "dense phase" by some authors, is independent of the superficial gas velocity and remains equal to the average bed voidage at minimum fluidization conditions. A large number of studies have shown, however, that this was not always true [11][12][13][14]. Deviations from the predictions of the two-phase theory are specially large for powders belonging to group A of Geldart's classification.

Abrahamsen and Geldart [13] correlated the flowrate of gas going through the dense or emulsion phase in fluidized beds of such powders. The superficial velocity U_D corresponding to that gas was correlated with:

$$U_D/U_{mf} = 0.770 \; z^{-0.244} \; (U_{mb}/U_{mf})^{0.71} \tag{14}$$

The deviation from the two-phase theory is thus a function of the ratio of the minimum bubbling velocity to the minimum fluidization velocity, which, as we saw earlier, can be used as a criterion to differentiate group A powders from group B powders. This correlation indicates that there will be almost no deviation from the two-phase theory for group B powders.

Abrahamsen and Geldart also correlated the gas voidage in the dense phase:

$$[(1-\varepsilon_{mf})/(1-\varepsilon_D)] = [H_D/H_{mf}] = [U_D/U_{mf}]^{0.178} \tag{15}$$

3. SIZE AND VELOCITY OF GAS BUBBLES

Gas bubbles are formed at the gas distributor, either directly at the distributor or from the break-up of the distributor jets.

They then rise through the bed and coalesce with neighboring bubbles on their way to the bed surface where they explode. Bubble coalescence also results from the fact that large bubbles rise faster and can thus catch and absorb smaller bubbles. Bubble coalescence thus results in an increase of bubble size with the height from the distributor. Bubble growth can be limited by the column size or bubble stability considerations. The rising velocity of a gas bubble increases as its size increases.

3.1 Bubble Size and Velocity in Freely Bubbling Beds

Since bubbles are not perfectly spherical, the "bubble diameter" d_b is the diameter of a sphere which has a volume equal to the bubble volume.

When the bed is not slugging and the maximum stable size has not been reached, the velocity of a single bubble rising through a bed at minimum fluidization conditions is given by:

$$U_{bmf} = K\,(g\,d_b)^{0.5} \qquad (16)$$

The value of the proportionality constant K depends on the particle size. A value of 0.711 as proposed by Davidson and Harrison [34] is the most commonly used. In the case of freely bubbling beds, an empirical correction factor equal to the difference between the superficial gas velocity and the minimum fluidization velocity is added to the bubble velocity to account for bubble interaction and solid circulation effects. The equation thus obtained was first proposed by Davidson and Harrison [34] (this equation is controversial):

$$U_b = (V_g - U_{mf}) + 0.711\,(g\,d_b)^{0.5} \qquad (17)$$

The numerous correlations available in the literature have been reviewed by Thonglimp et al. [15] and Corella and Bilbao [35]. Some are purely empirical correlations while others are based on bubble coalescence models. A correlation which agrees well with available experimental data was proposed by Darton [36]. It is based on a coalescence model and gives the bubble diameter at a height z above a perforated plate gas distributor:

$$d_b = 0.54\, g^{-0.2} (V_g - U_{mf})^{0.4} \left[z + (16A_o)^{0.5}\right]^{0.8} \qquad (18)$$

where A_o is the area of distributor plate associated with each hole.

3.2 Slugging Regime

As the superficial gas velocity is increased, the gas bubbles grow until they have either reached their maximum stable size or until their growth is limited by the column wall. In the latter case, the fluidized bed reaches the slugging regime and the gas bubbles become bullet-shaped axisymmetric slugs whose base is almost as large as the column diameter. With coarse solids and small diameter columns, flat-nosed slugs through which solid particles rain down can appear [14][37].

Stewart and Davidson (38) correlated slug velocity data (for axisymmetric slugs) with the following expression:

$$U_{slug} = 0.35\ (gD_c)^{0.5} + (V_g - U_{mf}) \tag{19}$$

A comparison of equations (17) and (19) shows that slugs are slower than free gas bubbles of the same size. The gas residence time will thus be higher in slugging beds that in freely bubbling beds. In addition, hydrodynamics are completely different in slugging beds. This indicates that reaction data obtained from small bench or pilot plant columns which can easily reach the slugging regime should be carefully scaled up to industrial columns which, due to their large size, do not usually operate in the slugging regime.

Several criteria must be satisfied for slugging to occur. The first criterion is that the maximum stable size of the bubbles must be larger than the column diameter. The second criterion is that the superficial gas velocity must be smaller than the value at which the turbulent regime is initiated [37] (the turbulent regime is characterized by small elongated gas voids which are continuously coalescing and splitting). The third criterion sets the minimum superficial velocity required for slugging. It was proposed by Stewart and Davidson [38] who assumed that each slug was separated by a dense phase plug which has a height equal to two column diameters:

$$V_g \geq U_{mf} + 0.07\ (gD_c)^{0.5} \tag{20}$$

The fourth slugging criterion considers that a minimum bed height is required for slugging [39]:

$$H_{bed} \geq 3.5\ D_c\left[1 - (N)^{-0.5}\right] \tag{21}$$

The third criterion assumes that, at the onset of slugging, gas slugs are separated by two column diameters. Since slug separations which were as large as 8 column diameters have been reported (40), it might be advisable to replace the third and fourth criteria by a simpler criterion which states that slugging will not occur unless the bubble diameter as calculated for a free bubbling bed (e.g. by equation (18)) is larger than $0.4D_c$.

When slugging occurs in a column, slugging seems to usually start at a height equal to about two column diameters and to be fully developed at a height equal to about 10 column diameters. At higher positions, the slugging regime is fully established.

3.3 Maximum Stable Bubble Size

There is little published data to confirm the existence of a maximum bubble size because the columns used in academic research are too small: bubble growth is then stopped by the column walls before the bubbles can grow to their maximum size. However, data from industrial columns seem to confirm the existence of a maximum bubble size [41]. Although there is some controversy about the mechanism which stops bubble growth, all interpretations postulate that solid particles penetrating into the bubbles are responsible for bubble breakup.

Harrison et al. [42] assumed that, as a gas bubble rises through a fluidized bed, the shear force exerted by the dense phase particles on the bubble induces a circulation of gas inside the bubble. As a first approximation, the velocity of this circulating gas can be assumed to be equal to the bubble velocity. When the velocity of the circulating gas becomes larger than the terminal velocity of the bed particles, particles are "sucked" into the bubble from its wake and destroy it. For multisize particles, one must decide which particle size to select to calculate the terminal velocity. Although he does not agree with Harrison's premises, Grace [37] suggests the use of a particle diameter equal to 2.7 d_p where d_p is the Sauter mean diameter of the bed particles. Harrison's theory was later modified [43] by stating that particles would be sucked into the bubble from its wake when the circulating gas velocity became higher than the saltation velocity of the particles.

A different approach starts with the assumption that bubble breakup is caused by "fingers" of particles raining through the roof of the gas bubble. Yates [14] reviewed several experimental studies for indirect clues as to which of these two approaches would be valid. He found indirect confirmations of both approaches. More direct experimental data are required to validate any of these theories.

3.4. Beds of Large Particles

Fluidized beds of large particles are of growing industrial importance (e.g. fluidized bed combustion of coal and fluidized bed drying). Bubbles in beds of large particles (i.e. with powders of Geldart's group D) are different from bubbles obtained with smaller particles. For example, since the superficial gas velocity is usually not much higher than the minimum fluidization velocity, the gas velocity in the emulsion phase is usually larger than the bubble velocity. Yates [14] gives a short review of this problem.

3.5 Effect of Bed Internals on Gas Bubbles

Bed internals such as diplegs, heat exchanger tubes, sampling probes and baffles can have an important effect on gas bubbles. The effects of vertical rods such as cyclone diplegs and vertical heat exchanger tubes have been reported by Yates et al. [45]. With arrays of vertical tubes, the bubbles will be split if the center-to-center tube spacing is between 33 and 90% of the bubble size. Otherwise, the gas bubbles will be stabilized by the rods and will grow bigger and rise at a higher velocity. Horizontal tubes will split bubbles which are significantly larger than the tube diameter. However, the bubbles will recombine quickly by coalescence unless there is a succession of several rows of tubes stacked vertically. This is usually the case with arrays of heat exchanger tubes which thus have a beneficial effect on fluidized bed operation by reducing the bubble size. However, Zenz [22] does not advise the use of horizontal heat exchanger tubes which do not perform well in industrial fluidized beds (for example, they impede axial mixing of the solids, resulting in large temperature gradients in the bed). Other fixed internals such as horizontal perforated plates and modified versions [37][45][46] are used to reduce the bubble size. Although they are used successfully in many industrial fluidized beds, under some conditions they can be quickly eroded, warped or plugged. Free-floating bubble breakers, which do not have these problems, can also be used [47].

4. ENTRAINMENT FROM GAS-SOLID FLUIDIZED BEDS.

The flux of the solids entrained above the TDH can be measured either by isokinetic sampling or by catching the solids recovered by the cyclones. For the flux of solids entrained below the TDH, isokinetic sampling or a movable gas exit connected to a cyclone can be used (54).

4.1 Transport Disengaging Height (TDH)

As solids recovery equipment such as cyclones and electrostatic separators is expensive, the flux of solids entrained into the recovery train must be minimized. The column exhaust must thus be located above the TDH. On the other hand, the column must be kept as short as possible to minimize capital costs. In practice, when internal cyclones are used, the exact position above the TDH of the column exhaust to the cyclones will be set by the dipleg pressure balance.

Since no model has managed to successfully interpret experimental data, empirical correlations must be used. Zenz [49] correlated the TDH with the difference between the superficial gas velocity and the minimum bubbling velocity for various diameters of the bubbles exploding at the surface of the bed. Wen and Chen [50] reviewed the literature data and correlations. They suggested the following correlation:

$$\mathrm{TDH} = 0.25\ \ln\left[100(F_o - F_\infty)/F_\infty\right] \tag{22}$$

where F_o is the flux of particles ejected from the bed surface and F_∞ is the flux of particles entrained above the TDH.

4.2 Entrainment Above the TDH

Wen and Chen [50] reviewed the correlations available in the literature. However, as demonstrated by Zenz [22], these correlations are very unreliable outside the narrow range of experimental conditions for which they were developed. Moreover, they cannot predict the size distribution of entrained solids.

Zenz and Weil [51] proposed a model the basic assumption of which is that the flux of solids entrained above the TDH is equal to the maximum solids flux which could be carried in a pneumatic transport line operating at the same superficial gas velocity as the fluidized bed column. Thus, for each particle size cut, the choking flux for pure monosize particles G_i is calculated from correlations developed for pneumatic transport and the flux of particles of that size entrained above the TDH is given by:

$$F_i = x_{bi} G_i \tag{23}$$

where x_{bi} is the weight fraction of this particle size cut in the bed solids. The total flux of solids entrained above the TDH is then given by:

$$F = \Sigma_i F_i \tag{24}$$

and the size distribution of the entrained material is given by:

$$x_i = F_i / F \tag{25}$$

Briens and Bergougnou [52] tested this model with available experimental data and found that in most cases, the predicted entrainment flux was within one order of magnitude of the actual experimental flux. This model could not accurately predict the size of entrained particles above the TDH.

Gugnoni and Zenz [53] proposed a model which proceeds in two steps. The first step is the calculation of the size distribution of the solids entrained above the TDH with the Zenz-Weil model. The mean diameter of the particles entrained above the TDH is then calculated. In the second step, the total flux of solids entrained above the TDH is calculated with an empirical formula. Testing of this model [52] showed that it consistently overestimated the entrained flux and that its predictions were within one order of magnitude of the experimental values. It could not accurately predict the size distribution of the entrained particles.

Briens and Bergougnou [52] proposed a model which is based on three assumptions:

1) For each particle size, the flux of solids entrained above the TDH is limited by choking, i.e. it cannot be larger than the fraction of the total choking load which is attributed to that particle size when the column above the TDH is assumed to behave as a pneumatic transport line.

2) For each particle size, the flux of solids entrained above the TDH cannot be larger than the flux of solids ejected from the bed surface,

3) Assumptions 1 and 2 would be sufficient to calculate the entrained flux for monosize particles. In the case of multisize particles, an extra assumption is required. It is, thus, assumed that the size distribution of the entrained particles will be such that it will maximize the entrained flux, namely, the gas will entrain preferentially the smallest particles.

All values predicted with this model did not differ from the experimental values by more than a factor of 2. Moreover, it successfully interpreted the results obtained by Large et al. [54] who reduced the entrained flux by two orders of magnitude by removing the smallest 1 wt% of the bed particles.

The Briens-Bergougnou model demonstrates that three approaches can be used to reduce the amount of solids entrained above the TDH:

1) The choking flux can be reduced by reducing the superficial gas velocity. This can be achieved, for example, by adding a large diameter section at the top of the column.

2) The flux of solids ejected from the bed surface can be reduced by reducing the bubble size or by placing some baffles just above the bed surface.

3) Minute changes in the size distribution of the bed solids can sometimes result in large reductions of the flux of solids entrained above the TDH [54]. Special attention should thus be given to the size distribution of fresh solids, attrition of bed solids and cut-points of the cyclones.

4.3 Solids Concentration Below the TDH

In fluidized bed reactors, a significant part of the reaction can take place in the freeboard (e.g. some catalytic reactors, fluidized bed combustors). The concentration or hold-up of solids in the freeboard must then be evaluated.

There are two ways to express the solids concentration:

1) The flowing density: $\rho_F = F / V_g$ (26)

2) The static holdup: $\rho_h = F / U_p$ (27)

They are not identical because the solid particles do not travel as fast as the gas.

Several models can predict the flux of solids entrained below the TDH [55],[56]. It is, however, preferable to use the empirical correlation proposed by Large et al. (54):

$$F = F_{\infty} + (F_{o} - F_{\infty}) \exp(-a\ H) \qquad (\text{with } a \approx 4\ m^{-1}) \qquad (28)$$

4.4 Flux of Solids Ejected from the Bed Surface

Two types of models have been proposed to predict the flux of particles ejected from the bed surface, depending on whether the ejected solids are assumed to come from the bubble nose or the bubble wake [57]. In all cases, reducing the size of the gas bubbles exploding at the surface of the bed will result in a significant reduction of the amount of ejected solids.

5. GAS DISTRIBUTORS

5.1 Jetting and Bubbling at Distributor Orifices

To prevent severe erosion by jet impingement of internals such as heat exchanger tubes, one must know how far a vertical jet issuing from a distributor such as a perforated plate will penetrate into a fluidzed bed. This knowledge is also required to prevent piercing by jets of the surface of shallow beds. Blake et al. [58] reviewed the correlations and data available in the literature for the penetration depth of upward jets into fluidized beds. They proposed the following correlation:

$$\frac{L}{d_h} = 87.\left[\frac{V_h^2}{g\ d_h}\right]^{.29}\left[\frac{\rho_g}{\rho_p}\right]^{.48}\left[\frac{\rho_g\ V_h\ d_p}{\mu_g}\right]^{-.17} \qquad (29)$$

If tuyeres with horizontal jets are used as gas distributors, they must be positioned so that they will not be eroded by jets issuing from adjacent tuyeres. Zenz [22] gives a correlation to calculate the penetration depth of horizontal jets into fluidized beds. He also gives a correlation to calculate the penetration depth of downward jets. This correlation would also be used to ensure that downward jets issuing from a pipe ring with downward-facing holes would not erode the support plate.

To simplify, two regimes of bubbling behavior can be considered. When the distributor pressure drop is low, the momentum of the gas entering the bed through tuyeres or orifices is low and bubbles form directly at the grid holes. When the pressure drop is high, jetting can be expected and the bubbles form at the tip of the jets with a diameter equal to about half the jet penetration depth [22]. Numerous intermediate regimes exist between these two extreme regimes. For example, bubbling has been reported to occur by intermittent formation of "bubble chains". There is also some controversy as to whether stable jets can exist and whether "jets" are not actually formed of very fast succeeding bubbles [14]. In the case of bubbles forming directly at the orifice, the following correlation can be used to calculate their initial size [59]:

$$d_{bo} = g^{-0.2} \left[6 \; A \; (V_g - U_{mf})/(\pi N)\right]^{0.4} \quad (30)$$

The bubble size at any height in the bed will depend on its initial size and thus on the grid design. This is especially important for shallow beds.

Zenz [22] assumed that bubbles forming at an orifice are surrounded by a shell of downward moving particles with a diameter 50% larger than the bubble diameter. Thus the center-to-center spacing between the distributor holes should be larger than (1.5 d_{bo}) to avoid premature coalescence of bubbles at the orifice.

Fakhimi [60] noted that at low gas rates, some of the distributor holes were not bubbling gas and were thus inactive. This phenomenon appeared to be particularly noticeable with low pressure drop plates. He developed a model based on the assumption that this was caused by particle bridging. Whitehead [61] observed the same phenomenon with multi-tuyere distributors.

In a commercial fluidized bed, the presence of defluidized or semi-fluidized particles on the distributor in-between the gas inlet points is very undesirable since with "sticky" or "tacky" materials, these zones will grow and eventually plug the distributor. With exothermic catalytic reactions, these defluidized zones would lead to the formation of hot spots on the distributor. Fakhimi [60] developed a model which predicts the formation of defluidized zones and Zenz [22] suggested a distributor made of a honeycomb of nearly touching cones or hexagons to eliminate these stagnant zones.

5.2 Spatial Distribution of Bubbles and Solids Circulation Patterns

Uneven spatial distribution of bubbles within a fluidized bed affects the efficiency of a fluidized bed reactor by causing a broadening of the residence time distribution of the gas. In regions where more bubbles flow, they will grow by coalesence and rise faster. They will also induce circulation currents or "Gulf-Streaming" of the emulsion phase [62]. Such induced currents will increase gas backmixing as well as affect heat transfer from immersed surfaces. Solids circulation and associated benefits or problems have been reviewed by Whitehead [63]. It should be noted that with curved distributor plates, an uneven hole spacing is required to offset uneven distribution of the gas bubbles.

5.3 Prevention of Solids Attrition

With solids which are easily attrited, the gas velocity at the gas-solids interface must be reduced without decreasing the distributor pressure drop which would adversely affect the gas distribution. This can be achieved with shrouded grids [22].

5.4 Prevention of Grid Leakage

Substantial backflow of particles through the grid (also called weeping or grid leakage) is highly desirable because it can lead among other things, to distributor erosion or plugging and particle attrition. Briens et al. [64] showed that grid leakage was caused by pressure fluctuations due to bubble formation at the grid and sloshing in the fluidized bed. This led to a mathematical model for grid leakage which was verified experimentally. Leakage was also found to vary with bed height and a sharp maximum of the grid leakage was observed for a bed height to diameter ratio of about 0.5. Wave breakers and cone-shaped grid holes were found to reduce leakage drastically. Selecting the proper grid thickness reduced grid leakage by up to 4 orders of magnitude. Properly designed tuyere distributors can practically eliminate grid leakage [65].

5.5 Effect of Grid Design on Heat and Mass Transfer

Grid jets have been shown to be important in gas-solids heat transfer. In the case of fast chemical reactions, the grid region accounts for a large fraction of the total conversion. Grid design can thus affect dramatically the overall conversion obtained in a fluidized bed reactor by affecting mass transfer at the distributor level [6].

Notation

a constant (equation 28), m^{-1}
A bed cross-section area, m^2
A_o area corresponding to each distributor hole, m^2
Ar Archimedes number (see equation 2)
C_1 constant in equation (3)
C_2 constant in equation (3)
D_c column diameter, m
d_b bubble diameter, m (diameter of a sphere with a volume equal to the bubble volume).
d_{b0} bubble diameter at the distributor, m
d_h hole diameter, m
d_p particle diameter (Sauter mean diameter in the case of multisize particles), m
d_{pmax} maximum particle diameter, m
d_{pmin} minimum particle diameter, m
F flux of entrained solids, $kg/(s.m^2)$
F_o flux of solids ejected from bed surface, $kg/(s.m^2)$
F flux of solids entrained above the TDH, $kg/(s.m^2)$
F_i flux of solids entrained above TDH for size cut i, $kg/(s.m^2)$

g gravity constant, m/s
G choking load for pure, monosize particles of size d_{pi}, kg/(s.m^2)
H bed height, m
H_{bed} bed height, m
H_D height of bed corresponding to the dense phase only, m
H_{mb} bed height at minimum bubbling conditions, m
H_{mf} bed height at minimum fluidization conditions, m
K proportionality constant
L penetration depth of the jet, m
M_{bed} mass of the bed, kg
N total number of gas distributor holes
n exponent in Richardson-Zaki type equation
Q_b bubble gas flowrate, m/s
Re_{mf} Reynolds number at minimum fluidization conditions
TDH transport disengaging height, m
U^* dimensionless gas velocity, V_g/U_{mf}
U_b bubble velocity m/s
U_{bmf} velocity of a single bubble at minimum fluidization conditions, m/s
U_D superficial gas velocity through the dense phase, m/s
U_{mb} minimum bubbling velocity, m/s
U_{mf} minimum fluidization velocity, m/s
U_p particle velocity, m/s
U_{slug} slug velocity, m/s
U_t particle terminal velocity, m/s
U_t' gas velocity corresponding to ε=1 (see eq.12),m/s
V_g superficial gas velocity, m/s
V_h gas velocity through the distributor holes, m/s
W_{bed} bed weight, N
W_f weight fraction of fines (i.e. with d_p < 45 μm) in the bed
x_{bi} weight fraction of particles of size cut i in the bed
x_i weight fraction of particles of size cut i (above the TDH)
z vertical position above the gas distributor, m
ΔP_{bed} pressure drop through the bed, Pa
ε bed voidage at operating conditions
ε_D voidage in the dense phase
ε_{mb} bed voidage at minimum bubbling conditions
ε_{mf} bed voidage at minimum fluidization conditions
Φ_s particle sphericity = ratio of the surface of a sphere of same volume as the particle over the actual particle surface
μ_f fluid dynamic viscosity, Poiseuilles
μ_g gas dynamic viscosity, Poiseuilles
ρ_{bed} average bed density, kg/m^3
ρ_F flowing density
ρ_f fluid density, kg/m^3
ρ_g gas density, kg/m^3

ρ_h static hold-up
ρ_p particle density, kg/m^3
$\sum$ summation sign (summation over all entrainable size cuts)

REFERENCES

1. Kunii, D. and O. Levenspiel. "Fluidization Engineering". John Wiley and Sons, New York, 1969.
2. Zenz, F.A., Othmer, D.F., "Fluidization and fluid-particle systems". Reinhold Publishers, New York, 1960.
3. Briens, C., M.A. Bergougnou and C.G.J. Baker. "Leakage of Solids (Weeping, Dumping) at the grid of a 0.6m diameter gas fluidized bed". Proceedings of the Second Engineering Foundation Conference on Fluidization. Cambridge University Press, Cambridge University, England, 1978, 32-37.
4. Behie, L.A., M.A. Bergougnou, C.G.J. Baker and T.E. Base. "Further studies on momentum of grid jets in a gas fluidized bed". CJChE 49, (1971), 557-561.
5. Behie, L.A., M.A. Bergougnou and C.G.J. Baker (1975). "Heat Transfer from a grid jet in a large gas fluidized bed", CJChE, 53, (1975) 25-30.
6. Behie, L.A., M.A. Bergougnou and C.G.J. Baker. "Mass Transfer from a grid jet in large gas fluidized bed". Proceedings of the First Engineering Foundation Conference on Fluidization. Asilomar, California, U.S.A. 1, (1976) 261-278.
7. Mayer, I. and M.A. Bergougnou. "Apparatus for Fluidized Solid Systems". U.S.P. 3,910,769, assigned to EXXON Research and Engineering Company, Linden, N.J., U.S.A.
8. Guigon, P., J.F. Large, M.A. Bergougnou and C.G.J. Baker. "Particle interchange through thin and thick baffle plates in multistage gas fluidized beds". Proceedings of the Second Engineering Foundation Conference on Fluidization, Cambridge, England. Cambridge University Press (1978), 134-139.
9. Newby, R.A., S. Katta and D.L. Keairns. "Regeneration of Calcium based SO_2 sorbents for fluidized bed combustion: engineering evaluation". U.S.E.P.A. - 600/7-78-039. March, 1978, 48-63.
10. Matsen, J.M. "Some characteristics of large circulation systems". Proceedings of the First Engineering Foundation Fluidization Conference. Hemisphere Publishing Corporation, Washington, D.C. (1976), 135-139.
11. Grace, J.R. and R. Clift, "On the two-phase theory of fluidization", Chem. Eng. Sci., 29, (1974) 327-334.
12. Rowe, P.N., Santoro, L., Yates, J.G., "The division of gas between bubble and interstitial phases in fluidized beds of fine powders", Chem. Eng. Sci., 33 (1978) 133-140.
13. Abrahamsen, A.R. and Geldart D., "Behaviour of gas-fluidized beds of fine powders", Powder Tech., 26, (1980) 47-55.

14. Yates, J.G., "Fundamentals of Fluidized-bed Chemical Processes", Butterworths, 1983.
15. Thonglimg, V., Hiquily, N., Laguerie, C., "Vitesse minimale de fluidisation et expansion des couches fluidisées par un gaz", Powder Technology, 38 (1984) 233-253.
16. Leva, M., "Fluidization", McGraw Hill, 1959.
17. Grewal, N.S. and Saxena, S.C., "Comparison of commonly used correlations for minimum fluidization velocity of small solid particles", Powder Techn., 26, (1980) 229-234.
18. Ergun, S., "Fluid flow through packed columns", Chem. Eng. Prog., 48, (1952) 89-94.
19. Richardson, J.F. and St. Jeronimo, M.A., "Velocity-voidage relations for sedimentation and fluidization", Chem. Eng. Sci., 34 (1979) 1419-1422.
20. Wen, C.Y., Yu Y.H., "Mechanics of fluidization", Chem. Eng. Prog. Symp. Series, 62, (1966) 100-111.
21. Botterill, J.S.M., "The effect of operating temperature on the velocity of minimum fluidization, bed voidage and general behaviour", Powder Tech., 31, (1982) 101-110.
22. Zenz, F.A., "State of the art review and report on critical aspects and scale-up considerations in the design of fluidized bed reactors", Chapter 4, final report on phase 2 of contract number DE-AC 21-80 MC 14141, U.S. Department of Energy, 1982.
23. Chitester, D.C., Kornosky, R.M., Fan, L.S., Danko, J.P., "Characteristics of fluidization at high pressure", Chem. Eng. Sci., 39, (1984) 253-261.
24. Yamazaki, R., Hong, G.F., Jimbo, G., "The behavior of gas-solid fluidized bed at elevated temperature", in "4th International Conference on Fluidization", Tokyo, Kunii, D. and Toei, R. eds, Engineering Foundation, N.Y. 1983.
25. Geldart, D., "Types of gas fluidization", Powder Tech., 7, (1973) 285-292.
26. Chen, P., Pei, D.C.T., "Fluidization characteristics of fine particles", Can. J. Chem. Eng., 62, (1984) 464-468.
27. Vanacek, V., "Fluidized Bed Drying", Leonard Hill, London, 1966.
28. Margiatto, C.A., Siegell, J.H., "Determination of porous particle density", Powder Tech., 3 , (1983) 105-106.
29. Abrahamsen, A.R., Geldart, D., "Behaviour of Gas-fluidized beds of fine powders", Powder Tech., 26, (1980) 35-46.
30. Harriott, P., Simone, S., Chapter 3 in "Handbook of Fluids in Motion", Cheremisinoff N.P., Gupta R. eds., Ann Arbor Science, 1983.
31. Zenz, F.A., Othmer, D.F., "Fluidization and Fluid-Particle Systems", Reinhold Publishing, 1960.
32. Geldart, D., Wong, A.C.Y., "Fluidization of powders showing degrees of cohesiveness", Chem. Eng. Sci., 39, (1984) 1481-1484.

33. Avidan, A.A., Yerushalmi, J., "Bed expansion in high velocity fluidization", Powder Tech., 32, (1982) 223-232.
34. Davidson, J.F., Harrison, D., "Fluidized Particles", Cambridge University Press, 1963.
35. Corella, J., Bilbao, R., "The effect of variation of the gas volume and of the bubble size on the conversion in a fluidized bed", Int. Chem. Eng., 24, (1984) 302-310.
36. Darton, R.C., "A bubble growth theory of fluidized bed reactors", Trans. Inst. Chem. Engrs., 57, (1979) 134-138.
37. Grace, J.R., Chapter 8 in "Handbook of Multiphase Systems", Hetsroni G. ed., Hemisphere Publishing Corporation, 1982.
38. Stewart, P.S.B., Davidson, J.F., "Slug flow in fluidized beds", Powder Tech., 1, (1967) 61-80.
39. Darton, R.C., Lanauze, R.D., Davidson, J.F., Harrison, D., "Bubble growth due to coalescence", Trans. Inst. Chem. Eng., 55, (1977) 274-280.
40. Thiel, W.J., Potter, O.E., "Slugging in fluidized beds", IEC Fund., 16 (1977) 242-247.
41. Matsen, J.M., "Evidence of a maximum bubble size in a fluidized bed", AIChE Symp. Ser. 69, No. 128 (1973) 30-33.
42. Harrison, D., Davidson, J.F., de Kock, J.W., "On the nature of aggregative and particulate fluidization", Trans. Inst. Chem. Eng., 39, (1961) 202-211.
43. Guedes de Carvalho, J.F.R., Harrison, D., "Fluidization under pressure", Inst. of Fuel Symp. Ser. No. 1 (1975) paper B1.
44. Yates, J.G., Cheesman, D.J., Mashingaidze, T.A., Howe, C., Jefferis, G., "The effect of vertical rods on bubbles in gas fluidized beds", in "Fluidization" (Proc. 4th Int. Conf.), Tokyo, Kunii, D., Toei, R., eds., (1983) 103-110.
45. Loew, O., "Particle and bubble behaviour and velocities in a large-particle fluidized bed with immersed obstacles", Powder Technology, 22, (1979) 45-57.
46. Jin, Y., Yu, Z.Q., Zhang, L., "Pagoda-shaped internal baffles for fluidized bed reactors", Int. Chem. Eng., 22, (1982) 269-279.
47. Gbordzoe, E., "Hydrodynamics of floating contactors in a fluidized bed", M.E.Sc. thesis, Engineering Sc., University of Western Ontario, London, Canada, 1979.
48. Ismail, S., Chen, J.C., "Volume fraction of solids in the freeboard region of fluidized beds", AIChE Symp. Ser. No. 234, 80, (1984) 114-118.
49. Zenz, F.A., "Particulate solids: the third phase in chemical engineering", Chem. Eng., 90, No. 24, (1983) 61-67.
50. Wen, C.Y., and L.H. Chen, "Fluidized bed freeboard phenomena: entrainment and elutriation", AIChE J., 12, (1982) 117-128.
51. Zenz, F.A., and N.A. Weil, "A theoretical-empirical approach to the mechanism of particle entrainment from fluidized beds", AIChE J., 4, (1958) 472-479.

52. Briens, C., and M.A. Bergougnou, "New model for entrainment from fluidized beds" AIChE J. to be published (1985).
53. Gugnoni, R.J., and F.A. Zenz, "Particle entrainment from bubbling fluidized beds", Grace J.R., Matsen, J.M. ed. Plenum Press, New York (1980) 501-508.
54. Large, J.F., Y. Martinie, and M.A. Bergougnou, "Interpretative model for entrainment in a large gas fluidized bed", International Powder and Bulk Solids Handling and Processing Conference, May 1976.
55. Kunii, D., and O. Levenspiel, "Fluidization Engineering", Krieger Publishing Co., New York, Chapter 10 (1977).
56. Morooka, S., K. Kawazuishi, and Y. Kato, "Holdup and flow pattern of solid particle in freeboard of gas-solid fluidized bed with fine particles", Powder Technology, 26, (1980) 75-82.
57. Pemberton, S.T., "Entrainment from fluidized beds", Ph.D. Thesis, Trinity College, Cambridge, U.K. (1982).
58. Blake, T.R., Wen C.Y., Ku, C.A., "The correlation of jet penetration measurements in fluidized beds using nondimensional hydrodynamic parameters", AIChE Symp. Ser. No. 234, 80 (1984) 42-51.
59. Geldart, D., "The expansion of bubbling fluidized beds", Powder Tech., 1, (1968) 355-368.
60. Fakhimi, S., Harrison, D., "Multi-orifice distributors in fluidized beds: a guide to design", Chemeca 70, Session 1, Inst. Chem. Engrs., (1970) 29-46.
61. Whitehead, A.B., Dent, D.C., "Behavior of multiple tuyere assemblies in large fluidized beds", in "Proc. Int. Symp. on Fluidization", Netherlands University Press (1967) 802-820.
62. Werther, J., Molerus, O., "The local structure of gas fluidized beds, II" Int. J. Multiphase Flow, 1, (1974) 123-138.
63. Whitehead, A.B., Dent, D.C., McAdam, J.C.H., "Fluidization studies in large gas-solid systems. Part V. Long and short term pressure instabilities", Powder Tech., 18, (1977) 231-237.
64. Briens, C., Bergougnou, M.A., Baker, C.G.J., "Grid leakage (weeping, dumping, particle backflow) in gas fluidized beds: the effect of bed height, grid thickness, wave breakers, cone-shaped grid holes and pressure drop fluctuations", in "Fluidization", Grace J.R., Matsen, J.M., eds. Plenum Press, New York, (1980) 423-420.
65. Tyagi, A., "Grid pressure drop and solid leakage in a large gas-solid fluidized bed", M.E.Sc. thesis Engineering Science, The University of Western Ontario, London, Canada, 1984.

THREE-PHASE FLUIDIZED BED REACTORS

H. de Lasa and S.L.P. Lee

Faculty of Engineering Science, The University of Western Ontario, London, Ontario, Canada N6A 5B9

INTRODUCTION

The particular characteristics of three-phase fluidized bed reactors have been covered in several recent reviews by Ostergaard [1], Wild [2], Epstein [3], Baker [4] and Muroyama and Fan [5]. Epstein [3] distinguished in particular the difference between three-phase fluidized beds and slurry reactors. In slurry reactors the size of the solid particles is normally smaller than 0.1 mm while in three-phase fluidization the particle diameter is bigger than 0.2 mm. The volumetric solid fraction is another significant difference, being 10% or below for slurry reactors and between 20-40% in three-phase fluidized bed units. In three-phase fluidized beds the particles are supported by the liquid and/or the gas while in slurry reactors the solid particles are suspended by the momentum transferred from the gas bubbles to the liquid and from the liquid to the solids. In slurry reactors the solid particles are normally carried into and out of the unit by the liquid stream. In three-phase fluidized beds the solids are not transported out of the unit by the liquid stream, they are fed and withdrawn independently of the liquid stream [3]. Epstein [3] introduced an interesting classification for three-phase reactors and particularly proposed four modes of operation for three-phase fluidized beds: Mode I: Cocurrent upflow circulation of gas and liquid with the liquid as a continuous phase, Mode II: cocurrent upflow circulation of gas and liquid with the gas as a continuous phase, Mode III: Countercurrent circulation of gas (upflow) and liquid (downflow) with the liquid as the continuous phase, Mode IV: Countercurrent circulation of gas (upflow) and liquid (downflow) with the gas as the continuous phase.

The most common of three-phase fluidized bed is Mode I. Some recent contributions [6],[7],[8] have covered Mode III of operation normally known as inverse fluidization. The particles of the bed have a density lower than that of the liquid; the liquid is circulated downward and the gas is introduced countercurrently to the liquid.

This review will essentially cover the cocurrently gas dispersed phase operation (Mode I) considering that it corresponds to the most successful industrial application of three-phase fluidized beds - the H-Coal process for coal liquefaction [9] and the H-Oil process for hydrocracking and hydrodesulfurization of heavy crudes [10]. Some other applications such as the catalytic methanation of CO and H_2 [11], the production of calcium bisulphite [12] and the production of zinc hydrosulphite [13] are also considered as potential processes for Mode I of three-phase fluidized operation.

Important characteristics of Mode I of operation closely relate to the bubble flow patterns which are coalesced bubble flow, dispersed bubble flow, slug flow, and transitional flow. The exact definition of these regimes is rather subjective and is frequently the result of visual observations. These flow patterns determine many of the properties of cocurrent three phase fluidized beds such as: porosity, bubble characteristics, mixing, heat and mass transfer. The specific values of these properties, their changes and their interdependence with respect to the flow patterns is covered in the following sections of this review.

Liquid-Solid Fluidized Beds

In order to describe the behaviour of three-phase fluidized beds it is very important to achieve a good understanding of the characteristics of liquid-solid fluidization. This is because the particles in three-phase beds are essentially fluidized by the liquid. An important question then is to define the conditions of minimum fluidization in liquid-solid systems. At minimum fluidization the pressure drop in the dense bed, frequently represented by the Ergun equation for fixed beds, becomes equal to the buoyant weight of the particles. The following identity then results.

$$\Delta p_B = H_{mf}(1-\varepsilon_{Lmf})(\rho_S-\rho_G)\frac{g}{g_c} = 150\left(\frac{1-\varepsilon_{Lmf}}{\varepsilon_{Lmf}^{3}}\right)\left(\frac{\mu_L V_{Lmf}}{d_p^{2}}\right)\frac{H_{mf}}{g_c} +$$

$$+ 1.75\left(\frac{1-\varepsilon_{Lmf}}{\varepsilon_{Lmf}}\right)\rho_L\frac{v_{Lmf}}{d_p}\,H_{mf} - \frac{1}{g_c} \qquad (1)$$

Multiplying both sides of this identity by $\rho_L d_p{}^3/\mu_L^2\ (1-\varepsilon_{Lmf})$ and after the appropriate simplifications it is possible to establish the following relationship between the Galileo number (Ga), the Reynolds number for the particle at the condition of minimum fluidization (Re_{pmf}) and the porosity of the bed at minimum fluidization (ε_{Lmf}).

$$Ga = 150\left(\frac{1-\varepsilon_{Lmf}}{\varepsilon_{Lmf}}\right) Re_{pmf} + \frac{1.75}{\varepsilon_{Lmf}^{3}} Re_{pmf}^{2} \qquad (2)$$

The bed porosity, ε_{Lmf}, can thus be expressed as a function of the Re_{pmf} and the Ga numbers.

$$\varepsilon_{Lmf} = f(Re_{pmf}, Ga) \qquad (3)$$

Because of the smooth expansion of liquid-solid fluidized beds eq. (3) is applicable for liquid-solid fluidized beds well beyond the condition of minimum fluidization. The following correlation can then be proposed:

$$\varepsilon_L = A\ Re_p^{\alpha}\ Ga^{\beta} \qquad (4)$$

This type of correlation certainly applies for the limiting condition $\varepsilon_L \to 1$ which corresponds to the case where the liquid velocity approaches the particle terminal settling velocity or $Re_p \to Re_{pt}$.

$$1 = A(Re_{pt})^{\alpha}\ Ga^{\beta} \qquad (5)$$

Equating (4) and (5) it is possible to establish a relationship between the velocity in the liquid phase, the terminal velocity and ε_L.

$$\frac{u_L}{v_{Lt}} = \varepsilon_L^{\ n} \qquad (6)$$

This relationship, known as the Richardson-Zaki equation, has been modified in order to provide a better correlation of the experimental data [14].

$$\frac{u_L}{v_{Li}} = \varepsilon_L{}^n \tag{7}$$

where $\log v_{Li} = \log v_t - d_p/D$

and the power n is defined as a function of the particle Reynolds number [9].

$$n = 4.65 + 20\ d_p/D \qquad Re_{pt} < 0.2 \tag{7a}$$

$$n = (4.4 + 18\ d_p/D\ )\ Re_{pt}{}^{0\ 03} \qquad 0.2 < Re_{pt} < 1 \tag{7b}$$

$$n = (4.4 + 18\ d_p/D\ )\ Re_{pt}{}^{0\ 01} \qquad 1 < Re_{pt} < 200 \tag{7c}$$

$$n = 4.4\ Re_{pt}{}^{0\ 1} \qquad 200 < Re_{pt} < 500 \tag{7d}$$

$$n = 2.4 \qquad Re_{pt} > 500 \tag{7e}$$

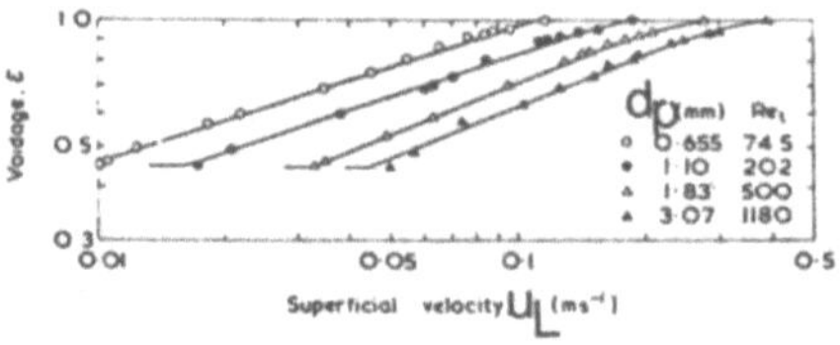

Figure 1. Change of the bed porosity with u_L [15]

Reprinted with permission from Ind. Eng. Chem., Proc. Des. Dev., 16,2 (1977). Copyright 1977 American Chemical Society.

It is important to mention as well (Figure 1) that for the higher values of Re_{pt} there is an increasing curvature in the log ε-log u_L plot and this is particularly noticeable for porosities of about 0.9. This behaviour certainly represents a limitation to eq. (7). In spite of this limitation an alternative to eqs.(7a) to (7e) was proposed by Garside ad Al-Dibouni [15] to unify in a single equation the ε_L predictions:

$$\frac{5.09 - n}{n - 2.73} = 0.104\ Re_{pt}^{0.877} \tag{8}$$

Three-Phase Fluidized Beds

Bed Porosity. The overall bed porosity is defined as the fraction of the bed occupied by the gas and the liquid and is represented by ($\varepsilon_G + \varepsilon_L$). Several authors have described the influence of various operating parameters such as superficial gas velocity, particle size, particle density, fluid density, surface tension and viscosity on ($\varepsilon_G + \varepsilon_L$).

An interesting phenomenon is the bed expansion or contraction upon injecting gas into a liquid-solid fluidized bed. For example, with large particles the bed height increases monotonically with gas velocity. Nevertheless, an initial decrease of bed height can be observed if small particles are used (Figures 2 and 3). Stewart and Davidson [16] advanced an explanation for this phenomenon indicating that when an air bubble rises through a water-fluidized bed, the gas bubble is followed by a liquid wake almost free of particles. The combined air/liquid velocity becomes much higher than the average velocity of the water circulating through the interstices of the bed. Therefore, the liquid rising through the three phase system with the air bubbles has a much smaller residence time than that of the liquid circulating through the bed interstices in the absence of air. Consequently the bed may contract rather than expand. A similar argument was given by Ostergaard [17]. Rigby and Capes [18], who essentially agreed with the views described above, postulated that the wakes

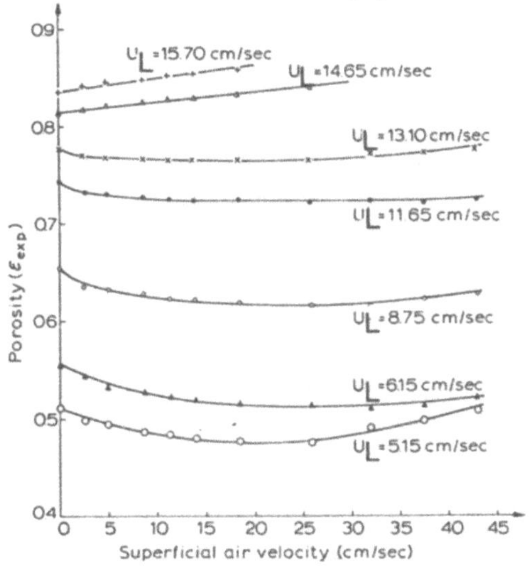

Figure 2. Effect of the superficial air velocity on the bed expansion for different u_L. Particles: 2-mm glass beads [22].

show a stable portion and vortices shed by the bubbles. It was concluded by Rigby and Capes [18] that the vortices could strongly affect the process of initial bed contraction and that this was particularly true for small particles.

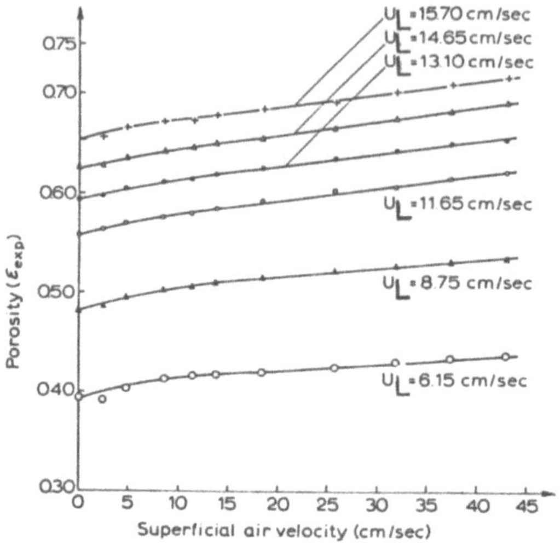

Figure 3. Effect of the superficial air velocity on the bed expansion for different u_L. Particles:4-mm glass beads [22].

The question of the prediction of the conditions of initial bed contraction was analyzed by Epstein and Nicks [19] who derived a criterion for predicting the condition of initial bed contraction using the following equations.

$$\varepsilon_G + \varepsilon_L + \varepsilon_S = 1 \quad (9)$$

$$\varepsilon_L = \varepsilon_W + (1 - \varepsilon_G - \varepsilon_W)\, \varepsilon'_{LP} \quad (10)$$

$$u_G = v_G\, \varepsilon_G \quad (11)$$

$$u_L = \varepsilon_W v_G + (1 - \varepsilon_G - \varepsilon_W)\, \varepsilon'_{LP}\; v_{LP} \quad (12)$$

$$v_{LP} = (\varepsilon'_{LP})^{n-1}\; v_t \quad (13)$$

Eq. (12) assumes that the liquid divides itself between the solids-free wakes and the particulate region of the fluidized bed. On the other hand, Eq. (13) considers that the interstitial velocity of the liquid can be represented by an equation similar to the one proposed by Richardson and Zaki [14].

An algebraic combination of eqs. (10) and (12) leads to

$$\varepsilon_L + \varepsilon_G = \frac{u_L - u_G k}{v_{LP}} + \frac{u_G(1+k)}{v_{LP} + v_{BL}} = f(v_{LP}, v_{BL}, k, u_L, u_G) \quad (14)$$

At the same time combining eqs. (12),(13) and (14), an expression for u_G as a function of v_{LP}, v_{BL}, k, v_t, n, u_L is obtained [19].

$$u_G = \{u_L(v_t/v_{LP})^{1/(n-1)} - v_{LP}\} / \{k(v_t/u_{LP})^{1/(n-1)} - v_{LP}(1+k)/(v_{LP}+v_{BL})\} \quad (15)$$

or

$$u_G = f(v_{LP}, v_{BL}, k, v_t, n, u_L) \quad (16)$$

It can be assumed that the various parameters v_{BL}, k, v_t, n, u_L are known which is equivalent to considering that the injection of gas takes place in a liquid-solid fluidized bed of specified properties and expansion. Then eqs. (14) and (15) giving u_G and $(\varepsilon_L + \varepsilon_G)$ can be considered only as a function of v_{LP} only. The differential velocity v_{BL} is determined using the appropriate equation for the gas and liquid phases. The bubble size and velocity were specified parameters [19].

Eqs. (14) and (15) can be differentiated with respect to v_{LP}. The ratio of these two derivatives defines $d(\varepsilon_L + \varepsilon_G)/du_G$ at the condition $u_G \to 0$.

$$\lim_{u_G \to 0} \frac{d(\varepsilon_L + \varepsilon_G)}{du_G} = \lim_{u_G \to 0} \frac{d(\varepsilon_L + \varepsilon_G)}{dv_{LP}} \Big/ \lim_{u_G \to 0} \frac{du_G}{dv_{LP}} \quad (17)$$

or

$$\lim_{u_G \to 0} \frac{d(\varepsilon_L + \varepsilon_G)}{du_G} = \frac{(\frac{n}{n-1} + k)\, v_{LP} - ((1+k)\, u_L + \frac{k v_{BL}}{n-1})}{(\frac{n}{n-1})\, v_{LP}\, (v_{LP} + v_{BL})} \quad (18)$$

The magnitude and sign of the change of $(\varepsilon_L+\varepsilon_G)$ with u_G can be evaluated with this equation. Because n is always bigger than 1 the denominator of eq. (18)is always a positive term. Then in the process of assigning a sign to eq. (18) it is necessary to consider only the numerator of that equation. A positive numerator indicates bed expansion while a negative numerator bed contraction. Almost simultaneously with the development of eq. (18) by Epstein and Nicks [19], Darton and Harrison [21] proposed another expression to predict the transition point between a fluidized bed that shows initial contraction and one that shows initial expansion.

$$(1+k)(1-\varepsilon_L+\frac{\varepsilon_L}{n})-\frac{\varepsilon_L\, k[V_b+(u_L/\varepsilon_L)]}{n\, u_L}=0 \qquad (19)$$

If both terms of eq. (19) are multiplied by $n\, u_L/(n-1)\varepsilon_L$ and if at the same time the appropriate algebraic transformations are performed the following equation results [20].

$$(\frac{n}{n-1}+k)v_{LP}-(1+k)\,u_L-\frac{kV_b}{n-1}=0 \qquad (20)$$

It should be mentioned that eq. (20) and the numerator of eq. (18) become for the bubble flow the same equation because v_{BL}, the relative velocity between the bubble and the liquid, is very close to V_b.

Equation (19) provides information about the factors that lead to the condition of initial bed contraction or, in other words, to the following inequality.

$$(1+k)\left[1-\varepsilon_L+\frac{\varepsilon_L}{n}\right]<\varepsilon_L\frac{k}{n}\frac{V_{br}}{u_L} \qquad (21)$$

This expression shows that initial bed contraction is favoured by: a) a high bubble velocity, V_b, b) at a given hold-up by a smaller liquid flow rate and c) a high volumetric ratio, k. It can also be seen from the criterion given by eq. (21) that the bed height will always expand monotonically if the liquid velocity is high enough. This prediction was observed experimentally by Bruce and Revel-Chion [22] for 4mm, 6mm and 8 mm glass particles.

El-Temtamy and Epstein [23] reassessed, however, the question of bed contraction or expansion including the potential influence of solid hold-up in the bubble wake. The question becomes of particular significance for solids smaller than 1-2 mm and particle densities that do not exceed 3000 kg/m^3. For these systems Batia and Epstein [24] proposed a generalized wake model with a parameter x that incorporates the solid content in the wake. Both eq. (10) and eq. (12) were modified as follows:

$$\varepsilon_L = \varepsilon'_{LP}(1 - \varepsilon_G - \varepsilon_W + x\,\varepsilon_W) + \varepsilon_W(1 - x) \qquad (22)$$

and

$$u_L = \varepsilon'_{LP}\, v_{LP}(1 - \varepsilon_G - \varepsilon_W) + \varepsilon_W v_G(1 - x) \qquad (23)$$

where the term x represents the ratio of solid hold-ups in the bubble wakes to the solid hold-up in the liquid particulate fluidized region. The parameter x was allowed to vary between 0 and 1.

Under these conditions eq. (18) becomes:

$$\lim_{u_G \to 0} \frac{d(\varepsilon_L+\varepsilon_G)}{du_G} = \frac{\left[\frac{n}{n-1} + k(1-x)\right]v_{LP} + \frac{xk(v_{LP}+v_{BL})}{n-1} - \left[(1+k)u_L + \frac{k(1-x)v_B}{n-1}\right]}{\left(\frac{n}{n-1}\right) v_{LP}\,(v_{LP} + v_{BL})} \qquad (24)$$

Because the denominator of eq. (24) is always a positive term, the conditions of bed expansion or contraction when $u_G \to 0$ can be predicted by observing the sign of the numerator of eq. (24).

The predictions of eq. (24) were tested with glass beads (0.456 mm and 1.08 mm) and lead shots (2.18 mm) for liquid superficial velocities of water and water solutions of polyethylene glycol in the 0.4 - 38 cm/s range [23]. The k parameter was estimated by using the spherical cap assumption and ε_L from previous research studies. It was proved that the sign of the numerator of eq. (24) correctly predicts the initial contraction or expansion of the bed.

An interesting concept developed by El-Temtamy and Epstein [23] concerns the critical x value, x_c, above which it could be predicted that expansion takes place and below which it could be estimated that contraction occurs. This x_c value can be calculated setting the numerator of eq. (24) equal to zero and solving for $x=x_c$. Another important observation concerns the influence of surface tension or solid wettability on the initial expansion of the fluidized bed [23]. In the case of low surface tension liquids and/or non-wettable solids the solids tend to rain through the gas bubbles increasing x and consequently increasing the numerator of eq. (24) and hence the tendency to expand when gas is introduced in the bed. In this respect Bhatia et al. [25] proved experimentally using 1 mm glass beads in an air-water fluidized bed the crucial importance of wettability on the initial expansion or contraction of three-phase fluidized beds. Glass particles (wettable solids) showed initial contraction while the same glass particles coated with teflon (rendered in this way non wettable) lead to an expansion of the bed from the lowest gas velocities onward.

For estimating the bed porosity in three-phase fluidized beds ($\varepsilon_G + \varepsilon_L$), Ostergaard [17] proposed a semiempirical method. It was assumed in the Ostergaard model that the bed consists of a liquid particulate phase, a bubble phase and a wake phase. It was considered as well that the wake moves with the bubble velocity and their porosity was identical to that of the liquid fluidized phase. Under these assumptions, the following equation was proposed to correlate u_L with ε_G, ε_W, V_b, ε'_{LP}, u'_{LP}.

$$u_L = u'_{LP}(1-\varepsilon_G - \varepsilon_W) + V_b\ \varepsilon_W\ \varepsilon'_{LP} \qquad (25)$$

where u'_{LP} and ε'_{LP} are the superficial liquid velocity and liquid porosity, respectively are based on the liquid particulate phase.

At the same time Ostergaard proposed [17] a questionable correlation, derived from data reported by Nicklin [26], to correlate V_b and u_G and u_L

$$V_b = 21.7 - 4.6\ \ln u_G + u_L \qquad (26)$$

With these two equations, an expression of the Richardson--Zaki type applied to the particulate phase and a correlation function between ε_W, ε_G and u_L an iterative method of calculation of ε was proposed. The ($\varepsilon_G + \varepsilon_L$) results predicted with this method and the experimental values differed significantly.

Dakshinamurty et al. [27] studied ($\varepsilon_G + \varepsilon_L$) in beds consisting of different types of solid particles (rockwool shots, glass balls, glass beads, iron shots, sand), two liquids (kerosene, water) and air. An attempt was made to use the semi-empirical method proposed by Ostergaard [17]. It was found that Ostergaard's method could correlate the data for particles of 1 mm. and 3.3 mm in diameter but could not correlate the data if the particle sizes were bigger than 3.3 mm. Because of these problems Dakshinamurty et al. [27] develop an empirical correlation between ε, the bed porosity, and two dimensionless groups (u_L/v_t) and ($\mu_L u_G/\sigma$)

$$\varepsilon = a\ (u_L/v_t)^b\ (\mu_L u_G/\sigma)^c \qquad (27)$$

for $Re_p < 500$ $b = 0.41$ $c = 0.08$ $a = 2.65$

$Re_p > 500 \qquad b = 0.6 \qquad c = 0.08 \qquad a = 2.12$

Satisfactory agreement between the measured bed porosities and the predicted values was found [27]. An average deviation of 3.7% - 5.6% between eq. (27) and the measured ε value was obtained. The constant 2.65 on eq. (27) for $Re_p < 500$ was later on revised by Dakshinamurty et al. [28] to 2.85. Eq. (27) shows, however, a significant limitation because it fails to predict the porosity for zero gas velocity conditions. In fact when $u_G \to 0$, the bed porosity is not zero and it should certainly approach the predictions of Richardson and Zaki's equation [14].

Bruce and Revel-Chion [22] measured the bed porosity, $(\varepsilon_G+\varepsilon_L)$, in three-phase fluidized beds consisting of glass-air-water using the following relationship.

$$(\varepsilon_G + \varepsilon_L)_{exp} = 1 - \varepsilon_S \quad \text{where} \quad \varepsilon_S = \frac{W}{\rho_S A\, H} \tag{28}$$

These authors compared the $(\varepsilon_G+\varepsilon_L)_{exp}$ with the predictions of eq. (27). The size of the glass particles ranged between 2 mm and 8mm, the liquid velocity between 5 and 15.7 cm/s, and the gas velocities between 2.5 and 43.0 cm/s. It was observed that only the correlation developed by Daksinamurty et al. [27] provided reliable results for the lower gas velocities.

An empirical correlation similar to the one proposed by Dakshinamurty et al. [27] was derived by Begovich and Watson and applied to 2381 data points coming from ten contributions [29].

$$\varepsilon_G+\varepsilon_L = a\, u_L^b\, u_G^c\, (\rho_S - \rho_L)^e\, d_p^e\, \mu_L^f\, D^g \tag{29}$$

where $a = 0.371 \pm 0.017$ $\qquad e = 0.268 \pm 0.010$

$b = 0.27 \pm 0.011$ $\qquad f = 0.055 \pm 0.008$

$c = 0.041 \pm 0.005$ $\qquad g = -0.033 \pm 0.013$

$d = -0.316 \pm 0.011$

The limitation of Doraiswamy's correlation also exists for eq. (29). This equation is not appropriate for predicting the bed porosity for the case of low gas velocity.

Kim et al. [30], [31] studied extensively the bed porosity and its changes with various operating parameters such as gas velocity, liquid viscosity and surface tension in a two dimensional bed. The gas velocity was found to have relatively little influence on (ε_L+ε_L). In some cases, as expected from eq. (21), when the gas velocity increased the bed porosity slightly decreased (initial bed contraction). It was observed by Kim et al. [30],[31] that the bed porosity remained relatively constant while the surface tension was varied from 40 dynes/cm to 80 dynes/cm. The liquid viscosity had, however, a much more significant effect on (ε_G+ε_L). For instance, the influence of viscosity changes in the 1 to 50 cp range on (ε_G+ε_L) for 1 mm. glass beads, 2.6 mm. gravel and 6 mm glass beads is shown in Figure 4 . On the basis of these results the following two correlations were developed to estimate (ε_G + ε_L) [31]. These two correlations were expressed in terms of two key dimensionless numbers We and Fr. They were conceived, at the same time, to account for the important condition of initial bed expansion or contraction:

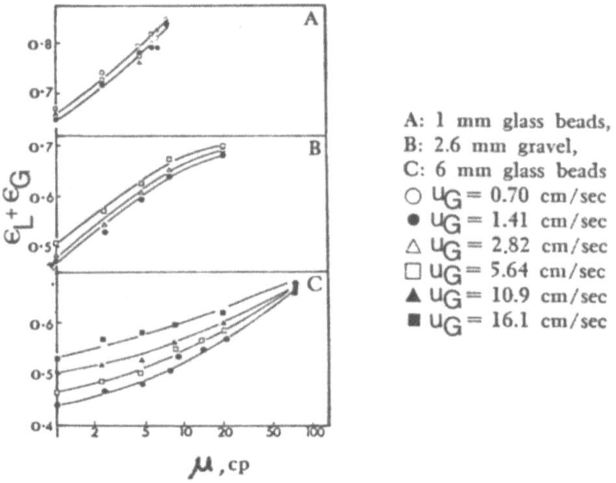

Figure 4. Effect of the liquid viscosity on bed porosity in three-phase fluidized bed (u_L = 5.4 cm/s) [31].

- For beds which initially expand when gas is introduced.

$$(\varepsilon_G + \varepsilon_L) = 1.40\ Fr_L^{0.178}\ W_e^{0.078} \tag{30}$$

- For beds which initially contract when gas is introduced

$$(\varepsilon_G + \varepsilon_L) = 1.301\ Fr_L^{0.128}\ We^{0.073}\ \exp\{0.031(u_L/u_G)\ \varepsilon_L^*\} \tag{31}$$

where ε^*_L is the liquid holdup in the corresponding liquid-solid fluidized bed which is given by

$$\varepsilon_L^* = 1.353\ Fr_L^{0.206}\ Re_L^{-0.10}$$

Another significant contribution to predicting bed expansion is based on the generalized wake model developed by Bhatia and Epstein [24]. This model, already described in previous sections of this review can be applied for the prediction of ($\varepsilon_G+\varepsilon_L$) combining eqs. (13),(22) and (23)

The following expression then results:

$$\varepsilon_G+\varepsilon_L = \left[\frac{u_L-u_G k(1-x)}{v_t(1-\varepsilon_G-k\varepsilon_G)}\right]^{1/n} (1-\varepsilon_G(1+k-kx)) + \varepsilon_G(1+k-kx) \qquad (32)$$

For the particular situation when $x \to 0$, the case of the solid-free wake model, eq. (32) reduces to the expression proposed by Efremov and Vakhrushev [33]. In general, when $x = 0$, three parameters have to be predicted, - k, x and ε_G - to effectively use eq. (32). Rigby and Capes [18], Darton and Harrison [21] tried to relate, the k parameter, ratio of the liquid wake to the bubble volume, to the volume of a single bubble. El-Temtamy and Epstein [32] suggested that k of single bubbles could be estimated by assuming spherical cap bubbles and a sphere-completing wake. However, given the difficulty of predicting bubble sizes, Efremov and Vakhrushev [33] suggested the average value of k in the bed be related with u_G, u_L and particle properties. For instance, the following empirical expression to predict k was proposed [21].

$$k = 1.4\,(u_L/u_G)^{0.33} - 1 \qquad (33)$$

This equation suggests that $k = 0$ at $u_L/u_G = 0.4$, a condition which seems to correspond to the onset of slugging. Epstein [20] questioned, however, the adequacy of defining k with eq. (33) when $u_L/u_G \to 0$. At the same time El-Temtamy and Epstein [32] tested critically the predictions of five different correlations observing a common trend - the increase of k with u_L.

The parameter x can be estimated by means of a correlation that stresses the importance of particle characteristics, particle-liquid interaction and gas-liquid relative velocity [20].

$$0 < v_{Li}/v_{BL} < 1.14 \qquad x = 1 - 0.877(v_{Li}/v_{BL}) \qquad (34)$$

$$v_{Li}/v_{BL} > 1.14 \qquad x = 0 \qquad (35)$$

Finally the ε_G parameter can for instance, be related to ε_L as proposed by Darton and Harrison [21]. To this end, the concept of drift flux, volumetric flux of gas relative to a surface moving at the average velocity, is used. It is considered that in the case of the uniform bubbling regime the drift flux becomes very close to the product of ε_G and the rising velocity of an isolated bubble in a stagnant medium. It should be pointed out that an alternative expression for the drift flux is required in the case of the churn-turbulent regime. Then, once k, x and ε_G are assessed the generalized wake model, eq. (32), provides an implicit expression to determine ε_L and $(\varepsilon_G + \varepsilon_L)$.

In order to measure the individual contribution of the liquid phase and gas phase to the $(\varepsilon_G + \varepsilon_L)$ term, the following equation was considered [31].

$$-\frac{\Delta P_B}{H} = (\varepsilon_L \rho_L + \varepsilon_G \rho_G + \varepsilon_s \rho_s) \frac{g}{g_c} \quad (36)$$

This equation assumes that the acceleration of the gas and liquid flows and the frictional losses on the walls are minimal [31]. Recently, Moroyama and Fan [5] claimed that eq. (36) could also be applied to the inverse three-phase fluidized bed system.

A combination of eqs (9), (36) and (28) provides the following relationships for determining experimentally ε_G and ε_L.

$$\varepsilon_G = \frac{-\frac{\Delta P_B}{H}\frac{g_c}{g} + (1 - \frac{W}{A\,H\,\rho_S})\rho_L + \frac{W}{A\,H}}{(\rho_L - \rho_G)} \quad (37)$$

and

$$\varepsilon_L = \frac{\frac{\Delta P_B}{H}\frac{g_c}{g} - \rho_G (1 - \frac{W}{A\,H\,\rho_s}) - \frac{W}{A\,H}}{(\rho_L - \rho_G)} \quad (38)$$

This method involves the definition of ΔP_B, H and W. In catalytic processes the total weight of solids W is normally a constant or close to a constant value. The total height of the bed, H can be determined visually or using the pressure profile along the bed axis. This axial pressure profile is frequently recorded using a series of manometers attached to the wall of the vessel. Due to the fact that bubbles are in a discontinuous phase, the measured temperature profiles could oscillate to a certain degree and these fluctuations will depend on the gas flow rates. In other words, local parameters such as the cross-sectional averaged pressure are fluctuating temporally and could significantly differ from their medium value. One possible approach is to photograph the manometer board "freezing" the manometer readings. Another more suitable alternative is to use a pressure transducer with a Scanivalve mechanism recording the pressure at various bed levels in several seconds (Figure 5). To obtain the total height of bed, H, the pressure profiles determined in the dense bed and in the freeboard region are extrapolated towards the bed interface. The point of intersection of these two quasilinear pressure profiles define H.

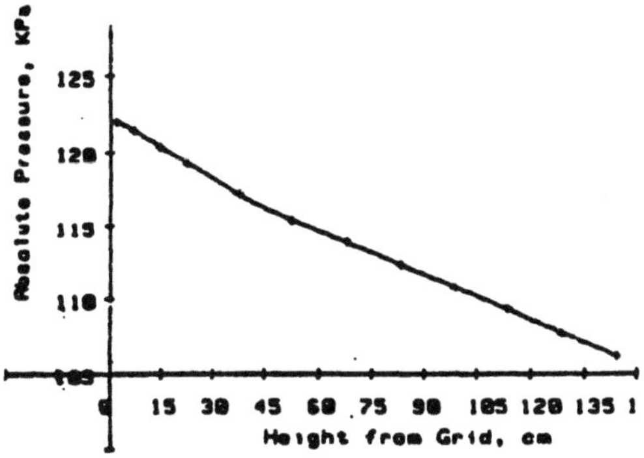

Figure 5. Change of the pressure with the height above the grid.

El-Temtamy [34] suggested a simple method for assessing the various hold-ups. The technique involved the measurement of pressure gradients for the three-phase bed and for the corresponding two-phase solid-liquid bed having the same total height (and therefore the same ε_s). Consequently, combining eq. (36), eq. (9) and a third equation (eq. (36) with $\varepsilon_G = 0$) the ε_G, ε_L and ε_s hold-ups were determined.

Kim et al. [31] correlated empirically the ε_L values derived with eq.(38) in terms of the Froude, Reynolds and Weber numbers using the following correlation:

$$\varepsilon_L = 1.504\ Fr_L^{0.234}\ Fr_G^{-0.086}\ Re_L^{-0.082}\ We^{0.092} \qquad (39)$$

Alternative methods for predicting the individual hold-ups have been proposed stressing the fact that the estimation of ε_S using eq. (28) involves the definition of the total height of the bed either visually or by extrapolation of the pressure gradients measured in the three-phase region and in the two-phase region above the bed, respectively. Both methods could lead to significant errors particularly at the conditions of high fluid flow rates or for beds of fine particles. In those cases the width of the transition region increases substantially and the determination of H either visually or by extrapolation becomes unsatisfactory.

To eliminate these problems several alternative approaches have been proposed to determine at least one of the three hold-ups (ε_L, ε_G, or ε_S) generating independently in this way a third equation that could be used instead of eq.(28) involving H. Epstein [3] described for instance a method consisting of the use of two quicky closing valves, located at the inlet and outlet of the fluidized bed section. Once the two valves were closed, the volume occupied by the gas was assessed. A similar method was proposed by Kim et al.[30] who proved that the gas and liquid hold-ups can be measured by simultaneous shutting-off the liquid and gas flows.

Several other proposals followed this idea of measuring at least one of the hold-ups independently. In some cases local hold-up values were also obtained. For instance, Rigby et al. [35], Darton and Harrison [37] used an electrical resistance probe to assess ε_G, Vasalos et al. [36] employed a radioactive tracer to determine ε_G, Ishida and Tanaka [38] suggested that the use of two fiber optics , a flat-end fiber and one ending in a prismatic shape would provide characteristic signals for bubbles and solids, de Lasa et al. [39] showed that a U-shaped single cord fiber has potential for determining ε_G and ε_S. Begovich and Watson [40] proposed the direct measurement of ε_L using an electroconductivity technique originally developed by Achwal and Stepanek [41] for gas-liquid systems. Begovich and Watson [40], Dhanuka and Stepanek [42] confirmed that in the case of three-phase fluidized beds the electroconductivity of the liquid was proportional to the ion concentration, to the cross-sectional area of the conducting liquid and it was inversely proportional to the length of the path between electrodes. Under these conditions, the conductivity is proportional to the liquid hold-up, ε_L.

$$\varepsilon_L = \frac{\gamma}{\gamma_o} \qquad (40)$$

Combining this result with eqs. (9) and (36) Begovich and Watson [40] defined the hold-ups of the various phases. The overall hold-ups ε_L, ε_S and ε_G measured with the electroconductivity technique and the pressure gradient method were compared. Deviations between the ε_L, ε_G and ε_S values resulting from the electroconductivity technique and the pressure gradient method were significant in the higher liquid velocity range (5 - 10 cm/s). Because the electroconductivity probe used by Begovich and Watson consisted of two platinum electrodes, each 1.4 cm^2, attached to a movable plexiglass ring, local measurements of ε_L at different levels of the fluidized bed were obtained. These local ε_L values and eqs. (9) and (36) allowed the assessment of hold-ups in the case of electroconductive liquids. In some situations, however, such as prototypes and simulators, this handicap can partially be overcome using, instead of the fluid of the process, an electroconductive liquid with similar fluid dynamic properties.

Kato et al. [43] used the pressure profile method and the electroconductivity probe in a bidimensional column. The liquid hold-ups obtained by the two methods agreed well except near the top of the fluidized bed. Because this study involved experimentation with air, aqueous solutions of carboxymethyl cellulose at different concentrations and glass particles of different sizes (0.42 mm, 0.66 mm, 1.2 mm, 2.2 mm) the effects of superficial liquid velocity, liquid viscosity and particle size on the liquid hold-up were considered. With this data a correlation similar to the one derived by Richardson and Zaki for liquid-solid fluidized beds was proposed [43].

$$\frac{u_L}{v_t} = \left(\frac{\varepsilon_L}{\varepsilon_L^*}\right)^n \qquad (41)$$

$$\frac{a-n}{n-b} = c\, Re_{pt}^{0.9} \qquad (42)$$

and ε^*_L is the extrapolated value of the liquid hold-up at $u_L = v_t$.

The various parameters involved in eq. (42) were evaluated by Kato et al. [43] from experimental data. The proposed correlations for a,b,c, and K follow:

$$a = 5.1(1+16.9\ K^{0.285}),\ b = 2.7,\ c = 0.1(1+4.43\ K^{0.165}) \qquad (42a)$$

and

$$K = \rho_L u_G^4/g\sigma \qquad (42b)$$

Substitution of these parameters into eq. (42) gives an implicit relation between n, K and Re_t. It was proved by Kato et al. [43] that eq. (41) provided a good representation of the experimental data obtained in that study. In other words, as can be visualized from eq. (41) the prediction of the liquid hold-up is less complex than that of $(\varepsilon_G + \varepsilon_L)$ or ε_G. The hold-up of the liquid increases monotonically with u_L and it decreases with u_G.

Due to the fact that ε_G is significantly affected by the flow regime it would seem that no satisfactory expression is yet available to predict ε_G. The Vail et al. [44] correlation is only a first attempt to assess ε_G. With the purpose of assessing ε_G, Morooka et al. [46] further applied the electroconductivity technique and the pressure gradient method to determine the various hold-ups in three-phase fluidized beds composed of air, water and glass particles. In particular changes of the gas hold-up with radial position were analyzed. It was shown that the changes of ε_G with radial position were well correlated with the following equation.

$$\frac{\varepsilon_G}{(\varepsilon_G)_{av}} = 2[1 - (\frac{r}{R})^2] \qquad (43)$$

The gas hold-up values obtained using the electroconductivity probe method and the static pressure method were compared by Morooka et al. [46]. The values agreed very well in the freeboard region.

In spite of the difficulties in predicting ε_G, the technical literature indicates that for glass particles smaller than 2.5mm in diameter the gas hold-up is significantly reduced by the presence of solid particles. Kim et al. [30] defined this regime as a bubble-coalescing condition. This is attributed to the fact that small particles promote bubble coalescence which results in higher rising velocities. The opposite is also true - larger particles tend to cause break up of bubbles. This condition is named the bubble disintegration regime [30]. The important effect of solid concentration on gas hold-up was analyzed by Kato et al. [46] with 0.08 - 0.15 mm particles. This work showed that an increase in solid hold-up generally decreased ε_G. The effect becomes insignificant at high gas velocities (greater than 10 to 20 cm/s) and at low gas velocities ($u_G < 3$ cm/s). Concerning the effect of liquid velocity on ε_G it seems that the free settling velocity of the particulate phase plays an important role. In beds of large and heavy particles the gas hold-up increases with u_L, whereas in beds of smaller particles gas hold-up was found to decrease with u_L or to be independent of the liquid velocity. [47],[30].

Bubble Characteristics

Studies concerning bubbles characteristics in three-phase fluidized beds can be grouped into three categories. The first category includes the analysis of the relationship between bubble size and the velocity of single bubbles. The second group considers the characteristics of bubbles in freely bubbling beds.

Finally, in the third category the main aim is to determine the mechanisms of bubble break-up [49].

Results for single rising bubbles of a known volume were reported by Massimilla et al. [49]. The apparatus consisted of a fluidized bed column 3.5in I.D. Several types of particles were used: - glass beads 0.79mm, 1.09mm; - iron sand 0.2mm; - silica sand 0.22mm. Water and air were employed as the liquid and gas phases, respectively. It was found that the relationship between the diameter of a single bubble and the bubble velocity was affected by the state of expansion of the liquid-solid bed and was almost independent of particle size and density.

Henrikson and Ostergaard [50] measured the rising velocity of single air bubbles in a two dimensional fluidized bed containing glass particles of 0.2mm, 1mm, 3mm in diameter. The liquids used were water, methanol and glycerol solutions. It was found that the bubble velocity, corrected for the wall effect, was well correlated by an equation involving the radius of a circular cap. The equation proposed was similar to the one used by Davies and Taylor for bubbles rising in inviscid liquids.

$$V_b = K'(gR)^{n'} \qquad (44)$$

In fact, it was observed that, except for the 3mm particles V_b was proportional to the square root of the radius of the circular cap. However, K was not the same for the various systems investigated. Because K should be the same for all inviscid liquids El-Temtamy and Epstein [51] modified eq. (44) expressing the bubble velocity as a relative velocity with respect to the liquid phase rather than relative to the column wall.

$$V_{br} = v_{BL} = V_b - \frac{u_L}{\varepsilon_L} \qquad (45)$$

This redefinition of the bubble velocity involved in eq. (45) made it possible to show that the two parameters n' and K' were constant for the various operating conditions and close to the values reported for pure liquids. Deviations from this trend were observed, however, for the smaller bubbles (in the 1.5 cm range) and for the lowest bed voidages. The discrepancy at low bed voidages was attributed to a non-Newtonian behaviour of the suspension [51]. It was also observed that the included angle of the bubble was affected by the viscosity of the liquid.

Rigby et al. [35], Park et al. [52] studied the behaviour of bubbles in a 10 cm diameter fluidized bed using an electro-resitivity probe made of two 0.1 cm Co-Ni-Fe alloy wires. The bed contained 0.120mm, 0.290mm, and 0.47mm glass particles and 0.775 mm sand particles. Both bubble frequencies and average bubble velocities were measured in a bubbling bed. It was stressed in their paper that a relationship between the bubble velocity and the bubble characteristic dimension (axial bubble cord) involve a $(V_b-u_G-u_L)$ velocity term following a model introduced by Nicklin [26]. With this concept Ribgy et al. [35] proposed the following correlation stressing the influence of bed expansion on bubble rising velocity.

$$(V_b-u_G-u_L)\left(\frac{1-\varepsilon}{\varepsilon}\right)^2 = 32.5\ell^{1.53} \qquad (46)$$

Considering at the same time the relationship between ℓ and R, $\ell \simeq 1.14R$, eq. (46) shows that the bubble velocity term depends on the bubble curvature radius to the 1.53 power. This functionality, R to the 1.53 power, was much stronger than the one suggested by Henrikson and Ostergaard [50] and El-Temtamy and Epstein [51].

Kim et al. [49] performed an extensive investigation in order to describe the characteristics of bubbles in freely bubbling beds containing 1mm and 6mm glass particles and 2.6mm irregular gravel. The fluidizing liquids were water, water-acetone, solutions of sugar and solutions of carboxymethyl cellulose. The study was based on the analysis of movie film taken in a two dimensional bed. Two correlations were developed using a multiple regression analysis, one for the equivalent bubble diameter, d_b, and a second one for the relative bubble velocity V_{br}.

$$d_b = 13.4\ u_L^{0.052}\ u_G^{0.248}\ \upsilon^{0.008}\ \sigma^{0.034} \qquad (47)$$

and

$$V_{br} = 83.1\ u_L^{0.065}\ u_G^{0.039}\ \upsilon^{0.025}\ \sigma^{0.179} \qquad (48)$$

and

$$V_{br} = 18.0\ d_b^{0.989} \qquad (49)$$

where u_L, u_G, V_{br}, are in mm/s, d_b is in mm, σ is in dyne/cm and ν in $mNsec^n/m^2$.

Eqs. (47) and (48) describe, for the conditions considered by Kim et al. [49], (u_G = 2.0-50.0 cm/s) the significant effect of u_G on V_{br} and d_b. Moreover, the same equations also show the relatively small and more complex dependance of d_b and V_{br} on u_L, ν and σ parameters. The bubble velocity observed by Kim et al. [49] was about 50-60% higher than the velocities that could be predicted using eq. (44) for the same bubble sizes. It was claimed that this difference resulted from the fact that bubble clouds rise considerably faster than a single isolated bubble in a fluidized bed.

However, it is highly possible that error was introduced in the measure of bubble sizes. In a bidimensional column, 25mm thick, parabolic shaped bubbles (side view) can be expected. Therefore, the bubble diameters measured should be smaller than their actual sizes. This effect is more significant for the smaller bubbles. Consequently, the apparent higher bubble velocities could in fact correspond to a situation where the observed V_{br} are related to bubbles that were, as a result of the measurement technique, assessed with a reduced size.

Recently, de Lasa et al. [53], [54], Lee et al. [55], [56], developed a novel U-shaped fiber optic probe to study bubble properties and hold-up in three-phase fluidized beds. This probe has the potential of being applicable to high temperature and high pressure coal liquefaction reactors and heavy oil hydrocracking units. Several of these probes were employed in a multiprobe configuration to define the following bubble characteristics: bubble velocity, axial bubble length, bubble size distribution, bubble frequency, bubble included angle. Each one of the probes was made from a single core optical fiber (400μm). The fibers were bent in a U-shape with a 0.5mm radius according to the following principles: the radius of curvature of the U has to be large enough so that the angle of incidence at the turning point is bigger than the angle of total reflection when the same fiber is exposed to the gas phase. If this condition is met the source light is conserved in the fiber. In addition, the U has to be curved in such a way that the radius is small enough to secure an angle of incidence at the turning point smaller than the angle of total reflection for the same fiber dipped in the liquid. With this design of the U the source light is totally or partially lost in the liquid and conserved in the probe when exposed to the bubbles.

Using the U-shaped probes correlations were developed to predict how V_{br} changed with ℓ, the axial bubble length [56] (Figure 6). These correlations were developed from the data obtained in a cylindrical plexiglass column 30 cm diameter containing 335μm glass beads.

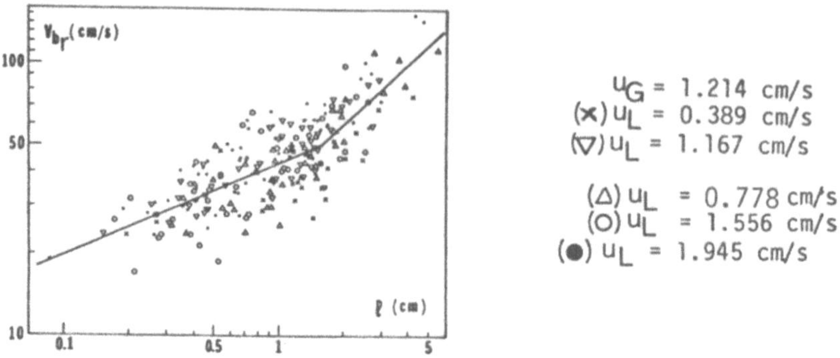

Figure 6. Change of the bubble velocity with respect to the axial bubble length [56].
Reproduced by permission of the American Institute of Chemical Engineers

$$V_{br} = (43.3 \pm 1.82)\ell^{0.342 \pm 0.059} \qquad \ell < 1.56 \qquad (50)$$

$$V_{br} = (36.44 \pm 7.4)\ell^{0.732 \pm 0.21} \qquad \ell > 1.56 \qquad (51)$$

The observation of two different regimes is not surprising and could reveal a similarity between gas-liquid flow and fluidized beds. In fact, it was indicated by Clift et al. [57] that in gas-liquid systems the velocities of small ellipsoidal bubbles are less sensitive than those of large spherical cap bubbles to the bubble length. Using the U-shape multiprobe configuration device the shape of larger bubbles was also investigated. This was achieved through the definition of the bubble included angle. The coexistence of slug, disk-like and spherical cap bubbles in three phase fluidized beds was also observed [56].

The introduction of solid particles into a two phase flow system (gas-liquid) could result in an enhanced bubble growth or in bubble splitting. Fine particles favour bubble coalescence while coarse solids bubble breaking. For glass beads or sand fluidized by air and water the transition particle diameter from bubble coalesence to bubble breaking regime is about 3 mm. Ostergaard [58] analyzed the growth of air bubbles emerging from a single orifice (3mm diameter) in a bed of small sand particles (0.64mm) fluidized by water recording the bubble frequency and its changes with bed position. The bubble frequency was measured using an electrical resistance probe at both the orifice injector and the bed surface. The change between these two frequencies provided information about bubble coalesence. It was observed that bubble coalescence was influenced by bed porosity. In beds of small particles, a reduction of the bed porosity favours bubble coalescence [58]. These results - coalescing in beds of small particles - are in agreement with those obtained by Rigby et al. [35] in beds of 0.120 mm, 0.290 mm and 0.470 mm glass particles and 0.775 μm sand particles. They also agreed with the observations of Massimilla et al. [49] in beds of 0.790 mm, 1 mm glass particles, 0.200 mm silica sand particles and 0.220 mm iron particles.

Page and Harrison [59] analyzed the process of bubble coalescence in air-water-0.5 mm sand particles taking pictures of the bubbles at the top surface of the dense bed. It was observed that at the lower liquid velocity of 0.85 cm/s the average bubble size decreased with the gas flow rate. With this experimental evidence Page and Harrison [59] argued that the average bubble size and the bubble size distribution in a fluidized bed is the result of competing effects - bubble splitting and bubble coalescence. Moreover, qualitative criteria were proposed to define the conditions of coalescence and splitting: - bubbles can split by turbulent eddies which are favoured by higher gas velocities, - the rate of coalescence is higher in liquids of higher viscosity, - the apparent viscosity of the bed decreases with an increase of u_L which in turn favours bubble splitting, - bubbles can be split by single particles and/or by groups of particles.

In beds of coarse glass particles, 6 mm diameter, the observed bubble sizes are normally quite small, (2mm of bubble equivalent diameter) [60]. Consequently, the bubbles emerging from the beds are quite uniform and small compared with the ones for gas-liquid systems containing no solids. This seems to show the significant extent of bubble break-up taking place in those systems. In order to explain this phenomenon two different mechanisms have been proposed. A first mechanism suggests that bubble break-up is due to penetration by particles into the bubble. A second mechanism considers that bubble break-up is a result of the Taylor instability of the bubble roof. To explore these two possible mechanisms, Henriksen and Ostergaard [60] observed the effect of 5mm steel spheres and 3-6 mm glass particles on air bubbles having 2 cm equivalent diameters. In none of the experiments did the bubbles disintegrate when spheres fell through them. It was postulated, using the Taylor instability analysis that only for particles exceeding 8 mm or in the case of bombardment by successive particles this effect could be of significance [60]. Bhatia [61] indicated that the critical diameter (8mm) could be even higher if the liquid fluidized bed viscosity was considered in the instability analysis. In this respect, Henrikson and Ostegaard [60] argued, following a previous study by Clift and Grace [62], that roof instability was more likely as the origin of bubble break-up. This effect is a consequence of turbulence in the liquid phase and resulted in a "finger" of liquid projected down from the roof. This liquid inside the bubble grew in size and eventually divided it.

Mixing Within Phases

The conditions of mixing in three-phase fluidized bed, are certainly the result of the behaviour and the interactions of the gas, liquid and solid phases.

The degree of mixing of the gas phase is normally quite low. Peclet numbers in the 40-50 range have been reported. For instance, Ostergaard and Michelsen [63] measured the Peclet number using a tracer technique involving pulses of argon-41 and following these perturbations with gamma ray detectors. The detectors were located at two different axial positions and fixed on the wall of the column. Because of the finite solubility of argon-41 in water, a correction to the measured signal was made to take into account the influence of absorption-desorption of the tracer in the liquid phase. For the analysis of the argon pulses the piston flow with axial dispersion model was adopted. The transfer function of the input-output perturbations or the weighted moments of the same signals were used instead of the conventional statistical moments (Laplace parameter = 0). With these techniques, a smaller influence of tail errors and consequently more reliable Pe_G parameters were observed [63]. Michelsen and Ostergaard [64] also used the pulse perturbations of argon-41 to trace mixing in the gas phase. It was mentioned by these authors that the axial dispersion model was not suitable in systems with rapid bubble coalescence. More recently Vasalos et al.[65], employed the same techniques developed by Ostergaard and Michelsen in an H-Coal fluidized bed reactor simulator. Extrudate desulfurization catalysts (1.8 mm diameter, 5.1 mm length) were fluidized in a glass column of 15 cm. diameter. Mineral oil and coal fines-kerosene slurry were used as fluidizing media. The viscosity of these liquids is similar to H-Coal liquids. These authors reported that considerable gas mixing existed in that unit.

Mixing in the liquid phase of three-phase fluidized beds is much more important than in the gas phase. Peclet numbers in the 0.005 - 0.5 range have been reported. It seems that if the fluidized bed operates in the bubble-coalescing regime axial mixing is augmented while in the bubble-disintegrating regime axial mixing is decreased. These observations are relative to the mixing conditions in a solid-free system operating at the same gas and liquid flow rates as for the two cases described above. The increase of mixing in the bubble-coalescing regime seems to be a consequence of larger and faster moving bubbles that stir the liquid vigorously. At the same time the decrease of axial mixing is the result of smaller and slower moving bubbles. These bubbles, that usually develop in the bubble-disintegrating regime, stir the liquid to a much smaller extent.

Ostergaard [67] used an unsteady tracer technique employing gamma-ray emitting ammonium bromide solution, to assess the liquid mixing in a 21.6 cm diameter column containing 1.1 mm, 3 mm and 6 mm glass ballotini. It was observed that for the 1.1 mm and 3 mm particle the axial dispersion significantly changed with u_G in the 0.5 - 20 cm/s range. The liquid velocity, u_L, had only a moderate effect on the mixing of the liquid. For 6 mm particles only at the higher gas velocities were the axial dispersion coefficients affected.

Vail et al. [68] used a steady state technique involving steady injection of aqueous NaCl solution at a given level of the column, to monitor the axial mixing of the liquid. The fluidized bed contained air, water and 0.87 mm spheres with a density of 2700 kg/m^3.The concentration profiles of tracer measured below the injection point provided information about axial dispersion coefficients and the characteristic mixing length (ratio of the axial dispersion with respect to the fluid velocity). It was observed that the presence of solids and the increase of the gas and liquid velocities were all factors promoting axial mixing of the liquid phase.

Kim et al. [30] used an unsteady pulse and an unsteady step injection of potassium permanganate solution in beds containing 2 mm glass beads and 2.6 mm irregular gravel. It was observed that the height of mixing unit parameter (HMU = $2H/Pe_L$) proportional to the axial mixing coefficient, increased with u_L and u_G. An empirical correlation was proposed to relate HMU with the Reynolds and Froude numbers for the gas and liquid phases.

$$\frac{HMU}{H_o} = \frac{2H}{Pe_L} = 0.068 \left(\frac{Fr_{pL}\rho_S}{\rho_L}\right)^{0.128} \left(\frac{Fr_{pG}\rho_S}{\rho_G}\right)^{0.168} \left(Re_{pL}Re_{pG}\right)^{0.120} + \left(\frac{HMU}{H_o}\right)_{u_G=0} \qquad (52)$$

El Temtamy et al. [69] determined the mixing in the liquid phase in a 5 cm. diameter fluidized bed using a steady state tracer technique similar to the one by Vail et al. [68]. Ammonium chloride solution was injected into the fluidized bed at a given level. When a steady state tracer distribution was attained, samples of the liquid upstream from the point of injection were collected. With this information the axial dispersion coefficients were determined in beds of 0.45 mm, 0.96 mm, 2 mm and 3 mm glass beads. No backmixing was detected for 3mm particles.

It was observed, however, that axial dispersion in the liquid phase was significant for 0.96 - 2 mm particles. This result may be due to the increase of bubble size. Similar results were obtained by Michelsen and Ostergaard [64]. However, in the Michelsen and Ostergaard case [64] for similar operating conditions the effect of u_G was much more pronounced and this was attributed to the use of a bigger column, 15.2 cm instead of 5 cm. Concerning the influence of particle size on axial mixing of the liquid phase it was observed that this effect was quite moderate [69]. Moreover El-Temtamy et al. [70], using the same set-up, injected the tracer at a point instead of along a plane, and analyzed the downstream concentration of tracer. In this way, both axial and radial dispersion coefficients were obtained. It was observed that the axial mixing coefficients were about 10-40 times bigger than the values of the radial mixing coefficients.

Joshi [71] using energy balances, proposed a unified model for estimating the liquid mixing. This model under the specific conditions of three-phase fluidized beds provided the following correlation where D_L, the axial dispersion coefficient, is given as a function of the various operating parameters u_G, u_L, ε_S, ε_L and ε_G.

$$D_L = 0.29\ (u_c + u_L)D \qquad (53)$$

where

$$u_c = 1.31\left\{gD(u_G + u_L)\frac{\rho_L u_L}{\varepsilon_S\rho_S + \varepsilon_L\rho_L} - v_{st}\,\varepsilon_S\left(\frac{\rho_S}{\rho_S\varepsilon_S + \rho_L\varepsilon_L} - 1\right) - \varepsilon_G V_{b\infty}\right\}^{1/3} \qquad (53a)$$

The correlation proposed by Joshi [71] was tested using about 80% of the liquid axial mixing coefficients reported in the literature by Kato et al.[46], Michelsen and Ostergaard [64], Ostergaard [67], and Vail et al. [68]. For the selected data points, the model conditions of 20% dissipation of input energy, was fulfilled. A 16% deviation between the experimental D_L values and the model predictions was observed.

Because most of mixing studies in three-phase fluidized beds have been carried out using water as the liquid phase, Kim and Kim [72] reported results of studies where the viscosity and surface tension of the liquid phase was changed over a wide range. Solutions of glycerol, methanol, alkyl polyester sulfonate and carboxymethyl cellulose in water were used. The tracer was NaCl and the tracer concentrations were monitored by seven conductivity probes. The data were analyzed using the axial dispersion plug flow model. It was observed that the effect of changing the viscosity from 1 - 30 cp and the surface tension from 40-70 dyne/cm was marginal. Kim and Kim [72] suggested that this was a consequence of the turbulent nature of the flow regimes in three-phase fluidized beds. A correlation was then proposed to predict Pe_L from knowledge of dp, D, u_L and u_G.

$$Pe_L = 20.19 \left(\frac{d_p}{D}\right)^{1.66} \left(\frac{u_L}{u_L + u_G}\right)^{1.03} \quad (54)$$

The mixing of solid particles in a three-phase fluidized bed was analyzed by Evstrop'eva et al. [73] taking cine-camera films of the particle motion in bidimensional beds of 2.5 mm and 3.9 mm glass beads. Two types of tracer particles were employed:
- coloured glass spheres (heavy tracers) having the same dimension and density as the other particles in the bed, - polystyrene particles (light tracers) having similar dimensions but a smaller density (1000 Kg/m^3) than the rest of the solid phase. The use of these two types of tracers allowed Evstrop'eva et al. [73] to analyze the particle motion for solids having a density very close to the solid phase and to the liquid phase, respectively. Histograms showing the distribution of particle velocites for light and heavy particles were defined. With this experimental data both the average velocities and the pulsation components were derived. It was observed that the average and the pulsation components increased with air flow rates and liquid velocity. The effect of particle diameter on particle velocities did not show a clear trend.

Concerning the mixing of particles, interesting observations were reported by Epstein et al. [74] who analyzed the various factors affecting the on-set of stratification or mixing in a fluidized bed consisting of a light solid (polypropylene or nylon spheres) and a heavy liquid (KI aqueous solution). It was shown that the condition of incipient mixing or stratification was essentially the result of the acting body forces. Using this concept, ρ_{mix} (gas plus liquid and solids) should be compared with ρ_s to define the on-set of mixing or stratification. If $\rho_{mix} < \rho_S$ then mixing should be achieved. For the opposite, if $\rho_{mix} > \rho_S$ then stratification should result. Deviation from this condition was significant at the higher gas flow rates and therefore the criterion should be taken as a conservative estimate of the on-set of mixing.

Because in several applications of three-phase fluidized beds a size distribution is commonly encountered, Fan et al. [75] analyzed the conditions of particle mixing and particle stratification using a binary mixture of solids. Using a 7.62 cm. diameter plexiglass column and two types of binary solid mixtures - 3mm and 4 mm glass particles, and 3 mm and 6 mm glass particles three possible mixing states were observed. These three states defined as: - complete segregation, - partial intermixing, - complete intermixing. The small particles used in this study were colored black while the large particles were white. It was shown by Fan et al. [75] that a strong correlation existed between the particular condition of particle mixing and the flow regime. For instance, it was observed (Figure 7) that complete mixing occurred in the coalescence bubble regime and slugging regime but only slightly in the transition regime (condition between slug and dispersed flow). Partial intermixing took place mainly in the dispersed bubble regime, the transition regime and only slightly in the bubble coalescing regime. Finally, complete segregation occurred solely in the dispersed bubble regime.

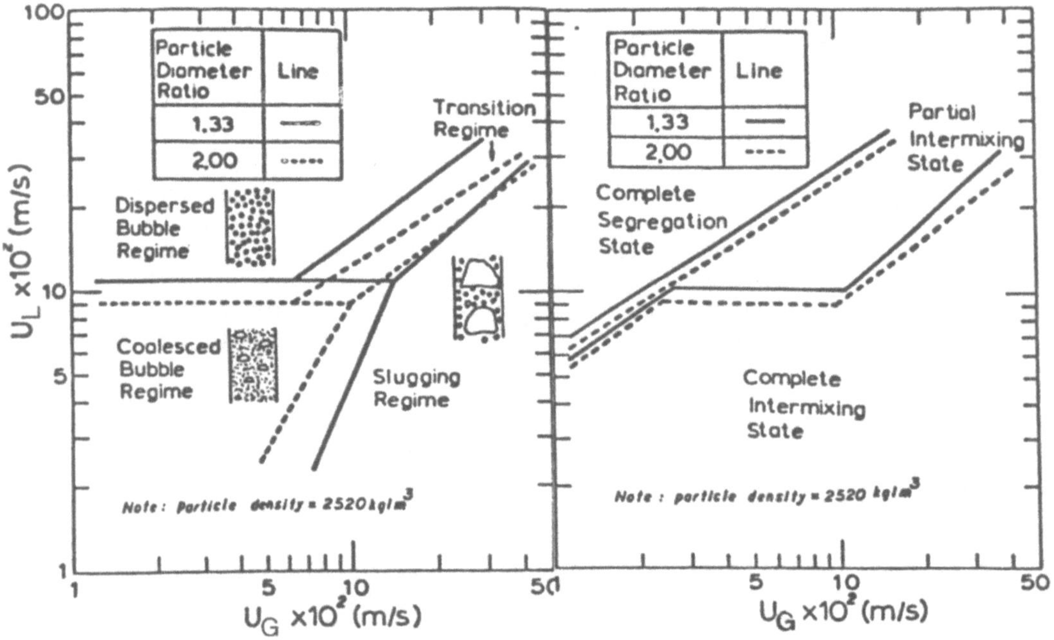

Figure 7. Flow regimes and solid mixing for binary mixtures with particle diameter ratios of 1.33 and 2. [75].
Reproduced by permission American Institute of Chemical Engineers.

Heat Transfer. Viswanathan et al. [76] measured wall-to-bed heat transfer in a jacketed fluidized bed containing by air, water and quartz particles (0.928 mm and 0.649 mm). Steam was circulated through the jacket. The amount of condensate collected and the sensible heat gained by the liquid phase were used to assess the heat transferred to the fluidized bed. The wall-to-bed heat transfer coefficient, h, and its dependence upon the ratio of air to liquid flow are shown in Figure 8. It was found that h increased first, reached a maximum and then decreased. It was speculated that the ascending portion of the curve up to the maximum point was probably due to the increased turbulence in the vicinity of the wall created by rising bubbles. It was argued that the decrease of h after the maximum could be the result of having the heat transfer surface partially covered with air which has a small thermal conductivity. In Figure 8 it can be observed

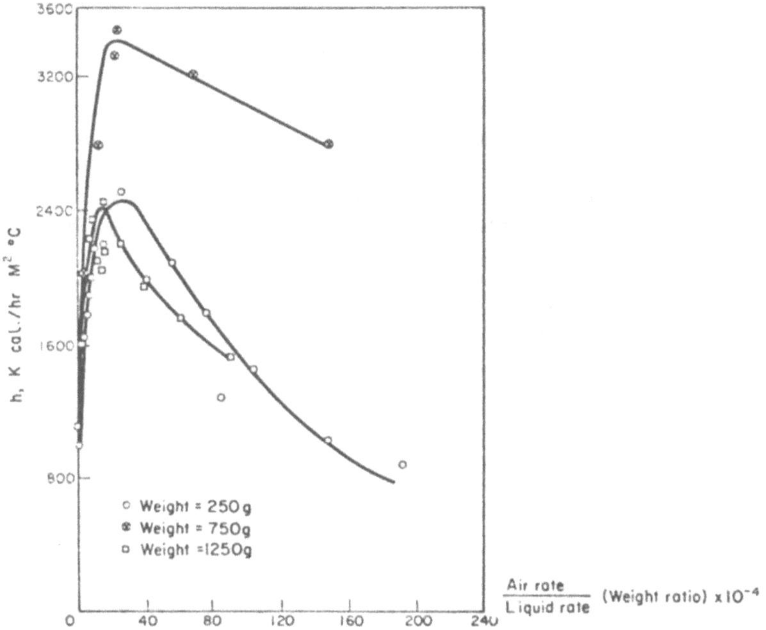

Fig.8. Heat Transfer data in three-phase fluidized beds of quartz sand, dp - 0.928 mm [76]. Reprinted with permission from Chem.Eng.Sci., 20 (1964). Copyright 1964 Pergamon Press.

that for a given particle size, h is dependant on the total weight of solids. It was shown by Viswanathan et al. [76] that there is an optimum solid weight which gives maximum heat transfer. Increase of h with weight of solids is possibly related to the fact that a higher concentration of particles assisted the process of reducing the thickness of the liquid film that develops near the wall. Decrease of h beyond a critical weight was probably due to an increased tendency of the solids to settle in the bed.

Armstrong et al. [77], Baker et al. [78] measured heat transfer coefficients between a heat transfer surface and a three-phase fluidized bed (air-water-glass beads). The fluidized bed was 24 cm in diameter and 274 cm in height. The heat transfer surface was concentrically located inside the column and consisted of a brass cylinder (6.35 cm in diameter by 25.4 cm long) housing four electrical heating elements. The heat transfer coefficients were assessed considering the power input to the heaters, the surface area of the brass cylinder and the temperature difference between the bed and cylinder surface. Experiments were conducted with four particle sizes 0.5 mm, 1 mm, 3 mm and 5 mm. The h values and their changes with u_G, u_L and d_p were analyzed and compared with h in gas-liquid and liquid-solid systems. Armstrong et al. [77], Baker et al. [78] found that h increased with u_G. The increase was initially quite rapid. However, at the higher gas flow rates, this effect became less marked and approached a constant value. It was mentioned that the introduction of gas into a liquid-solid fluidized bed did not affect drastically the bed porosity. It is, in fact, the higher turbulence promoted by the gas who gives the higher h values observed in three-phase fluidized beds. Change of h with u_L was studied and an optimum u_L was found. The effect of solid particles on h was also analyzed by Armstrong et al. [77], Baker et al. [78]. It was shown that the introduction of solids into a gas-liquid bed has in general the effect of increasing the heat transfer coefficient. An exception to this were the beds of small particles (dp = 0.15 mm) subjected to the higher superficial velocities (u_G = 11.9 - 23.7 cm/s for u_L cm/s). Changes of h with particle size were small for these conditions(Figure 9).

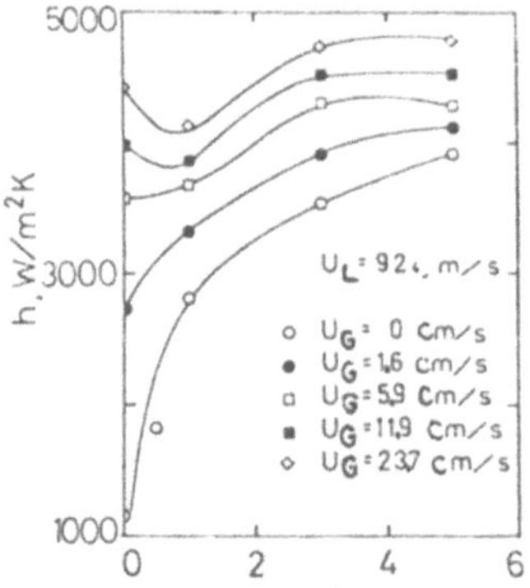

Figure 9. Change of the wall-to-bed heat transfer coefficient with particle size, u_L = 9.2 cm/s [78].

Using this data, the following correlation was proposed

$$h = 1977\, u_L^{0.070}\, u_G^{0.059}\, d_p^{0.106} \qquad (55)$$

where h is in w/m^2°K, u_L and u_G in mm/s and dp in mm.

Kato et al. [43] measured wall heat transfer coefficients in two fluidized bed columns having diameters of 5.2 cm and 12 cm, respectively. Water and aqueous solutions of carboxymethyl cellulose were used as the liquid phase and air was employed as the gas phase. The diameters of the glass beads were 0.42 mm, 0.66 mm, 1.2 mm and 2.2 mm. An important aspect of this contribution was the assessment of the influence of liquid viscosity on h. An electrically heated copper section fixed at the wall of the column was used as the source of heat. The heat transfer coefficients were calculated from the heat flux and temperature difference between the wall and the bed. Kato et al. [43] correlated h in terms of a modified Nusselt number, a modified Reynolds number, the Prandtl number and Fr_G. The equation predicts Nu' as an additive effect of the liquid-solid and gas phases.

$$Nu' = 0.044\,(Re'\,Pr_L)^{0.78} + 2.0\,Fr_G^{0.17} \qquad (56)$$

where $Nu' = \frac{h\,d_p\,\varepsilon_L}{k_L(1-\varepsilon_L)}$ and $Re' = \frac{d_p\,U_L\,\varepsilon_L}{\mu_L(1-\varepsilon_L)}$

The correlation of Kato et al. [43] was tested with their own results and with data previously reported by Baker et al. [43] and Fukuma et al. [79]. The calculated values agreed with the experimental results within ± 30%.

Mass Transfer. An important question in the process of assessing the gas-liquid mass transfer in three-phase fluidized bed is given by the fact that the gas-side mass transfer coefficient can generally be neglected. In other words, the overall gas-liquid mass tansfer coefficient is dominated by the liquid-side mass transfer coefficient, k_La_L. Ostergaard and Suchozebrshi [80] investigated the absorption of carbon dioxide in three phase fluidized beds of 1 and 6 mm glass ballotini and water. The volumetric mass transfer coefficient or k_La_L increased with gas flow rate but was uninfluenced by variation of the liquid velocity. Particle size was observed to have a pronounced effect on the absorption rate (Figure 10). The k_La_L values for 6 mm particles were approximately ten times larger than the coefficients for the 1 mm particles. The coefficients in solid-free bubble columns were of intermediate magnitude. The difference between the k_La_L values for beds of 1 mm and 6 mm particles corresponded to the different hydrodynamic state of the bed. Beds of small particles are characterized by bubble coalescence and by low gas-liquid interfacial areas. On the other hand, beds of large particles (6 mm) are characterized by bubble disintegration and have large gas-liquid interfacial areas.

Ostergaard and Fosbol [81] reassessed the conditions of gas-liquid mass transfer from the gas phase to the liquid phase in fluidized beds of 1 mm and 6 mm particles by measuring the rate of

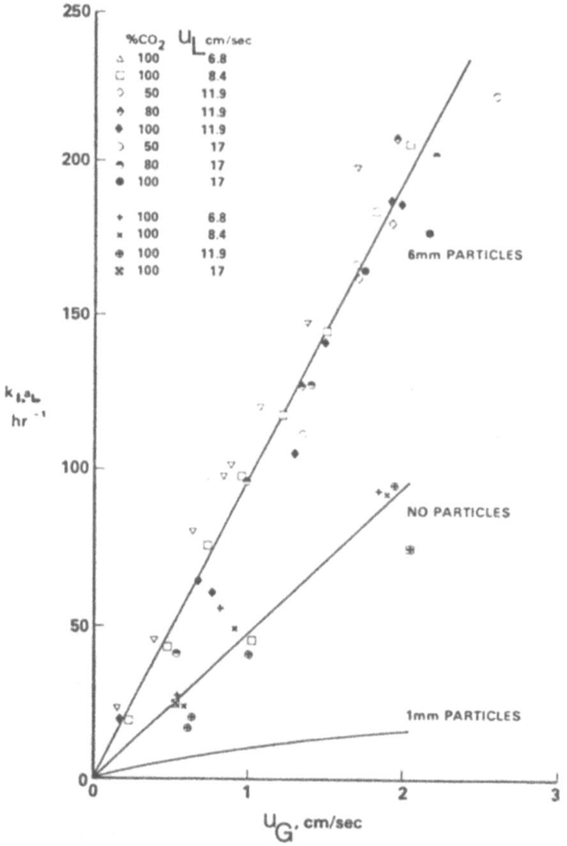

Figure 10. Volumetric absorption coefficient for beds of 1-mm particles, - no particles at different u_L [80].

absorption of oxygen in water. The use of oxygen as the absorbed gas is advantageous when compared with carbon dioxide because of the low oxygen solubility and correspondingly low depletion of the gas phase during its passage through the bed. Liquid samples were withdrawn at the inlet, the outlet and at four different levels along the column. The oxygen concentration of the liquid phase was determined by titration. The k_La_L coefficients were calculated under the assumption that fluid flow could be described as piston flow without axial mixing. Ostergaard and Fosbol [81] claimed that this assumption was adequate for beds of 6mm particles showing low mixing in the liquid phase. However, this hypothesis is questionable in bubble columns and in beds of 1 mm particles in which the liquid mixing is significant. The results obtained by Ostergaard and Fosbol [81] with the oxygen absorption technique were in accordance with previous observations. The superficial liquid velocity had no significant effect on the absorption rate in beds of 6 mm and in bubble columns. However, in beds of 1 mm particles an increase of liquid velocity caused a marked increase in k_La_L. It was argued that the change in u_L affected bubble coalescence and consequently the bed mass transfer. It was also observed by Ostergaard and Fosbol [81] that k_La_L decreased with increasing bed height. It was apparent that k_La_L varied markedly with axial position in the bed. In beds of 1 mm particles k_La_L decreased steadily with increasing distance whereas in beds of 6 mm particles an initial increase to a maximum value was followed by a decrease in k_La_L. Several factors were advanced to explain these results. However, the dominant effect seems to be the influence of bubble coalescence in beds of 1 mm particles that tends to reduce k_La_L with height. It was suspected that for 6 mm particles the bubble disintegration regime was reduced close to the gas distributor, the increased bubble size increasing k_La_L. Further up the column k_La_L diminished when bubble size stabilized and slip velocity decreased.

Dakshinamurty et al [82] determined volumetric mass transfer coefficients by absorbing oxygen and CO_2 in water in a three-phase fluidized bed consisting of particles with a size between 1.06 and 6.84 mm and densities between 2400-11170 Kg/m^3. The glass column was 5.6 cm in diameter and 68 cm high. The gas and liquid velocities were changed over the range 0.5 to 8 cm/s and 1.72 to 18 cm/s, respectively. It was assumed that the fluid flow pattern could be described with the piston flow model without axial mixing. These investigators confirmed the results of other authors [80], that the k_La_L values were increased by the presence of the solid particles. The only exception to this was for particles with a size smaller than 1 mm.

Lee and Worthington [83] investigated the absorption of CO_2 into water and into a sodium sulphonate solution in a bed of 6 mm particles. In addition to the measurements of k_La_L they also measured the interfacial area, a_L, at the top of the dense bed using a light transmission technique and assumed that a_L was virtually the same throughout the bed. Lee and Worthington [85] found a similar effect of u_G and u_L on k_La_L as observed by Ostergaard et al. [80],[81]. Fakushina [84] took the experimental data of Lee and Worthington [83] and proposed an empirical equation to correlate a_L with the Re_L, Re_G, dp, D and ε. At the same time the k_L coefficient was correlated using an equation similar to one used for packed bed columns with cocurrent flow.

Strumillo and Kudra [85] used absorption with chemical reaction,- CO_2 in NaOH solution - to determine a_L in a three phase fluidized bed. They employed the method proposed by Danckwerts which enables k_L and a_L to be determined independently.

Dhanuka and Stepanek [85] used a method that involved the simultaneous measurement of the volumetric liquid mass transfer coefficients and the interfacial area. The importance of this approach measuring independently k_L and a_L was stressed considering that k_L and a_L depend on different parameters. For instance, it is expected that k_L depends on the diffusivity and the viscosity while a_L is affected by surface tension. Dhanuka and Stepanek [86] measured oxygen desorption into carbon dioxide and nitrogen. The liquid samples were withdrawn at one bed position (62 cm above the grid) and the dissolved oxygen concentration in the sample was measured with an oxygen meter. It was claimed that the k_La_L predictions were not significantly affected if liquid backmixing was ignored. The experimental k_La_L values as a function of the superficial velocities were reported for glass ballotini particles of 1.98 mm, 4.08 mm and 5.86 mm diameter respectively. For 1.98 mm particles k_La_L increased with u_L and u_G. For beds of 4.08 mm and 5.86 mm solids the liquid velocity had no significant effect on k_La_L and the volumetric mass transfer coefficient was mainly a function of the gas velocity. The influence of the particle size on k_La_L confirmed trends reported previously [80], [81]. The beds of 1.98 mm particles had k_La_L values slightly lower than in bubble columns while the k_La_L values in beds of 4.08 mm and 5.86 mm particles were much higher than for a bubble column. The bed of 5.86 mm particles was the most efficient contactor of the three tested. To determine the gas-liquid interfacial area, absorption of CO_2 in $NaCO_3$ - $NaHCO_3$ buffer solution with hypochlorite as a catalyst was employed. The results are shown in Figures 11 and 12. For 4.08 mm and 5.86 mm particles, a_L increased fairly linearly with u_G. It can be observed that the changes of a_L and the volumetric mass transfer coefficients with u_L and u_G follow similar trends. These results also indicate the important influence of a_L on the k_La_L group.

Alvarez Cuenca et al. [87], [88] studied oxygen mass transfer in three-phase fluidized beds stressing the need of a two zone model. This was done to overcome the apparent inability of the piston flow model and the axial dispersion model to represent the experimental data. The k_La_L values in each zone were calculated from averaged volumetric mass transfer coefficients between the grid and a particular axial position. Because the k_La_L parameters in each zone - grid region and bulk zone - were calculated from averaged parameters instead of using the local volumetric mass transfer coefficient the conclusions should be reassessed using an appropriate method of data analysis.

Recently Catros et al. [89] measured oxygen concentration profiles in a 17.2 cm plexiglass column and a bed containing 0.508 mm particles. It was observed that axial dispersion was limited

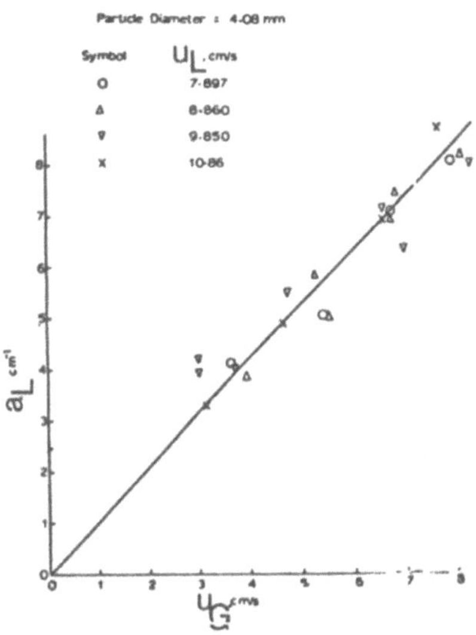

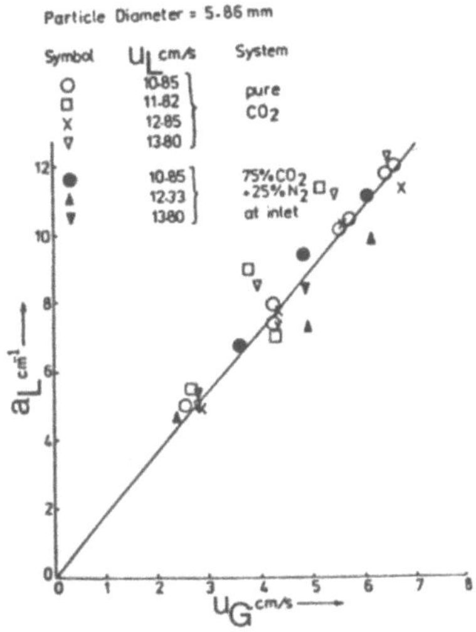

Figure 11 Effect of gas and liquid velocities on a_L dp = 4.08 mm [86]

Figure 12 Effect of gas and liquid velocities on a_L dp = 5.08 mm [86]

Reproduced by permission of the American Institute of Chemical Engineers

in the dense bed. However, axial mixing became significant in the freeboard region as well as in experiments run with the column free of solids (bubble column). The effect of horizontal baffles on the mass transfer was also evaluated in this study [89].

Finally, the conditions of liquid-solid mass transfer were explored by Prakash et al.[66] and Arters and Fan [90] as reported by Muroyama and Fan [5]. Arters and Fan [90] employed cylindrical particles of benzoic acid (diameter: 1.5 mm, length: 4 mm) fluidized by air and water. Results showed that the liquid-solid mass transfer in a three-phase fluidized bed was higher than that of a two-phase (liquid-solid system) fluidized bed operating at the same liquid velocity.

<u>Notation</u>

a	parameter eq. (27), eq. (29), eq. (42).
a_L	specific interfacial area
A	crossectional area of the bed (cm^2), parameter eq. (4)
b	parameter eq. (27), eq. (29), eq. (42).
c	parameter eq. (27), eq. (29), eq. (42).
d	parameter eq. (29).
d_b	equivalent bubble diameter (cm)
d_p	particle diameter (cm)
D	diameter of the column (cm)

D_L axial dispersion coefficient (cm^2/s)
f parameter eq. (29)
Fr_p dimensionless Froude Number for the particle = U/dp g
g gravity acceleration (cm/s^2) and constant in eq. (29).
gc gravity acceleration ($Kg/Kgf.cm/s^2$)
Ga dimensionless Galileo number = $(\rho_s-\rho_L)\rho_L g\ d_p{}^3/\mu^2$
h wall-to-bed heat transfer coefficient
H total height of the dense bed (cm)
HMU height of mixing unit parameter (cm)
k ratio of the wake volume to bubble volume (cm^3wake/cm^3bubble)
k' thermal conductivity of the liquid
k_L liquid side mass transfer coefficient (cm/s)
K parameter eq. (42).
ℓ axial bubble cord (cm)
n parameter of the Richardson-Zaki equation, (eq. 7)
n' parameter eq. (44).
Pe Dimensionless Peclet number
r radial position in the column (cm)
R radius of curvature (cm), radius of the column (cm)
Re_p Dimensionless Reynolds number for the particle $\rho u d p/\mu$
u superficial velocity (cm/s)
u' superficial velocity based in the particulate phase (cm/s)
v linear velocity (cm/s)
v_{Li} modified linear velocity used in eq. (7)(cm/s)
v_{BL} relative velocity of the gas phase (bubbles) with respect to the liquid phase (cm/s)
v_{LP} linear velocity of the liquid in the particulate phase (cm/s)
V_b velocity of the bubble with respect to the walls of the column (cm/s)
V_{br} relative velocity of the bubble with respect to the liquid phase (cm/s)
$V_{b\infty}$ velocity of an isolated bubble (cm/s)
W total weight of solid particles in the bed (Kg)
We Dimensionless Weber number
x volumetric fraction of the wake occupied by the solid particles (cm^3 particles/cm^3 wake)
x_c critical volumetric fraction of the wake occupied by the solids above which it could be predicted that the initial bed expansion takes place and below which it could be estimated that contraction occurs.
ε volumetric fraction based on the bed volume
ε' volumetric fraction based on the particulate phase volume
α constant parameter eq. (4)
β constant parameter eq. (4)
γ electroconductivity in the bed
γ_o electroconductivity in the liquid alone

ρ density (Kg/m^3)
μ viscosity (g/cm.s)
σ surface tension (dyne/cm)
Δp pressure drop (Kgf/cm^2)
υ generalized viscosity constant (mNs^n/m^2)

Subscripts

av average
B bed
exp experimental
G gas phase
L liquid phase
L_{mf} liquid phase at minimum fluidization
LP particulate phase
Lt liquid phase at terminal velocity condition
mix mixture = gas + liquid + solid
mf at minimum fluidization condition
pmf particle at minimum fluidization condition
pt particle at terminal velocity condition
S solid
t at terminal velocity condition
W wake

REFERENCES

1. Ostergaard, K., "Three-Phase Fluidization" in Fluidization eds. J.F. Davidson and D. Harrison, 751-780 (New York, Academic Press, 1971).
2. Wild, G., "Les Reacteurs à lits fluidisé gaz-liquide-solide. Etat de l'art et perspectives industrielles", Entropie, 106 (1983), 3-37.
3. Epstein, N., "Three Phase Fluidization: Some Knowledge Gaps", Can. J. Chem. Eng. 59 (1981), 649-657.
4. Baker, C., "Three-Phase Fluidization" in Multiphase Chemical Reactors, eds. A. Rodrigues, J. Calo and N. Sweed (The Netherlands, Sithoff and Noordhoff, 1981).
5. Muroyama, K. and L.S. Fan, "Fundamentals of Gas-Liquid-Solid Fluidization", AIChE Journal 31 (1985) 1-34.
6. Chern, S.H., K. Muroyama and L.S. Fan, "Hydrodynamics of Constrained Inverse Fluidization an Semifluidization in a Gas-Liquid-Solid System". Chem. Eng. Sci. 38(1983) 1167-1174.
7. Fan, L.S., K. Muroyama and S.H. Chern, "Hydrodynamics of Inverse Fluidization in Liquid-Solid and Gas-Liquid-Solid Systems", Proceedings 2nd World Chem. Engng. Conference, Montreal, Canada, 3(1981)41-44.
8. Fan, L.S., K. Muroyama and S.H. Chern, "Hydrodynamics Characteristics of Inverse Fluidization in Liquid-Solid and Gas-Liquid-Solid Systems". Chem. Eng. J. 24 (1982) 143-150.

9. Schuman, S.C., R.H. Wolk and M.C. Chervanak, "Hydrogenation of Coal" U.S. Patent 3301393 (1967).
10. Adlington, D. and E. Thompson, "Desulfurization in Fixed and Fluidized Bed Catalyst Systems". Proc. 3rd. Euro. Chem. React. Eng., 203-213 (Pergamon Press, Oxford, 1965).
11. Blum, D.B. and J.J. Toman, "Three-Phase Fluidization in a Liquid Phase Methanator", AIChE Symp. Ser. 73, No. 161 (1977), 115-120.
12. Volpicelli,G. and L. Massimlla, "Three-Phase Fluidized Bed Reactors, An Application of the Production of Calcium Bisulphite Acid Solutions", Chem. Eng. Sci, 25 (1970) 1361-1372.
13. Sastri, N.V.S., N. Epstein, A. Hinrata, I. Koshijima and M. Izumi, "Zinc Hydrosulphite by Three Phase Fluidization: Experiments and Model". Can. J. Chem. Eng. 61 (1983) 635-646.
14. Richardson, J.F. and W.N. Zaki, "Sedimentation and Fluidization: Part I". Trans. Inst. Chem. Engrs. 32 (1954) 35-53.
15. Garside, J. and M. Al-Dibouni, "Velocity-Voidage Relationships for Fluidization and Sedimentation in Soid-Liquid Systems". Ind. Eng. Chem., Process Des. Dev. 16,2(1977).
16. Stewart, P.S.B. and J.F. Davidson, "Three Phase Fluidization: Water, Particles and Air". Chem. Eng. Sci, 19 (1964) 319-322.
17. Ostergaard, K., "On Bed Porosity in Gas-Liquid Fluidization". Chem. Eng. Sci., 20 (1965) 165-167.
18. Rigby, G.R. and C.E. Capes, "Bed Expansion and Bubble Wakes in Three-Phase Fluidization". Can. J. Chem. Engng. 48 (1970) 343-348.
19. Epstein, N. and D. Nicks, "Contraction or Expansion of Three-Phase Fluidized Beds". Fluidization Technology. ed. D. Keairns 389-397 (Washington, Hemisphere Publishing Corporation 1976).
20. Epstein, N, "Criterion for Initial Contraction or Expansion of Three-Phase Fluidized Bed" Can. J. Chem. Engng. 54 (1976) 259-263.
21. Darton, R.C. and D. Harrison, "Gas and Solid Hold-up in Three-Phase Fluidization", Chem. Eng. Sci. 30(1975) 581-586.
22. Bruce, P.N. and L. Revel-Chion, "Bed Porosity in Three-Phase Fluidization, Powder Tech. 10(1974) 243-249.
23. El-Temtamy, S.A., N. Epstein, "Contraction or Expansion of Three-Phase Fluidized Beds Containing Fine/Light Solids". Can. J. Chem. Engng. 57(1979)520-522.
24. Bhatia, V.K. and N. Epstein, "Three-Phase Fluidization: A Generalized Wake Model". Proceedings of the International Symposium on Fluidization and Its Applications 380-392 (Toulouse, Cepadues-Editions, 1974).
25. Bhatia, V.K., K.A. Evans and N. Epstein. "Effects of Solids Wettability on Expansion of Gas-Liquid Fluidized Bed" IEC Process Des. Develop 11(1972)151-152.

26. Nicklin, D.J., "Two Phase Bubble Flow" Chem. Eng. Sci, 17 (1962)693-702.
27. Dakshinamurty, P., V. Subrahmanyam and J.N. Ras "Bed Porosities in Gas-Liquid Fluidization" IEC Proc. Des. Dev., 10(1971)322-328.
28. Dakshinamurty P., D.V. Rao, R.V. Subbaraju and V. Subratimanyan, "Bed Porosities in Gas-Liquid Fluidization" IEC Proc. Des. Develop. 11 (1972) 318-319.
29. Begovich, J.M. and J.S. Watson, "Hydrodynamic Characteristics of Three-Phase Fluidized Beds" in Fluidization eds. Davidson and Keairns, 190-195 (Cambridge, Cambridge University Press, 1978).
30. Kim, S.D., C.G.J. Baker and M.A. Bergougnou, "Hold-up and Axial Mixing Characteristics of Two and Three-Phase Fluidized Beds", Can. J. Chem. Engng. 50 (1972)695-701.
31. Kim, S.D., C.G.J. Baker and M.A. Bergougnou, "Phase Hold-up Characteristics of Three-Phase Fluidized Beds". Can. J. Chem. Engng., 53(1975)134-139.
32. El-Temtamy, S.A. and N. Epstein, "Bubble Wake Solids Content in Three-Phase Fluidized Beds". Int. J. Multiphase Flow 4(1978)19-31.
33. Efremov, G.I. and I.A. Vakhrushev, "A Study of the Hydrodynamics of Three-Phase Fluidized Beds", Intern. Chem. Engng. 10(1970)37-41.
34. El-Temtamy, S.A., Ph.D. Thesis, Cairo Univ. (1974).
35. Rigby, G.R., G.P. van Blockland, W.H. Park and C.E. Capes. "Properties of Bubbles in Three Phase Fluidized Beds as Measured by an Electroresistivity Probe". Chem. Eng. Sci., 25(1970)1729-1741.
36. Vasalos, I.A. et al., "Study of Ebullated Bed Fluid Dynamics for H-Coal". Final Report, U.S. Dept. of Energy under Contract DE-AC05-10149(1980).
37. Darton, R. and D. Harrison, "The Rise of Single Gas Bubbles in Liquid Fluidized Beds", Trans Inst. Chem. Engrs. 12(1974)301.
38. Ishida, M. and H. Tanaka, "An Optical Probe to Detect Both Bubbles and Suspended Particles in a Three-Phase Fluidized Bed". J. Chem. Eng. Japan, 15(1982)389-391.
39. de Lasa, H., S.L.P. Lee and M.A. Bergougnou. "Multiprobe System for Measuring Bubble Characteristics, Gas Hol-up, Liquid Hold-up and Solid Hold-up in a Three-phase Fluidized Bed". Patent Application for Canada and USA (1985).
40. Begovich, J.M. And J.S. Watson, "An Electroconductivity Technique for the Measurement of Hold-ups in Three-Phase Fluidized Beds". AIChE J. 24(1978)351-354.
41. Achwal, S. and J. Stepanek. "An alternative method of determining hold-up in gas-liquid systems" Chem. Eng. Science, 30(1975) 1443-1444.

42. Dhanuka, V.R. and J.B. Stepanek, "Gas and Liquid Hold-up and Pressure Drop Measurements in a three-Phase Fluidized Bed" in Fluidization, eds. Davidson and Keairns, 179-183, (Cambridge University Press, 1978).
43. Kato, Y., K. Uchide, T. Kago and S. Morooka, "Liquid Hold-up and Heat Transfer Coefficient Between Bed and Wall in Liquid-Solid and Gas-Liquid-Solid Fluidized Beds", Powder Tech., 28(1981)173-179.
44. Vail, Y.K., N. Manakov and V.V. Manshilin, "The gas contents of three-phase fluidized beds". Int. Chem. Engng. 10(1970)244-247.
45. Morooka, S., K. Uchida and Y. Kato, "Recirculating Turbulent Flow of Liquid in a Gas-Liquid-Solid Fluidized Bed", J. Chem. Eng. Japan, 15(1982) 29-34.
46. Kato, Y., A. Nishiwaki, T. Fukuda and S. Tanaka, J. Chem. Eng. Japan 5(1972) 112 -
47. Kara, S., B.G. Kelkan, Y.T. Shah and N. Carr "Hydrodynamics and Axial Mixing in a Three-Phase Bubble Column". IEC Proc. Des. Dev. 21(1982) 584-594.
48. Kim, S.D., C.G.J. Baker and M.A. Bergougnou, "Bubble Characteristics in Three-Phase Fluidized Beds" Chem. Eng. Sci., 32(1977) 1299-1306.
49. Massimilla, L.A., Solimando and E. Squillace, "Gas Dispersion is Solid-Liquid Fluidized Beds", Brit. Chem. Eng. 6(1961) 232-239.
50. Henrikson, H.K. and K. Ostergaard, "Characteristics of Large Two-Dimensional Air Bubbles in Liquids and in Three-Phase Fluidized Beds". Chem. Eng. J. 7(1974) 141-146.
51. El-Temtamy, S.A. and N. Epstein, "Rise Velocities of Large Single Two-Dimensional and Three-Dimensional Gas Bubbles in Liquids and in Liquid Fluidized Beds". Chem. Eng. J., 19(1980) 153-156.
52. Park, W.H., W.K. Kang, C.E. Capes and G.L. Osberg, "The Properties of Bubbles in Fluidized Beds of Conductive Particles as Measured by an Electroresistivity Probe", Chem. Eng. Sci, 24(1969) 851-865.
53. de Lasa, H., S.L.P. Lee and M.A. Bergougnou, "Application of a Photometric Technique to the Study of Three-Phase Fluidized Beds". Proceedings 33rd Can. Chem. Engng. Conference, 2(1983) 503-508.
54. de Lasa, H., S.L.P. Lee and M.A. Bergougnou, "Measurement in Three-Phase Fluidized Beds using a U-Shaped Optical Fiber". Can. J. Chem. Engng. 62(1984) 165-169.
55. Lee, S.L.P., H. de Lasa and M.A. Bergougnou, "A U-Shaped Fiber Optic Probe to Study Three-Phase Fluidized Beds". Accepted for publication Power Tech. (1985).
56. Lee, S.L.P., H. de Lasa and M.A. Bergougnou. "Bubble Phenomena in Three-Phase Fluidized Beds as Viewed by a U-Shaped Fiber Optic Probe". Accepted for publication AIChE Symp. Series (1985).

57. Clift, R., J.R. Grace and M.E. Weber, "Bubbles, Drops and Particles". (New York, Academic Press, 1978).
58. Ostergaard, K., "On the Growth of Air Bubbles Formed at a Single Orifice in a Water Fluidized Bed", Chem. Eng. Sci.21(1966) 470-472.
59. Page, R.E. and D. Harrison, "The Size Distribution of Gas Bubbles Leaving a Three-Phase Fluidized Bed". Powder Tech. 6(1972) 245-249.
60. Henriksen and K. Ostergaard, "On the Mechanism of Break-up of Large Bubbles in Liquids and Three-Phase Fluidized Beds", Chem. Eng. Sci, 29 (1974) 626-629.
61. Bhatia, V.K., "Stability of Bubbles in Fluidized Beds", Ind. Eng. Chem. Fundam. 15 (1976) 86.
62. Clift, R. and J.R. Grace, "The Mechanism of Bubble Break-up in Fluidized Beds", Chem. Engng. Sci. 27(1972) 2309-2310.
63. Ostergaard, K. and M.L. Michelson, "On the Use of the Imperfect Tracer Pulse Method for Determination of Hold-up and Axial Mixing". Can. J. Chem. Engng. 47(1969)107-112.
64. Michelsen, M.L. and K. Ostergaard, "Hold-up and Fluid Mixing in Gas-Liquid Fluidized Beds" Chem. Engng.J. 1(1970) 37.
65. Vasalos, I.A., D.N. Rundell, K.E. Megiris and G.J. Tjatjopoulos, "Hold-up Correlations in Slurry-Solid Fluidized Beds". AIChE Journal 28(1982) 346-348.
66. Prakash, A., C. Briens and M.A. Bergougnou, "Mass Transfer Between Solid Particles and Liquid in a THree-Phase Fluidized Bed". Proceedings 34th Can. Chem. Engineering Conference, 1(1984) 478-481.
67. Ostergaard, K., "Hold-up, Mass Transfer and Mixing in Three-Phase Fluidization", AIChE Symposium Series 74, No. 176(1978) 82-86.
68. Vail, Y.K., N. Manakov and V.V. Manshilin, "Turbulent Mixing in a Three-Phase Fluidized Bed". Intern. Chem. Engng. 8(1968) 293-296.
69. El-Temtamy, S.A., Y.O. El-Sharnoubi and M. El-Halwagi, "Liquid Dispersion in Gas-Liquid Fluidized Beds, Part I: Axial Dispersion. The Axially Dispersed Plug-Flow Model". Chem. Eng. J. 18(1979) 151-159.
70. El-Temtamy, S.A., Y.O. El-Shamoubi and M. El-Halwagi "Liquid Dispersion in Gas-Liquid Fluidized Beds Part II: Axial and Radial Dispersion. The Dispersed Plug-Flow Model" 18(1979) 161-168.
71. Joshi, J.B., "Axial Mixing in Multiphase Contactors. A Unified Correlation". Trans. I. Chem. Eng. 58 (1980) 155-165.
72. Kim, S.D. and C.H. Kim, "Axial Dispersion Characteristics of Three-Phase Fluidized Beds". J. of Chem. Engng. of Japan 16(1983) 172-178.
73. Evstrop'eva I.P., I.N. Taganov and P.G. Romankov, "Experimental Study of the Rates of Motion of the Phases in a Three-Phase Solid-Liquid-Gas System". Theor. Found. Chem. Eng. 6(1972) 545-549.

74. Epstein, M., D.J. Petrie, J.H. Linehan, G.A. Lambert and D.H. Cho. "Incipient Stratification and Mixing in Aerated Liquid-Liquid or Liquid-Solid Mixtures". Chem. Eng. Sci., 36(1981) 784-787.
75. Fan, L.S., S.H. Chern and K. Moroyama, "Qualitative Analysis of Solids Mixing in a Gas-Liquid-Solid Fluidizee Bed Containing a Binary Mixture of Particles". AIChE J. 30(1984) 858-860.
76. Vismanathan, S., A.S. Kakar and P.S. Murti, "Effect of Dispersing Bubbles into Liquid Fluidized Beds on Heat Transfer Hold-up and hold-up at constant bed expansion". Chem. Eng. Sci. 20(1964) 903-910.
77. Armstrong, E.R., C.G.J. Baker and M.A. Bergougnou. "Heat Transfer and Hydrodynamic Studies on Three-Phase Fluidized Beds". Fluidization Technology, Ed. D.L. Keairns, 453-457, (Washington, Hemisphere Publishing, 1976).
78. Baker, C.G.J., E.R. Armstrong and M.A. Bergougnou. "Heat Transfer in Three-Phase Fluidized Beds". Powder Tehc. 21(1978) 195-204.
79. Fukuma, M., K. Muroyama, N. Yasunishi, M. Matsuura and S. Sawa, Preprints of the 45th Annual Meeting of the Soc. of chem. Engrs., Osaka, Japan (1980)665.
80. Ostergaard, K. and W. Suchozebrshi. "Gas-Liquid Mass Transfer in Gas-Liquid Fluidized Beds". Proc. Eur. Symp. Chem. React. Eng. 21 (Oxford, Pergamon Press, 1969).
81. Ostergaard, K. and P. Fosbol, "Transfer of Oxygen across the Gas-Liquid Interface in Gas-Liquid Fluidized Beds" Chem. Eng. J. (1972)105-111.
82. Dakshinamurty, P., C. Chiranjevi, V. Subrahmanyam and P.K. Rao, "Studies of Gas-Liquid Mass Transfer in a Gas-Liquid Fluidized Bed". Fluidization and its applications 429-439 (Toulouse, Cepadues-Editions, 1973).
83. Lee, J.C. and H. Wortington, "Gas-Liquid Mass Transfer in Three-Phase Fluidized Beds". Int. Chem. Eng. Symp. Series, 38, Multi-Phase Flow Systems I (London, The Inst. Chem. Engrs. 1974).
84. Strumillo, C. and T. Kudra, "Interfacial Area in Three-Phase Fluidized Beds". Chem. Eng. Sci., 32(1977) 229-232.
85. Fukushima, S., "Gas-Liquid Mass Transfer in Three-Phase Fluidized Bed". J. Chem. Eng. Japan 12(1979) 489-491.
86. Dhanuka, V.R. and J.B. Stepanek, "Simultaneous Measurement of Interfacial Area and Mass Transfer Coefficient in Three-Phase Fluidized Bed". AIChE J. (1980) 1029-1038.
87. Alvarez-Cuenca, M. and M.A. Nerenberg, "The Plug Flow Model for Mass Transfer in Three-Phase Fluidized Beds and Bubble Columns". Can. J. Chem. Engng. 59(1981)739-765.
88. Alvarez-Cuenca, M., C.G.J. Baker, M.A. Bergougnou and M.A. Nerenberg. "Oxygen Mass Transfer in Three-Phase Fluidized Beds Working at Large Flow Rates". Can. J. Chem. Eng. 61(1983) 58-63.

89. Catros, A., J.R. Bernard, C.L. Briens and M.A. Bergougnou. "Generalized Model for Gas-Liquid Mass Transfer in Three-Phase Fluidized Beds with and Without Horizontal Baffles". To be published (1985).
90. Arters, D. and L.S. Fan, "Liquid-Solid Mass Transfer in a Gas-Liquid-Solid Fluidized Bed". AIChE Meeting, San Francisco (1984).

LIQUID PHASE MASS TRANSFER COEFFICIENTS AND INTERFACIAL AREA IN THREE PHASE FLUIDIZATION

U.C. Tosyalı and B.Z. Uysal

Department of Chemical Engineering
Middle East Technical University
Ankara, Turkey

INTRODUCTION

Various aspects of three-phase fluidization have been the subject of numerous investigations due to its high potential for industrial applications in areas such as catalytic processes, hydrocracking and desulphurization of petroleum products, coal liquefaction, hydrogenation of unsaturated fats, and production of calcium bisulfite liquor. Extensive studies have been published concerning the hydrodynamics of three-phase fluidized beds, such as expansion [1-8], pressure drop [9], gas and liquid hold up [10-14], minimum fluidization velocity [12] and axial mixing [8, 10, 14-17].

Particle size and density have been determined to have a marked effect upon the size of gas bubbles and, thus, on the hydrodynamic characteristics and performance of three-phase fluidized beds [4, 8, 18-20]. Bubble coalescence becomes important when the particles are smaller than some critical size, about 2mm, at low liquid rates and high gas rates, and bubble disintegration gains importance in beds of large particles at high liquid rates and low gas rates.

Mass transfer across the gas-liquid interface is the rate-controlling step in most of the applications of three-phase fluidization and, in particular, the volumetric mass transfer coefficient, $k_L^o a$, has a profound effect on bed performance. Knowledge of the individual liquid side mass transfer coefficient, k_L^o, and the interfacial area, a, and the effect of various parameters on these terms, rather than the volumetric mass transfer coefficient, $k_L^o a$, are more useful to set the process conditions to obtain high yields. There have been however, more studies on $k_L^o a$ than the determination of k_L^o and a separately.

Massimilla et al. [21] measured the rate of absorption of CO_2 in water using 3 sizes of silica and glass particles smaller than 1mm. They concluded that the absorption rate increased with increasing liquid velocity and decreased with increasing particle size.

Ostergaard and Suchozebrski [22] also investigated the absorption of CO_2 in water using 1mm and 6mm particles. These investigators found that the volumetric absorption coefficient increased with increasing gas flow rate but was not affected by liquid velocity. Particle size was, however, determined to have a very important effect. The volumetric absorption coefficients in beds of 6mm particles were much greater than the coefficients in beds of 1mm particles. The authors found that the coefficients in solid free bubble columns were of intermediate magnitude. The difference between the absorption coefficients in beds of small and large particles were explained by bubble behaviour, i.e. beds of small particles are characterized by bubble coalescence and hence low interfacial area, whereas beds of large particles are characterized by bubble disintegration and hence, have a large interfacial area. Ostergaard and Fosbol [23] reported the results of oxygen absorption by tap water in beds of 1mm and 6mm glass particles. Volumetric mass transfer coefficients decreased with the bed height in beds of 1mm particles and bubble columns. In beds of 6mm particles the mass transfer coefficients were observed to pass through a maximum with changing bed height.

Ostergaard [14] investigated CO_2 absorption in a carbonate/ bicarbonate buffer and offered some correlations for volumetric and area-based mass transfer coefficients, as well as interfacial area, for 1mm, 3mm and 6mm particles. The interfacial area was found to depend only on gas velocity for all particle sizes. Liquid phase mass transfer coefficients were found to be affected only by gas velocity for the smallest and largest particles, but both gas and liquid rates affected mass transfer coefficients for 3mm particles.

Dhanuka and Stepanek [24] measured volumetric and area-based mass transfer coefficients and interfacial areas in a three-phase fluidized bed using the technique suggested by Robinson and Wilke [25]. The chemical system used was 1.01 M sodium carbonate-sodium bicarbonate solution with a buffer ratio of one, containing 0.01 M sodium hypochlorite to enhance the reaction rate. Pure CO_2 and CO_2 with N_2 as the gas phase and three sizes (1.98, 4.08, and 5.86mm) of glass ballotini as the solid phase were used. Their measurements showed that the particle size has a profound effect on $k_L^o a$ and a, while no difference in k_L^o values were observed for different particle sizes. Further, it was reported that both $k_L^o a$ and a increased with gas rate for all particle sizes. Area based mass transfer coefficients were found to increase with increasing gas rate for the smallest particle size, to decrease for intermediate sizes, and not to be influenced for the largest size. No effect of liquid velocity was observed on $k_L^o a$ and a in beds of the two larger particles, while both $k_L^o a$ and a increased with liquid rate for the smallest particles. Liquid rate did not affect k_L^o for all particle sizes.

EXPERIMENTAL AND MATHEMATICAL TECHNIQUE

In the present study, the technique suggested by Robinson and Wilke [25] was used. Experiments were performed by simultaneous chemical absorption of CO_2 in NaOH solution and physical desorption of dissolved oxygen from the liquid phase. Therefore, physical desorption and chemical absorption rates were measured under truly identical physico-chemical and hydrodynamic conditions.

Danckwerts' Surface Renewal Theory was employed in the development of the model for the process. The stoichiometry of the reaction considered in this work was $A + 2B \rightarrow$ Products. The absorption of CO_2 into strong hydroxide solutions has been successfully treated as absorption accompanied by an irreversible second order reaction between CO_2 and OH^- ions.

The mechanism of chemical absorption is as follows,

$$CO_2\ (g) \rightarrow CO_2\ (l)$$

$$CO_2 + OH^- \rightleftarrows HCO_3^- \qquad \text{reaction (1)}$$

$$HCO_3^- + OH^- \rightleftarrows CO_3^= + H_2O \qquad \text{reaction (2)}$$

Reaction (1) is a fast reaction and reaction (2) is instantaneous. Therefore, the rate of CO_2 removal by reaction with OH^- ions is only determined by the rate of reaction (1) which is then the rate-controlling step. In relatively strong hydroxide solutions, pH > 10, the equilibrium concentration of HCO_3^- ions can be neglected and the overall reaction is

$$CO_2 + 2OH^- \rightarrow CO_3^= + H_2O \qquad \text{reaction (3)}$$

The reaction rate is given by $k_2 C_A C_B$, where A refers to CO_2, B refers to OH^-, and k_2 is the second order rate constant of reaction (1).

The physical situation for the absorption of gas A followed by a practically irreversible reaction with liquid phase reactant B can be represented by a pair of time dependent diffusion equations.

$$\frac{\partial C_A}{\partial t} = D_A \frac{\partial^2 C_A}{\partial x^2} - k_2 C_A C_B \tag{1}$$

$$\frac{\partial C_B}{\partial t} = D_B \frac{\partial^2 C_B}{\partial x^2} - 2\, k_2 C_A C_B \tag{2}$$

The initial and boundary conditions are taken to be

$$\text{at } t=0,\ x>0 \quad ; \quad C_B = C_{B_o} \quad , \quad C_A = 0 \tag{3a}$$

$$t>0,\ x=0 \quad ; \quad \frac{\partial C_B}{\partial x} = 0 \quad , \quad C_A = C_{A_i} \tag{3b}$$

$$t>0,\ x\to\infty \quad ; \quad C_B = C_{B_o} \quad , \quad C_A = 0 \tag{3c}$$

Equations (1) and (2) can be solved numerically with boundary conditions given by Equations (3). The mass transfer coefficient with chemical absorption, based on Danckwerts' surface renewal theory, is then given by

$$k_L = \frac{1}{C_{A_i}} \int_o^\infty s\, e^{-st}\, R(t)\, dt \tag{4}$$

where R(t) is the rate of gas absorption per unit area obtained by solution of the differential equations.

It is often convenient to present the effect of a chemical reaction in terms of an enhancement factor (E) which is defined as the ratio of the average mass flux with chemical reaction to the average mass flux without chemical reaction, but with the same driving force.

$$E = k_L a \,/\, k_L^o a = k_L \,/\, k_L^o \tag{5}$$

Since according to Danckwerts' surface renewal theory

$$k_L^o = \sqrt{D_A\, s} \tag{6}$$

the enhancement factor can be calculated numerically using Equations (4), (5) and (6).

DeCoursey [26] offered an approximate analytical integration of differential equations (1) and (2) and suggested the following expression for the enhancement factor,

$$E_D = - \frac{M}{2(E_a-1)} + \left(\frac{M^2}{4(E_a-1)^2} + \frac{E_a M}{(E_a-1)} + 1 \right)^{1/2} \quad (7)$$

where E_a is the asymptotic value of the enhancement factor for an instantaneous reaction as given by Danckwerts [27] and expressed by

$$E_a = 1 / \text{erf } B \quad (8)$$

where B is given by the implicit relation [28]

$$\frac{1}{2} \left(\frac{C_{B_o}}{C_{A_i}}\right) \left(\frac{D_B}{D_A}\right)^{1/2} \text{erf } B \exp B^2 = \text{erfc} \left[B \left(\frac{D_A}{D_B}\right)^{1/2} \right] \exp \left[B^2 \frac{D_A}{D_B} \right] \quad (9)$$

and

$$M = D_A \, k_2 \, C_{B_o} \, / \, (k_L^o)^2 \quad (10)$$

A comparison of estimates by this relatively simple relation for the enhancement factor with the results of numerical calculations showed very good agreement over a wide range of variables. It was found that the maximum discrepancy in the enhancement factor was less than 7% [28]. Thus, because of the simplicity and accuracy of Equation (7) its use can be recommended for enhancement factor predictions.

Experimental enhancement factor and the DeCoursey formula enable evaluation of the surface renewal rate and physical area-based mass transfer coefficient, k_L^o. Interfacial area can then be determined from the measured values of the volumetric mass transfer coefficient and the predicted value of the area-based mass transfer coefficient:

$$a = k_L^o a \, / \, k_L^o \quad (11)$$

The procedure adapted can therefore be written as follows:

i) Since the concentration of CO_2 in the bulk liquid is zero due to chemical reactions (1) and (2), the chemical absorption coefficient for CO_2 is calcualted from

$$(k_L a)_{CO_2} = \frac{\overline{R_A a}}{\bar{C}_{A_i}} \quad (12)$$

where $\overline{R_A a}$ is the mean absorption rate per unit volume and can be experimentally determined from inlet and outlet CO_2 concentrations of the gas, i.e.

$$\overline{R_A a} = \overline{U}_G \, \Delta C_A / Z_T \tag{13}$$

were $\overline{U}_G$ is the average superficial gas velocity.

ii) As the concentration of oxygen in the bulk gas is practically zero, its interfacial concentration at the gas side is also zero. Interfacial mole fraction of oxygen at the liquid side which is supposed to be in equilibrium with gas side interfacial concentration can thus be safely taken as zero. Therefore,

$$(k_L^o a)_{O_2} = \frac{U_L}{Z_T} \ln \frac{x_{in}}{x_{out}} \tag{14}$$

iii) The mass transfer coefficient for physical absorption of CO_2 can be obtained from the mass transfer coefficient for physical desorption of oxygen after correction for different diffusivities. Since k_L^o is proportional to $D^{½}$;

$$(k_L^o)_{CO_2} = (k_L^o)_{O_2} \left(\frac{D_{CO_2}}{D_{O_2}} \right)^{½} \tag{15}$$

or

$$(k_L^o a)_{CO_2} = (k_L^o a)_{O_2} \left(\frac{D_{CO_2}}{D_{O_2}} \right)^{½} \tag{16}$$

iv) The experimental enhancement factor is evaluated from

$$E = (k_L a \, / \, k_L^o a)_{CO_2} \tag{5}$$

v) The parameter B, a constant value, is estimated by Equation (9).

vi) E_a is determined from Equation (8).

vii) The experimental enhancement factor is equated to E_D, and k_L^o is estimated from Equations (7) and (10) using the Newton-Raphson or Wegstein method.

viii) The surface renewal rate, s, can be predicted from Danckwerts model, i.e.

$$k_L^o = \sqrt{D_A s} \tag{6}$$

ix) The interfacial area, a, is determined from the measured value of $k_L^o a$ and the predicted value of k_L^o.

$$a = (k_L^o a \, / \, k_L^o)_{CO_2} \tag{11}$$

The assumptions involved in this procedure are as follows:

i) The surface renewal rate is identical for both physical and chemical absorption. This can be accomplished using simultaneous chemical absorption and physical desorption under the same hydrodynamic conditions.

ii) Temperature and the values of the physico-chemical quantities such as solubilities, diffusivities and reaction rate constant remain uniform and constant.

iii) Axial liquid mixing can be ignored. It has been shown to be small for particle sizes used in this research [10]; therefore, appreciable effect of axial mixing on absorption rate is not expected.

iv) The most important assumption is that the effective area for physical desorption is the same as for chemical absorption.

To ensure this last assumption, Danckwerts' parameter (γ) is chosen such that it is equal to one, where γ is defined as the ratio of the increase in absorption capacity of the liquid to the enhancement factor.

$$\gamma = C/E \tag{17}$$

When, $\gamma = 1$, the effective area for chemical absorption is equal to that for physical desorption and when $\gamma >> 1$, the effective area is significantly greater for chemical absorption than for physical desorption. Use of a second-order reaction such as reaction (3) gives values of γ near unity. For example, γ is about 0.8 for 0.2 M NaOH for reaction (3) [29]. Rapid first-order reactions usually give values of γ greater than 10.

EXPERIMENTAL

The technique involving simultaneous absorption with chemical reaction and physical desorption was employed to determine mass transfer coefficients with and without chemical reaction under identical hydrodynamic conditions. The gas phase consisted of CO_2 and N_2, and the liquid phase consisted of 0.2M NaOH solution containing dissolved oxygen. 3mm and 4mm glass beads were used as the solid phase.

CO_2 and N_2 concentrations were kept about 90% and 10%, respectively, in the inlet gas stream. The reason for introducing N_2 to the gas stream was to improve the accuracy of gas analysis made by gas chromatography as its concentration remained practically the same during passage through the three phase fluidized bed.

Keeping the CO_2 concentration as high as 90% was however required in order to be able to neglect the gas side resistance to mass transfer.

A schematic diagram of the experimental set-up is shown in Figure 1. The three-phase fluidized bed was contained in a 42mm ID glass column. The height of the active zone of the bed was 120mm and the total height of the column was 400mm. The bed was equipped with a wire screen to support the particles. Static bed height was varied from zero to 84mm. The liquid phase was introduced to the bottom of the column and after passing through a calming section entered the active zone. The gas was introduced through a ring distributor containing six equispaced 1mm diameter holes. The ring was placed just below the wire screen at the bottom of the column.

The liquid tank was equipped with an impeller. A cooling coil was inserted into the tank to remove the heat evolved during preparation of the NaOH solution. Sodium hydroxide was dissolved in distilled water to provide an initial feed concentration of about 0.2 M. Oxygen from a cylinder was bubbled through the tank.

Experiments were performed at room temperature to study the effect of liquid and gas velocities, particle diameter and static bed height on the mass transfer coefficient and interfacial area. Gas and liquid flow rates were varied from 1.03×10^{-4} to 1.82×10^{-4} m^3/s and 3.3×10^{-6} to 25×10^{-6} m^3/s, respectively.

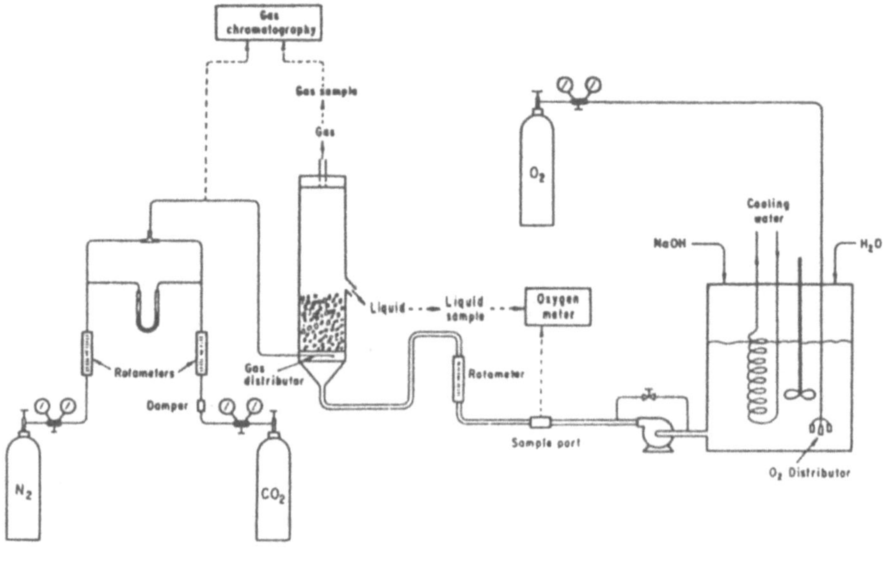

Figure 1 Experimental set-up

Gas and liquid samples were analyzed by gas chromatography and using an oxygen meter, respectively. The rate of CO_2 absorption with chemical reaction was obtained from the gas phase anlaysis. The liquid side volumetric physical mass transfer coefficient was determined from the desorption rate of oxygen. Detailed description of the experimental set up, procedure and analysis of data is given by Tosyalı [30]. Methods of estimating the interfacial CO_2 concentration, diffusivities of CO_2 and OH^- in the liquid phase, reaction rate constant, which are all required in data analysis, can be found elsewhere [31, 32].

RESULTS AND DISCUSSION

The effect of particle diameter, superficial liquid velocity, superficial gas velocity and static bed height on the rate of mass transfer across the gas-liquid interface in a three-phase fluidized bed were investigated.

At the beginning of the experiments, the effect of the initial concentration of dissolved oxygen in the liquid phase on the volumetric physical mass transfer coefficient was checked. Results of experiments showed that there was no effect of this kind. Therefore, no effort was made to keep the initial oxygen concentration constant for the rest of the experiments.

Volumetric chemical and physical mass transfer coefficients, k_La and k_L^oa, evaluated using the experimental results for a superficial gas velocity of 131.4mm/s and a static bed height of 84 mm are presented as a function of the superficial liquid velocity for 3 and 4mm particle sizes in Figures 2 and 3. It should be noted here that volumetric and area based physical mass transfer coefficients in Figure 3 and in similar following figures are plotted for carbon dioxide. These two figures show that the volumetric mass transfer coefficients obtained using simultaneous absorption with chemical reaction and physical desorption increase with increasing liquid velocity. This observation is in agreement with the findings of Massimila et al. [21] and the results of Dhanuka and Stepanek [24] for small particles but not in agreement with the results for 4.08mm particles reported by later authors. The present observations also confirm the results reported by Ostergaard [14] for 3mm particles. The next step was to investigate whether this increase in volumetric mass transfer coefficient with increasing liquid velocity was primarily due to the effect of this parameter on k_L^o or on a. Area based physical mass transfer coefficients, k_L^o, and interfacial areas, a, evaluated for the same conditions, are shown in Figures 4 and 5. Inspection of these figures reveals that increase in the volumetric mass transfer coefficient was effectively due to the steep increase in interfacial

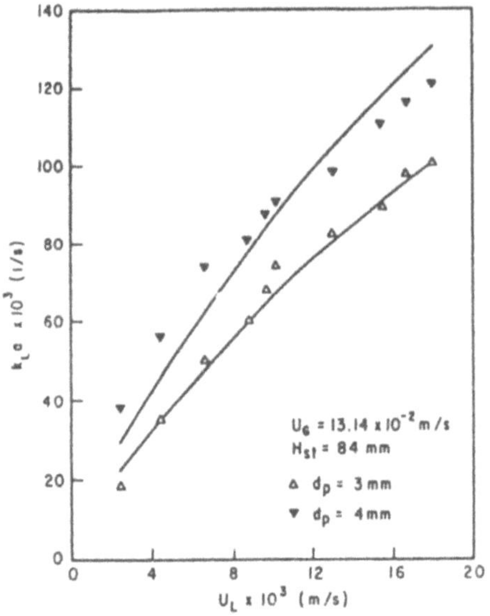

Figure 2. Variation of volumetric chemical mass transfer coefficient with liquid velocity and effect of particle size.

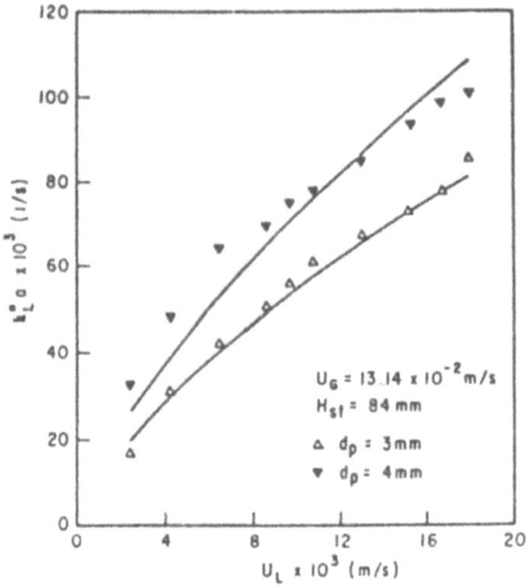

Figure 3. Variation of volumetric physical mass transfer coefficient with liquid velocity and effect of particle size.

area with increasing liquid velocity. Area based physical mass transfer coefficient, k_L^o, decreases with liquid velocity as shown in Figure 4. However, the interfacial area increases more rapidly with liquid velocity than k_L^o decreases. So the net effect of U_L is to increase $k_L^o a$. The positive effect of liquid velocity on the interfacial area probably arises from the increase in liquid hold up and the reduction of bubble coalescence with increasing liquid velocity [10].

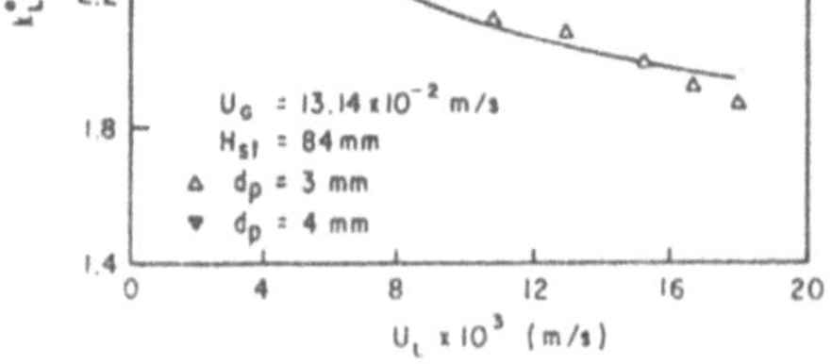

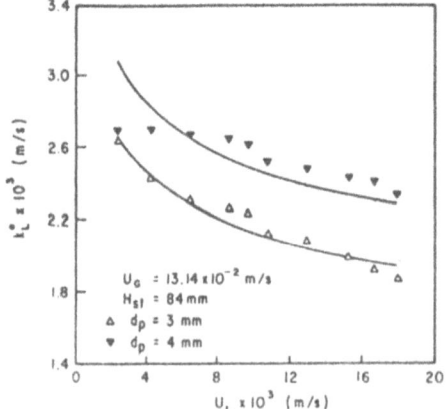

Figure 4. Variation of area-based physical mass transfer coefficient with liquid velocity and effect of particle size.

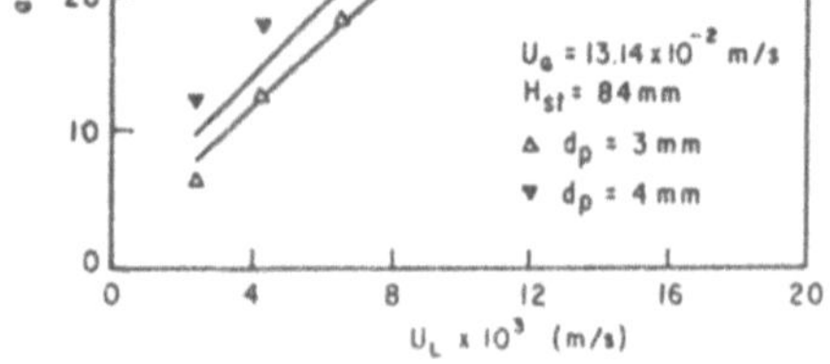

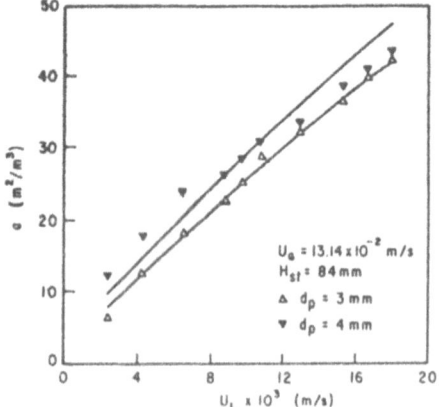

Figure 5. Variation of interfacial area with liquid velocity and effect of particle size.

It can be concluded from quantitative comparison of the results for 3mm and 4mm particles in Figures 2-5 that particle size has an important effect on all k_L^o, a, k_L^oa and k_La. Larger values of these parameters were obtained as particle size was increased. This increase in total mass transfer rate with increasing particle size confirms previous studies by Ostergaard and Suchozebrski [22] and by Dhanuka and Stepanek [24]. The similarity of effect of operating parameters on mass transfer coefficients and interfacial area for both 3 and 4mm particle sizes and comparison of these results with those reported in the literature imply that bubble disintegration dominates in three-phase fluidization for all the cases of this study.

When the effect of gas velocity is considered, k_La, k_L^oa and a increase while k_L and k_L^o decrease with increasing gas velocity (Figures 6, 7, 8). Again the effect of gas velocity on the interfacial area dominates in determining the effect of gas velocity on the volumetric physical and chemical mass transfer coefficients. This effect of gas velocity on interfacial area is probably because of the increase in gas hold up with gas velocity. Similar effects of gas velocity on volumetric mass transfer coefficients and interfacial area were reported by Dhanuka and Stepanek [24]. Volumetric mass transfer coefficient values obtained in the present study are, however, a little smaller than those reported by these authors. Since the liquid velocities employed in their research were higher, this difference can be regarded as reasonable. On the other hand, interfacial area values found by Dhanuka and Stepanek are about one order of magnitude greater than the values of the present work. This considerable discrepancy can be partly attributed to lower liquid rates in the present work and may be explained in part by the effect of increase in the absorption capacity due to a fast first order reaction in buffer solutions leading to high values of γ and thus high values of the interfacial area as suggested by Joosten and Danckwerts [29]. The trend for the effect of gas velocity on mass transfer coefficients and interfacial area also agrees with the results presented by Ostergaard [14]. However, present values of the mass transfer coefficients are about 2-5 times greater than those reported by Ostergaard. Agreement between interfacial area values is better. The decrease in k_L^o values with gas rate, which was also observed by Dhanuka and Stepanek [24] for 4.08mm particles, may be explained by the reduction in surface renewal rate associated with uniform distribution of smaller bubbles due to bubble disintegration [33].

Having set the factual effects of liquid velocity, gas velocity and particle size on mass transfer coefficients and interfacial area for H_{st}=84mm, correlations were attempted for these characteristic parameters by non-linear regression analysis of the experimental results. Empirical correlations in S.I. units, obtained for

CO_2, are presented below.

$$k_L^o a = 956.6\ U_L^{0.70}\ U_G^{0.37}\ d_p \tag{18}$$

$$k_L^o = 0.01134\ U_L^{-0.15}\ U_G^{-0.38}\ d_p^{0.54} \tag{19}$$

$$a = 84356.3\ U_L^{0.85}\ U_G^{0.75}\ d_p^{0.46} \tag{20}$$

Multiple regression coefficients for Equations (18), (19), and (20) were found to be 0.96, 0.92, and 0.96, respectively, which indicate fits that are quite satisfactory. Although Equations (18), (19) and (20) were developed using estimated values of $k_L^o a$, k_L^o and a

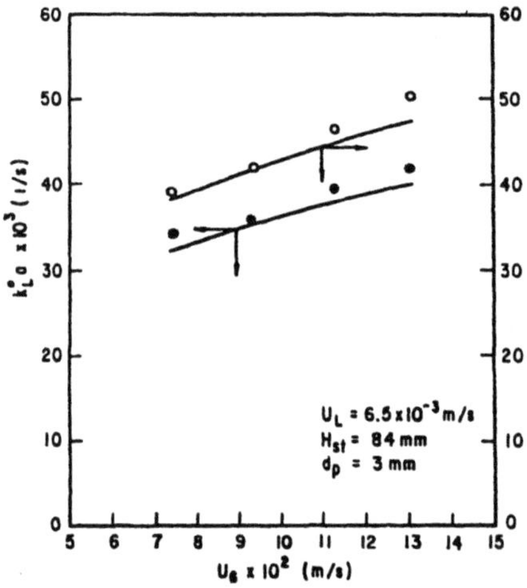

Figure 6. Effect of gas velocity on volumetric physical and chemical mass transfer coefficients.

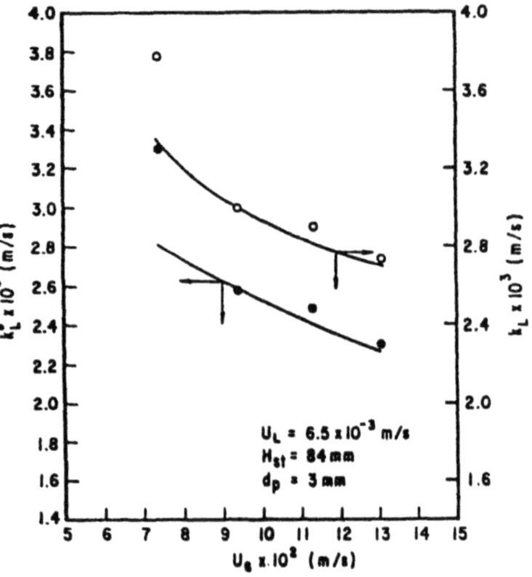

Figure 7. Effect of gas velocity on area-based physical and chemical mass transfer coefficients.

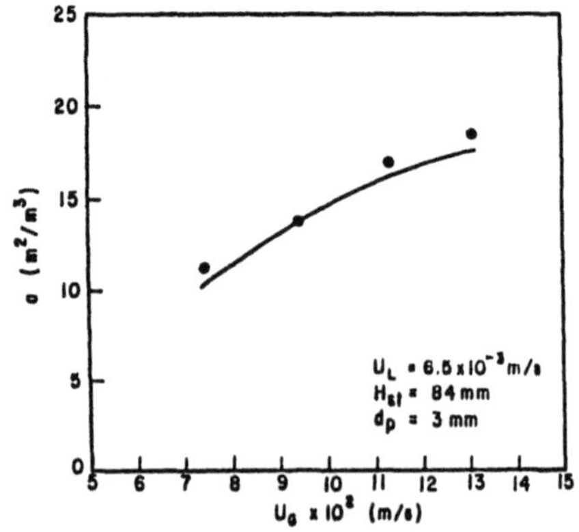

Figure 8. Effect of gas velocity on interfacial area.

separately, it may be noted that the product of Equations (19) and (20) gives Equation (18). Figures 9 and 10 also show the adequacy of the fit of the proposed correlations to the experimental data. It should also be stated here that solid lines in Figures 2-8 were drawn using the appropriate correlations given by Equations (18), (19) and (20).

Correlations for chemical mass transfer coefficients, k_L and k_La, were not regarded as necessary as they differ for different chemical systems and reactions employed. On the other hand, correlations for k_L^oa and k_L^o can be adapted to other physical systems by just making a diffusivity correction. Mass transfer coefficients assessed with absorption and chemical reaction can then be estimated using Equations (18) and (19) together with the enhancement factor calculated specifically for the chemical system of interest.

The effect of static bed height was investigated by performing a separate set of experiments. Static bed heights greater than 84mm might lead to maldistribution of solid particles because of the small diameter of the column used. Therefore, smaller static bed heights were used. Results of these experiments are presented in Figures 11, 12, and 13. It was found that volumetric mass transfer coefficients and interfacial areas increased with increasing aspect ratio which was defined as the ratio of static bed height to column diameter. However, area based mass transfer coefficients decreased as static bed height increased. As the general evaluation of all the results implies that bubble disintegration dominates for the conditions of the present study, increase in interfacial area with bed height is reasonable. This finding agrees with the results obtained by Ostergaard and Fosbol [23] for large particles.

When the experimental values for different static bed heights in these figures are compared with the results obtained from the experiments with no solid particles (bubble column) which are also given on the ordinate, it may be concluded that introduction of solid particles improved the volumetric mass transfer coefficients as well as interfacial area by a factor of at least 2. k_L^o and k_L seem to be negatively affected by use of solid particles in the bed. These results then show that interfacial area increases very appreciably with addition of solid particles and this increase in interfacial area causes the increase in volumetric mass transfer coefficients resulting in higher mass transfer capacities for three-phase fluidization in comparison with bubble columns.

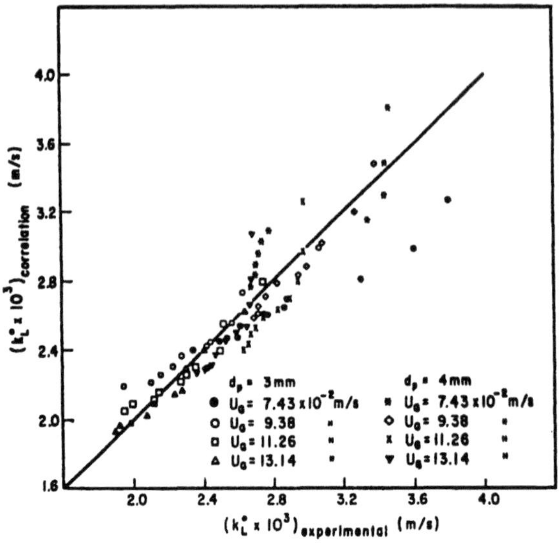

Figure 9. Comparison of experimental and estimated values of area-based physical mass transfer coefficient.

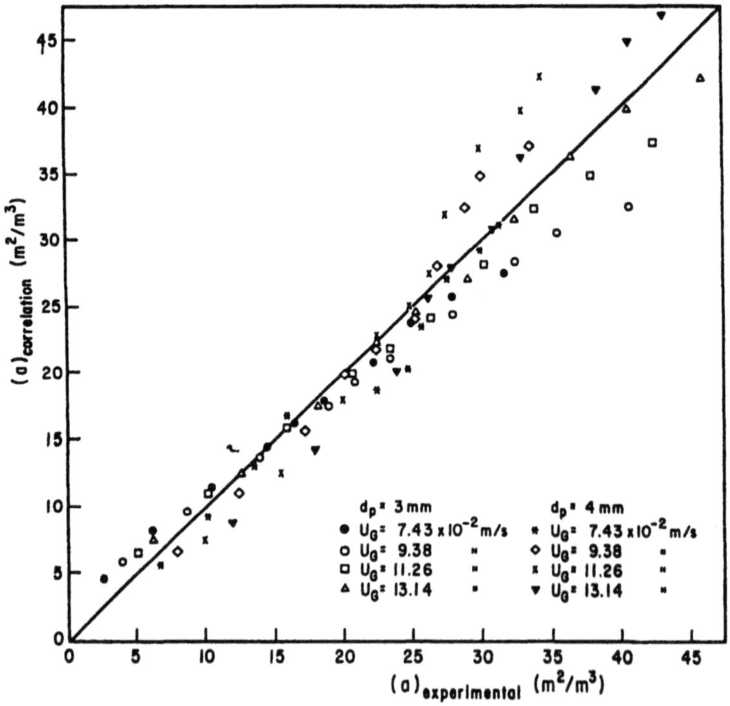

Figure 10. Comparison of experimental and estimated values of interfacial area.

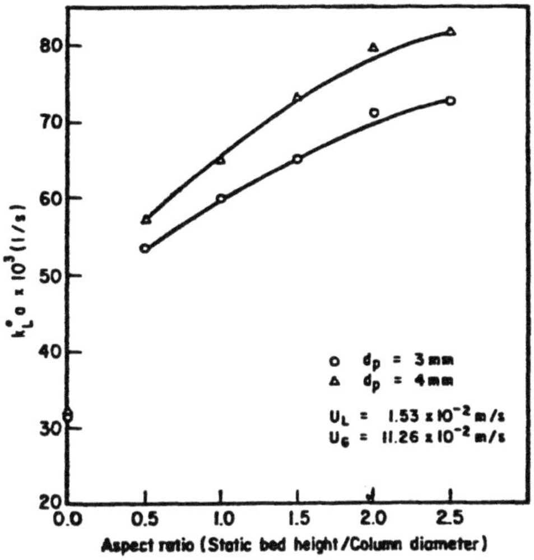

Figure 11. Effect of static bed height on volumetric physical mass transfer coefficient.

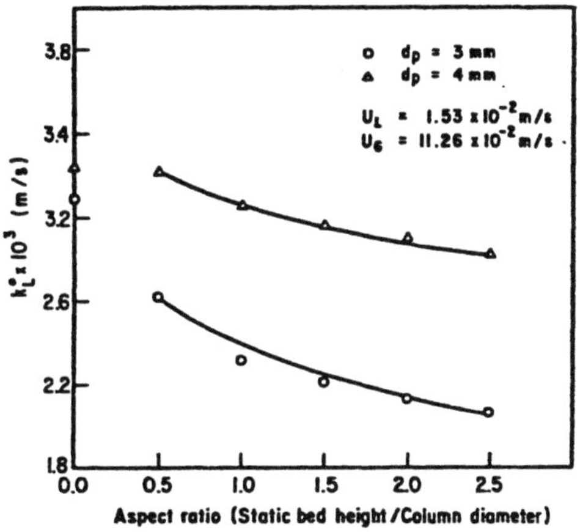

Figure 12. Effect of static bed height on area-based physical mass transfer coefficient.

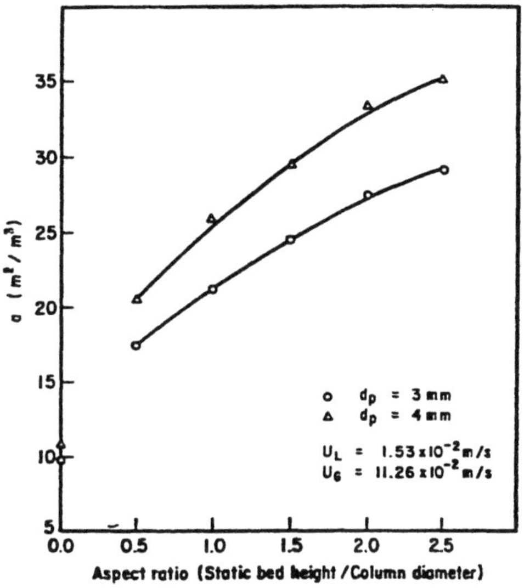

Figure 13. Effect of static bed height on interfacial area.

NOTATION

a	:	Interfacial area, m^2/m^3
B	:	Constant determined from Equation (9)
C_A	:	Concentration of dissolved gas (CO_2), $kmol/m^3$
C_{Ai}	:	Interfacial concentration of dissolved gas (CO_2), $kmol/m^3$
C_B	:	Concentration of liquid phase reactant (OH^-), $kmol/m^3$
C_{B_o}	:	Concentration of liquid phase reactant (OH^-) in the bulk, $kmol/m^3$
ΔC	:	Concentration driving force, $kmol/m^3$
D_A	:	Diffusivity of CO_2, m^2/s
D_B	:	Diffusivity of liquid phase reactant (OH^-), m^2/s
E	:	Enhancement factor
E_a	:	Asymptotic value of enhancement factor, 1/erf B
E_D	:	Enhancement factor predicted by DeCoursey formula
H_{st}	:	Static bed height, m
k_L^o	:	Area-based physical mass transfer coefficient, m/s
k_L	:	Area-based chemical mass transfer coefficient, m/s
$k_L^o a$	:	Volumetric physical mass transfer coefficient, 1/s
$k_L a$	:	Volumetric chemical mass transfer coefficient, 1/s
k_2	:	Second order reaction rate constant, 1/s
M	:	Generalized variable for a second-order reaction, $D_A k_2 C_{B_o}/k_L^{o\,2}$.
R_A	:	Rate of gas absorption per unit area, $kmol/m^2.s$
$\overline{R_A}$	:	Mean absorption rate per unit area, $kmol/m^2.s$
s	:	Surface renewal rate, 1/s
t	:	Time, s
U_G	:	Superficial gas velocity, m/s
U_L	:	Superficial liquid velocity, m/s
x	:	Distance, m
x_{in}	:	Mole fraction of oxygen at the inlet
x_{out}	:	Mole fraction of oxygen at the outlet
Z_T	:	Total height, m
γ	:	Danckwerts parameter, i.e. ratio of the increase in capacity of the solution to absorb a gas to the enchancement factor.

Superscript

–	:	Average value.

REFERENCES

1. Turner, R., "Fluidization", 42-58, (London, Soc. Chem. Ind., 1964).

2. Stewart, P.S.B., and Davidson, J.F., "Three Phase Fluidization: Water, Particles and Air", Chem. Eng. Sci.,19 (1964) 319-322.
3. Adlington, D., and Thompson, E., "Desulfuriziation in Fixed and Fluidized Bed Catalyst Systems", Proc. 3rd. European Symp. Chem. React. Eng., 203-210,(Oxford, Pergamon Press, 1965).
4. Ostergaard, K., "Fluidization", 58-60, (Society of Chem. Ind., London,1964).
5. Ostergaard, K., "On Bed Porosity in Gas-Liquid Fluidization", Chem. Eng. Sci., 20 (1965) 165-167.
6. Ostergaard, K., and Theisen, P.I., "The effect of Particle Size and Bed Height on the Expansion of Mixed Phase (Gas-Liquid) Fluidized Beds", Chem. Eng. Sci., 21 (1966) 413-417.
7. Sherrard, A.J., "Three Phase Fluidized Beds", Dissertation, University of College of Swensea, 1966.
8. Kim, S.D., Baker, C.G., and Bergougnou, M.A., "Hold Up and Axial Mixing Characteristics of Two and Three Phase Fluidized Beds", Can. J. Chem. Eng., 50 (1972) 659-701.
9. Lee, J.C., and Al-Dabbagh, N., "Onset of Fluidization at high gas rates", Fluidization, 184-189 (Cambridge University Press, 1978).
10. Michelsen, M.L., and Ostergaard, K., "Hold Up and Fluid Mixing in Gas Liquid Fluidized Beds", Chem. Eng. J., 1 (1970) 37-45.
11. Kim, S.D., Baker, C.G., and Bergougnou, M.A., "Phase Hold Up Characteristics of Three Phase Fluidized Beds", Can. J. Chem. Eng., 53 (1975) 134-139.
12. Begovich, J.M., and Watson, J.S., "Hydrodynamic Characteristics of Three Phase Fluidized Beds", Fluidization, Eds. Davidson and Keairns, 190-195 (Cambridge, Cambridge University Press, 1978).
13. Dhanuka, V.R., and Stepanek, J.B., "Gas and Liquid Hold Up and Pressure Drop Measurements in a Three Phase Fluidized Bed", Fluidization, Eds Davidson and Keairns, 179-183 (Cambridge, Cambridge University Press, 1978).
14. Ostergaard, K., "Hold Up, Mass Transfer, and Mixing in Three Phase Fluidization", AIChE Symp. Ser. 176, 74 (1978) 82-86.
15. Vail, Yu. K., Manakov, N. Kh., and Manshilin, V.V., "Turbulent Mixing in A Three Phase Fluidized Bed", Int. Chem. Eng., 8 (1968) 293-296.
16. Vail, Yu. K., Manakov, N. Kh., and Manshilin, V.V. "The Gas Contents of Three Phase Fluidized Beds", Int. Chem. Eng. 10 (1970) 244-247.
17. Kim, C.H., and Kim, S.D., "Liquid Phase Dispersion in Three Phase Fluidized Beds". Fluidization, 195-199 (Cambridge, Cambridge University Press, 1981).
18. Massimilla, L., Solimando, A., and Squillace, E., Brit. Chem. Eng., 6 (1961) 232-239.

19. Kim, S.D., Baker, C.G., and Bergougnou, M.A., "Bubble Characteristics in Three Phase Fluidized Beds", Chem. Eng. Sci., 32 (1977) 1299-1306.
20. Henriksen, H.K., and Ostergaard, K., "Characteristics of Large Two Dimensional Air Bubbles in Liquids and in Three Phase Fluidized Beds", Chem. Eng. J., 7 (1974) 141-146.
21. Massimilla, L., Majuri, N., and Signorini, P., 1959. Cited in "Fluidization" by Davidson, J.F., and Harrison, D., (London and New York, Academic Press, 1977).
22. Ostergaard, K., and Suchozebrski, W., "Gas-Liquid Mass Transfer in Gas-Liquid Fluidized Beds", Proc. 4th Int. Sym. Chem. React. Eng. (Pergamon Press, Oxford, 1969).
23. Ostergaard, K., and Fosbol, P., "Transfer of Oxygen Across The Gas-Liquid Interface in Gas Liquid Fluidized Beds", Chem. Eng. J., 3 (1972) 105-111.
24. Dhanuka, V.R., and Stepanek, J.B., "Simultaneous Measurement of Interfacial Area and Mass Transfer Coefficient in Three Phase Fluidized Beds", AIChE J., 26 (1980) 1029-1038.
25. Robinson, C.W., and Wilke, C.R., "Simultaneous Measurement of Interfacial Area and Mass Transfer Coefficients for Well-Mixed Gas Dispersion in Aqueous Electrolyte Solutions", AIChE J., 20 (1974) 285-294.
26. DeCoursey, W.J., "Absorption With Chemical Reaction, Development of a New Relation for The Danckwerts Model", Chem. Eng. Sci., 29 (1974) 1967-1872.
27. Danckwerts, P.V., "Gas Liquid Reactions", (New York, McGraw-Hill Book Co., 1970).
28. Matherson, E.R., and Sandall, O.C., "Gas Absorption Accompanied by a Second Order Chemical Reaction Modeled According to The Dankcwerts Surface Renewal Theory", AIChE J., 24 (1978) 552-554.
29. Jooesten, G.E.H., and Danckwerts, P.V., "Chemical Reaction and Effective Interfacial Areas in Gas Absorption", Chem. Eng. Sci., 28 (1973) 453-461.
30. Tosyalı, U.C., "Mass Transfer Coefficients and Interfacial Area in Three Phase Fluidization", M.S. Thesis, Middle East Tech. Univ., Ankara, 1984.
31. Nijsing, R.A., Hendriksz, R.H., and Kramers, H., "Absorption of CO_2 in Jets and Falling Films of Electrolyte Solutions, With and Without Chemical Reaction", Chem. Eng. Sci., 10 (1958) 88-104.
32. Hikita, H., Asai, S., and Takatsuka, T., "Absorption of Carbondioxide into Aqueous Sodium Hydroxide and Sodium Carbonate-bicarbonate Solutions", Chem. Eng. J., 11 (1976) 131-141.
33. Matheron, E.R., and Sandall, O.C., "Effective Interfacial Area Determination by Gas Absorption Accompanied by Second Order Irreversible Chemical Reaction, AIChE J., 25 (1979) 332-338.

DESIGN AND SIMULATION OF BUBBLE COLUMN REACTORS

WOLF-DIETER DECKWER

TECHNISCHE CHEMIE, UNIVERSITÄT OLDENBURG
D-2900 OLDENBURG, F. R. GERMANY

1. INTRODUCTION

Bubble column reactors (BCR) are contactors in which a discontinuous gas phase moves in the form of bubbles relative to a continuous liquid phase. In the liquid, a reactive solid or catalyst may be fluidized. BCR can be single staged or multi staged, batch or continuous with regard to the liquid (or slurry) phase. Continuous BCR can be operated cocurrently or countercurrently.

Fig. 1 shows various types of BCR. Compared to other multiphase reactors some of the main advantages of BCR are:
- less maintenance is necessary and sealing is no problem due to the absence of moving parts
- high rates of mass transfer can be obtained
- solids can be handled without significant erosion or plugging problems
- little floor space is required and BCR are not expensive.

Considerable backmixing of both phases, high pressure drop and coalescence of bubbles are, however, some of the disadvantages. Therefore, BCR have been modified in many ways to suit particular applications (see Fig. 1). If low backmixing is required sectionalized (multi channel or multi staged) BCR must be used. Loop reactors are used in biotechnology as, for instance, the ICI pressure cycle fermenter and its deep shaft modification [1]. For certain reactions like air oxidations where high gas component conversion is not necessary and low pressure drop is needed, a horizontally sparged BCR can be used [2]. When complete gas phase conversion and hence a long gas residence time is required, downflow BCR can be used. Examples where such reactors are used include phosgene decom-

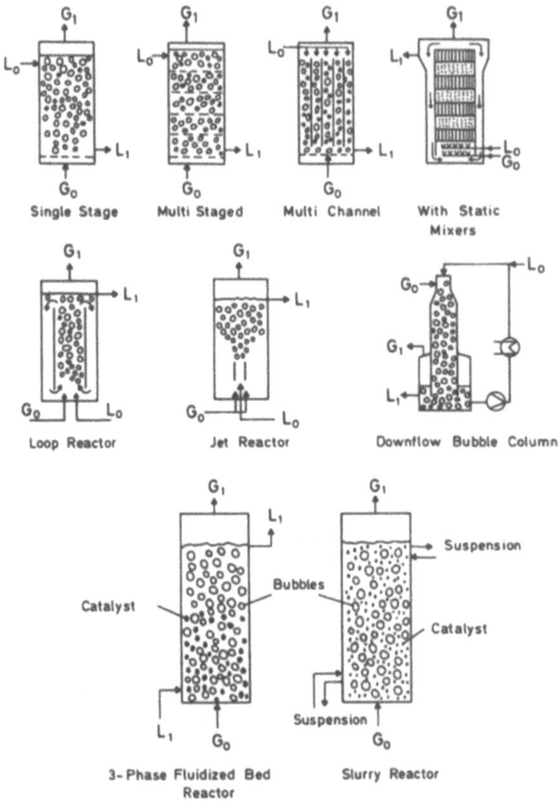

Fig. 1: Various types of BCR [5]
Reproduced by permission of the Am. Inst. of Chem. Engs.

position [3] and water ozonization [4]. Vertically sparged, single staged BCR are most commonly used in industry and hence, in this contribution, the main emphasis is given to this type. Applications range from very mild conditions in biotechnology (such as production of single cell protein and cultivation of animal cells) to conditions prevailing in the chemical industry (oxidations, hydrogenations, chlorinations, alkylations) to high temperature and high pressure processes like coal liquefaction, Fischer-Tropsch synthesis, and coal gasification in a molten iron bath. Summaries about BCR applications have been given by Shah et al. [5] and Deckwer [6]. Bubble column slurry reactors have been treated by Chaudhari and Ramachandran [7, 8].

To achieve a quantitative description of BCR mathematical models are formulated as usual in chemical reaction engineering. In general, mathematical models are applied, firstly, to evaluate and interprete experimental data and, secondly, to calculate conversions, selectivities, performances and reactor sizes. In industrial

practice one encounters two main tasks:
- The reactor configuration is given and can be modified only slightly (for instance, by introducing phase recycle, installing additional heat transfer area or changing the gas sparger). The task is to calculate conversions, selectivities, etc. as function of the operating conditions and thus to find the optimum performance of the given setup.
- The inlet conditions and the desired production rate is known. It is then required to find the best type of reaction vessel and its size, i.e., that geometrical arrangement of the reactor which yields the desired production under consideration of appropriate boundary conditions (selectivity, separation of products, etc.).

Though BCR and their various modifications are simple with regard to construction, modeling and reliable design and scaleup is not an easy task. Fig. 2 summarizes the major phenomena involved in BCR modeling. First of all, there are the processs specific factors like stoichiometry, gas solubility, heats of solution and the kinetic model which has to comprehend all the main reactions of the

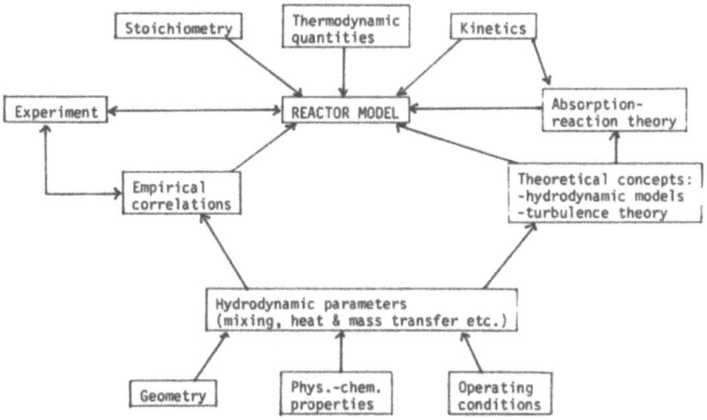

Fig. 2: Phenomena involved in BCR modeling

chemical system under consideration. In addition, we have as a major influence the fluid dynamic parameters such as phase holdups and their local distributions, mixing phenomena, heat and mass transfer, etc. It is not possible to fix these parameters by external means, instead they are a result of the combined action of physical properties, the geometrical dimensions of the vessel, the phase distributing devices, and the operating conditions. Indeed, the acessibility of the fluid dynamic parameters involved in the model equations is one of the serious problems and a difficult task in the design and simulation of BCR.

It is still common practice to estimate the fluid dynamic properties from empirical correlations. Such correlations are usually developed from "cold flow" measurements which are often not properly designed and evaluated. It is understood that use of empirical correlations is of limited value and their predictions may lead to serious errors. This is particularly valid for those quantities which characterize interfacial properties like mass transfer coefficients, interfacial areas and phase holdups. It is now obvious that properties like density, viscosity, and surface tension are not always sufficient to describe fluid dynamic and interfacial phenomena. This will be illustrated by two examples:

- Tarmy et al. [9] measured gas holdup under various cold flow conditions in order to predict the holdup for the EDS coal liquefaction process. These authors observed rather low holdup values and churn-turbulent flow in their cold flow studies while, surprisingly, under the actual conditions of the EDS process bubbly flow prevailed and gas holdup was as high as 0.5.
- In the chlorination of toluene extremely low overall mass transfer coefficients were observed [10]. Even after accounting for gas side resistances, the liquid side mass transfer coefficients were smaller by one order of magnitude than those predicted from correlations [11, 12]. A speculative explanation of this discrepancy could be the formation of a Cl_2-toluene charge transfer complex which decelerates diffusion of Cl_2 from the gas/liquid interface into the bulk phase. Of course, empirical correlations do not account for such an additional interfacial resistance.

To a large extent, such uncertainties are due to the occurrence of different flow regimes in BCR. Bubbly flow or quiescent bubbling with almost uniformly sized bubbles with similar rise velocities is found at low gas flow rates. At higher gas velocities the homogeneous or churn-turbulent flow regime is characterized by large bubbles. Different dependencies of the fluid dynamic properties are found in different flow regimes. These dependencies are not yet clearly understood and, additionally, no reliable flow charts for BCR are available.

As a general rule of thumb, caution and reserve is recommended when applying empirical correlations to new processes. They may predict reliable data in many cases while, in others, they fail completely. In recent years, new hydrodynamic models for BCR have been developed [13 - 16]. Summaries can be found in papers by Shah and Deckwer [17] and Deckwer and Schumpe [18]. As a main result, these models predict liquid mixing and circulation flow rates and confirm dependencies upon various parameters found empirically. It is understood that the hydrodynamic models refer to the bulk fluid flow and are therefore not able to predict typical interfacial parameters and microphenomena within the bulk liquid phase like micromixing and segregational effects. Newly developed hydrodynamic models will certainly have a strong influence on the formulation of BCR models in the near future. This will be discussed later in section 5.

The various empirical and semitheoretical correlations for estimating design parameters of BCR have been reviewed and critically evaluated by Shah et al. [5] and Deckwer [6]. Also specific recommendations for their use have been given by these authors. Therefore, this problem needs no further consideration here. This contribution concentrates on the present status and future trends of the design and simulation of BCR on the basis of mathematical models. In what follows, it is assumed that the reader is familiar with the basic principles of absorption-reaction theory as outlined in chemical reaction engineering text books and standard reference monographs [19, 20].

2. CLASSIFICATION OF BCR MODELS

Models of BCR can be developed on the basis of various view points. The mathematical structure of the model equations is mainly determined by the residence time distribution of the phases, the reaction kinetics, the number of reactive species involved in the process, and the absorption-reaction regime (slow or fast reaction in comparison to mass transfer rate). One can anticipate that the gas phase as well as the liquid phase can be either completely backmixed (CSTR), partially mixed, as described by the axial dispersion model (ADM), or unmixed (PFR). Thus, it is possible to construct a model matrix as shown in Fig. 3. This matrix refers only to the gaseous key reactant (A) which is subjected to interphase mass transfer and undergoes chemical reaction in the liquid phase. The mass balances of the gaseous reactant A are the starting point of the model development. By solving the mass balances for A alone, it is often possible to calculate conversions and space-time-yields of the other reactive species which are only present in the liquid phase. Heat effects can be estimated, as well. It is, however, assumed that the temperature is constant throughout the reactor volume. Hence, isothermal models can be applied.

In the model matrix of the gaseous reactant a classification is also given with respect to the absorption-reaction regime. In the models <11> to <33>, it is supposed that the gaseous reactant is present in the bulk of the liquid phase ($c_{AL} > o$). Hence, the process takes place in the absorption regime with slow reaction. If the absorption is fast or instantaneous then the gaseous reactant undergoes complete reaction in the liquid side film at the gas/liquid interface. In this case, the mixing behavior of the bulk liquid phase is without any influence ("o"), provided temperature and concentration of liquid phase reactant B are locally independent. The models with absorption accompanied by fast chemical reaction are denoted by <41>, <42> and

<43> in Fig. 3. There are also applications whereby the gas phase composition is approximately constant. This is observed in hydrogenations and oxidations, particularly, when high recycle flow rates are employed. In such cases the gas phase balance needs no consideration as indicated by "o". The absorption rate and space-time-yield can be calculated from the liquid phase balance, alone.

In spite of the fact that in larger diameter BCR gas phase dispersion coefficients are high, it can be anticipated that the Bodenstein number (Bo) of the gas phase which characterizes the overall dispersion effects is significantly larger than Bo of the liquid phase:

$$Bo_L = \frac{L\ u_L}{\varepsilon_L\ E_L} << Bo_G = \frac{L\ u_G}{\varepsilon_G\ E_G} \tag{1}$$

This has to be attributed to higher gas throughputs ($u_G > u_L$) and lower gas holdups ($\varepsilon_G < \varepsilon_L$). Therefore, when modeling BCR it is not realistic to assume higher mixing in the gas phase than in the liquid phase. Hence, models <21>, <31> and <32> will not be considered any further. Also the assumption of plug flow in the liquid phase only applies to a few cases as, for instance, packed bubble columns and three-phase fluidized beds where high liquid flow rates must be applied to fluidize larger particles.

Liquid-phase	Gasphase			
	CSTR	ADM	PFM	"o"
CSTR	<11> +	<12> +	<13> +	<14> +
ADM	<21> -	<22> +	<23> +	<24> +
PFM	<31> -	<32> -	<33> +	<34> +
"o"	<41> +	<42> +	<43> +	<44> +

Fig. 3: Matrix of BCR models based on mixing properties of gas and liquid phase [6]

Owing to low gas solubility in liquids the concentration of the gaseous reactant in the liquid phase is usually much smaller than that of the liquid reactant B. As mixing in BCR is usually large, particularly if their diameter is large, and hence the values of Bo_L are low, say $Bo_L < 0.2$, it is often valid to assume complete mixing of the liquid phase with regard to the liquid phase reactants. The same assumption applies with regard to temperature. However, the Bodenstein number, i.e., the ratio of convective transport to dispersive transport (see eq. (1)) only applies to those components remaining in one phase. With respect to the gaseous reactant which takes part in interfacial mass transfer, the Bodenstein number is not an appropriate criteria to judge on liquid phase mixing. A better criteria is the ratio of the mixing time to a characteristic mass transfer time, i.e.,

$$\Phi = \frac{\text{mixing time}}{\text{mass transfer time}} = \frac{L^2/\varepsilon_L E_L}{1/K_L a} . \qquad (2)$$

Typical values of Φ for reactions taking place in the slow reaction-absorption regime are given in table 1.

Only if the mixing time is not considerably larger than the mass transfer time, the assumption of backmixed liquid is also justified for the dissolved gaseous reactant. For instance, in the case of the Fischer-Tropsch synthesis, the liquid phase synthesis gas concentration profiles are only rather flat if $\Phi < 10$. Such a low value of Φ can be realized only in large diameter reactors which are not too high. For tall reactors, the ADM has to be applied for the liquid phase. On the other hand, in chlorinations of aromatic compounds a considerable and unusually high gas and liquid side mass transfer resistance occurs. Hence, the mass transfer time is large giving low values of Φ. For such reactions, complete liquid phase mixing with regard to the dissolved gaseous reactant is justified even in tall BCR.

Table 1: Typical values of Φ (ratio of mixing time and mass transfer time)

Reaction	d_R/L in BCR, cm 20/200	100/1000
Isobutene hydration from C_4 cuts [34]	18	47
Fischer-Tropsch synthesis in slurry phase [37]	53	141
Chorination of toluene [10]	0.2	0.5

Besides the absorption regime and residence time distribution, other phenomena modify the structure of the model equations and mathematical efforts to solve them. One of these phenomena is the height dependency of the gas velocity. On the one hand, the mole flow rate of the gas phase decreases usually as a result of absorption and reaction while, on the other hand, the gas may expand due to decreasing hydrostatic head. Of course, expansion is only important in tall reactors operated at low pressure, i.e., atmospheric pressure. In general, both effects, contraction and expansion, should be considered properly, for instance, by balancing the inerts in the gas phase which introduces the gas velocity as an additional variable. Thus, the model equations become nonlinear even for linear reaction kinetics.

3. BCR IN THE SLOW REACTION REGIME

BCR are particularly well suited to carry out reactions in the slow reaction regime of absorption. Due to the high liquid holdup BCR provide for a large liquid volume where the reaction can take place. Also, in slurry reactors where the reaction takes place at the surface of the solid catalyst particles belong to the slow reaction regime. Only a few exceptions are known where absorption enhancement due to the slurry phase reaction has been observed [6, 20 - 22]. Strictly speaking, enhancement and hence transition to the fast reaction regime can only be expected if the diameters of the particle fines are considerably less than the liquid film thickness at the gas/liquid interface.

Of course, the residence time distribution of the liquid phase is of importance for calculating reactor performance in the slow reaction regime. In the matrix of Fig. 3 the models <11> through <33> are thinkable in this regime. However, for reasons given before only models <11>, <12>, <13>, <22>, and <23> will be considered in what follows.

3.1 Complete liquid phase mixing

Various authors [7, 8, 10, 12, 23, 24] have proposed models for BCR and bubble column slurry reactors (BCSR) which are based on the assumption of a backmixed liquid and slurry phase, respectively. However, it should be emphasized once again, that this assumption is only justified for the gaseous reactant if mixing time and mass transfer time are of the same order of magnitude, i.e., the value of Φ in eq. (1) is in the range of one or less. The advantage of assuming backmixed liquid is that for a number of kinetic laws closed solutions and explicite expressions for the space-time-yield achievable in BCR and BCSR can be developed. This has been shown by Chaudhari and Ramachandran [7] for the case of slurry reactors with plug flow of the gas phase. Their procedure can be applied to BCR, of course, and can be extended to dispersed flow in the gas phase, i.e., model <12>. A restriction in the treatment of Chaudhari and Ramachandran is, however, that isobaric conditions and constant gas velocity are assumed.

As in Chaudhari and Ramachandran [7] the gas phase balance of reactant A in model <13> (gas phase in PF) is given by

$$-u_G \frac{dc_A}{dx} - K_L a \, (c^*_{AL} - c_{AL}) = 0 \, . \tag{3}$$

Introducing Henry's law

$$p_A = He \, c^*_{AL} = c_A \, RT \tag{4}$$

and the Stanton number

$$St = K_L a \frac{L}{u_G} \frac{RT}{He} \tag{5}$$

it follows

$$-\frac{dc_A}{dz} - St \, (c_A - c_{AL} \, He/RT) = 0 \, . \tag{6}$$

As the liquid phase is assumed to be completely mixed, c_{AL} is constant and eq. (6) can be integrated

$$\frac{c_A - c_{AL} \, He/RT}{c_{Ao} - c_{AL} \, He/RT} = e^{-St \, z} \, . \tag{7}$$

The gas phase outlet concentration is given by

$$c_{A1} = c_{Ao} e^{-St} + c_{AL} \frac{He}{RT} (1 - e^{-St}) \, . \tag{8}$$

The specific absorption rate which is equivalent to space-time-yield follows from

$$R_A a = \frac{1}{V_R} (\dot{V}_{Go} \, c_{Ao} - \dot{V}_{G1} \, c_{A1}) \, . \tag{9}$$

If the gas volume flow is constant ($\dot{V}_G = \dot{V}_{Go} = \dot{V}_G$) one obtains by considering eq. (7)

$$R_A a = \frac{He}{RT} \frac{\dot{V}_G}{V_R} (1 - e^{-St}) \left(c_{Ao} \frac{RT}{He} - c_{AL}\right) \, . \tag{10}$$

Introducing the liquid phase concentration in equilibrium with gas inlet concentration

$$c^*_{ALo} = c_{Ao}\ RT/He \tag{11}$$

leads to

$$R_A a = Q\ (c^*_{ALo} - c_{AL}) \tag{12}$$

with

$$Q = \frac{He}{RT}\frac{\dot{V}_G}{R_T}\ (1 - e^{-St})\ . \tag{13}$$

The constant liquid phase concentration c_{AL} can be eliminated by considering the kinetic law of the reaction taking place in the bulk liquid. For a first order law in A the specific absorption rate in terms of reaction volume (gas-in-liquid dispersion) is given by

$$R_A a = k_1 \varepsilon_L c_{AL}\ . \tag{14}$$

Introducing eq. (14) in eq. (12) one obtains for the overall specific absorption rate

$$R_A a = \frac{c^*_{ALo}}{\frac{1}{Q} + \frac{1}{k_1 \varepsilon_L}}\ . \tag{15}$$

Chaudhari and Ramachandran [7] derived expressions for $R_A a$ for other kinetic laws, as well. A selection is given in Table 2.

For the other models (<11> and <12>) to be dealt with in this section the procedure is quite analogous. Again, at first the gas phase balance is solved giving eq. (12) with a modified expression for Q. This is followed by elimination of c_{AL} using appropriate kinetic laws which leads again to the same expressions for $R_A a$ in Table 2. In model <11> with backmixed gas phase the equation for Q is as follows

$$Q = \left(\frac{RT}{He}\frac{V_R}{\dot{V}_G} + \frac{1}{K_L a}\right)^{-1}\ . \tag{16}$$

The mass balance of the gas phase in model <12> with axial dispersions is

$$\varepsilon_G E_G \frac{d^2 c_A}{dx^2} - u_G \frac{dc_A}{cx} - K_L a\ (c^*_{AL} - c_{AL}) = o\ . \tag{17}$$

Table 2: Relations for specific absorption rates of models <11>, <12> and <13> with backmixed liquid phase and constant gas flow (Absorption with slow reaction, Q is given in text)

Type of reaction	Kinetic law r_A in mol l^{-1} s^{-1}	Specific absorption rate $R_A a$, mol l^{-1} s^{-1}
Order 1	$k_1 c_{AL}$	$\left[\frac{1}{Q} + \frac{1}{k_1 \varepsilon_L}\right]^{-1} c^*_{ALo}$
Order 2	$k_2 c_{AL}^2$	$\frac{Q^2}{4k_2\varepsilon_L}\left[1 - \sqrt{1 + 4\frac{k_2\varepsilon_L}{Q} c^*_{ALo}}\right]^2$
Order 1/2	$k_{1/2} c_{AL}^{1/2}$	$\frac{(k_{1/2}\varepsilon_L)^2}{2Q}\left(\sqrt{1+4\left(\frac{Q}{k_{1/2}\varepsilon_L}\right)^2 c^*_{ALo}}-1\right)$
Order 0	k_o	$k_o \varepsilon_L$
Michaelis-Menten	$\frac{\mu c_{AL}}{K_M + c_{AL}}$	$\frac{1}{2}\alpha_o \left(1 - \sqrt{1 - 4\, Q\varepsilon_L \mu c^*_{ALo}/\alpha_o^2}\right)$ $\alpha_o = Q(c^*_{ALo} + K_M) + \mu\varepsilon_L$
Reaction of 2 gaseous species in liquid phase	$k_2 c_{AL} c_{DL}$	$\frac{1}{2}\beta_o \left(\beta_1 - \sqrt{\beta_1^2 - 4/(z\beta_o)}\right)$ $\beta_o = Q_A Q_D\, c^*_{ALo} c^*_{DLo}$ $\beta_1 = \frac{1}{\nu Q_A c^*_{ALo}} + \frac{1}{Q_D c^*_{DLo}} + \frac{1}{\nu k_2\, c^*_{ALo} c^*_{DLo}}$
$A \rightleftharpoons E$	$k_1\,(c_{AL} - c_{AL,Gl})$	$(c^*_{ALo} - c_{AL,Gl})\left[\frac{1}{Q} + \frac{1}{k_1\varepsilon_L}\right]^{-1}$
$2A \rightleftharpoons E$	$k_2\,(c_{AL}^2 - c_{AL,Gl}^2)$	$Q\,\gamma\left(1 - \sqrt{1 - (c^{*2}_{ALo} - c^2_{AL,Gl})/\gamma^2}\right)$ $\gamma = c^*_{ALo} + \frac{Q}{2k_2\varepsilon_L}$

Considering eq. (1), (4), and (5) and introducing

$$\hat{c}_A = c_A - c_{AL} \frac{He}{RT} \tag{18}$$

gives

$$\frac{1}{Bo_G} \frac{d^2\hat{c}_A}{dz^2} - \frac{d\hat{c}_A}{dz} - St\ \hat{c}_A = o\ . \tag{19}$$

The boundary conditions are

$$z = o \qquad \hat{c}_A = \hat{c}_{Ao} - \frac{1}{Bo_G} \frac{d\hat{c}_A}{dz} \tag{20}$$

$$z = 1 \qquad \frac{d\hat{c}_A}{dz} = o \tag{21}$$

and solution for the reactor outlet (z = 1) gives

$$\frac{c_{A1} - c_{AL}\ He/RT}{c_{Ao} - c_{AL}\ He/RT} = \frac{4}{N} q \exp(Bo_G/2) \tag{22}$$

with

$$q = \sqrt{1 + 4\ St/Bo_G} \tag{23}$$

and

$$N = (1 + q)^2 \exp(Bo_G/2) - (1 - q)^2 \exp(Bo_G/2)\ . \tag{24}$$

Calculating the absorption rate as in eq. (9) leads to eq. (12) with

$$Q = \frac{\dot{V}_G}{V_R} \frac{He}{RT} (1 - \frac{4}{N} q \exp(Bo_G/2)) \tag{25}$$

This gives a uniform description for the models <11> to <13>. The space-time-yield (specific absorption rate) can be calculated from the expressions given in Table 2 whereby the appropriate equations for Q are found from eq. (16) (model <11>), eq. (25) (model <12>), and eq. (13) (model <13>). The entire treatment can be extended to catalytic slurry reactors (BCSR) by considering the liquid/solid mass transfer resistance and pore diffusional effects. In this case the value of Q in the absorption rate expressions of Table 2 has to be replaced by Q* which is defined by

$$Q^* = (\frac{1}{Q} + \frac{1}{k_s a_s})^{-1}\ . \tag{26}$$

Possible pore diffusional resistances are accounted for in the usual manner by introducing effectiveness factors in the rate law

[7, 8].

In model <14>, the gas phase concentration is constant and hence c^*_{AL} is constant as well. The steady state absorption rate is given by

$$R_A a = K_L a\,(c^*_{AL} - c_{AL}) = K_L a\,(c_A \frac{RT}{He} - c_{AL}) \,. \tag{27}$$

As usual, this equation is used to eliminate c_{AL} in the appropriate kinetic expression and solved for $R_A a$.

On the basis of the analytical solutions, i.e. eq. (12) incorporating the appropriate expressions for Q, the influence of gas phase mixing on reactor performance can be studied. For a first order reaction it can be shown that the ratio of absorption rates and conversions, respectively, of the mixed (m) and unmixed (p) model is given by

$$\frac{(R_A a)_m}{(R_A a)_p} = \frac{X_m}{X_p} = \frac{St\,(Da\gamma + 1 - e^{-St})}{(1 - e^{-St})(Da\gamma + \gamma\,Da\,St + St)} \tag{28}$$

with

$$Da = k_1 \varepsilon_L \frac{V_R}{\dot{V}_G} \tag{29}$$

$$\gamma = \frac{RT}{He} \,. \tag{29a}$$

The ratio $(R_A a)_m/(R_A a)_p$ is plotted versus γ Da in Fig. 4 with St as a parameter. At low values of Da and relatively high St numbers the ratio approaches one as the mixed liquid phase is nearly saturated and, hence, gas phase mixing is unimportant. If, on the other hand, Da $\gg$ St, one approaches the diffusional regime, where the bulk liquid phase concentration is nearly zero and gas phase mixing has its largest effect. It is interesting to note that $(R_A a)_m/(R_A a)_p$ passes a minimum value in dependence on St. For $St > 1$ and Da $\gg$ St it follows from eq. (28) that the Stanton number which leads to a minimum value of the ratio (i.e. the maximum difference between the two models) is given by $\exp(St) = 1 + St + St^2$. Solution leads to $St = 1.8$ and $(X_m/X_p)_{min} = 0.773$. Therefore, the maximum reduction due to gas phase mixing is given by $X_m \geq 0.773\,X_p$. For obtaining the same conversions in reactors with backmixed and unmixed gas phase the volume of the backmixed reactor must be larger. For the volume ratio the following relation can be derived

$$\frac{V_{Rm}}{V_{Rp}} = \frac{(\gamma\,Da)_p + St_p}{(\gamma\,Da)_p\,St_p}\,\frac{X}{1 - X} \tag{30}$$

where the index p refers to the values of γ, Da, and St which give the conversion X in the unmixed reactor. As a rule of thumb it can be concluded from various calculations that $V_{Rm}/V_{Rp} \leq 2.2$.

The expressions for the absorption rate in Table 2 were derived without explicite consideration of an influence of the liquid phase reactant B. If A undergoes a reaction of the type

$$A + \nu\ B \rightarrow \text{products} \tag{31}$$

in liquid phase the following relations are valid for continuous throughput of liquid phase

$$R_B a = \nu\ R_A a \tag{32}$$

$$\dot{V}_L (c_{Bo} - c_B) = \nu\ V_R\ R_A a \tag{33}$$

$$X_B = \frac{\nu\ \dot{V}_G}{\dot{V}_L} \frac{c_{Ao}}{c_{Bo}} X_A\ . \tag{34}$$

If the reaction in the bulk liquid phase follows the kinetic law

$$r_A = k_{mn}\ c_A^{\ m}\ c_B^{\ n} \tag{35}$$

then $R_A a$ has to be evaluated with

$$k_m = k_{mn}\ c_B^{\ n}\ . \tag{36}$$

Rearrangement of eq. (33) to

$$c_B = c_{Bo} - \frac{\nu\ R_A a}{1/\tau_L} \tag{37}$$

with $\tau_L = V_R/\dot{V}_L$ and introduction in eq. (36) leads to

$$k_m = k_{mn}\ (c_{Bo} - \frac{\nu\ R_A a}{1/\tau_L})^n\ . \tag{38}$$

Use of eq. (38) in the relations given in Table 2 yields implicite expressions for $R_A a$ which can be solved by iteration methods. If $m = n = 1$ and hence

$$R_A a = k_2'\ \varepsilon_L\ c_{AL}\ c_B \tag{39}$$

c_{AL} and c_B can be introduced from eq. (12) and (37) which gives

$$\frac{R_A a}{k_2' \varepsilon_L} = (c_{Bo} - \frac{\nu R_A a}{1/\tau_L}) (c^*_{ALo} - \frac{R_A a}{Q}) \qquad (40)$$

and

$$R_A a = \frac{1}{2} \beta_o (\beta_1 - \sqrt{\beta_1^2 - 4/(\nu\beta_o)}) \qquad (41)$$

with

$$\beta_o = c^*_{ALo} c_{Bo} Q/\tau_L \qquad (42)$$

and

$$\beta_1 = \frac{1}{c_{Bo}/\tau_L} + \frac{1}{\nu Q c^*_{ALo}} + \frac{1}{\nu k_2' \varepsilon_L c^*_{ALo} c_{Bo}} . \qquad (43)$$

In many industrial applications BCR are operated in a semi-batch manner. Only the gas is in continuous flow while the liquid is batch. As shown by Chaudhari and Ramachandran [7], the reaction time needed for a desired conversion of B can be calculated easily if the assumption underlying models <11>, <12>, and <13> are fulfilled. The variation of B follows from

$$-\frac{dc_B}{dt} = R_B a = \nu R_A a \qquad (44)$$

or

$$t = -\frac{1}{\nu} \int_{c_{Bo}}^{c_B} \frac{dc_B}{R_A a} . \qquad (45)$$

Assuming that eq. (39) applies to the reaction kinetics one obtains with consideration of eq. (12) for the batch reaction time

$$t = -\frac{1}{\nu c_{Ao} RT} \int_{c_{Bo}}^{c_B} He (\frac{1}{Q} + \frac{1}{k_2' \varepsilon_L c_B}) dc_B . \qquad (46)$$

Provided that Q ($K_L a$), ε_L and He are independent of c_B integration yields

$$\frac{c_{Bo} - c_B}{Q} + \frac{1}{k_2' \varepsilon_L} \ln \frac{c_{Bo}}{c_B} = \nu c_{Ao} \frac{RT}{He} t . \qquad (47)$$

Also for other cases the batch reaction time can be calculated [7], cf. Table 2 of this paper.

It should be pointed out once again that the treatment of BCR given here for the models <11> to <13> is based on the assumption of a constant gas flow rate. This assumption can be justified if the concentration of the absorbing and reacting gaseous component is small, for instance, when removing impurities from gas streams. Of course, assuming constant gas flow rate is correct for such reactions where the same amount of gas is generated as is consumed. An example is the chlorination of aromatics where HCl is liberated. As indicated by the value of Φ given in Table 1, this type of chlorination reactions can really be treated with the assumption of complete liquid phase mixing (even with regard to the dissolved chlorine). Therefore, the mathematical treatment given here for models <11>, <12> and <13> should particularly be applicable to chlorinations of aromatics. This is demonstrated in the following example.

Example 1: Chlorination of Toluene

The batch chlorination of toluene (with continuous Cl_2 flow) was studied in a BCR by Lohse et al. [10]. At low catalyst concentrations ($FeCl_3$) and in the presence of trace impurities of water the reaction takes place in the slow reaction regime. The chlorination of toluene (T) and the consecutive chlorination of monochloro toluene to dichloro toluene

$$\underset{(T)}{C_7H_8} \xrightarrow[-HCl]{Cl_2} \underset{(MCT)}{C_7H_7Cl} \xrightarrow[-HCl]{Cl_2} \underset{(DCT)}{C_7H_6Cl_2} \qquad \text{(E 1.1)}$$

follows a kinetic law which is first order in Cl_2 and aromatic compound for given concentrations of catalyst and H_2O:

$$- r_T = k_2 \, c_{Cl_2} \, c_T \qquad \text{(E 1.2)}$$

$$- r_{MCT} = k_2' \, c_{Cl_2} \, c_{MCT} - k_2 \, c_{Cl_2} \, c_T \, . \qquad \text{(E 1.3)}$$

The governing transient balance equation of Cl_2 in the gas phase is given by

$$\frac{\varepsilon_G L}{u_{Go}} \frac{\partial c_G}{\partial t} = - \frac{\partial c_G}{\partial z} - \frac{K_L a L}{u_{Go}} (c_L^* - c_L) \qquad \text{(E 1.4)}$$

which can be rendered dimensionless

$$\frac{\partial Y}{\partial \Theta} = -\frac{\partial Y}{\partial z} - St_G(Y-X) \ . \qquad \text{(E 1.5)}$$

The dimensionless quantities in eq. (E 1.5) are defined as follows

$$Y = c_G/c_{Go} \qquad \text{(E 1.6)} \qquad\qquad X = c_L/c^*_{Lo} \qquad \text{(E 1.7)}$$

$$\Theta = t\ u_{Go}/(\varepsilon_G L) \qquad \text{(E 1.8)}$$

$$St_G = \frac{K_L a L}{u_{Go}} \frac{RT}{He} . \qquad \text{(E 1.9)}$$

It can be shown that the gas phase concentration responds quickly to changes in liquid phase composition. Therefore, the gas phase can be treated as quasi-stationary and eq. (E 1.5) reduces to

$$\frac{dY}{dz} = - St_G(Y-X) . \qquad \text{(E 1.10)}$$

As X is constant, solution with consideration of Y(o) = 1 gives

$$Y = X + (1-X)e^{-St_G z} . \qquad \text{(E 1.11)}$$

The transient balance for Cl_2 in the liquid is given by

$$\frac{dX}{dt} = \frac{K_L a}{L} \int_0^1 (Y(z)-X)dz - k_2\ c_{To}\ XZ_1 - k_2'\ c_{To}\ XZ_2 \qquad \text{(E 1.12)}$$

where

$$Z_1 = c_T/c_{To} \qquad \text{(E 1.13)} \qquad\qquad Z_2 = c_{MCT}/c_{To} \qquad \text{(E 1.14)}$$

Introducing the logarithmic mean of the driving concentration difference calculated from eq. (E 1.11), the liquid phase balance of Cl_2 is as follows

$$\frac{dX}{dt} = \frac{K_L a}{\varepsilon_L St_G} [1 - X - (1-X)e^{-St_G}]$$

$$- k_2\ c_{To}\ XZ_1 - k_2'\ c_{To}\ XZ_2 \qquad \text{(E 1.15)}$$

The balance equations of the liquid organic compounds are given by

$$T: \quad \frac{dZ_1}{dt} = - k_2\ c^*_{Lo}\ XZ_1 \qquad \text{(E 1.16)}$$

$$\text{MCT:} \quad \frac{dZ_2}{dt} = k_2\, c^*_{Lo}\, XZ_1 - k'_2\, c^*_{Lo}\, XZ_2 \qquad \text{(E 1.17)}$$

$$\text{DCT:} \quad \frac{dZ_3}{dt} = k'_2\, c^*_{Lo}\, XZ_2 \qquad \text{(E 1.18)}$$

where

$$Z_3 = c_{DCT}/c_{To} \;. \qquad \text{(E 1.19)}$$

Using the appropriate data of mass transfer coefficients, Cl_2 solubilities and kinetic constants which were all determined independently, numerical solution of eq. (E 1.15) to (E 1.18) gives an excellent description of the experimental chlorination data [10]. An example is given in Fig. E 1. It is worthwhile to mention that the striking agreement shown in Fig. E 1 could only be reached by accounting properly for the temperature rise at the gas/liquid interface. This temperature increase results from Cl_2 absorption and reaction and reduces the Cl_2 solubility at the interface.

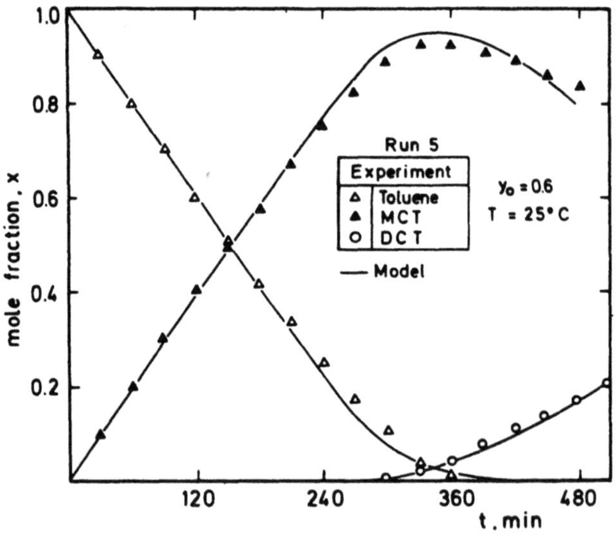

Fig. E 1: Description of experimental data of batch chlorination of toluene in BCR ($c_{cat}(FeCl_3) = 2.96 \times 10^{-3}$ mol/l, Cl_2 inlet mol fraction = 0.6, T = 298 K)[10]. Reprinted with permission from Chem. Engng. Sci. 38, Lohse M., E. Alper and W.-D. Deckwer, Modeling of Batch Catalytic Cheorination of Toluene. Copyright (1976), Pergamon Press

Usually, in chemical industry there are many applications where high concentrations of gaseous reactants are used, and simultaneously high conversions are desired. In such cases, the assumption of constant gas flow cannot be maintained. Consideration of gas flow variations increases the mathematical efforts. Even for first order reaction and the most simple models, i.e. <11> and <13>, the design equations are considerably more complicated.

Under isobaric operating conditions (hydrostatic head ≪ operating pressure) the gas volume flow at outlet for complete mixing of all phases (model <11>) is given by

$$\dot{V}_{G1} = \dot{V}_{Go} - R_A a \, V_R \, V_M \, . \tag{48}$$

With this relation the outlet gas concentration and c^*_{AL} follows from eq. (9)

$$c_{A1} = c^*_{AL} \frac{He}{RT} = \frac{\dot{V}_{Go} \, c_{Ao} - R_A a \, V_R}{\dot{V}_{Go} - R_A a \, V_R \, V_M} \tag{49}$$

Introducing eq. (49) in

$$R_A a = K_L a \, (c^*_{AL} - c_{AL}) \tag{50}$$

and assuming first order reaction (i.e. eq. (14)) gives

$$R_A a = K_L a \, \left(\frac{RT}{He} \frac{\dot{V}_{Go} \, c_{Ao} - R_A a \, V_R}{\dot{V}_{Go} - R_A a \, V_R \, V_M} - \frac{R_A a}{k_1 \, \varepsilon_L}\right) . \tag{51}$$

Rearrangement and solving for $R_A a$ leads to

$$R_A a = \frac{1}{2} \frac{\dot{V}_{Go}}{V_R \, V_M} \alpha_1 \alpha_2 \left(1 - \sqrt{1 - 4 \frac{V_R V_M c^*_{ALo}}{\dot{V}_{Go} \, \alpha_1^2 \alpha_2}}\right) \tag{52}$$

with

$$\alpha_1 = \frac{1}{K_L a} + \frac{V_R \; RT}{\dot{V}_{Go} \; He} + \frac{1}{k_1 \varepsilon_L} = \frac{1}{Q} + \frac{1}{k_1 \varepsilon_L} \tag{53}$$

$$\alpha_2 = \left(\frac{1}{K_L a} + \frac{1}{k_1 \varepsilon_L}\right)^{-1} \tag{54}$$

In eq. (52) only the negative sign of the root is reasonable. This can be confirmed by calculating absorption rates from eq. (52) and comparing for the limiting case, i.e. the diffusional regime ($1/k_1\varepsilon_L \rightarrow o$).

In model <13> the gas phase is in plug flow and the absorption rate and c^*_{AL} are locally dependent. Therefore, eq. (48) and (49) which are valid for the lumped model only, cannot be used. The gas phase balance with consideration of a variable gas flow is given by

$$- \frac{d}{dx} (u_G \, c_A) - K_L a \, (c^*_{AL} - c_{AL}) = o \ . \tag{55}$$

Introducing the contraction or expansion factor, respectively, which is defined as

$$\zeta = \frac{\dot{V}_G(X = 1) - V_{Go}}{\dot{V}_{Go}} \tag{56}$$

Then the variation of the gas velocity can be expressed as function of the conversion

$$u_G = u_{Go} \, (1 + \zeta X) \tag{57}$$

with

$$X = 1 - \frac{u_G \, c_A}{u_{Go} \, c_{Ao}} \ . \tag{58}$$

Making use of these relations eq. (55) can be rearranged to

$$\frac{dX}{dz} = St \left[\frac{1 - X}{1 + \zeta X} - \frac{c_{AL}}{c^*_{ALo}} \right] \tag{59}$$

which gives on integration for the reactor outlet

$$- \frac{\zeta}{\psi} X - \frac{1 + \zeta}{\psi^2} \ln(1 - \frac{\psi X}{1 - c_{AL}/c^*_{ALo}}) = St \tag{60}$$

where

$$\psi = 1 + \zeta \, c_{AL}/c^*_{ALo} \ . \tag{61}$$

Eq. (60) corresponds with eq. (7) for constant gas flow. As usual, it is now necessary to eliminate c_{AL} by accounting for the reaction rate in the backmixed phase. Assuming m-th order reaction of A in the backmixed liquid phase one obtains

$$\frac{c_{AL}}{c^*_{ALo}} = \frac{1}{\gamma} \left(\frac{X}{Da_m}\right)^{1/m} \tag{62}$$

with

$$Da_m = k_m \, c_{Ao}^{m-1} \, V_R \, \varepsilon_L / \dot{V}_G \, . \tag{63}$$

Combining eq. (60), (61) and (62) one obtains the following implicit relation for the conversion of model <13> with variable gas flow and for m-th order reaction

$$(St \{1 + \frac{\zeta}{\gamma} (\frac{X}{Da_m})^{1/m}\} + \zeta X) \, (1 + \frac{\zeta}{\gamma} (\frac{X}{Da_m})^{1/m})$$

$$= (1 + \zeta) \, (\ln [1 - \frac{1}{\gamma} (\frac{X}{Da_m})^{1/m}]$$

$$- \ln [1 - X - (1 + \zeta X) \frac{1}{\gamma} (\frac{X}{Da_m})^{1/m}]) \tag{64}$$

For model <13> with constant gas flow or $\zeta = o$, one can derive the following explicit equation for the conversion from the relations given in Table 2

$$m = 1: \qquad X = \frac{\gamma \, Da_1 \, (1 - e^{-St})}{1 + \gamma \, Da_1 - e^{-St}} \tag{65}$$

$$m = 2: \qquad X = \frac{(1 - e^{-St})^2}{4 \, Da_2 \, \gamma^2} \, (1 - \sqrt{1 + 4 \, Da_2 \gamma^2 / (1 - e^{-St})})^2 \tag{66}$$

$$m = 1/2: \qquad X = \frac{Da_{1/2}^2 \, \gamma}{1 - e^{-St}} \, (\sqrt{1 + 4(1 - e^{-St})^2 / (Da_{1/2}^2 \, \gamma)} - 1) \tag{67}$$

In order to demonstrate the effect of gas flow variations due to absorption and reaction, Fig. 5 compares the conversion of model <13> with constant and with variable gas velocity for various values of ζ. It is observed that the discrepancy between both models depends only in a small extent on the individual values of Da, γ, St, and m as long as $|\zeta| \leqq 0.5$. Therefore, the curves shown in Fig. 5 present average values. Only for $\zeta = -1$ a larger difference was found. However, no clear dependence on the individual parameters could be discerned. Therefore, the area is shaded in

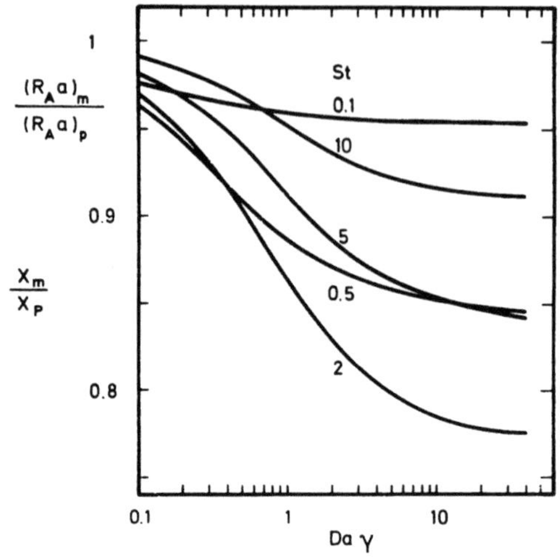

Fig. 4: Effect of gas phase mixing on conversion and absorption rate for first order reaction - Comparison of models <11> and <13> both with backmixed liquid phase

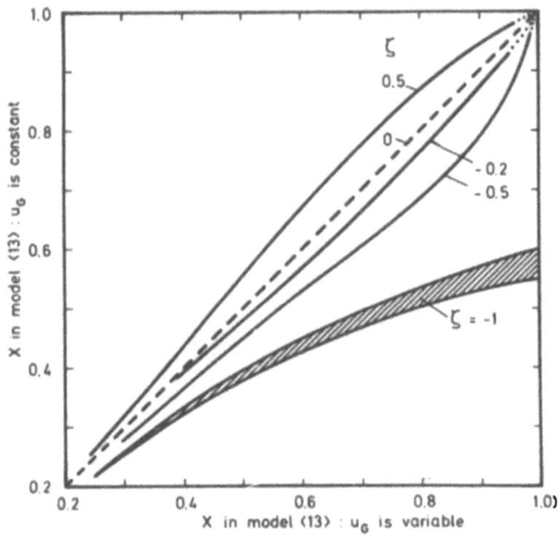

Fig. 5: Effect of gas flow variations (given by ζ, eq. (56)) on conversion in model <13>

Fig. 5. For little contraction, i.e. ζ = - o.2, the effect is moderate. The largest conversion differences are about o.o4. Therefore, application of constant gas flow models can be justified for many cases of industrial importance like separation of impurities or desired products from gas streams. If ζ increases the deviations increase remarkably and are larger for contraction than for expansion as shown for $|\zeta|$ = o.5 in Fig. 5. It is understood that constant flow models should not be applied for larger values of ζ. In the case of contraction which is more frequent in chemical reaction engineering, such values can give considerable overdesign. Only for some oxidation and hydrogenation reactions with low conversion per pass and by application of gas phase recycle may constant gas flow models still work as a useful approximation.

3.2 Dispersed plug flow in liquid phase

In models <22>, <23>, and <24> of the matrix in Fig. 3, finite mixing in the liquid phase is taken into account. This is done by applying the axial dispersion model (ADM) to both phases (model <22>). Making use of the ADM for modeling multiphase reaction systems with largely different phase velocities is certainly only a rough approximation. Particularly, in view of new hydrodynamic models proposed in recent years models which resemble more realistically the fluid flow in BCR can be constructed. Such models are still under development and some aspects will be briefly discussed later. At present, it appears only reasonable to employ the ADM where the various mixing and dispersive phenomena are lumped into one single parameter, e.g. the axial dispersion coefficient E. The ADM has been found to be a reliable and reasonable basis to interprete mass transfer and reaction data from various BCR [25]. The applicability of the ADM to design and simulation of BCR is therefore evident.

Dispersion models for two-phase systems were proposed by Miyauchi and Vermeulen [26, 27] for the case of continuous liquid-liquid contactors. For such systems, Hartland and Mecklenburgh [28] systematically developed and analyzed possible analytical solutions for a variety of limiting cases. If the mathematical treatment used for extractors is applied to BCR, the closed solutions are more complicated [29, 30] and only numerical simulations can be recommended to detect the influence of a certain parameter [30, 31]. Only for the case of model <24> with constant gas phase mole fraction (pressure can be variable), an easy to handle analytical solution is available which can successfully be used for evaluation of experimental profiles, for instance, in oxygen mass transfer measurements [25, 32]. Usually, the assumption of constant gas concentration is not at all permitted, and if gas flow variations are also considered, the balance equations are nonlinear even for first order rate processes. Therefore, only numerical solution procedures are appropriate to solve the resulting nonlinear boundary value problem. In the following, the balance equations are developed for models <22> and <23> under consideration of gas flow and pressure variations. This is done only for the gaseous reactant A which undergoes first order reaction in the liquid phase which is the most common case.

The gas phase balance in the slow reaction regime for model <23> is

$$- \frac{d}{dx} (u_G c_A) - k_L a \, (c^*_{AL} - c_{AL}) = o \; . \tag{68}$$

To account for the dependency of u_G on x a balance over the inerts i is considered, i.e.,

$$\frac{d}{dx} (u_G \; c_i) = o \tag{69}$$

giving

$$u_G \; P \; (1 - x_A) = u_{Go} \; P_o (1 - x_{Ao}) \; . \tag{70}$$

With eq. (70) it follows from eq. (68)

$$-u_{Go} \; (1 - x_{Ao}) \; \frac{P_o}{RT} \frac{d}{dx} \left(\frac{x_A}{1 - x_A}\right) - k_L a \; (c^*_{AL} - c_{AL}) = o \; . \tag{71}$$

The liquid phase balance is given by

$$\varepsilon_L \; E_L \; \frac{d^2 c_{AL}}{dx^2} + a^* \; u_L \; \frac{d \; c_{AL}}{dx} + k_L a \; (c^*_{AL} - c_{AL}) - k_1 \varepsilon_L c_{AL} = o \tag{72}$$

where $a^* = -1$ for cocurrent flow and $a^* = 1$ for countercurrent flow of the phases. If ε_G is constant the hydrostatic head in the column decreases linearly with x, hence

$$P = P_T + \rho g \; \varepsilon_L \; (L - x) \; . \tag{73}$$

Using Henry's law in the form

$$P x_A = He \; c^*_{AL} \tag{74}$$

and introducing dimensionless quantities, i.e.,

$$z = x/L \tag{75}$$ $$\bar{x}_A = x_A / x_{Ao} \tag{76}$$

$$\alpha = \rho g \varepsilon_L \; L / P_T \tag{77}$$ $$\beta = 1 + \alpha(1 - z) \tag{78}$$

$$Bo_L = u_L L / \varepsilon_L E_L \tag{79}$$ $$Da = k_1 \varepsilon_L L / u_L \tag{80}$$

$$St_L = k_L a\ L/u_L \qquad (81) \qquad St_G = k_L a\ LRT/(u_{Go}\ He) \qquad (82)$$

$$\bar{y}_A = \frac{c_{AL}}{c^*_{AL}\ (z = o)} = \frac{c_{AL}\ He}{P_T\ (1 + \alpha)\ x_{Ao}} \qquad (83)$$

eqns. (71) and (72) yield by differentiation and rearrangement

$$-\frac{d\bar{x}_A}{dz} = \frac{St_G\ (1 - x_{Ao}\ \bar{x}_A)^2}{(1 + \alpha)\ (1 - x_{Ao})} \left|\beta\ \bar{x}_A - (1 + \alpha)\ \bar{y}_A\right| \qquad (84)$$

$$\frac{1}{Bo_L} \frac{d^2\ \bar{y}_A}{dz^2} + a^* \frac{d\ \bar{y}_A}{dz} = -\frac{St_L\ \beta}{1 + \alpha} \bar{x}_A + (St_L + Da)\ \bar{y}_A\ . \qquad (85)$$

These equations are subject to the following boundary conditions for concurrent flow ($a^* = -1$)

$$z = o \qquad \bar{x}_A = 1 \qquad \bar{y}_A = \frac{1}{Bo_L} \frac{d\ \bar{y}_A}{dz} \qquad (86)$$

$$z = 1 \qquad \frac{d\ \bar{y}_A}{dz} = o \qquad (87)$$

and for countercurrent flow

$$z = o \qquad \bar{x}_A = 1 \qquad \frac{d\ \bar{y}_A}{dz} = o \qquad (88)$$

$$z = 1 \qquad \bar{y}_A = -\frac{1}{Bo_L} \frac{d\ \bar{y}_A}{dz}\ . \qquad (89)$$

Eqns. (84) to (89) present the dimensionless steady state isothermal balance equations of model <23> [33]. In model <22>, dispersion in gas phase is considered. The governing gas phase balance equation is derived as usual [6, 33] and leads to

$$\frac{1}{Bo_G} \frac{d^2\ \bar{x}_A}{dz^2} - \left(\frac{\alpha}{Bo_G\ \beta} + \bar{u}_G\right) \frac{d\ \bar{x}_A}{dz} = St_G\ \bar{x}_A\ (1 - x_{Ao}\ \bar{x}_A)$$

$$- St_G \frac{1 + \alpha}{\beta} \bar{y}_A\ (1 - x_{Ao}\ \bar{x}_A)\ . \qquad (90)$$

The dimensionless gas velocity $\bar{u}_G = u_G/u_{Go}$ is obtained from an entire balance over the gas phase giving

$$\frac{d\,\overline{u}_G}{dz} = \frac{\alpha}{\beta}\,\overline{u}_G - St_G\, x_{Ao}\,(\overline{x}_A - \overline{y}_A\,\frac{1+\alpha}{\beta}) \quad . \tag{91}$$

Eqns. (90) and (91) and the liquid phase balance eq. (85) are the basic equations of model <22> which must be solved with the conditions

$$z = o \qquad \overline{u}_G = 1 \qquad \overline{x}_A = 1 + \frac{1}{Bo_G}\,\frac{d\,\overline{x}_A}{dz} \tag{92}$$

$$z = 1 \qquad \frac{d\,\overline{x}_A}{dz} = o \tag{93}$$

and the liquid phase boundary conditions, eq. (86) to (89) for co-current or countercurrent flow, respectively.

Under isobaric conditions, i.e., the pressure at reactor top P_T is large compared to the hydrostatic pressure of the liquid phase in the column and hence $\alpha = o$ and $\beta = 1$, it follows from eq. (90) and (91)

$$\frac{1}{Bo_G}\,\frac{d^2\,\overline{x}_A}{dz^2} - \frac{d}{dz}\,(\overline{u}_G\,\overline{x}_A) - St_G\,(\overline{x}_A - \overline{y}_A) = o \ , \tag{94}$$

which accounts still for gas flow variations by the absorption process. If such variations are also neglected, model <22> reduces for co-current flow ($a^* = 1$) to

$$\frac{1}{Bo_G}\,\frac{d^2\,\overline{x}_A}{dz^2} - \frac{d\,\overline{x}_A}{dz} - St_G\,(\overline{x}_A - \overline{y}_A) = o \tag{95}$$

$$\frac{1}{Bo_L}\,\frac{d^2\,\overline{y}_A}{dz^2} - \frac{d\,\overline{y}_A}{dz} + St_L\,(\overline{x}_A - \overline{y}_A) - Da\,\overline{y}_A = o \ . \tag{96}$$

Numerical simulations of model <22>, i.e. eqns. (85), (90), and (91) are shown in Figs. 6 to 10 [33]. The conversion X as function of St_G for various values of St_L is shown in Fig. 6. It can be seen that X increases St_G, but decreases with increasing St_L. This behaviour is understood if one realizes that small St_L means high liquid velocities and high St_G implies preferentially high gas solubilities. The effect of the pressure ratio α on X is depicted in Fig. 7. Relevant values of α encountered in industrial processes as well as in waste water treatment and fermentations lie in

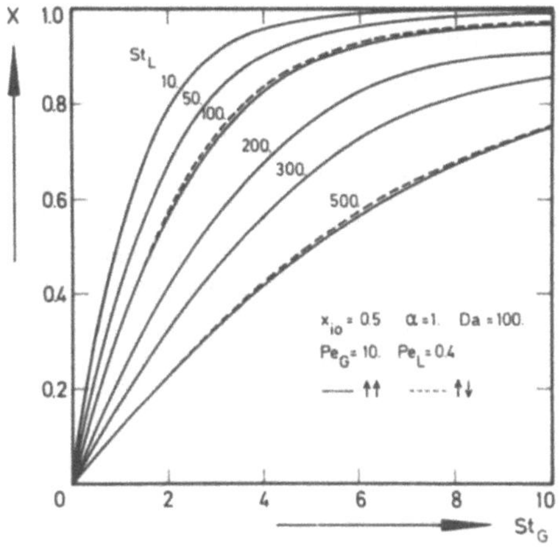

Fig. 6: Conversion as function of St_G for various St_L [33] Reprinted with permission Chem. Engng. Sci., 31. Copyright (1976) Pergamon Press.

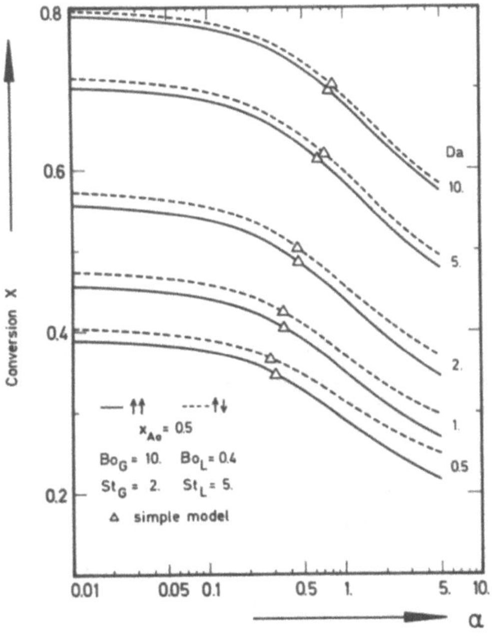

Fig. 7: Conversion vs. pressure ratio α [33]. Reprinted with permission Chem. Eng. Sci.,31. Copyright (1976) Pergamon Press.

the interval: o.25 ≦ α ≦ 1.5. In this region of α, X distinctly depends on α for all values of Da. The decrease of X is caused by gas expansion as a result of reduced pressure. Thus the gas velocity is increased giving a lower gas phase residence time. With increasing α the gas expansion cannot be balanced by gas shrinkage resulting from absorption. However, it is understood that the effect of α depends on the amount of inerts present in the inlet gas. Curves of X vs. x_{Ao} for countercurrent flow (↑↓) and Da = 5 are given in Fig. 8. It is seen that with increasing x_{Ao} the conversion increases, too. Particularly, under isobaric conditions (α = o) X increases

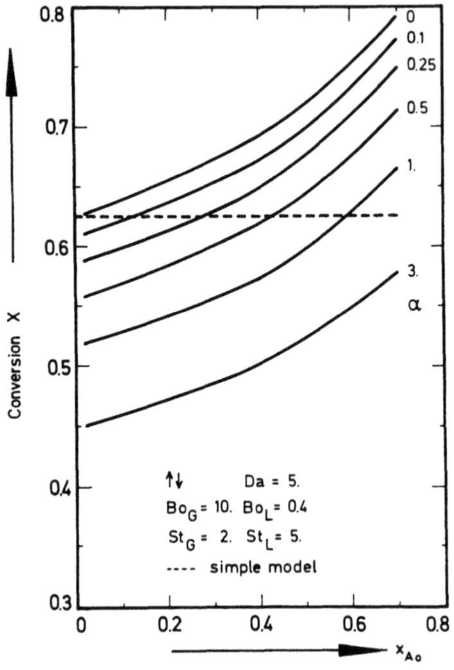

Fig. 8: Conversion as function of inlet mole fraction for various values of α [33] . Reprinted with permission Chem. Engng. Sci., 31. Copyright (1976) Pergamon Press.

remarkably with x_{Ao}. The triangles given in Fig. 7 and the dotted line in Fig. 8 refer to a simplified version of model <22> given by eqns. (95) and (96). The simulations shown in Figs. 7 and 8 clearly demonstrate that BCR design must be based on models which account for gas velocity changes. Only for isobaric conditions and x_{Ao} ≦ o.1 can the use of simpler models be justified.

The effect of dispersion in both phases on BCR performance operating in the slow reaction regime is shown for countercurrent flow in Fig. 9. The effect of Bo_L on X depends largely on gas phase

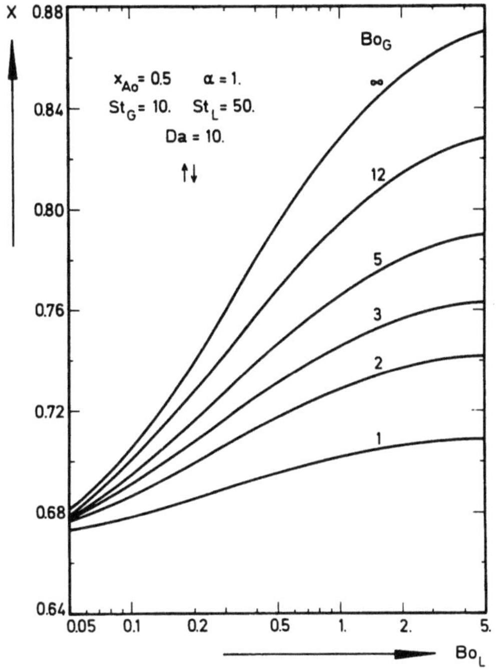

Fig. 9: Effect of dispersion in both phases on X for countercurrent flow [33]. Reprinted with permission Chem. Eng. Sci., 31, Copyright (1976) Pergamon Press.

dispersion (Bo_G). It is very pronounced for high values of Bo_G. If the reaction rate in the liquid phase is increased and St_L is reduced, liquid phase concentration is low ($\bar{y}$ < o.2). As shown in Fig. 10, the conversion is little affected by variations of Bo_L under such conditions. However, the influence of Bo_G on X is again very pronounced. Therefore, one can conclude as a rule of thumb that gas phase dispersion has a major impact on BCR performance while the influence of liquid phase dispersion depends on the concentration level in the liquid phase.

The applicability of model <23> as formulated here with consideration of gas flow and pressure variations has been demonstrated by Deckwer [34] for absorption and hydration of isobutene in sulfuric acid and by Deckwer et al. [35, 36] for CO_2 mass transfer. The physical absorption of CO_2 in water proved to be a very useful tool to simulate conditions which often prevail in chemical processes, i.e. large amounts of gas are absorbed in a liquid under drastic changes of gas volume flow. As an example Fig. 11 shows CO_2 gas phase profiles measured in a BC of 20 cm diameter and 7.2 m height. As shown in Fig. 11, the profiles computed on the basis of model <23> excellently describe the measured data. It is understood that

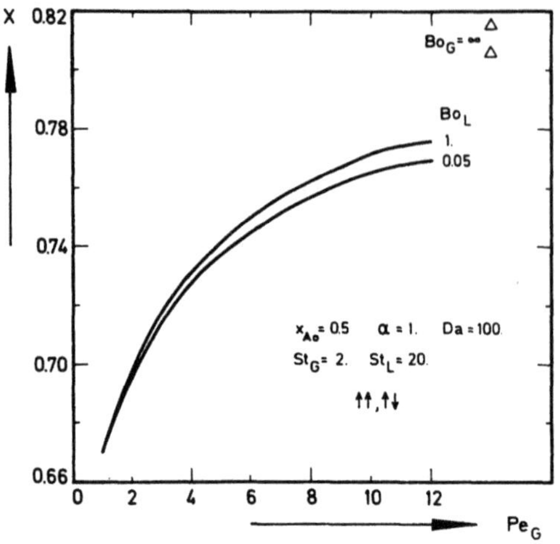

Fig. 10: Effect of dispersion on X at low concentrations in liquid phase [33] . Reprinted with permission Chem. Engng. Sci., 31, Copyright (1976) Pergamon Press.

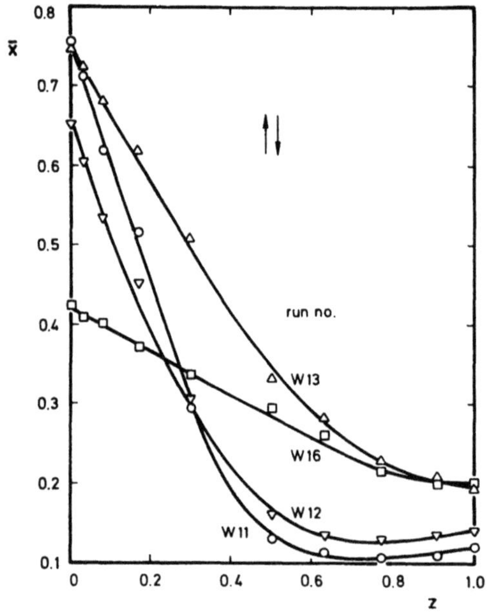

Fig. 11: Experimental and computed gas phase profiles for countercurrent CO_2 absorption [35]

simpler models (with constant gas flow and pressure) are not able to describe the experimental data.

On the basis of the assumptions of model <22> and <23> the Fischer-Tropsch synthesis in a slurry phase BCR has been modeled [37, 38]. As this hydrocarbon synthesis from synthesis gas ($CO + H_2$) is accompanied by considerable volume contraction, it is clear that gas flow variations have to be accounted for. The developed models are useful to evaluate experimental data from bench scale units and to simulate the behavior of larger scale Fischer-Tropsch slurry reactors. Though only simplified kinetic laws were applied, the predictions of the model are in reasonable agreement with data reported from 1.5 m diameter demonstration plant. Fig. 12 shows computed space-time-yields (STY) as a function of the inlet gas velocity. As the Fischer-Tropsch reaction on suspended catalyst takes place in the slow reaction regime, it is understood that STY passes through a maximum in dependence of u_{G0}. The predicted maximum is in striking agreement with experimental observations [37].

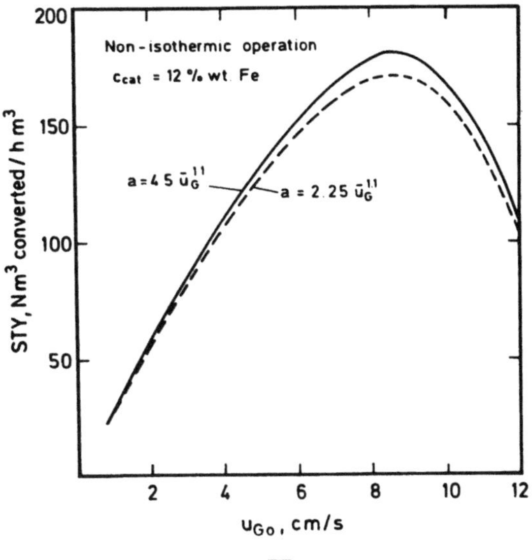

Fig. 12: Space-time-yield vs. gas velocity under the conditions of the Fischer-Tropsch synthesis in slurry phase
Reprinted with permission from Ind.Eng.Chem.Proc.Des.Dev.21 (1982) 222-241 copyright (1982) American Chemical Society.

4. FAST REACTION REGIME

If the reaction is fast compared to mass transfer the gaseous species A is completely consumed in the liquid film at the gas/ liquid interface. Therefore, the concentration of A in the bulk liquid is zero. This is also the case in the so-called diffusional regime which is a limiting range of the slow reaction regime. In the diffusional regime the bulk liquid concentration is zero as well, but an absorption enhancement does not occur. For zero liquid concentration the mixing behavior of the liquid phase does not play any role. Hence, the models <41> to <44> apply.

In model <41> and <44> the gas phase concentration is constant as a result of complete mixing in model <41> and due to low conversion and differential reactor behavior in model <44>. In both cases the results of the absorption-reaction theory [19, 20] can be applied directly to calculate absorption rates, space-time-yields and conversion. Therefore, no further treatment is required here.

Various rate expressions resulting from different kinetic laws have been treated in model <42> and <43> by Mashkar [39] and Juvekar and Sharma [40]. Closed solutions for a number of cases were developed [40]. In what follows, only the more important case of first order kinetics with respect to A will be considered. For model <43> with consideration of gas flow and pressure variations the governing balance equation follows from eq. (84) by setting $\overline{y}_A$ = o and St_G = St_{GR}

$$- \frac{1 - x_{Ao}}{\overline{x}_A \, (1 - x_{Ao} \, \overline{x}_A)^2} \, d \, \overline{x}_A = St_{GR} \frac{\beta}{1 + \alpha} \, dz \tag{97}$$

where

$$St_{GR} = K_{AR} a \, \frac{L}{u_{Go}} \, \frac{RT}{He} \; . \tag{98}$$

Here K_{AR} is a generalized absorption-reaction parameter. Appropriate expressions for K_{AR} are given in Table 3. Integration of eq. (97) and introducing the conversion gives

$$x_{Ao} \, X - (1 - x_{Ao}) \ln(1 - X) = \frac{1 + o.5 \, \alpha}{1 + \alpha} \, St_{GR} \; . \tag{99}$$

This simple relation can be used for design calculations.

In model <42> gas phase dispersion is taken into account. If isobaric conditions ($\alpha \to o$) are assumed and gas flow variations are neglected ($x_{Ao} \to o$; equimolar absorption and desorption, etc.), the balance equation is

$$\frac{1}{Bo_G} \frac{d^2 \, \overline{y}_A}{dz^2} - \frac{d \, \overline{y}_A}{dz} - St_{GR} \, \overline{y}_A = o \; . \tag{100}$$

Table 3: Relations for K_{AR} in eq. (98). First order reaction in A. $M = Da\ k_{1n}\ c_B^n/k_L^2$

K_{AR}, cm/s	Criteria for application
k_L or $(\frac{1}{k_L} + \frac{1}{k_G\ He})^{-1}$	$\sqrt{M} < 1$, $k_L a \ll \varepsilon_L k_{1n}\ c_B^n$
$k_L \sqrt{1 + M}$	$\sqrt{M} \simeq 1$, $k_G\ He \gg k_L \sqrt{1 + M}$
$(\frac{1}{k_G\ He} + \frac{1}{k_L \sqrt{1 + M}})^{-1}$	$\sqrt{M} \simeq 1$, $k_G\ He \simeq k_L \sqrt{1 + M}$
$k_L \quad M$	$\sqrt{M} \gg 1$, $k_G\ He \gg k_L \sqrt{M}$
$(\frac{1}{k_G\ He} + \frac{1}{k_L \sqrt{M}})^{-1}$	$\sqrt{M} \gg 1$, $k_G\ He \simeq k_L \sqrt{M}$

In addition: $\sqrt{M} \ll (c_B/\nu c^*_{AL}) \sqrt{D_B/D_A}$

This corresponds to the one-phase ADM, the solution of which for $z = 1$ is given by

$$\bar{y}_A = \frac{4\ q\ \exp(Bo_G/2)}{(1 + q)^2 \exp(q\ Bo_G/2) - (1 - q)^2 \exp(-q\ Bo_G/2)} \tag{101}$$

with

$$q = \sqrt{1 + 4\ St_{GR}/Bo_G} \quad . \tag{102}$$

If $Bo_G \geq 2$, the second term in the denominator of eq. (101) can be neglected with an error less than 10^{-3} in $\bar{y}_A$. Under this condition, an explicite equation for the reactor length can be derived from eq. (101)

$$L = \frac{2E_G\ \varepsilon_G}{u_G\ (1 - q)} \ln\left[\frac{1 - X}{4q}(1 + q)^2\right] \ . \tag{103}$$

q does not involve L as

$$St_{GR}/Bo_G = K_{AR}a \frac{\varepsilon_G E_G}{u_G^2} \frac{RT}{He} . \tag{104}$$

Example 2: Calculation of Absorber Height

Carbon dioxide has to be scrubbed from synthesis gas into carbonate-bicarbonate buffer solution using arsenite as a catalyst. Assume the subsequent data apply:

Diameter of unstaged BCR, m	1.0
Temperature, K	333
Pressure, MPa	5.06
Gas velocity, cm/s	5.5
Gas hold-up ε_G	0.17
Mass transfer coefficient $k_L a$, cm/s	0.04
Enhancement factor E	3.2
Inlet mol fraction of CO_2, y_0	0.038
Henry's constant, MPa l/mol	20
Gas phase dispersion coefficient, cm^2/s	1.7×10^4

For the sake of simplicity eq. (103) is used to calculate the height of the chemical absorber. From the given data it follows

$$\frac{St_{GR}}{Bo_G} = \frac{k_L a\ E}{u_G} \frac{RT}{He} \frac{\varepsilon_G E_G}{u_G} = 1.69$$

and q = 2.787 and thus from eq. (103) L = 1206 cm. Calculation of the gas phase Bodenstein number gives $Bo_G = 2.3$. Hence applicability of eq. (103) is justified and the absorber height will be 12 m.

For the more general case with variable gas velocity due to the counteracting effects of absorption and expansion and with consideration of a fast reaction of order m in A and n in B the governing balance equation of the gas phase is given by

$$\varepsilon_G E_G \frac{d}{dx} (c_G \frac{dx_A}{dx}) - \frac{d}{dx}(u_G c_A) - K_{AR}a\ c_{AL}^{*m'} = 0 \tag{105}$$

with

$$m' = (m + 1)/2 \tag{106}$$

$$K_{AR} = \sqrt{\frac{2}{m+1} D_A \, k_{mn} \, c_B^n} \quad . \tag{107}$$

With the dimensionless quantities used in model <22> and <23> eq. (105) can be rearranged to

$$\frac{1}{Bo_G} \frac{d^2 \bar{x}_A}{dz^2} - \left(\frac{\alpha}{Bo_G \, \beta} + \bar{u}_G\right) \frac{d \bar{x}_A}{dz} - \bar{x}_A \frac{d \bar{u}_G}{dz} + \frac{\alpha}{\beta} \bar{u}_G \bar{x}_A$$

$$- St^*_{GR} \, (\beta \, x_{Ao})^{m'-1} \, \bar{x}_A^{\,m'} = o \tag{108}$$

where

$$St^*_{GR} = K_{AR} a \frac{L}{u_{Go}} \frac{RT}{He} (P_T/He)^{m'-1} \quad . \tag{109}$$

An equation for $\bar{u}_G$ is obtained again from the entire balance giving in dimensionless form

$$\frac{d \bar{u}_G}{dz} = \frac{\alpha}{\beta} \bar{u}_G - St^*_{GR} \, \beta^{m'-1} \, x_{Ao}^{\;m'} \, \bar{x}_A^{\,m'} \quad . \tag{110}$$

With consideration of eq. (110) the component balance, eq. (108), can be simplified to

$$\frac{1}{Bo_G} \frac{d^2 \bar{x}_A}{dz^2} - \left(\frac{\alpha}{\beta \, Bo_G} + \bar{u}_G\right) \frac{d \bar{x}_A}{dz} - St^*_{GR} \, (\beta \, x_{Ao})^{m'-1} \, \bar{x}_A^{\,m'} (1 - x_{Ao} \bar{x}_A) = o . \tag{111}$$

The coupled eqns. (110) and (111) must be solved numerically with consideration of the proper boundary conditions.

When discussing model <22> it has been pointed out that gas phase dispersion largely influences BCR performance. This conclusion is also valid for the fast reaction regime. For the important case of $m' = m = 1$ design charts similar to that developed by Levenspiel and Bischoff can be constructed in order to estimate the effect of Bo_G, St_{GR}, x_{Ao}, and α on the conversion [6]. An example of such a design diagram is shown in Fig. 13 for $x_{Ao} = 0.6$. The diagram can particularly be used to estimate the additional volume demand caused by increasing dispersion as the ratio of Stanton numbers St_{GR} (ADM) (used in model <42>) and St_{GR} (PFM) (used in model <43>) corresponds with the volume ratio. Diagrams as shown in Fig. 13 are particularly useful as the Stanton number of model <43>

(without gas phase dispersion) required for a certain conversion can simply be calculated from eq. (99).

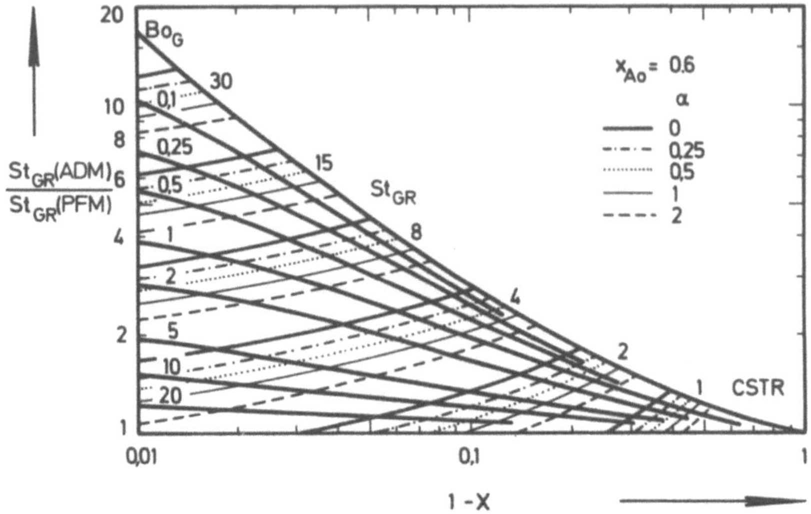

Fig. 13: Effect of Bo_G, St_{GR}, and α on conversion and volume ratio (St_{GR} (ADM)/St_{GR} (PFM)) for x_{Ao} = 0.6 and fast first order reaction [6]

Example 3: Design of an Isobutene Absorber

Isobutene is separated from C_4 crack fractions by absorption and reaction in sulfuric acid solutions [34]:

$$\ce{C4H8} + H_2O \xrightarrow[H_2SO_4]{k_1} \ce{C4H9OH}$$

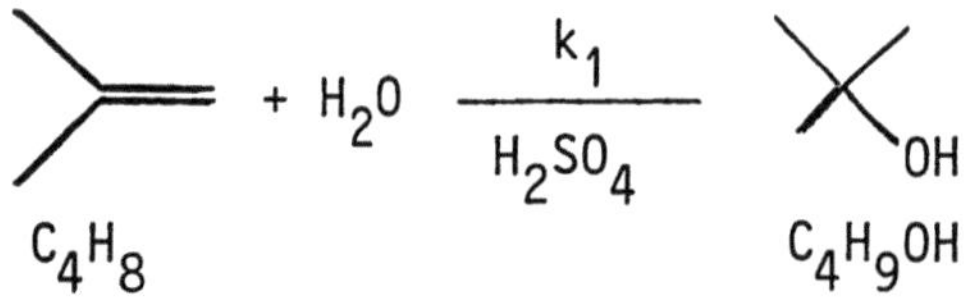

Calculate the length of a 1.2 m diameter BCR required for 95 per cent conversion. Assume that the following data apply:

Gas velocity u_{Go}, cm/s	5
Gas hold-up ε_G	0.23
Interfacial area a, cm^{-1}	12
Pressure at BCR top P_T, kPa	200
Inlet mol fraction x_{Ao}	0.6
Reactor diameter d_R, m	1.2
Density of liquid P, g/cm^3	1.26
Henry's constant, kPa cm^3/mol	2×10^7
Mass transfer coefficient k_L, cm/s	0.01

Rate constant k_1, s^{-1}	1.34
Temperature T, K	303
Diffusivity D, cm^2/s	3×10^{-6}

First, it has to be checked in which absorption-reaction regime this process takes place. The enhancement factor for first order reaction gives

$$E = (1 + k_1 D/k_L^2)^{1/2} = 1.023 \ .$$

Hence absorption enhancement by fast chemical reaction is negligible. However, the criteria for the diffusional regime [19, 20]

$$\frac{k_L^2 \ (k_1 + 1/\tau_L)}{k_1 D \ (k_1 + k_L a + 1/\tau_L)} \gg 1 \gg \frac{k_L a}{k_L a + k_1 + 1/\tau_L}$$

gives for the present reaction system with consideration of $1/\tau_L \ll k_1, k_L a$

$$18.6 \gg 1 \gg 0.104 \ .$$

Therefore, the process occurs in the diffusional regime and a liquid phase has not to be considered. Due to the large diameter of the BCR dispersion in gas phase cannot be neglected. In addition, gas flow variations must be considered. Hence, the mass balances for this particular example are given by eq. (110) and (111). Instead of solving the nonlinear boundary value problem eq. (99) is used to calculate firstly the height of the isobutene absorber with neglect of gas phase dispersion. With this value known the actual required length can be estimated by making use of the design chart, i.e. Fig. 13.

For the present data the LHS of eq. (99) gives 1.768, thus it follows

$$1.768 = \frac{1 + 0.5\ \alpha}{1 + \alpha}\ St_{GR} \ .$$

α and St_{GR} are functions of the reactor length. Introducing

$$\alpha^* = \alpha/L = \rho_L \ g\varepsilon_L/P_T$$

$$St_{GR}^* = St_{GR}/L = k_L a \ \frac{RT}{He \ u_{Go}}$$

gives the following quadratic equation in L:

$$L^2 + L\,\frac{2(St_{GR} - 1.768\alpha^*)}{St^*_{GR}\ \alpha^*} - \frac{2 \times 1.768}{St^*_{GR}\ \alpha^*} = 0$$

with $\alpha^* = 4.75 \times 10^{-4}$ cm^{-1} and $St^*_{GR} = 3.02 \times 10^{-3}$ cm^{-1} the height of the BCR with plug flow in the gas phase follows as

$$L\ (PFM) = 665\ \text{cm}\ .$$

The gas phase dispersion coefficient E_G is estimated from the correlation of Mangartz and Pilhofer [5]

$$E_G = 5 \times 10^{-4} \left(\frac{u_G}{\varepsilon_G}\right)^3 d_R^{1.5} = 6752\ \text{cm}^2/\text{s}\ .$$

Assuming a height of 10 m for the axial dispersed reactor a Bodenstein number can be estimated

$$Bo_G = \frac{u_{Go}\ L}{\varepsilon_G\ E_G} = \frac{5 \times 1000}{0.23 \times 6752} = 3.2$$

For this value of Bo_G and $x_{Ao} = 0.6$ the design chart gives for $\alpha \approx 0.5$ at $X = 0.95$ a value of $St_{GR}(ADM)/St_{GR}(PFM)$ of about 1.8. Hence, the length of the BCR with consideration of gas phase dispersion is

$$L(ADM) = L(PFM) \times 1.8 = 1190\ \text{cm}.$$

5. REACTOR MODELS VERSUS HYDRODYNAMIC MODELS

In recent years, new hydrodynamic models for the churn-turbulent flow regime in BCR have been developed [13 - 17]. Though these models use different approaches, they lead to equations describing liquid velocity profiles, slip velocities, circulation flow, etc. The ADM and its various modifications are generally thought to be a rather realistic description of BCR. However, the ADM presupposes the existence of turbulent eddies which are small in comparison to characteristic dimensions of the reactor. In view of the conclusions drawn from hydrodynamic BCR modelling one has seriously to question, whether

- the ADM which involves only one global parameter (E) can really be applied to comprehend the entire liquid circulation flow pattern of churn-turbulent flow in BCR
- the ADM is appropriate to describe the gas phase residence time distribution which results from the occurrence of bubble classes with different sizes and rise velocities.

The existence of different bubble classes in churn-turbulent flow can be shown easily by dynamic gas disengagement measurements. Simplified evaluation of such measurements yields a splitting of the bubble phase into two bubble classes, i.e., one class of small bubbles with low rise velocity and another class of large bubbles with high rise velocity. Two bubble classes with such properties were also the starting point of a BCR model proposed by Joseph and Shah [41]. Experimental techniques and devices are now available to measure the entire bubble spectrum with regard to size and rise velocity. This information could form the basis of more realistic models of the gas phase flow in BCR.

With regard to liquid flow the results of the more sophisticated models have not yet been fully incorporated in BCR modelling. It is particularly believed that the circulation cell model derived by Joshi and Sharma [15] on the basis of the energy balance method will stimulate BCR modelling. Cell models are very flexible and can easily be adapted to various flow patterns. Two examples are given in Fig. 14 and 15.

6. SOME PRACTICAL DESIGN ASPECTS AND SCALEUP RULES

In this last section an attempt will be made to derive some rules of thumb which are helpful when operating and designing BCR. It will be demonstrated that on the basis of very simple considerations and models relations between important operating variables and useful conclusions can be obtained.

6.1 Relations between conversion and space-time-yield

Let us consider a reaction of gaseous reactant A with liquid phase reactant B: $A + \nu B \rightarrow C$ which follows first order kinetics in A and B

$$r_A = \frac{1}{\nu} r_B = k_2 \, c_A \, c_B \, . \tag{112}$$

Assuming (1) complete mixing in the gas and liquid phase, (2) constant gas flow rate, (3) absorption with slow reaction and (4) neglecting convective transport of A ($u_L \, (c_{ALo} - c_{AL}) \rightarrow 0$) the following relation between the conversion of A and B can be derived

$$\frac{X_A}{1 - X_A} \, \frac{He}{RT \, St} = \frac{1}{1 + \frac{St/Da}{1 - X_B}} \tag{113}$$

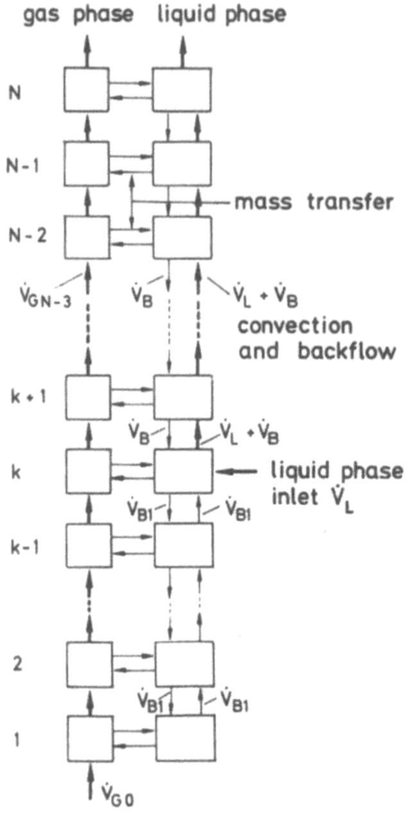

Fig. 14: Back flow cell model of BCR with arbitrary liquid inlet [36]

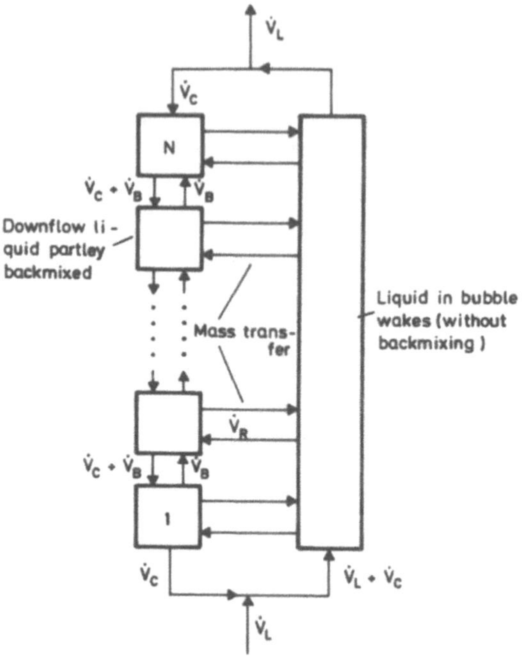

Fig. 15: Model of the liquid phase with zones of different flow

with

$$St = k_L a\ V_R/\dot V_G \tag{114}$$

$$Da = k_2\ c_{Bo}(1 - \varepsilon_G)\ V_R/\dot V_G\ . \tag{115}$$

Eq. (113) shows that X_A and X_B behave in opposite ways, e.g., if $X_B \to 0$, then X_A has its maximum value, namely

$$\frac{X_{A\ max}}{1 - X_{A\ max}} = St\ \frac{RT}{He}\ \frac{Da}{Da + St}\ . \tag{116}$$

If, in addition, Da ≫ St, the diffusional regime as the limiting case of the slow reaction regime is obtained

$$X_{A\,max} = \frac{St\ RT/He}{1 + St\ RT/He} \;. \tag{117}$$

On the other hand, if $X_A \to 0$, it follows from eq. (113) that $X_B = X_{B\,max} = 1$ which, however, can only be valid if the liquid phase residence time is high enough. Consideration of the balance on B

$$\dot{V}_L(c_{Bo} - c_B) = \nu\, k_2\, c_B\, c_{AL}\, (1 - \varepsilon_G)\, V_R \tag{118}$$

leads to

$$\frac{X_B}{1 - X_B} = Da_B \left[1 - X_A \left(1 + \frac{He}{St\ RT}\right)\right] \tag{119}$$

with

$$Da_B = \nu\, k_2\, (1 - \varepsilon_G) c^*_{ALo}\, \tau_L \;. \tag{120}$$

Now, if $X_A \to 0$, it follows

$$X_B = X_{B\,max} = \frac{Da_B}{1 + Da_B} \;. \tag{121}$$

Therefore, high conversions of A can only be realized in the diffusional regime, i.e., if Da and c_{Bo}, respectively, are large. On the other hand, high conversions of B can be obtained if the saturation concentration of A in the liquid phase is high, i.e., if Da_B is large. Relatively high conversions of A and B can be realized simultaneously only in large reactor volumes, if both St, Da and Da_B are large. However, this is at the expense of the space-time-yield which will be very low. The specific absorption rate which is equivalent to reactor performance and space-time-yields is given by

$$R_A a = k_L a\, (c^*_{AL} - c_{AL}) = k_2\, (1 - \varepsilon_G)\, c_{AL}\, c_B \;, \tag{122}$$

and can be referred to the maximum value of $R_A a$, i.e.,

$$R_A a_{max} = k_2\, (1 - \varepsilon_G)\, c^*_{ALo}\, c_{Bo} \;. \tag{123}$$

Then it can be shown that the relative space-time-yield is given by

$$\psi = \frac{R_A a}{R_A a_{max}} = \frac{(1 - X_A)(1 - X_B)}{1 + (Da/St)(1 - X_B)} . \tag{124}$$

The maximum value $\psi = 1$ can only be reached if $St \gg Da$ and $X_A = X_B = 0$. For arbitrary values of St and Da a maximum value $\psi = \psi_{max}$ follows, if $X_A = X_B = 0$. Fig. 16 is a plot of X_A vs. X_B for various values of $\psi < \psi_{max}$ and Da = St. It can be seen that high conversions of A and B can only be obtained at low space-time-yields. For a given space-time-yield, high conversions of B can only be achieved on the expense of low conversions of A and vice versa. The same conclusions are obtained if the treatment given here is extended to the absorption regime with fast reaction [6].

6.2 Optimum reactor performance in the slow reaction regime

By elimination of c_{AL} from eq. (122) the following relation can be derived

$$R_A a = \frac{k_2 c_B (1 - \varepsilon_G) k_L a}{k_2 c_B (1 - \varepsilon_G) + k_L a} c^*_{AL} \tag{125}$$

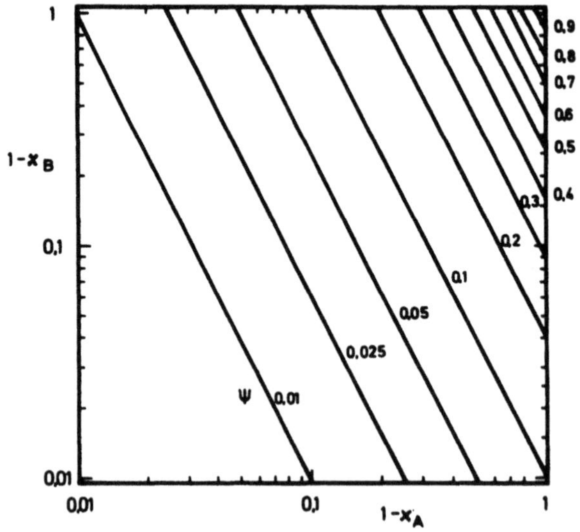

Fig. 16: Relation between X_A, X_B, and relative space-time-yield ψ. [6]

which shows that for given values of $k_2\ c_B$, k_L, and c^*_{AL} the specific absorption rate depends only on a and ε_G. Both of these quantities are solely functions of the gas velocity for a given set-up and gas/liquid system. Increasing u_G increases a and hence the mass transfer rate, while the effective reaction volume $(1 - \varepsilon_G)$ decreases. Therefore, $R_A a$ passes through a maximum value as function of u_G. Assuming the following dependencies

$$\varepsilon_G = \varepsilon^*_G\ u_G^m \tag{125a}$$

$$a = a^*\ u_G^n \tag{125b}$$

where ε^*_G and a^* are the value of ε_G and a at u_G = 1 cm/s, eq. (126) can be derived for the optimum gas velocity which corresponds to the maximum value of $R_A a$

$$u^{m+n}_{G,opt}\ \frac{k_L a^*\ m\ \varepsilon_G{}^*}{k_2\ c_B\ n} - u^{2m}_{G,opt}\ \varepsilon^{*2}_G + 2\ u^m_{G,opt}\ \varepsilon^*_G - 1 = 0\ . \tag{126}$$

As a rough approximation it can often be assumed that m = n = 1 . Then eq. (126) reduces to a quadratic equation. Solution of this equation and using only the negative root leads to the following simple relation

$$u_{G,opt} = (1/\varepsilon^*_G)\left(1 + \sqrt{\frac{k_L a^*}{k_2\ c_B\ \varepsilon^*_G}}\right)^{-1} \tag{127}$$

In Fig. 17 the dependence of $R_A a$ on u_G is shown for few values of $k_2\ c_B$. It can be seen that the maximum is very flat, hence, in this range $R_A a$ is not sensitive to u_G. As high gas velocities imply low conversions of gaseous reactant, operation of the system at lower gas flow rates is more favorable. From the curves of Fig. 17 one can discern that 90 % of the maximum space-time-yield can be reached at only half the optimum gas velocity ($u_G = 0.5\ u_{G,opt}$). However, this conclusion cannot be generalized (see Fig. 12). It is, therefore, recommended that each individual reaction system be analyzed separately.

Example 4: Optimum Gas Velocity for Fischer-Tropsch Slurry Reactor

The Fischer-Tropsch synthesis (FTS) can be carried out on Fe catalysts suspended in inert hydrocarbons. This process mode offers a number of advantages. The stoichiometry of the FTS can be simplified by

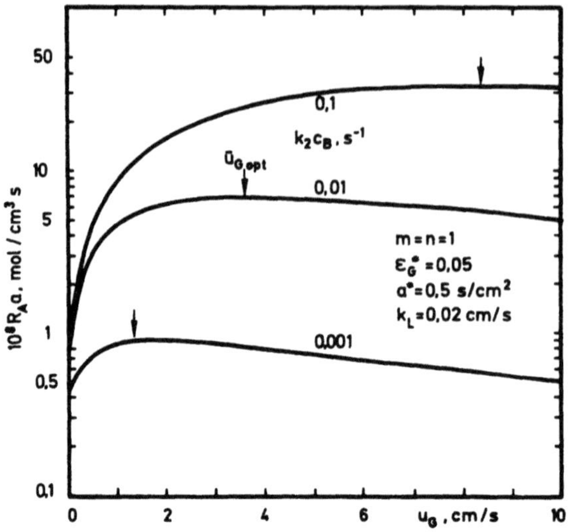

Fig. 17: Specific absorption rate as function of gas velocity

$$CO + (1 + \frac{n}{2})\ H_2 \xrightarrow{Fe} CH_n + H_2O$$

where CH_n presents a mixture of various hydrocarbons. Under certain conditions it is possible to approximate the kinetics of the FTS by a first rate law in H_2. Estimate the gas velocity which maximizes the space-time-yield ($STY = R_A a$) for a conversion of about 90 per cent. Assume that model <11> and the following data apply [30, 38]:

$k_1 = 0.512\ s^{-1}$
$\varepsilon_G = 0.053\ u_G^{1.1} = a_1\ u_G^{1.1}$
$k_L a = 0.045\ u_G^{1.1} = a_2\ u_G^{1.1}$
$\zeta = -0.5$ (volume contraction factor)
(u_G in cm/s, $k_L a$ in s^{-1})

With this data eq. (126) can be written as

$$u_{G,opt}^{2.2}\ (\frac{a_1 a_2}{k_1} - a_1^2) + u_{G,opt}^{1.1}\ 2a_1 - 1 = 0\ .$$

Solution gives

$$u_{G,opt}^{1.1} = \frac{1}{a_1\left(1 + \sqrt{\frac{a_2}{k_1 a_1}}\right)} = 8.25\ (\mathrm{cm/s})^{1.1}$$

$$u_{G,opt} = 6.81\ \mathrm{cm/s}.$$

This presents an average gas velocity which is defined by

$$\bar{u}_G = u_{G,opt} = u_{Go}\ (1 + 0.5\ \zeta\ X_{CO+H_2})$$

with $\zeta = -0.5$ and $X_{CO+H_2} = 0.9$ one obtains for optimum inlet velocity of synthesis gas ($CO+H_2$)

$$u_{Go,opt} = 8.8\ \mathrm{cm/s}.$$

By comparison with Fig. 12 it can be seen that the estimate on the basis of model <11> agrees well with calculations based on the more sophisticated model <22>. It should also be mentioned that the FTS was operated in the Rheinpreussen-Koppers demonstration plant under optimal conditions with an inlet gas velocity of $u_{Go} = 9.5$ cm/s.

6.3 Scaleup rules

As outlined in the introduction, there is no doubt that modelling and, hence, scaling up of BCR should be based on mathematical models. It should be pointed out, however, that the computational efforts can only be justified if the fluiddynamic parameters, the heat and mass transfer properties, kinetic data etc. are known with sufficient accuracy. This is often not the case and their determination may be tedious and time-consuming. Therefore, in industrial practice another approach must be applied to quickly fulfill the scaleup task without knowing the above details. A common procedure is to measure space-time-yields and conversion in laboratory and bench scale unit and, on the basis of such data, to scale up to the final design. When using this approach, one must be aware of those quantities and phenomena which are strongly dependent on the scale of the reactor. A summary of guidelines is given in Table 4. In addition, both experience and fortune may contribute to solving the design task safely and economically. Often the optimum conditions are only found during longer operation of the BCR.

Table 4: Influences on scaling up of BCR

Quantity effected by geometrical size of BCR	Absorption regime with slow reaction	Absorption regime with fast reaction
Interfacial area and gas holdup (bubble coalescence, generation of large bubbles and bubble slugs)	Important, but both $k_L a$ and r_A determine $R_A a$, search for u_G, which gives maximum value of $R_A a$	Very important; $R_A a$ is proportional to a
Gas phase dispersion	Important at high absorption rates and high conversion of gaseous reactant, respectively.	Very important, comparable with one-phase reactors
Liquid phase dispersion	Important for conversion of B, little influence on conversion of A	No pronounced influence on conversion of A, little effect on B conversion
Catalyst settling and dispersion	Large effects are possible, dependent on d_{cat} and ρ_{cat}	Only possible if $d_{cat} << \delta_{eff}$, in this case catalyst is uniformly distributed
Flow regime (change of flow regime: bubbly ↔ churn-turbulent ↔ slugs)	Little effect, not very pronounced with respect to $R_A a$	Drastic effects

7. CONCLUSIONS

In this paper a reasonable classification of BCR models has been given. As criteria for such a classification the mixing behavior of the phases and the absorption-reaction regime has been used. The mathematical efforts to solve the balance equations of the various models differ largely. However, this does not present a serious problem in view of the available numerical solution methods and computational techniques. Versatile analytical solutions for a variety of rate laws could be developed for the fast and slow reaction regime provided complete liquid phase mixing can be presupposed in the last case.

BCR of industrial size should be designed on the basis of model <22> and <42>, respectively, under consideration of pressure effects and gas flow variations. It is, however, also shown that it is often sufficient to apply simpler models in order to obtain significant conclusions for many practical purposes. One should always keep in mind that the mathematical effort involved in solving the model equations is only worthwhile, if all the hydrodynamic parameters, the mixing and mass transfer properties and the kinetic and solubility data as well are known with sufficient accuracy. In this regard parameter studies on the basis of various models can be very helpful.

There is no doubt that more sophisticated models than those presented here can be constructed. However, it is believed that theory is advanced far enough. What is really needed are careful experimental studies on chemically reactive systems of industrial significance.

Nomenclature

a	specific gas/liquid interfacial area, cm^{-1}
a_S	specific liquid/solid interfacial area, cm^{-1}
Bo_G	gas phase Bodenstein number, $u_G L/\varepsilon_G E_G$
Bo_L	liquid phase Bodenstein number, $u_L L/\varepsilon_L E_L$
c_A	concentration of gaseous reactant in gas phase, mol/cm^3
c_{AL}	concentration of gaseous reactant in liquid phase, mol/cm^3
c^*_{AL}	equilibrium concentration of gaseous reactant in liquid phase, mol/cm^3
$\hat{c}_A$	concentration variable defined by eq. (18)
Da	Damköhler number defined by eq. (29)
E	axial dispersion coefficient, cm^2/s
g	gravitational acceleration, cm/s^2
He	Henry's constant, p_A/c^*_{AL}, Pa cm^3/mol
k_L	liquid side mass transfer coefficient, cm/s
k_n	n-th order reaction rate constant, $s^{-1}(cm^3/mol)^{n-1}$
k_s	liquid/solid mass transfer coefficient, cm/s
K_L	overall liquid mass transfer coefficient, cm/s
K_M	Michaelis-Menten constant
L	reactor length, cm
m	reaction order with regard to A
n	reaction order with regard to B
p	partial pressure, Pa
P	total pressure, Pa
P_T	pressure at BCR top, Pa
Q	parameter defined by eq. (13), (16), or (25), respectively
r	reaction rate, $mol/(cm^3\ s)$
R	gas constant, Pa $cm^3/(mol\ K)$
$R_A a$	specific absorption rate, space-time-yield, $mol/(cm^3\ s)$
St	Stanton number defined by eq. (5)

St_G	Stanton number defined by eq. (82)
St_{GR}	Stanton number defined by eq. (98)
St_L	Stanton number defined by eq. (81)
t	batch reaction time, s
T	temperature, K
u	linear velocity, cm/s
$\underline{u}_G$	dimensionless linear gas velocity, u_G/u_{Go}
$\dot{V}$	volume throughput, cm^3/s
V_M	molar volume, cm^3/mol
V_R	reactor volume, cm^3
x	axial coordinate, cm
x_A	gas phase mole fraction
$\underline{x}_A$	normalized mole fraction, x_A/x_{Ao}
X_A	conversion
$\underline{y}_A$	dimensionless liquid phase concentration, defined by eq. (83)
z	dimensionless axial coordinate, x/L
α	pressure parameter defined by eq. (77)
β	parameter defined by eq. (78)
γ	parameter, RT/He
ε	phase holdup
ζ	contraction-expansion factor defined by eq. (56)
μ	specific growth rate, $g/(cm^3\ s)$
ν	stoichiometric coefficient
ρ	density, g/cm^3
ψ	relative space-time-yield defined by eq. (124)
τ	average residence time, s

Indices

A	gaseous key reactant
B	liquid key reactant
G	gas phase
L	liquid phase
m	mixed reactor
p	plug flow reactor
0	reactor inlet
1	reactor outlet

References

1. Hines, D.A. The Large Scale Pressure Cycle Fermenter Configuration. Dechema Monographs 82 (1978) 55-64.
2. Joshi, J.B. and M.M. Sharma. Mass Transfer Characteristics of Horizontal Sparged Contactors. Trans. Instn. Chem. Engrs 54 (1976) 42-53.
3. Herbrechtsmeier, P. and R. Steiner. Untersuchungen an einem Blasensäulen-Abstromreaktor. Chem. Ing. Tech. 50 (1978) 944-947.
4. Herbrechtsmeier, P., H. Schäfer and R. Steiner. Gas Absorption in Downflow Bubble Columns for the Ozone-Water System. Ger. Chem. Eng. 4 (1981) 258-264.
5. Shah, Y.T., B.G. Kelkar, S.P. Godbole and W.-D. Deckwer. Design Parameters Estimations for Bubble Column Reactors. AICHE J. 28 (1982) 353-379.
6. Deckwer, W.-D. Reaktionstechnik in Blasensäulen (Frankfurt a.M., Verlag Sauerländer und Salle, 1985).
7. Chaudhari, R.V. and P.A. Ramachandran. Three-Phase Slurry Reactors. AICHE J. 26 (1980) 177-201.
8. Chaudhari, R.V. and P.A. Ramachandran. Three-Phase Catalytic Reactors (New York, Gordon and Breach Science Publ., 1983).
9. Tarmy, B., M. Chang, C. Coulaloglou and P. Ponzi. The Three-Phase Hydrodynamic Characteristics of the EDS Coal Liquefaction Reactors: Their Development and Use in Reactor Scale up. Proc. ISCRE 8. Inst. Chem. Engrs. Symp. Ser. No. 87 (1984) 303-317.
10. Lohse, M., E. Alper and W.-D. Deckwer. Modeling of Batch Chlorination of Toluene in Bubble Column. Chem. Eng. Sci. 38 (1938) 1399-1409.
11. Calderbank, P.H. and M. Moo-Young. The Continuous Phase Heat and Mass Transfer Properties of Dispersions. Chem. Eng. Sci. 16 (1961) 51-64
12. Akita, K. and F. Yoshida. Gas Holdup and Volumetric Mass Transfer Coefficient in Bubble Columns. Ind. Eng. Chem. Proc. Des. Dev. 12 (1973) 76-80.
13. Ueyama, K. and T. Miyauchi. Properties of Recirculating Turbulent Two-Phase Flow in Gas Bubble Columns. AICHE J. 25 (1979) 258-266.
14. Kojima, E. et al. Liquid Phase Velocity in a 5.5 m Diameter Bubble Column. J. Chem. Eng. Japan 13 (1980) 16-21.
15. Joshi, J.B. and M.M. Sharma. A Circulation Cell Model for Bubble Columns. Trans. Instn. Chem. Engrs. 57 (1978) 244-251.
16. Zehner, P. Impuls-, Stoff- und Wärmetransport in Blasensäulen. Chem. Ing. Tech. 54 (1982) 248-251.
17. Shah, Y.T. and W.-D. Deckwer. Handbook of Fluids in Motion. (Eds. N.P. Cheremisinoff and R. Gupta, Ann Arbor, Ann Arbor Science Publ., 1983).
18. Deckwer, W.-D. and A. Schumpe. Preprints Technik der Gas/flüssig und der Dreiphasen-Strömung (Düsseldorf, GVC-VDI-Gesellschaft, 1984).

19. Danckwerts, P.V. Gas-Liquid Reactions. New York, McGraw-Hill, 1970).
20. Doraiswamy, L.K. and M.M. Sharma. Heterogeneous Reactions. Vol. 2. Fluid-Fluid Reactions (New York, J. Wiley & Sons, 1984).
21. Alper, E., B. Wichtendahl and W.-D. Deckwer. Gas Absorption Mechanism for Catalytic Slurry Reactors. Chem. Eng. Sci. 35 (1980) 217-222.
22. Pal, S.K., M.M. Sharma and V.A. Juvekar. Fast Reactions in Slurry Reactors: Catalyst Particle Size smaller than Film Thickness: Oxidation of aqueous Sodium Sulphide Solutions with activated Carbon as Catalyst at elevated Temperatures. Chem. Eng. Sci. 37 (1982) 327-336.
23. Satterfield, C.N. and G.A. Huff. Effects of Mass Transfer on Fischer-Tropsch Synthesis in Slurry Reactors. Chem. Eng. Sci. 35 (1980) 195-202.
24. Bukur, D. Some Comments on Models for Fischer-Tropsch Reaction in Slurry Bubble Column Reactors. Chem. Eng. Sci. 38 (1983) 441-446.
25. Deckwer, W.-D., K. Nguyen-tien, B. Godbole and Y.T. Shah. On the Applicability of the Axial Dispersion Model in Bubble Columns. AICHE J. 29 (1983) 915-922.
26. Miyauchi, T. and T. Vermeulen. Longitudinal Dispersion in Two-Phase Continuous-Flow Operations. Ind. Eng. Chem. Fundam. 2 (1963) 113-126.
27. Miyauchi, T. and T. Vermeulen. Diffusion and Back-Flow Models for Two-Phase Axial Dispersion. Ind. Eng. Chem. Fundam. 2 (1963) 304-310.
28. Hartland, S. and J.C. Mecklenburgh. A comparison of differential and stage-wise countercurrent extraction with backmixing. Chem. Eng. Sci. 21 (1966) 1209-1221.
29. Langemann, H. Axiale Diffusion und Durchmischung in stationären Zweiphasen-Gegenstrom-Reaktoren. Chem. Ztg.-Chem. App. 92 (1968) 391-406
30. Langemann, H. Stationäre Konzentrationsverteilungen in isothermen Gas-Flüssigphase-Gegen-und-Gleichstrom-Reaktoren mit linearem Druckgefälle. Chem. Ztg.-Chem. App. 92 (1968) 845-859.
31. Cichy, P.T., J.S. Ultman and T.W.F. Russel. Two-Phase Reactor Design Tubular Reactors - Reactor Model Development 61 (1969) (No. 8) 6-26.
32. Deckwer, W.-D., K. Ngyen-tien, A. Schumpe and S. Serpemen. Oxygen Mass Transfer into Aerated CMC Solutions in a Bubble Column. Biotechn. Bioeng. 24 (1982) 461-481.
33. Deckwer, W.-D. Non-Isobaric Bubble Columns with Variable Gas Velocity Chem. Eng. Sci. 31 (1976) 309-317.
34. Deckwer, W.-D. Absorption and Reaction of Isobutene in Sulfuric Acid III. Considerations on the Scale up of Bubble Columns. Chem. Eng. Sci. 32 (1977) 51-57.
35. Deckwer, W.-D., I. Adler and A. Zaidi. A Comprehensive Study on CO_2-Interphase Mass Transfer in Vertical Cocurrent and Countercurrent Gas-Liquid Flow. Can. J. Chem. Eng. 56 (1978) 43-55.

36. Deckwer, W.-D., J. Hallensleben and M. Popovic. Exclusion of Gas Sparger Influence on Mass Transfer in Bubble Columns. Can. J. Chem. Eng. 58 (1980) 190-197.
37. Deckwer, W.-D., Y. Serpemen, M. Ralek and B. Schmidt. Modeling the Fischer-Tropsch Synthesis in Slurry Phase. Ind. Eng. Chem. Proc. Des. Dev. 21 (1982) 222-241.
38. Kuo, J.C.W. Slurry Fischer-Tropsch/MOBIL Two-Stage Process of Converting Syngas to High Octane Gasoline. Final DOE-Report Contract No. DE-AC22-80PC 300 22 (Mobil Res. & Dev. Corp., 1983).
39. Mashkar, R.D. Effect of Backmixing on the Performance of Bubble Column Reactors. Chem. Eng. Sci. 29 (1974) 897-905.
40. Juvekar, V.A. and M.M. Sharma. Some Aspects of Process Design of Gas/Liquid Reactors. Trans. Instn. Chem. Engrs. 55 (1977) 77-91.
41. Joseph, S. and Y.T. Shah. A Two-Bubble Class Model for Churn Turbulent Bubble Column Slurry Reactors. ACS Symp. Series 27 (1984) 149-167.

SLURRY REACTORS, FUNDAMENTALS AND APPLICATIONS

A.A.C.M. Beenackers*) and W.P.M. van Swaaij**)

*) Chemical Engineering Laboratories, University of Groningen, Nijenborgh 16, Groningen, The Netherlands
**) Laboratory of Chemical Engineering, Twente University of Technology, P.O. Box 217, Enschede, The Netherlands

1. INTRODUCTION

Slurry reactors are applied nowadays in a wide range of chemical processes, both on a laboratory scale and in industrial practice. Applications range from catalytic hydrogenation of edible oils, coal hydrogenation, Fischer- Tropsch reactions to biological reactors (e.g. single cell proteins) and polymerization reactors. Doraiswamy and Sharma [1] identified over 50 different processes in which slurry reactors have been used and more applications are to be expected. A lot of experimental work has been devoted to slurry systems. Overviews have been given in textbooks on three phase reactors (see Doraiswamy and Sharma [1] and Shah [2]). Recent review articles are: Chaudhari and Ramachandran [3], Deckwer and Alper [4], Hofmann [5], Shah et al. [6]. Even more information is available on three phase fluidization. This regime is related to slurry reactors and sometimes no distinction is made with slurry reactors. In the present lecture series there is a separate contribution on three phase fluidization and therefore it will not be discussed in this paper. After a discussion of the different properties of slurry reactors as compared to other three phase systems and the different modifications of slurry reactors, hydrodynamic aspects are considered: minimum suspension criteria, hold-up fractions and axial mixing. Then regimes of mass transfer in relation to chemical reactors will be treated in more detail. Subsequently, developments in reactor modelling will be discussed and progress in application of such models to real life systems will be summarized. The paper concludes with prospective new developments and a comprehensive list of conclusions.

2. PROPERTIES OF SLURRY REACTORS

In slurry reactors an attempt is made to realize intensive and intimate contact between a gas phase component, usually to be dissolved in the liquid phase, a liquid phase component and a finely dispersed solid. With respect to this purpose slurry reactors are related to packed bed reactors with the different gas/ liquid flow regimes that can be realised (such as trickle flow, pulsed flow, dispersed bubble flow etc.) Also, there is a lot of similarity with three phase fluid bed systems. These latter systems share many properties with slurry reactors but the main difference is the fact that in fluid beds with upward fluid flows the drag force acting on the solids by the gas and liquid flow is on the average balanced by the net weight of the particles, while in slurry reactors the overall liquid-solid slip velocities are practically zero and particles remain suspended by the action of the turbulence in the liquid phase. This usually implies that somewhat larger particles are applied in three phase fluid beds allowing for a more or less restricted height of the expanded particle bed under the action of the gas and liquid flow only, thus creating a liquid/gas freeboard (in slurry reactors usually d_p < 100-200 μm [7] and no clear freeboard exists). Typical properties of slurry reactors, and for comparison packed bed co-current downflow trickle flow reactors, are summarized in Table I. Most properties indicated for slurry reactors also hold for the three phase fluidized beds. An important difference is related to the somewhat larger particle sizes normally used in three phase fluidization allowing for a simple separation between fluid and particles. The properties indicated in Table I can be advantageous or disadvantageous depending on the application. We will shortly discuss the different properties in the light of examples of industrial applications.

Table I. Typical properties of three phase reactors

Property	Slurry reactors	Co-current downflow trickle flow reactors
macro-mixing liquid	intermediate to ideally mixed	close to plugflow at high liquid rates
macro-mixing gas phase	usually intermediate to ideally mixed	close to plugflow at high liquid rates
solids mixing	usually intermediate to ideally mixed	no mixing
solids replacement	continuous	usually intermitted (difficult)
particle size	small to very small	restricted by pressure drop
pressure drop	high (hydrostatic pressure)	low
wetting	complete	intermediate to complete
radial heat transport	very fast	slow
heat removal/addition	easy	difficult (interstag cooling or cold shot)
danger of plugging	small	can be high
gas/liquid mass transfer rate constant	intermediate	high
liquid solids mass transfer rate constant	very high	high
liquid solids separation	difficult, costly	easy

2.1. Macro-mixing of liquid. Normally this reduces the average liquid phase reactant concentration which results in most cases in a decrease in reaction rate. Sometimes this effect can be beneficial. Satterfield and Huff [8] showed that in Fischer Tropsch synthesis in a slurry reactor, due to the intense mixing, a much higher effective H_2/CO ratio can be present within the liquid phase of the reactor than in the feedstock at a given excess H_2/CO ratio in the feedstock over the usage ratio. This has important advantages in minimizing carbon formation and catalyst deactivation due to CO disproportionation. If overall low axial mixing in the liquid phase is desired cascades of slurry reactors should be applied. On the other hand the degree of mixing in a trickle flow reactor can be increased by applying liquid recycle.
2.2. Macro-mixing of the gas phase is an important problem only if high gas phase conversions are desired and specially if countercurrent operation is aimed for. Examples are absorption of acid gases (CO_2, NO_x, SO_2) in lime slurries. A single stage slurry reactor is not specially suited for these conditions and if a high gas phase conversion is aimed for, multiple stage or slurry tray columns may be a solution if slurry operation is to be selected.
2.3. Solids mixing and solids replacement are important in cases where the solids have a short lifetime. This can be the case if the solid phase is to be converted such as in coal hydrogenation or if the catalyst is rapidly deactivating as in hydrodesulfurization and demetallation of oil residues. For these applications slurry reactors have a definite advantage over trickle flow reactors, specially if the residence time distribution can be reduced, e.g., by putting slurry vessels in series. Then also countercurrent operation is possible. Recent modifications of the trickle flow reactor as applied by Shell involve bunkerflow of the solids allowing for continuous removal and addition of catalyst (see [9], [10]).
2.4. Particle size in slurry reactors can be small to very small, even down to the (sub)micron range. This allows for high particle effectiveness factors even at high reaction rates. Slurry reactors therefore can have high conversion rates per unit slurry volume. Other advantages can be related to the small particle size, such as the large external surface as, e.g., in edible oil hydrogenation, and also in the case of poremouth plugging, as, e.g., in the (auto) catalytic deposition of metals (Ni, V) in the pore mouths of catalysts for desulfurization of residues (see [11]).
2.5. Pressure drop in slurry reactors is usually more or less independent of the gas flow and close to the hydrostatic pressure. Of course, there is also the pressure drop required for the gas distributor. In trickle flow the pressure drop is strongly influenced by the gas flow rate and because of the generally larger particle size the pressure drop will often be lower than in slurry reactors. In three phase reactors where large gas flows are to be processed and where pressure drops are an important cost factor, such as in lime scrubbing of flue gases simple slurry reactors are avoided and special contactors allowing for a low pressure drop are preferred

(see e.g. Coca and Diaz [12]).

2.6. Particle wetting in slurry reactors is always complete while in trickle flow, specially at low specific liquid flow rates, wetting can be uncertain and a problem. This is nearly always a disadvantage as partial wetting implies dead zones, hot spots and channeling. Only in cases where part of the fluid to be processed is a gas, and part a liquid under reactor conditions, partial wetting may allow for the gas phase to be in better contact with the solids. This happens, e.g., in hydroprocessing of light oil fractions (such as hydrodesulfurization).

2.7. Radial heat transport may be a problem in trickle flow but is never a problem in slurry reactors. Also the heat removal/addition in slurry reactors is much more easy because high heat transfer rates to cooling surfaces are possible. In trickle flow reactors interstage cooling, cold shot techniques and cooling with an evaporating solvent are possible; and the latter technique can also be applied with slurry reactors.

2.8. In case plugging and/or fouling is a problem, trickle flow clearly has a disadvantage over slurry reactors. A well known phenomenon is the pressure drop built-up during operation of a trickle flow hydrodesulfurization reactor limiting the operating period. Similar problems do not occur in slurry reactors.

2.9. Gas/liquid mass transfer depends on many factors, as will be described below. Related to the lower liquid hold-up the interfacial area per unit volume liquid in trickle flow columns can be very high which is an important advantage in mass transfer.

2.10. Liquid solids separation is not a problem in trickle flow but can be a difficult and costly operation in slurry reactors, especially if very fine particles have to be removed from (viscous) liquids, e.g., by filtration. As already stated, it is in this respect that three phase fluid bed reactors, which have properties rather similar to slurry reactors, are markedly different.

3. DIFFERENT TYPES OF SLURRY REACTORS

Slurry reactors can be classified according to the phases where the reactants are present. Table II gives an overview. The most important distinction is whether the solid phase is a reactant or a catalyst. In principle, the solids could also be inert and only present to increase mass transfer between phases as is often the case, e.g., in trickle flow reactors. In slurry reactors the introduction of solids for this purpose only is not worthwhile, with the exception of solids like zeolites and activated carbon for enhancement of mass transfer or improvement of selectivity [21, 22] but in such a system the solid is not really inert. Another example is the turbulent contactor in which large but light balls are moved by a gas flow and irrigated by a liquid phase. However, this regime falls outside the scope of the present presentation. If the solid is a reactant as well as the gas phase and liquid phase, the situation becomes rather complex; nevertheless, it corresponds to many practical situations (see e.g. Shah [2]). A rather exceptional

Table II. Slurry reactors classified according to the chemical system

	Typical examples	ref.	remark
gas liquid and solids all reactants	- thermal coal liquifaction - CO_2 absorption in a lime suspension - single cell protein	[12] [13] [14]	
gas and solids are reactants	- hydride formation and decomposition in a slurry	[15]	liquid to improve heat transport properties, avoid dust entrainment, allowing to operate an absorber-desorber continuously
gas phase is reactant solid is a catalyst liquid is inert	- Fischer-Tropsch process - hydrogenation of ethylene using a suspended Raney nickel catalyst	[8] [16]	liquid phase used to suspend catalyst and improve heat transport
gas phase and liquid phase are reactants solid is a catalyst	many industrial reactions: - hydrogenation of edible oil - hydro-desulphurization - oxygen consuming reactions in activated carbon slurries	 [17] [18] [19]	active carbon may also enhance oxygen transfer (see D2)

situation is that the liquid phase is inert and gas phase and solid phase are reactants. A recently published example is the continuous absorption/desorption of hydrogen in hydridable metal/oil slurries (Ptasinski et al. [16]). Of course, there must be a good reason to introduce a liquid phase here as it adds to the total transport resistances. In the case of the hydride slurries the liquid phase allows for rapid heat removal, prevents metal or hydride dust particles from being entrained with the gas and makes it possible to circulate the solids continuously between an absorber and desorber.

For similar reasons an inert liquid phase is often used if the solid is a catalyst, heterogeneously catalyzing a desired gas phase reaction. Apart from enhancing the heat transport rates, the fine dispersion of catalyst and the control of the reactants (re)distribution over the available catalyst can be important factors.

In a large class of slurry reactors both the gas phase and liquid phase are reactants and the solid is a catalyst. Much research and development work has been carried out and is still going on for this class of reactors.

A different classification of slurry reactors can be based on the contacting pattern and the mechanical devices applied to influence the contacting patterns and mass transfer. Fig. 1 gives an overview. We will not consider three phase fluidization. The three phase sparged slurry reactor is in fact the most simple slurry reactor. The liquid and solid phase can be operated batchwise or continuously. For continuous operation the intense mixing of the liquid/solid emulsion may be a disadvantage because it renders the macromixing close to a single ideal mixer operating at outlet conditions. This can be counteracted by putting more slurry spargers

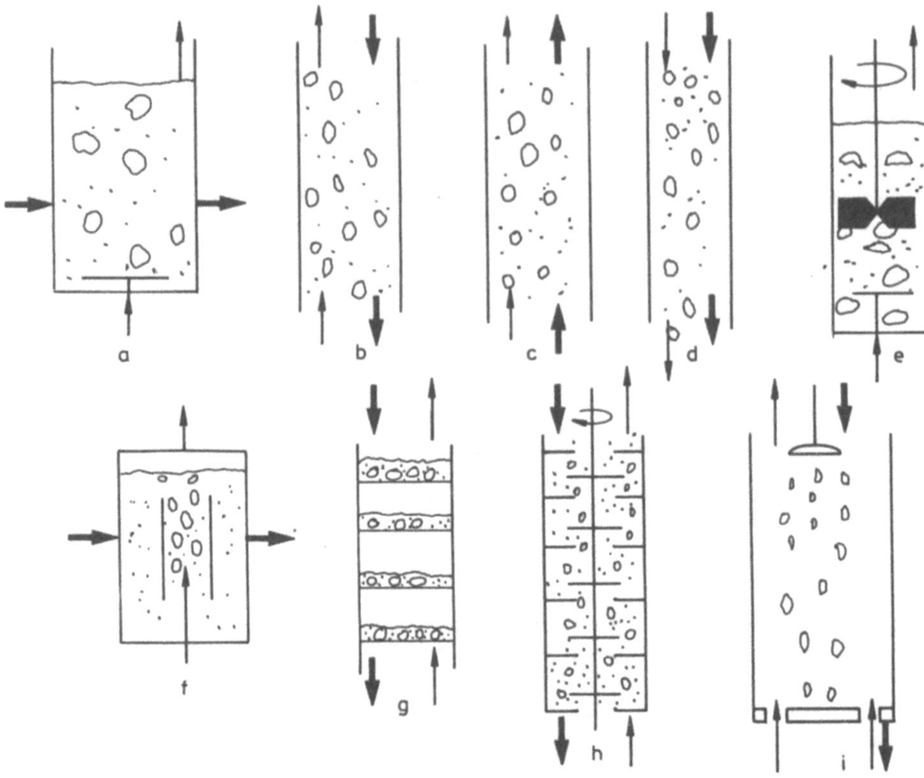

Fig. 1 Slurry reactors as classified by the contacting pattern and mechanical devices
(a) three-phase sparger
(b) countercurrent column
(c) co-current upflow
(d) co-current downflow
(e) stirred vessel
(f) draft tube reactor
(g) tray column
(h) rotating disc contactor
(i) three phase spray column
liquid flow gas flow

in series. If the particles are removed between the stages, the contacting pattern of liquid and solid can also be manipulated (e.g. countercurrent contact) which may be advantageous in relation to catalyst aging (e.g. in hydrodesulfurization of residues). If the contact time of liquid/solid suspension with the gas phase is relatively short and also in case of reactors of high L/d_t values, it may be useful to distinguish a co- and countercurrent contacting pattern and, specially in relation to the gas hold-up, upflow and downflow. To improve the solids suspension and/or to improve mass or heat transfer in many cases a stirrer is added to the system. In some cases a stirrer can be avoided and yet the solid can be kept in suspension by a relatively small amount of gas by means of a draft tube placed in the slurry reactor. Other possibilities are external slurry recycling which can be combined with jet mixing. If staging and countercurrent operation are essential a slurry tray column and a rotating disk column have been suggested (Ghim et al. [23]) to shape a slurry reactor towards a countercurrent contactor. In flue gas purification where a low pressure drop is essential, lime or $CaCO_3$ slurries are used in spray contacting with the gas phase (see [12]). Sometimes it is meant to evaporate the water phase but also simple slurry scrubbing is envisaged.

4. HYDRODYNAMICS OF SLURRY REACTORS

Hydrodynamics of slurry reactors includes the study of minimum gas velocity or power input to just suspend the particles (or to fully homogeneously suspend the particles), bubble dynamics and the hold-up fractions of gas, solids and liquid phases. A complicating problem is the large number of slurry reactor types in use (see fig. 1) and the fact that most correlations available are at least partially of an empirical nature. We will therefore restrict ourselves to sparged slurry columns and slurries in stirred vessels. A second problem is the difference with three phase fluidization. To avoid too much overlap we will only consider those cases where superficial liquid velocities are so low that its contribution to suspension of the particles is relatively unimportant.

4.1. Minimum suspension criteria

For the design of slurry reactors, whether agitated only by the flowing gas or assisted by one or more stirrers, the conditions at which the particles are just suspended are very important. If the overall contacting efficiency with the solids is considered, it increases rapidly if more and more solids are suspended but once the condition of minimum suspension is passed and no solids are laying on the bottom of the vessel, further increase of the energy input for suspension only moderately increases the contacting efficiency. Another aim could be to reach for complete homogeneous suspension. This condition is generally difficult to reach for many applications and an arbitrary criterium should then be introduced. Therefore, generally only a minimum suspension criterium is considered.

In sparged vessels with a stagnant liquid medium a minimum suspension gas velocity can be defined at which all solid particles are just suspended. Three different experimental techniques have been used to obtain an operational definition of minimum suspension gas velocities:

1. Particles do not remain on the bottom of the vessel for more than say one or two seconds. This has to be checked by visual observation through a transparent bottom.
2. The lowest gas velocity at which the pressure drop over the liquid phase indicates that all particles are just suspended (see Fig. 2).

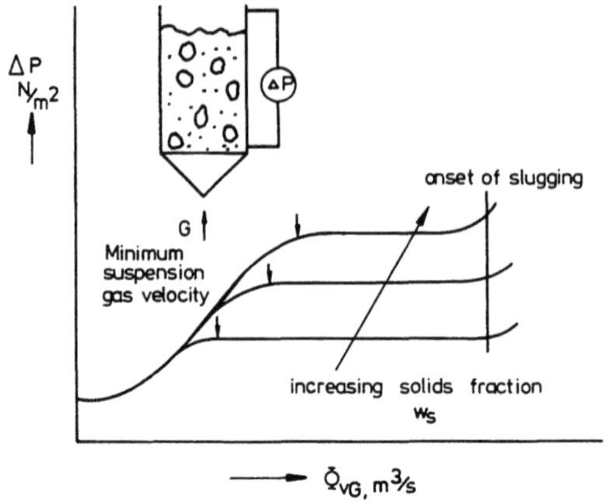

Fig. 2 Experimental technique to find the minimum gas velocity for particle suspension in gas-liquid-solids sparged columns.

3. Observing the solids concentration at a given position above the bottom of the vessel where it passes through a maximum value or a discontinuity [24].

Narayanan et al. [25] used a visual observation technique and have given relations for the minimum gas velocity to suspend the particles. To obtain a theoretical basis, they compared a pick-up velocity previously derived [25] on the basis of a force balance:

$$v_i = (2g(\rho_S-\rho_L)[\frac{d_p}{\rho_L} + \frac{w_S L_{SL}}{\rho_S+w_S\rho_L}])^{1/2} \tag{1}$$

which was taken equal to the liquid velocity in upward direction for which it was derived [25]:

$$v_u = u_G + 1/3\ (2gL_{SL}\varepsilon_G(\frac{\rho_L-\rho_G}{\rho_L}))^{0.5} \tag{2}$$

An empirical relation was introduced to relate the bubble hold-up with the superficial gas velocity u_G.

$$\varepsilon_G = 6.2\ u_G \quad \text{for} \quad u_G < 0.067\ \text{m/s}$$
$$\varepsilon_G = 0.765 u_G^{0.38} \quad 0.067 < u_G < 0.22\ \text{m/s} \tag{3}$$

Combining equations (1), (2) and (3) results in a minimum gas velocity necessary to suspend the particles: u_G (min, Theor.). However, it was clear from comparison with the experimental values that a correction factor should be introduced for particle concentration and vessel diameter, d_t:

$$u_G(\text{min, actual}) = 4.3(\frac{d_t}{0.0508})^n e^{-10w_S} u_G\ (\text{min, Theor.})(w_S < 0.1) \tag{4}$$

$$u_G\ (\text{min, actual}) = 1.25\left(\frac{d_t}{0.0508}\right)^n e^{-3w_S}\ u_G\ (\text{min, Theor.})\ (w_S > 0.1)$$

in which $n = 0.2$ if $d_p < 100$ μm and $n = 0.5$ if $d_p > 200$ μm.

These correlations were found to be valid up to a liquid viscosity of 0.006 Ns/m^2 with column diameters of 0.0508, 0.114 and 0.141 m. Roy et al. [26], applying the pressure drop technique, studied a large variety of gas solid and liquid systems including non-aqueous systems and particles with different degrees of wetting (quartz, coal, catalyst pow- ders). Results were presented as critical hold-ups which correspondends to the hold-up of particles that can be just suspended at a given gas velocity. Kato et al. [27] showed that in many cases there will remain a solids profile in the column rendering the completely suspension criterium doubtful for tall slurry columns. A pressure drop technique was also applied by Imafuku et al. [28]. From this work it also became clear that with a draft tube applied in the system a much smaller volumetric flow rate of gas was required to suspend a same amount of solids.
In stirred vessels usually the stirrer provides most of the energy required to keep the particles suspended. The minimum suspension criteria for a three phase stirred tank are usually derived from those obtained in the absence of the gas phase. It has been shown by Chapman et al. [29] that there is a considerable influence of the gas phase which depends on the geometry of the system. In the absence of a gas phase the stirrer speed at minimum suspension was studied by Zwietering [30] who investigated a wide range of materials and experimental conditions. Although this work is already relatively old (1958) and resulted in a purely empirical expression, its results are still generally used for design purposes and to check new theoretical approaches. Zwietering proposed:

$$\omega_{min} = C1\ \nu^{0.1}\ d_p^{0.2}\ \left(g\ \frac{\Delta\rho}{\rho_L}\right)^{0.45}\ w_S^{0.13}\ \ d_s^{-0.85} \tag{5}$$

The exponents were found to be independent of impeller type, vessel size, impeller clearance and impeller to tank diameter ratio. The dimensionless constant, C1, accounted for variations in system geometry (e.g. on d_t/d_s). This would indicate that the basic mechanism leading to minimum suspension may be the same for rather different stirrer geometries. Table III, which is an updated version of the one given by Nienow [31], indicates different exponents found in a few other investigations. A recent and interesting study of this phenomena has been made by Chapman et al. [24] who found exponents close to those of Zwietering. Baldi et al. [32] made a relatively successful theoretical approach. It has been assumed in their work that the suspension of particles is mainly due to eddies of a certain critical scale. This scale would be of the order of magnitude of the particle size. From an energy balance it follows:

Table III. Exponents found by different authors for the Zwietering type of equations

Authors		w_S	ν	$\Delta\rho$	d_p	d_s (at constant d_t)
Zwietering	[30]	0.13	0.1	0.45	0.2	-2.35 (R), -1.67 (A)
Nienow	[31]	0.12	-	0.43	0.21	-2.25 (R)
Pavlushenko	[51]	-	0.2	-	0.4	-2.5 (A)
Oyama	[189]	~ 0	-	1.0	1.3	-1.67 (?)
Kneule	[52]	0.17	-	0.5	0.17	-1.67 (M)
Narayanan	[53]	0.22	-	0.5	0.5	-2.00 (R)
Weisman	[54]	0.17	- terminal velocity concept			-1.67 (R)
Kolar	[55]	0.10	- terminal velocity concept			-2.24 (A,M)
Baldi	[32]	0.125(*)	0.17(*)	0.42(*)	0.14(*)	theoretical function
Chapman	[24]	0.12	0	0.4	0.15	-2.25 (R), -1.5 (A)

(*) for d_s/(clearance height) = 1 only
R = radial flow impeller
A = axial flow impeller
M = mixed downflow impeller

$$\rho_L v'^2 \div d_p \Delta\rho g \qquad (6)$$

v' is the fluctuating velocity of the critical eddies. Because the scale of these eddies is generally much larger than of those that dissipate their energy by viscous forces (Kolmogoroff scale), v' can be expressed as [32]:

$$v' \div (e_b d_p)^{1/3} \qquad (7)$$

in which e_b is the power input per kg liquid close to the bottom of the stirred vessel. The latter factor is difficult to evaluate. In fact, the energy dissipation within a stirred vessel is far from homogeneous nor is it isotropic. There is supportive evidence for the use of equation (7) for the average turbulent velocity [32]. The average value of the energy dissipation $\bar{e}$ can be found from:

$$\bar{e} = \frac{4\phi\ \omega^3 d_s^5}{\pi d_t^3} \qquad (8)$$

assuming $e_b \div \bar{e}$; a constant power number (which is reasonable in the turbulent regime), and constant ratio d_s/d_t, it is found for the minimum suspension stirring speed:

$$\omega_{min} \div d_p^{1/6}(g\ \frac{\Delta\rho}{\rho_L})^{0.5} d_s^{-2/3} \qquad (9)$$

The exponents in equation (9) are close to those observed by Zwietering (eq. 5) although no influence of particle concentration nor viscosity influence is predicted. Baldi et al. [32] have observed that if the ratio of $\bar{e}$ and e_b has been properly accounted for and allowing for an influence of particle concentration, the exponents given in Table III are obtained which are rather close to those found by Zwietering. It should be noted, however, that this similarity only holds for a specific clearance height of the stirrer. In a recent extensive study, Chapman [24] concluded that for some geometries the relations of Baldi et al. appeared to work well but overall the Zwietering relation is still the soundest basis for design. A notable exception is the prediction of "the scaling up". In fact from the equation of Zwietering, Chapman derived $\bar{e}_{min} \div d_t^{-0.55}$, while from their own results obtained with tank diameters up to 1.83 m they found $e_{min} \div d_t^{-0.28}$ in agreement with those mentioned for use in industrial practice $e_{min} \div d_t^{-0.25}$, [33]. The theoretical approach of Baldi et al. [32] has the potential of extension to other geometries (e.g. jet stirring or gas/solid/liquid spargers).

<u>Influence of the gas flow rate on the minimum suspension criterion in stirred vessels</u>

Little information is available about hydrodynamics of three phase stirred systems. Most data concern two phase gas/liquid stirred vessels which fall outside the scope of the present work. Although often much less power is required to disperse the gas phase then to suspend particles, it was noted already by Arbiter et al. [34] that the introduction of a gas phase in a stirred particle-liquid suspension may cause settling of part of the suspended particles corresponding with a decrease in power consumption. Similar conclusions may be drawn from the work of Queneau et al. [35], Wiedmann et al. [36] and Subbarao and Taneja [37]. It was shown by Wiedman et al. [36] that the minimum mixing requirement for the gas phase itself was independent of particle conditions (density difference, volume fraction, d_p etc.). Chapman et al. [29] presented data for a wide range of vessels (up to 1.83 m diameter) with different impellers. The power input necessary to cause suspension under aerated conditions was found to be higher than for unaerated conditions demonstrating that the presence of the gas phase has an additional effect probably in damping local turbulence and velocities near the vessel base. Slightly lower exponents were found for the influence of most of the variables as compared to equation (5); the most significant being the influence of the particle liquid density difference. To summarize:

$$\omega_{min} \text{ (gassed conditions)} \div \nu^{0.0}\, d_p^{0.12}\, \Delta\rho^{0.22} w_S^{0.12} \qquad (10)$$

Chapman et al. [29] proposed an extremely simple relation from which the increase in stirrer speed required to suspend all particles under aerated conditions over unaerated conditions is given for disc turbines of diameter $d_t/2$ and clearance height of $d_t/4$:

$$\Delta\omega = 56.4\ \Phi_{v,G}/V_L \qquad (11)$$

where $\Delta\omega$ is the increase in impeller speed to maintain suspension (rev/s)

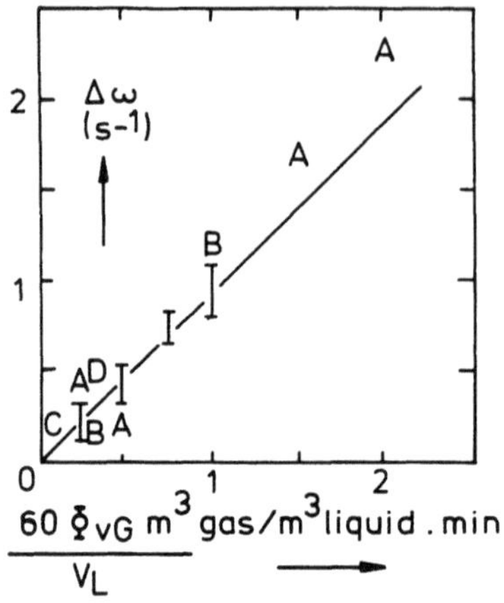

Fig. 3 Increase in stirrer speed necessary to keep the particles suspended if gas is fed to the stirred vessel (Chapman et al. [29])
Disk turbines of diameter $d_t/2$ and clearance $d_t/4$
A. d_t = 0.29 m. D. d_t = 1.83 m.
B. d_t = 0.30 m. Barred line ranges for d_t = 0.56 m.
C. d_t = 0.91 m. at different conditions

Similar relations are given for other geometries. The relation (11) is shown in fig. 3 for disk turbines of half the vessel diameter with clearance $d_t/4$ tested on different scales and with different particles. This geometry was recommended as the safest impeller type for three phase agitators with respect to amount of data available and stability in large vessels, although other geometries (like upward pumping impellers) may have advantages in special situations.

4.2. Hold-up fractions in slurry reactors

Sparged columns

The average hold-up fractions of gas, solids and liquids should satisfy the equation:

$$\varepsilon_G + \varepsilon_L + \varepsilon_S = 1 \tag{12}$$

In contrast to three phase fluid beds where the relation between the three phase hold-ups can be rather complex, in slurry reactors with the much smaller particles and slip velocities the relation between ε_L and ε_S is often simple as it is fixed by the feed ratio of solids and liquid phases, or liquid and solids volumes are constant (in batch systems). The bubble hold- up is much more difficult to predict, first of all because of the different regimes that might prevail both in stirred vessels and slurry sparger columns. These regimes are shared with two phase gas liquid systems (see fig. 4). In a three phase sparger the regimes are (see Shah [6]):

- uniform bubbling at low gas velocities: $\varepsilon_G \div u_G^n$ n = 0.7-1.2
- churn-turbulent flow at higher gas velocities with a mixture of large and small bubbles: $\varepsilon \div u_G^n$; n = 0.4-0.7
- slugging in small diameter columns where the largest bubbles are comparable to the column diameter

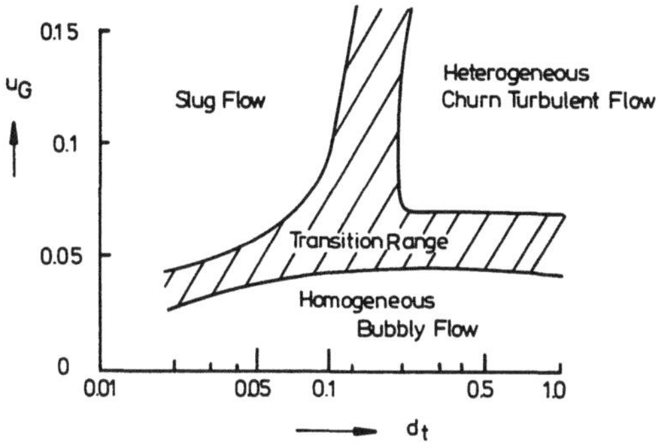

Fig. 4 Approximate position of flow regimes for water and diluted aqueous solutions. (Shah et al. [6])

Darton [38] recently proposed for the churn-turbulent regime in three phase fluidization to use the equation of Akita and Yoshida [39] written in an explicit form and using liquid/solid suspension bulk properties instead of liquid properties.

$$\varepsilon_G = 1/8 \ \ln\left[1+8Cg^{7/24}\left(\frac{\mu_B}{\rho_B}\right)^{-1/6}\left(\frac{\sigma}{\rho_B}\right)^{-1/8} u_G\right] \tag{13}$$

$0 < \varepsilon_G < 0.4$
C = 0.2 for normal systems
C = 0.25 for aqueous systems of electrolytes

In this equation μ_B is not the liquid viscosity but the liquid/solid suspension viscosity given by [38]

$$\mu_B = \mu_L \ \exp(36.15 \ \varepsilon_S^{2.5}) \tag{14}$$

A similar approach will probably be valid for churn-turbulent flow slurry sparged reactors and also for the bubbling regime using existing relations for gas-liquid systems while replacing the liquid properties by slurry bulk properties. This would be of great help in the design because in two phase systems much more information is available. For recent reviews in this field see e.g. Shah et al. [6] and Heynen and van 't Riet [40]. Generally the presence of solids with $d_p > 40$ µm does not influence the gas hold-up very much as follows also from the data of Ying et al. [41] who found the two phase correlation of Akita and Yoshida [39] to be also valid for three phase systems. At low gas velocities (Figure 5) there is a slight decrease in gas hold-up for smaller particles at increasing concentration probably due to an increasing bubble coalescence rate, but the effect becomes insignificant for larger gas velocities ($u_G > 0.1$ m/s) (Kato et al. [27] and Ying et al. [41]).

Gas hold-up in stirred vessels

For stirred vessels the gas hold-up in a slurry system can be related to the gas hold-up in the corresponding gas-liquid systems.

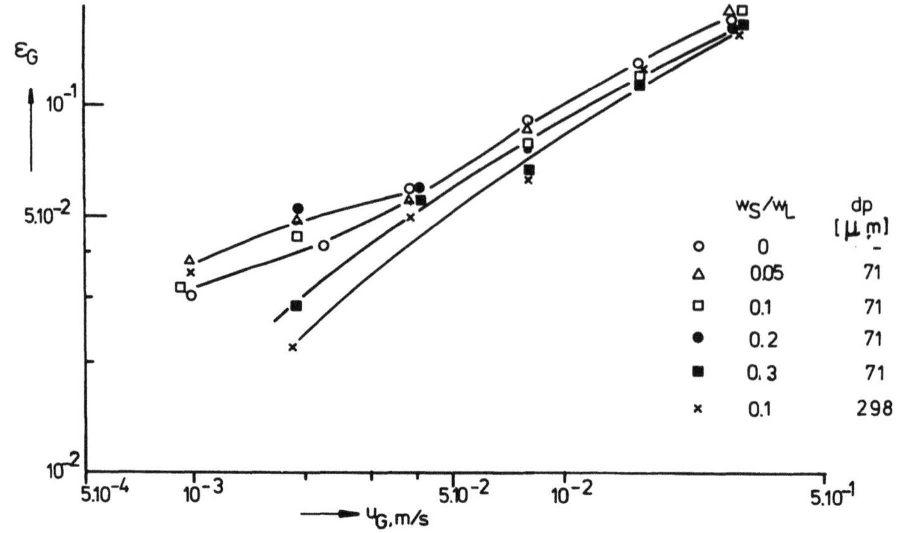

Fig. 5 Influence of particles on gas hold-up in a stirred vessel (Wiedmann et al. [62]). Water-air-glass spheres; 6 blade disc turbine: $d_s = 1/3\ d_t$; Re_S=200,000; d_t= 0.45 m.; clearance is 0.5 d_t.

Generally as it is shown in Figure 5 the effects of the solids at higher gas velocities are not very important except for large particles and high concentrations (Wiedmann [36]) and probably for very small particles, $d_p < 10\ \mu m$.

4.3. Axial mixing in slurry reactors

We can distinguish axial mixing in the three different phases G-L-S and we will consider stirred vessels and sparged columns separately. In stirred vessels the liquid mixing will in most cases be rather close to one completely mixed stage and both the macro and micro mixing problems will not be much different from those of a two phase stirred vessel or even single phase stirred vessels and fall outside the scope of the present work. On gas phase mixing of stirred slurry reactors to the authors knowledge no data are available.

In gas/liquid/solid sparged columns the situation is somewhat more complicated. Just like in the case of gas-liquid systems different regimes will have to be distinguished. Here the most important variables are d_t and u_G. If it is assumed:

$$D_l \div d_t^p\ u_G^q \qquad (15)$$

it is recommended for p a range from 1-1.5 and for q values from 0-2. For instance Rice et al. [42] concluded that the wide range for q was due to the different regimes and recommended for:
chain bubbling q = 2,
bubbling flow q = 1,
churn turbulent flow q = 1/3,
slug flow q = 0.

The effects of liquid velocity (at least at low velocities), direction of flow and liquid properties on axial mixing are only minor for Newtonian fluids. Recent correlations on gas liquid columns were given by Joshi and Sharma [43]. Field and Davidson [44] measured the dispersion in a large industrial column (d_t = 3.2 m, H = 19 m) and found, as shown in Table IV, agreement with the correlations of Deckwer et al. [45] and Joshi and Sharma [43].

Table IV. Liquid phase axial dispersion correlations for sparged columns

Deckwer et al. [45]	$D_l = 0.68\, d_t^{1.4}\, u_G^{0.33}$
Baird and Rice [57]	$D_l = 0.35\, d_t^{1.33}\{u_G[\frac{\epsilon_G}{1-\epsilon_G}]g\}^{1/3}$
* Joshi [43]	$D_l = 0.29\, d_t(v_c+u_L)$ v_c = circulation velocity which is a complex function [43]
* Kato et al. [27]	$Pe_L = 13\sqrt{Fr}/(1+8(\sqrt{Fr})^{0.85})$
Towell and Ackermann [49]	$D_l = 1.225\, d_t^{1.5} u_G^{0.5}$
Hikita and Kikukawa [56]	$D_l = (0.15+69\, u_G^{0.77}) d_t^{1.25} (\frac{0.001}{\mu_s})^{0.12}$

* based on three phase column

The influence of particles on axial mixing can be expected to be small at least for low concentrations and small particles. This is confirmed by early experiments of Kato et al. [27,46]. For particle sizes ranging from 63 to 177 µm and concentrations up to 200 kg/m^3 the Pe_L is essentially a function of the Fr number:

$$Pe_L = \frac{u_G d_t}{D_l} = 13(\sqrt{Fr})/(1+8Fr^{0.425}) \qquad (16)$$

$$\sqrt{Fr} = \frac{u_G}{\sqrt{g d_t}} \qquad (17)$$

A similar situation of a Pe_L function of Fr was observed for gas liquid systems:

$$Pe_L = \frac{u_G d_t}{D_l} = 13(\sqrt{Fr})/(1+6.5(\sqrt{Fr})^{0.8}) \qquad (18)$$

Fig. 6 shows the changes of D_l with u_G in the liquid phase of slurry reactors. Recent data of Kara et al. [47] together with the predictions of two correlations are reported.
The situation in three phase fluid beds should be clearly distin-

guished from the slurry reactors. Here the axial dispersion depends much more on the superficial velocity of the liquid (see e.g. Michelson and Ostergaard [48]) but this regime falls outside the scope of the present work.

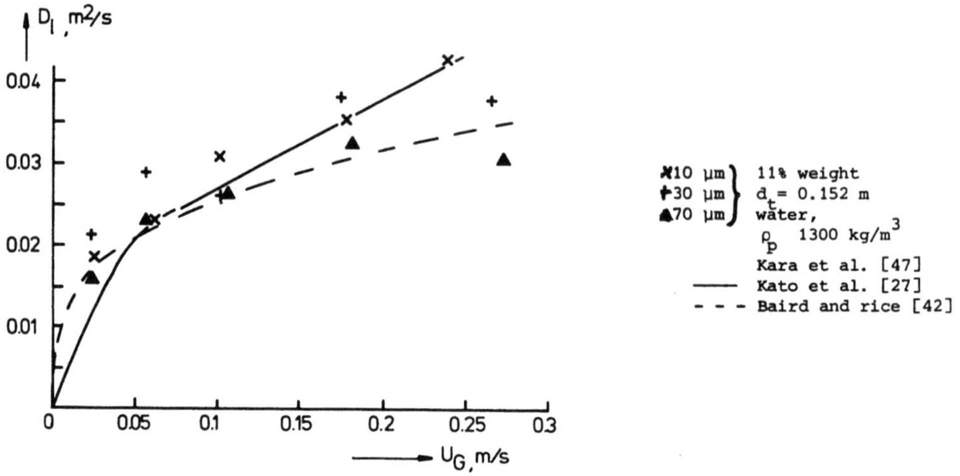

Fig. 6 Axial dispersion in the liquid phase of a slurry column. Comparison of data of Kara et al. [47] with correlations

Data on gas phase dispersion in three phase sparged columns are scarce. For two phase systems a few correlations are available [44,48,49,50], these correlations are shown in Table V. However the scatter in the data is quite large. The influence of the column diameter varies from d_t^2 (Towell and Ackerman [49]) to d_t^1 (Diboun and Schügerl [50]) although in most recent correlations $d_t^{1.5-1.33}$ is taken. The correlation of Field and Davidson [44] has been tested for large scale equipment and should therefore be recommended:

Table V. Gas phase axial dispersion in sparged columns correlations

Towell and Ackermann [49]	$D_{1,G} = 1.97\ d_t^2\ u_G$
Diboun and Schügerl [50]	$D_{1,G} = 5\ u_r d_t$
Mangartz and Pilhofer [60]	$D_{1,G} = 50\ d_t^{1.5}\ (u_G/\varepsilon)^3$
Field and Davidson [44]	$D_{1,G} = 56.4\ d_t^{1.33}\ (u_G/\varepsilon)^{3.56}$

$$D_{1,G} = 56.4\ d_t^{1.33}\ \left(\frac{u_G}{\varepsilon_G}\right)^{3.56} \qquad (19)$$

To the knowledge of the authors the data on gas phase axial mixing

in three phase systems only refers to three phase fluidization (Schügerl [58] and Michelson and Ostergaard [48]). Since the cause of axial dispersion can probably be traced back to rising velocity differences in bubble swarms, in case of lack of data equation (19) should be used, except probably for very concentrated slurries.
The dispersion of the solids phase has been found to be nearly equal to that of the liquid phase at least for small particle diameters as general prevailing in slurries [27].
The axial dispersion of solids can be measured using a steady state diffusion sedimentation technique. This method was applied e.g. by Kato et al. [27] and recently by Brian and Dyer [59]. At zero liquid rates, if the particles are only suspended by the gas flow, a solid concentration profile will be established:

$$W_S/W_{S_0} = \exp\left(-\frac{v_p z}{D_{1,S}}\right) \tag{20}$$

Provided the particle settling velocities v_p are known, this equation allows the calculation of $D_{1,S}$. Usually experiments at nonzero liquid rates are used to evaluate v_p and $D_{1,S}$ separately. A similar situation as used in this experimental set-up might occur in practice if slurry column reactors are operated close to the conditions given by the minimum suspension criterium. In this case reactor calculations should take the solids concentration profiles into account. A recommended correlation for the solids dispersion coefficient for small particles is given by Kato et al. [27]:

$$Pe_S = \frac{13\sqrt{Fr_G}}{1+8Fr_G^{0.425}} \tag{21}$$

$$Pe_S = \frac{u_G d_t}{D_{1,S}} \tag{22}$$

which is identical to the equation for liquid phase dispersion.

5. MASS TRANSFER WITH CHEMICAL REACTION

The average conversion rate per m^3 reactor of a gaseous reactant A absorbing in the liquid present in a well stirred slurry semi-batch reactor and then reacting according to an irreversible first order kinetics at the external surface of non porous catalyst particles, is given by [61]. (Figure 7).

$$J_A a = \bar{c}_{AG} / \left\{ \underset{(R_1)}{\frac{1}{k_G a}} + \underset{(R_2)}{\frac{1}{m k_L a}} + \underset{(R_3)}{\frac{1}{m k_{LS} a_p}} + \underset{(R_4)}{\frac{1}{m k_1' a_p}} \right\} \tag{23}$$

provided $\bar{c}_{AG}$ is constant all over the reactor. R_1, R_2, R_3, and R_4 are the so-called resistances that control the overall conversion rate, with:

$R_1 = 1/k_G a$ = the resistance to mass transfer in the gas phase
$R_2 = 1/mk_L a$ = the resistance to mass transfer in the liquid phase at the gas liquid interface
$R_3 = 1/mk_{LS} a_p$ = the resistance to mass transfer in the liquid phase

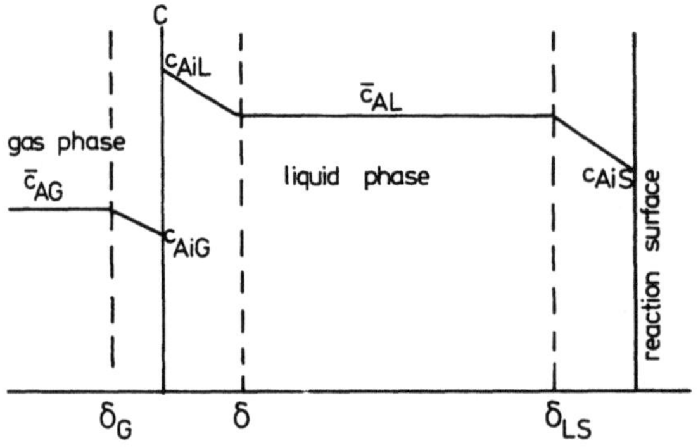

Fig. 7 Concentration profiles in mass transfer and reaction in series; slurry reactor with non porous particles; film theory. (From Westerterp et al. [61])

at the liquid-solid interface

$R_4 = 1/mk_1''a_p$ = the resistance to reaction at the external catalyst surface

Let us first focus on R_4. If the catalyst is of a porous type, which is the rule rather than the exception, then R_4 in Equation (23) has to be replaced by:

$$R_4' = d_p/(6\ m\ a_p D_i \phi\ \tanh\ \phi) \tag{24}$$

R_4' is the overall resistance for a first order reaction with internal diffusion within the porous catalyst particles. For kinetics different from first order, the relations for both R_4 and R_4' will be different and, as a rule, will still contain a concentration term. For instance for n-th order kinetics Equation (24) is still valid in good approximation provided ϕ is defined according to:

$$\phi = \frac{d_p}{6}\ \left(\frac{2}{n+1}\ k_n'' a_S (c_{AiS})^{n-1}/D_i\right)^{0.5} \tag{25}$$

For more complex kinetics, as e.g. Langmuir-Hinshelwood, see e.g. [61]. The numerical value of $\bar{c}_{AiS}$ in Equation (25) will, in general, depend on the values of $\bar{c}_{AG}$, R_1, R_2 and R_3. From the above we may conclude that R_4 and R_4' are functions of the intrinsic reaction kinetics and the particle properties (a_p, a_S, D_i, d_p) whilst for non-first order kinetics also $\bar{c}_{AiS}$ plays a role. Consequently, R_4 and R_4' are independent of the hydrodynamics within the slurry reactor except indirectly via R_1, R_2 and R_3 in case of non-linear intrinsic reaction kinetics.

Concerning R_2, the relation given in Equation (23) only holds if reaction and/or adsorption within the film for mass transfer around the bubbles (film theory, $\delta = D/k_L$) can be neglected relative to the conversion taking place in the bulk of the slurry phase. In some systems this is not the case and then the numerical value of R_2 can be reduced due to enhancement of the mass transfer process at the gas-liquid interface by the occurrence of fast chemical reactions or physical adsorption processes parallel with the gas to liquid mass transfer process. We then have:

$$R_2 = 1/(mk_L aE_A) \tag{26}$$

where E_A not only depends on reaction and/or adsorption kinetics but, via k_L, also may depend on the hydrodynamics in the reactor.

From the above analysis we conclude that for an adequate description of mass transfer with chemical reaction in slurry reactors we need reliable data on the following two types of parameters:

(a). parameters which are specific for slurry reactors (k_G, k_L, a, E_A)

(b). parameters which are not specific for the type of reactor applied (intrinsic reaction kinetics and particle properties as a_p, a_S, d_p and, to some extent, also D_i)

In this review we will limit our scope to the parameters of type (a). only. Below, we summarize new information on each parameter as collected from open literature published since 1980. However, we start with a short discussion of the product of k_L and the specific contact area, a, (k_La) because this often is the overall rate controlling step. Fortunately, this parameter is also relatively easy to measure; easier than k_Ga and k_L and a separately. Therefore it is always essential to estimate as early as possible in the progress of the reactor design which resistance(s) might be rate controlling because that dictates the information necessary for a sound design and the type of experiments to be done if the available information turns out to be insufficient.

5.1. The volumetric liquid side mass transfer coefficient at the gas-liquid interface (k_La)

Often, the gases used are only sparingly soluble in the liquid ($k_G/mk_L \gg 1$ or $R_1 \ll R_2$); the reactions involved not extremely fast ($E_A = 1$ in Equation (26)) but still fast enough to ascertain a bulk concentration of reactant A in the slurry ($\bar{c}_{AL}$) to be effectively zero, also because nearly always $a_p \gg a$. Then, also R_3 and R_4 both $\ll R_2$ and Equation (23) simplifies to:

$$J_A a = k_L a \; m \; \bar{c}_{AG} \tag{27}$$

For a stationary operating semi-batch stirred tank reactor the mass balance gives:

$$J_A a = -\Phi_A/V_r = (\Phi_{vG0}c_{AG0} - \Phi_{vG1}c_{AG1})/V_r \tag{28}$$

which shows a conversion rate ($-\Phi_A$, kmol A/s) to be proportional to k_La and independent of both intrinsic reaction kinetics and intra particle diffusion processes. On the other hand, such a reaction system can be used to measure k_La directly via the following equation:

$$k_La = -\Phi_A/(m\bar{c}_{AG1}V_r) \tag{29}$$

Ideally, the model reaction is the same as the reaction under investigation. The difficulty here is to establish the conditions under which $R_2 >> R_1$, R_3 and R_4. Alternatively, also one of the many well known gas-liquid model reactions may be applied for the purpose of measuring k_La but then the problem is to carry out that model reaction in a slurry with properties as close as possible to the actual slurry under investigation both with respect to the liquid and the solids. Probably most popular nowadays are transient techniques, often using physical absorption only [62,63] but also with reaction (see e.g. [64,65]). In fact, there exists quite a variety of methods to measure k_La and we refer to the methods of Laurent et al. [66] and Sharma et al. [67,68] which concentrate on gas-liquid systems but are equally well applicable to gas-slurry systems. As long as the presence of the solids does not affect k_L and a, the existing relations for k_La in two phase gas-liquid reactors may be applied to predict k_La for slurry reactors too. For reviews summarizing the available information for gas liquid reactors see [69-79].

Crucial, of course, is the question: under what conditions do solids not interfere with k_La? According to Joosten et al. [80], particles with diameters 50 $\mu m \leq d_p < 200$ μm do not affect k_La appreciably, provided the volumetric solids fraction remains below 15%. However, as we know now, even then the process of mass transfer can be enhanced if the particles are very reactive or if the volumetric absorption capacity of the particles with respect to A is much larger than the solubility of A in the liquid [4,63,81,82]. With increasing solids concentrations above 20%, k_La rapidly decreases depending on particle properties. Joosten tried to correlate his data with the effective slurry viscosity but this attempt was not completely successful [80]. Recently, new data have been reported by Sada et al. [83] measured with CO_2 absorption in an aqueous slurry containing typically 2 μm $Mg(OH)_2$ particles. Despite the reaction between CO_2 and $Mg(OH)_2$, enhancement could be excluded due to the low solubility of $Mg(OH)_2$. The measurements of Joosten et al. [80] and Sada et al. [83] are compared in Figure 8. We see that Joosten's conclusion for particles > 50 μm is not valid for fine particles with diameters smaller than the film thickness for mass transfer ($\delta=D/k_L$) and we conclude that more data are needed, particularly for particle sizes below 50 μm. But also for larger part-

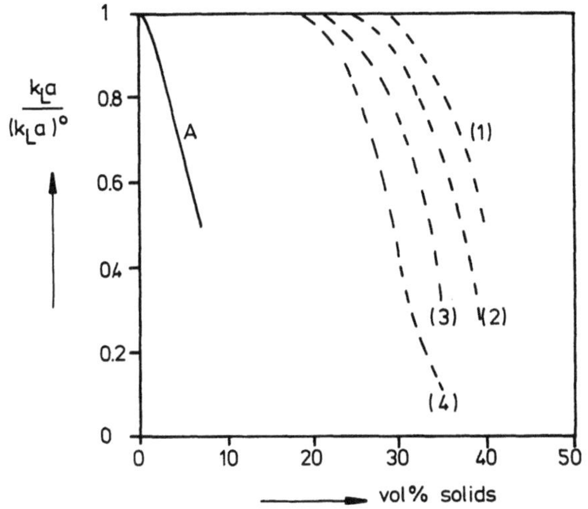

Fig. 8 Relative decrease of k_La as a function of solids concentration
A: 2 μm $Mg(OH)_2$ solids in bubble column [83]
1-4: 50-250 μm solids in stirred tank [80]

(1)	glass	88 μm
(2)	glass	53 μm
	polyprop.	53-105 μm
(3)	sugar	74-105 μm
(4)	polyprop.	250 μm

icles an equation predicting k_La from solids and fluid properties is still lacking. Probably, Figure 8 illustrates the existing uncertainty in the appropriate gas-liquid mass transfer correlations in real life slurry reactors, which usually operate also at increased temperature and pressure where virtually no experimental data on k_La are available. Even for one of the most interesting applications and probably one of the most intensely studied slurry systems since 1980, the Fischer-Tropsch synthesis, there is still uncertainty, not only on the actual values of k_La but also on the fact whether either R_2 or R_4 is the rate controlling step. A recent debate has been taken place concerning this matter between Satterfield, Deckwer and others [84-91]. Such uncertainty is the more astonishing because a rate limitation in the transport of hydrogen from the gas phase to the liquid phase not only greatly reduces the conversion rate realized in the reactor but, for this reaction, significantly influences product selectivity, as has been pointed out by Satterfield and Huff [8] (Figure 9). Main reason for such uncertainties still is a lack of consistent relations predicting both k_L and the specific contact area, a, as function of observable parameters as power consumption, specific gas load and gas, liquid and solids properties. In this respect it is interesting that Satterfield and Huff [8] only report that their catalyst is smaller than 53 μm diameter. Figure 8, however, shows that the actual particle diameter might greatly have influenced the appropriate value of k_La. Additionally, the limited experimental data available, have been measured with aqueous systems at ambient temperatures and atmospheric pressure. Still less is known for organic liquid mixtures at higher temperatures and particularly at increased operating pressures. It is these conditions which make experiments expen-

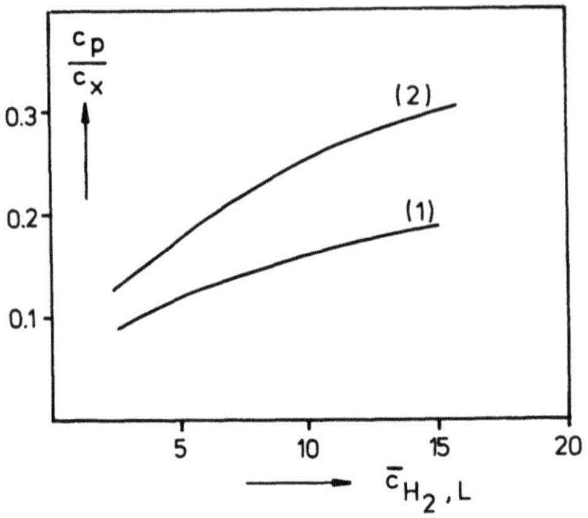

Fig. 9 Experimentally obtained product ratio as a function of $\bar{c}_{H_2,L}$ in Fischer-Tropsch synthesis at T = 248 °C, P = 0.3-1.1 MPa and for various H_2-CO feed ratios .
(1) P = propane, X = propene
(2) P = n-heptane, X = 1 and 2 heptenes
Data from Satterfield and Huff [8]

sive and difficult but which are relevant, not only for Fischer-Tropsch but also for methanol synthesis in a slurry phase, most hydrogenations including coal liquefaction, etc.

5.2. The volumetric gas side mass transfer coefficient (k_Ga)
Measurement techniques for k_Ga are reviewed by Sharma et al. [67, 68,92] and by Laurent et al. [66]. With systems in which the gas phase resistance prevails, k_Ga is determined analogously to the determination of k_La in a slurry reactor. In case the gas phase is perfectly mixed, it follows, analogous to Equation (29):

$$k_Ga = -\Phi_A/(\bar{c}_{AG1}V_r) \tag{30}$$

This relation holds for both gas-liquid and gas-solid systems provided the gas phase resistance is completely controlling the rate of the transfer process. Also a chemical reaction can be chosen for which k_G/mk_LE_A is in the order of one, whilst both R_3 and $R_4 << R_1$. Then, from Equations (23 and 25) it follows:

$$J_Aa = \frac{\bar{c}_{AG}}{\frac{1}{k_Ga} + \frac{1}{mk_LaE_A}} \tag{31}$$

and for an ideal tank reactor, with a mass balance, analogous to Equation (29):

$$\frac{1}{k_Ga} + \frac{1}{mk_LaE_A} = \frac{\bar{c}_{AG1}V_r}{-\Phi_A} \tag{32}$$

By varying the average volumetric reaction rate constant, e.g. with

a catalyst or by changing the concentration of a second reactant already present in the liquid phase, E_A may be varied in such a way that the range $k_G/(mk_LE_A) < 1$ to $k_G/(mk_LE_A) > 1$ is completely covered. For $E_A = \phi$, the values of k_Ga and the specific contact area, a, can be found simultaneously by plotting $V_r\bar{c}_{AGl}/(-\Phi_A)$ against $1/k_L\phi$. Alternatively, by applying an instantaneous bimolecular reaction in the liquid ($E_A = E_{A\infty}$), k_Ga and k_La can be found simultaneously by plotting $V_r\bar{c}_{AGl}/(-\Phi_A)$ against $1/E_{A\infty}$. However, these methods are time consuming because a series of experiments with varying ϕ is necessary to obtain one set of k_Ga, a or k_Ga, k_La respectively, that is at one particular combination of gas and liquid load. Therefore we suggest the simultaneous absorption of two gases, one gas reacting chemically giving k_La (for E_A=1 or E_A=$E_{A\infty}$) or a (for $E_A = \phi$) and the other gas giving k_Ga at exactly the same hydrodynamic conditions. Such a method, originally developed by Robinson and Wilke [93] and Beenackers and van Swaaij [94] for a slightly different purpose, see next section, can be easily modified for measuring both k_G and a in one single experiment. A possibly prospective system might be the simultaneous absorption of H_2S and CO_2 in aqueous solutions of amines with inert solids to simulate actual slurry conditions. For CO_2, the conditions can be chosen such that $E_A = \phi$. With this reaction the specific contact area, a, can be obtained. The absorption rate of H_2S is, under certain conditions, limited by k_G. If so, the product k_Ga will be obtained from the observed H_2S conversion in the same experiment and k_G follows from $k_G = k_Ga/a$.
Despite the availability of the experimental techniques, virtually no information is available on k_Ga in slurry reactors [5,95]. One might argue that such information is not important, because for sparingly soluble gases, k_Ga usually is not rate controlling while for very soluble gases, conversion will usually be complete, whatever reactor design is selected, due to the large numerical value of k_Ga, making such processes extremely fast. The latter part of this reasoning is not always correct, however. If selectivity is affected by liquid concentrations of gaseous reactants, accurate information on k_G is vital for designing a process unless the possibility of any gas phase rate controlling effect can be excluded with certainty.

5.3. The true gas-liquid specific contact area

Specially for fast reactions, where enhancement at the gas-liquid interface occurs, knowledge of the true gas liquid specific contact area, a, is desired rather than knowledge of the product k_La only. One method is the measurement of both gas hold-up and the Sauter mean bubble diameter $\bar{d}_b$:

$$\bar{d}_b = \sum (n_i d_b^3) / \sum (n_i d_b^2) \qquad (33)$$

Then, the specific contact area follows from:

$$a = 6\varepsilon_g/\bar{d}_b \tag{34}$$

See for reviews on this technique [79,96]. This method has been followed by Deckwer et al. [97,98] for determining the specific contact area, a, in Fischer-Tropsch reactors. The advantage of the method is that it is applicable for any system under the relevant reactor conditions. Disadvantages of the method are that usually only local values of the specific contact area, a, are obtained, that it is time consuming and that the bubble size distribution might be disturbed by the sample technique. A second method is the so-called chemical absorption technique which has been introduced by Westerterp [99,100] and Yoshida and Yoshiharu [101] and has found a wide application since then. General review articles have been published by Linek and Vacek [102], Sharma et al. [67,68,93] and by Laurent et al. [66]. The principle of the method is simple: if a component is transferred from a non-reacting phase into a reacting phase in which it reacts chemically enhanced according to $J_A a = k_L\phi a c_{Ai}$, then the product $k_L\phi$ is a function of kinetics only and is independent of k_L. Under these conditions the contact area, a, follows straightforwardly from the observed conversion rate and the kinetics, e.g., for a mixed tank reactor with no gas phase limitation,

$$a = \frac{J_A a V_r}{k_L\phi m\bar{c}_{AG1}} = \frac{-\Phi_A}{k_L\phi m\bar{c}_{AG1}} \tag{35}$$

With a second reactant B, already present in the liquid, we have, for kinetics of order (1,1) and $E_{A,\infty} >> \phi > 2$:

$$k_L\phi = \sqrt{(k_{1,1}D_A\bar{c}_B)} \text{ and} \tag{36}$$

$$a = \frac{-\Phi_A}{m\bar{c}_{AG1}((k_1 D_A\bar{c}_B))^{1/2}} \tag{37}$$

For a steady state column reactor the specific contact area, a, can be determined by solving

$$\frac{d(u\bar{c}_A)}{dz} - D_1\varepsilon\frac{d^2\bar{c}_A}{dz^2} + a\ m\ \bar{c}_{AG}\sqrt{(k_{1,1}D_A\bar{c}_B)} = 0 \tag{38}$$

with the appropriate boundary conditions. In this relation $\bar{c}_A$, u, ε and D_1 refer to the (dispersed) non-reaction phase. The specific contact area, a, is found by trial and error. In practice, many complications occur, e.g.

- the reaction influences the specific area [103]
- gas phase resistance is not negligible
- the condition $2<\phi<<E_{A\infty}$ is not fulfilled in a part of the reactor
- the measured area depends on the ratio:
  factor by which reaction in creases the capacity of the reaction phase)/ (enhancement factor)

The latter effect has been observed in packed columns [104].

Most common model reactions for the measurement of a are the absorption of CO_2 in hydroxide solution (Danckwerts [115]) and the absorption of oxygen in aqueous sodium sulphite solutions catalysed by CO^{2+}, (Reith and Beek [105], and Linek and Vacek [102]). The latter system is sensitive to the sulphite kinetics, the cobalt and the water quality. An induction period before the reaction starts cannot be excluded [67]. The suitability of a great number of gas-liquid reactions as model reactions has been discussed by Sharma et al. [67,68]. The numerical value of the specific contact area is often related to the power consumption. The relation between these two parameters depends on the reactor type, (Nagel et al. [106]). Reviews, covering conventional two-phase gas-liquid reactors have been presented by Laurent and Charpentier [69], Charpentier [70] and van Landeghem [71]. For bubble columns we refer to the reviews of Gestrich et al. for liquid side mass transfer coefficients [72], specific contact area [73] and hold up [74] and further to Mashelkar [75]. For gas-liquid stirred tank reactors we refer to the reviews of Sideman et al. [76], Reith [78], Reith and Beek [77] and van Dierendonck et al. [79]. From the above reviews it appears that relations are available for predicting the value of specific contact areas, particularly in aqueous systems and at atmospheric pressure. Which one is the best, usually will depend on the actual reactor conditions. It is always risky to use a relation outside the range of conditions under which the underlying experiments have been carried out. Table VI gives the relations of Akita and Yoshida [107]; Table VII the relations of van Dierendonck for bubble columns only. For stirred tank reactors, the reader is referred to the literature [79,108]. Note the difference in bubble diameter between pure liquids and liquids containing electrolytes or surfactants. Recently a new relation has been proposed by Hammer et al. for pure liquids [109,110]. It has been broadly tested with respect to viscosity and density of the liquid but is limited with respect to u_G. The paper also shows how mixtures may give substantially higher values of contact area than each of the components separately. If data are needed far away from ambient temperatures at increased pressures and with non-aqueous systems, still measurements are necessary at actual reactor conditions because the general relations found in the literature may lack the desired accuracy or fail completely. This has been shown [97,98] by Deckwer for Fischer-Tropsch reactor systems. He therefore measured the area in Fischer-Tropsch reactors experimentally. His result:

Table VI. Specific contact area in bubble columns according to Akita and Yoshida [107] $10 \le T\ [°C] \le 40$; $790 \le \rho_L\ [kg/m^3] \le 1165$; $0.58*10^{-3} \le \mu_L [Ns/m^2] \le 21*10^{-3}$; $60*10^{-3} \le \sigma\ [N/m] \le 74*10^{-3}$; $0.26*10^{-9} \le D_L[m^2/s] \le 4.2*10^{-9}$

$$d_b/d_t = 26(Bd)^{-0.50}(Ga)^{-0.12}(Fr)^{-0.06}$$

$$\frac{\varepsilon}{(1-\varepsilon)^4} = 0.20(Bd)^{1/8}(Ga)^{1/12}(Fr)^{0.5}$$

$$a = 6\varepsilon/d_b$$

Table VII. Specific contact area in G-L bubble columns ($d_G > 0.15$ m) according to van Dierendonck [79,108]; $T = 20$ °C; $0.4*10^{-3} \le \mu\ [Ns/m^2] \le 2.6*10^{-3}$; $18*10^{-3} \le \sigma\ [N/m] \le 75*10^{-3}$; $700 \le \rho_L\ [kg/m^3] \le 1230$.

relation	pure liquids	solutions with electrolytes or surfactants
ε	$\varepsilon = 1.2(Fr^3\ Bd^3/\ Ga)^{1/8}$ (for $\varepsilon < 0.45$ and $u_G < 0.4$ m/s)	ibid.
d_b	$d_b/d_t = (0.95 \pm 0.05)\ Bd^{-0.5}$ (for $u < 0.03$ m/s)	ibid.
	$d_b/d_t = 2.50\ (Bd^3FrGa)^{-1/8}$ (for $0.03 < u_G < 0.3$ m/s)	$d_b/d_t = 1.45\ (Bd^3FrGa)^{-1/8}$ (for $0.03 < u_G < 0.3$ m/s)
a	$6\varepsilon/d_b$	ibid.

$$a = 710\ u_G^{1.1} \tag{39}$$

with u_G in m/s and a in m^{-1}. On the other hand, Brown [111] successfully applied the equations of Akita and Yoshida [107] for k_L and the contact area, a, in bubble columns to design a pressurized slurry reactor for methanol synthesis. Also here the reactor conditions are far away from the conditions used in the experiments of Akita and Yoshida. Brown [111] realized, however, that k_La is strongly influenced particularly by the gas hold-up ε_G. Therefore, Brown [111] measured ε_G under actual reactor conditions. This experimental value was substituted in the k_La relation of Akita and Yoshida [107] rather than the relation of Akita and Yoshida [107] for ε_G. Though Brown [111] still had to carry out experiments, his approach is attractive because measurements of ε_G usually are simpler and cheaper than measurements of k_La.

Data on the influence of solids on the value of the contact area, a, are very scarce and do not show a uniform pattern. In Figure 10

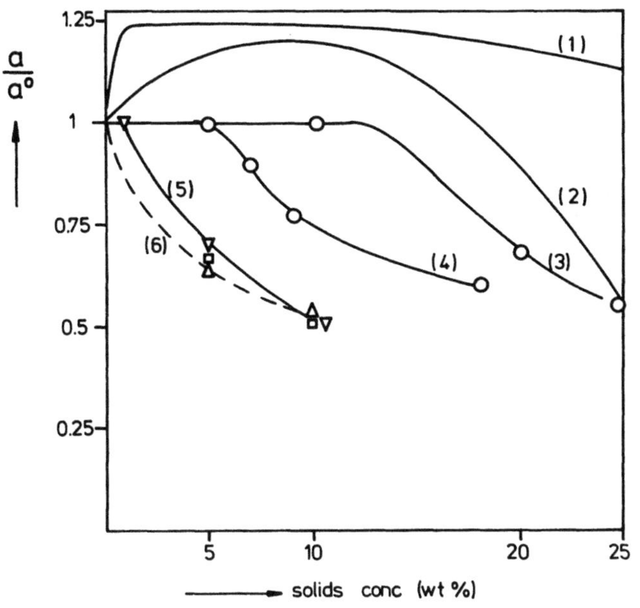

Fig. 10 Relative change of gasliquid interfacial area as a function of solids concentration (1) 7 μm $Ca(OH)_2$ particles in BC [83]
(2) 2 μm $Mg(OH)_2$ particles in BC [83]
(3) (●) 0-10 μm activated carbon in BC; u_G = 8 cm/s [112]
(4) (○) 0-10 μm Kieselguhr in BC; u_G = 8 cm/s [112]
(5) (▼) 0-10 μm Al_2O_3 in BC; u_G = 8 cm/s [112]
(6) 100 μm glass bedds in STR [113]
▲: $\dot{W}/V_r$ = 0.1 kW/m^3, ■ : $\dot{W}/V_r$ = 1 kW/m^3
BC = bubble column, STR = stirred tank reactor

we summarize data from the open literature [83,112,113]. All have been measured by a chemical method with aqueous systems. For relatively large particles in a stirred tank reactor (d_p = 100 μm), the measurements of Kürten and Zehner [113] (Figure 10, line 6) suggest a rapid decrease of area with increasing solids concentration. Also with small particles in the 1-15 μm range the data of Quicker et al. [112] suggest a rapid decrease of interfacial area with increasing solids concentration, though not before a certain, minimum solid content has been exceeded. However, a unique correlation that predicts this decrease in area quantitatively has not been found yet. Quicker et al. tried a relation of the type:

$$a = c u_G^n \mu_{eff}^m \tag{40}$$

with μ_{eff} an effective viscosity of the liquid-solids suspension. They indeed found an unique relation for kieselguhr and activated carbon particles:

$$a = 5.83\ u_G^{0.81}\mu_{eff}^{-0.22} \qquad \text{(S.I.-units)} \quad (41)$$

but a completely different relation for alumina particles:

$$a = 2.05\ 10^{-8}\ u_G^{0.65}\mu_{eff}^{-3.09} \qquad (42)$$

suggesting that an equation of type of eq (40) does not give all parameters influencing the interfacial area.

Different from other authors, Sada et al. [83] recently reported a significant increase of the specific contact area upon adding small particles (curves 1 and 2, Figure 10). So far, no explanation has been presented of this different behaviour but we guess that the increase might have been caused by a change of the system from a pure liquid to an ionic system. Sada et al. [83] measured the area by absorbing CO_2 which reacts with the hydroxide dissolved and, probably, to some extent, also with the particles undissolved. The system becomes ionic which, according to the relations of van Dierendonck (Table VII), results in much smaller bubbles and consequently in larger values of the interfacial area. The data of Quicker et al. [112] can be replotted to check whether the relative decrease of interfacial area only depends on the amount and type of solids added or that the relative decrease also varies with the superficial gas velocity u_G in the bubble contactor. Figure 11 shows that a/a^0 may decrease with u_G but not always and that no unique relation has been found yet. We conclude that, though a few interesting papers have been published recently, we are still far from a unique correlation predicting the interfacial area directly from a system and for given operating parameters. Much work still has to be done in this respect.
With no data available, Hoffmann [5] suggests as a scale up rule for design purposes that the same interfacial area will require the same total power input per unit volume of liquid. This still is a useful suggestion but both Figure 10 and the paper of Nagel et al. [106] indicate that its accuracy is limited.

5.4. The true liquid side mass transfer coefficient at the gas-liquid interface (k_L)

Numerical values of the mass transfer coefficient k_L can be derived from the previously discussed measurements of the contact area, a, and of the product k_La via $k_L = k_La/a$. However, in fluid-fluid reactions, the hydrodynamics of the system may be influenced by the chemical reaction, causing a change in both k_L and a with a change of the reaction rate. This has been shown by Linek [103,114,115] and Brian et al. [116]. This disadvantage can be overcome by using the so-called Danckwerts method [117] in which carefully selected chemical reactions of a moderate fast rate are applied with $\phi \simeq 1$. Sharma and Danckwerts reviewed suitable reaction systems for both

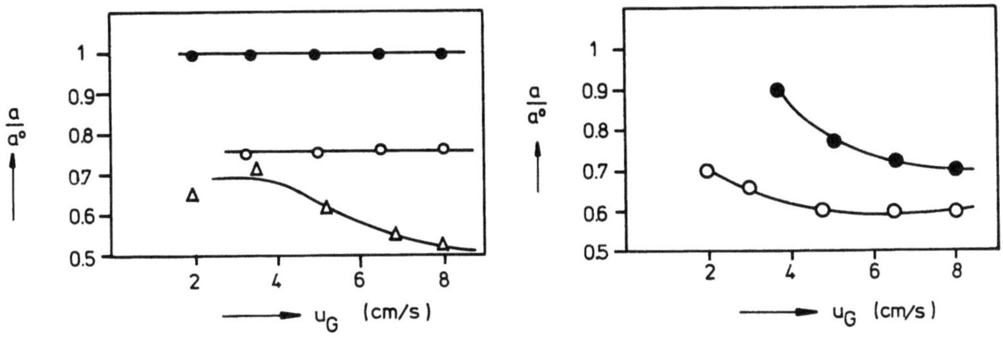

Fig. 11 Relative decrease of interfacial area in a bubble column due to the addition of fine solids (0-15 μm) as a function of superficial gas velocity
(a) solids concentration 10 wt % (char 9.2)
(b) solids concentration 20 wt % (char 18.4)
● kieselguhr (ρ = 2360)
o activated char (ρ = 1800)
▲ alumina (ρ = 3180)
Data taken from Quicker et al. [112]

gas-liquid and liquid-liquid reactors [67]. According to the Danckwerts penetration model for kinetics (1,1), no diffusion limitation of B and $\bar{c}_{AL}$ zero, J_A follows from:

$$J_A = mc_{AG} \sqrt{(D_A k_{1,1} \bar{c}_B + k_L^2)} \qquad (43)$$

so that, e.g., for a stirred tank reactor:

$$\frac{\Phi_A}{m\bar{c}_{AG} V_r} = a\sqrt{(D_A k_{1,1} \bar{c}_B + k_L^2)} \qquad (44)$$

Now $k_{1,1}\bar{c}_B$ is varied from $\phi \ll 1$ to $\phi \gg 1$ by changing $\bar{c}_B$ or, preferably, by changing $k_{1,1}$ by varying the catalyst concentration. In case k_L and the contact area, a, do not change with the reaction rate a straight line is obtained in a plot of $(\Phi_A/m\bar{c}_{AG}V_r)^2$ against $k_{1,1}\bar{c}_B$. From the slope, $a^2 D_A$, and the intercept, $(k_L a)^2$, both k_L and a are obtained. Unfortunately, this method is very time consuming because every (k_L,a)-combination calls for a whole series of experiments. Therefore Robinson and Wilke [93] and Beenackers and Van Swaaij [94] introduced the method of simultaneous mass transfer of two components. One component reacts chemically enhanced according to $E_A = \phi$ and gives information on the value of the interfacial area, the other component is inert and gives information on the value of the product $k_L a$ under exactly the same hydrodynamic conditions. In this way k_L and the contact area, a, are obtained by one experiment only. Concerning numerical values of k_L in two phase gas-liquid reactors, we refer to the same reviews mentioned in the previous section with respect to interfacial area. For conventional reactors k_L usually can be predicted from the existing correlations with much more accuracy than the contact area, a. We suggest the Calderbank - Moo Young [118] equations as interpreted by van Dierendonck for pure liquids [108] (see Table VIII). For bubble columns Akita and Yoshida [107] found:

$$k_L = 0.5\left(\frac{g^5 D_L^4 \rho_L^3 d_b^4}{\sigma^3}\right)^{1/8} \tag{45}$$

With respect to the influence of solids on the numerical value of k_L, again, the data are scarce. Firstly, one has to distinguish between two types of particles: those with no or negligible adsorption capacity for the component transferred and those with, preferably, a high adsorption capacity. From the latter type of particles it is known now, and also understood, that it may enhance the mass transfer process. It is confusing, however, to ascribe this to an increase in the mass transfer coefficient k_L. According to our view, this effect should be described in the form of an enhancement factor, analogous to the established theories on mass transfer with

Table VIII. Mass transfer coefficients in two phase gas-liquid reactors according to van Dierendonck [108]

system	k_L
bubble column	$k_L = 0.42\left(\frac{\mu_L g}{\rho_L}\right)^{1/3}\left(\frac{D_L}{\nu_L}\right)^{1/2}$ $(d_b > 2\ mm)$
both ionic and non-ionic	$k_L = k_{L(d_b \geq 2\ mm)}\ 500\ d_b$ $(d_b < 2\ mm)$
liquids; stirred	with: minimum k_L:
tank with pure liquids and	$k_L = 0.28\left(\frac{\mu_L g}{\rho_L}\right)^{1/3}\left(\frac{D_L}{\nu_L}\right)^{2/3}$
solutions with non-ionic solutes	
stirred tank with solutions	$k_L^{-1} = k_{L,o}^{-1}\left[1.2 + 260\frac{\mu_L}{\sigma} v_r\right]$
of ionic solutes	$k_{L,o} = 1.13\left(\frac{D_L v_b}{d_{b,o}}\right)^{0.5}$; $d_{b,o} = 0.8\left(\frac{\sigma}{\rho_L g}\right)^{0.5}$
	$v_b = 1/18\ \rho_L g\ d_b^2/\mu_L$; $(Re_b < 1)$
	$v_b = 1/4\ d_b\left(\frac{\Delta\rho^2 g^2}{\rho_L \mu_L}\right)^{1/3}$; $(30 < Re_b < 10^3)$
	$v_b = 1.76\left(\frac{\Delta\rho g d_{b,o}}{\rho_L}\right)^{0.5}$; $(Re_b > 10^3)$

chemical reaction (see next section). As a consequence, reliable measurements on the influence of particles on k_L should be carried out with non absorbing particles. Alper et al. [82] measured the influence of quartz particles (d_p<5 μm) and of oxyrane-acrylic beads (d_p = 15-20 μm) on k_L in a stirred cell reactor and found no effects. Also Sada [119] et al. found no effect of 0-20 wt % $Ca(OH)_2$ fine solids. Consequently, the conclusion that k_L is not affected as long as the solids concentration remains below 20 wt % seems to be valid.

On the other hand, Sada et al. [83] recently reported a substantial decrease of k_L with solids concentration in a bubble column (Figure 12). This is not necessarily conflicting with the findings of Alper et al. [82] because in a bubble column k_L may be influenced indirectly through a change in the interfacial area (and therefore a change in the bubble diameter d_b) due to the presence of particles. However, addition of 20 wt % $Mg(OH)_2$ solids apparently results in a decrease of both the contact area, a, and k_L (Figures 10 and 12). Now a decrease in the contact area, a, would suggest larger bubbles and therefore we would expect k_L to remain either constant or, if the increasing d_b causes a change in bubble regime from stagnant to mobile, even an increase in k_L. Here, the data seem to be inconsistent. In studying the effects of solids on k_L and the contact area, a, not only the average particle diameter but also the possible influence of particle size distribution should be investigated.

5.5. Solids-liquid mass transfer

The rate of mass transfer between the liquid phase and the solid particles can be a limiting factor in slurry reactors. Therefore a reliable relation for the prediction of this transfer rate is desired. Two different approaches are used to find a theoretical

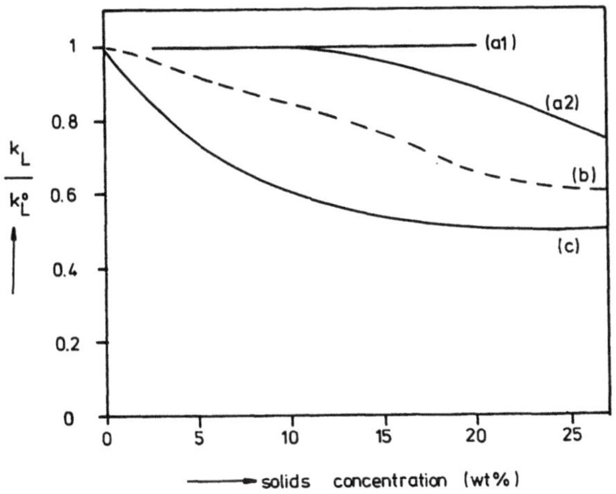

Fig. 12 Relative decrease of the true mass transfer coefficient (k_L/k_L^o) due to the addition of fine solids to the liquid.
(a) in stirred cell contactor with "fine" (10 μm?) $Ca(OH)_2$ particles (a1) [119]
(a2) [120]
(b+c) in bubble column [83] with 7 μm $Ca(OH)_2$ particles (b) and 2 μm $Mg(OH)_2$ particles (c)

basis for correlating mass transfer data. For a single particle in a stagnant infinite diffusion field the Sherwood number is given by:

$$Sh = \frac{k_{LS}d_p}{D} = 2 \tag{46}$$

In slurry systems, of course, the mass transfer is enhanced by convective currents originating from bulk convection and turbulence. In the first approach main emphasis is put on bulk convection and mass transfer related to the particle (terminal) slip velocity v_T e.g. via the modified Frössling equation:

$$Sh = 2 + m\, Re_p^{0.5} Sc^{1/3} \tag{47}$$

$$Re_p = \frac{d_p v_T}{\nu} \tag{48}$$

For a turbulent flow field m is not equal to the constant given by Frössling : (m = 0.6). Harriot [121, 122] and Shah [2]. showed that m values in slurries may vary from 1-4 times the values obtained for terminal velocities for a freely falling particle. A fundamental difficulty is the estimation of the slip velocity which should be related to the global system parameters [123,124]. Therefore although this type of approach could be applied to three phase fluid beds where slip velocities can be more directly related to liquid velocities in slurry reactors it is not to be recommended. Moreover, the concept of slip velocity in relation to the terminal velocity completely fails if the densities of liquid and particles are the same. Chapman et al. [125] nevertheless recommended mass transfer relations based on circulation velocities in stirred slurry reactors because of the more direct relation of the mass transfer to flow patterns in the tank, which is obscured by taking energy input as a basis. Energy input per unit mass of liquid is the principle factor in the second approach to the liquid/solids mass transfer problem, on the Kolmogoroff's theory of local isotropic turbulence. This theory leads to a Re number based on the velocity of the critical eddies:

$$Re \div \left(\frac{\bar{e} d_p^4}{\nu^3}\right)^{1/3} \tag{49}$$

The Re number, for energy dissipation by the gas phase, e.g., in sparged columns $\bar{e} = u_G g$, can be used in an equation similar to the Frössling equation (Sano et al. [126], Sänger and Deckwer [127]). The latter authors proposed:

$$Sh = 2 + 0.545\, Sc^{1/3} \left(\frac{\bar{e} d_p^4}{\nu^3}\right)^{0.264} \tag{50}$$

Equation (50) is recommended for sparged gas/solids/liquid contactors and has been tested over a wide range of conditions:

$$10^{-3}<(\frac{\bar{e}d_p^4}{\nu_L^3})<10^{-7} \text{ and } 175<Sc<50.000.$$

s indicated by Sano et al. [126], who presented an overview of the correlations for solid-liquid stirred vessels, the same relation can be used for stirred vessels with an energy input given by the total energy input supplied by the gas phase and the stirrer. This procedure has been criticized by Chapman et al. [125] because several phenomena related to particle suspension cannot solely be described by a specific energy input and a lot of interesting details on stirrer qualities would be lost. More work will be needed to clarify this point.

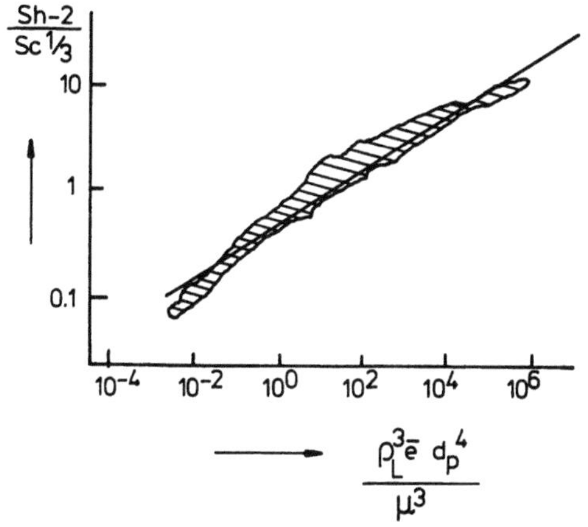

Fig. 13 Correlation of liquid-solid mass transfer data in bubble columns (Sänger and Deckwer [127]). Shaded area: experimental points 137<Sc<50,000

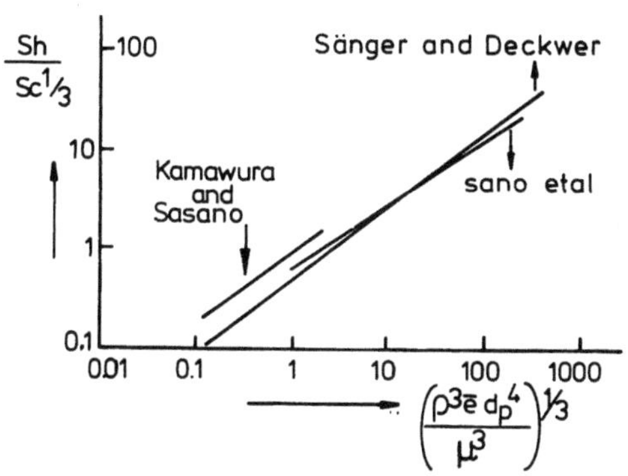

Fig. 14 Correlations of liquid-solid mass transfer coefficients in bubble columns (Sänger and Deckwer [127])

5.6. Measurement of intrinsic reaction kinetics

Measuring intrinsic reaction kinetics is not an easy procedure and in any case asks for skilled reaction engineers to avoid the many pitfalls that may hamper a sensible analysis of the experimental findings. The problem always is to find conditions for which mass transfer limitation effects ideally are absent or at least can be separated from the intrinsic kinetics. Therefore, in most reactors a form of intense mixing is applied which, in combination with a low solids concentration, may result in absence of gas-liquid mass transfer effects. Both continuous and semi-batch stirred tank slurry reactors are in use. Here the problem still is that the mass transport to the particles is only slightly influenced by the mixing effect of the stirrer, particularly with small particles. For this reason, reactors with internal recycle, as originally developed for catalytic gas-solid reactions (spinning basket), also have been applied for measuring kinetics in three phase systems. Recently, both Ohta et al. [128] and Njiribeako et al. [129] measured the intrinsic kinetics in the catalytic oxidation of aqueous phenol solutions with such a reactor. For the same system also a well-stirred slurry semi-batch reactor has been used [130]. Operating and design data of such a slurry reactor, when used for kinetic measurements, can be found in the paper of Huff and Satterfield [131]. Additional advantages of basket type model reactors are:

- less problems with catalyst attrition
- easier sampling because no danger of particle clogging of the particles in the sampling pipes

These advantages must be counter-balanced by such disadvantages as risk of insufficient catalyst wetting due to centrifugal forces at high rotation speeds and the fact that the particle diameter usually can not be as small as for slurry reactors because d_p is limited by the screen size of the basket. If so, elimination of mass transfer effects still may provide a problem. For an entry on advantages and disadvantages of the various model reactor types presently in use for measuring intrinsic kinetics in three phase systems we refer to the recent review of Shah and Smith [132] and to the book of Shah [2].

A new development may be the use of transient techniques in slurry model reactors. Weng and Smith [133] recently presented a model for the reactant concentration in the gas outlet if a pulse of reactant was introduced in the gas feed. Mathematically they showed the possibility of distinguishing between the adsorption rate and the true surface reaction rate. A slightly different mathematical treatment is presented by Datta and Rinker [64]. We do not know of experimental work on measuring intrinsic kinetics using such transient techniques.

5.7. Enhancement of gas-liquid mass transfer

In slurry reactors, enhancement of the gas-liquid mass transfer process can occur due to both a fast chemical reaction and the

Table IX. Types of enhanced mass transfer at gas liquid interface in slurry reactors

type no.	system type	references
I	mass transfer followed by physical adsorption on small particles (shuttle mechanism; "grazing factor")	Kars et al. [81] Alper et al. [82] Kralj & Sincic [63]
II	mass transfer followed by fast homogeneous reaction in the liquid (particles are inert)	Numerous; see e.g. [61]
III	mass transfer followed by fast reaction in liquid with reactant originating from small solving particles	Sada et al. [83,119,120, 134], Alper and Deckwer [135], Ramachandran & Sharma [136], Uchida et al. [137,138]
IV	ibid, but with heterogeneously reacting small particles	Alper and Deckwer [135] Ptasinski et al. [16,139]
V	mass transfer followed by fast catalytic reactions in small particles	Kut et al. [140] Alper et al. [141] Sada et al. [142,143] Sharma et al. [21,130] Wimmers & Fortuin [144, 145]

presence of small porous particles with a high adsorption capacity for the components transferred. It may result in a change of R_2 in Equation (23) according to Equation (26). However, in deriving expressions for the enhancement factor, every system asks for a special analysis, because the mechanism causing the enhancement is intrinsically different for each case. In Table IX we classify the various types of slurry systems deserving an individual treatment. Below, each system is discussed separately.

Type I: Mass transfer with physical adsorption on small particles
Much confusion on the influence of particles on k_L and k_La has been created in the open literature by not properly distinguishing between the type of particles used. Various authors report a substantial increase of k_La or the contact area, a, by adding small quantities of fine activated carbon particles with diameters in the order of the film thickness or smaller. As has been shown by Kars et al. [81] and more recently also by Alper et al. [82], this effect is caused by the high adsorption capacity of activated carbon for the component transferred. The effect completely disappears if non-porous particles with an equally small diameter are used instead (Figure 15). The statement, often found in the literature dealing with activated carbon particle slurries, that the presence of such particles causes an increase in k_La or in k_L is confusing and should be avoided. According to our view, both k_La and k_L should be defined in such a way that these parameters are independent, not only of the physical absorption capacity in two phase

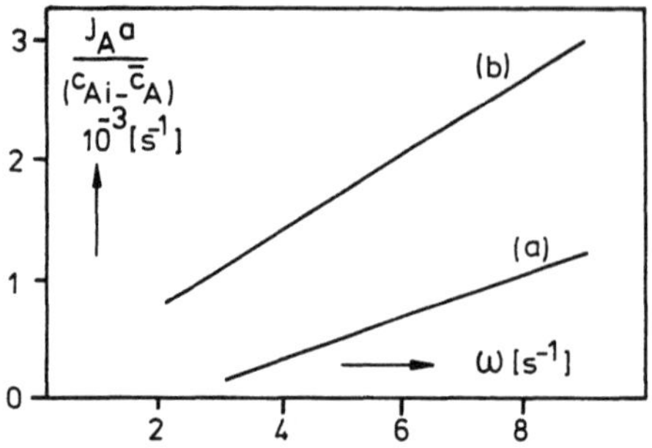

Fig. 15 Rate of physical mass transfer of oxygen in aqueous solutions in a stirred cell reactor
(a) no particles or quartz (d_p < 5 μm) or oxyrane acrylic beads (d_p ≃ 15-20 μm) added
(b) with 0.1-3 wt % activated carbon added
Adapted from Alper et al. [82]

fluid-fluid systems, but as well of the physical adsorption capacity in two phase fluid-solid systems, and of the reaction kinetics in reactive systems. This is common sense, but k_La should also be independent of the physical adsorption capacity of any solids present which, untill now, has not been generally accepted. For reactive systems the concept of the enhancement factor, E_A, has been developed:

$$E_A = \frac{J_{A,\ \text{with reaction}}}{J_{A,\ \text{without reaction}}} \tag{51}$$

with both fluxes based on the same driving force for mass transfer $(c_{Ai}-\bar{c}_A)$ where A is the reactant transferred into the reaction phase. It is our opinion that a similar enhancement factor should be defined in the case of mass transfer with physical adsorption on small particles in slurry reactors. We propose, by analogy with Equation (51):

$$E_{A,phys} = \frac{J_{A,\ \text{with adsorbing particles present}}}{J_{A,\ \text{with similar non-adsorbing particles present}}} \tag{52}$$

with both fluxes based on the same driving force for mass transfer $(c_{Ai}-\bar{c}_A)$. The relations presented in Figure 15 have been measured in a stirred cell with a flat and therefore a constant gas-liquid interface so that $E_{A,phys}$ can be calculated directly by dividing the upper line values by the corresponding lower line values. The result is shown in Figure 16, line b. Notice that the independent variable ω^3 is proportional to the specific power input. Recently, additional data have been published by Kralj and Cincic [63]. These authors measured both k_La and the gas hold-up as a function of specific power input in a stirred tank reactor both with and without small activated carbon particles in the liquid. These findings with 0.25 wt % activated carbon particles could be correlated according to the following equation

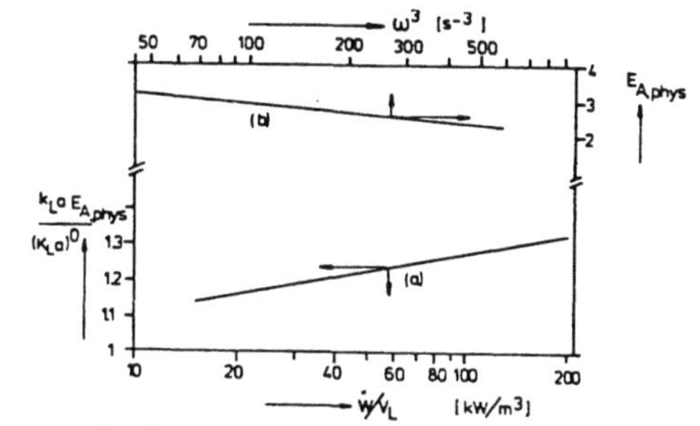

Fig. 16 Enhancement of mass transfer in physical absorption of oxygen in oxygen in aqueous solutions due to the addition of activated carbon particles as a function of specific power input
(a) 0.25 wt % in stirred tank reactor [63]
(b) 1-30 wt % in stirred cell; $\bar{d}_p$ < 5 μm [82]

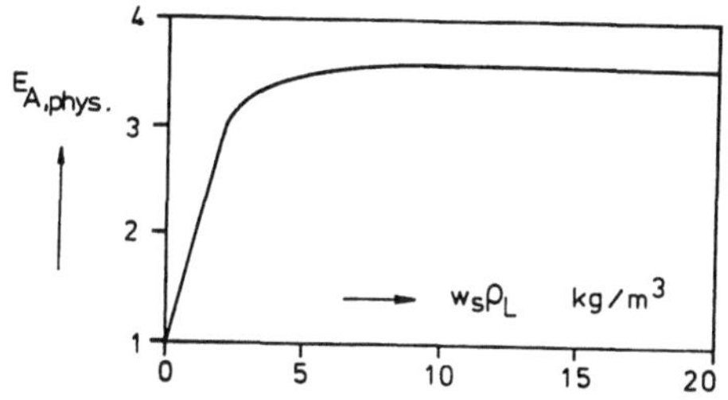

Fig. 17 Enhancement factor with physical absorption of CO_2 in aqueous solutions as a function of the activated carbon concentration in a stirred cell with ω = 1.33 s^{-1} and $\bar{d}_p$ < 5 μm.
Adapted from Alper et al. [82]

$$\frac{k_L a\ E_A}{(k_L a)^o} = 0.6158\ (\frac{\bar{W}}{V_L})^{0.058}\ u_G^{-0.038} \qquad (53)$$

Equation (53) is also shown in Figure 16 (line (a)). At a first glance the data of Alper et al. (line (b)) look inconsistent with the data of Kralj and Cincic (line (a)) but this is not necessarily so. Firstly, Kralj and Cincic used a lower carbon concentration of 0.25 wt %, and for such a low value Alper et al. report a lower enhancement factor (Figure 17). Secondly, in the experiments of Kralj and Cincic [63] the value of the interfacial area probably decreased for the lower values of $\bar{W}/V_L$ and increased for the higher values of $\bar{W}/V_L$ after addition of the particles, both relative to the conditions with no particles present. This, at least, is suggested by the results of their hold-up measurements. Kralj and Cincic [63] do not report the values of the interfacial areas realized in their experiments. It prevents an unambiguous comparison with the data reported by Alper et al. [82]. Measurements of Kars et al. [81] suggest (Figure 18) that enhancement is already possible for d_p substantially larger than the film thickness. In their experiments $k_L \simeq 2.5*10^{-4}$ m/s, $D_L = 1.2*10^{-9}$ m²/s and consequently in experiments with their smallest particle diameter (34 μm) $d_p/\delta = k_L d_p/D_L = 7$. With increasing d_p, however, the enhancement rapidly decreases.

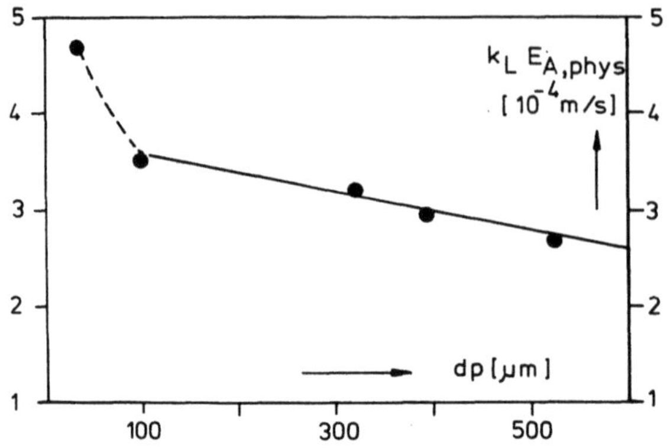

Fig. 18 Decrease of enhancement of mass transfer in the physical absorption of propane in slurries of activated carbon in water with increasing average particle diameter. (1 wt % activated carbon; $D/k_L \simeq 5$ m) From Kars et al. [81]

Concerning the theory on predicting E_A, virtually nothing has been published as yet. Only Kars et al. [81] mention an asymptotic value of E_A for $d_p \to 0$.

$$E_{A,phys}\ (d_p \to 0) = (1 + (1-\varepsilon_s)m_{LS})^{0.5} \tag{54}$$

with: m_{LS} = the distribution coefficient of the absorbing component between solids and liquid (kmol/m^3 solids)/(kmol/m^3 liquid). For the measurements reported in Figure 18, Equation (54) predicts for $d_p \to 0$ the asymptotic value of $E_{A,phys}$ = 4.6 which is very close to the value reported for d_p = 34 μm. As follows from Figure 17, after a certain minimum amount of particles have been added, no further increase of the enhancement is found with increasing solids content. Similar observations have been reported for mass transfer with reactions of type III and V (Table IX). As first pointed out by Alper et al. [82] for catalytic reactions, this might be caused by the fact that with increasing E_A the penetration depth of A decreases and may become of the same order of magnitude as the particle diameter. Such a mechanism may also play a role in the physical adsorption processes discussed here. It suggests a maximum enhancement in the order of:

$$E_{A,max} = \alpha \frac{D_L}{k_L d_p} \tag{55}$$

In the experiments reported in Figure 17, the particle diameter, d_p, is smaller than 5 μm [82] while δ was larger than 50 μm [135]. With the experimentally observed $E_{A,max} \simeq 3.5$, the data suggest: $\alpha \leq 0.35$. However, with no more data available, eqn (55) remains speculative. For intermediate values of d_p in the order of 0.5-10 D_L/k_L (= δ in film theory) equation (54) is too simplistic. Derivation of a realistic theory in this area might prove to be rather complicated as probably such effects as the movement of particles right through the boundary layer for mass transfer, whilst picking-up adsorbate there, have to be taken into account. Kars et al. [81] introduce the term "grazing" by the solids. We note the analogy between this type of enhancement in physical adsorption and the so-called shuttle mechanism in e.g. absorbing carbon dioxide in an aqueous mixture with secondary and tertiairy amines [61].

A fascinating application of enhancing mass transfer by physical adsorption on activated carbon recently has been published by Janakiraman and Sharma [21]. These authors studied the role of addition of activated carbon particles with $d_p << D_L/k_L$ in alkaline hydrolysis of tridecyl formate and benzyl benzoate (liquid-liquid systems) in alkaline hydrolysis of phenyl benzoate and ethyl p-nitro benzoate and in oximation of cyclododecanone. With carbon particle loadings below 0.02 kg/m^3 (to prevent agglomeration) the conversion rates could be enhanced by a factor 1.2 to 5. All systems were analysed theoretically and good agreement with experimental findings was obtained.
Enhancement of the type as discussed here is also possible with small droplets instead of particles. Recently, such effects indeed have been observed (Fig. 19) and explained by a model [146]. The effect is of interest in biochemical reaction engineering where the low solubility of oxygen in the aqueous solutions often is the rate limiting step. Addition of small droplets of e.g. fluoro hydrocarbons, in which oxygen is much more soluble than in water, may increase the rate of oxygen transfer substantially.

Type II: Mass transfer with fast homogeneous reaction in the slurry: particles are inert

This case is basically similar to mass transfer and reaction in two phase systems which is well described in the literature.

Type III: Mass transfer with fast reaction in the liquid with a reactant originating from small dissolving particles

This system can be described as: $A(G) \rightarrow A(L)$
$B(S) \rightarrow B(L)$

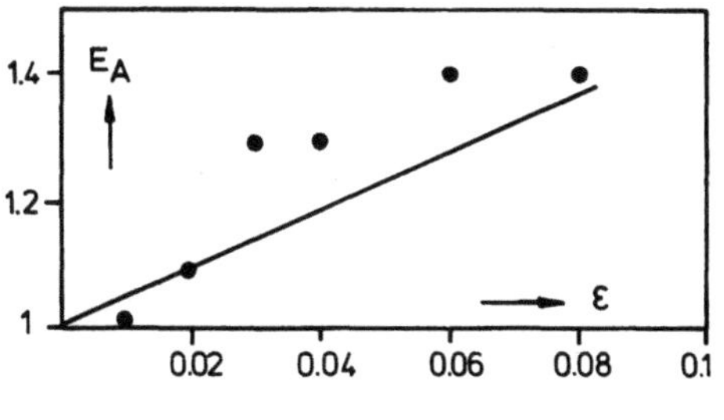

Fig. 19 Enhancement of oxygen mass transfer in aqueous solution by small droplets of hexadecane
ε vol % hexadecane
• experimental data
— model prediction
From Bruining et al. [146]

Table X. Mass transfer with fast reaction in the liquid with a reactant originating from small particles

SO_2 in $Mg(OH)_2$ slurries	Sada et al. [134]
CO_2 in $Ca(OH)_2$ slurries	Sada et al. [83,119]
SO_2 in $Ca(OH)_2$ slurries	Sada et al. [119,120]
CO_2 in $Mg(OH)_2$ slurries	Sada et al. [83]
SO_2 in melamine slurries	Uchida et al. [137]

$$A(L) + B(L) \rightarrow P(L)$$

A number of such reactions have been studied in detail during the last decade. We give an overview in Table X. For references before 1975 see [1].

As has been shown by Uchida and Wen [138] six different asymptotic cases of mass transfer interfering with instantaneous chemical reaction can be distinguished (Figure 20 a and b). In cases 1a to 1c the dissolution of the solids in the film at the G-L interface can be neglected relative to the dissolution in the bulk. Mass transfer of A then is in series with the dissolution process. The criterion under which this assumption is valid orginally was derived by Ramachandran and Sharma [136]:

$$Ha = \left(\frac{k_{LS} a_p D_A}{k_L^2 D_B}\right)^{1/2} < 0.2 \tag{56}$$

For: $\frac{k_{LS} a_p c_{Bi}}{k_L\ a\ c_{Ai}}$ larger than 1 (case 1a and 16) the controling resistance is at the gas-liquid rather than the liquid-solid interface and the gas-liquid mass transfer will be enhanced whilst the solids dissolution will be enhanced if that group is smaller than 1 (case 1c). In all three cases the enhancement is caused by the presence of dissolved B in either the gas-liquid (1a-1b) or the liquid-solid (1c) film and calculation of the enhancement factors is straightforward from the theory of mass transfer with homogeneous reaction in the liquid phase (Uchida and Wen [138]). The calculation of E_A is still straightforward for Ha > 2 and $(k_{LS} a_p c_{Bi})/(k_L a c_{Ai}) < 1$ (case 2c). A problem exists for both Ha > 2 and $(k_{LS} a_p c_{Bi})/(k_L a c_{Ai})$ > 1. We then have particles smaller than the film thickness for mass transfer at the gas- liquid interface and the absorption of A may be extra enhanced due to the dissolving of the particles within the film. A solution based on the film model originally was presented by Ramachandran and Sharma [136]. In their treatment they neglect any effect of possible enhanced dissolution of the particles in the film due to the chemical reaction. The latter effect has been included in a more recent model, published by Uchida et al. [147,148]. Figure 21 describes for concentration profiles at the liquid-solid and the gas-liquid interface according to these

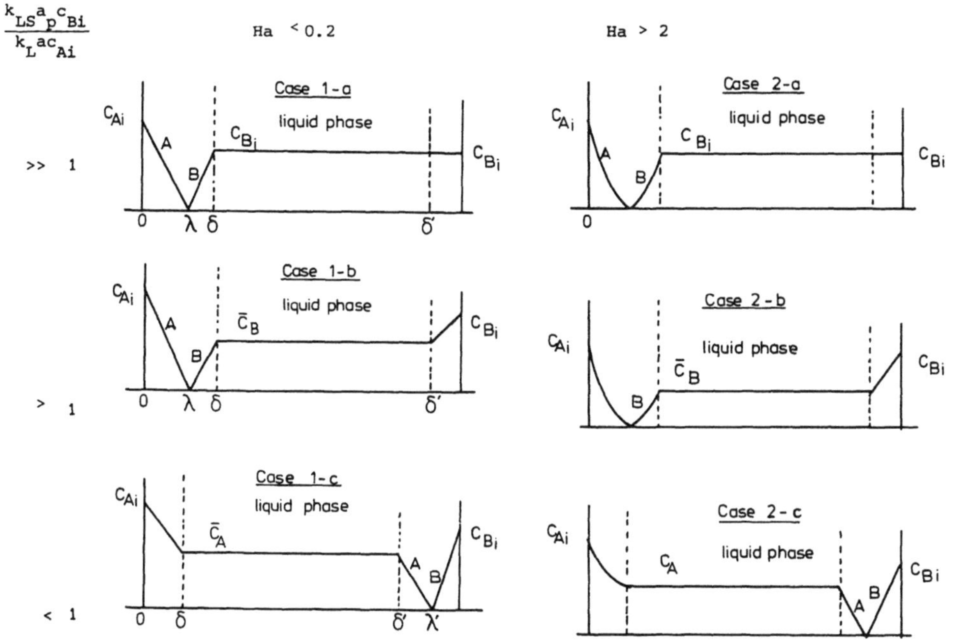

Fig. 20 The various cases for mass transfer with instantaneous reaction in the liquid with a reactant originating from solving particles

(a) Solid dissolution in the liquid film next to the gas-liquid interface is not important. Case 1-a, transfer of B is much faster than that of A so that the reaction plane is in the liquid film near the gas-liquid interface and the concentration of B in the bulk liquid phase is saturation concentration of B. Case 1-b, transfer of B is faster than that of A so that the reaction plane is in the liquid film near the gas-liquid interface. Case 1-c, transfer of A is faster than that of B so that the reaction plane is in the liquid film around the solid particles in the bulk liquid phase.
Adapted from Uchida and Wen [138]

(b) Solid dissolution in the liquid film next to the gas-liquid interface is important. Case 2-a, transfer of B is much faster than that of A so that the main reaction plane is in the liquid film near the gas-liquid interface and the concentration of B in the bulk liquid phase is saturation concentration of B. Case 2-b, transfer of B is faster than that of A so that the main reaction plane is in the liquid film near the gas-liquid interface. Case 2-c, transfer of A is faster than that of B so that the main reaction plane is in the liquid film around the solid particles in the bulk liquid phase.

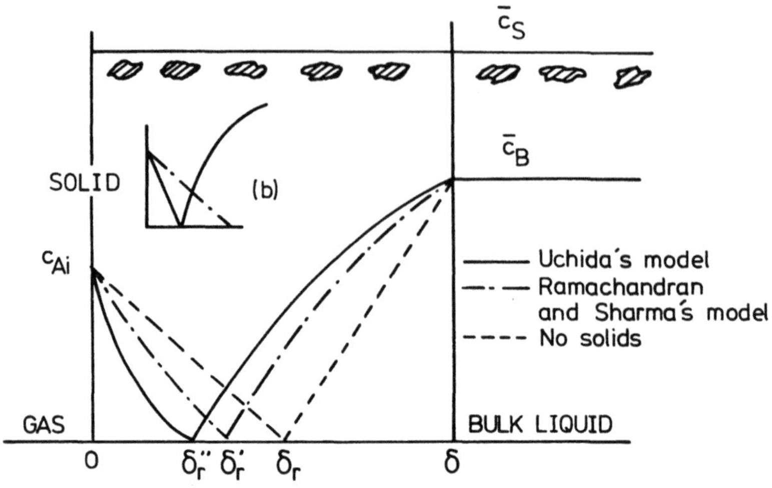

Fig. 21 Concentration profiles of dissolved gas A and sparingly soluble fine particles B (Infinitely fast reaction; no depletion of solid particles at gas-liquid interface)

models. Both models have been extended for finite reaction rates by Sada et al. [119] and experimentally found enhancement factors for the absorption for both CO_2 and SO_2 in $Ca(OH)_2$ slurries could be satisfactorily explained by the extended Uchida model [119]. There remains a problem however, film models by nature are stationary models and with the "diffusivity" of particles of 1-10 μm being very low (particularly in stirred cells where the diffusion of these particles with a higher density than the liquid must be in an upward direction) depletion of solids in the film can easily occur. In fact in such a case, stationary concentration profiles in the film will not occur before the film is empty of particles. Strictly speaking, film models should always predict an enhancement factor of 1 due to absence of soluble solids in the film under stationary conditions. For practical purposes there is no objection in taking the solids concentration in the film equal to the bulk concentration as long as the solids depletion time is large relative to the surface renewal time in terms of penetration theory. We did such a calculation for the example presented by Doraiswamy and Sharma [1, page 262-263] and found, assuming a molecular weight of the solids of 50 kg/kmol, an initial rate directly at the interface of $k_{LS}a_p$ $(c_{Bi}-0)$ = 750 kg/m^3 s;
The following parameters were used in this calculation: total solids present in bulk: 100 kg/m^3; penetration contact time from τ =

$4D/\pi k_L^2 = 0.1$ s.
Consequently there is substantial depletion of solids near the interface.
Recently, Sada et al. [120,149] reported absorption experiments for SO_2 and CO_2 in aqueous calcium hydroxide slurries in which they found much lower absorption rates than predicted by any of the models discussed above. They corrected their results by assuming a solids empty zone close to the interface. Following the reasoning of Alper et al. [82] according to which no particles can be suspended in a zone of thickness λ close to the interface, with λ of the order of the particle diameter, they took λ proportional to d_p and thus found a reasonable agreement between theory and experiments. This theory has been disputed [150] and therefore it might be worthwile to set up a rigourous numerical model based on penetration theory and taking solids depletion into account. Figure 22 illustrates the qualitative effect of particle depletion on concentration profiles. Finally it should be mentioned that a system belonging to class 1b (Figure 20) recently has been successfully modelled by Uchida et al. [137] (SO_2 in melamine slurry).

Type IV: Mass transfer followed by heterogeneous reaction in the particles present in the slurry

This type of mass transfer with reaction plays a role in various chemical slurry processes as in the absorption of CO_2 in MgO and CaS suspensions, SO_2 in $CaCO_3$ suspensions, hydrogenation of poly-

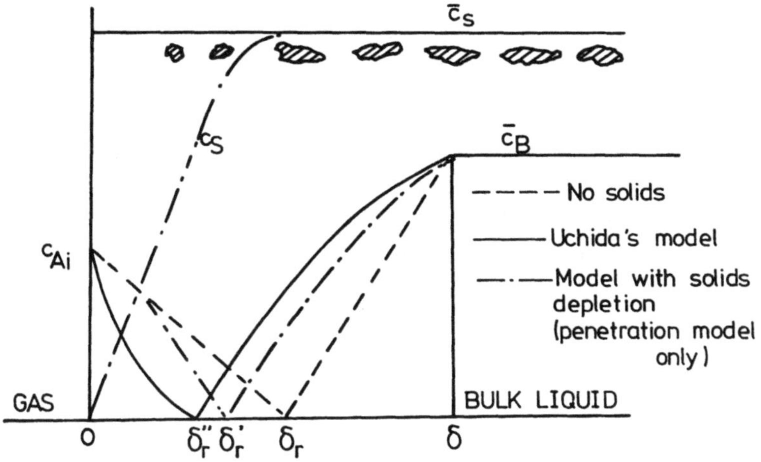

Fig. 22 Concentration profiles of dissolved gas A and sparingly soluble fine particles B (Infinitely fast reaction, both with and without depletion of solid particles at gas-liquid interface)

mers in suspension, oxidation of nylon in suspension etc. An extended list of examples with references is presented by Doraiswamy and Sharma [1]. Basically there are two different cases to be distinguished, i.e., where the product is soluble (shrinking particles) and where the particles retain their shape.

As far as we know mass transfer with chemical reaction in these systems has been analysed only for mass transfer with reaction at the gas-liquid interface in series, that is for conversion mainly taking place at the particles present in the bulk of the liquid. The description is straight forward from coupling the well established theory on mass transfer with reaction in series at the gas liquid interface with a mass balance over the bulk in which an overall rate expression is used in the form of
$R_A = f(\bar{c}_A, a_p, \text{kinetics}, E_{AS})$ where E_{AS} follows from the equally well established theory of mass transfer with reaction in dissolving or in porous particles, whichever mechanism occurs. In the latter case various particle conversion mechanisms are possible such as shrinking core, grain model etc. [61].
A comprehensive analysis dealing with the various asymptotic cases of gas-liquid mass transfer in series with various particle conversion mechanisms in the bulk has been presented recently by Doraiswamy and Sharma [1]. Such a model has been successfully applied to oxydesulfurization of coal [151]. As far as we know, no analysis has been presented as yet for the case where the particles are small with respect to the gas-liquid film for mass transfer, and consequently may enhance the gas-liquid mass transfer process. According to our experience it plays a role in a new process that we recently developed for the concentration of hydrogen from lean gas mixtures with a slurry containing finely hydridable metal particles [16,139,152-154], (Fig. 23).
A possible criterium for the occurrence of such enhancement at the gas- liquid interface is: $(K_{LS} a_p D_A)/(k_L^2) >> 1$ with $1/K_{LS} = 1/k_{LS} + 1/k_l''$, if the product is soluble, and $\frac{1}{K_{LS}} = \frac{1}{k_{LS}} + \frac{d_p}{6 D_{Ai} E_{AS}}$ for particles which retain their shape. Here E_{AS} follows from the conven-

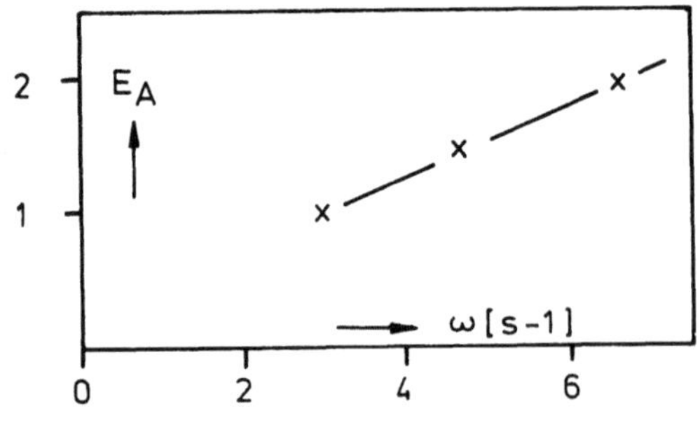

Fig. 23 Experimentally observed enhancement in absorption of hydrogen in a slurry of fine hydridable $LaNi_5$ particles ($w_s = 0.4$) suspended in an inert silicon oil (stirred cell, 70 °C, data derived from [16]

tional theory on mass transfer with reaction in porous reactive particles (shrinking core, grain model and others).
For no depletion of reactant close to the gas-liquid interface such a model can be developed rather straightforwardly along the lines as followed by, e.g., Uchida for sparingly soluble particles [147]. Of course, the differential equations describing local transport of A and conversion of B around and in the single particles will be different. However, in case depletion of solid B may occur close to the gas-liquid interface, we feel that film models are not realistic. Intuitively, we expect depletion of solids at the interface to be significant for: $(D_p\langle\bar{c}_B\rangle)/(D_A c_{Ai} E_A) \ll 1$
Now $\langle\bar{c}_B\rangle/c_{Ai}$ is $\gg 1$ but $D_p/D_A \simeq 0$ if not negative in a stirred cell with $\rho_S/\rho_L > 1$, so that solids depletion at the gas-liquid interface might be the rule, rather than the exception.

If depletion occurs then a numerical penetration model taking all effects into account is desired to predict enhancement factors and to check any forthcoming simplified model as soon as it comes-up. Such a model is under development by Holstvoogd et al. [155].

Type V: Mass transfer followed by fast catalytic reactions in small particles

Catalytic slurry reactors nowadays are the most important class of slurry reactors by far. Doraiswamy and Sharma [1] presented a list of 34 systems of commercial interest and hundreds of papers have been published in this area. In most applications, solids concentrations are relatively low and as a consequence there is no enhancement of the mass transfer at the gas-liquid interface due to the presence of the catalyst. The reactors can be modelled by the

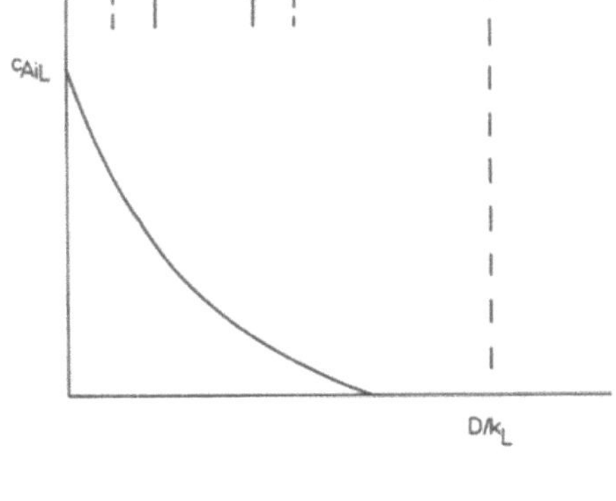

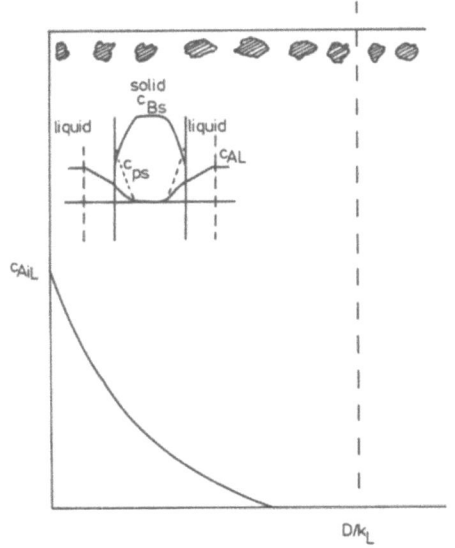

Fig. 24 Concentration profiles of dissolved gas A at gas-liquid interface and around insoluble porous reactive particles, remaining their shape; no depletion of solids near gas-liquid interface.

concept of mass transfer with reaction in series, starting from equation (23). This concept is almost classical by now and the reader is referred to the various handbooks [1,2,61].

With increasing catalyst concentrations and particle diameters small with respect to the film thickness, enhancement at the gas-liquid interface may occur and it is here that a review of the latest developments is justified. We agree, however, with Alper and Deckwer [141] that often it is not economic to carry out the reactions in the enhanced gas-liquid mass transfer mode. This asks for relatively high catalyst concentrations and, with typical film thickness in commercial slurry reactors in the order of $D_L/k_L \simeq 10$ μm, this may be even more important for very small particles in the one micron range. Such particles are difficult to remove from the product. Also, such small particles may easily coagulate. That may facilitate the separation process but unavoidably also kills the enhancement effect originally looked for. However, as Alper et al. [82] pointed out, there are a number of catalytic hydrogenation and partial oxidation reactions which, for selectivity reasons, must be carried out at nearly zero bulk concentration of dissolved gas which can be realized by only gentle agitation and then enhancement may become of importance. Strictly speaking, this statement is correct, but often it works out in exactly the wrong way. Many of these reactions are of a consecutive type:

$$A(G) \rightarrow A(L) \quad (57)$$
$$A(L) + B(L) \rightarrow P(L) \quad (58)$$
$$A(L) + P(L) \rightarrow X(L) \quad (59)$$

with P the desired and X the undesired product. With reaction (58) of lowest order with respect to A, concentration of A should be as low as possible wherever reaction takes place. But unless reaction (59) is zero order with respect to P, enhancement in the absorption of A can spoil the selectivity because, depending on the level of $\bar{c}_B/c_{Ai}$ and $\bar{c}_P/c_{Ai}$, it may cause substantial depletion of B and accumulation of P in the film (Figure 25). Both effects promote the undesired reaction (59) relative to reaction (58). For those systems the best selectivity is obtained for Al $\phi^2 >> 1$ but still $\phi < 0.2$. In other words only gently stirring is alright to keep the product $k_L a$ low but enhancement should be avoided. Only if both $\bar{c}_B/c_{Ai}$ and $\bar{c}_P/c_{Ai} >> E_A$ then enhancement promotes conversion rates without spoiling the selectivity. For a review of the effects of chemical enhancement on selectivity we refer to [61].
Another area where enhancement at the gas-liquid surface may become important is in viscous slurries where k_L is reduced due to the viscosity. A theory on possible enhancement assuming a uniform-distributed catalyst with particle diameter reduced to the molecular scale is straight forward; e.g. for first order kinetics and $\bar{c}_A$

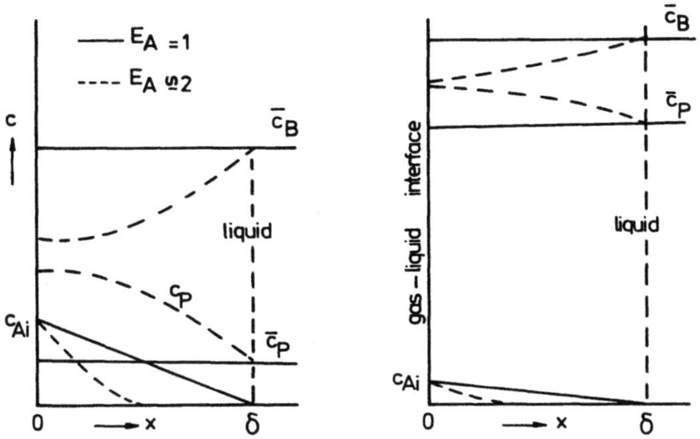

Fig. 25 Influence of enhancement of gas-liquid mass transfer on selectivity in consecutive system A(G) → A(L); A(L) + B(L) → P(L); A(L) + P(L) → X(L). Concentration profiles near gas-liquid interface.
(a) $\bar{c}_P/c_{Ai} \leq 1$; enhancement promotes conversion rate but spoils selectivity
(b) $\bar{c}_B/c_{Ai} \gg 1$; enhancement promotes conversion with only slight influence on selectivity.

= 0 (film theory):

$$E_A = \phi/\tanh \phi \tag{60}$$

$$\text{where } \phi = \frac{1}{k_L} \sqrt{\left(\frac{\varepsilon_S k_1}{D_A}\right)} \tag{61}$$

with k_1 the volumetric first order rate constant within the particle.

In the case of complete mass transfer limitation around the particles, the same is valid with:

$$\phi = \frac{1}{k_L} \sqrt{\left(\frac{k_{LS} a_p}{D_A}\right)} \tag{62}$$

which results in E_A being inversely proportional to d_p for $\phi > 2$.

A more refined theory taking intraparticle diffusion limitation into account has been developed by Sada et al. [142] while Pal et al. [130] consider the bimolecular reaction: A(G) → A(L)
A(L) + B(L) → P(L)
which is already fast with no catalyst and which could be substan-

tially enhanced by small carbon particles on which the reaction was so fast that external mass transfer to the particles controlled the conversion rate. For contribution of the catalyst being dominant and no diffusion limitation of B(L) they proved the set of eqs. (60 and 62) to be valid.

In order to check the theory, only limited experimental evidence is available with catalyst particles other than activated carbon or catalyst impregnated on activated carbon. With activated carbon, two effects usually play a role simultaneously: enhancement due to physical adsorption of the reactant (enhancement of type I) and enhancement due to chemical reaction. These effects should be separated as shown by Alper et al. [82]. They measured the enhanced oxygen absorption into glucose solutions containing Pt deposited on activated carbon as a catalyst. By changing the Pt catalyst concentration without changing the particle content these authors proved true chemical enhancement due to reaction. Their results are presented in Figure 26. Two aspects are noteworthy from this figure:

- Operating in the diffusional regime (R_2 rate controlling), an increase in catalyst concentration of a factor of 50 is necessary before any enhancement is observed. It is this effect that makes operation in the enhanced mode usually uneconomic.
- The enhancement is limited to 1.8.

The authors propose that E_A can not increase any further as soon as the effective reaction zone has been reduced, due to the enhancement, to a dimension in the order of the particle diameter. With an effective thickness of the reaction zone, according to film theory, given by $\delta_r = \delta/E_A$, this reasoning results in: $E_{A,max} = \delta/d_p$. Indeed, measurements with carbon dioxide in carbonate-bicarbonate buffer solutions containing carbonic anhydrase immobilized on solid

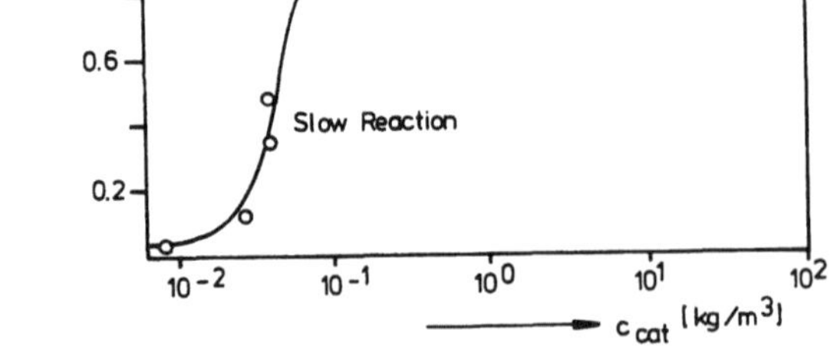

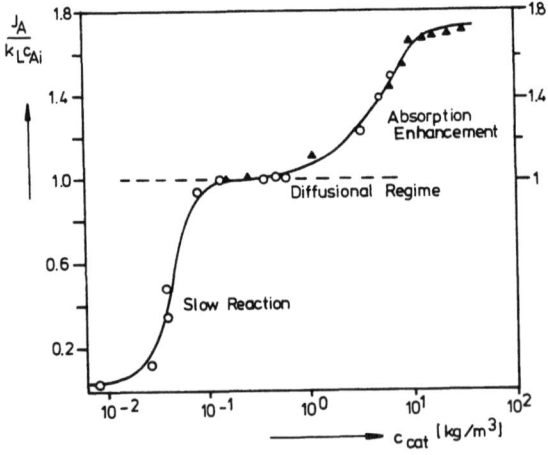

Fig. 26 Chemical enhancement in oxidation of glucose with Pt on carbon, measured in a stirred cell reactor;
▲ - total carbon = 32.8 kg/m^3;
o - total carbon = 9.8 kg/m^3.
From Alper et al. [82]

particles support this reasoning with reasonable accuracy [141]. Pal et al [130] obtained enhancement factors as high as 14 in oxidizing sodium sulphide solution with activated carbon as a catalyst. Their value of D_L/k_L was of the order of 7 μm. Particle size was as low as 1.7 μm which would suggest a maximum E_A according to the criterion of Alper and Deckwer of only 4.5, or alternatively: $E_{A,max} = 3\delta/d_p$. For enhancement due to physical adsorption we estimated (5.7 type I) $E_{A,max} = 0.35\ \delta/d_p$. So we may conclude that the question of $E_{A,max}$ asks for further research.

Recently Sada et al. [143] dealt with the problem of simultaneous absorption of two gases (SO_2 and O_2) in a slurry containing small activated carbon particles enhancing the absorption by catalysis:

$$SO_2(G) \rightarrow SO_2(L)$$
$$O_2(G) \rightarrow O_2(L)$$
$$SO_2(L) + 1/2O_2(L) + H_2O(L) \rightarrow H_2SO_4$$

Theoretical analysis based on the film model is rather straight forward but the experimental interpretation is ambiguous, possibly due to uncontrolled particle attrition. However, the possible effect of enhancement due to physical adsorption on the particles was not taken into account.

Wimmers et al. [144,145] measured the enhancement of absorption from a stationary bubble into a stagnant liquid catalyzed by fine activated carbon and arrived, both experimentally and theoretically, at:

$$E_A = 1 + 0.5\ d_b \sqrt{(k_1/D_A)} \qquad (63)$$

for stationary conditions. Their method deserves attention, both for its accuracy and for the fact that the complicating effect of possible co-enhancement due to physical adsorption can be excluded in this way.

6. NEW DEVELOPMENTS IN REACTOR MODELLING

Several textbooks are available which describe in detail how to design and scale-up slurry reactors, e.g., the book by Shah, which appeared in 1979 [2] and the new book (1984) of Doraiswamy and Sharma [1]. For modelling catalytic three phase reactors, we also refer to the papers of Ramachandran and Chaudhari [3,156]. At a previous NATO ASI Conference (Cesme-Izmir, 1981) Hofmann [5] presented a description of the reactor models governing the design and scale-up of slurry reactors. We therefore will concentrate on some new developments only.

6.1 Calculation of batch time

Ramachandran and Chaudhary [157] derived an equation for calculation of the batch time of a semi batch three phase slurry reactor with a catalytic reaction of type:

$A(G) \rightarrow A(L)$

$A(L) + \nu_B B(L) \rightarrow \nu_P P(L); \quad R_A = -k_{1,1} c_A c_B$

allowing for resistances at the gas-liquid interface ($1/k_L a$), at the liquid-particle interface ($1/k_{LS} a_p$) and a combined mass transfer plus reaction resistance within the porous particles. Their result is:

$$\tau_{batch} = \frac{\bar{c}_{Bo}}{\nu_B c_{Ai}} \left[\zeta_B \left\{\frac{1}{k_L a} + \frac{1}{k_{LS} a_p}\right\} + \frac{d_p^2 \rho_S I}{12 w_S \rho_L D_{iA}}\right] \tag{64}$$

with

$$I = \int_{1-\zeta_B}^{1} \frac{dy}{3\phi_o y^{0.5} \coth(3\phi_o y^{0.5}) - 1} \tag{65}$$

I is given on Fig. 27 in the form of a chart and ϕ_o is the Thiele modulus at t=0:

$$\phi_o = \frac{V_p}{A_p} \left(\frac{k_{1,1} \bar{c}_{Bo}}{D_{i,A}}\right)^{0.5} \tag{66}$$

For $\phi_o < 0.2$, an asymptotic solution is obtained (no diffusion limitation inside the particles). Then:

$$\tau_{batch} = \frac{\bar{c}_{Bo} \zeta_B}{\nu_B c_{Ai}} \left(\frac{1}{k_L a} + \frac{1}{k_{LS} a_p} - \frac{1}{k_{1,1} \bar{c}_{Bo}} \frac{\ln(1-\zeta_B)}{\zeta_B}\right) \tag{67}$$

In deriving the design-equation (64), c_{Ai} and solids distribution were assumed to be constant and homogeneous, respectively. This should always be checked, particularly in bubble columns.

6.2. Model for zero order and Langmuir-Hinshelwood type catalytic reactions

Some catalytic slurry systems show effective zero order kinetic behaviour. Design equations for such slurry systems are relatively easy to obtain, starting from:

$$R_A = -k_o \varepsilon_S. \tag{68}$$

However, for $c_{AiS} < \frac{k_o d_p^2}{24 D_{iA}}$, with c_{AiS} the concentration of A close to the particle external surface, no conversion in a certain region in the centre of the particles will occur due to absence of A in

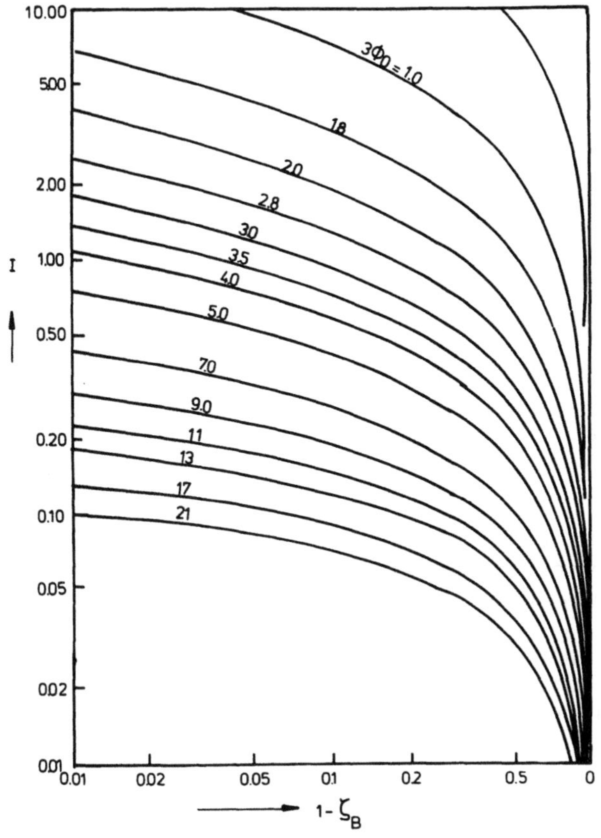

Fig. 27 Chart of integral I vs $(1 - \zeta_B)$ for various values of $3\Phi_0$. From Ramachandran and Chaudhari [157]

that region. Then, the observed conversion rate will be lower than calculated according to (68). To check this criterion we need a relation for c_{AiS}. For constant gas phase and bulk liquid phase concentrations, c_{AiS} follows from the mass balance: $J_A a = J_{A,S} a_p = -R_A$. The result is:

$$c_{AiS} = m\bar{c}_{AG} - k_o \varepsilon_S \left[\frac{1}{K_L a} + \frac{1}{k_{LS} a_p}\right]^{-1} \tag{69}$$

For conditions where this criterion and Equations (69) are satisfied, Chaudhari and Ramachandran recently published design equations taking the internal particle diffusion effects into account [158]. They applied this model to the hydrogenation of DNT, the ethynylation of formaldehyde and the oxidation of cyclohexene. Fig. 28 shows the influence of pressure on the conversion rate. The decrease in $(-R_A)$ below a critical pressure indeed could be predicted by this model within 23%. A complication not accounted for

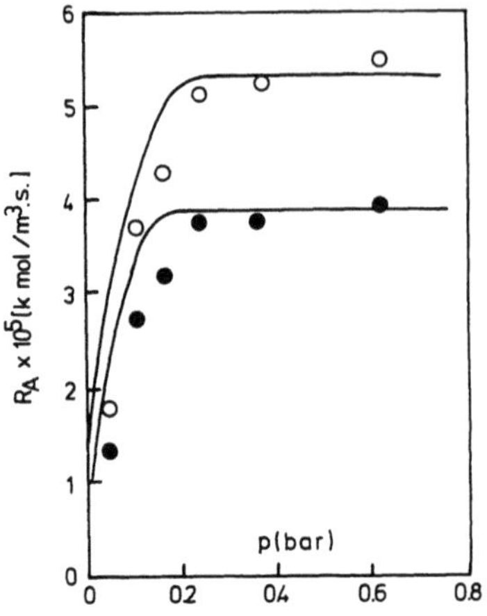

Fig. 28 Influence of pressure on rate of ethynylation of formaldehyde (B) in a slurry reactor explained by a model for mass transfer with zero order reaction.
—— theoretical
o experiments with $\bar{c}_B = 4$ kmol/m^3
• experiments with $\bar{c}_B = 3$ kmol/m^3
(From Chaudhari and Ramachandran[158])

in the model of Chaudhari and Ramachandran [158] is the fact that zero order reactions never remain zero order for reactant concentrations approaching zero. Usually a Langmuir-Hinshelwood type of kinetics of the form: $R_A = -k_1 c_A/(1+Kc_A)$ is more realistic. It shows that the kinetics approach first order for $c_A << 1/K$ and zero order for $c_A >> 1/K$. A slurry reactor model for non-linear kinetics of such a type also has been published by Ramachandran and Chaudhari [159].

6.3. Effect of axial dispersion in the gas phase

The effect of the residence time distribution of the gas bubbles on slurry reactor performance has been studied theoretically by Chang and Smith [160] using a tank-in-series model. The effect is maximal for intermediate solubilities, i.e. for: $mk_L a\, \tau_G \simeq 1$. Chang and Smith use a slightly different absorption parameter Ab defined as:

$$Ab = \frac{mk_L aV_L}{\Phi_{vG}} = \frac{mk_L a\tau_G}{\varepsilon} \tag{70}$$

and find maximum deviations Δ,

$$\Delta = \frac{(c_{G,1})_{N=N} - (c_{G,1})_{N=\infty}}{c_{G,o} - (c_{G,1})_{N=\infty}}, \tag{71}$$

in the gas outlet concentrations relative to plug flow for N = 1 and $\bar{c}_{AL} = 0$. It then can be in the order of 25% for Ab ≃ 1.8. For intermediate $\bar{c}_{AL} \simeq 1/2 c_{AiL}$ the effect is slightly lower (Fig. 29). We may safely conclude from the above analysis that the assumption of plug flow of the gas phase in bubble column slurry reactors usually will not result in substantial overestimates of the performance of the slurry reactor.

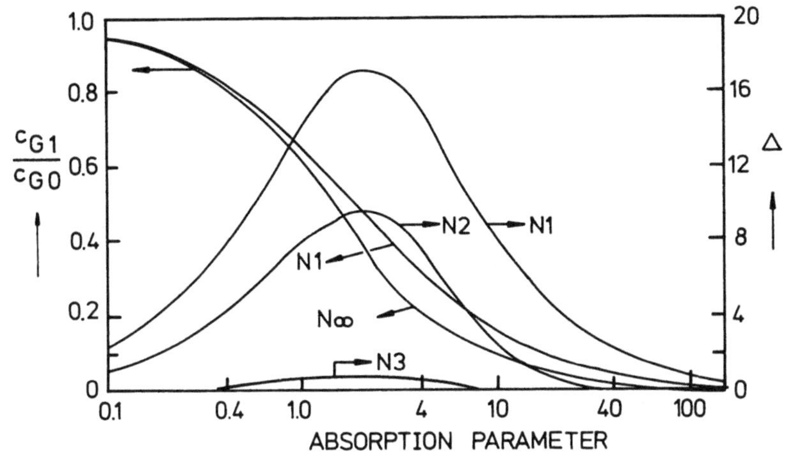

Fig. 29 Effect of RTD on exit gas concentrations in a slurry bubble column for Da = 1; $Da = ((1/\tau_L + k')/k_La$; $k' = k_1 \varepsilon_S \eta/(1 + k_1 \varepsilon_S \eta/k_{LS}a_p)$
From Chang and Smith [160]

6.4. Two zone model for the gas phase

The above conclusion does not imply that models assuming one single k_La value over the whole column always result in realistic models. For low gas and liquid velocities it may describe experimental findings satisfactorily but, particularly for high gas and liquid velocities, the mass transfer close to the grid can be much more effective than further upwards in the bubble column [161]. This naturally leads to a two-zone model for the mass transfer from the gas phase: a zone close to the grid where k_La is relatively large and the remaining part of the reactor volume above the grid zone where k_La is relatively low. Such a model has been developed by Alvarez-Cuenca and Nerenberg [161]. It is a three parameter model ($(k_La)_{grid}$, $(k_La)_{bulk\ zone}$, volume ratio of the two zones). It indeed gives an improved fit of their experimental findings for $u_L \approx 0.1$ m/s and $u_G \approx 0.2$-0.3 m/s but no relations for a priori predictions of the three parameters are given. More work may be useful in this area to get such relations. We note the similarity with Darton's model for gas-solid fluidized beds [162,163] where, from the grid upwards, bubbles grow rapidly until, at a more or less predictable height, bubble growth and break-up are at equilibrium resulting in a constant average d_b further upwards in the bed.

6.5. Effect of axial dispersion in the liquid and the solid phase

Where effects of axial dispersion of the gas phase can usually be neglected in slurry reactor design, the effects of mixing in the slurry phase can be quite significant, particularly for smaller column diameters, at lower catalyst loadings and larger particle sizes. New studies on dispersion-sedimentation models taking such effects into account have been presented recently by Govindarao and Chidambaram [164,165] and by Serpemen and Deckwer [166]. The new developments here are that settling effects with a first order surface reaction have been extended from a semi-batch reactor [167] to a continuous bubble column where catalyst accumulation effects can be significant [164,165] while the model of Parulekar and Shah for settling and accumulation effects in cocurrent bubble columns [168] has been extended to include mass transfer effects in and around porous particles both for first order, half and second order kinetics [166]. Govindarao and Chidambaram [164,165] use the equation of Kato et al. [27] to describe the backmixing in the continuous upflow slurry reactor:

$$\frac{u_G d_t}{D_{1,S}} = \frac{13(\sqrt{Fr})}{1 + 8\ Fr^{0.425}} (1 + 0.009\ Re(\sqrt{Fr})^{-0.8}) \qquad (72)$$

with $Re = v_p d_p \rho_L / \mu_L$, and predict increasing solids hold-up in particularly the bottom part of the reactor, relative to the solids hold-up in the slurry feed, with increasing particle size and decreasing gas and liquid velocity. This sedimentation and accumulation of particles is caused by the effective slip velocity of the particles and counteracted by the axial dispersion coefficient of the slurry phase. The effects on conversion can be significant as has been illustrated in Figs 30 and 31 for a first order catalytic surface reaction of the component transferred from the gas phase. For a constant solids load, $w_S \rho_L$, the conversion usually will decrease with increasing d_p due to decreasing a_p via $a_p = 6\varepsilon_S/d_p$. However, due to increasing sedimentation with increasing d_p, ε_S will increase with increasing d_p. As it works out, for small d_p, the specific solids area decreases with increasing d_p while for larger d_p, the sedimentation effect becomes more dominant resulting

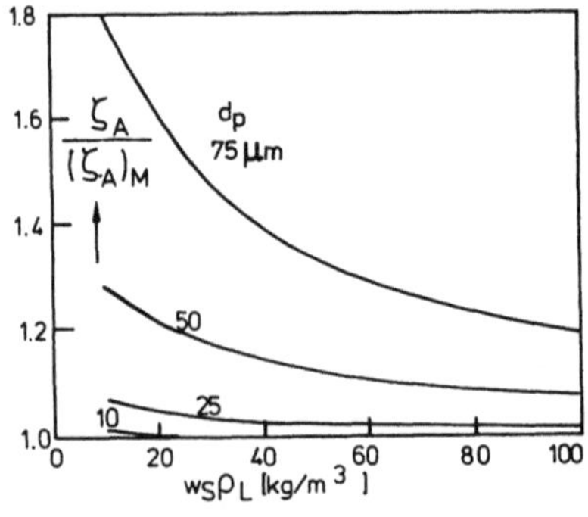

Fig. 30 Conversion enhancement $\zeta_A/(\zeta_A)_M$ with finite dispersion in the slurry phase (relative to the conversion with a completely backmixed slurry) vs. feed catalyst loading; first order catalytic surface reaction of gaseous component in upflow bubble column slurry reactor, u_G = 0.075 m/s; u_L = 0.025 m/s; ρ_S/ρ_L = 3.9; d_t = 0.2 m; k_1'' = 4 10^{-4} m/s.
(adapted from Govindarao and Chidabaram [165])

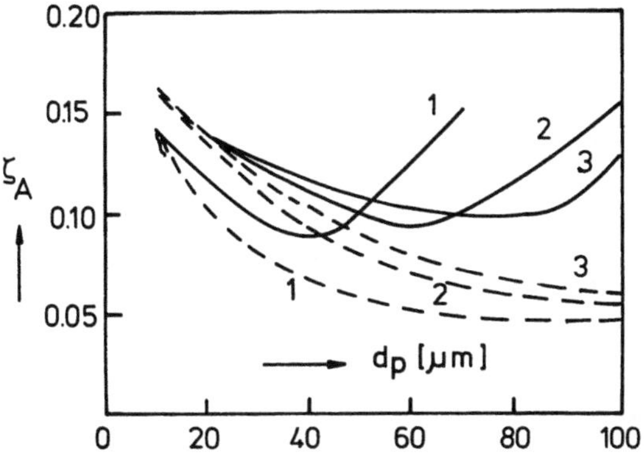

Curve	ρ_S/ρ_L [-]	u_L [m/s]
1	7.9	0.025
2	3.95	0.025
3	3.95	0.040

Fig. 31 Effect of ρ_S and u_L on the conversion vs particle diameter. First order catalytic surface reaction in bubble column upflow slurry reactor; $k_1'' = 4\ 10^{-4}$ m/s; $w_S\rho_L = 10$ kg/m^3; $u_G = 0.075$ m/s; $d_t = 0.2$m;
--- complete backmixed slurry
—— sedimentation effects taken into account via D_1 from eqn 72.
Adapted from Govindarao and Chidambaram [165].

in an increase of a_p with increasing d_p. As a result, particles of 100 μm may give better conversions relative to particles of say 40 μm, at the same solids load in the feed (Fig. 31). If so, the larger particles will be preferred for economic reasons. For small columns the above conclusions are, at least qualitatively, in agreement with the analysis of Serpemen and Deckwer [166]. These authors conclude, however, that the settling effects rapidly diminish with column diameter so that for columns with diameters above 30 cm catalyst settling effects can be neglected for solid particle diameters up to 200 μm (Fig. 32).

6.6. Modelling of a Multi Staged Agitated Column Reactor

Some authors have noted that a type of rotating disc contactor, originally designed as a liquid-liquid extractor, after some modifications, may have excellent properties to carry out slurry reactions [169,170]. In such a reactor both gas and liquid hold-up can be controlled independently, while backmixing can be greatly suppressed for all phases present. First data on axial mixing, hold-up and mass transfer in such a reactor have been published by Elenkov and Vlaev [169].

7. APPLICATION OF REACTOR DESIGN MODELS

The book by Doraiswamy and Sharma [1] contains an overwhelming list of slurry processes studied. Here, we restrict ourselves to the latest developments only (see Table XI). Until now, most studies have been dedicated to model reactors either to get at the intrinsic kinetics or to show overall performance of the system without

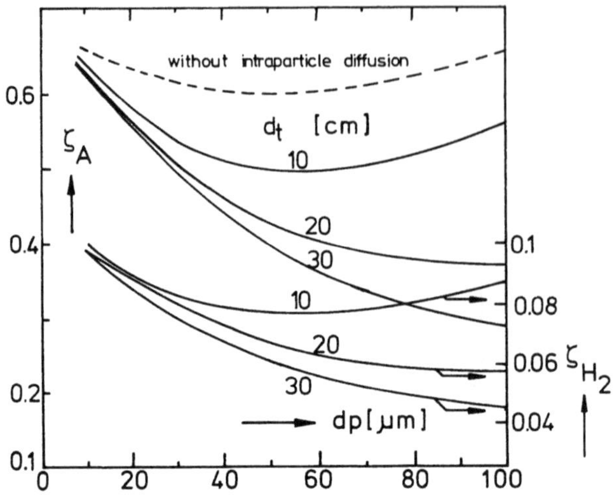

Fig. 32 Influence of catalyst settling in continuous hydrogenation of butynedial (A) in a bubble column slurry reactor as a function of d_p for various reactor diameters d_t
u_G = 0.04 m/s, $u_L = 10^{-4}$ m/s, $k_La = 8\ 10^{-2}\ s^{-1}$.
(adapted from [166])

Table XI. Recent applications of reactor models to slurry reactions

system	reference
methanol synthesis	Ledakowicz et al. [173] Brown [111] Kafarov et al. [174]
Fisher Tropsch synthesis	Stern et al. [172] Deckwer et al. [86,88,89,166] Bukur [171,175] Satterfield et al. [84,87,91, 131]
hydrogenation of phenolic compounds	Kut et al. [140,176]
oxidation of phenol	Ohta et al. [128]
coal liquefaction	Shah and Parulekar [177]
oxydesulfurization of coal	Joshi et al. [151]
hydrogenation of glucose	Turek et al. [178]
hydrogenation of DNT	Chaudhari & Ramachandran [158]
ethynylation of formaldehyde	ibid.
oxydation of cyclohexene	ibid.
methanation of CO	Serpemen & Deckwer [166]
hydrogenation of butynediol	ibid.

having a detailed knowledge of the interference of mass transfer limitations and intrinsic kinetics. The latter in particular holds for the older literature in the field. Information on successful modelling of real life commercial slurry reactors is still scarce in the open literature.

7.1. Particularly over the last 5 years and probably catalyzed by the oil crisis most attention has gone to Fisher-Tropsch slurry reactors; see for a review [97]. Satterfield et al. conclude mass transfer limitations dominate in real life reactors [84,87] which then may have an effect on product selectivity. Deckwer et al. have studied the reaction in a 3.8 cm bubble column [86] and presented a design model for a large scale reactor [89] (d_t = 1.29 m, L = 7.7 m, u_{Go} = 0.095 m/s) using a two phase dispersion model, taking into account heat effects, change of gas flow rate with conversion, gas-liquid and liquid-solid mass transfer and catalyst settling effects. They concluded that the latter two effects can be neglected in larger diameter columns. Also gas-liquid mass transfer limitations are small according to their analysis as long as conventional catalysts and favourable gas velocities are applied. They found a satisfactory agreement between their analysis and the Rhein-Preussen-Koppers demonstration plant. As discussed above, the difference in the results of Satterfield et al. and Deckwer et al. originate from different relations for predicting the interfacial area. It is generally accepted that more experimental work on this topic is necessary. Some authors use experimental data obtained by others in their own analysis [84,171,172]. That remains always risky.

7.2 Another area where attention has been focussed during the last 5 years is the synthesis of methanol in a bubble column slurry reactor and in a gas-liquid-solid fluidized bed. This process, being invented by Chem. Systems [179,181] meanwhile has been further developed by Air Products up to a 5 tons/day pilot unit at Laporte, Texas [111]. Recently three studies have been published, modelling this process, all based on experimental experience, see Table XII. Brown [111] applied a tank-in-series model with backmixing and concluded that for a balanced feed transport resistances do not play a role. This conclusion has been confirmed by Ledakowicz et al. [173]. These authors also conclude that sedimentation effects can be neglected. Kafarov et al. [174] assumed the slurry to be perfectly mixed and conclude mass transfer resistances play a role. This probably is caused by the fact that the particles used by these authors are much larger, which also makes the assumption of perfect slurry mixing questionable. Sedimentation effects probably should be considered here. For non-balanced CO-rich gas, the the experiments of Brown also indicate diffusion limitation effects but here the analysis is still ambiguous. More data are needed, particularly on the gas hold-up in the pilot plant under actual operating conditions.

7.3. Competitive, consecutive hydrogenation reactions of type:

$H_2(G) \rightarrow H_2(L) \rightarrow H_2(S)$
$B(L) \rightarrow B(S)$

Table XII. Modelling the Methanol Synthesis in a bubble column slurry reactor (catalyst CuZnAl)

	p [bar]	T [°C]	d_p [μm]
Brown [111]	35-70	200-270	5-10
Ledakowicz et al. [173]	20-50	220-250	< 53
Kafarov et al. [174]	40-150	220-280	2000

$$B'(L) \rightarrow B'(S)$$
$$2H_2(S) + B(S) \rightarrow P(S) \xrightarrow{H_2(S)} X(S)$$
$$2H_2(S) + B'(S) \rightarrow P'(S) \xrightarrow{H_2(S)} X'(S)$$
$$P(S) + P'(S) + X(S) + X'(S) \rightarrow P(L) + P'(L) + X(L) + X'(L)$$

in a stirred tank slurry reactor have been modelled by Kut et al. [140,176] assuming Langmuir-Hinshelwood kinetics. As known from theory on mass transfer with chemical reaction [61], any limitation in transport of P and P' from the catalyst will result in a decrease of selectivity with respect to the intermediate products. With phenolic compounds as substrate the authors could show that mass transfer limitations could be avoided by using sufficiently small catalyst particles (< 10 μm).

7.4. A recent overview of the various models applied to coal liquefaction and oxydesulfurization in slurry reactors has been presented by Shah and Gopal [182]. Concerning coal liquefaction, the most complete model up to date pro- bably is the one of Shah and Parelukar [177] for an adiabatic bubble column slurry reactor. It includes:

- pseudo first order kinetics in hydrogen
- mass transfer resistance from gas to liquid
- dispersion-solids sedimentation according to Eqn. (72)
- liquid dispersion according to [27], Eqn. (18)
- heat dispersion coefficient equal to mass dispersion coefficient for each phase

The actual temperature profiles observed in the reactor could be well predicted by the model, provided a fit value for k_La was taken to match the

observed hydrogen consumption: $k_La = 0.3\ d_t^{0.17}\varepsilon^{1.1}$ and heat dispersion coefficients for both the liquid and the solids phase were reduced by a factor of 3, relative to the predictions from Eqns (18) and (72), (Fig. 33).

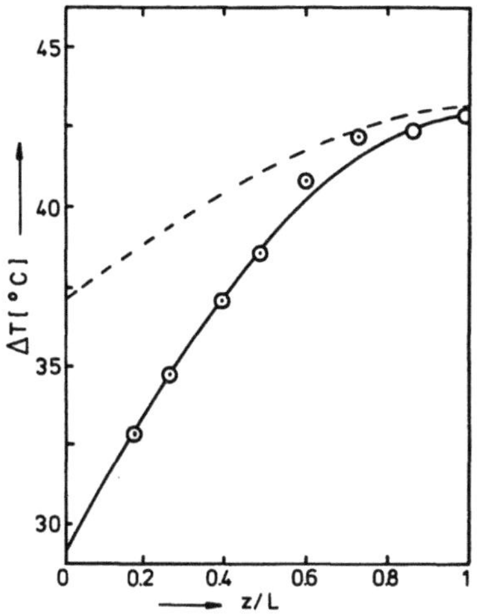

Fig. 33 Actual temperature profile in coal liquefaction slurry column compared with the model of Shah and Parulekar (from [177])
⊙ plant data
--- heat dispersion according to Kato et al.[108]
—— model predictions with heat dispersion coefficient reduced by a factor 3

d_t = 0.6 m, L = 9 m, $w_S \rho_L$ = 100 kg/m^3, u_L = 0.003 m/s, u_G = 0.03 m/s, d_p = 100 μm, T = 418 °C, P = 150 bar.

7.5. A fairly complete model of the oxydesulfurization of coal in bubble column slurry reactors has been presented by Joshi et al. [151]. It includes a dispersion-solids sedimentation model to describe the RTD of the reactive particles but solids accumulation is not included, hence: $D_{1,S} = D_{1,L}$. The equation used is taken from Joshi [43]:

$$D_{1,S} = D_{1,L} = 0.38 d_t^{1.33} g(u_G - \varepsilon_G v_b) \tag{73}$$

with v_b the terminal rise velocity of the bubbles. The reaction is assumed to be controlled by one single step of a shrinking core model, so either

- mass transfer around the particles, or
- diffusion through the ash layer, or
- chemical reaction at the surface of the unreacted core

It does not include particle size distribution, gas to liquid transfer resistance and gas contraction due to conversion of oxygen. The model fairly well (within 25%) predicts the actual conversions of pyritic sulfur of about 95% obtained at 175-200 °C, 67-69 bar and solids residence time typically 1800 seconds in a 22.2 mm bubble column, assuming the chemical reaction at the surface of the unreacted core to be rate controlling. For a recent overview on this process, see Shah and Albal [188].

8. PROSPECTIVE NEW DEVELOPMENTS IN SLURRY PROCESS APPLICATIONS

8.1. Methanol synthesis

After successful completion of the 5 tons/day pilot plant development program of Air Products [111], a slurry phase methanol synthesis process may reach the commercial market. The prospects are encouraging if all the advantages claimed can be demonstrated. These are:

- A better temperature control and no hot spots due to the large heat capacity of the inert liquid.
- A larger conversion per pass: 35-43% relative to typically 12% in a conventional fixed bed. This results in a reduction of recycle compression costs and in a larger flexibility with respect to feed gas composition, which could make it particularly attractive for CO rich synthesis gas from partial oxidation processes without using a shift reactor.
- A larger heat recovery efficiency relative to the ICI process of 98% vs 86%, due to external steam raising. This advantage probably will be much less pronounced if compared to the Lurgi process, as also Ledakowicz et al. [173] point out.
- A reactor of simple construction.
- Reduced catalyst deactivation rates.
- Constant catalyst activity due to continuous withdrawal of catalyst while simultaneously introducing new catalyst.
- Less catalyst attrition.

In fact, many of these advantages have been demonstrated already for the Fisher-Tropsch synthesis in slurry reactors compared to the classical fixed bed and entrained fluidized bed processes [97].

8.2. Combination of Fisher-Tropsch synthesis and Mobil process

In principle, as has been pointed out by Deckwer [97], it seems attractive to combine the slurry phase Fisher-Tropsch with the Mobil process in one single slurry reactor. This way CO rich synthesis gas, originating from coal gasifiers, might be converted directly, without using a shift reactor, to yield high octane number C_5-C_{11} hydrocarbons with probably 70% selectivity. It is a fascinating option requiring still much development work to prove its viability.

8.3. Hydrogen recovery from lean gas mixtures

Much work has been done over the past two decades on storage of hydrogen in packed beds of hydridable metal alloys as Mg_2Ni, FeTi and $LaNi_5$ aiming at the introduction of hydrogen as a clean motor fuel. The method can be used also for hydrogen recovery in packed beds by a pressure or temperature swing technique [183,184]. This technology still provides many problems as:

- poor heat transfer in packed beds
- embrittlement of the particles which may lead to blocking and entrainment of fines
- large forces on the reactor wall during adsorption
- discontinuous operation with four or more beds necessary

Recently, we started the development of a slurry process to overcome all these drawbacks [152-154]. Here, an inert oil containing small hydridable metal alloy particles is continuously recycled over an absorber and a desorber column. In this way, a continuous stationary process is obtained with excellent heat transfer properties, a low pressure drop and improved reactivity by the possibility of using small particles in the micron range. In fact, the conversion rate can be higher than in a packed bed due to the absence of heat transfer resistances and reduced intraparticle diffusion effects. As shown in Fig. 34, complete metal conversion can be obtained in only a few minutes, leading to a slurry with a hydrogen concentration comparable to liquid hydrogen. Any further decripitation of the(se) particles will not result in the problems mentioned for the packed bed method and there is also no problem of expansion forces on the reactor wall. The method also compresses hydrogen without the need of a compressor by operating the desorber at a higher temperature relative to the absorber. Bench scale experiments have been carried out successfully and a continuous pilot plant is in the start-up phase at Twente University of Technology [16,139].

8.4. Activated carbon as a mass transfer promotor

Application of small particles of activated carbon with $d_p \ll D_L/k_L$ to enhance the mass transfer process in multiphase reactions has been shown to be very prospective [21], particularly if the mutual solubility of the reactants is limited. It may find a wide range of applications if downstream particle separation problems can be solved.

8.5. Biochemical reaction engineering

Bioreactors using free cells are basically slurry reactors. Also if the biocatalyst is immobilized, the slurry reactor is one of the options. With the present up-swing of biotechnology the use of such slurry reactors may rapidly increase. One of the problems here might be the limited rate of oxygen mass transfer to the aqueous phase due to the low solubility and foaming problems in many bio-

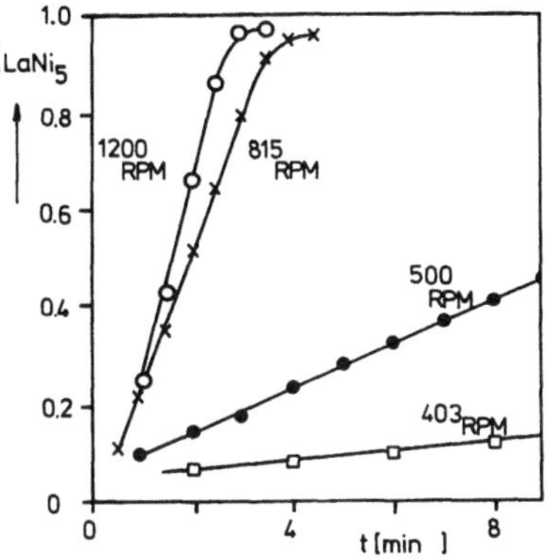

Fig. 34 $LaNi_5$ conversion vs absorption time in a semibatch stirred slurry tank reactor for various stirrer speeds

$H_2(G) \rightarrow H_2(L) \rightarrow H_2(S)$

$3H_2 + LaNi_5 \rightarrow LaNi_5H_6$

$T = 25\ ^oC$, $P_{H_2} = 26$ bar, $w_S = 28\%$.

The gas-liquid surface remains flat up to 400 RPM. Adapted from [16]

chemical systems. Here, the use of a second dispersed fluid phase with a relatively high solubility of oxygen might find application [146]. We studied a four-phase slurry reactor (octene phase, water, cells, air) to produce octene epoxide with Pseudomonas Oleverans and found that design rules for such complicated systems are virtually lacking [185].

8.6. Selective separation by using zeolites and activated carbon particles in a slurry

Selective absorption by using continuous slurry processes instead of packed bed swing techniques is under development [22,186,187]. Some advantages are of a similar kind as those outlined above for the hydrogen recovery process with hydridable metal particles in a slurry though the heat effects involved usually are less pronounced. Fig. 35 shows typical results as obtained by Tinge et al. [22]. Specific liquid absorption capacity is lower than for packed beds but this may be outweighed by the advantages of a continuous process with no or strongly reduced attrition problems and high absorption rates due to small particles, meanwhile maintaining good selectivity or getting improved selectivity if the liquid acts as a filter for one of the components.

9. CONCLUSIONS

A clear and operational boundary should be established between slurry reactors and three phase fluid beds. Particle diameter alone (e.g. d_p < 200 μm for slurry reactors) is not sufficient and boundaries should be based on hydrodynamic regime, settling velocities and turbulence parameters.

9.1. Minimum suspension criteria

Minimum suspension criteria can be based on fluid velocity or on energy input in the fluid phases. The energy input per unit volume fluid phase is generally applicable to widely different geometries.

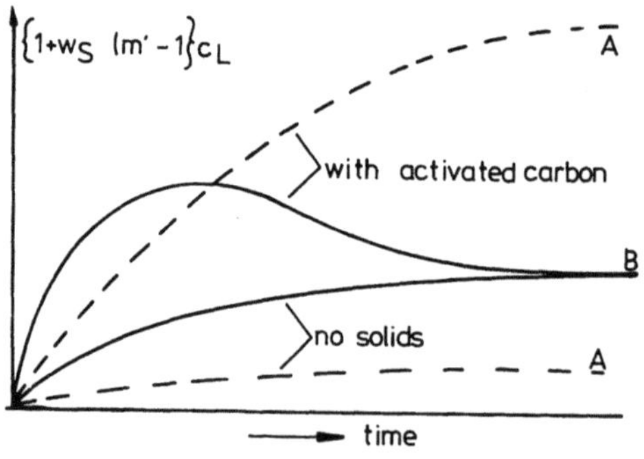

Fig. 35 Application of activated carbon for selective absorption in a slurry. Stirred cell experiments for Tinge et al. [22]

However, the concept fails to predict the influence of practically important parameters like vessel bottom geometry while in stirred vessels it does not correctly take the influence of the gas phase into account.

Minimum suspension criteria for solids in three phase stirred vessels can be related to those of the two phase stirred vessels. Data on three phase systems are relatively scarce, however. The same holds for three phase sparged columns. Moreover, in laboratory equipment with a large value of L/d_t it may be difficult to obtain a state of homogeneous particle distribution. More data on the influence of scale of operation on this phenomenon is required.

9.2. Gas hold-up

The gas hold-up in three phase sparged vessels depends on the gas velocity, but, just as in liquid-gas systems, different regimes must be distinguished (uniform bubbling, churn-turbulent flow, slugging etc.). Data with high slurry concentrations are scarce. If two phase relations will have to be applied for high slurry concentrations then the fluid properties should probably be replaced by liquid solids suspension properties.

9.3. Axial mixing

At lower particle concentrations axial mixing in the liquid phase of three phase sparged columns can be derived from those of the two phase system, taking into account the appropriate hydrodynamical regime. The axial dispersion coefficients for the solids phase are almost equal to those of the liquid phase. Specially for small scale columns at high L/d_t values, the settling tendency of the particles may have to be taken into account when applying the axial mixing coefficient to reactor performance calculations.

9.4 .Volumetric liquid side mass transfer coefficient (k_La)

More data are needed on the influence of pressure on k_La, also in two-phase systems. More data are also needed on the influence of solids on k_La, particularly for particles smaller than 50 μm, but also for larger particles at solids concentration above 20 vol %.

9.5. Volumetric gas-side mass transfer coefficient (k_Ga)

More data are needed on k_Ga for very reactive systems where k_Ga may influence selectivity by affecting concentrations in the slurry phase.

9.6. Specific gas liquid contact area

For reactive systems where capacity depends on the contact area, a, rather than k_La, more data are needed on the influence of pressure and liquid mixture composition on the value of the specific contact area. In case of uncertainty on the contact area, the second best alternative to measuring the area is measuring gas hold-up and combine it with the existing relations for d_b according to a =

$6\varepsilon/d_b$. Even small amounts of fine solids as low as 1 wt % may substantially change the value of the contact area but no consistent theory explaining unambiguously the measured effects has been presented yet.

9.7. Liquid side mass transfer coefficient

Usually either the product k_La or the contact area alone (a) affect specific reactor capacity. Moreover, uncertainty in predicting k_L seems to be less than in k_La and a, particularly in two-phase systems.

As a rule k_L is lowered by solids concentration but probably merely indirectly through a change in bubble size. No unique relation on the influence of solids on k_L is available yet.

Reported increase of k_L with solids concentration usually is caused by enhancement of mass transfer due to physical adsorption on activated carbon. Such an effect should be described by the concept of enhancement through adsorption rather than by increase of k_L by solids.

9.8. Solids-liquid mass transfer coefficient

Solid/liquid mass transfer in both sparged vessels and stirred vessels can be successfully described with the energy input concept. Direct relation with fluid-particle slip velocities could reveal influence of geometrical factors not covered by the energy input concept. Probably, in stirred vessels the influence of the gas phase can not be completely accounted for by the total energy input/m^3 liquid concept. However, data on this aspect are relatively scarce and more work is needed to clarify this aspect.

9.9. Measuring techniques

Simultaneous absorption of two gases in a reactive liquid with different regimes of mass transfer with reaction for each gaseous component may be a more efficient tool in measuring k_G, k_L and the contact area, a, relative to the more classical techniques.

Transient techniques, already common in measuring mass transfer parameters may gain interest in measuring kinetics in slurry reactors as well.

Basket type reactors have gained interest in measuring kinetics in slurry reactors and the advantages and disadvantages of this model reactor in slurry applications by now are well understood.

9.10. Enhancement of mass transfer through reaction or through physical adsorption on solids

The enhancement of mass transfer by physical adsorption on activated carbon particles smaller than the film thickness is qualitatively understood but more work is needed on predicting the maximum

possible value of E_A and on predicting E_A for particle sizes in the order of the film thickness (D_A/k_L).
Multiphase reactions between mutually sparingly soluble reactants can be enhanced substantially by adding solids with a great adsorption capacity and $d_p << D_L\ k_L$. Though probably not applied in practice yet, it has a definite potential provided downstream separation problems can be solved. A similar statement holds for adding fluorohydrocarbons as a finely dispersed oxygen carrier to the aqueous phase in bio slurry reactors.

Since 1980 significant progress has been made in describing mass transfer with reaction for one of the reactants originating from soluble particles. Also here, there is a limit in the maximum enhancement, depending on d_p. Theories assuming no particles present at the interface within a distance d_p have been shown to be useful but we call for developing penetration models taking particle depletion into account.

Theory on mass transfer with reactive particles is well described as long as mass transfer and reaction are in series. Also here, enhancement effects with small particles recently have been observed but the theory by and large still has to be developed.

Theory on mass transfer with catalytic reaction in catalyst particles suspended in a slurry is well established as long as mass transfer and reaction are in series. In commercial reactors the latter will be the rule rather than the exception. Again, small particles ($d_p << D_L/k_L$) can enhance mass transfer up to a maximum which is still not well understood. However, significance for industrial engineering practice is limited. In studying such enhancement effects the use of activated carbon particles is tricky because enhancement by adsorption may interfere with enhancement by reaction.

9.11. New developments in reactor modelling

Design models for complex kinetics either can be found in the literature or can be easily derived from by now standard theory on mass transfer with reaction in slurries.

The effect of axial dispersion of the gas phase on conversion in slurry reactors usually is a second order effect only but more attention to the mass transfer parameters close to the grid can be worthwile.

Successful application of dispersion-solids sedimentation models in continuous bubble column reactors has been achieved but it also has been shown that the solids sedimentation effects rapidly decrease with reactor diameter. For columns larger than 30 cm, solids settling effects usually can be neglected for particle sizes up to 200 μm.

Further development of staged slurry column reactors can be attractive.

9.12. Application of reactor design models

Publications on successful application of realistic design models to commercial scale slurry processes are relatively scarce. Nevertheless, progress has been made, particularly in modelling slurry reactors for coal liquefaction, Fisher-Tropsch synthesis, methanol synthesis, oxydesulfurization of coal and selective hydrogenation where intermediates are the desired product. The result is encouraging, taking the lack of reliable mass transfer data at actual reactor conditions into account.

9.13. Prospective new developments in slurry process applications

Slurry reactors have a bright future. New applications can be expected in methanol synthesis, in combining the Fisher-Tropsch with the Mobil process, in hydrogen recovery from lean gas mixtures, in biochemical reaction engineering, in using activated carbon either to promote mass transfer in chemical reactors or to increase selectivity in separation processes, to mention a few.

LIST OF SYMBOLS

A_p	external surface area of particle	m
a	interfacial area of bubbles per unit volume reactor	m^2/m^3
a_p	external interfacial area of particles per unit volume reactor	m^2/m^3
a_S	internal surface area per unit porous particle volume	m^2/m^3
Al	Hinterland coefficient $(1-\varepsilon)/a\delta = (1-\varepsilon)k_L/(aD_{AL})$	-
Bd	Bond number $(g\ d_t^2\ \rho_L/\ \sigma)$	-
$\underline{c}_J$	molar concentration of species J	$[kmol/m^3]$
$\bar{c}_J$	molar concentration in the bulk of the liquid or gas phase	$[kmol/m^3]$
d	diameter	m
D	molecular diffusion coefficient	m^2/s
D_i	intra particle diffusion coefficient	m^2/s
D_l	axial dispersion coefficient	m^2/s
D_p	diffusion coefficient for particles	m^2/s
$D_p\langle\bar{c}_B\rangle$	flux of particles to surface	$kmol/m^2.s$
e_b	specific power input per kg liquid close to the vessel bottom	W/kg.s
$\bar{e}$	average specific power input per kg liquid	W/kg.s

$\bar{e}_{min}$	average power input per kg liquid at minimum suspension criterium	W/kg.s
E_J	enhancement factor of component J; defined by Eqn 51	
E_{JS}	ibid, with respect to transport into porous particles	-
$E_{J\infty}$	ibid, for instantaneous reaction	-
$E_{A,phys}$	enhancement factor due to addition of activated carbon	-
Fr	Froude number (u_G^2 / g d_t)	-
g	gravity constant	m/s^2
Ga	Galilei number (g d_t^3 / ν_L^2)	-
Ha	Hatta number	
J	molar flux	$kmol/m^2s$
K	Langmuir adsorption coefficient	$m^3/kmol$
K_{LS}, K_L	overall mass transfer coefficient	m/s
k_G, k_L	individual mass transfer coefficients at G-L interface	m/s
k_{LS}	ibid, at liquid-solid interface	m/s
k_n,$k_{n,m}$	homogeneous reaction velocity constant of order n or n,m	$\frac{m^{3(n+m-1)}}{kmol^{(n+m-1)}s}$
k_n''	surface reaction rate constant of order n	$\frac{m^{3(n-1)+1}}{kmol^{(n-1)}s}$
L	length of reactor	m
L_{SL}	height of the slurry	m
M_J	molecular weight comp J	kg/kmol
m	distribution coefficient e.g. m = c_L/c_G at equilibrium	-
N	number of tanks in series	
n	exponent in Eqn (4)	
Pe_L	liquid/gas Peclet number	u_Gd_t/D_l
Pe_S	solid/gas Peclet number	$u_Gd_t/D_{l,s}$
R_1,R_2	resistances against mass transfer	
R_3,R_4	defined below Eqn (23)	s
R_J	molar rate of production per unit volume	$kmol/m^3s$
Re_p	= d_pv_T/ν	
Sh	Sherwood number	$k_{LS}d_p/D$
Sc	Schmidt number ν/D	
T	temperature	°C or K
u	superficial velocity	m/s
u_r	relative velocity G-L	m/s
V_p, V_r, V_L	volume of particle, reactor and liquid	m^3
v_b	bubble rise velocity	m/s
v_p	particle terminal velocity	m/s
v_r	effective tip velocity of stirrer	m/s
v_i	minimum pick-up velocity (eqn 1)	m/s

v_u	upward component of liquid velocity (eqn 2)	m/s
v'	fluctuating velocitiy of critical eddies	m/s
W	power input	W
w_S	mass of solids per kg liquid	-
z	coordinate in direction of flow	m
α	proportionality constant	
δ	film thickness for mass transfer (film theory); without an index in the liquid at G-L interface	m
δ_r	depth of penetration in mass transfer with instantaneous reaction	m
$\Delta\rho$	$\rho_L - \rho_G$ or $\rho_s - \rho_L$	kg/m^3
ε	gas hold up per m^3 reactor	
ε_L, ε_S	hold up of liquid and solids	
μ	dynamic viscosity	$N.s/m^2$
ν	kinematic viscosity	m^2/s
ν_J	stoichiometric coeff. of species J	
ζ_J	degree of conversion of species J	
η	effectiviness factor	
ρ	density	kg/m^3
σ	surface tension	N/m
Φ_J	production rate of J	kmol/s
Φ_v	volumetric flow rate	m^3/s
ω	stirring speed	s^{-1}
ω_{min}	minimum suspension stirring speed	s^{-1}
τ	contact time in penetration theory; also residence time	s
ϕ	reaction or Thiele modulus	-
ψ	power number	-

Subsripts

A,B,	reactant A, B etc.
b	bubble
B	liquid-solids suspension
G,L.S	gas, liquid, solids phase
p	particle
P	product P
s_t	stirrer reactor
0	initial or feed condition
1	in reactor outlet

Superscripts

0	without solids present

LITERATURE

[1] Doraiswamy, L.K. and Sharma, M.M., Heterogeneous reactions, Vol 2, Wiley (1984).

[2] Shah, Y.T., Gas-Liquid-Solid Reactor Design, McGraw-Hill inc, New York (1979).

[3] Chaudhari, R.V. and Ramachandran, P.A., AIChE J. 26 (1980) 177-201.

[4] Deckwer, W.D. and Alper, E., Chem. Ing. Tech., 52 (1980) 219-228.

[5] Hofmann, H., in: E. Alper (Ed.), Mass transfer with chemical reaction in multi phase systems, vol II, NATO ASI Series No. 73, Martinus Nijhoff Publishers, The Hague (1983) 171-197.

[6] Shah, Y.T., Kelkar, B.G., Godbole, S.P. and Deckwer, W.D., AIChE J., 28 (1982) 353-379.

[7] Muroyama, K. and Fan, L., AIChE J., 31 No 1 (1985) 1-35.

[8] Satterfield, Ch.N. and Huff, G.A., ACS Symp. Ser. 196 (1982) 225-236.

[9] Van Zijll Langhout, W.C., Ouwerkerk, C. and Pronk, K.M.A., Oil Gas J., Dec. 1, 1980, 120-126.

[10] Dautzenberg, F.M. and George, S.E., Advances in Catalytic Chemistry II, Symposium Salt Lake City, Utah U.S.A., May 1982.

[11] Dautzenberg, F.M., van Klinken, J., Pronk, K.M.A., Sie, S.T. and Wijffels, J.-B., ACS Symposium Series (1978), 65, 254-267.

[12] Coca, J. and Diaz, J.M., Dechema Monograph, 86 (1980) 517-528.

[13] Franck, H.G. and Knop, A., Kohleveredelung Chemie und Technologie, Springer Verlag, Berlin, New York (1979).

[14] Juvekar, V.A. and Sharma, M.M., Chem. Eng. Sci., 28 (1973) 825-837.

[15] Sittig, W., Faust, U., Präve, P. and Scholderer, J., Chemische Industrie, 30 (1978) 713.

[16] Ptasinsky, K.J., Beenackers, A.A.C.M. and van Swaaij, W.P.M., in: W. Palz and D. Pirrwitz (Eds), to be published in: Energy from Biomass Series E No 5.

[17] Calderbank, P.H., Evans, F., Farley, R., Jepson, G. and Poll, A., Proc. Symp. on Catalysis in Practice, Inst. Chem. Engrs., London (1963) 66-74.

[18] Baltes, J., Cornils, B. and Frohning, C.D., Chem. Ing. Tech., 47 (1975) 522-532.

[19] Joshi, J.B., Abichandani, J.S., Shah, Y.T., Ruether, J.A. and Ritz, H.J., AIChE J., 27 (1981) 937-945.

[20] Lefers, J., Koetsier, W.T. and van Swaaij, W.P.M., Chem. Eng. J. 15 (1978) 111-120.

[21] Janakiraman, B. and Sharma, M.M., Chem. Eng. Sci. 40 (1985) 235-247.

[22] Tinge, J.T., Mencke, K. and Drinkenburg, A.A.H., to be published.

[23] Ghim, Y.S., Kim, M.J. and Chang, H.N., Proc. 3rd Pac. Chem.

Eng. Congr. (1983) 164-168.
[24] Chapman, C.M., Nienow, A.W., Cooke, M. and Middleton, J.C., Chem. Eng. Res. Des., 61 (1983) 71-81.
[25] Narayanan, S., Bhatia, V.K. and Guha, D.K., Can. J. Chem. Eng. 47 (1969) 360-364.
[26] Roy, N.K., Guha, D.K. and Rao, N.M., Chem. Eng. Sci. 19 (1964) 215-225.
[27] Kato, Y., Nishiwaki, A., Fukuda, T. and Tanaka, S., J. Chem. Eng. Jpn., 5 (1972) 112-118.
[28] Imafuku, K., Wang, T.Y., Koide, K. and Kubota, H. J. Chem. Eng. (Japan), 1 No 2 (1968) 153-158.
[29] Chapman, C.M., Nienow, A.W., Cooke, M. and Middleton, J.C., Chem. Eng. Res. Des., 61 (1983) 167-181.
[30] Zwietering, T.N., Chem. Eng. Sci., 8 (1958) 244-253.
[31] Nienow, A.N., Chem. Eng. Sci., 23 (1968) 1453-1459.
[32] Baldi, G., Conti, R. and Alaria, E., Chem. Eng. Sci. 33 (1978) 21-25.
[33] Gates, L.E., Morton, J.R. and Fondy, P.L., Chem. Eng., 83 (1976) May 24, 144-150.
[34] Arbiter, N., Harris, C.C. and Yap, R.F., Trans. AIME 244 (1969) 134-148.
[35] Queneau, P.B., Jan, R.J., Rickard, R.S. and Lowe, D.F., Metalurg. Trans, B (1975), 6B (1), 149-157.
[36] Wiedmann, J.A., Steiff, A. and Weinspack, P.M., Chem. Eng. Commun., 6 (1980) 245-256.
[37] Subbarao, D. and Taneja, V.K., 3rd Euro.Conf. Mixing (BHRA, Cranfield) Vol. 1, (1979) 229-240.
[38] Darton, R.C., Proc. XVIth ICHMT Symp. Heat and Mass transfer in fixed and Fluidized beds, Dubrovnik, Yugoslavia (to be published).
[39] Akita, K. and Yoshida, F., Ind. Eng. Chem. Process Des. Dev., 12 (1973) 76-80.
[40] Heynen, J.J. and van 't Riet, K., Chem. Eng. J. 28 (1984) B21-B42.
[41] Ying, D.H., Givens, E.N. and Weimer, R.F., Ind. Eng. Chem. Proc. Des. Dev., 19 (1980) 635-641.
[42] Rice, P.G., Tuppurainen, P. and Hedge, R.M., Paper presented at ACS Meeting, Las Vegas, August 1980.
[43] Joshi, J.B., Trans. Inst. Chem. Engrs., 58 (1980) 155-165.
[44] Field, R.W. and Davidson, J.F., Trans. Inst. Chem. Engrs., 58 (1980) 228-236.
[45] Deckwer, W.D., Burckhart, R. and Zoll, G., Chem. Eng. Sci., 29 (1974) 2177-2188.
[46] Kato, Y., Kagaku Kogaku, 27 (1), (1963) 7-11.
[47] Kara, S., Kelkar, B.G. and Shah, Y.T., Ind. Eng. Chem. Process Des. Dev., 21 (1982) 584-594.
[48] Michelsen, M.L. and Ostergaard, K., Chem. Eng. J., 1 (1970) 37-46.
[49] Towell, G.D. and Ackermann, G.H., Proc. 5th European Symp. Chem. Reaction Eng. B1, Amsterdam (1972).

[50] Diboun, M. and Schügerl., K., Chem. Eng. Sci. 22 (1967) 147-160.

[51] Pavlushenko, I.S., Kostin, N.M. and Matveev, S.F., Zhur. Priklad. Khim., 30 (8) (1957) 1160-1169.

[52] Kneule, F., Chem. Ing. Tech., 28 (1956) 221-225.

[53] Narayanan, S., Batia, V.K., Guha, D.K. and Rao, M.N., Chem. Eng. Sci., 24 (1969) 223-230.

[54] Weisman, J. and Efferding, L.E., AIChE J., 6 (1960) 419-426.

[55] Kolar, V., Coll. Czech. Chem. Comm., 32 (1967) 526-534.

[56] Hikita, H. and Kikukawa, H., Chem. Eng. J., 8 (1974) 191-197.

[57] Baird, M.H.I. and Rice, R.G., Chem. Eng. J., 9 (1975) 171-174.

[58] Schügerl, K., Proc. Int. Symp. on Fluidization, Eindhoven, A.A.H. Drinkenburg (Ed.), Neth. Univ. Press, Amsterdam (1967) 782.

[59] Brian, W.B. and Dyer, P.N., ACS Symp. Ser. 237 (1984) 107-124.

[60] Mangartz, K.H. and Philhofer, T., Verfahrenstechnik, 14 (1980) 40-44.

[61] Westerterp, K.R., van Swaaij, W.P.M. and Beenackers, A.A.C.M., Chemical Reactor Design and Operation, Wiley, Chichester, (1984).

[62] Wiedmann, J.A., Steiff, A and Weinspack, P.M., Verfahrenstech. 14 (1980) 93-98.

[63] Kralj, F. and Sincic, D., Chem. Eng. Sci., 39 (1984) 604-607.

[64] Datta, R. and Rinker, R.G., Chem. Eng. Sci., 39 (1984) 893-901.

[65] Ivory, C.F. and Bratzler, R.L., Chem. Eng. Commun. 10 (4-5) (1981) 293-305.

[66] Laurent, A., Prost, C. and Charpentier, J.C., J. Chim. Phys., 72 (1975) 236-244.

[67] Sharma, M.M. and Danckwerts, P.V., Brit. Chem. Eng., 15 (4) (1970) 522-528.

[68] Sridharan, K. and Sharma, M.M., Chem. Eng. Sci. 31 (1976) 767-774.

[69] Laurent, A. and Charpentier, J.C., Chem. Eng. J. (Lausanne), 8 (1974) 85-101.

[70] Charpentier, J.C., in Luss, D. and Weekman (jr), V.W. (Eds), ACS Symp. Ser., 72 (1978) 223-61.

[71] van Landeghem, H., Chem. Eng. Sci., 35 (1980) 1912-1949.

[72] Gestrich, W., Esenwein, H. and Kraus, W., Chem. Ing. Tech., 48 (1976) 399-407.

[73] Gestrich, W. and Krauss, W., Chem. Ing. Tech., 47 (1975) 360-367.

[74] Gestrich, W. and Räse, W., Chem. Ing. Tech., 47 (1975) 8-13.

[75] Mashelkar, R.A., Brit. Chem. Eng., 15 (1970) 1297.

[76] Sideman, S., Hortacsu, O. and Fulton, J.W., Ind. Eng. Chem.

(7), 58 (1966) 33-47.
[77] Reith, T. and Beek, W.J., Chem. React. Eng., Proc. Eur. 4th, Brussels (1968); Pergamon, 1971 p. 191-204.
[78] Reith, T., Brit. Chem. Eng., 15 (1970) 1559-1563.
[79] Van Dierendonck, L.L., Fortuin, J.M.H. and Venderbos, D., Chem. React. Eng., Proc. Eur. Symp., 4th, 1968, Pergamon, 1971, 205-215.
[80] Joosten, G.E.H., Schilder, J.G.M. and Janssen, J.J., Chem. Eng. Sci., 32 (1977) 563-566.
[81] Kars, R.L., Best, R.J. and Drinkenburg, A.A.H., Chem. Eng. J., 17 (1979) 201-210.
[82] Alper, E., Wichtendahl, B. and Deckwer, W.D., Chem. Eng. Sci., 35 (1980) 217-222.
[83] Sada, E., Kumuzawa, H. and Lee, C.H., Chem. Eng. Sci., 38 (1983) 2047-2052.
[84] Satterfield, Ch.N. and Huff, J.A., Chem. Eng. Sci., 35 (1980) 195-202.
[85] Deckwer, W.D., Serpemen, Y., Ralek, M. and Schmidt, B., Chem. Eng. Sci., 36 (1981) 765-771.
[86] Deckwer, W.D., Serpemen, Y., Ralek, M. and Schmidt, B., Chem. Eng. Sci., 36 (1981) 791-792.
[87] Satterfield, Ch.N. and Huff (Jr), G.A., Chem. Eng. Sci., 36 (1981) 790-791.
[88] Deckwer, W.D., Lehmann, H.J., Ralek, M. and Schmidt, B., Chem. Ing. Tech., 53 (1981) 818-819.
[89] Deckwer, W.D., Serpemen, Y., Ralek, M. and Schmidt, B., Ind. Eng. Chem., Proc. Dev., 21 (1982) 231-241.
[90] Van Vuuren, D.S., Chem. Eng. Sci., 38 (1983) 1365-1367.
[91] Satterfield, Ch.N. and Huff (Jr), G.A., Chem. Eng. Sci., 38 (1983) 1367-1368.
[92] Sharma, M.M. and Mashelkar, R.A., I. Chem. Eng. Symp. Ser., Institute of Chemical Engineers, London, 28 (1968) 10-21.
[93] Robinson, C.W. and Wilke, C.R., AIChE J., 20 (1974) 285-294.
[94] Beenackers, A.A.C.M. and van Swaaij, W.P.M., Chem. React. Eng., Proc. Eur., 6th, Intern. Symp., 4th, 1976, Dechema, Frankfurt (M), (1976) VI-260 - VI-270.
[95] Charpentier, J.C.H., Adv. Chem. Eng., 11 (1981) 1-133.
[96] Towell, G.D., Strand, C.P., and Ackerman, G.H., AIChE-I Chem. E., Symp. Ser., No 10, Institute of Chemical Engineers, London, 1965, 97-105.
[97] Deckwer, W.D. in: E. Alper (Ed), Mass transfer with chemical reaction in multi phase systems, Vol II, NATO ASI Series, No. 73, Martinus Nijhoff Publishers (1983) 287-349.
[98] Deckwer, W.D., Louisi, Y., Zaid, A. and Ralek, M., Paper No 76a, presented at AIChE 72nd Annual Meeting, San Francisco, Nov. 1979.
[99] Westerterp, K.R., Eng. Dr. Thesis, Delft, 1962.
[100] Westerterp, K.R., van Dierendonck, L.L. and de Kraa, J.A., Chem. Eng. Sci., 18 (1963) 157-176.
[101] Yoshida, F. and Yoshiharu, M., Ind. Eng. Chem., Proc. Des.

Dev., 2 (1963) 263-268.
[102] Linek, V. and Vacek, V., Chem. Eng. Sci., 36 (1981) 1747-1768.
[103] Linek, V., Collect. Czech. Chem. Commun., 34 (1969) 1299-1301.
[104] Joosten, G.E.H. and Danckwerts, P.V., Chem. Eng. Sci. 28(2), (1973) 453-461.
[105] Reith, T. and Beek, W.J., Chem. Eng. Sci., 28 (1973) 1331-1339.
[106] Nagel, O., Kurten, H. and Sinn, R., Chem. Ing. Tech., 44 (1972) 899-903.
[107] Akita, K. and Yoshida, F., Ind. Eng. Chem., Proc. Des. Dev., 13 No 1 (1974) 84-90.
[108] Van Dierendonck, L., Thesis, Enschede, 1970.
[109] Hammer, H., Int. Chem. Eng., 21 (1981) 173-178.
[110] Hammer, H., Küsters, W., Schrag, H.J. Soemarno, A., Sahabi, U., Schönau, H. and Napp, W. in: L.K. Doraiswamy (Ed), Recent advances in the engineering analysis of chemically reacting systems, Wiley Eastern, New Delhi (1984) 379-395.
[111] Brown, D.M., ISCRE VIII, Edindurgh (1984) Preprints, Inst. Chem. Eng., Symp. Ser. 87 (1984) 699-708.
[112] Quicker, G., Schumpe, A. and Deckwer, W.D., Chem. Eng. Sci., 39 (1984) 179-183.
[113] Kürten, H. and Zehner, P., Ger. Chem. Eng., 2 (1979) 220.
[114] Linek, V., Chem. Eng. Sci., 27 (1972) 627-637.
[115] Linek, V. and Mayrhoferova, J., Chem. Eng. Sci., 24 (1969) 481-496.
[116] Brian, P.L.T., Vivian, J.E. and Matiatos, D.C., AIChE J., 13 (1967) 28-36.
[117] Danckwerts, P.V., Gas-Liquid reactions, McGraw-Hill, London (1970).
[118] Calderbank, P.H. and Moo Young, M.B., Chem. Eng. Sci., 16 (1961) 39-54.
[119] Sada, E., Kumazawa, H. and Butt, M.A., Chem. Eng. Sci., 32 (1977) 1165-1170.
[120] Sada, E., Kumazawa, H. and Lee, C.H., Chem. Eng. Sci., 39 1984) 117-120.
[121] Harriott, P., AIChE J., 8 (1962) 101-102.
[122] Harriott, P., AIChE J., 8 (1962) 93-101.
[123] Levins, D.M. and Glastonburg, J.R., Chem. Eng. Sci., 27 (1972) 537-543.
[124] Schwartzberg, H.G. and Treybal, R.E., Ind. Eng. Chem. Fund., 7, No 1 (1968) 6-12.
[125] Chapman, C.M., Nienow, A.W., Cooke, M. and Middleton, J.C., Chem. Eng. Res. Des. 61 (1983) 182-185.
[126] Sano, Y., Yamaguchi, N. and Adachi, T., J. Chem. Eng. Japan, 7 (1974) 255-261.
[127] Sänger, P. and Deckwer, W.D., Chem. Eng. J., 22 (1981) 179-186.
[128] Ohta, H., Goto, S. and Teshima, H., Ind. Eng. Chem., Fund. 19

(1980) 180-185.
[129] Njiribeako, A., Silverston, P.L. and Hudgins, R.R., Prepr. Can. Symp. Catal. 1977, 5th, 170-181.
[130] Pal, S.K., Sharma, M.M. and Juvekar, V.A., Chem. Eng. Sci., 37 (1982) 327-336.
[131] Huff, G.A. and Satterfield, Ch.N., Ind. Eng. Chem. Fundam., 21 (1982) 479-483.
[132] Shah, Y.T. and Smith, D., Proc. 8th Int. Symp. om Chem. React. Eng., ISCRE VIII, Edinburgh, September 1984; to be published by the Inst. Chem. Engrs, U.K.
[133] Weng, H.S. and Smith, J.M., Chem. Eng. J., 28 (1984) 115-124.
[134] Sada, E., Kumazawa, H., Butt, M.A. and Sumi, T., Chem. Eng. Sci., 32 (1977) 972-974.
[135] Alper, E. and Deckwer, W.D. in: E. Alper (Ed), Mass transfer with chemical reaction in multiphase systems, vol II, NATO ASI Series, No 73, Martinus Nijhoff Publishers (1983) 199-224.
[136] Ramachandran, P.A. and Sharma, M.M., Chem. Eng. Sci., 24 (1969) 1681-1686.
[137] Uchida, S., Miyazaki, M. and Masumoto, S., Chem. Eng. Sci., 39 (1984) 1527-1528.
[138] Uchida, S. and Wen, C.Y., Chem. Eng. Sci., 32 (1977) 1277-1281.
[139] Ptasinsky, K.J., Beenackers, A.A.C.M., van Swaaij, W.P.M. and Holstvoogd, R.; to be published in Energy from Biomass, Series E No 7, Reidel, Dordrecht (1985).
[140] Kut, O.M., Gut, G., Buehlmann, T. and Lussy, A., in: E. Alper (Ed), Mass transfer with chemical reaction im multiphase systems, vol II NATO ASI Series E, No 73, Martinus Nijhoff, The Hague (1983) p. 225- 237.
[141] Alper, E. and Deckwer, W.D., Chem. Eng. Sci., 36 (1981) 1097-1099.
[142] Sada, E., Kumazawa, H. and Butt, M.A., Chem. Eng. Sci., 32 (1977) 970-972.
[143] Sada, E., Kumazawa, H. and Hashizume, I., Chem. Eng. J., 26 (1983) 239-244.
[144] Wimmers, O.J. and Fortuin, J.M.H., preprints ISCRE 8, Inst. Chem. Eng. Symp. Ser., 87 (1984) 195-204.
[145] Wimmers, O.J., Paulussen, R., Vermeulen, D.P. and Fortuin, J.M.H., Chem. Eng. Sci., 39 (1984) 1415-1422.
[146] Bruining, W., Joosten, G., Hofman, H. and Beenackers, A.A.C.M., to be published.
[147] Uchida, S., Koide, K. and Shindo, M., Chem. Eng. Sci., 30 (1975) 644-646.
[148] Uchida, S., Miyachi, M. and Ariga, O., Can. J. Chem. Eng., 59 (1981) 560-562.
[149] Sada, E., Kumazawa, H. and Hashizume, I., Chem. Eng. Sci., 36 (1981) 639-642.
[150] Pasiuk-Bronikovska, W. and Ziajka, J., Chem. Eng. Sci., 17

(1982) 1823-1824.
[151] Joshi, J.B., Abichandani, J.S., Shah, Y.T., Ruether, J.A. and Ritz, H.J., AIChE J., 27 (1981) 937-945.
[152] Beenackers, A.A.C.M. and van Swaaij, W.P.M., Eur. Pat. Appl. No 83200660.5 dd 06.05.1983.
[153] Beenackers, A.A.C.M. and van Swaaij, W.P.M. in: G. Grassi and W. Palz (Eds), Energy from Biomass, Series E, vol 3, Reidel, Dordrecht (1982) 201.
[154] Beenackers, A.A.C.M. and van Swaaij, W.P.M., Neth. Patent Appl. No 8201885 (1982).
[155] Holstvoogd, R., Ptasinsky, K.J. and van Swaaij, W.P.M., to be published.
[156] Ramachandran, P.A. and Chaudhari, R.V., Chem. Eng., Dec. 1980, 74-85.
[157] Ramachandran, P.A. and Chaudhari, R.V., Chem. Eng. J., 20 (1980) 75-78.
[158] Chaudhari, R.V. and Ramachandran, P.A., Ind. Eng. Chem., Fund., 19 (1980) 201-206.
[159] Ramachandran, P.A. and Chaudhari, R.V., Can. J. Chem. Engng., 58 (1980) 412-415.
[160] Chang, H. and Smith, J.M., AIChE J., 29 (1983) 699-700.
[161] Alvarez-Cuenca, M. and Nederenberg, M.A., Proc. Int. Fermentation Symp. 6th, Advances in Biotechnology (1981) Vol 1, 477-503.
[162] Darton, R.C., in: La Nauze, R.D., Davidson, J.F. and Harrison, D., Trans. Inst. Chem. Engrs., 55 (1977) 274-280.
[163] Darton, R.C., Trans. Inst. Chem. Engrs., 57 (1979) 134-138.
[164] Govindarao, V.M.H. and Chidambaram, M., AIChE J., 30 (1984) 842-845.
[165] Govindarao, V.M.H. and Chidambaram, M., Chem. Eng. J., 27 (1983) 29-36.
[166] Serpemen, Y. and Deckwer, W., in: Alper, E. (Ed), Mass transfer with chemical reaction in multiphase systems, vol II: Three-Phase Systems, NATO ASI Series No 73, Martinus Nijhoff, The Hague, (1983) 239-255.
[167] Govindarao, V.M.H., Chem. Eng. J., 9 (1975) 229-240.
[168] Parulekar, S.J. and Shah, Y.T., Chem. Eng. J., 20 (1980) 21-33.
[169] Elenkov, D. and Vlaev, S.D. in: E. Alper (Ed), Mass transfer with chemical reaction in multiphase systems, vol II, Three-Phase Systems, NATO ASI Series No 73, Martinus Nijhoff (1983) 257-266.
[170] Brauer, H., Ger. Chem. Eng., 3 (1980) 66.
[171] Bukur, D.B., ChemEng. Sci., 38 (1983) 441-446.
[172] Stern, D., Bell, A.T. and Heinemann, H., Chem. Eng. Sci., 38 (1983) 597.
[173] Ledakowicz, S., Fang Li-Kang, Kotowski, W. and Deckwer, W.D., Erdöl und Kohle-Erdgas-Petrochemie vereinigt mit Brennstoff-Chemie, 37 (1984) 462-465.
[174] Kafarov, V.V., Ivanov, V.A. Brodskii, S.Ya. and Leonov, V.E.,

Theor. Found. Chem. Techn., (1982) 16(5), 644-649 (Russian); Engl. Translation: Plenum Publ. (1983) 0040-5795/82/1605-0437.
[175] Bukur, D.B. and Gupte, K.M., Chem. Eng. Sci., 38 (1983) 1363-1364.
[176] Kut, O.M. and Gut, G., Chimia, 34 (1980) 250-253.
[177] Shah, Y.T. and Parulekar, S.J., Chem. Eng. J., 23 (1982) 15-30.
[178] Turek, T., Chakrabarti, R.K., Lange, R., Geike, R. and Flock, W., Chem. Eng. Sci., 38 (1983) 275-283.
[179] Frank, M.E. and Sherwin, M.B., Hydrocarbon Processing, 122-124 (11) (1976).
[180] Frank, M.E., Proc. Intersoc. Energy Convers. Eng. Conf., 15th (2), (1980) 1567-1572.
[181] Sherwin, M.B. and Blum, D., EPRI Report AF-693, Project 317-2, May 1978.
[182] Shah, Y.T. and Gopal, J., in: E. Alper (Ed), Mass transfer with chemical reactions in multiphase systems, vol II: Three-Phase Systems, Martinus Nijhoff, The Hague, (1983) 267-286.
[183] Cholera, V. and Gidaspow, D., Proc. Intersoc. Energy Conv. Eng. Conf. (1977) 12, (1), 981-986.
[184] Cholera, V. and Gidaspow, D., Chem. Eng. Symp. Ser., 54 (1978) 21-32.
[185] Van der Meer, A.B., Beenackers, A.A.C.M. and Stamhuis, E.J., to be published.
[186] Maslan, F. (to Escambia Chemical Corp.) US Patent 2.823.763, US Patent 2.823.765 and US Patent 2.823.766 Appl. Feb. 1958.
[187] Astakov, V.A., Lukin, V.D., Romankov, P.G. and Tan, V.A., Zh. Prikl. Khim., 44 (2) (1971) 319-323.
[188] Shah, Y.T. and Albal, R.S., in E. Alper (Ed), Mass transfer with chemical reaction in multiphase systems, vol II, Three-Phase Systems, NATO ASI Series No 73, Martinus Nijhoff (1983) 351-364.
[189] Oyama, Y. and Endoh, K., Chem. Eng. Tokyo, (1956) 20:66.

HYDRODYNAMICS OF TRICKLE-BEDS. THE PERCOLATION THEORY

M. CRINE

Research Associate NFSR
Laboratoire de Génie Chimique
Université de Liège
Institut de Chimie - B6 - Sart Tilman
4000 LIEGE BELGIUM

1. INTRODUCTION

The performance of a chemical reactor, i.e., the relation between output and input, is affected by the properties of the reactional system (kinetics, thermodynamics,...) and by the contacting pattern. The importance of this contacting pattern is quite obvious in the case of multiphase reactors like trickle-bed reactors. Reactants are indeed present in both fluid phases and reactions occur at the contact of the catalytic solid phase. Unfortunately the description of the fluid flow pattern is very difficult because of the high number of intricate mechanisms that can control this pattern. The situation is so complex that, in many cases, we do not even know the essential hydrodynamic parameters that may affect the performance of the reactor.

Fluid flow pattern may be described in two ways : a fluid mechanisms approach and a global phenomenological approach based on the Residence Time Distribution (R.T.D) concept. The fluid mechanics approach tries to determine the velocity, concentration and temperature profiles within the reactor on the basis of fundamental equations of fluid flow hydrodynamics. This approach, when successful, leads to complex mass and heat balance equations requiring cumbersome numerical computations and yielding too detailed informations when a macroscopic description of the process is required by the chemical engineer.

The RTD concept first introduced by Danckwerts (1) is now quite classical and widely applied to describe liquid flow in chemical reactors. The distribution of residence times obtained by tracer techniques is directly affected by the liquid flow pattern. The shape of RTD curves must consequently reflect in some way the nature of this fluid flow pattern. Phenomenological models simulating these RTD curves allow us to determine how the actual fluid flow deviates from ideal situation such as plug flow, complete mixing,... The parameters contained in these phenomenological RTD models are thus quantifications of deviations from ideal fluid flows. When analyzing liquid flow in trickle-bed reactors, the situation becomes so complex that either the RTD models remain too simple to correctly represent the actual flow pattern, or these models are so complex and the number of parameters so high that they cannot be determined accurately and related to the operating conditions. This is clearly a limit of this second approach which, otherwise, is quite efficient to derive mass and heat balance equations for a trickle-bed reactor.

The two approaches are not conflictive but rather complementary. The fluid mechanics description may give a physical and theoretical meaning to the phenomenological parameters introduced in the RTD models and an indication on the way these parameters are affected by the operating conditions. In turn, the RTD models give a more accurate global description of fluid flow pattern. In this paper, we will present an application of this twosided approach to the analysis of the liquid flow hydrodynamics and the RTD models in trickle-bed reactors.

RTD models for trickle-bed reactors are quite numerous. They are reviewed in part 2 in order to evidence the main fluid flow characteristics that have been considered by the authors developing these models. The fluid mechanics description is based on percolation concepts. The main implications of these concepts are analyzed in part 3 whereas part 4 is devoted to the development of a percolation model describing the liquid flow distribution in a trickle-bed reactor. This model is then applied to derive correlations for the wetting efficiency and the dynamic liquid holdup (part 5) and, finally, for the axial dispersion coefficient (part 6) : a classical example of a phenomenological parameter introduced in a RTD model. It is worth noting that the wetting efficiency is actually a hydrodynamic parameter characterizing the liquid distribution within the packed bed. The difficulty to determine experimentally this parameter is however such that it is often considered as a phenomenological parameter determined by means of an appropriate RTD model (2).

2. R.T.D. MODELS

The R.T.D. models are based on two idealized states of mixing : the complete mixing and the plug flow i.e. the absence of mixing. The actual fluid flow is simulated by adding some simple transport processes to these idealized flow representations and/or by decomposing the reactor volume into different staged regions in which the fluid flow is represented by an idealized state of mixing. Each of these added transport mechanisms is characterized by 1 or 2 phenomenological parameters. The number of parameters is definitely limited to a practical value of two or three because of the limited amount of information available in experimental RTD curves as well as by the accuracy of the numerical technique used to estimate these parameters. This means that the description of the fluid flow remains oversimplified. Actually, these models should not be considered as describing the actual fluid flow hydrodynamics but rather as evaluating the deviations from plug flow or complete mixing in order to write correctly the mass and heat balance equations.

RTD models have been developed to characterize mixing in both gas and liquid phases. Actually, mixing in the gas phase has received very little attention because its influence on the behaviour of trickle flow reactors is rather weak. Gas commonly used as oxygen or hydrogen are scarcely soluble in most liquid reactants so that mass transfer phenomena may be neglected in this phase. Moreover the variations in gas phase composition between the inlet and outlet of the reactor are generally so small that mass balance has to be written in the liquid phase only.

The numerous RTD models developed for trickle bed reactors may be grouped into two classes :

1. Differential models in which the mass balance is applied to a volume element, i.e. represented by differential equations. The basic model in this class is the plug flow model;
2. Stagewise models in which the mass balance is applied to staged regions, i.e. represented by algebraic equations. The basic model in this class is the perfectly mixed cell model.

The main differential R.T.D. models are schematized in Figure 1.

- The plug flow model (Figure 1a) in which each volume element is assumed to have an equal residence time;
- the plug flow with dispersion (P.D.) model (Figure 1b) in which the observed residence time dispersion is represented by a single Fick's law-type mechanism (parameter Bo_L) superimposed onto the plug flow (3);
- the piston exchange model (Figure 1c) in which mass exchange (k_s) between the flowing zone (h_d) and a stagnant one (h_s) is superimposed onto the plug flow (4);
- the time delay models (Figures 1d and 1e) which assume a plug flow region (h_d) with fluid elements randomly delayed (t_D) in stagnant zones (h_s) perfectly or not perfectly (m) mixed (5, 6);
- the piston dispersion exchange model (Figure 1f) similar to the piston exchange model (figure 1c), except that axial dispersion is superimposed onto the plug flow in the flowing zone (7);
- the pulsed flow model (Figure 1g) in which the liquid

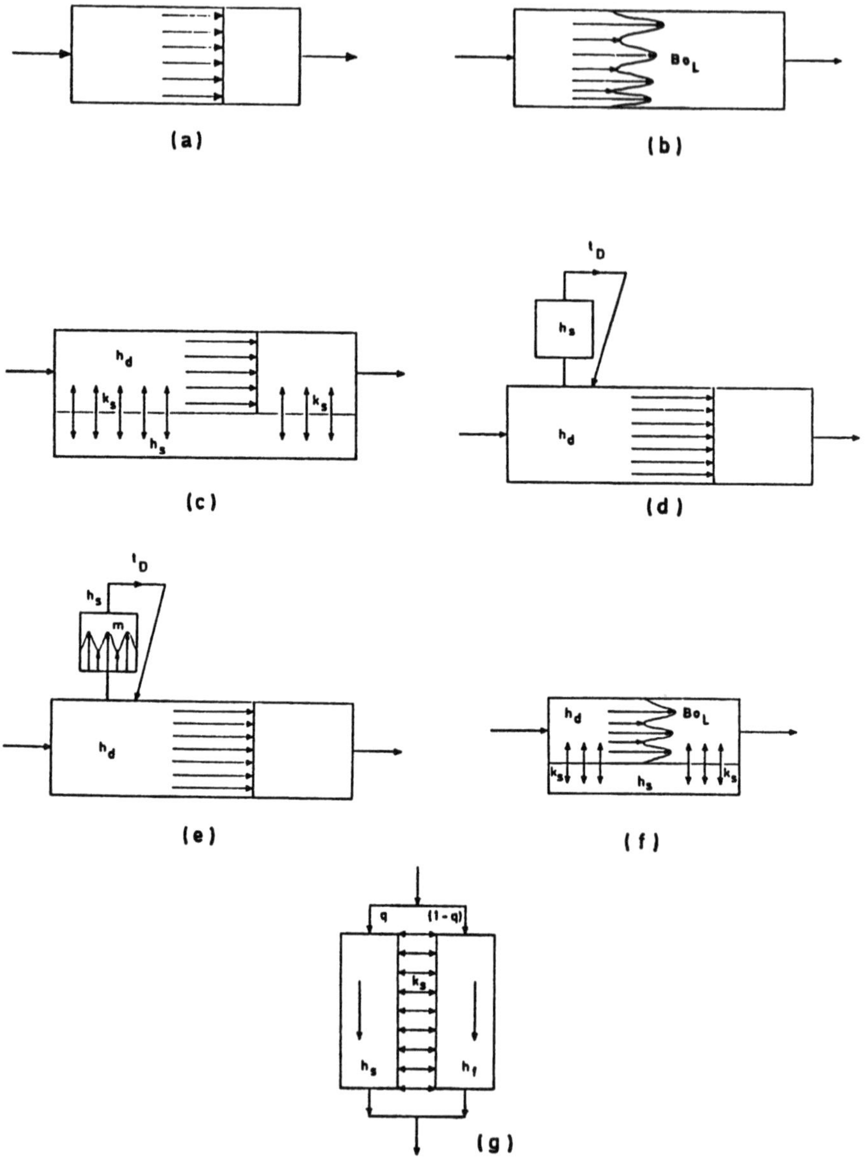

Figure 1
Differential RTD models

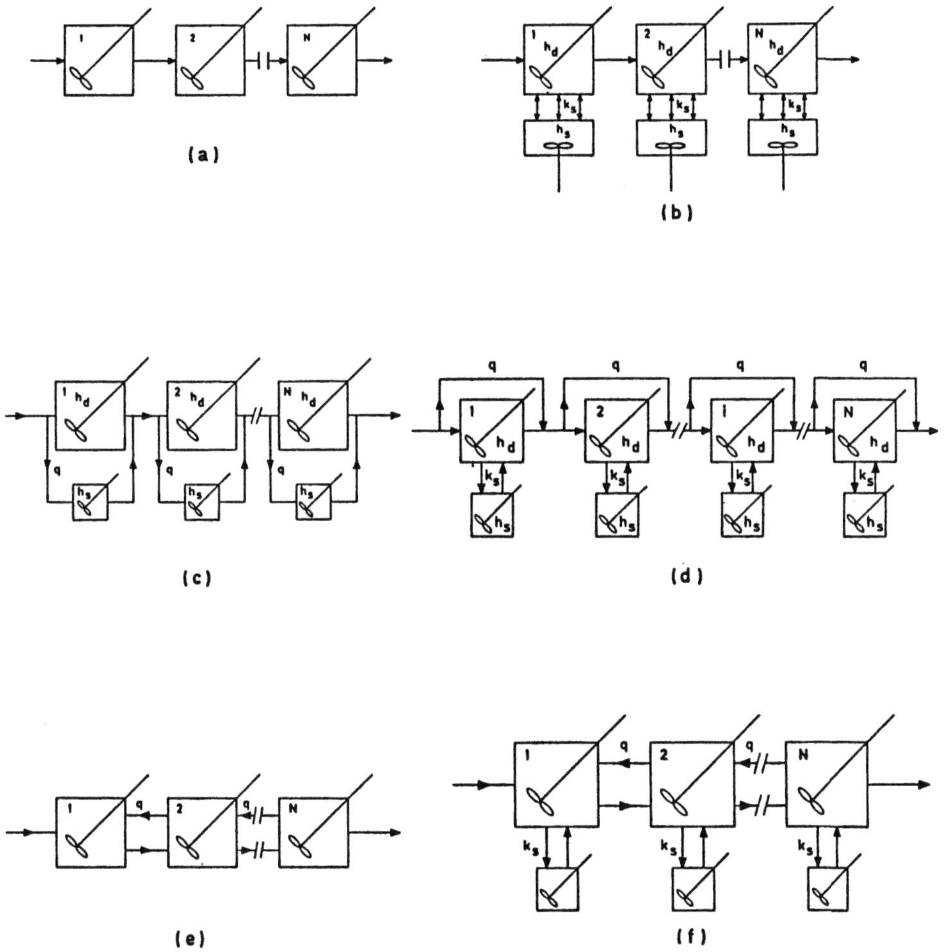

Figure 2
Stagewise RTD models

flow is distributed (q) between two piston flow regions (h_s and h_f) with continuous mass transfer (k_s) between them (8).

The main stagewise R.T.D. models are schematized in Figure 2.

- The cell in series model (Figure 2a) in which the parameter is the number N of cells connected in series (9);
- the cell exchange model (Figure 2b) in which mass exchange (k_s) between the flowing cells (h_d) and stagnant zones (h_s) is added (10);
- the parallel cell model (Figure 2c) in which each stage is made of two cells (h_d, h_s) in parallel between which the liquid flow is distributed (q) (7);
- The cell exchange by-pass model (Figure 2d) similar to a cell exchange model to which a by-pass (q) is added (11);
- the backflow cell model (Figure 2e) in which a backflow (q) is superimposed to the net flow through each cell;
- the backflow cell exchange model (Figure 2f) which consists in a cell exchange model to which backflow (q) is superimposed (12).

The stagewise models reported above assume lumped parameters within each of the staged regions. It is also possible to consider distributed parameters within these regions. Each region is then represented by a differential model (e.g. plug flow or P.D. model). That is the way followed by some researchers who tried to visualize the local fluid flow hydrodynamics. A staged region is assumed to correspond roughly to a packing element. The number N of staged regions may then be estimated by the ratio Z/dp between the reactor length and the packing dimension. The local liquid flow is represented by a trickling film over the surface of the particles and a mixing region at the contact point between particles (Figure 3). The models of Michell and Furzer (13) and Rao and Varma (14) belong to this category. The first model (Figure 3a) represents the local liquid flow by a plug flow in the trickling film and an imperfect mixing at the contact points visualized by a by-pass (q). The second model (Figure 3b) represents the liquid flow by a PD model in the trickling film (Bo_L) and an imperfect mixing visualized by a mass exhange (k_s) between a flowing cell (h_d) and a stagnant zone (h_s). Both Michell and Furzer (13) and Rao and Varma (14) compared their models with experiments and concluded that - because of the high value of the ratio Z/dp (i.e. the number N of stages) - the dispersion is mainly affected by the mechanisms introduced within each stage (by-pass q, Bodenstein number Bo_L,...) rather than by the discretization into a finite number of stages.

This remark introduces the problem of the comparison between differential and stagewise models. The cell in series model (Figure 2a) and the axial dispersion model (Figure 1b)

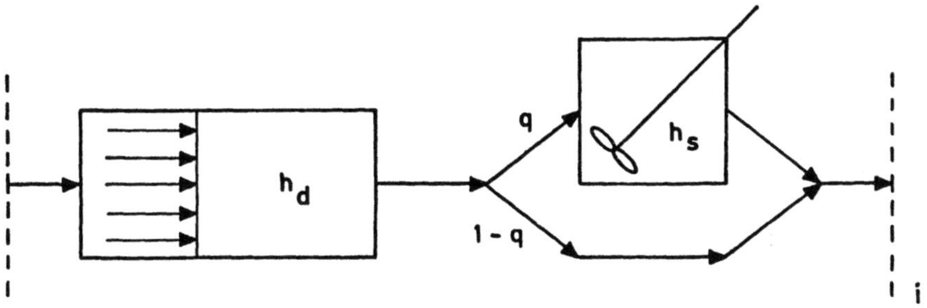

Figure 3a
Local liquid flow vizualisation
by Michell and Furzer (13)

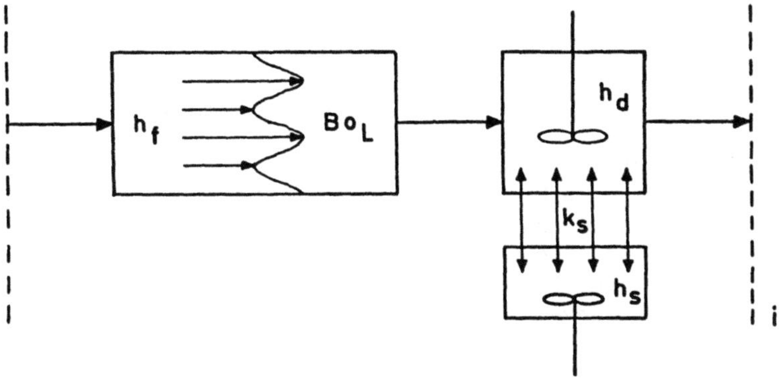

Figure 3b
Local liquid flow vizualisation
by Rao and Varma (14)
Reproduced by permission American Institute of Chemical Engineers.

are often compared because they have rather similar RTD curves in the usual range of operating conditions. They are however based on completely different dispersion mechanisms. Mixing in the PD model is assumed to occur at a very small scale when compared with the size of the reactor, e.g. at the molecular scale or the particle scale. In the staged model, mixing occurs at the scale of the stages. In the first case, the intensity of mixing is limited but mixing has access to the whole bed whereas, in the second case, mixing is complete but delimited to the staged regions. Of course, when the number of stages is very large, i.e. their size is very small, both models become nearly identical.

Besides the distinction between differential and stagewise models, the analysis of Figures 2 and 3 shows two main hypotheses concerning the nature of mixing. In models like the PD model or the cells in series model, mixing is assumed to result from some homogeneous turbulence. The PD model is certainly the most currently used and a lot of correlations have been proposed for its basic parameter - the Bodenstein number. A comparison between some of these correlations, carried out by Goto et al. (15) indicates large discrepancies. Actually many workers (e.g. 13, 16) have shown that this model cannot interpret satisfactorily the whole shape of an experimental RTD curve especially at high time values. This explains the discrepancies between different Bodenstein correlations which strongly depend on the nature of the numerical technique used to estimate the parameter. This disagreement has motivated different investigators to introduce a new concept in modeling liquid flow hydrodynamics - the heterogeneity of the liquid flow. This is the concept on which models like the piston exchange or the cell exchange models are based. In such models, the liquid is assumed to be distributed among two zones of flow - a flowing zone (plug flow, perfectly mixed cell,...) and a stagnant zone. Different models may be derived according to the law ruling the mass exchange between these zones. More elaborate models could also be developed assuming more than two zones of flow in order to represent more realistically the actual liquid flow in trickle-bed reactors. Such models would however have too many parameters to be estimated from experimental RTD curves.

Actually, the RTD models reported in this section evidence some questions that cannot be answered by RTD experiments but rather by a detailed analysis of the liquid flow hydrodynamics :

- the scale of mixing (e.g. the number of stages in a stagewise model)
- the intensity of mixing (e.g. the Bodenstein number in a differential model)
- the nature of mixing (homogeneous turbulence or liquid flow maldistribution).

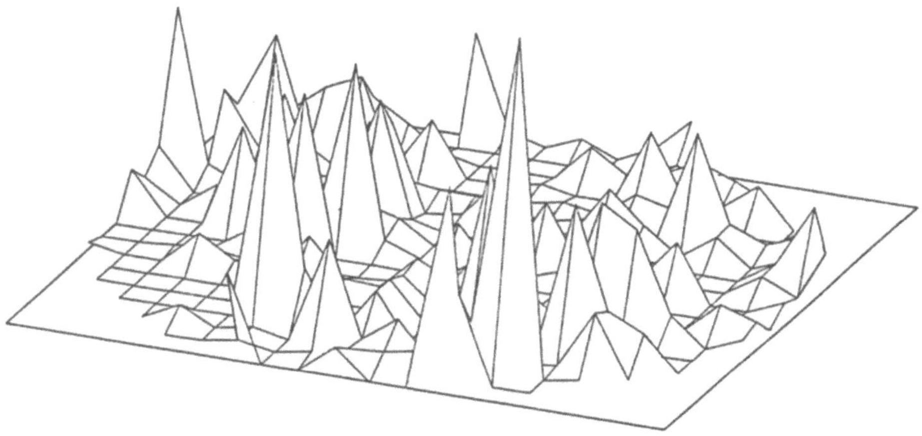

Figure 4
Liquid flow maldistribution at the outlet of a trickling filter (media : Filterpak CR-50; specific area : 220 m^{-1}; liquid flowrate : 0.19 kg m^{-2} s^{-1}) (19)

3. THE DISTRIBUTION OF LIQUID : A DIFFUSION OR PERCOLATION PROCESS?

3.1. Influence of the fluid flow regime

The RTD models reviewed in the preceeding section may be considered as black box models. They simulate RTD curves measured at the outlet of the reactor, adopting some simplified description of the fluid flow hydrodynamics. The amount of information that can be extracted from an RTD experiment is quite limited so that the level of description of the liquid flow remains very simple. For the sake of simplicity, these models adopt a homogeneous picture of fluid flows. For example, the diffusional mechanism adopted to describe mixing in the PD model is uniformly extended from a volume element to the whole packing. The same concept is adopted in the PE model in which the mass transfer process between flowing and stagnant zones is uniformly distributed along the axis of the reactor. Such homogeneous or pseudo-homogeneous models are phenomenologically incorrect. Gas and liquid flow maldistributions have indeed been observed in trickle-bed reactors (17, 18) as well as in a similar apparatus, the trickling filter used in biological wastewater treatment. An example of liquid flow distribution measured at the outlet of this latter type of apparatus is depicted in Figure 4 (19). Obviously the liquid flow is far from being homogeneous. It is rather puzzling to know whether the diffusional picture adopted in classical RTD models may be maintained, or not. As a matter of fact, mixing may always be represented by a diffusional mechanism provided the scale of mixing i.e. the level of description is kept low enough relative to the bed scale (e.g. the molecular scale). The question is then to know how this diffusional picture is modified when moving from the lowest level of description to the bed scale. This question should be answered by applying volume averaging techniques to the fundamental equation of the PD model. It is out of the scope of this paper to describe in detail these techniques which may be a very powerful tool to analyse heterogeneous flows. We will nevertheless consider two limiting cases of practical importance - the case of very high interactions and the case of very low interactions between gas and liquid flows. These two cases correspond to different flow regimes as defined by Charpentier and Favier (20) or Gianetto et al. (21). High interactions are induced in the regime of pulsing and/or foaming flow (depending on the fluid properties) and the regime of spray flow. Low interactions are characteristic of the trickling flow regime. The range of operating parameters corresponding to these regimes may be determined from a flow regime map developed, e.g. by Charpentier and Favier (20) and reproduced in Figure 5.

In the case of high gas-liquid interactions, fluid flow is essentially governed by these fluid-fluid interactions. If, in order to simplify the problem, we assume that fluid flow is governed solely by these interactions, we obtain a scattering process whose each elementary step is determined by the local random properties of the flowing species. This is typically a diffusion process (22). In consequence, the diffusional picture of mixing as well as PD or related homogeneous models may be kept up from the molecular to the bed scale. Obviously, the molecular diffusivity introduced at the molecular scale must be replaced by an eddy diffusivity characterizing turbulence or mixing phenomena. A theoretical determination of this new diffusivity as well as its dependence versus the operating conditions (i.e. the effect of hydrodynamics) would require the use of volume averaging techniques as mentioned above. Nevertheless, in the absence of such a theoretical development, one may recommend the use of well established RTD models with adequate parameter correlations. It must also be recalled that axial dispersion in the regime of high interactions is very weak (23) so that an approximate determination of this parameter may be quite sufficient.

In the case of low gas-liquid interactions - i.e. the trickling flow regime - fluid flow is no more governed by these fluid-fluid interactions but rather by fluid-solid interactions.

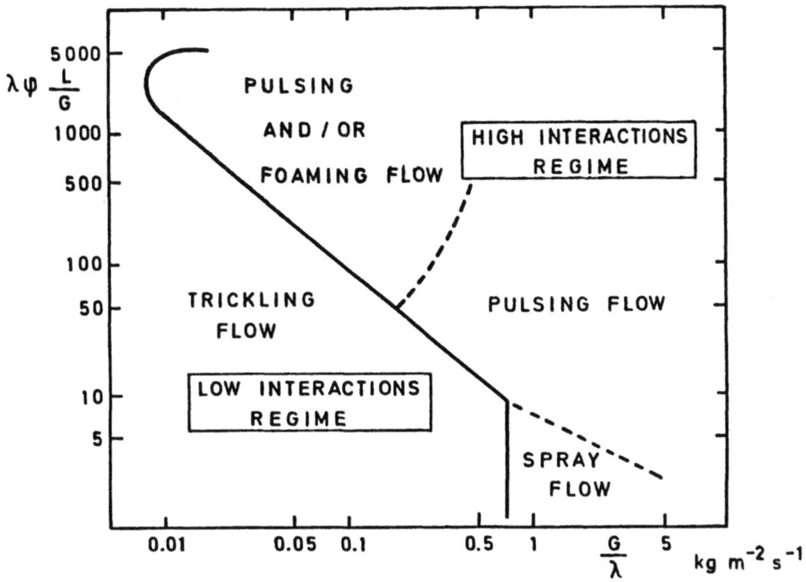

Figure 5
Map of fluid flow regimes (20)

$$\lambda = [(\rho_G/\rho_{air})\ (\rho_L/\rho_{water})]^{0.5}$$

$$\varphi = (\sigma_{water}/\sigma_L)\ [\mu_L/\mu_{water})\ (\rho_{water}/\rho_L)^2]^{0.33}$$

Reproduced by permission American Institute of Chemical Engineers.

The liquid trickling through the packed bed is divided into different channels or rivulets, each of which wets a certain area of the packing surface. The number, the size and the direction of the channels leaving contact points between particles depend on properties of the packing (geometrical, surface properties, etc...) governing the local accessibility of the packing. The clustering of all these elementary channels generates at the bed scale liquid flow structures or preferential flow paths which are characteristic of the liquid flow maldistribution. This representation of the liquid flow implies the scattering process to be governed at each elementary step by a local property (accessibility) of the scattering medium. Such a process is called a percolation process. It exhibits a memory effect related to the local properties of the packing whereas a diffusion process does not exhibit such characteristics. This memory effect explains the existence of preferential flow paths stable in time. In this case, mixing processes are governed by transport mechanisms occurring not only at the molecular and particle scale (i.e. very small scales) but also at the scale of the preferential flow paths - the bed scale. A diffusional picture is thus highly questionable. To answer this question, it is necessary to consider the problem from a theoretical standpoint, applying volume averaging techniques to the basic equation of a diffusional mechanism. That is the purpose of section 4, the remaining of the present section being devoted to a general description of percolation processes.

What is a percolation process ?

Percolation provides an intuitively appealing and transparent model for dealing with the unruly geometries which occur in many strongly disordered materials such as randomly packed beds. Percolation processes were introduced in the mathematical literature by Broadbent and Hammersley in 1957 (22). The growing importance of the subject may be illustrated by a review of some of the articles written in the preceedings years (25, 26, 27, 28). The term percolation itself was first introduced by Hammersley. He had already in mind at that time the passage of a fluid through an interconnected network of channels, with some of the channels being blocked (at random). This image corresponds rather well to the situation prevailing in a coffee percolator. Some channels of the network are blocked or disconnected - they correspond to pores of the packing which cannot be irrigated, i.e. which are not accessible to the liquid flow. This concept of random accessibility is essential in a percolation process; it opens the way to a mathematical description of the liquid flow heterogeneity.

Percolation theory has been developed initially within the context of regular lattices. To account for the unruly geometry of the scattering medium, a non geometric two-state property (e.g. the accessibility to liquid flow), which carries the statistical character of the problem, is randomly assigned to each bond (bond percolation) or site (site percolation) of the lattice. The trickling problem is best represented by a bond percolation process. The accessibility of the packed bed to the liquid is indeed determined by the local states of pore irrigation i.e. by the local states of connection between two neighbour contact points.

Figure 6 illustrates some aspects of bond percolation on a square lattice. This figure shows the changes occurring within one portion of the lattice as the fraction p of unblocked bonds is increased from 0.1 to 0.4 and 0.6. Several clusters formed by contiguous unblocked bonds are indicated (n denotes cluster size). At $p = 0.1$ and $p = 0.4$ all clusters are finite but at $p = 0.6$, infinite clusters (actually represented on the figure by a cluster connecting opposite sites of the sample)

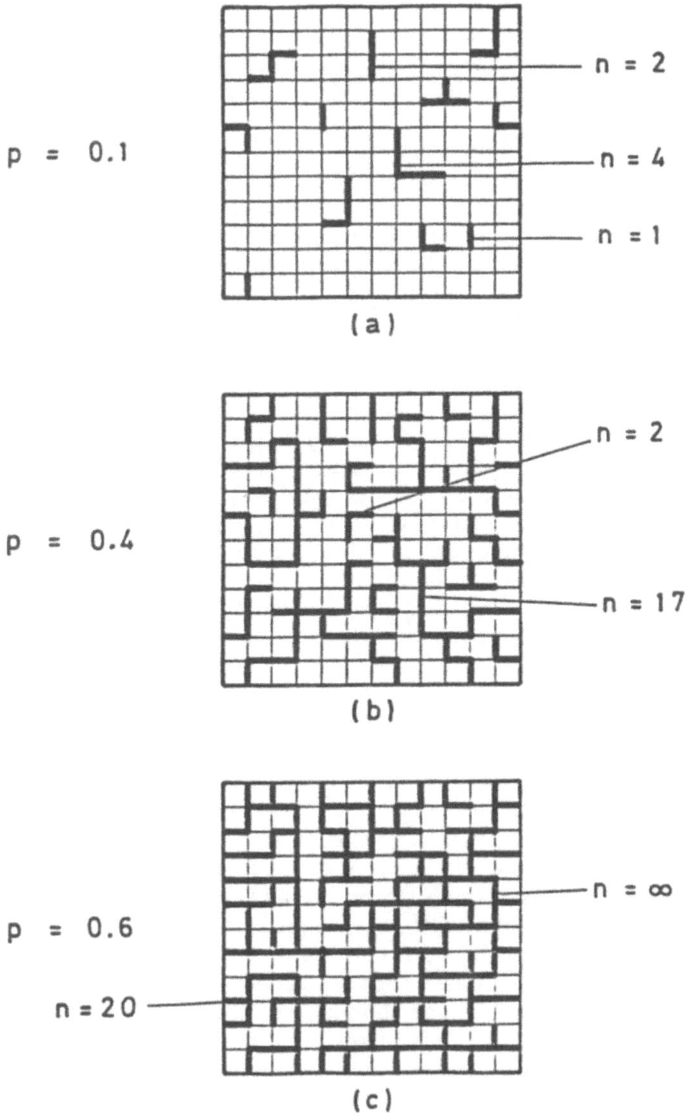

Figure 6
Bond percolation in a square lattice

are present in addition to finite ones. Between Figures 6b and 6c, a critical threshold (equal to 0.5 in a square lattice) has been crossed below which there are only finite clusters and above which both finite and infinite clusters coexist. As indicated in Figures 6a and 6b, small clusters dominate the scene as p is small, and larger clusters appear as p increases. As p approaches the critical threshold the average cluster size diverges. Considerable research has been devoted to the study of the phenomena of percolation. The evolution of interest in this field has been spurred by the relationship between the percolation problem and the more general one of phase transitions. All these works have led to a number of crucial results in the understanding of the behaviour in the neighbourhood of the critical threshold. There exists, however, another problem particular to percolation which is more closely related to the mechanisms of gas-liquid flow through a packed bed - the transport processes through percolation structures.

An easy way to analyze this problem consists in simulating by Monte Carlo methods random walks within the percolation structures. A simple example will illustrate the influence of the percolation structures on the random walk which otherwise would reduce to a diffusion problem. Let us consider (Figure 7) a regular 1D-lattice. Particles move randomly stepwise in this lattice with equal probability either to the right or to the left. At each step n, the square of the displacement $R^2(n)$ relative to the local origin is computed. When the various values of $R^2(n)$ are averaged over a large number of simulations, we obtain the mean-square displacement $\langle R^2(n)\rangle$, characterizing the random walk problem. In the case of a random walk ruled by a diffusional mechanism, the local direction of displacement is a property of the moving particles. A few steps of this type of random walk are illustrated in the lefthand side of Figure 7. This diffusional random walk is characterized by a mean-square displacement $\langle R^2(n)\rangle$ equal to the number of steps n as reported in Figure 8. In the case of a random walk within percolation structures, the local directions of displacement are attributed randomly to the bonds of the lattice. Particles move inside the lattice, following these local directions. A few random walk steps are illustrated in the righthand side of Figure 7. Actually, after a few steps the particle remains trapped between two bonds of opposite directions. This phenomena of trapping or reduction of accessibility is a crucial property of a percolation process. The mean-square displacement $\langle R^2(n)\rangle$ must saturate at a value $\langle R^2(\infty)\rangle$ for a sufficiently large number of steps as reported in Figure 8. This very simple example may be generalized to 2D and 3D lattices. We would observe the same phenomenon of saturation below the critical threshold (actually the critical threshold in a 1D lattice equals 1). Above the critical threshold, $\langle R^2(n)\rangle$ may tend to infinity but with a slope determined by the extent of the per-

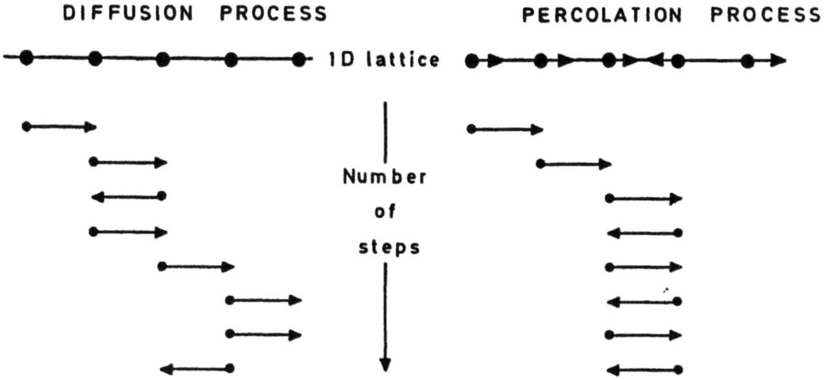

Figure 7
Random walk in a 1D lattice

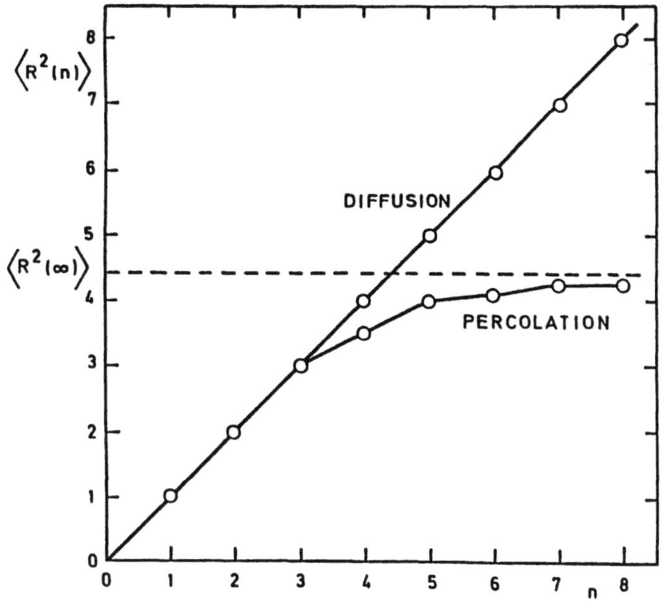

Figure 8
Mean square displacement versus the number of steps in a 1D lattice

colation structures.

These few examples of percolation problems evidence the influence that percolation structures may have on fluid flow hydrodynamics when fluid-solid interaction is the factor determining fluid flow through packed beds (i.e. at low gas-liquid interactions).

4. THE PERCOLATION MODEL

4.1. Percolation structures

The unruly geometry of the packed bed is represented by randomly assigning to each bond of a regular lattice a two-state property representing the local accessibility to liquid flow. A typical correspondence between the packing and such a lattice is illustrated in Figure 9. The sites of the lattice correspond to contact points between particles whereas the bonds correspond to pores or volume elements joining two neighbour contact points. The juxtaposition of unblocked or accessible bonds generates clusters, some of which have an infinite length, as observed in the simple example of part 3 of this paper. These infinite length clusters may contribute to a transport property - like liquid flow - provided dead-end trajectories have been first eliminated (29). The resulting clusters represent what we will call the percolation structures.

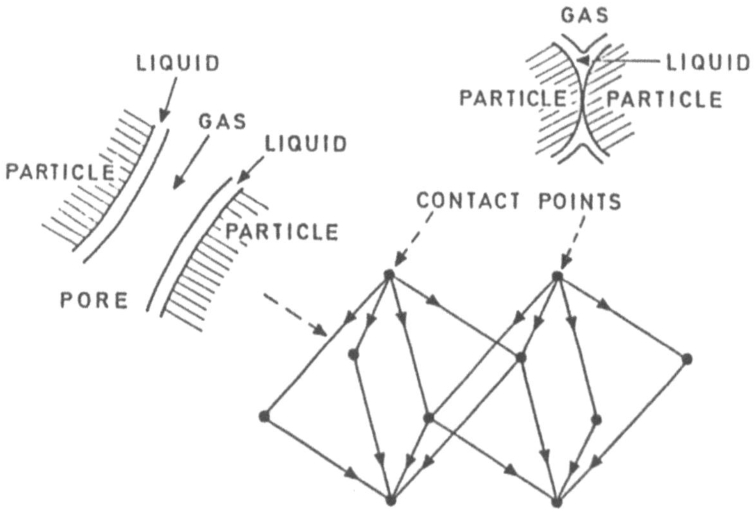

Figure 9
Analogy between packing and lattice

This lattice representation provides a powerfull tool for numerical simulations using e.g. Monte Carlo methods. While somewhat out of the scope of this paper, these numerical simulations are interesting to interpret because they evidence the type of liquid flow properties it is possible to account for by means of the percolation concept. Figure 10 illustrates a typical result of these simulations. It represents for different cases, the distribution of irrigated zones in a cross section of the packing. The black squares show the intersections of the liquid flow paths with the cross section.

Before interpreting these numerical simulations, it is interesting to compare them with some experimental results represented in figure 11 in a similar way. These results have been obtained by locally measuring the liquid velocity at the outlet of a trickling filter (19). The agreement between experiments and numerical simulations seems to be qualitatively statisfying and suggest us to analyze in further details flow-path distributions represented in Figure 10.

These flow path distributions are fundamentally stochastic and must be quantified by some characteristic parameters. The first one and certainly the most useful is the fraction of bonds belonging to percolation structures. It is graphically represented by the fraction of black squares in figure 10. This fraction corresponds also to the fraction of irrigated packing i.e. the wetting efficiency f_w.

For small values of f_w (f_w = 0.18, Figure 10a), the structures of the flow are threadlike and isolated from one another. They consist of independent rivulets, the cross sections of which are shown by isolated black squares. For values of f_w which are a little larger (f_w = 0.36; Figure 10b), small aggregates of very cut-off forms appear. They correspond to expansions and contractions of the threadlike structures. This remains a rivulet flow, since each trajectory is bounded at least partially by non irrigated zones (white squares), i.e. they conserve some degree of freedom for spreading. For still higher values of f_w, the ramification of the structures increases and above a certain threshold, a transition is observed (f_w = 0,79, Figure 10d). The size of the aggregates is such that many flow trajectories are completely bounded by irrigated zones, i.e. they lose any degree of freedom for spreading. These trajectories are representative of a film flow. There is a distribution between the rivulets and films, with the proportion of films increasing with f_w. The existence of rivulets and films is phenomenologically correct. It has been observed by many investigators and has provided, e.g. the basis for the development of the liquid holdup model derived by Charpentier et al. (30).

Figure 10 represents examples of bond irrigation distributions. It would be possible to draw similar pictures for site irrigation distributions. The black squares would corres-

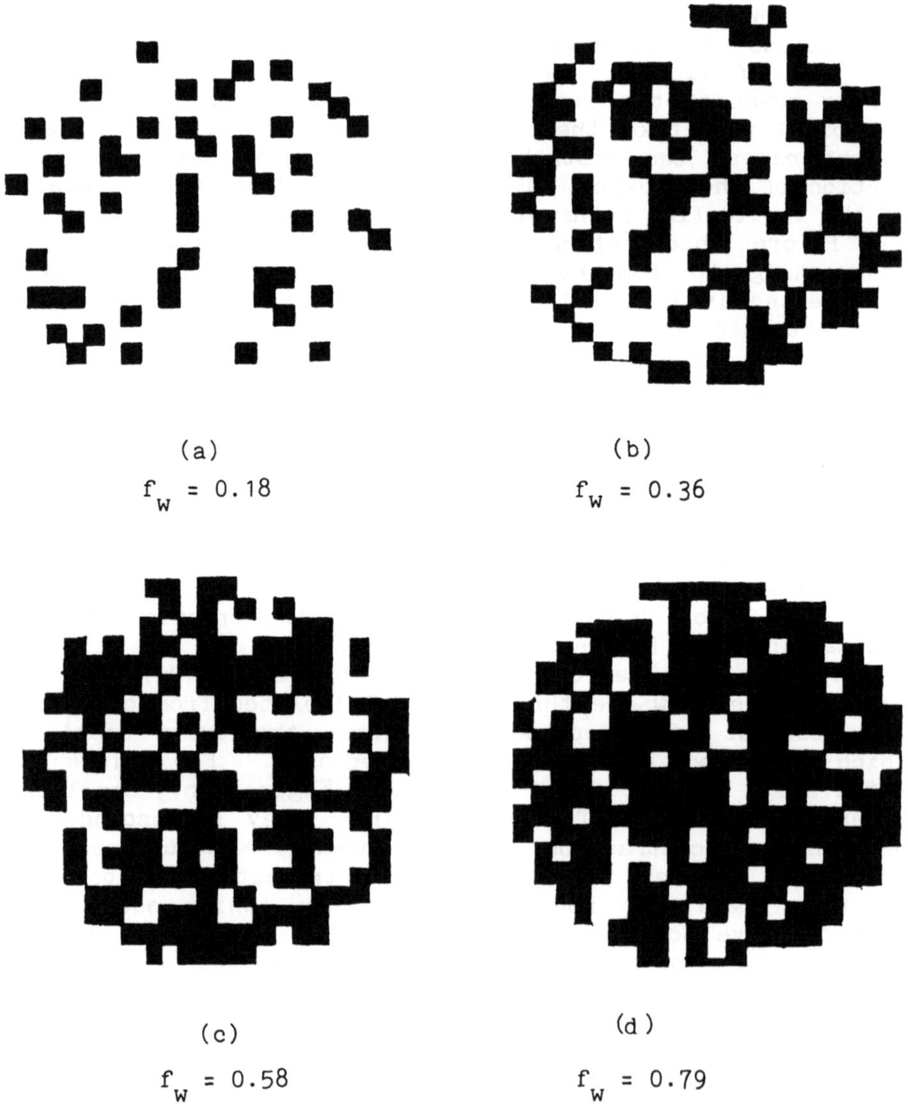

Figure 10
Numerical simulations of percolation structures in a centered cubic lattice

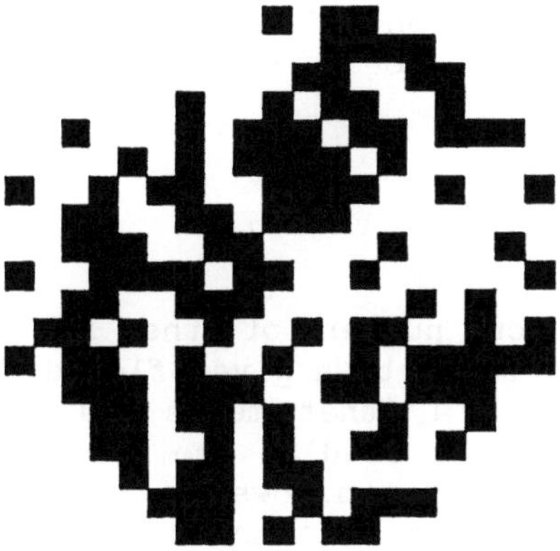

(a)

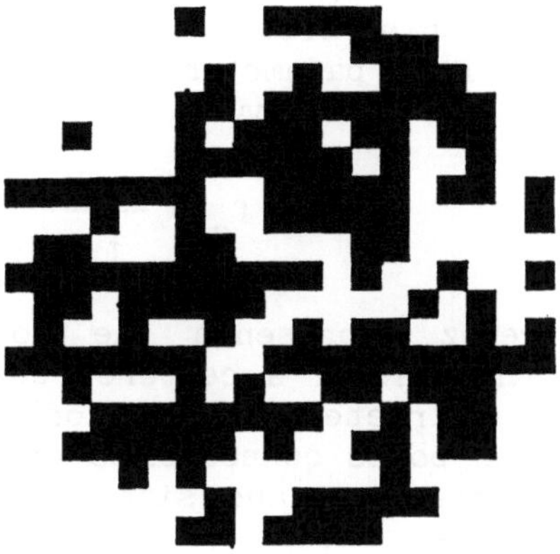

(b)

Media : Filterpak CR-50

Specific area : 220 m^{-1}

Liquid velocity

(a) : 0.1 kg m^{-2} s^{-1}

(b) : 0.2 kg m^{-2} s^{-1}

(c) : 0.5 kg m^{-2} s^{-1}

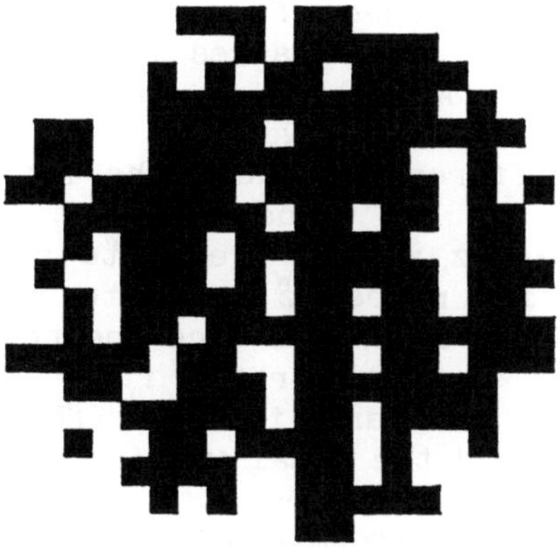

(c)

Figure 11
Experimental determinations of the liquid distribution at the outlet of a trickling filter (uniform liquid distribution at the top of the column)

pond to irrigated sites. The characteristic parameter quantifying this distribution is the fraction of irrigated sites f_s. This parameter may be correlated to f_w on the basis of numerical simulations (31) by

$$f_s = \frac{z\, f_w}{1 + (z-1)\, f_w^{\,z/z-1}} \qquad (1)$$

where z represents the coordination number of the lattice. In the case of a centered cubic lattice with a given flow direction (directed percolation), z equals 4 (that means only half of the bonds connected to one site may be used). The comparison of these bond and site flow distributions allows us to introduce the concept of local wetting. This wetting will be said to be complete when all the usable bonds leaving a site (4 in the case of the above example) are irrigated. The partial local wetting will in turn be defined as the ratio f_w/f_s and related to f_w by

$$\frac{f_w}{f_s} = \frac{1 + (z-1)\, f_w^{\,z/z-1}}{z} \qquad (2)$$

The dependence of this ratio versus f_w is depicted in figure 12. For high values of f_w, both quantities are nearly identical. Actually, in this case, all the sites are irrigated and the local wetting equals the global wetting efficiency. A difference between the ratio f_w/f_s and f_w appears nevertheless for low values of f_w. The ratio tends indeed to a finite value 1/z when f_w tends to zero. The role played by the coordination number z is quite interesting. It defines the accessibility of the lattice and consequently of the packing :

- For z = 1, the ratio f_w/f_s is constant and equals 1; all the flow paths are isolated from one another whatever the value of f_w and the local wetting is complete.
- For $z = \infty$, the ratio f_w/f_s is always equal to f_w; all the packing is uniformly (f_s = 1) but partially wetted whatever the value of f_w.

It is clear from these two limiting cases, that the accessibility of the lattice increases with the coordination number z. This later parameter is not easy to determine experimentally. It may be estimated by assimilating it to the coordination number of the packing, i.e. the mean number of contact points between particles. While not fully rigourous, this approximation seems to be logical. Indeed, the greater the coordination number of the packing, the greater the number of trajectories into which a flow path may be divided and the easier the spreading of the liquid within the packing. The mean number of contact points between particles may be correlated to the porosity of the packing, providing thus a way to estimate the value of z (32).

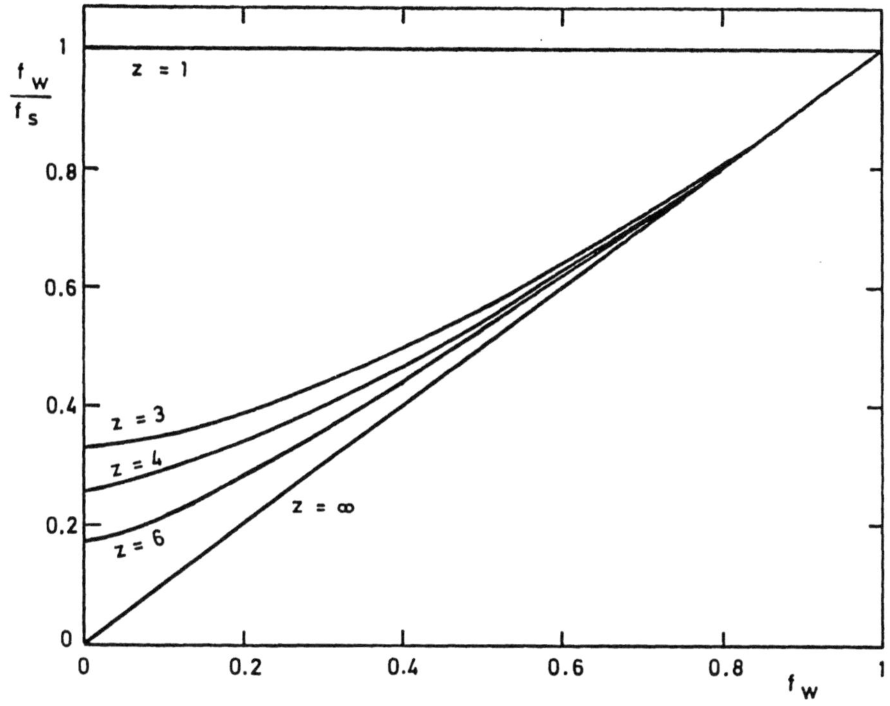

Figure 12
Partial local wetting f_w/f_s versus
the wetting efficiency f_w

4.2. Stochastic description of the liquid flow

Due to the random and intricate nature of the percolation structures, streamtubes of the liquid flow divide and rejoin repeatedly at each intersection point. This results in a liquid flow distribution which is controlled by a stochastic process. It may thus be analyzed in terms of maximum entropy. This concept is indeed a very elegant route for the estimation of the most probable configuration of a stochastic process. It assumes that this most probable configuration is obtained when the entropy of the process is a maximum. For the case we are interested in, the entropy is essentially configurational. It corresponds to the number of different flow configurations that may be adopted to achieve a liquid flow distribution.

The local liquid flowrate in the packing is represented by a density of connections attributed to each bond of the lattice. Let us characterize these bonds by densities i which can take whole number values 1, 2, ..., ∞ . The value 0 obviously corresponds to bonds which do not relate to percolation structures. This flow representation implies two main assumptions :

- the liquid flowrate takes only discrete values
- the liquid flowrate varies between 0 and infinity.

These two assumptions - leading obviously to important simplifications of the mathematical treatment - are however not too restricting. The discretization is a way to replace integrals by summations and will not affect significantly the results of computation provided the step of discretization remains small relative to the mean value of the flow density. Infinite or very large values of the flow density are obviously quite unrealistic but it will be shown (see Eq. 7) that the probability of these levels is negligible.

Let us represent the number of bonds with a density i by l_i. The number C of different flow configurations l_o, l_1, l_2,... is given by

$$C = \frac{l_t!}{\prod_{i=o}^{\infty} l_i!} \qquad (3)$$

where l_t represents the total number of bonds in the lattice. The configurational entropy H is given by

$$H = \ln C \qquad (4)$$

The entropy may be maximized taking two constraints into account :

- the total number of bonds is fixed

$$\sum_{i=o}^{\infty} l_i = l_t \qquad (5)$$

- the mean density of connection $\langle i \rangle$ is fixed

$$\sum_{i=o}^{\infty} i\, l_i = \langle i \rangle\, l_t \qquad (6)$$

Such a problem of maximization may be elegantly treated using indeterminate Lagrangian multipliers (31).

The resulting most probable distribution of densities of connection is then

$$\frac{l_i}{l_t} = \frac{1}{\langle i \rangle + 1} \left[\frac{\langle i \rangle}{\langle i \rangle + 1}\right]^i \qquad (7)$$

Eq. 7 exhibits an exponential decrease with the local density of connection. This exponential decrease tends to zero (uniform distribution) as the mean density of connection tends to infinity (i.e. as $\frac{\langle i \rangle}{\langle i \rangle + 1}$ tends to unity).

4.3. The minimum wetting velocity.

The discrete distribution described by Eq. 7 is characterized by a minimum non zero value of the density of connection - it equals 1. The actual velocity distribution observed in a trickle-bed reactor has a similar characteristic. There exist a minimum liquid velocity u_m below which the liquid film trickling over a solid surface becomes unstable. The smaller this minimum liquid velocity, the better the ability to spread over the packing surface. This parameter gives a physical meaning to the concept of packing accessibility. Accounting for the proportionality between the liquid flow velocity and the density of connection, Eq. 7 may be transformed in an actual liquid velocity distribution (Eq. 8). $\langle i \rangle$ is proportional to $\langle u_L \rangle$ and i is proportional to u_{Li}, the minimum liquid velocity u_m being the proportionality factor. This gives :

$$\alpha_i = \frac{u_m}{\langle u_L \rangle + u_m} \left[\frac{\langle u_L \rangle}{\langle u_L \rangle + u_m}\right]^{\frac{u_{Li}}{u_m}} \qquad (8)$$

where α_i represents the fraction of the packing surface covered by a liquid film with a velocity u_{Li}.

The dependence of the parameter u_m versus the operating conditions may be analyzed in two ways. It may be considered as a wetting parameter which has to be estimated for each experimental set of data. This will be the approach to be adopted in the next sections. u_m may also be considered as the liquid velocity that provides the more stable flow configuration in an isolated rivulet. This situation of isolated rivulets is indeed approached when f_w tends to zero (see e.g. Figure 10a) i.e. when the liquid velocity in each rivulet tends to its minimum value u_m. The velocity that yields the more stable flow configuration corresponds also to a minimum of energy dissipations. It may thus be determined by minimizing the sum of all the sources of energy dissipation related to the creation of the liquid film (33). The number, the nature and the importance of these sources depend of course on the local flow representation adopted. In the case of a laminar film trickling over planar surfaces with negligible shear stress at the gas-liquid due to the gas flow (low pressure drop), we have to consider two sources of dissipations :

- the viscous drag whose importance is determined by the liquid viscosity;
- the surface energy dissipated to replace a gas-solid interface by gas-liquid and liquid-solid interfaces.

The minimization of the energy dissipations leads to the following proportionality relation (32, 34)

$$u_m \div a \left[\frac{\mu_L}{\rho_L^3 (\rho_L g + \delta_{LG})}\right]^{1/5} E_s^{3/5} \qquad (9)$$

This relation shows that u_m increases (i.e. the wettability or accessibility of the packed-bed decreases) when increasing the liquid viscosity μ_L or the surface energy as well as when decreasing the pressure drop δ_{LG}. The surface energy depends not only on surface physico-chemical properties (surface tension, contact angle) but also on composition or temperature heterogeneities within the liquid film. This means that E_s is affected by mass and heat transfer across the film, i.e. by the chemical reaction rate. This dependence has been verified experimentally in the case of hydrogenation of maleic acid into succinic acid in a trickle-bed reactor (34). Such a dependence between the accessibility to liquid of a packed bed and the chemical reaction rate explains why the hydrodynamics may be affected by the chemical transformation involved in a trickle-bed reactor

5. MODELING HYDRODYNAMIC PROCESSES

5.1. General

Fluid flows or any related hydrodynamic quantities in a packed bed may be analyzed at various levels each one leading to completely different observations. Both the discretization adopted in the lattice representation and the nature of a packed bed suggest us to consider the following fundamental levels of observation :

- the particle scale which corresponds to the level of a bond of the lattice;
- the bed scale which corresponds to the level of the whole lattice.

Hydrodynamic processes must be modeled at these two levels. The first one introduces the effects of the local gas-liquid-solid interactions characterized by a local flow hydrodynamics model. The second level introduces the effect of the liquid flow maldistributions occurring at the bed scale and of course corresponds to the level at which we can carry out experiments.

The methodology for modeling hydrodynamic processes will then be as follows :

- modeling at the particle scale leading to a relation between a hydrodynamic quantity x and the local superficial liquid velocity u_{Li};
- modeling at the bed scale by averaging the relation $x(u_{Li})$ using an appropriate weighting factor directly related to the velocity distribution defined by the percolation model.

Most of the hydrodynamic quantities are extension properties so that the averaging formula reduces to :

$$\langle x \rangle = \sum_{i=o}^{\infty} \alpha_i \; x(u_{Li}) \qquad (10)$$

where α_i represents the fraction of packing surface covered by a liquid film with a velocity equal to u_{Li} (Eq.9). This

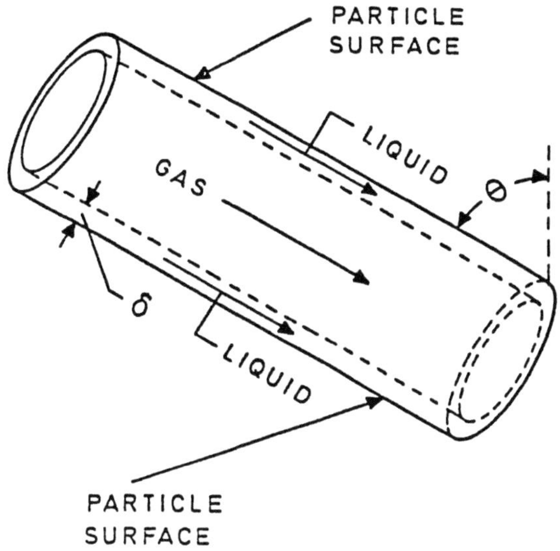

Figure 13
Pore flow model

methodolgy will be applied - as an example - to the modeling of two different hydrodynamic quantities - the wetting efficiency and the dynamic liquid holdup.

5.2. Modeling at the particle scale

The modeling at this level requires a representation of the local gas-liquid flow. We will consider here a rather simple example of pore flow model - the cocurrent gas-liquid flow through a straight pore (Figure 13). The curvatures of gas-liquid and liquid-solid interfaces are assumed to be negligible. This implies the liquid film thickness to be small relative to mean particle size. A laminar liquid flow is assumed without shear stress at the gas-liquid interface. This latter assumption was also adopted to derive Eq. 11 for the minimum liquid velocity u_m and corresponds to the regime of low gas-liquid interactions (35).

The wetting efficiency f_w is quite easily described by assuming a complete local wetting when the liquid flow through the pore is different from zero.

$$f_w \quad \begin{array}{ll} = 1 & \text{for } u_{Li} \geqslant u_m \\ = 0 & \text{for } u_{Li} = 0 \end{array} \qquad (11)$$

The dynamic liquid holdup h_d on planar inclinated surfaces is given by Crine et al. (36)

$$h_d = \left(\frac{3}{\cos^2\Theta}\right)^{1/3} a^{2/3} \left(\frac{\mu_L}{\rho_L g + \delta_{LG}}\right)^{1/3} u_{Li}^{1/3} \qquad (12)$$

The angle Θ characterizing the mean inclination to the vertical may be estimated, e.g. by assuming a random distribution of each individual pore inclination.
In this case :

$$\cos\Theta = \frac{2}{\pi} = 0.6366 \qquad (13)$$

which leads to :

$$h_d = 1.95 \; a^{2/3} \left(\frac{\mu_L}{\rho_L g + \delta_{LG}}\right)^{1/3} u_{Li}^{1/3} \qquad (14)$$

5.3. Modeling at the bed scale

The hydrodynamic quantities $<f_w>$ and $<h_d>$ are estimated at the bed scale by specifying in Eq. 10 the various parameters in terms of u_{Li}.

$$<f_w> = \frac{u_m}{<u_L> + u_m} \sum_{i=o}^{\infty} \left[\frac{u_L}{<u_L> + u_m}\right]^{u_{Li}/u_m} f_w\,(u_{Li}) \qquad (15)$$

Noticing that $f_w\,(u_{Li})$ is always equal to 1 excepted when u_{Li} equals zero, one may deduce

$$<f_w> = \frac{<u_L>}{<u_L> + u_m} \qquad (16)$$

Similarily, for h_d, one obtains

$$<h_d> = 1.95\, a^{2/3} \left(\frac{<\mu_L>}{\rho_{Lg} + \delta_{LG}}\right)^{1/3} \frac{<u_L>}{<u_L>+<u_m>} \sum_{i=o}^{\infty} \left[\frac{<u_L>}{<u_L>+u_m}\right]^{u_{Li}/u_m} u_{Li}^{1/3} \qquad (17)$$

Approximating the summation in Eq. 17 by an integral leads to (36) :

$$<h_d> = 1.74\, a^{2/3} \left(\frac{<\mu_L>}{\rho_L g + \delta_{LG}}\right)^{1/3} \left[\frac{<u_L>}{<u_L>+u_m}\right]^{2/3} <u_L>^{1/3} \qquad (18)$$

Eqs. 16 and 18 may be presented in a dimensionless form

$$<f_w> = \frac{Re_L}{Re_L + Re_m} \qquad (19)$$

$$<h_d> = 1.74 <f_w>^{2/3} Re_L^{1/3}\, Ga_L^{*\,-1/3}\, (adp)^{2/3} \qquad (20)$$

where Re_m represents the Reynolds number in which the liquid velocity equals u_m and Ga_L^* represents a modified Galileo number in which $\rho_L g$ is replaced by $\rho_L g + \delta_{LG}$ in order to account for the effects of the pressure drop.

Eq. 19 is an increasing function Re_L, reaching asymptotically unity for very large values of Re_L. It is worth analyzing this dependence in terms of the apparent log-slope because most of the empirical correlations of $<f_w>$ are based on power laws of the Reynolds number. Eq. 19 may be approximated by the following proportionality relation :

$$\langle f_w \rangle \div \left(\frac{Re_L}{Re_m}\right)^{1-\langle f_w \rangle} \qquad (21)$$

Eq. 20 may be similarily approximated

$$\langle h_d \rangle \div Re_L^{1-2/3 \langle f_w \rangle} \; Re_m^{2/3 (\langle f_w \rangle - 1)} \qquad (22)$$

For very low values of $\langle f_w \rangle$, both $\langle h_d \rangle$ and $\langle f_w \rangle$ are proportional to Re_L. In this case, each flow trajectory is independent from each other : they are completely free to spread on the packing surface (see figure 10a). This means that an increase of the liquid velocity is accompanied by a proportional increase of the wetted surface, the film thickness remaining at its minimum value. This implies $\langle h_d \rangle$ and $\langle f_w \rangle$ to vary proportionally to Re_L. When the liquid velocity increases, $\langle f_w \rangle$ increases too and the flow trajectories become more and more constrained by their neighbours. The increase of wetted surface is no more proportional to Re_L as indicated by Eq. 21. The film thickness corresponding to the local liquid velocity is no more constant and increases progressively. The combined effects of the variations of wetted area and film thickness are represented by an apparent log-slope equal to $1 - \frac{2}{3} \langle f_w \rangle$ in Eq. 22. For nearly complete wetting, we have a pure film flow : the wetted area remains constant and the film thickness varies proportionally to the power 1/3 of the Reynolds number. Eq. 21 and 22 are thus in complete agreement with the phenomenon of transition between rivulet ($\langle f_w \rangle \ll 1$) and film ($\langle f_w \rangle \simeq 1$) flow structures observed when analyzing graphically the percolation structures (Figure 10). The exponent of Re_m in Eqs 21 and 22 is always negative. This implies both $\langle h_d \rangle$ and $\langle f_w \rangle$ to decrease when increasing Re_m i.e. when decreasing the wettability or accessibility of the packing. This influence is at its maximum for very small values of the liquid Reynolds number ($Re_L \ll Re_m$).

Different attempts to correlate empirically $\langle f_w \rangle$ and $\langle h_d \rangle$ have been reported in the literature. These correlations are often based on a dimensional analysis suggesting the use of dimensionless groups such as Re_L, We_L, Fr_L and a packing geometrical factor. Mills and Duduković (37) proposed the following relation based on experimental data consisting exclusively of conditions of flow, particle size and bed porosity typical of trickle flow operations :

$$\langle f_w \rangle = \tanh \left[1.09 \, Re_L'^{\,0.28} \, Fr_L'^{\,0.21} \, We_L'^{\,-0.16} \left(\frac{a d_p}{\varepsilon}\right)^{0.03}\right] \qquad (23)$$

Re_L', Fr_L' and We_L' are modified dimensionless groups in which the particle size dp is replaced by the inverse of the packing specific area a. The form of this function is suggested by the observed asymptotic approach to a constant value of $\langle f_w \rangle$. Eq. 23 predicts that $\langle f_w \rangle$ should decrease as the particle size

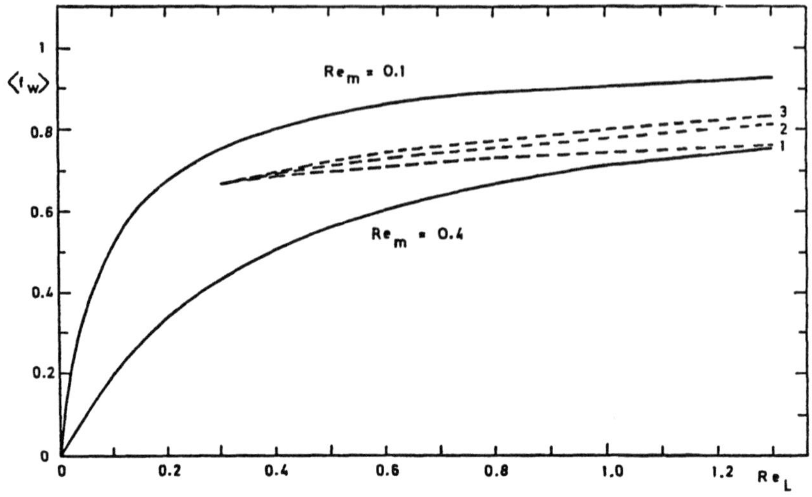

Figure 14
Comparison of Eq. 19 (solid lines) with empirical correlations of $\langle f_w \rangle$ (dotted lines) : $\langle f_w \rangle = a\, Re_L^b$
1 : b = 0.097; 2 : b = 0.137; 3 : b = 0.148 (37)
a = 0.747 a = 0.783 a = 0.803

decreases. Such a dependence implies that, in Eq. 16 or 19, the minimum velocity u_m and consequently the Re_m decreases also with the particle size. This decrease which is not formally predicted by Eq. 9 may in fact be explained by the effect of capillary forces which reduce the energy necessary to wet the solid surface, i.e. E_s. Eqs. 19 and 23 are compared in figure 14 together with power law correlations also suggested by Mills and Dudukovic (37) for some specific sets of experimental data. These empirical correlations and Eq. 19 give rather similar results when Re_m is varied between 0.1 and 0.4. It should be noted that the exponents of the power law correlations agree rather well with the mean value of 1 - $\langle f_w \rangle$ (i.e. roughly 0.2) predicted by Eq. 21.

For the dynamic liquid holdup, Specchia and Baldi (35) proposed a correlation specifically valid for the trickling flow conditions (low gas-liquid interactions)

$$\langle h_d \rangle = 3.86\, \varepsilon^{0.35}\, Re_L^{0.545}\, Ga_L^{*-0.42}\, (adp)^{0.65} \qquad (24)$$

The exponents of the dimensionless groups Ga_L^* and adp in Eqs 20 and 24 are very close. The exponent of the Reynolds number in Eq. 24 may be interpreted on the basis of Eq. 22. This yields a mean value of $\langle f_w \rangle$ close to 0.7. It should also be noted that other correlations of $\langle h_d \rangle$ not specific of trickling flow conditions have been reported in the literature (38, 39).

They generally suggest an increase of the Reynolds number exponent when using less wettable packings. This is well accounted for by Eq. 22 which predicts an exponent equal to 1-2/3 f_w i.e. increasing with Re_m. This role of wetting parameter played by Re_m, controlling the accessibility of the packed-bed to the liquid flow will be confirmed when analyzing its influence on the axial dispersion coefficient.

6. MODELING AXIAL DISPERSION

6.1. General

The most conventional RTD model describing axial dispersion in trickle flow reactors consists of a plug flow to which a diffusional mechanism is superposed. The phenomenological parameter quantifying the dispersion is an effective diffusivity or its related dimensionless group, the Bodenstein number. This parameter reflects the contribution of different mechanisms which may be grouped according to the scale at which they occur :

- molecular level : corresponding to the scale of a few molecules;
- particle scale : at which the hydrodynamics is described by a pore flow model (see figure 13);
- bed scale : at which the hydrodynamics is described by the percolation model.

At the molecular level, dispersion is controlled by the molecular diffusivity D_m, i.e by collisions between molecules. At the particle scale, dispersion is controlled by the local hydrodynamics, i.e. by velocity profiles, turbulence,... in the liquid films. At the bed scale, dispersion results from velocity distribution within the percolation structures.

The use of the axial dispersion model give rise to the problem of knowing whether a diffusional picture can be adopted to describe dispersion at each of these levels, or not. We will consider the modeling of axial dispersion at the particle and bed scales, discussing the necessary assumptions to adopt such a diffusional picture.

6.2. Modeling at the particle scale

In the trickling flow regime, the liquid flow is generally laminar. At the particle scale, the dispersion is caused by the laminar flow, i.e. it results from the combined effects of velocity profiles in the liquid films which create transverse concentration profiles and of transverse molecular diffusion which tends to reduce these profiles. Two limiting cases are generally considered depending on the predominant effect. When the residence time in the liquid film is large enough, transverse concentration profiles are completely flattened out by molecular diffusion. This case reduces to Taylor's diffusion (40, 41). When dealing with liquid films trickling over planar surfaces, it may be demonstrated (13) that the axial dispersion coefficient D_L is given by

$$D_L = 0.057 \frac{v^3 \mu_L}{D_m \rho_L g} \tag{25}$$

where v_L represents the interstitial liquid velocity (i.e. the superficial liquid velocity u_L divided by the liquid holdup). The usual range of operating conditions for trickle-bed reactors (small liquid holdup, small particles) are however such that the residence time between two contact points is rather small and, consequently, the conditions of applicability of Taylor's diffusion are not satisfied.

The opposite limiting case leads to neglect of the effect of transverse molecular diffusion. Each streamtube in the liquid film are independent of each other. In this case, the axial dispersion D_L may be related to the intersticial liquid velocity v_L as follows (36)

$$D_L = D_m + 1.307\, v_L\, dp \tag{26}$$

when a planar trickling film is considered (i.e. the same pore flow model as the one used for the modeling of the dynamic liquid holdup).

Eqs. 25 and 26 differ by their dependence upon the liquid velocity. The convective term of Eq. 26 is proportional to v_L whereas Taylor's model predicts an exponent equal to three for this dependence. Furthermore, the convective term of Eq. 26 is unaffected by molecular diffusion. Experimentally, some investigators (42) have shown that the RTD curves are nearly independent of the molecular diffusivity of the tracer used. Eq. 26 seems consequently to be the most plausible one and should be adopted to model particle scale dispersion in trickle flow reactors.

6.3. Modeling at the bed scale

6.3.1. Averaging formula. Two basic mechanisms drive dispersion at this level. The first one is static - streamtubes of the flow structures divide and rejoin repeatedly at the intersections of flow passages. This results in a variation of length and orientation of flow paths which induces dispersion of tracer. The second mechanism is dynamic - the residence time within a flow passage depends on the different local velocities encountered. The dispersion phenomenon depends thus also on the liquid velocity distribution.

To avoid having to treat unnecessary mathematical complications, we will adopt the simplest type of model that allows us to relate residence time and velocity distributions. It consists in an array of independent parallel channels of equal lengths L_c as schematized in figure 15. Each channel is representative of a particular flow path between two successive intersections. The description of the static contribution to the dispersion phenomenon is obviously quite oversimplified.

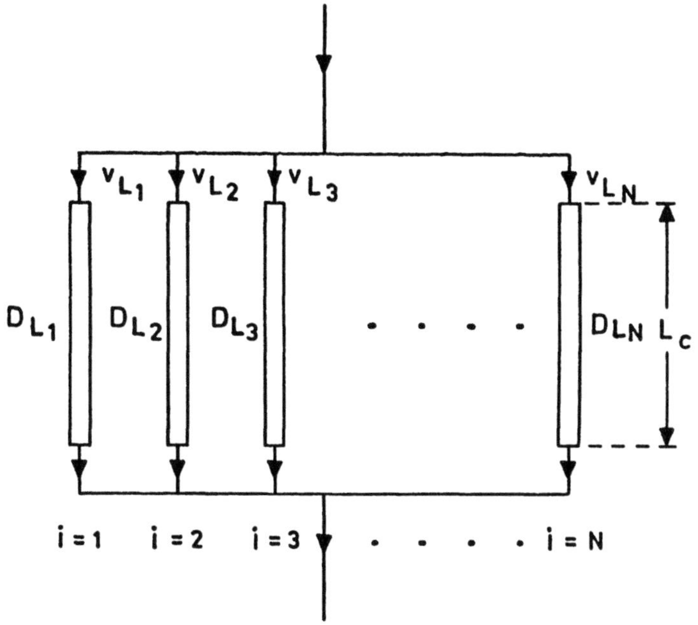

Figure 15
Array of parallel independent channels

The free path between two intersection points along the flow trajectories is assumed to be constant and equal to L_c. Actually in a trickle bed, the free path varies randomly. This simplification will require to relate somewhat empirically L_c to the operating conditions (see Eq. 33).

Each channel of the array is characterized by a velocity v_{Li} and a concentration C_i. If we assume that the dispersion phenomenon within each channel is governed by a diffusional mechanism, it follows

$$\frac{\partial C_i}{\partial t} + v_{Li} \frac{\partial C_i}{\partial Z} = D_{Li} \frac{\partial^2 C_i}{\partial Z^2} \tag{27}$$

D_{Li} is the axial dispersion coefficient describing the residence time distribution in channel i. It represents the contribution of the particle scale (microscopic level) mechanisms and may thus be determined from Eq. 26.

The dispersion at the bed scale may be determined by averaging Eq. 27 over the whole set of channels and introducing variables $\langle v_L \rangle$, $\langle C \rangle$ and $\langle D_L \rangle$. These new variables cannot be related exactly by a relation similar to Eq. 27. This means clearly

that the dispersion phenomenon at the bed scale is not ruled by a diffusional mechanism.
The solution of the following equation

$$\frac{\partial C}{\partial t} + \langle v_L \rangle \frac{\partial C}{\partial Z} = \langle D_L \rangle \frac{\partial^2 C}{\partial Z^2} \qquad (28)$$

may nevertheless be an approximation of the exact solution having the same zeroth, first and second moments along the bed axis provided $\langle D_L \rangle$ is determined as follows (43)

$$\langle D_L \rangle = \sum_{i=1}^{N} w_i D_{Li} + \langle v_L \rangle^2 t \sum_{i=1}^{N} w_i \frac{v_{Li}}{\langle v_L \rangle} \left[\frac{v_{Li}}{\langle v_L \rangle} - 1\right] \qquad (29)$$

w_i is the fraction of the relative amount of tracer in channel i. According to Carbonell (43), w_i is equal to the relative local volumetric flow rate i.e.,

$$w_i = \frac{v_{Li}}{\langle v_L \rangle} \frac{A_i}{A_t} \qquad (30)$$

where A_i is the cross-section of channel i and A_t the total cross-section of the whole set of channels.

Eq. 29 is an averaging formula of $\langle D_L \rangle$ analogous to Eq. 10 for the averaging of extension hydrodynamic quantities. The first term represents the contribution of each individual channel. The second term is the contribution to dispersion from differences in liquid velocities among the different channels. The differences cause the tracer to spread and the amount of spread to be proportional to the residence time t. This implies Eq. 27 to have a more pronounced tail than the classical diffusion-like equation. The main consequence of the time dependence exhibited by Eq. 29 is that the spread of the RTD curves will differ according to whether it is computed along the bed axis at a fixed time or along the time axis at a fixed position in the bed. This provide further evidence that the dispersion process is not diffusive (44). The exact solution may still be further approximated by keeping $\langle D_L \rangle$ constant with time, i.e., by replacing the time t by the mean residence time t_m in the array of channels

$$t_m = \frac{L_c}{v_L} \qquad (31)$$

It follows

$$\langle D_L \rangle = \sum_{i=1}^{N} w_i D_{Li} + \langle v_L \rangle L_c \sum_{i=1}^{N} w_i \frac{v_{Li}}{\langle v_L \rangle} \left[\frac{v_{Li}}{\langle v_L \rangle} - 1\right] \qquad (32)$$

Eq. 32 shows clearly that the solution is an intermediate case between a purely diffusive mechanism which would be represented by the first term of Eq. 32 (independent of L_c but a function of the dispersion at the particle scale) and a purely convective mechanism which would be represented by the second term of Eq. 32 (proportional to L_c but independent of the dispersion at the particle scale). It should however be kept in mind that this second term is a time-averaged approximation. It is justified provided the length L_c of the array of channels remains small enough relative the total length of the reactor. This means that the scale of mixing must be relatively small relative to the bed scale.

6.3.2. Correlation of the axial dispersion coefficient. Eq. 32 allows us to derive a bed scale correlation of the axial dispersion coefficient $< D_L >$ provided the different parameters appearing in this equation may be related to the operating conditions by means of the percolation model.

The mean length L_c corresponds to the mean free path along the flow trajectories of the percolation structures. In the absence of a detailed analysis of the frequency distribution of intersection points along these trajectories, we will assume the distribution to be a geometrical one. That means that the mean frequency should be proportional to the density of flow trajectories i.e. the irrigation rate $<f_w>$. The mean length L_c (i.e. the period of the distribution) is thus inversely proportional to $<f_w>$. Noticing L_c is also proportional to the characteristic size dp of the particles, it follows

$$L_c = b \frac{dp}{<f_w>} \tag{33}$$

The proportionality factor b must be introduced at least for two reasons :

- the minimum distance between two intersection points is probably not exactly equal to the characteristic size dp;
- the oversimplification introduced in Figure 15 leads us to neglect the contribution of the channel length distribution to the dispersion process; for this reason the parameter L_c appearing in Eq. 32 is somewhat empirical and justifies the introduction of the proportionality factor b.

The distribution of liquid velocities in the array of channels may be represented by the velocity distribution in the percolation structures (Eq. 9). The weighing factors ω_i are easily related to α_i by

$$w_i = \frac{\alpha_i \, v_{Li} \, \delta_i}{\Sigma \alpha_i \, v_{Li} \, \delta_i} \tag{34}$$

where δ_i represents the local liquid film thickness. In the case of a laminar flow on planar surface of mean inclination

Θ, δ_i is related to v_{Li} by

$$\delta_i = \left(\frac{3\mu_L}{\rho_L g \cos\Theta} v_{Li}\right)^{1/2} \qquad (35)$$

It is possible to express the summations in Eq. 32 in terms of moments δ^n of the film thickness distribution

$$\delta^n = \Sigma\, \delta_i^n \qquad (36)$$

Approximating these summations by integrals allows us to derive the following equation (36)

$$\langle D_L\rangle = D_m + 2.22 \langle v_L\rangle\, dp + 2.183 \langle v_L\rangle\, L_c \qquad (37)$$

These variables are directed along the local flow direction which is assumed to have a mean inclination Θ. Eq. 37 may be rewritten in terms of axial variables by a change of axis

$$\langle D_L\rangle = D_m \cos^2\Theta + 2.220 \cos\Theta \langle v_L\rangle\, dp + 2.183 \cos\Theta \langle v_L\rangle L_c \qquad (38)$$

This expression involves three contributions :

- the contribution of the molecular diffusivity (dispersion at the molecular scale) which is usually negligible;
- the contribution of the particle scale dispersion which is proportional to the characteristic particle size dp;
- the contribution of the bed scale dispersion which is proportional to the mean length L_c.

L_c may be replaced in Eq. 38 by its dependence on the Reynolds number (Eqs 19 and 33) while Eq. 38 may be presented in a dimensionless form

$$Bo_L^{-1} = Bo_m^{-1} \cos^2\Theta + (2.220 + 2.183\, b) \cos\Theta + 2.183\, b \cos\Theta\, Re_m\, Re_L^{-1} \qquad (3$$

The Reynolds numbers (Re_L and Re_m) are superficial variables whereas the Bodenstein numbers (Bo_L : axial dispersion and Bo_m : molecular diffusivity) are interstitial variables.

Two limiting cases may be distinguished

- $Re_L \ll Re_m$ (very small irrigation rate; $\langle f_W\rangle = \frac{Re_L}{Re_m}$)

$$Bo_L \simeq \frac{Re_L}{2.183 \cos\Theta\, b\, Re_m}, \qquad (40)$$

the Bodenstein number is proportional to the superficial Reynolds number.

- $Re_L \gg Re_m$ (nearly complete irrigation; $\langle f_W\rangle \simeq 1$)

$$Bo_L \simeq \frac{1}{(2.220 + 2.183\, b) \cos\Theta}, \qquad (41)$$

the Bodenstein number is nearly constant

Between these two limiting cases, the apparent log-slope of Bo_L versus Re_L varies continuously between unity and zero when increasing the liquid velocity. This agrees fairly well with empirical correlations whose apparent log-slope ranges about 0.2-0.5 (15, 23, 45). Eq. 37 predicts also that there is a linear relationship between the inverses of the Bodenstein and Reynolds numbers and that the slope of this straight line is proportional to Re_m. This representation will be adopted to compare Eq. 39 with some experimental published in the literature.

Buffham and Rathor (46) used air and aqueous glycerol solutions flowing in a 3.8 x 10^{-2} m column packed with 3 x 10^{-3} m by 3 x 10^{-3} m non-porous ceramic Raschig rings. The liquid viscosity was varied between 10^{-3} and 7.5 x 10^{-3} Ns m^{-2}. In this range, the authors observed rather small variations of the static surface tension and density (less than 6% and 15%, respectively). The inverse of the Bodenstein number estimates they obtained is plotted in figure 16 against the inverse of the superficial Reynolds number. Actually, the authors reported experiments carried out with three different bed lengths. Owing to the relatively large scatter between the different results, only averaged values for the three lengths are reported in figure 16.

Three different liquid viscosities were used : 10^{-3}, 4.5 x 10^{-3} and 7.5. x 10^{-3} Ns m^{-2}. Eq. 39 was fitted separately on these three sets of data.

The agreement between experiments and theory is rather satisfying. The experimental points are indeed fairly well situated on three straight lines whose slope varies with the liquid viscosity. The proportionality factor b was kept constant for the three sets of data. According to the theory, this parameter is indeed a geometrical factor characteristic of the packing used and thus independent of the fluid properties. The parameter Re_m which characterizes the slope of the straight lines decreases rather slowly as the liquid viscosity increases. This causes u_m to increase roughly proportionally to a power 0.4 of the liquid viscosity. In Eq. 9, u_m increases with a power 0.2. The agreement with experiments is rather satisfying owing to the relatively scatter of data and to the inaccuracy of the parameter estimates.

Crine et al. (36) used air and water flowing in a 3.8 x 10^{-2} m glass column packed with 3 x 10^{-3}m glass spheres. The fluid properties were kept constant (μ_L = 10^{-3} Ns m^{-2}) but two different particle wettabilities were obtained by means of a surface treatment. The inverse of the Bodenstein number estimates are reported in figure 17 against the inverse of the liquid superficial Reynolds number for different gas superficial velocities ranging between 2 x 10^{-2} and 5 x 10^{-1} kg m^{-2} s^{-1}. In view of the experimental scatter, no definite influence of the gas flowrate can be observed. Eq. 39 was

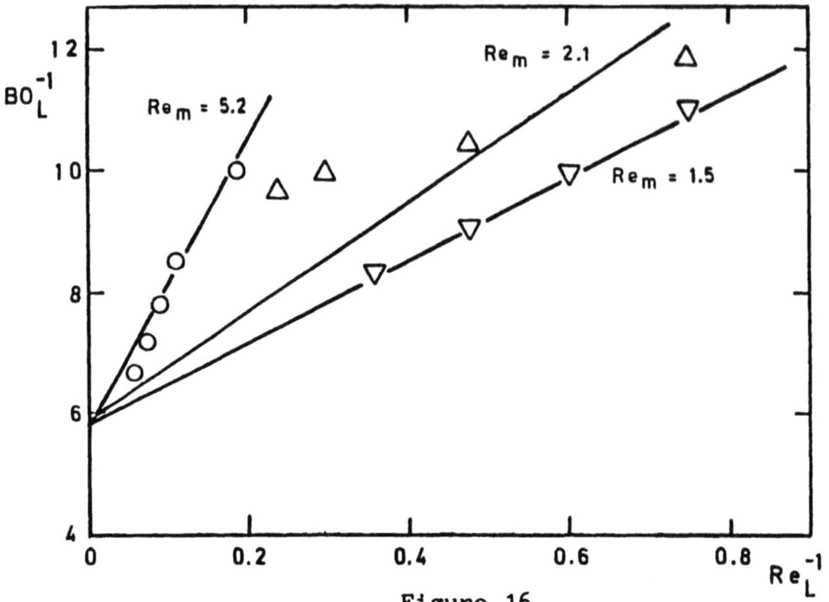

Figure 16
Bodenstein number as a function of the Reynolds number (Buffham and Rathor experiments; ▽ : μ_L = $7.5 10^{-3}$ Nsm^{-2}, △: μ_L = 4.5 10^{-3} Nsm^{-2}; ○: μ_L = 10^{-3} Nsm^{-2})

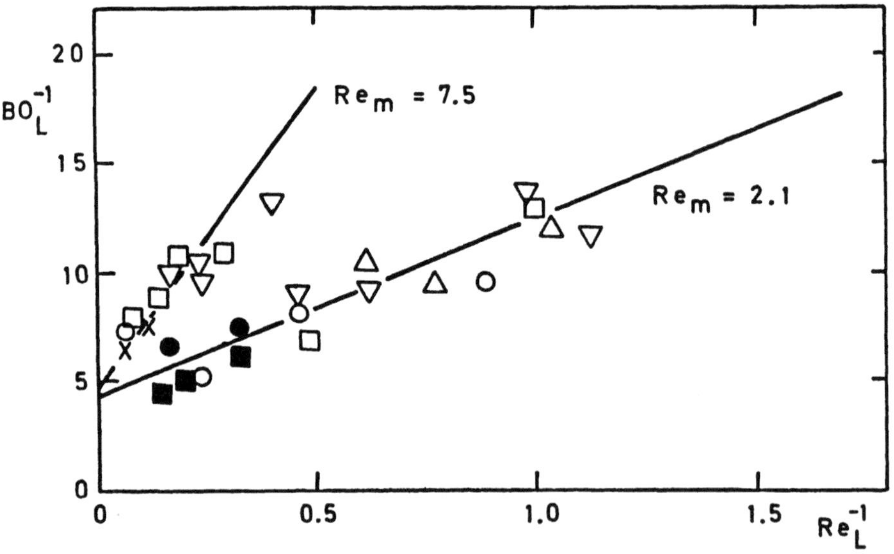

Figure 17
Bodenstein number as a function of the Reynolds number (Crine et al. experiments; gas flowrates : △: 0.0 ms^{-1}; ▽: 0.05 ms^{-1}; □: 0.1 ms^{-1}; ○: 0.15 ms^{-1}; X: 0.25 ms^{-1}; ●· 0.4 ms^{-1}·■· 0.51 ms^{-1})

fitted separately on the two sets of data (poor and good particle wettabilities). The agreement between experiments and theory is once again satisfying. The experimental points are more or less situated on two straight lines whose slopes differ according to the particle wettability. Re_m and consequently u_m are smaller for the wettable particles confirming thus the rôle of wetting parameter played by these variables.

7. CONCLUSIONS

The gas-liquid flow through a random packing may be represented by a percolation process. The percolation theory allow us to account for the random nature of the packing by introducing the essential concept of packing accessibility. This accessibility is represented by the lattice coordination number in the percolation model. It is directly related to texture properties of the packing.

An "entropy" analysis of the liquid flow through the percolation structures allow us to derive a theoretical expression of flow distribution. This expression may be used as the basis of averaging formula of various hydrodynamic mecanisms. The resulting models involve both parameters characterizing the mechanism modeled at the particle scale and a parameter defining the effective solid wettability, i.e. the minimum liquid velocity u_m. The various models analysed in this paper and compared with experiments yield logical variations of the parameter u_m with the operating conditions (solid wettability, liquid viscosity).

The percolation approach opens also the way to some statistical hydrodynamics of liquid flow through random media.

NOMENCLATURE

- a specific area of the particles (L^{-1})
- A_i cross-section of channel i (L^2)
- A_t total cross-section of the set of channels (L^2)
- b proportionality constant introduced in Eq. 33 (-)
- Bo_L Bodenstein number of liquid, $\langle v_L\rangle\, dp/\langle D_L\rangle$ (-)
- C number of flow configurations (-)
- D_L axial dispersion coefficient (L^2T^{-1})
- D_m molecular diffusivity (L^2T^{-1})
- dp equivalent particle diameter (L)
- E_s surface energy dissipated by the creation of a liquid film (MT^{-2})
- Fr'_L modified Froude number $u_L^2\, a/g$ (-)
- f_s fraction of irrigated sites (-)
- f_w^s fraction of irrigated bonds or wetting efficiency (-)
- g acceleration of gravity (LT^{-2})
- G_* gas superficial mass flow rate $(ML^{-2}T^{-1})$
- Ga_L modified Galileo number, $dp^3\, \rho_L\, (\rho_L g + \delta_{LG})/\mu_L^2$ (-)
- H configurational entropy (-)

h_d dynamic liquid holdup or fraction of flowing zones in RTD models (-)

h_s fraction of stagnant zones in RTD models (-)
i density of connection in channel i (-)
k_s mass transfer coefficient between flowing and stagnant zones in RTD models (LT^{-1})
L liquid superficial mass flow rate ($ML^{-2}T^{-1}$)
l_c mean length of channel (see Figure 15) (L)
l_i number of bonds with a density of connection i (-)
l_t total number of bonds (-)
m mixing parameter in Figure 1e (-)
n cluster size in a percolation process or number of steps in a random walk (-)
N number of cells in series in a RTD model or number of channels in Figure 15 (-)
p fraction of active bonds in a percolation process (-)
q fraction of bypass flow (Figures 1g and 2d) or backflow (Figure 2e) (-)
Re_L Reynolds number of liquid, $\rho_L \langle u_L \rangle dp/\mu_L$ (-)
Re_L^* modified Reynolds number of liquid, $\rho_L \langle u_L \rangle /a\mu_L$ (-)
Re_m Reynolds number associated with the minimum liquid velocity u_m, $\rho_L u_m dp/\mu_L$ (-)
$R^2(n)$ square of the displacement as a function of the number of steps n in a random walk process (-)
t residence time (T)
t_D delay time in Figures 2d and 2e (T)
t_m mean residence time in the array of channels (T)
u_L superficial liquid velocity (LT^{-1})
u_m minimum liquid velocity (LT^{-1})
v_L intersticial liquid velocity (LT^{-1})
We_L' modified Weber number of liquid ($\rho_L \langle u_L^2 \rangle /\sigma_L a$) ($LT^{-1}$)
w_i weighting factor in Eq. 29 (-)
x general denotation of a hydrodynamic quantity in Eq. 10
z coordination number of lattice (-)
Z total length of bed (L)

Greek letters

α_i fraction of packing surface covered by a liquid film with a superficial velocity equal to u_{Li} (-)
δ film thickness (L)
δ_{LG} pressure drop per unit of bed length ($ML^{-2}T^{-2}$)
ε bed porosity (-)
μ_G gas viscosity ($ML^{-1}T^{-1}$)
μ_L liquid viscosity ($ML^{-1}T^{-1}$)
ρ_G gas density (ML^{-3})
ρ_L liquid density (ML^{-3})
Θ mean inclination of a pore (-)
σ_L surface tension of liquid (MT^{-2})

REFERENCES

1. Danckwerts, P.V., Chem. Eng. Sci., 2 (1953) 1.
2. Colombo, A.J., Baldi, G. and Sicaroli, S., Chem. Eng. Sci., 31 (1976) 1101.
3. Sater, V.E. and Levenspiel, O., Ind. Eng. Chem. Fund., 5 (1966) 86.
4. Hochmann, J.M. and Effron, E., Ind. Eng. Chem. Fund., 3 (1969) 63.
5. Buffham, B.A. and Gibilaro, L.G., Chem. Eng. Sci., 1 (1970) 31.
6. Oorts, A.J. and Hellinckx, L.J., Chem. Eng. J., 7 (1974) 147.
7. Van Swaaij, W.P., Charpentier, J.C. and Villermaux, J., Chem. Eng. Sci., 24 (1969) 1083.
8. Lerou, J.J., Glasser, D. and Luss, D., Ing. Eng. Chem. Fund., 19 (1980) 66.
9. Deans, H.A. and Lapidus, L., A.I.Ch.E.Jr., 6 (1960) 656.
10. Deans, H.A., Soc. Petrol. Eng. J., 3 (1963) 49.
11. Raghuraman, J. and Varma, Y.B., Chem. Eng. Sci., 28 (1973) 585.
12. Popovic, M. and Deckwer, W.D., Chem. Eng. J., 11 (1976) 67.
13. Michell, R.W. and Furzer, I.A., Trans. Instn. Chem. Eng., 50 (1972) 334.
14. Rao, V.G. and Varma, Y.B., A.I.Ch.E.Jr., 22 (1976) 612.
15. Gato, S., Levec, J. and Smith, J.M., Catal. Rev. Sc. Eng., 15 (2), (1977) 187.
16. Charpentier, J.C., Bakos, M. and Le Goff, P., Proceedings 2nd Conference on "Quelques Applications de la Chimie Physique", Vedzprem, Hungary (1971) 31.
17. Porter, K.E. and Templeton, J., Trans. Instn. Chem. Eng., 46 (1968) 86.
18. Bemer, G.G. and Zuiderweg, F.J., Chem. Eng. Sci., 33 (1978) 1637.
19. De la Cal, J.C., Missa, A., Crine, M. and L'Homme, G., Proceedings 32d Mediterranean Congress on Chemical Engineering, Barcelona, Spain (1984) 189.
20. Charpentier, J.C. and Favier, M., A.I.Ch.E.Jr., 21 (1975) 1213.
21. Gianetto, A., Baldi, G., Specchia, V. and Sicardi S., A.I.Ch.E.Jr., 24 (1978) 1087.
22. Broadbent, S.R. and Hammersley, J.M., Proc. Cam. Phil. Soc., 53 (1957) 629.
23. Herskowitz, M. and Smith, J.M., A.I.Ch.E.Jr., 29 (1983) 1.
24. Frisch, H.L. and Hammersley, J.M., J. Soc. Ind. Appl. Math., 11 (1973) 894.
25. Shante, V.K. and Kirkpatrick, S., Adv. Phys., 20 (1971) 325.
26. De Gennes, P.G., La Recherche, 7 (1976), 919.

27. Stauffer, D., Phys. Reports, 45 (1979) 1.
28. Blanc, R. and Guyon, E., in "Percolation structures and processes" (Jérusalem : The Israel Physical Society, 1983), pp. 229-250.
29. Crine, M. and Marchot, P., Int. Chem. Eng., 23 (1983) 663.
30. Charpentier, J.C., Prost, C., Van Swaaij, W.P. and Le Goff, P., Chimie Ind. Génie Chimique, 99 (1968) 803.
31. Crine, M. and Marchot, P., Int. Chem. Eng., 24 (1984) 53.
32. Suzuki, M., Makino, K., Yamada, M. and Iinoya, K., Int. Chem. Eng., 21 (1981) 482.
33. Crine, M., Doctorat en Sciences Appliquées, Université de Liège (1978).
34. Ruiz, P., Crine, M., Germain, A. and L'Homme, G., in M.P. Dudukovic and P.L. Mills, ed., "Chemical and catalytic reactor modeling" (Washington : ACS Symposium Series, N° 237) pp 15-36.
35. Specchia, V. and Baldi, G., Chem. Eng. Sci., 32 (1977) 515.
36. Crine, M., Asua, J.M. and L'Homme, G., Chem. Eng. J., 25 (1982) 183.
37. Mills, P.L. and Dudukovic, M.P., Proceedings 2nd World Congress on Chemical Engineering, Montreal, Canada (1981) 143.
38. Gardner, G.C., Chem. Eng. Sci., 5 (1956) 101.
39. Warner, N.A., Chem. Eng. Sci., 11 (1959) 149.
40. Taylor, G.I., Proc. Roy. Soc. (London), A 219 (1953) 186.
41. Taylor, G.I., Proc. Roy. Soc. (London), A 223 (1954) 446.
42. Kan, K.M. and Greenfield, P.F., Ind. Eng. Proc. Des. Dev., 18 (1979), 140.
43. Carbonell, R.G., Chem. Eng. Sci., 34 (1979) 1031.
44. Sahimi, M., Davis, H.T. and Scriven, L.E., Chem. Eng. Comm., 23 (1983) 329.
45. Shah, Y.T., "Gas-Liquid Reactor Design" (New York : Mc Graw-Hill, 1979).
46. Buffham, B.A. and Rathor, M.N., Trans. Instr. Chem. Eng., 56 (1978) 266.
47. Michell, R.W. and Furzer, I.A., Chem. Eng. J., 4 (1972) 53.

ENGINEERING ASPECTS OF TRICKLE BED REACTORS

R. M. Koros

Exxon Research and Engineering Company
Florham Park, NJ 07932

This review deals with the engineering aspects of trickle-bed reactors as it applies to general design, scale-up, pilot plant design and operation, and reactor troubleshooting. Also, research needs in trickle-bed technology are discussed.

1. GENERAL DESIGN PRINCIPLES

Mixed-phase downflow fixed bed "trickle bed" reactors have been in use commercially for several decades for hydrotreating hydrocarbons which are liquid at reaction conditions. These reactors are used to hydroprocess such feeds as solvents, gas oils, lubricants, atmospheric resids and vacuum resids. The basic elements of the reactor (Figure 1.1) are: a top inlet; a distributor tray with multiple distribution elements; a catalyst bed; and a supporting grid or an outlet collector. Gas and liquid feeds enter at the top and are distributed evenly over the surface of the bed by the tray. The liquid flows down over the catalyst in the bed generally wetting the catalyst and thereby contacting it. The gas flows cocurrently down the bed along with the liquid contacting the catalyst mostly through first dissolving in the liquid. In this review, progress in the rational design, scale up and troubleshooting of these reactors will be described.

1.1 Considerations in Design

In mixed phase downflow fixed bed reactors, the principal step is generally the contacting between the liquid phase and the catalyst. Gaseous reactants can also directly react on the catalyst, but usually they transfer from the gas to the liquid before

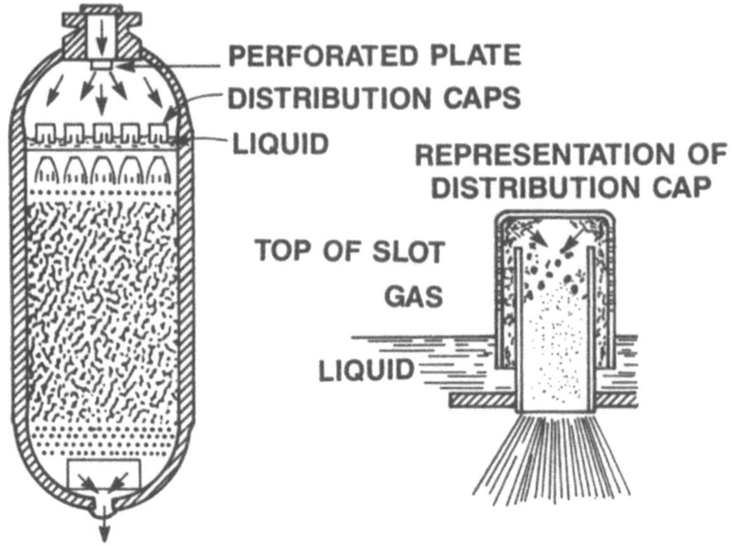

Figure 1.1. Schematic of mixed-phase fixed bed reactor (after B. L. Tarmy [75]).

reacting on the catalyst. Hence, in the design of mixed phase downflow fixed bed reactors both the flow dynamics and the mass transfer from the fluids to the catalyst are important issues. For most reactions, heat of reaction must also be dissipated. Hence, heat transfer between the liquid phase and the catalyst, and radially across the bed can also be important issues. In the flow of gases and liquids in the reactor, gravity plays an important role but usually pressure drop must be expended to move the fluids through the reactor. Reactor pressure drop can vary widely depending on the flow rates and the flow regimes existing in the reactor. Lastly, the design engineer must consider the radial and axial distribution patterns in the catalyst bed. This information along with the contacting patterns are required to define the efficiency of a trickle bed reactor. Important issues with respect to these considerations will be discussed in the next sections.

1.2 Flow Dynamics

Mixed phase downflow fixed bed reactors can operate in several flow regimes. These bear resemblance to flow regimes in two phase flow in pipes. These flow regimes are the gas continuous ("trickle bed"), bubble flow, pulse flow and gas continuous blurring regime or spray flow. Many maps have been proposed in the literature to distinguish the boundaries between these various flow regimes. One such map by Sato et al [62] (Figure 1.2.1) is given in terms of the gas mass velocity and liquid mass velocity. At low gas and liquid mass velocities, the flow is in the gas continuous trickle regime. Increasing the liquid mass velocity at low gas mass velocity brings the flow into the dispersed bubble regime. At intermediate gas and liquid mass velocities, pulse flow is obtained. The gas-continuous blurring regime or spray flow is obtained at low liquid mass velocities and very high gas mass velocities.

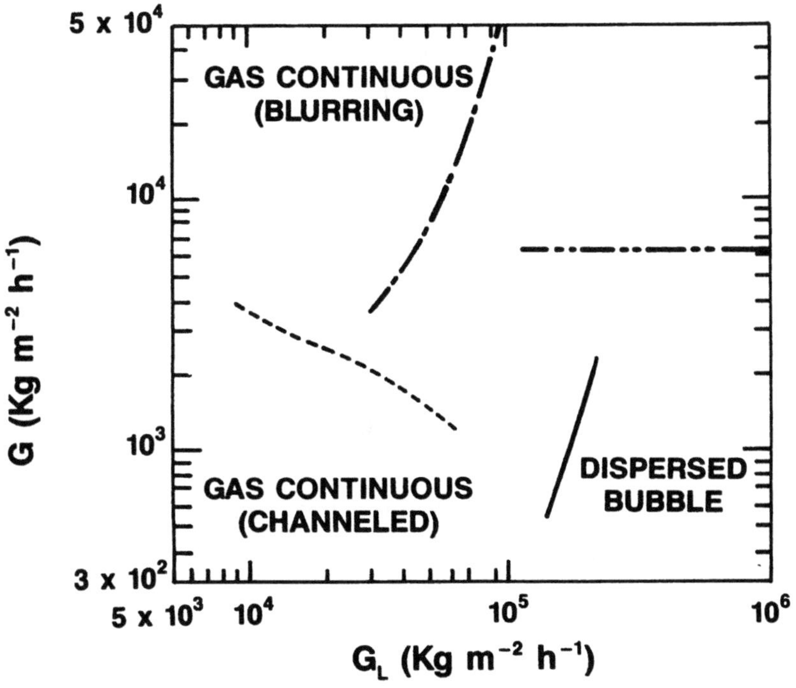

Figure 1.2.1. Flow regimes in mixed-phase downflow fixed beds (after Y. T. Sato, et al [62]).

The position of the boundaries is a function of liquid and gas properties as shown by Chou, Worley and Luss [6]. For example, properties such as foaminess was shown by both Talmor [73] and Charpentier and Favier [4] to also affect the position of the boundaries. Tosun [76] tested the available flow maps with a variety of data he obtained and concluded that the map by Charpentier et al [4] fits all the data best. The data contains observations made using both aqueous and organic non-foaming fluids. This map, shown in Figure 1.2.2, distinguishes only between a high and low interaction regimes. The high interaction regime includes both pulsed and bubbly flow. The coordinates of the map are $L\lambda\psi/G$ and G/λ.

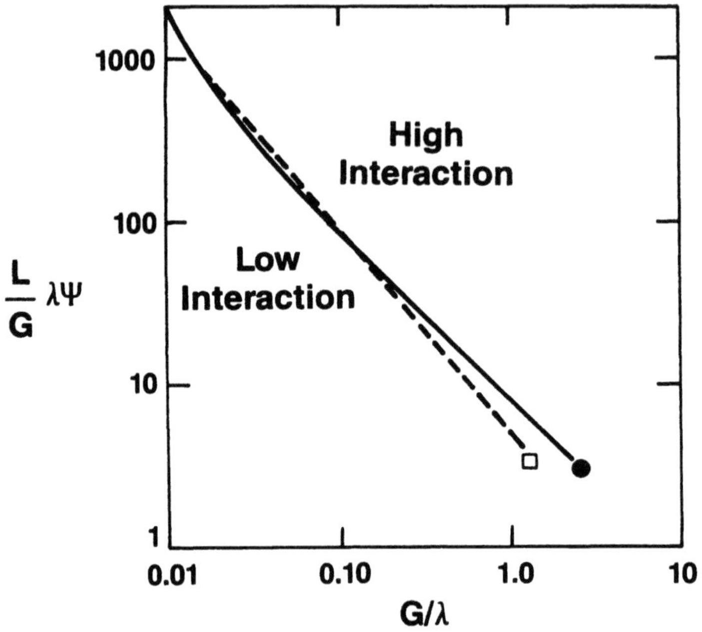

Figure 1.2.2. Boundary line between high and low interaction regime: (●) Tosun [76]; (□) Charpentier and Favier [4] (after Tosun, G. [76]).

where L, G - are the liquid and gas mass velocities.

$$\Psi = \left(\frac{\sigma_W}{\sigma_L}\right)\left[\frac{\mu_L}{\mu_W}\left(\frac{\rho_W}{\rho_L}\right)^2\right]^{1/3}$$

$$\lambda = \left[\left(\frac{\rho_a}{\rho_W}\right)\left(\frac{\rho_L}{\rho_a}\right)\right]^{1/2}$$

σ_W, σ_L = surface tension of water and the liquid

μ_W, μ_L = viscosity of water and the liquid

$\rho_W, \rho_a, \rho_L, \rho_G$ = densities of water, air, the liquid and the gas

Few diagnostics exist to determine the nature of the flow regime in a reactor. Specifically, pulsed flow can be distinguished from the other regimes by the time variations in pressure drop. However, Tosun [77] observed that pulsed flow is visually observed before the onset of pressure fluctuation. The bubble flow regime has characteristically high pressure drops.

Generally accepted means for determining the nature of the flow regime in a reactor do not exist and is a research need, in particular because the flow regime in the pilot plant may very well differ from that in the commercial unit. Pilot plants in general operate at low liquid and gas mass velocities and are, therefore, well in the gas continuous "trickle-bed" regime. Commercial units are usually at intermediate or high mass velocities and are, therefore, away from that region of the map. Contacting behavior and pressure drop may differ quite widely on scale-up. Hence, it is important to verify in a pilot plant the flow regime projected for the commercial unit using the actual feed at the design conditions.

A diagnostic to distinguish between the foaming and non-foaming regime was proposed by Charpentier et al [5] in which the propensity for foam to collapse was correlated to pressure drop. This foam collapse data is obtained in a special apparatus in which the propensity for two gas bubbles to collapse is measured in the liquid in question. This technique requires that the foam collapse data be obtained at reactor conditions.

1.3 Pressure Drop and Dispersion

Because of the low gas and liquid mass velocities in pilot plant reactors, pressure drop is low and generally difficult to measure. In commercial reactors, however, it is usually important and computations of pressure drop are usually needed to set pump and compressor designs. Several pressure drop correlations are available in the literature. One popular one is that by Larkins, White and Jeffrey [39] modified by Reiss [58] which applies to both the trickle gas continuous regime and the pulse flow regime. The correlation is represented below.

$$\log \left[\frac{(\Delta P/\Delta L)_{LG}}{\delta_L + \delta_G}\right] = \frac{0.416}{0.666 + [\log^2 \chi]} \qquad (1)$$

$$\chi = \left(\frac{\delta_L}{\delta_G}\right)^{0.5} \qquad (2)$$

In this equation, $(\Delta P/\Delta L)_{LG}$ is the mixed phase pressure drop per unit bed length and δ_L and δ_G, are the liquid in gas pressure drop per unit bed length as calculated using the Ergun equation for each of the liquid and gas flowing separately as single phases through the reactor. Weekman and Myers [86] extended this correlation to foamy systems by redefining the parameter χ as shown below:

$$\chi^1 = \frac{\delta_L}{\delta_G}^{0.5} \qquad (3)$$

Specchia and Baldi [70] proposed an alternate equation of the form first suggested by Turpin and Huntington [82] using the pressure drop as a function of a friction factor for gas-liquid flow in the bed which modifies a gas kinetic energy term, $\rho_G U_G^2$, with ρ_G, as the gas density and U_G as the superficial gas velocity in the bed. The friction factor is a function the gas and liquid Reynolds numbers and ψ as given in Section 1.2.

Clements and Schmidt [9] proposed that $(\Delta P/\Delta L)_{LG}$ depends on the particle size (d_p), the liquid viscosity (μ_L), the gas and liquid Reynolds number (Re_G, Re_L) and the gas Weber number (We_G):

$$[(\Delta P/\Delta L)_{LG}/\delta_G)] \quad \alpha \quad [\mu_L d_p \left(\frac{\varepsilon}{1-\varepsilon}\right)^3 \left(\frac{Re_G\ We_G}{Re_L}\right)] \qquad (4)$$

This relationship was shown to be valid in both the gas continuous and in the pulse flow regimes. Clements and Schmidt [9] and Tosun [76] observed that there is a smooth transition from one regime to the other.

For non-foaming liquids, Tosun [77] compared the correlations of Weekman and Myers [86], Midoux et al [46], Sato et al [62], Larkins et al [39] with the new data reported in the publication (Figure 1.3.1) and found that the correlation below gave the best overall fit:

$$\phi_L = [(\Delta P/\Delta L)_{LG} \;/\; (\delta_G]^{1/2} = 1 + \frac{1}{\chi} + \frac{1.424}{\chi^{0.576}} \qquad (5)$$

where $\chi = [\delta_L/\delta_G]^{1/2}$

None of the available correlations account for possible variations attributable to small impurities in the gas or liquid feeds nor do they predict foaminess. Because of this uncertainty, pressure drop data should be obtained when handling an unfamiliar feed. Both the feed and product material should be tested at liquid and gas mass velocities which are projected for the commercial unit.

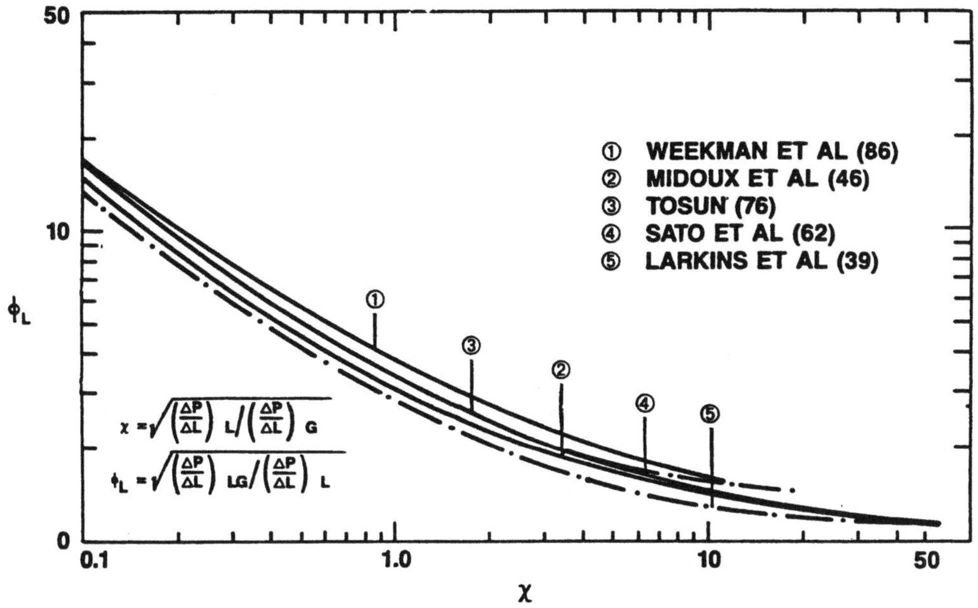

Figure 1.3.1. Comparison of several proposed pressure drop correlations (after Tosun, G. [76]).

1.4 Flow Distribution

The importance of good flow distribution of the liquid in a mixed phase downflow fixed bed reactor was first pointed out by Ross [60]. Ross compared the performance of a pilot plant and a commercial reactor and explained the difference in the performance by differences in the distribution in each of these reactors. The performance of the pilot and the commercial reactors were compared through measurement of the conversion of the hydrodesulfurization reaction. Flow distribution in the commercial reactor was quantified through tracer measurements. It was shown that those commercial reactor designs in which the tracer data showed most uniform distribution in the reactor also gave the best performance. Unfortunately, this publication did not describe the changes in the distributor trays which led to differences in distributions and performances. A recent report by Krambeck [36] discusses the use of a multicomponent tracer with wide boiling range to trace the vapor and liquid flows in a trickle bed to help confirm and/or assess the extent of maldistribution.

Laboratory studies are available dealing with radial dispersion from single points and other geometries which can be used to quantify distribution patterns from specific distributor tray designs. The underlying principle in the interpretation of these laboratory studies, is that dispersion in a trickle bed reactor is through dispersion of rivulets of liquid over the catalyst bed. These rivulets act within the bed much like raindrops on a pane of glass. As the liquid flows over catalyst packed in a bed, the flow can move radially in random directions causing radial dispersion. It is generally believed that the flow tends to an equilibrated condition down the bed for each specific superficial velocity in the bed. Thus, typically, laboratory measurements report the axial downflow distance that is required to achieve this equilibrated flow condition.

One such study by Hoftyzer [29] proposed a characteristic dimensionless parameter lz/y^2 to characterize this dispersion. "l" is the lateral dispersion coefficient, z is the axial distance down the bed required to achieve the equilibrated condition, and y is the radial distance for which this equilibrated uniform flow is found. The dispersion coefficient can be quantified by using the data from several workers [24], [52], [72], [89] as shown by Koros [35]. Thus, a K_{crit} value can be obtained which gives the combination of the dimensions that are required to achieve good dispersion of liquid. This is given below:

$$K_{crit} = \frac{y^2}{d_p^{0.5} z} \leq 4 \quad (6)$$

Where d_p is the diameter of the catalyst particle. For values of K_{crit} (in $cm^{0.5}$) less than 4, it can be assumed that the flow is uniform. This equation predicts that for a 2.54 cm diameter pilot plant using a catalyst 0.16 cm in size a single point distributor will give good distribution 4 cm down the bed. Hence, a short bed of inerts which is usually used in pilot plant reactors will be enough to rectify the potential maldistributions caused by a single point feed. For a commercial unit, it is quite apparent that multipoint distribution is required for adequate distribution without wasting large amounts of reactor straight side.

None of the available studies in the literature dealing with dispersion directly relate efficiency of catalyst utilization to dispersion. However, a conservative basis of design would be to assume that in the zone where a uniform distribution is being established, catalyst contacting efficiency is only partial.

Wall flow was also the subject of study. Herskowitz [24] established that 16 particle diameters are needed to avoid excessive wall flow. On the other hand, Koros [34] showed that only 1.4 particle diameters are required as long as the liquid mass velocity is above 10 kg/m^2s. To rectify possible flow maldistributions in a commercial reactor, Chou [8] proposed the insertion of redistribution internals which were claimed to bring wall flow into the main bed flow.

The effect of axial dispersion is more readily quantifiable than the effects of radial dispersion. Mears [44] derived an expression

$$\frac{Z}{d_p} > \frac{20n}{Pe_\ell} \ln \frac{C_o}{C_i} \tag{7}$$

relating the reactor height to the particle diameter ratio, Z/d_p, which is required to assure that the apparent reaction rate deviates from the actual reaction rate by less than 5% due to axial dispersion. This Z/d_p ratio is a function of the order of the reaction, n, the Peclet number, Pe_ℓ, and the conversion in the reactor as given by the logarithim of the ratio of the inlet to the outlet concentrations, C_i and C_o. Measurements of the Peclet number for liquids in trickle bed reactors by Hochman and Effron [28] showed that the Peclet number decreases with decreasing liquid velocity implying that small pilot plants should have large z/d_p ratios to minimize axial dispersion. Also, the Peclet number shows large variabilities implying that reactors with marginal Z/d_p ratios could be influenced by variabilities in the Peclet number from run to run. Hochman and Effron reported that startup conditions also affect the Peclet number. It is fairly well known phenomena that for very short pilot plant reactors large variabilities in reactor performance are observed. This effect is certainly one explanation of the observed variabilities. Later in this review, other potential causes will also be highlighted.

The Peclet number for gas varies inversely with liquid velocity and is approximately the same for gases and liquids for pilot plants. Hence, the same concerns for gas axial dispersion exists in the pilot plant as for liquids.

For the high velocities and large Z/d_p ratios in commercial reactors, gas and liquid phase axial dispersion is of no consequence.

1.5 Mass Transfer

In considering the possibility of mass transfer limitations in a mixed phase downflow fixed bed reactor, several mass transfer steps must be considered: from the liquid to the catalyst, from the flowing liquid to the stagnant liquid, from the gas to the liquid, and from the gas to the catalyst. One or several mass transfer steps may be limiting and, therefore, each must be considered.

Mass transfer correlations from liquid to solids are based on total interfacial area of the solids. Hence, the correlations include partial wetting effects. As expected, the mass transfer coefficient increases with increasing liquid flow rate. It is somewhat insensitive to the gas flow rate. The sensitivity to the gas rate increases in the pulse and dispersed bubble regime. Several correlations for the mass transfer coefficient, k_s, have been proposed; each of the form:

$$\varepsilon_b \, Sh \, Sc^{-1/3} = f\,(Re_L, Re_g) \tag{8}$$

where ε_b is the bed void fraction; Sh, the Sherwood number ($= k_s d_p/D$); Sc, the Schmidt number ($= \Upsilon/D$) and Re_L, Re_G, the liquid and gas Reynolds number ($= Vd_p/\Upsilon$); V, the fluid velocity; Υ, the kinematic viscosity; D, the diffusivity and d_p, the particle diameter. One such correlation by Yoshikawa et al (88) unifies the data from several workers for cocurrent gas/liquid downflow and upflow and for liquid-full operation, as follows:

$$\varepsilon_b Sh \, Sc^{-1/3} = (1+0.003\, Re_L Re_G)^{0.5}(0.765\, Re_L^{0.78} + 0.365\, Re_L^{0.614}); \tag{9}$$

valid for $0.46 < d_p < 1.3$mm, $0 < Re_G < 12$, $0.5 < Re_L < 50$. This correlation, shown in Figure 1.5.1 has data only for aqueous systems, so the basis for using dimensionless parameters is questionable; however, Herskowitz [27] presents mass transfer data taken during the catalytic hydrogenation of cyclohexene that agrees well with this correlation. In the high interaction regime, Herskowitz finds a good match with the correlations of Chou, Worley and Luss [7] and Reuther, Yang and Hayduk [59] but found disagreement with other workers. A two to three-fold range in mass transfer coefficients were found among the reported correlations. Using these correlations values for $k_s a$ (where a is the area/volume of the catalyst) from about 10^{-2} to 10^{-1} sec^{-1} are predicted for flow rates ranging from pilot plant to commercial units. These values are considerably higher than normal hydrotreating reaction rate constants as reported by de Bruijn [12], Van Dongen [85]. Hence, liquid-solid mass transfer is usually not limiting.

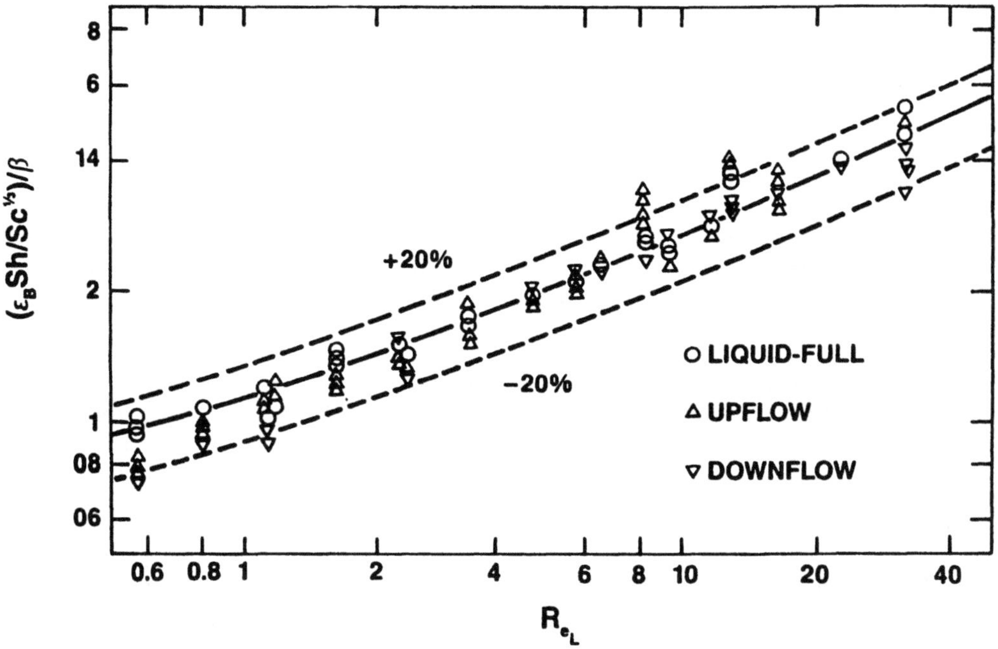

Figure 1.5.1. Unified liquid-solids mass transfer correlation for cocurrent downflow and upflow mixed-phase and liquid full fixed bed reactors (after Yoshikawa, et al [88]).

The mass transfer from the flowing liquid to the stagnant liquid is about 10 times larger than the liquid to solid mass transfer coefficient as shown by measurements of Hochman and Effron [28] and Kan and Greenfield [32]. Hence, this mass transfer rate is usually negligible.

Mass transfer coefficients for gas to liquid are usually smaller than for liquid to solid mass transfer in low velocity pilot plants but are higher at the higher velocities in commercial units. In either case, however, gas-liquid mass transfer is usually not limiting. As part of a study of formic acid oxidation in a laboratory trickle bed, Goto and Smith [19] showed that at low gas velocities (0-0.8 cm/s), the gas-liquid and liquid-solid mass transfer are comparable for a 0.5 mm particle. For a 2 mm particle, the gas-liquid coefficient is about three times the solid-liquid coefficient.

A recent dimensional correlation for organic liquids by Midoux [47] gives the gas-liquid mass transfer coefficient, $k_L a$

(in s^{-1}) as a function of the bed pressure drop, δ_{LG} (in m H_2O/m of bed) as follows:

$$\frac{k_L\, a}{\sqrt{D}} = 5.0 \times 10^8 \left(\frac{\varepsilon_b}{a_c}\, \delta_{LG}\right)^{1.63} \qquad (10)$$

where D is the diffusivity in m^2/s; ε_b, the bed porosity and a_c, the specific area of the packing, m^{-1}. This correlation predicts values ranging from ~10^{-2} to 1 s^{-1} for pressure drops of 0.2 to 2 m H_2O/m bed, a range which covers a large variety of small and large scale applications. This mass transfer rate is considerably higher than that required for normal hydrotreating; hence, it is usually not limiting.

Mass transfer directly between the gas and the wet catalyst particle is thought to be extremely high (Herskowitz and Moseri, [26]) though there are no direct confirming measurements. However, because of this high rate, interesting contacting phenomena can be observed (Section 2.1.2) when using highly active catalysts which are limited by mass transfer through the liquid phase.

Criteria for determining a priori the possibility of a mass transfer limitation was given by Satterfield [64]. The criteria assumes that mass transfer does not limit the reaction if the concentration in the bulk liquid phase, C_L, is less than 5% different from that at the surface of the catalyst. For a first order reaction,

$$C_L > \left(\frac{10\, R_p d_p}{3k_s}\right) \qquad (11)$$

satisfies the criteria. In the equation, d_p is the diameter of the catalyst pellet, k_s the mass transfer coefficient between the liquid and the solid and R_p, the reaction rate per unit pellet volume. This criteria was adapted by Koros [35] to be usable if the intrinsic activity, k_i, is known as follows:

$$\frac{k_s}{d_p k_i} < 3 \qquad (12)$$

Another criteria due to Satterfield [65] is one that is based on work by Petersen [56], as developed further by Froment and Bischoff [16], which has as a starting point the observation that the film mass transfer coefficient in a catalytic system cannot limit unless pore diffusion is also limiting. Hence, a mass transfer limitation on the outside of the porous catalyst pellet can only be important, for example, if the Weisz and Prater modulus Φ is greater than about 3 to 10 which corresponds to an effectiveness factor, η, of 0.3 to about 0.7. Thus, the criteria suggests that if the effectiveness factor of the catalyst in question in a reactor is close to unity mass transfer limitations cannot be important.

The experimental diagnostic suggested by Koros and Nowak [33] allows checking for the possible mass transfer limits in a particular reactor without altering or disturbing the fluid dynamics of the reactor. The diagnostic involves measuring the apparent reaction rate of a catalyst as a function of varying the intrinsic catalyst activity. The intrinsic catalyst activity is changed keeping the particle size and fluid dynamic conditions constant. This can be done by changing the loading of the active catalyst in the pellet. If the apparent reaction rate varies linearly with the intrinsic activity of the catalyst, the reaction is kinetically controlled. If a square root relationship is found, pore diffusion in the catalyst limits. If, on the other hand there is no change in the apparent reaction rate with catalyst activity, then outside mass transfer is limiting. An example of how the intrinsic activity of the catalyst in a hydrotreating trickle bed operation may be varied would be by varying the metal loading of say, a platinum on alumina catalyst. The generality of the diagnostic is discussed by Madon, O'Connell and Boudart [42].

1.6 Heat Transfer

Heat transfer in trickle bed reactors is not a well studied phenomena. Heat transfer coefficients reported in the literature vary widely and more studies are needed. There is general agreement, however, that due to the poor and irregular heat transfer in trickle bed reactors, there is a potential for hot spotting in this type of reactor.

Hot spots in trickle beds with highly exothermic reactions have been observed both commercially and in pilot plants and are believed to be a consequence of shadowing of the flow within the reactor by either internals or other flow disturbances. The low flow behind the obstruction persists due to the low radial dispersion of the flow. Very high temperatures can build up because of the localized lower space velocities and the low heat transfer and low radial dispersion of the locally generated heat. Jaffe [31] simulated the formation of the hot spots in a trickle bed reactor using parallel and series networks of stirred tank reactors as suggested by Dean and Lapidus [11]. The simulation was for a hydrogenation and the calculations were supported by kinetics obtained in a small pilot plant. Jaffe concluded that hot spotting due to shadowing from internals is a real concern and should be kept in mind in the design of internals for trickle bed reactors.

Eigenberger [15] reported that low flow transients coupled with shadowing in mixed phase reactors can cause even greater hot spotting. This amplification was ascribed to the localized consumption of liquid phase reactant stagnated behind an obstruction. With a localized flow interruption caused by a transitory flow maldistribution, little or low fluid would flow into or out of the shadowed dead zone. Thus, virtually all heat removal out of the zone by sensible heat flow or due heat transfer is lost. Eigenberger, et al [15] also used the Dean and Lapidus model to calculate temperature profiles. They simulated the catalytic hydrogenation of benzene to produce cyclohexane, a highly exothermic reaction.

For steady state conditions, they calculated a 100°C buildup behind a small obstruction with a radius of 0.085 m. in a reactor 6 m in diameter. Their calculation showed that at the conditions they chose, the hot spot would materialize at about 0.2 m downstream from the obstruction. The transient analysis further showed that temperature could rise as much as 500°C in a short time, approximately 600 seconds. This dramatic rise in temperature is accompanied by a depletion of the liquid phase and hydrogen. This hot spot temperature is several times higher than the adiabatic temperature rise given by the steady state value at the hot spot.

The authors report that they have observed extremely high temperatures in a commercial reactor. In particular, this observation was not coupled with a known change in reactor conditions. However, the temperature levels observed could be explained by reduction of hydrogen flow to about 1/4 of the normal which the authors surmise could have occurred due to a transitory partial localized plugging of the bed by dirt in the feed or by a flow obstruction caused by broken catalyst particles within the bed. Berkelew [2] also reported hot spotting due to localized random maldistributions in hydrotreaters.

In general, the recommendation for adiabatic trickle bed reactors is to avoid internals that obstruct flow and to try to prevent localized partial plugging of the bed through good flow distribution. For reactions that require heat transfer radially through a wall such as in tubular reactors, it is recommended that the pilot plant reactor be run at conditions similar to the commercial reactor. Tests at the low mass velocities typical for pilot plants may be too conservative to get an economic design. Due to the large disagreement amongst the available correlations, as reviewed by Herskowitz [26], scale up from a small scale pilot plant to a larger commercial unit is not recommended for highly exothermic reactions using heat transfer at the wall.

2. LABORATORY AND PILOT PLANT REACTOR DESIGN AND OPERATION

The behavior of several configurations of trickle-bed laboratory and pilot plant reactors have been studied and reported in the literature. In this section, they will be reviewed to compare their characteristics under various reactor conditions.

Commonly, trickle beds reported in the literature have diameters of approximately 2.5 cm, with short to medium length beds, about 30 to 200 cm. They are run under a variety of, but usually low, liquid and gas velocities. The catalyst is packed below a calming zone of inert particles. Alternatively, a short bed length is used, but with the catalyst diluted with inert particles such as fine sand. In another type, high liquid velocity is obtained through recycle. Both batch and continuous flow modes have been reported for this latter reactor configuration. Each of these types of reactors will be discussed separately with examples given from the literature to help characterize their performance.

2.1 Variable Liquid Velocity Downflow Reactors

In the analysis of the performance of these reactors, we have to distinguish between those which are limited by the transport of a liquid phase reacting component and those which are limited by a reacting gas phase component. The characteristics of these two types of reactors are somewhat different as demonstrated by the examples below. Table 1 gives a listing of selected examples from the literature for each of these reactor types.

2.1.1 Liquid phase component transport limited reactions. An industrially important reaction of this type is hydrotreating to remove sulfur, nitrogen and metals from gas oils and residual oils. These feeds are commonly liquid under reactor conditions. For example, Henry and Gilbert [23] reported studies using a vacuum gas oil in a 1000 cc reactor of 3.2 cm inside diameter containing a 0.6 cm thermowell. Catalyst beds of 1.6 mm extrudates, 114 cm and 325 cm long were used. These tests (Figure 2.1.1) showed that the global rate of reaction, as exemplified by the hydrocracking yield, was best correlated by the inverse liquid hourly space velocity to the 2/3 power and the reactor length to the 1/3 power. This correlation was based on the assumption of first order kinetics and that the contacting is proportional to liquid holdup. It was proposed that the performance of the reactor can be described by the equation below,

$$\mathrm{Log}\,[c_o/c_i] \;\alpha\; Z^{1/3}\,(SV)^{-2/3}\,d_p^{-2/3}\,\mu_L^{1/3} \qquad (13)$$

where c_o, c_i are the outlet and inlet concentrations of the key components; Z, the catalyst bed length; SV, the liquid hourly space velocity; d_p, the catalyst diameter; and μ_L, the liquid viscosity.

Table 1

EXAMPLES OF LABORATORY AND PILOT PLANT REACTORS USED FOR TRICKLE BED REACTIONS

Low and Variable Velocity Once-Through Mode

- o Liquid phase component transport limited reactions
  - Vacuum gas oil H/C, HDS, and HDN (Henry, [23])
  - Residua HDS (Montagna, [50], [51])
    (Paraskos, [54])
  - Hydrogen peroxide decomposition (Koros, [34])
  - Formic acid oxidation (Goto, [19])
  - Acetic acid oxidation (Levec)

- o Gas phase component transport limited reactions
  - Glucose oxidation (Tsukamoto, [78])
  - α-Methylstyrene hydrogenation (Herskowitz, [24])
  - Ethanol oxidation (Goto, [21])

High Liquid Velocity Recycle

- o Continuous
  - Residua hydrometallization (Van Dongen, et al [84])

- o Batch
  - Glucose oxidation (Tsukamoto et al, [78])
  - HDS of Residua (Papayannakos et al, [53])
  - Cyclohexene hydrogenation (Herskowitz et al, [26])

Low Velocity Downflow with Diluent Fine Sand Fill

- Vacuum gas oil HDS (De Bruijn, [12])
- Residua Hydrodemetallization and HDS (Garcia, [17])
- Vacuum gas oil HDN (Van Klinken et al, [85])
- Hydrogen peroxide decomposition (Koros, [34])

Comparison of Stirred Tank and Trickle Bed Reactors

- Hydrogen peroxide decomposition (Koros, [34])
- Alpha methyl styrene hydrogenation (Herskowitz, [27])
- Glucose oxidation (Tsukamoto et al, [78])
- Alpha methylstyrene hydrogenation in n-hexane (Mill, [48])
- Glucose hydrogenation (Turek et al, [80])

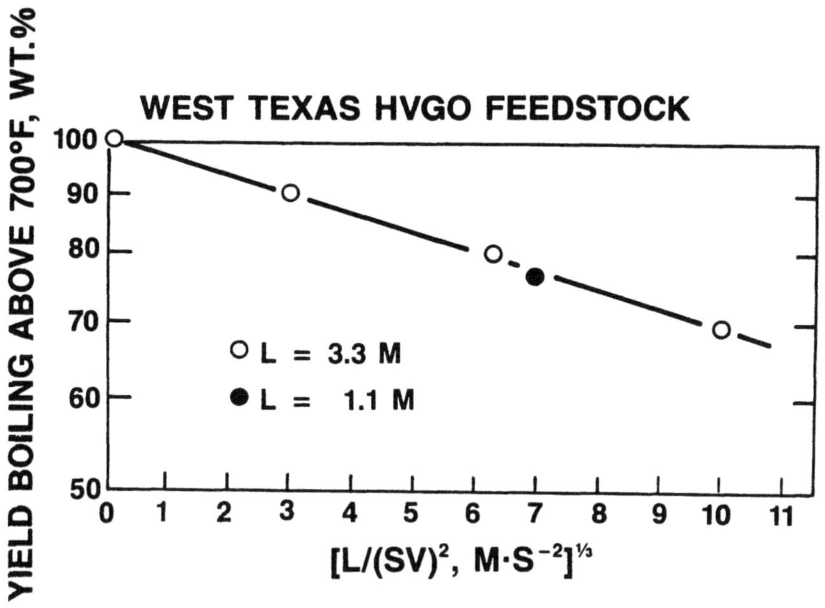

Figure 2.1.1. Effect of liquid space velocity and catalyst bed length on hydrocracking in trickle beds (after Henry, H. C. and J. B. Gilbert [23]).

Reprinted with permission from Ind. Eng. Chem. Process Des. Dev. 12(3),(1973),328-334. Copyright (1973) ACS.

On the basis of measurements for hydrocracking and hydrodenitrogenation, the authors concluded that contacting in normally used pilot plants run at liquid velocities between 5 x 10^{-3} to 0.3 cm/s are sensitive to liquid flow rate and contacting increases with increasing velocity. On the basis of the holdup model, they also suggested that the contacting efficiency should be a function of the diameter of the catalyst and the liquid viscosity as shown by the equation above.

In another process pilot plant study, Montagna and Shah [50] found similar effects using a series of atmospheric reduced crudes. Figure 2.1.2 shows a sample of the results in which the first order constant for desulfurization, denitrogenation, demetalization were shown to be a function of the reactor length demonstrating again that liquid velocity affects catalyst contacting. These studies were performed with 8 x 14 mesh and 20 x 30 mesh catalysts. The bed lengths were 10.5 to 84.2 cm and the catalysts were packed in a stainless steel reactor, 2.54 cm I.D with a 0.6 cm axial thermowell. Studies with a lighter residua, 53% reduced crude (Figure 2.1.3) did not show that length, i.e., liquid velocity, affected performance.

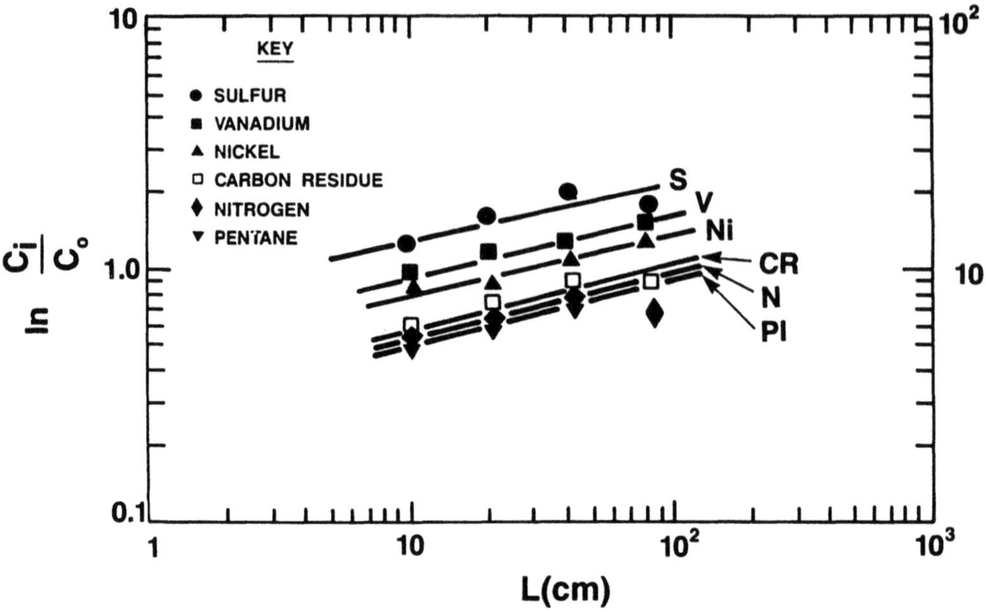

Figure 2.1.2. Effect of reactor length at constant space velocity on the reaction rate for HDS, HDN and demetallization of residua (36% KATB) (after Montagna, A. A. and Y. T. Shah [50]). Reprinted with permission from Ind. Eng. Chem. Proc. Des. Dev. 14 (1975) 479-483. Copyright (1975) American Chemical Society.

In a follow-up study, Montagna, Shah and Paraskos [51] studied the effect of particle size on the reactor performance. The same reactor was used but with 100 cc charge. Catalyst sizes ranged from about 0.4 to 3 mm. A heavy vacuum bottoms, 22% reduced crude, and a lighter atmospheric bottoms, 36% reduced crude, were used for these tests. Figure 2.1.4 shows that for the smaller catalyst, the first order kinetic constant remains constant with increasing particle size and then decreases as would be expected from the onset of pore diffusion limitations. The data was interpreted with the model shown below:

$$\ln [c_i/c_o] \;\alpha\; d_p^{0.82} \text{ or } \ln [c_i/c_o] \;\alpha\; \tanh (k^2 d_p)/d_p^{0.82} \qquad (13')$$

where k is the first order reaction constant.

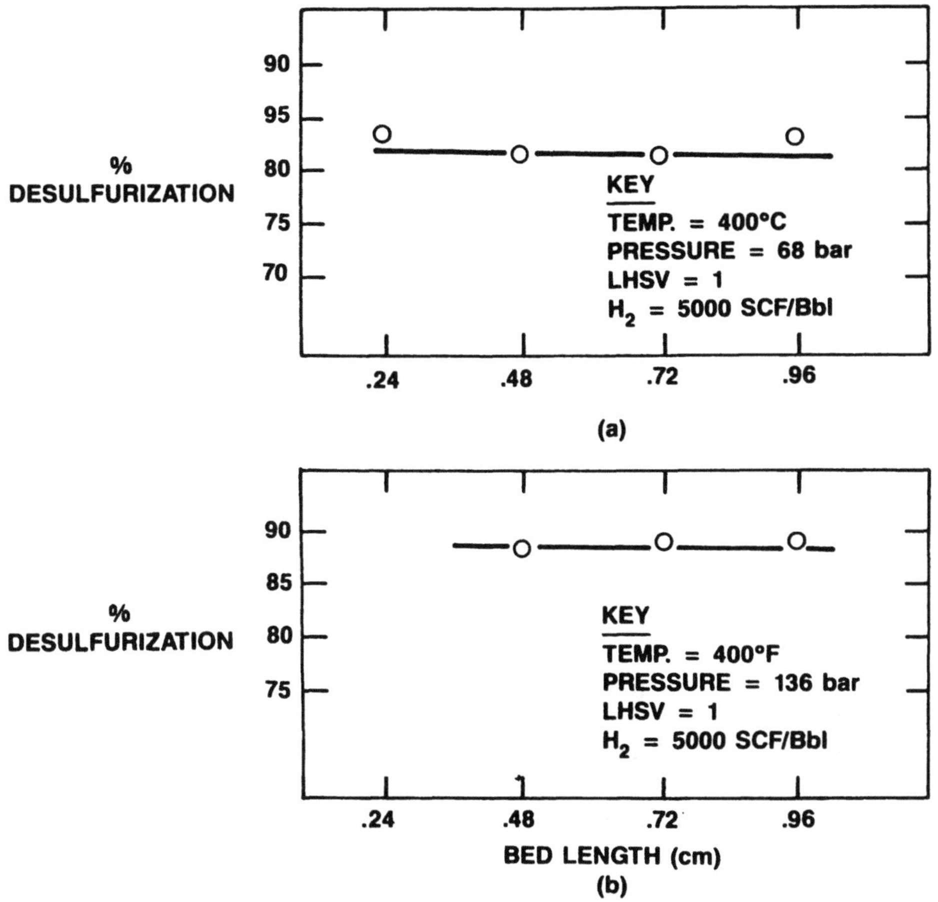

Figure 2.1.3. No effect of bed length found for hydrodesulfurization of 53% KATB residua at 68 and 136 bar (after Montagna, A. A. and Y. T. Shah [50]). Reprinted with permission from Ind, Eng. Chem. Proc. Des. Dev. 14 (1975) 479-483. Copyright (1975) ACS.

These workers point out that their data are best correlated with a model which has a particle size dependency, $d_p^{0.82}$, and not $d_p^{0.18}$ as suggested by the holdup model of Henry and Gilbert [23]. This higher exponent is obtained using a model which correlates the contacting efficiency with catalyst wetting. The wetted area correlation for packing of Puranik and Vogelpohl [57] was used for the model development.

Paraskos, Frayer and Shah [54] studied the hydrotreating of light gas oil, vacuum gas oils and reduced crudes. It was concluded that the effect of mass velocity depended on the nature of the feed. The slope of the first order reaction constant with reactor length varied from 0.05 to 0.55 depending on the temperature, pressure, liquid hourly space velocity and the nature of the reaction. Thus, they concluded that the holdup model of Henry and Gilbert [23] and of Mears [45] are not generally applicable.

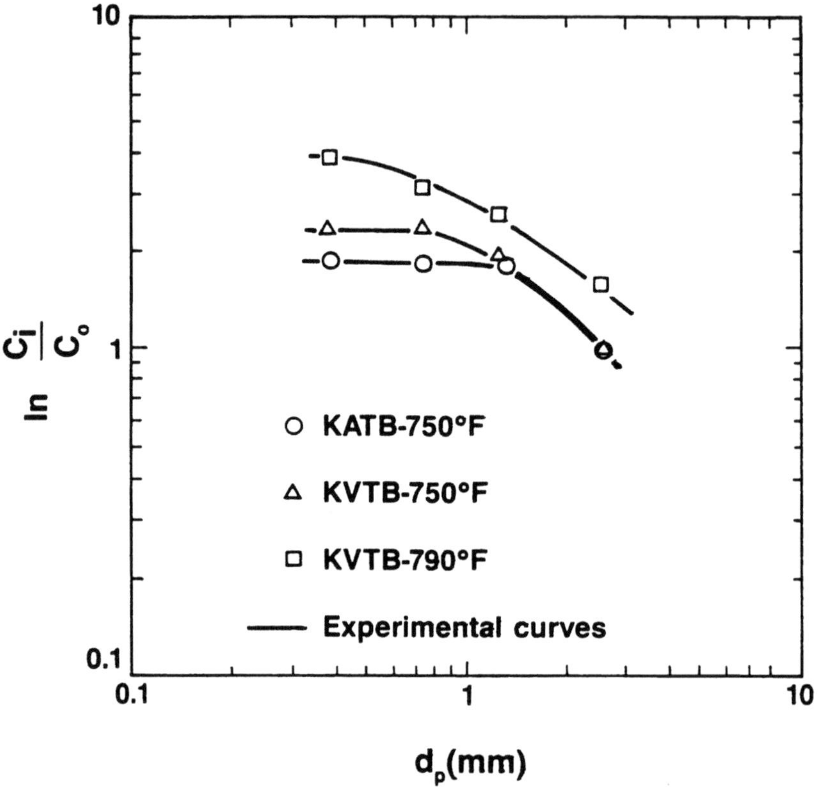

Figure 2.1.4. Effect of particle size on contacting in HDS of residua (after Montagna, A. P., et al [51]). Reprinted with permission from Ind.Eng.Chem.Proc.Des.Dev.16(1)(1977)152-155 Copyright (1977) American Chemical Society.

Koros [34] found sizeable liquid velocity effects for the rate of hydrogen peroxide decomposition on a charcoal catalyst in a variety of reactor diameters and lengths (0.46 cm I.D. by 122 cm, 0.78 cm I.D. by 122 cm and 2.54 cm I.D. by 183 cm) and catalyst sizes of varying sizes (6/14 mesh granules, 0.32 cm by 0.32 cm pils and 0.32 cm extrudates). These studies were performed over a wide range of liquid velocities from 0.41 to 8.2 kg/m^2.s (5.5 x 10^{-3} - 1.1 cm/s).

In this study, the catalyst was calibrated using a stirred tank reactor. It was found that the efficiency of catalyst utilization in the trickle bed varied widely with the liquid velocity, as shown in Figure 2.1.5. The dependence of catalyst utilization with liquid mass velocity varied depending on the reactor tube to catalyst diameter ratio with the lower ratios being more sensitive to liquid velocity as expected. In some cases, no effect of liquid mass velocity was observed. This was surmised to be due to irreproducibilities in the packing. In general, a critical mass velocity for maximum catalyst utilization appears to exist at about 3 kg/m^2s (0.4 cm/s). The activation energy for the reaction was measured to be 42 kg/mol both in a stirred tank and in the trickle bed reactor tests showing that the contacting inefficiencies are not due to purely mass transfer limitations. The first order kinetic reaction constant was 3 V/H/V, a rate which is comparable to normal hydrotreating reactions.

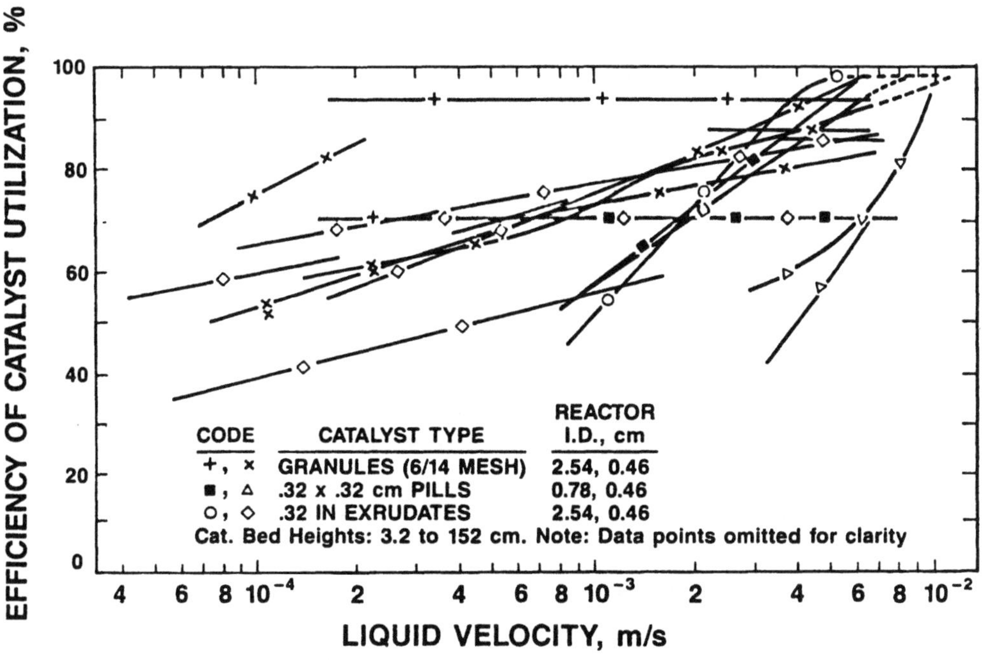

Figure 2.1.5. Effect of liquid velocity on the efficiency of catalyst utilization for the H_2O_2 decomposition over charcoal (after Koros, R. M. [34]).

Levec and Smith [41] studied the oxidation of dilute acetic acid solutions in a trickle bed reactor 2.54 cm I.D. diameter by 30 cm long operated at an elevated pressure of about 70 atmospheres. Catalyst sizes were 0.054 and 0.24 cm. Liquid flow rate varied from 0.075 to 0.26 cm/s. In this reaction oxygen is the limiting component; however, due to the high pressure, its concentration in the liquid is quite high. The results (Fig. 2.1.6) show that the liquid velocity has a very small effect on the global reaction rate and that the performance of the reactor is not affected by the catalyst size in the range studied. The reaction rate is slow since pore diffusion limitations were not exhibited. This reaction showed small contacting changes with liquid velocity.

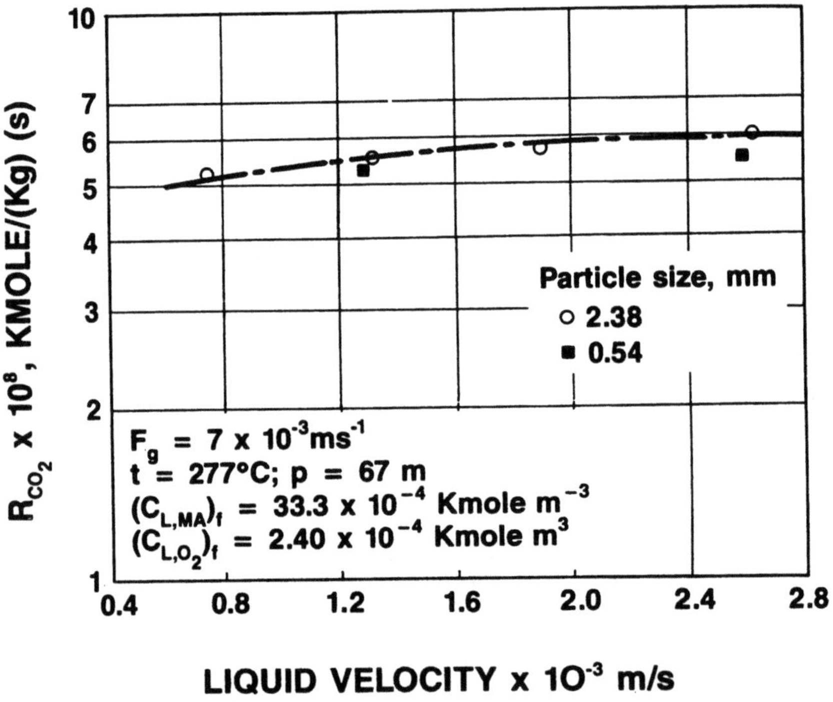

Figure 2.1.6. Effect of liquid velocity on acetic acid oxidation (after Levec, J. and J. M. Smith [41]). Reproduced by permission of the American Institute of Chemical Engineers.

For formic acid oxidation at 40 atmospheres, the reaction rate as reported by Goto and Smith [19] is about an order of magnitude higher than that for acetic acid oxidation. The results shown in Figure 2.1.7 were calculated using the intrinsic reaction kinetics determined by Baldi et al [1] together with mass transfer coefficients from Goto and Smith [19]. Reasonable match was obtained. A moderate effect of liquid flow rate on the global reaction rate was observed.

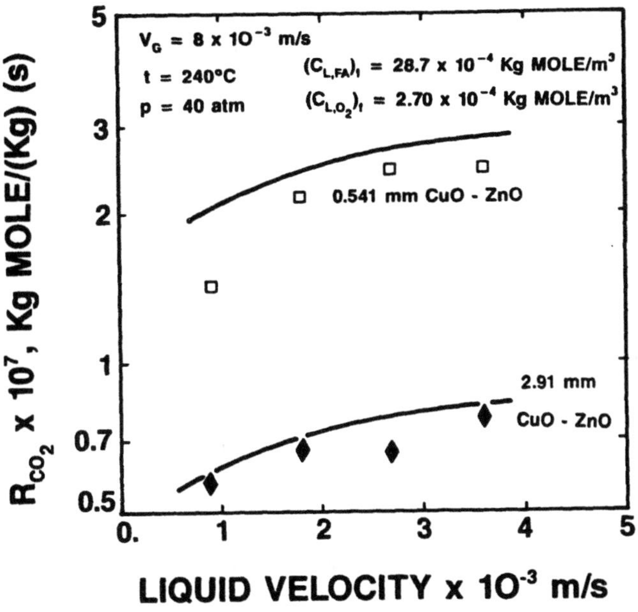

Figure 2.1.7. Effect of liquid velocity on the oxidation of formic acid at 40 atm., pressure, - calculated (after Goto, S. and J. M. Smith [19]). Reproduced by permission American Institute of Chemical Engineers.

These reactor studies showed that trickle bed reactors operating with limiting liquid phase reactants may exhibit mass velocity sensitivities. These sensitivities are a function of the nature of the reaction, catalyst size and the reactor tube to catalyst size ratio.

2.1.2 Gas phase component transport limited reactions. There are several studies in the literature representative of reactions limited by transport of the gas phase component. In these studies, the gas phase component must first dissolve in the liquid to be then transported to a catalytic site by diffusion into the catalyst pores. Alternatively, the gas phase reactant can contact the catalyst particle outer surface directly without first dissolving in the moving liquid stream. Because the direct contact of the gas with the catalyst reduces the mass transfer resistance, there is a potential for increased reaction rates for situations in which the catalyst is poorly wetted. Thus, minima in the observed global reaction rates are possible in this case. This is in contrast with reactions in which the limiting reactant is a liquid phase component where partial wetting due to low liquid velocity leads to reduced contacting or at best no change in contacting. Table 1 shows several examples in which the behavior of gas phase limited reactions was studied.

Herskowitz, Carbonell and Smith [24] studied the behavior of α-methylstyrene hydrogenation over a Pd/Al_2O_3 catalyst (Figure 2.1.8). Large changes in the global rate were observed as a function of changing liquid velocity including a minimum in the rate. The reaction rate here is about 100-fold higher than for acetic acid oxidation. Also, the concentration of hydrogen in the liquid is considerably lower than the oxygen in the oxidation being that this reaction was carried out at one atmosphere pressure instead of 70 atmospheres. The minimum shown by the more active 2.5% Pd/Al_2O_3 catalyst is explained by the transition of the controlling transport limit from direct gas phase transfer to a partially wetted catalyst to transport through the flowing liquid to the wetted catalyst. The lower activity 0.75% Pd/Al_2O_3 catalyst did not show this effect.

Herskowitz and Mosseri [26], using the same reaction but with a much more active catalyst, found an even greater sensitivity to the liquid flow rate and in addition also found a sensitivity to the gas rate. As shown in Figure 2.1.9, increasing gas flow rates increase the global rate of reaction at low liquid flow rates until a certain gas rate is reached. Then the effect reverses and the reaction rate decreases upon further increase of the gas rate. This observation is explained by suggesting that at the higher gas rates the liquid is driven off the surface of the catalyst generating dry inactive zones. The zero gas flow rate performance is well predicted by calculations using liquid-solid mass transfer coefficients reported by Goto and Smith [19]. Using this mass transfer correlation and the wetting efficiency model of Tan and Smith [74], a consistent correlation of the data was obtained. This figure shows that at zero gas flow rate full catalyst wetting was achieved over the entire range of liquid velocities. The global rate, however, decreased to about one third of the maximum at the lowest liquid velocity. At gas velocities higher than the maximum shown in this figure, which corresponds to 515 cm/s, it was found that the wetting efficiency would drop even further, but the prediction of this effect is beyond the scope of the Tan and Smith model.

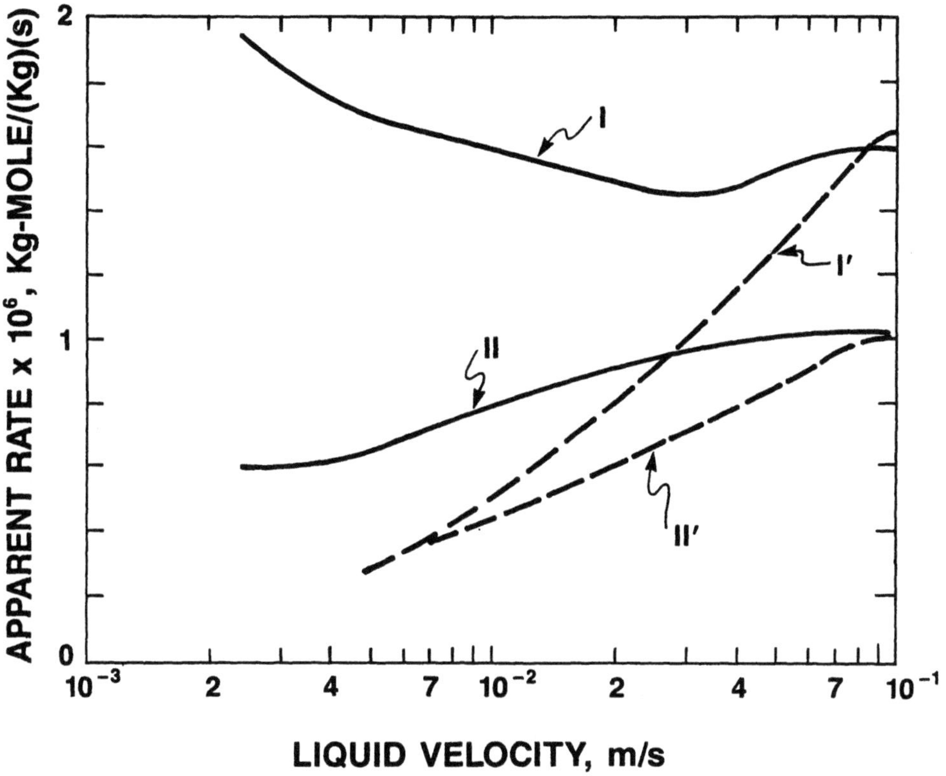

Figure 2.1.8. Effect of liquid velocity on the apparent reaction rate for a reaction in a fixed bed reactor limited by gas/liquid mass transfer. Hydrogenation of alpha methyl styrene on 2.5% Pd/Al_2O_3 (I) and 0.75% Pd/Al_2O_3 (II). Calculated apparent rate assuming liquid contacting/mass transfer as the only limitations (I', II') (after Herskowitz, M., R. G. Carbonell and J. M. Smith [24]. Reproduced by permission AIChE.

Another example of a reaction where the global rate had a minimum as a function of liquid velocity is the oxidation of ethanol in an aqueous solution (Goto and Mabuchi [21]). This reaction was carried out in a 5 cm inner diameter by 20 cm high reactor. The liquid velocities ranged from 0.3 to 0.4 cm/s. The catalyst size was 5.7 mm. This reaction system shows the minimum in global activity with increasing superficial liquid velocity in spite of the fact that the activity is relatively low. However, due to the large catalyst size, it was found that a very high liquid superficial velocity, i.e., 2.5 cm/s, was needed in the upflow mode to eliminate mass transfer limitations. Thus, this reaction system is indeed mass transfer limited confirming that the increased activity with decreased liquid rate is attributable to improve direct gas-solid mass transfer.

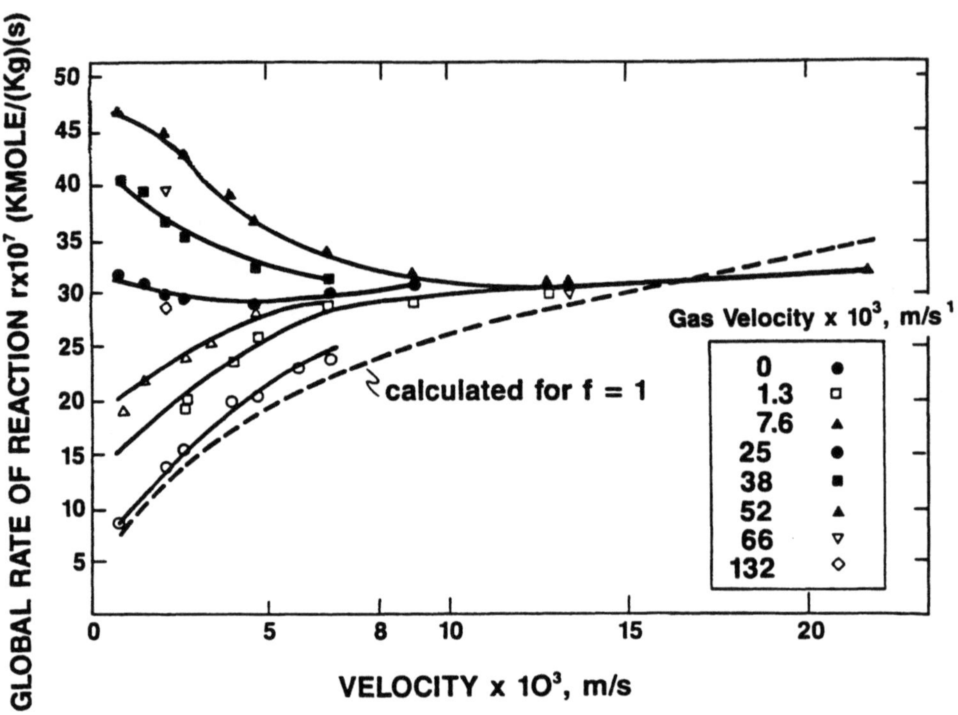

Figure 2.1.9. Effect of gas and liquid velocities using a high activity catalyst for α-methylstyrene hydrogenation (after Herskowitz, M. and S. Mosseri [26]).
Reprinted with permission from Ind.Eng.Chem.Fundam.22(1983) 4-6 Copyright (1983) American Chemical Society.

The global glucose oxidation rate in a low mass velocity trickle bed was found by Tsukamoto, Morita and Okada [78] to be higher than for a liquid full reactor in which the liquid was presaturated with oxygen. The catalyst was an activated charcoal impregnated with immobilized enzymes. The charcoal was 0.055 and 0.011 cm in size. The reactor was 2.1 cm I.D. and 27 cm long. The liquid velocity was varied between 0.044 to 0.337 cm/s and the gas velocity between 1.63 and 7.35 cm/s. Under these conditions the reaction is under pore diffusion limit control. Thus although a minimum in global rate was not reported, the unexpected better performance of the trickle bed in comparison with the fully wetted conditions was indicative of a parallel, more direct contacting path between the gas and the liquid.

Summarizing then, it has been shown that for at least three reactions, that under strong mass transfer control with a limiting gas phase component, liquid mass velocity effects on the global reaction rates showing a minimum can be obtained due to the countering effects of liquid velocity on gas/catalyst and liquid/catalyst contacting.

2.2 Low Mass Velocity Downflow Reactors with Fine Diluent

Improved contacting in laboratory trickle beds by the addition of a fine diluent to the catalyst bed have now been reported by several workers for a variety of reactions. Koros [34] showed that for the hydrogen peroxide decomposition in methanol, the introduction of 50 mesh sand into the interstices of the bed improved the contacting efficiency over a wide range of liquid velocities as shown in Figure 2.2.1. The effect was demonstrated by taking a bed of 3.2 mm extrudates in a 2.54 mm diameter reactor

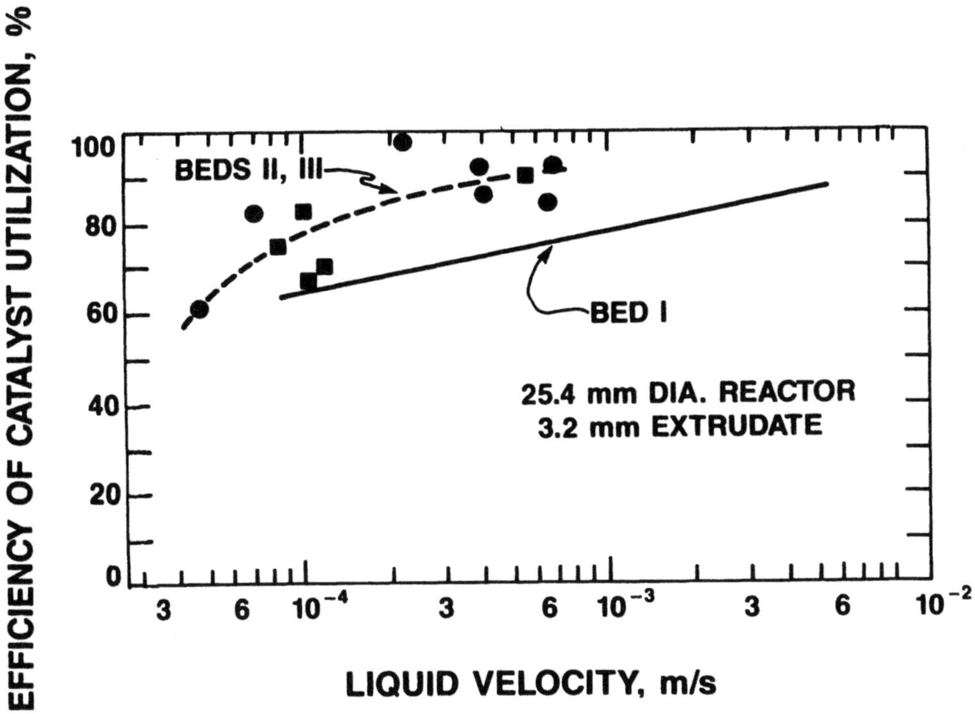

Figure 2.2.1. Effect of imbedding the catalyst bed in sand Bed I packed with vibration. Bed II (●) packed with interstices filled 50 mesh sand. Bed III (■) catalyst imbedded in a 50 mesh sand bed (after Koros, R. M. [21]).

and first determining the efficiency of catalyst utilization as a function of liquid mass velocity. The bed was then filled with 50 mesh inert sand in such a way that the interstices were filled without disturbing the bed. This is designated as "Bed II" in the figure. Efficiency was improved by about 30 to 40%. Comparable increase in catalyst efficiency was obtained when the same catalyst was repacked, embedded in a 50 mesh sand bed ("Bed II").

Garcia and Pazos [17] using a variety of diluents ranging in mesh sizes from 10/16 through 28/100 found that for the hydrodesulfurization of Venezuelan Jobo deasphalted oil, catalyst utilization was increased with the use of small diluent. A 1.6 mm extrudate catalyst was used. Table 2 shows the results of the tests. A bed length of 32 cm was used with the catalyst to bed ratio of 42:90 cm^3/cm^3. Apparent rate constants for the diluted beds were 32-170% higher than for the undiluted beds. The reactor used for these studies was 1.9 cm I.D. by 170 cm long with a thermowell measuring 0.6 cm O.D.

Table 2

INFLUENCE OF DILUENT PARTICLE SIZE AND BED POROSITY ON HDS AND HDN ACTIVITIES

Run	Particle Diameter (mm)	Bed Porosity	SV (h^{-1})	HDS (wt%)	HDV (wt%)
A	1.63	0.39	1.0	77.2	52.0
			2.0	62.3	36.4
B	0.74	0.34	1.0	85.9	69.7
			2.0	73.6	48.6
C	0.49	0.34	1.0	89.0	81.0
			2.0	78.0	64.9
D	0.45	0.25	1.0	83.0	70.9
			2.0	75.3	54.1

1/6 mm extrudate, bed length = 32 cm, catalyst/bed ratio = 40/90 cm^3/cm^3

(After Garcia, W. and J. M. Pazos [17]).

Similar effects on contacting were found by Van Klinken and Van Dongen [85]. They used a bench scale reactor 1 cm in length with an empty cross-section of about 2.5 cm^2 operated at a liquid velocity of 1.9 x 10^{-2} cm/s for the hydrodenitrogenation of a vacuum distillate. The bed was diluted 1:1 with a 0.2 mm silicon carbide powder. Both residence time tracers and denitrogenation studies were conducted. It was found the liquid holdup of the diluted bed was approximately twice that of the undiluted bed and that the first order rate constant for nitrogen removal increased between 3 to 10 times. It was also observed that the reproducibility of the data improved with dilution.

De Bruijn [12] used a 21 mm and a 38 mm inside diameter pilot plants both provided with axial thermowells of 0.6 mm outside diameter. It was found that upon dilution, the variation of the catalyst activity due to differences in average extrudate length were eliminated. The variations were observed in the undiluted 50 ml bed. Changes of almost a factor of two were found for a two-fold length change in the of average extrudate length. In the diluted bed, only a very small change was detected. Likewise, the effect of extrudate diameter on catalyst activity was found to be much more pronounced in the undiluted small reactor bed than in the diluted reactor. As a matter of fact, the diluted small reactor behaved exactly the same as the longer, higher velocity 1000 ml undiluted catalyst bed. Both contacting and bed length effects involving axial dispersion were proposed as explanations for these dilution effects. Fine sand diluent increases the number of particles per unit length of reactor that the flow has to pass through, thus decreasing the potential deleterious effect of axial dispersion on conversion for short beds as pointed out by Mears [44].

In summary, several workers using a variety of feedstocks and conditions have found that diluting small trickle bed reactors with fine grained inert material improves contacting as evidenced by measurements of increased apparent reaction rates. In addition, it has been found that the reaction rate measurements obtained with diluted beds, even when running at very low liquid velocities, are much more consistent than when run undiluted.

2.3 Recycle Trickle Beds - Batch and Continuous

Several workers have reported studies using trickle beds with a recycle loop operated in either batch or continuous flow mode. The inherent advantage of this set-up is that the liquid velocity to the bed can be set as high as needed to effect complete contacting independent of the length and space velocity used in the reactor. The risk in such an operation is that the liquid is exposed to a large holdup volume. Hence, if there are thermal reactions in addition to the catalytic ones, these will be accentuated.

Van Dongen et al [84] reported studies with hydrodemetalization of heavy residual oils using a continuous flow recycle trickle bed. The trickle bed was operated at a liquid velocity of 3 kg/m^2s (0.3 cm/s). The authors comment that this velocity is of the order of magnitude of commercial reactors so that good contacting can be assumed. Furthermore, they checked this assumption by finding that halving or doubling the liquid velocity did not affect the rate of sulfur and vanadium removal.

Papyannakos [53] used a batch recycle trickle bed reactor to study the kinetics of heavy desulfurization of petroleum residua. In this work, 200 gm of feedstock were used and recycled. No report is made of the liquid flow rates used. Van Dongen also ran in a batch wise mode, however, in this case the feed was passed through the reactor semi-continuously. Each pass was collected and rerun.

Herskowitz et al [26] and Tsukamoto et al [78] also used batch recycle reactors for their studies of trickle beds in cyclohexene hydrogenation and glucose oxidation, respectively. In the cyclohexene hydrogenation studies, liquid velocities as high as 2 cm/s were used. In glucose oxidation, velocities up to 0.35 cm/s were used.

None of these studies reported any negative aspects with respect to recycle operations. It is concluded, therefore, that this mode of operation is acceptable and can readily negate concerns with contacting. However, direct measurements of plug flow conversions with a deactivating catalyst are difficult to execute.

3. SCALEUP MODELS

A variety of scaleup models for trickle beds are presented in the literature. All the models assume that even distribution of the fluid over the entire top catalyst bed surface will be achieved with the proper design of inlet distributor hardware. But each model depicts differently the interaction of fluid dynamics, mass transfer and the intrinsic kinetics of the catalytic reaction.

The models can be classified according to the basic hydrodynamic premise underlying the model. The "holdup model" assumes that liquid-catalyst contacting is proportional to the amount of liquid that is held up in the bed. The "wetting model" presupposes that the key factor controlling catalyst utilization is the extent of wetting of the particles by liquid. The "partial wetting/mass transfer catalyst effectiveness model" assumes that contacting can be described by a conventional mass transfer model modified to take into account the fraction of the catalyst that is wet. The "stagnancy/catalyst effectiveness model" assumes that a key parameter controlling the contacting efficiency is the size of the stagnancies within the bed.

3.1 Holdup Scaleup Model

The "holdup" model assumes that contacting is proportional to the liquid holdup in the catalyst bed. This model, proposed by Henry and Gilbert [23], uses total holdup measurement as a basis and presupposes that each element of liquid hold-up is associated with an equivalent catalyst element and that all of these equivalences are of equal efficiency without respect to the nature of the reaction. Liquid velocity, particle size and fluid physical property affect contacting only as those parameters affect holdup.

This model was successfully used to correlate hydrocracking and hydrodesulfurization data that Henry and Gilbert obtained in a pilot plant reactor. A functionality with respect to reactor length of $Z^{0.33}$ and inverse with space velocity ($SV^{-0.67}$) predicted by the model was confirmed. Subsequent studies by Montagna, et al [50], [51] pointed out that these functionalities did not generally apply. Liquid holdup was found to be not specific enough to describe trickle bed contacting. However, the concept that a parameter akin to holdup, such as wetting, was established as a working basis for subsequent models.

3.2 Catalyst Wetting Scaleup Model

The "catalyst wetting" model assumes that the rate of reaction is proportional to the wetted surface of the catalyst and that this wetting can be described by correlations such as those given by Onda [52] for high liquid velocities and by Puranik and Vogelpohl [57] for low velocities. Mears [45] combined these expressions with a model for a first order reaction and obtained the relationships below for high liquid velocities:

$$\ln [c_i c_o] = \frac{k}{SV} [1 - \exp(-W\,U)^{0.4} (SV)^{0.4}] \qquad (14)$$

and for low liquid velocities,

$$\ln [c_i/c_o] \;\alpha\; \frac{Z^{0.32}}{SV^{0.68}} \; d_p^{\,0.18} \; \mu_L^{\,-0.05} \; \left(\frac{\sigma_w}{\sigma_L}\right)^{0.2} \qquad (15)$$

In these expressions C_i and C_o are the inlet and outlet concentrations of the key component; k, the first order kinetic constant; SV, space velocity; W, a constant; Z, the catalyst bed length; d_p, the catalyst particle diameter; μ_L, liquid viscosity and σ_w, σ_L, the surface tension of water and the liquid.

The high velocity expression has, because of the exponential, a saturation characteristic with increasing velocity, U, which is consistent with the experimental observations (see Section 2.1.1).

The low mass velocity expression suggests that for a given reactor length the logarithm of the ratio of the inlet to the outlet concentration of the key component should be inversely proportional to the 0.68 power of the space velocity. This, however, was not confirmed by subsequent studies by Paraskos et al [54], and Montagna et al [50], [51]). It does represent the data of Henry and Gilbert [23]. Hence, it can be concluded that the exponent on the bed length and space velocity varies with the feedstock and perhaps catalyst activity.

Gianetto et al [18] suggested that a more general expression would be one in which the exponents for the length and space velocity were treated as fitting parameters as shown in the equation below:

$$\ln \left[\frac{c_i}{c_o}\right] \; \alpha \; \frac{Z^m}{(SV)^{1-m}} \tag{16}$$

The value of zero for m would indicate there is no liquid velocity effect. Increasing values of m would indicate greater velocity effects on the conversion with a value of m=1 describing a situation in which the catalyst efficiency is proportional to the liquid velocity.

3.3 Partial Wetting/Mass Transfer/Catalyst Effectiveness Scaleup Model

This model type requires a good knowledge of the wetting efficiency of the reactor. It also requires the use of mass transfer coefficients which do not have a wetting factor in them. Wetting efficiencies are usually obtained from tracer experiments which in themselves have model assumptions imbedded in them. Also, very few mass transfer coefficients are available in which the interfacial area and mass transfer coefficient are measured independently. Hence, though sophisticated, at this time these models do not have very strong predictive capabilities. However, they do give a scaling structure for situations where the wetting efficiencies can be controlled on scaleup.

The basic premise of this model as proposed by Mills, Baudry and Dudukovic [48] is that the activity of a catalyst bed in trickle flow is a function of the intrinsic catalyst activity moderated by an overall efficiency of the trickle bed as shown below,

$$R_{TB} = \eta_{TB} \, R_i \tag{17}$$

$$\eta_{TB} = \frac{\eta_{CE}}{\frac{1}{\eta} + \frac{\Lambda^2}{Bi_{wo}}} + \frac{1 - \eta_{CE}}{\frac{1}{\eta} + \frac{\Lambda^2}{Bi_d}} \qquad (18)$$

where

R_{TB} = apparent trickle bed reaction rate

R_i = intrinsic reaction rate of the catalyst

η_{TB} = overall efficiency of a trickle bed.

η_{CE} = catalyst wetting efficiency

η = effectiveness factor of totally wetted catalyst.

Λ = $(V_p/S_e)\,(k_v/D_e)^{1/2}$

$Bi_{d,wo}$ = Biot number of the dry and wet catalyst, k_s (or k_L) $V_p/D_e\,S_e$

D_e = effective pore diffusivity of key reactant within the catalyst

k_v = first order volumetric kinetic constant

k_L, k_s = gas/liquid and gas/solid mass transfer coefficients

V_p/S_e = particle volume/wetted area

In this formulation, it is assumed that the overall reaction rate is the sum of the reactions that are occurring on the wetted and non-wetted portions of the catalyst pellet surface. Over the wetted surface, the reactants reach the surface of the catalyst through a liquid film which may be under mass transfer control. The reactants can also reach the surface directly through the gas phase to the unwetted proportion of the catalyst through a possible mass transfer limitation in the gas phase. This model simplifies to all the expected forms. For example, for the case of no mass transfer control but with moderate pore diffusion control, the formulation for η_{TB} becomes equal to $\eta_{CE}\,\eta$.

Mills, et al [48] tested this model with the hydrogenation of α-methylstyrene dissolved in n-hexane. A reactor 1.9 cm in diameter by 21 cm long was used packed with 1.6 mm cylindrical catalyst particles. The catalyst consisted of a 5% palladium on alumina catalyst. The effectiveness factor of the catalyst was obtained in a stirred tank slurry reactor and in a stirred basket reactor. Liquid superficial mass velocities of 0.14 to 2.4 kg/m^2s

and gas superficial mass velocities of hydrogen saturated with hexane vapor of 0.24 to 0.49 kg/m^2s were used. The catalyst wetting efficiency was calculated on the basis of dynamic tracer tests from a previous study. None of the available mass transfer correlations gave an appropriate fit to the data, however, using the form of a mass transfer correlation, a consistent set of fitted parameters were found that adequately represented the data for cocurrent downflow and upflow, and countercurrent flow.

A modified form of this model adapted by Zhenglu, Han-Yu and Smith [89] for a non-volatile reactant is shown below:

$$\eta_o = \frac{\frac{3}{\phi^1}\left(\frac{3}{\tanh \phi^1} - \frac{1}{\phi^1}\right)}{1 + \frac{\phi}{NU_{LS}}\left(\frac{3}{\tanh \phi^1} - \frac{1}{\phi^1}\right)} \qquad (19)$$

where

η_o = overall effectiveness factor

$\phi^1 = \frac{\phi}{f}$

$\phi = r_p\,(k_\nu/D_e)^{1/2}$

f = wetting efficiency

$NU_{LS} = \varepsilon_p\, k_s\, r_p/D_e$

ε_p = catalyst void fraction

The basis for this model is the same as the previous one. Only the term dealing with the wetted portion of the catalyst is handled and the Thiele modulus, Φ, is modified by the wetting efficiency, f, as suggested by Dudukovic [13]. Calculations using this model compare well with the overall effectiveness factors calculated using the cubic model of Herskowitz [24] as shown by Figure 3.3.1 for two values of the Nusselt number. These figures show that under severe pore diffusion control, i.e., large values of ϕ, the overall effectiveness factor is proportional to the wetting efficiency. For low values of ϕ, approaching no pore diffusion limitations, the overall effectiveness factor is unaffected by the wetting efficiency. Hence, this model predicts that liquid velocity effect should be most pronounced for high activity pore diffusion limited reactions, as experimentally observed.

The partial wetting model was adapted by Lee and Smith [40] to yield criteria for determining conditions under which negligible transport or partial wetting would obtain. To assure no pore diffusion limitations, it was found that the conventional criteria that the effectiveness factor, η, should approach unity also applies for potentially partially wetted systems. For negligible transport limitations, it was proposed that the following inequality should hold:

$$\left(\frac{R\ d_e\ \rho_p}{k_L\ C_L\ \varepsilon_P}\right) < \frac{\beta}{1-\beta} \qquad (20)$$

and

$$\beta = \frac{(\overline{M}\ \Delta\ H)\ D_e\ C_L}{\lambda_e\ T} \qquad (21)$$

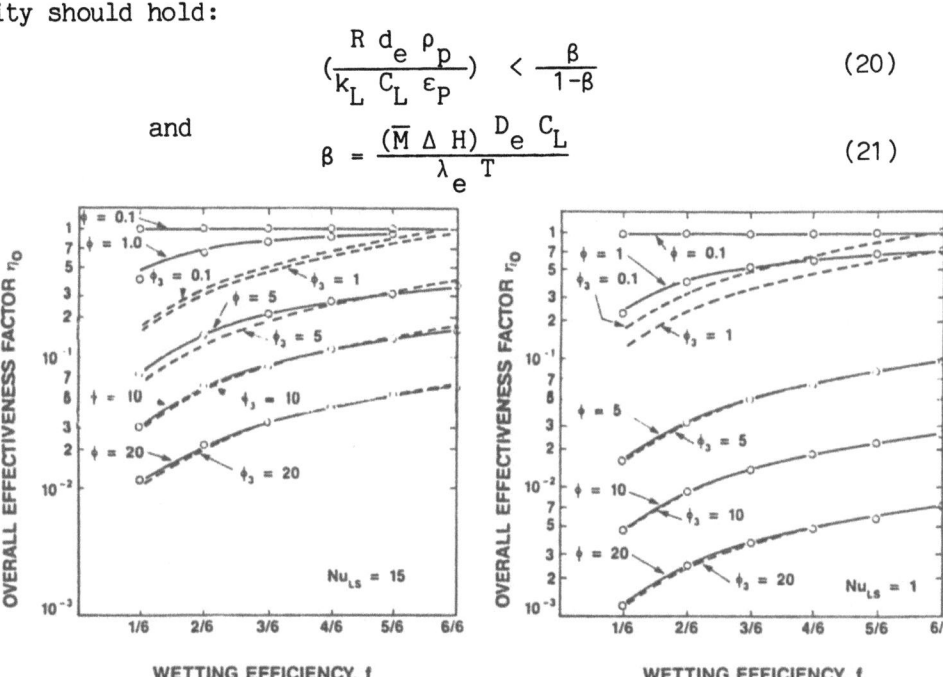

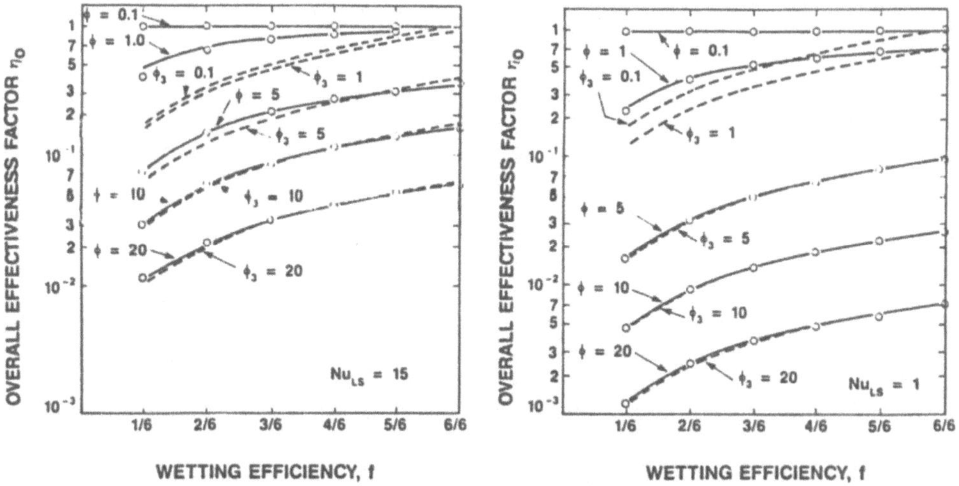

Figure 3.3.1. Partial wetting model predictions for reaction by a non-volatile component for NU_{LS} = 15 and 1. __ cubic model, __ x weighting-factor method, o Zhenglu model NU_{LS} = 15 and 1 (after Zhenglu, P. et al [90]). Reproduced by permission AIChE.

where R is the global reaction rate; d_e, the pellet dimension (volume/surface area); k_L, mass transfer coefficient across the liquid interface; C_L, the gas concentration in the liquid; ρ_p the pellet density; ε_p, the catalyst pellet void fraction; D_e^p, λ_e, the effective diffusivity and thermal conductivity; $-\Delta H$, the exothermic heat of reaction and T, the temperature. This criterion has the same form as a simplified one for isothermal conditions proposed by Satterfield (63), namely:

$$\frac{R\ d_e}{k_L\ C_L} < 0.3 \qquad (22)$$

For confirming partial wetting, Lee and Smith proposed the following inequalities:

$$\eta > \eta_L \text{ when } f < 1 \qquad (23)$$

$$\eta\ \phi^2 = \left(\frac{R\ d_e\ \rho_P}{D_e\ C_L}\right) > \frac{\phi \tanh \phi}{1 + \dfrac{\phi \tanh \phi}{Di}} \qquad (24)$$

$$\frac{R\ d_e\ \rho_p}{k_L\ C_L\ \varepsilon_p} > 1 \quad , \quad \frac{\phi\ \tanh\ \phi}{Bi_{ML}} \gg 1 \tag{25}$$

where

$\phi = d_e\ (\rho_p\ k_s/D_e)^{1/2}$; f is the fraction wetted and $Bi_{ML} = \varepsilon_p\ k_L\ d_e/D_e$

These criteria correctly predict the observation by Herskowitz et al [24] that for the hydrogenation of α-methylstyrene, partial wetting occurs below about 3 cm/s.

Turek [80] using a simplified partial wetting model was able to reconcile measurements from a slurry, two stirred "basket" reactors and a trickle bed. The reaction was glucose hydrogenation.

3.4 The Stagnancy/Catalyst Effectiveness Scaleup Model

This model Koros [35] assumes that the key factor determining the efficiency of catalyst utilization in a trickle-bed reactor is the size of the stagnancies in the catalyst bed. Several tracer studies have established the existence of stagnancies in trickle-beds. These stagnancies are associated with non-flowing liquid holdup; hence the catalyst particles within these stagnancies must be reached by diffusion from the outer periphery of the stagnancy. The exchange between the flowing streams and the stagnancies are quite high Hochman [28], and hence should not be limiting.

Tracer studies of Colombo et al [10] showed that the effective diffusivity of porous particles in low velocity trickle beds were considerably less than that for well-wetted particles at high velocities. They observed that the effective diffusivity of the particles increased monotonically with liquid velocity until a plateau was reached at a liquid velocity which is very close to that obtained for full utilization in other contacting studies (Satterfield, [65]; Koros, [34]). They surmised that the reason for the lowered effective diffusivity is that partially wetted particles or those adjacent to a stagnant zone would not allow ready access of the tracer to the pores of the particle. An alternate explanation for these observations is that the porous catalyst particle is bathed in a stagnant pool of liquid and that, in effect, the lower apparent diffusivity reflects the larger diffusion path into the pores of the clump of particles included in the stagnant zone. Increasing liquid velocity decreases the size of the stagnancy and thereby increases the effective diffusivity.

This extension of the stagnancy model (Koros, [35]) can be used to explain the liquid velocity behavior of catalyst contacting reported by Paraskos et al [54], Montagna et al [51] and Shah et al [52]. The argument goes as follows: Take a flowing stream as shown in Figure 3.4.1 meandering through a trickle bed. At high velocities, the rivulets will penetrate between each of the particles thoroughly and bathe each of the catalyst particles completely giving good access for reaction, i.e., 100% contacting efficiency. As the liquid velocity is lowered, the rivulets no longer bathe each of the particles and instead meander in such a way as to contact the edge of clumps of particles. Within these clumps the liquid is essentially stagnant.

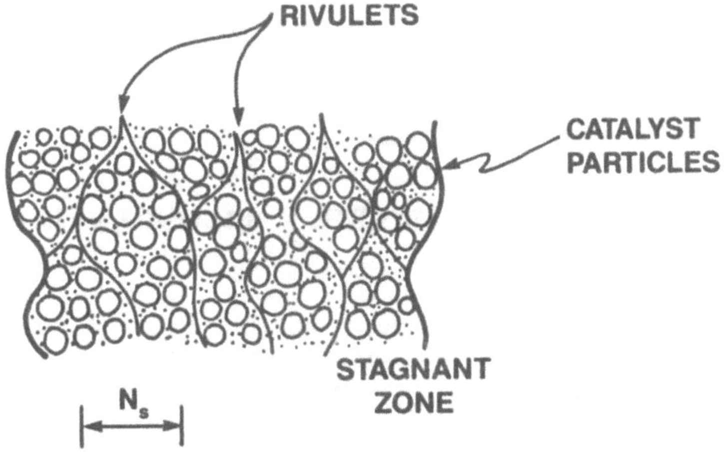

Figure 3.4.1. Stagnant zone model for contacting efficiency in a trickle bed (after Koros, R. M. [35]).

In effect, then, each of these clumps is like a virtual catalyst with a bimodal pore structure; that is, a few large pores interconnecting the particles plus the small pore network inherent in the catalyst particle. The effective diffusivity of such a clump will be a combination of the effective diffusivity of the porous particle and these large "macro-pores". But the effective size of the catalyst particle is no longer the size of the particle but instead is the size of the clump.

Some quantification of the model can be accomplished. With respect to the effect of liquid velocities, it can be assumed that the number of particles, n_s, in a clump will decrease with increasing velocity to the limit of unity for full utilization as shown in Figure 3.4.2. At low velocities, n_s ~3-5 based on various tracer study results (Hochman [28], Colombo [10]). On the basis that contacting and holdup is improved through dilution with small inert particles (Section 2.2), it can be surmised that as the particle size becomes smaller, the rivulet number density becomes larger and hence rivulet contacting of each particle will be enhanced. A sketch of such a functionality is shown in the Figure 3.4.2.

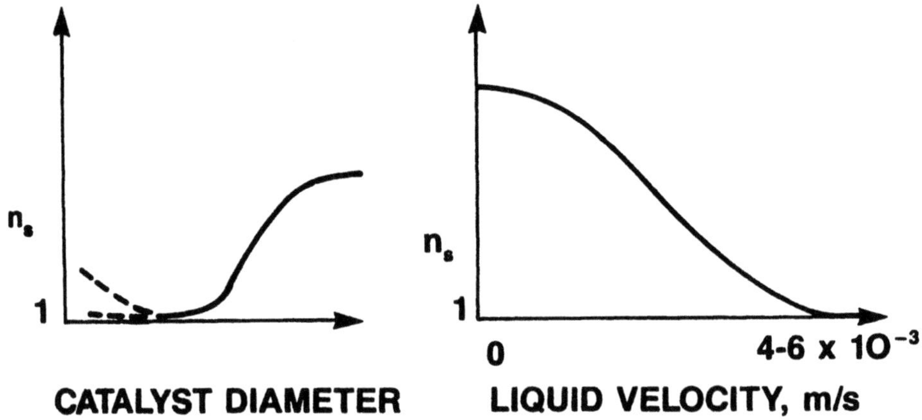

Figure 3.4.2. Conjectural relationships of the stagnant zone size and liquid mass velocity and catalyst diameter for the stagnant zone model for contacting efficiency in a trickle bed (after Koros, R. M. [35]).

These surmised functionalities of clump size with liquid velocity and particle size can then be used to analyse the behavior of catalyst particles in a trickle bed using a modified catalyst Weiz and Prater modulus, ϕ_{MV}. For a clump of particles in the bed,

$$\phi_{MV} = \frac{n_s^2 \, d_p^2}{36 \, D_e} k_{app} \tag{26}$$

where n_s is the number of catalyst particles in a clump; d_p, the catalyst particle diameter; D_e, the effective diffusivity of the catalyst and k_{app}, the apparent first order reaction constant. For $n_s = 3 - 5$ and $\phi_{MV} \ll 1$, there should be no liquid velocity effect on contacting for low liquid velocity reactors. For $\phi_{MV} > 1$, a liquid velocity effect is expected.

Because n_s is a function of liquid velocity and particle size, a catalyst particle which is pore diffusion limited will exhibit a sensitivity to liquid velocity and particle size above and beyond that expected in a single phase reactor. Thus, the model predicts that the sensitivity of the observed reaction rate to particle size for reactions which are diffusion limited will be larger for trickle bed reactors than for single phase reactions. Being that n_s increases with increasing d_p, the limiting relationship of the effectiveness factor will not be d_p^{-1} as for single phase reactions, but instead d_p^{-n} where n is greater than 1. Furthermore, this behavior will be exhibited even for reaction rates low enough so that liquid film mass transfer limitations do not exist. Thus, this model, as discussed by Koros [35], explains the liquid velocity effects found in relatively slow hydrotreating reactions such as reported by Paraskos et al [54] and Montagna et al [50], [51].

In summary, the stagnancy/catalyst effectiveness model predicts that liquid and/or gas velocity effects on the apparent reaction rate will be observed for catalysts which are at least marginally diffusion limited and run in a trickle bed reactor under low velocity conditions. The model predicts that for scaleup of reactions which are diffusion limited or at least marginally so, the pilot plant should be designed to run at elevated velocities which do not show sensitivity to liquid velocity. Conversely, if a pilot reactor is used for providing data for scaleup showing velocity effects, there is a good likelihood that the catalyst suffers from diffusion limitations.

To quantify this model further a more thorough knowledge of the physics of the flow in trickle beds is needed. However, at this time it can be used as a criteria for when to expect a liquid mass velocity effect and as a vehicle for checking the consistency of experimental data.

3.5 Summary of Scaleup Models

There has been an evolution of scaleup models for trickle beds over the past decade. All of the models provide insights into the operation of trickle bed reactors, but none provide an a priori vehicle for scaleup. The models provide means for diagnosing the behavior of pilot plants and to guide their improvement when needed. For liquid phase limited reactions, the models predict that scaleup from small scale on the basis of constant space velocity will generally give conservative results provided the larger unit is equipped with appropriate distributors. For gas-limited reactions, the models predict that the larger units may perform worse than the small units. To further quantify scaleup of trickle beds, a better understanding of the physics of the mixed phase flow in the bed is needed.

4. TROUBLESHOOTING

A very practical application of trickle-bed technology is in the troubleshooting of commercial hydrotreaters. Recently Edgar, Johnson, Pistorius and Varadi [14] summarized the guidelines for troubleshooting malfunctioning hydrotreaters. They used a fault analysis approach. Here we will discuss those elements of their fault analysis that relate to the analysis of the hydrotreater reactor performance. It will be presumed in the discussion that a hydrotreater is malfunctioning either as a recently built new unit or as one that has been operated as per design previously.

4.1 Check Operating Conditions

Normally reactors will be equipped with thermocouples within the catalyst bed and at the inlet and outlet. As a first step in troubleshooting, the temperature profiles should be plotted and checked for consistency with heat and material balance calculations. Likewise, pressure drop and pressure drop profiles should also be checked for consistency versus predictions for the actual conditions of the operation. Pretreatment of the catalyst such as sulfiding should be checked against normal or pilot plant procedures.

Unit startup details and run procedures should be reviewed to assess whether or not reactor conditions occurred which could have damaged the catalyst. For example, extensive reduction of the catalyst prior to sulfiding would be harmful to the activity of the catalyst. Likewise, inadvertent high temperature processing of olefinic material particularly with a fresh active catalyst could cause premature coking and loss of catalyst activity. Generally, catalyst pretreatment performed improperly would be expected to harm the apparent catalyst activity.

4.2 Low Initial Catalyst Activity Due to Poor Flow Distribution During Presulfiding.

One manifestation of hydrotreater malfunctioning is low conversion. This is usually noted on the basis of past history on the feedstock or from scaleup correlations from pilot plant data. Poor presulfiding, as mentioned above, can cause low catalyst activity. However, not only process malfunctions can cause problems in activity but also poor flow distribution during presulfiding.

One evidence of poor flow distribution during presulfiding is premature hydrogen sulfide breakthrough during the sulfiding procedure. This would be evidence of channelling. Additionally, non-uniform bed temperatures during presulfiding would also provide evidence for maldistribution. The sulfiding reaction is exothermic and rapid; therefore, in a properly functioning reactor the temperature wave should advance in the catalyst bed evenly across the entire diameter. As discussed in Section 1.6, radial heat transfer in trickle beds is poor, hence, radial variations in temperature due to maldistributions are expected to be sustained and readily detected.

The H_2S breakthrough should coincide with the completion of the temperature wave front in the reactor. If abnormalities are observed, checks should be made to ascertain that the minimum flow rates to assure good distribution within the bed were in use during the presulfiding procedure. Criteria for these will be discussed below.

4.3 Diagnostics for Hydrodynamic Problems

Hydrodynamic problems in hydrotreaters evidence themselves in several possible ways. Poor contacting can cause low apparent activity even if the presulfiding was done correctly. The reactor can show low temperature response. Also, the bed can exhibit high pressure drop. Lastly, there can be unexpectedly high response of conversion to small flow rate changes. Interpretation of each of these diagnostics will be discussed below.

- Low catalyst activity. Catalyst bypassing due to a defective distributor or due to low liquid velocity could cause low apparent activity. As discussed in Section 3, low catalyst utilization would be expected under these circumstances which would manifest themselves as low apparent catalyst activity.

- Low temperature response. Hydrodesulfurization reactions usually double in rate for about every 20°F increase in temperature. Hence, measurable changes in conversion should be attained for a change of 10°F. If the expected change in conversion does not occur, hydrogen starvation, bypassing or channelling may be the cause. Hydrogen starvation and bypassing can be determined by ascertaining if the correct flows are being brought to the reactor. That is, that there are no inadvertent leaks along the way or malfunctioning meters. Causes for channeling and their remedies will be dealt with below.

- High bed pressure drop. Abnormally high pressure drop can be caused by bed slumping following catalyst breakup or by the use of the wrong size inert bed topping or support. Bed slumping can be caused by poor packing of the bed. Slumping can sometimes occur after pressure surges. Hence, packing procedures and pressure and flow rate records should be investigated for evidences of malfunctions.

Fouling can be from several sources, such as corrosion products from equipment ahead of the reactor, particulate matter being carried in by the feed or by chemical reactions occurring on a catalyst.

Catalyst breakup can occur during shipping, catalyst loading or due to process surges. High pressure drop caused by the use of the wrong size inerts in packing the bed would be due to the migration of the catalyst through the inert support with subsequent plugging of outlet hardware and piping.

- High response to flow rate changes. Unexpected product quality changes in response to small feed rate changes suggest channelling. For instance, changes of feed rates of 25 to 50% can be used for this diagnostic. This diagnostic is useful; however, in most cases, commercial units do not have the flexibility to change flow rates over a wide enough range to use this method to search out and determine that channelling is the true cause for a malfunction.

4.4 Causes for Channelling

The principle causes for channelling are non-uniform catalyst loading, low mass velocity, poor or defective distributor and catalyst fouling. These are discussed below.

- Non-uniform catalyst loading. Catalyst is normally loaded into the reactor by means of a sock or through pneumatic means. If the void fraction of the packed catalyst is not monitored and controlled properly, non-uniform void fractions can be obtained resulting in channelling and poor or variable contacting as several laboratory studies have shown (Section 2).

- Low mass velocity. As suggested by laboratory studies discussed in Section 2, poor catalyst contact of liquid is to be expected for mass velocities below 1.4 $kg/m^2 \cdot s$ (0.2 cm/s). The preferred range is 2-7 $kg/m^2 \cdot s$ (0.3-0.9 cm/s). If during presulfiding or during processing the mass velocity is not in the preferred range, lowered apparent catalyst activity can be observed. In case the flow rates were low during presulfiding the catalyst may be damaged and may require regeneration and resulfiding under proper conditions. Low mass velocities during processing may be due to unplanned turndown conditions or poor distributor design.

- Poor or defective distributor. Particularly in an old unit with new feeds, it is possible that the distributor is hydrodynamically mismatched with the current operation. This should be checked. In addition, distributor tray leaks or tilt could cause malfunctioning of the distributor. Good distributor performance is essential to the operation of the reactor. Laboratory studies have shown that trickle beds fed with non-uniform flows will establish uniform flow with difficulty (Section 1.4).

- Internal obstructions. Studies with exothermic reactions have shown that internal obstructions such as thermowells, supports and other internal hardware that may obstruct the flow can cause channelling as discussed in Section 1.6.

- Catalyst Fouling. Channelling due to catalyst fouling can be caused by non-uniform deposition of foulants, such as by corrosion products surging through the system due to a thermal shock. Defective, poorly placed or designed trash baskets, or overloaded trash baskets on top of the bed can cause non-uniform flow to be initiated from the top of the bed resulting in channelling. As discussed above, pressure surges which cause bed movement can cause bed disturbance, and non-uniform bed heights and void fractions, that will result in channelling.

4.5 Troubleshooting Summary

If it is determined that poor performance of the reactor is tied to sulfiding issues, air burning and restarting is usually advised. If the investigation determines that channelling is the issue, then entrance into the reactors is required for inspection and confirmation of the cause for the channelling. If the causes are not associated with the malfunctions at the top of the bed, the catalyst may have to be removed and reloaded.

5. SUMMARY

Trickle bed reactor technology is widely applied in the petroleum and petrochemical industry. It is a mechanically simple reactor to build and operate. Scale-up is still not well understood, though for typical applications scale-up tends to be forgiving. Several scale-up models have been proposed, but none have been universally accepted in the literature. The physics of the flow through the bed have not been elucidated yet and this is a serious hindrance to the development of definitive design and scale-up models. Phenomenological correlations for pressure drop, mass and heat transfer are available but are generally not in good agreement with each other. Repeatability and reproducibility of these measurements are a serious issue. The literature has several examples now of consistent treatments of experimental data obtained in well stirred and trickle bed reactors, giving confidence that the goal of "a-priori" design of trickle beds based on intrinsic kinetics is reasonable.

NOMENCLATURE

a	=	Area/volume of interfacial contact, m^{-1}
a_c	=	Specific area of packing, m^{-1}
$Bi_{d,wo}$	=	Biot number of the dry and wet catalyst, k_S (or k_L) V_p/D_eS_e, dimensionless
c_i, c_o	=	Inlet and outlet concentrations of key components, kg mole · m^{-3}
C_L	=	Gas concentration in the liquid, kg mole · m^{-3}
d_e	=	Pellet dimension (volume/surface area), m^{-1}
d_p	=	Diameter of catalyst pellet, m
D	=	Molecular diffusivity, $m^2 \cdot s^{-1}$
D_e	=	Effective pore diffusivity of key reactant within the catalyst, $m^2 \cdot s^{-1}$
E_a	=	Axial dispersion coefficient in the bed, $m^2\ s^{-1}$
f	=	Wetting efficiency, dimensionless
F_{LG}	=	Friction factor for gas-liquid flow in the bed, dimensionless
G	=	Gas mass velocity, kg · $m^{-2}\ s^{-1}$
$-\Delta H$	=	Heat of reaction, kjoule · $kgmole^{-1}$
k	=	First order reaction constant, s^{-1}
k_{app}	=	Apparent first order reaction constant, s^{-1}
K_{crit}	=	Value of combination of bed dimensions within which good dispersion of liquid is achieved, $cm^{-0.5}$
k_i	=	First order constant of intrinsic activity, s^{-1}
k_L, k_S	=	Gas/liquid and gas/solid mass transfer coefficients, $m \cdot s^{-1}$
k_v	=	Volumetric first order reaction constant, s^{-1}
L	=	Liquid mass velocity, kg · $m^{-2} \cdot s^{-1}$
l	=	Lateral dispersion coefficient of liquid in a bed, cm
LHSV	=	Liquid hourly space velocity, h^{-1}
m	=	Fitting parameter exponent on bed length for contacting model, dimensionless
n	=	Reaction order, dimensionless
n_s	=	Number of particles in a catalyst clump within a stagnant zone, dimensionless
NU_{LS}	=	Nusselt number, = $(\varepsilon_p\ k_s\ r_p/D_e)$, dimensionless
$(\Delta P/\Delta L)_G$	=	Gas phase bed pressure drop, kPa · m^{-1}
$(\Delta P/\Delta L)_L$	=	Liquid phase bed pressure drop, kPa · m^{-1}
$(\Delta P/\Delta L)_{LG}$	=	Mixed phase pressure drop per unit bed length, kPa · m^{-1}
Pe_L	=	Peclet number for the liquid phase $(= \frac{V\ d_p}{E_a\ \varepsilon_L})$, dimensionless
r_p	=	Radius of catalyst pellet, m
R	=	Reaction rate, kg-moles · $m^{-3}\ s^{-1}$
Re_G	=	Gas Reynolds number (Ud_p/Υ), dimensionless
Re_L	=	Liquid Reynolds number (Ud_p/Υ), dimensionless
R_i	=	Intrinsic reaction rate of the catalyst, kg-mole · $m^{-3} \cdot s^{-1}$

R_p	=	Reaction rate per unit pellet volume, kg-mole · m^{-3} · s^{-1}
R_{TB}	=	Apparent trickle bed reaction rate, kg-mole · m^3 · s^{-1}
Sc	=	Schmidt number (= γ/D), dimensionless
S_e	=	Wetted area of particle, m^2
Sh	=	Sherwood number (= $k_s\ d_p/D$), dimensionless
SV	=	Space velocity, h^{-1}
T	=	Temperature, ^oC
U	=	Superficial gas velocity, m s^{-1}
V	=	Fluid velocity, m · s^{-1}
V_p	=	Particle volume, m^3
W	=	Fitting constant for catalyst wetting model, s · m^{-1}
We_G	=	Weber number ($V^2\ d_p\ \rho_L/\sigma_L$), dimensionless
y	=	Effective radial distance for dispersion from a point source, cm
z	=	Axial distance down the bed necessary to achieve even liquid distribution, cm
Z	=	Catalyst bed length, m

Greek Letters

β	=	Thermal catalyst modulus = $(\overline{M}\ \Delta\ H)\ D_e\ C_L/\lambda_e T)$, dimensionless
λ	=	Physical parameter grouping ratio used in flow maps, dimensionless
λ_e	=	Effective thermal conductivity, W · m^{-1} $^oC^{-1}$
Λ	=	Catalyst diffusion modulus = $(V_p/S_e)\ (k_v/D_e)^{1/2}$, dimensionless
γ	=	Kinematic viscosity, m^2 · s^{-1}
δ_L , δ_G	=	Liquid and gas pressure drop per unit bed length calculated using Ergun equation, kPa · m^{-1}
ε_L	=	Liquid holdup in bed porosity, dimensionless
ε_b	=	Bed void fraction, dimensionless
ε_p	=	Catalyst pellet void fraction, dimensionless
η	=	Effectiveness factor of totally wetted catalyst, dimensionless
η_{CE}	=	Catalyst wetting efficiency, dimensionless
η_o	=	Overall effectiveness factor, dimensionless
η_{TB}	=	Overall efficiency of a trickle bed, dimensionless
μ_w, μ_L	=	Viscosity of water and liquid, Pa · s
ρ_P	=	Pellet density, kg · m^{-3}
ρ_w, ρ_a, ρ_L, ρ_G	=	Densities of water, air, liquid and gas, kg · m^{-3}
σ_w, σ_L	=	Surface tension of water and liquid, mN · m^{-1}
Φ	=	Weisz and Prater Modulus, dimensionless

ϕ_L	=	Mixed phase pressure drop parameter = $[(\Delta P/\Delta L)_{LG}/\delta_G]^{1/2}$, dimensionless
ϕ_{MV}	=	Modified Weiz and Prater Modulus used to test for liquid velocity effects in trickle beds, dimensionless
ψ	=	Physical parameter grouping used in flow maps, dimensionless
χ	=	Mixed phase pressure drop parameter = $(\delta_L/\delta_G)^{1/2}$, dimensionless

REFERENCES

1. Baldi, G., S. Goto, C. K. Chow and J. M. Smith. Catalytic Oxidation of Formic Acid in Water. Ind. Chem. Eng. Proc. Des. Dev. 13 (1974) 447-452.
2. Barkelew, C. H. and B. S. Gambhir. Stability of Trickle Bed Reactors in Chemical and Catalytic Reactor Modeling. M. P. Dudukovic and P. L. Mills, ed. (ACS, 1984).
3. Capra, V., S. Sicardi, A. Gianetto and J. M. Smith. Effect of Liquid Wetting on Catalyst Effectiveness on Trickle-Bed Reactors. Can. J. of Chem. Eng. 60 (1982) 282-288.
4. Charpentier, J.-C., N. Favier. Some Liquid Holdup Experimental Data in Trickle-Bed Reactors for Foaming and Nonfoaming Hydrocarbons. AICHE J. 21 (1975) 1213-1218.
5. Charpentier, J.-C., N. Favier. Hydrodynamics of Two-Phase Flow Through Porous Media. Chemical Engineering of Gas-Liquid-Solid Catalyst Reactions, G. P. L'Homme Editor. Precedings of an International Symposium Held at the University of Liege, March 1978, 78-108.
6. Chou, T. S., F. L. Worley and D. Luss. Ind. Eng. Chem. Process Des. Dev. 16 (1977) 424-427.
7. Chou, T. S., F. L. Worley and D. Luss. Local Particle-Liquid Mass Transfer Fluctuations in Mixed-Phase Cocurrent Downflow Through a Fixed Bed in the Pulsing Regime. Ind. Eng. Chem. (Fundan) 18 (1979) 279-283.
8. Chou, T. S. Liquid Distribution in a Trickle-Bed with Redistribution Screens Placed in the Column. Ind. Eng. Chem. Process Des. Dev. 23 (1984) 501-505.
9. Clements, L. D. and P. C. Schmidt. Paper Presented at AICHE Sixty-Ninth Annual Meeting, November 1976, Chicago Illinois.
10. Colombo, A. J., G. Baldi and S. Sicardi. Chem. Eng. Sci. 31 (1976) 1101-1108.
11. Dean, H. A. and L. Lapidus. A Computational Model for Predicting and Correlating the Behavior of Fixed-Bed Reactors. AICHE J. 6 (1960) 656-663.
12. De Bruijn, A. Testing of HDS Catalysts in Small Trickle-Phase Reactors. Sixth International Congress of Cat. Proc. (Fundan) (1977) 951-964.
13. Dudukovic, M. P. Catalyst Effectiveness Factor and Contacting Efficiency in Trickle-Bed Reactors. AIChE J. 23 (1977) 940-944.
14. Edgar, N. D., A. D. Johnson, J. T. Pistorius, T. Varadi. Troubleshooting Made Easy, Hydrocarbon Processing (May 1984) 65-70.
15. Eigenberger, G. and V. Wegerle. Runaway in an Industrial Hydrogenation Reactor. (7th International Symposium of Chemical Reaction Engineering, Boston, 1982) A.C.S. Symp. Series No. 196, 133-143.

16. Froment, G. and K. B. Bischoff. Chemical Reactor Analysis and Design. (Wiley, New York, 1979).
17. Garcia, W. and J. N. Pazos. Hydrodynamic Effects in Trickle-Bed Laboratory Reactors for Hydrodesulphurization and Hydrodemetallization of Heavy Feeds. Chem Eng. Sci. 37 (1982) 1589-1591.
18. Gianetto, A., G. Baldi, B. Specchia and S. Sicardi. Hydrodynamics and Solid-Liquid Contacting Effectiveness in Trickle-Bed Reactors. AICHE J. 24 (1978) 1087-1104.
19. Goto, S. and J. M. Smith. Trickle-Bed Reactor Performance. Part 2. Reaction Studies. AICHE Journal 21 (1975) 714-720.
20. Goto, S. and J. M. Smith. Performance of Slurry and Trickle-Bed Reactors: Application to Sulfur Dioxide Removal. AICHE J. (1978) 286-293.
21. Goto, S. and K. Mabuchi. Oxidation of Ethanol in Gas-Liquid Cocurrent Upflow and Downflow Reactors. Can. J. Chem. Eng. 62 (1984) 865-869.
22. Hanika, J., V. Vosecky and V. Ruzicka. Dynamic Behavior of the Laboratory Trickle Bed Reactor. Chem Eng. J. 21 (1981) 109-114.
23. Henry, H. C. and J. B. Gilbert. Scale Up of Pilot Plant Data for Catalytic Hydroprocessing. Ind. Eng. Chem. Process Des. Dev. 12 (3) (1973) 328-334.
24. Herskowitz, M., R. G. Carbonell and J. M. Smith. Effectiveness Factors and Mass Transfer in Trickle-Bed Reactors. AICHE J. 25 (2) (1979) 272-282.
25. Herskowitz, M. and J. M. Smith. Trickle-Bed Reactors: A Review. AICHE J. 29 (1983) 1-18.
26. Herskowitz, M., and S. P. Mosseri. Global Rates of Reaction in Trickle-Bed Reactors: Effects of Gas and Liquid Flow Rates. Ind. Eng. Chem. Fundam. 22 (1983) 4-6.
27. Herskowitz, M. and M. Abuelhaija. Liquid-Solid Mass Transfer in a Trickle-Bed Reactor Measured by Means of a Catalytic Reaction. Chem. Eng. Sci. 40 (1985) 631-634.
28. Hochman, J. M. and E. Effron. Two-Phase Cocurrent Downflow in Packed Beds., Ind. Eng. Chem. Fundam. 8 (1969) 63-71.
29. Hoftyzer, P. J. Trans. Inst. Chem. Engrs. (London) 42 (1964) T109-T117.
30. Iannibello, A., S. Marengo and A. Guerci. Performance of a Pilot Trickle-Bed Reactor for Hydrotreating of Petroleum Fractions: Dynamic Analysis. Ind. Eng. Chem. Proc. Dev. 22 (1983) 594-598.
31. Jaffe, S. D. Hot Spot Simulation in Commercial Hydrogenation Processes. Ind. Eng. Process Des. Dev. 15 (1976) 410-416.
32. Kan, K. M. and P. F. Greenfield. A Residence-Time Model for Trickle-Flow Reactors Incorporating Incomplete Mixing in Stagnant Regions. AICHE J. 29 (1) (1983) 123-132.
33. Koros, R. M. and E. J. Nowak. A Diagnostic Test of the Kinetic Regime in a Packed Bed Reactor. Chem. Eng. Sci. 22 (1967) 470.

34. Koros, R. M. Catalyst Utilization and Mix-Phase Fixed Bed Reactors. Proceedings of Fourth International, Sixteenth European Symposium on Chemical Reaction Engineering. Heidleberg, Federal Rep. of Germany (April 6-8, 1976) 372-381.
35. Koros, R. M. Scale-up Considerations for Mixed Phase Catalytic Reactors. Multiphase Chemical Reactors Vol. II - Design Methods. A. E. Rodriguez, J. M. Calo and N. H. Sweed, eds. Sijthuff and Noordhoff (1981).
36. Krambeck, F. J. Computers and Modern Analysis in Reactor Design. The 8th Inf. Symp. on Chem. React. Eng. (EFCE Pub. Series 37 1984).
37. Krishnaswany, S. and J. R. Kittrell. Effective External Diffusion on the Deactivation Rates. AICHE J. 27 (1981) 125-131.
38. Kwanda, K., H. Takuuschi, N. H., Koyana, Kogaku. 32 (1976) 121.
39. Larkins, R. P., R. R. White, and P. W. Jeffrey. Two-Phase Concurrent Flow in Packed Beds. AICHE J. 7 (1961) 231-239.
40. Lee, H. H. and J. M. Smith. Trickle-Bed Reactors: Criteria of Negligible Transport Effects and of Partial Wetting. Chem. Eng. Sci. 37 (2) (1982) 223-227.
41. Levec, J. and J. M. Smith. Oxidation of Acetic Acid Solutions in a Trickle-Bed Reactor. AICHE J. 22 (1) (1976) 159-168.
42. Madon, R. L., J. P. O'Connell and M. Boudart. Catalytic Hydrogenation of Cyclohexene: Part II. Liquid Phase Reaction on Supported Platinum in Gradientless Slurry Reactor. AICHE J. 24 (1978) 904-911.
43. Marangozis, J. Effect of Catalyst Particle Size on Performance of a Trickle-Bed Reactor. Ind. Eng. Chem. Process Des. Dev. 19 (1980) 326-328.
44. Mears, D. E. The Role of Axial Dispersion in Trickle Flow Laboratory Reators. Chem. Eng. Sci. 26 (1971) 1361-1366.
45. Mears, D. E. Role of Liquid Holdup and Effective Wetting in the Performance of Trickle Bed Reactors. ACS Symp. Ser. 133 (1974) 218-227.
46. Midoux, N., M. Favier and J. C. Charpenteier. Flow Pattern, Pressure Loss and Liquid Hold-up Data in Gas-Liquid Downflow Packed Beds with Foaming and Non-foaming Hydrocarbons. J. Chem. Eng. Japan 9 (1976) 350-356.
47. Midoux, N., B. I. Morsi, M. Purwasamita, A. Laurent and J. C. Charpentier. Interfacial Area and Liquid Side Mass Transfer Coefficient in Trickle Bed Reactors Operating with Organic Liquids. Chem. Eng. Sci. 29 (1984) 781-794.
48. Mills, T. L., E. G. Beaudry and M. T. Dudukovic. Comparison and Prediction of Reactor Performance for Packed Beds with Two-Phase Flow: Downflow, Upflow and Countercurrent Flow. I. Kin. E. Symposium Ser. No. 87 (ISCRE 8) (1984) 527-534.

49. Morita, S. and J. M. Smith. Mass Transfer and Contact and Efficiency in a Trickle-Bed Reactor. Ind. Eng. Chem. Fundam. 17 (2) (1978) 113-120.
50. Montagna, A., Y. T. Shah. The Role of Liquid Holdup, Effective Catalyst Wetting, and Backmixing on the Performance of a Trickle-Bed Reactor for Residue Hydrodesulfurization. Ind. Eng. Chem. Proc. Des. Dev. 14 (1975) 479-483.
51. Montagna, A., Y. T. Shah, J. A. Paraskos. Effect of Catalyst Particle Size on Performance of a Trickle-Bed Reactor. Ind. Eng. Chem. Process Des. Dev. 16 (1) (1977) 152-155.
52. Onda, K., H. Takenuchi and Y. Kogama. Effect of Packing Materials on the Wetted Surface Area. Kagaku Kogaku 31 (1967) 126.
53. Papyannakos, N. and J. Marangosis. Kinetics of Catalytic Hydrodesulfurization of a Petroleum Residue in a Batch-Recycle Trickle Bed Reactor. Chem. Eng. Sci. 39 (1984) 1051-1061.
54. Paraskos, J. A., J. A. Frayer and Y. T. Shah. Effect of Holdup, Incomplete Catalyst Wetting and Backmixing During Hydroprocessing in Trickle-Bed Reactors. Ind. Eng. Chem. Process Des. Dev. 14 (1975) 315-322.
55. Pavko, A., D. N. Misic and J. Levec. Kinetics in Three-Phase Reactors. The Chem. Eng. J. 21 (1981) 149-154.
56. Petersen, E. E. Chemical Reaction Analysis, Prentice-Hall, Englewood Cliffs, NJ (1965).
57. Puranik, S. S. and A. Vogelpohl. Effective Interfacial Area in Irrigated Packed Columns. Chem. Eng. Sci. 29 (1974) 501-507.
58. Reiss, L. P. Cocurrent Gas-Liquid Contacting in Packed Volumes. Ind. Eng. Chem. Process Des. Dev. 6 (1967) 486-499.
59. Reuther, K. J., C. Yang, W. Hayduk. Particle Mass Transfer During Cocurrent Downward Gas Liquid Flow in Packed Beds. Ind. Eng. Chem. Proc. Des. Dev. 19 (1980) 103-107.
60. Ross, L. D. Performance of Trickle Bed Reactors. Chem. Eng. Prog. 61 (1965) 77-82.
61. Sato, Y. T., Hirose, F., Takahashi, M. and Y. Toda. Performance of Fixed Bed Catalyst Reactor with Coke Current Gas Liquid Flow. First Pacific Chemical Engineering Congress (Japan) Part 2 (1972) 187-196.
62. Sato, Y. T., Hirose, F., Takahasi, M., Toda, Y., Hashicuhi, J. Flow Patterns and Pulsation Properties of Cocurrent Gas-Liquid Downflow in Packed Beds. Chem. Eng. Japan 6 (1973) 315-319.
63. Satterfield, C. N., A. A. Pelosoff and T. K. Sherwood. Mass Tansfer Limitations in a Trickle-Bed Reactor. AICHE J. 27 (1969) 226-234.
64. Satterfield, C. N. Mass Transfer in Heterogeneous Catalysis. (The MIT Press, Cambridge, 1970).
65. Satterfield, C. N. Trickle-Bed Reactors. AICHE Journal 21 (1975) 209-228.

66. Satterfield, C. N., G. A. Huff, H. G. Stenger, J. L. Carter and R. L. Madon. A Comparison of Fischer-Tropsch Synthesis in a Vapor-Phase Fixed-Bed Reactor and in a Slurry Reactor. AICHE Annual Meeting, San Francisco (1984).
67. Satterfield, C. N., and F. Ozel. Direct Solid-Catalyzed Reaction of a Vapor in an Apparently Completely Wetted Trickle Bed Reactor. AICHE J. 19 (1973) 1259-1261.
68. Sicardi, S., G. Baldi and V. Specchia. Hydrodynamic Models for the Interpretation of the Liquid Flow in Trickle-Bed Reactors. Chem. Eng. Sci. 35 (1980) 1775-1782.
69. Sicardi, S., G. Baldi, A. Gianetto and V. Specchia. Catalyst Areas Wetted By Forming a Semi-Stagnant Liquid and Trickle-Bed Reactors. Chem. Eng. Sci. 35 (1980) 67-73.
70. Specchia, V., and G. Baldi. Pressure Drop and Liquid Holdup for Two-Phase Cocurrent Flow in Packed Beds. Chem. Eng. Sci. 32 (1977) 515-523.
71. Specchia, V., G. Baldi and A. Gianetto. Solid-Liquid Mass Transfer in Cocurrent Two-Phase Flow Through Packed Beds. Ind. Eng. Chem. Proc. Des. Dev. 17 (1978) 362-367.
72. Stanek, B. and J. Hanika. The Effect of Liquid Flow Distribution on Catalytic Hydrogenation of Cyclohexene in an Adiabatic Trickle-Bed Reactor. Chem. Eng. Sci. 37. (1982) 1283-1288.
73. Talmor, E. Two-Phase Downflow Through Catalyst Beds. Part I. Flow Maps. AICHE J. 23 (1977) 868-874.
74. Tan, C. S. and J. M. Smith. Catalyst Particle Effectiveness with Unsymmetrical Boundary Conditions. Chem. Eng. Sci. 35 (1980) 1601-1609.
75. Tarmy, B. L. Reactor Technology in Kirk-Othmer. Encyclopedia of Chemical Technology. Vol. 19, 3rd Ed. (John Wiley, New York, 1982).
76. Tosun, G. A Study of Cocurrent Down Flow of Non-foaming Gas - Liquid Systems in a Packed Bed. 1. Flow Regimes: Search for Generalized Flow Map. Ind. Eng. Can. Process Des. Dev. 23 (1984) 29-35.
77. Tosun, G. A Study of Cocurrent Down Flow of Non-Foaming Gas - Liquid Systems in a Packed Bed. 2. Pressure Drop: Search for a Correlation. Ind. Eng. Can. Process Des. Dev. 23 (1984) 35-39.
78. Tsukamoto, T., S. Morita and J. Okada. Oxidation of Glucose on Immobilized Glucose Oxidase in a Trickle-Bed Reactor: Effect of Liquid-Solid Contacting Efficiency on the Global Rate of Reaction. Ken. Eng. Farm. Bull. 30 (1982) (5) 1539-1549.
79. Tsukamoto, T., S. Morita and J. Okada. Oxidation of Glucose on Immobilized Glucose Oxidase: Trickle-Bed Reactor Performance. (1983).
80. Turek, F., K. Chakrabarti, R. Lange, R. Geike and W. Flock. On The Experimental Study and Scale-up of Three-Phase Reactors. Chem. Eng. Sci. 38 (2) (1983) 275-283.

81. Turek, F.and R. Lange. Mass Transfer in Trickle-Bed Reactors at Low Reynolds Number. Chem. Eng. Sci. 36 (1981) 569-579.
82. Turpin, J. L. and R. L. Huntington. Prediction of Pressure Drop for Two-Phase, Two-Component Concurrent Flow in Packed Beds. AICHE J. 13 (1967) 1196-1202.
83. Van Deemter, J. J. Trickle Hydrodesulfurization - A Case History. Third European Symposium on Chemical Reaction Engineering. (1964) 215-222.
84. Van Dongen, R. H., D. Bode, H. Vanderijk and J. Van Klinken. Hydrodemetallization of Heavy Residual Oils in Laboratory Trickle-Flow Liquid Recycle Reactors. Ind. Eng. Can. Process Des. Dev. 19 (1980) 630-635.
85. Van Klinken, J. and R. H. Van Dongen. Catalyst Dilution for Improved Performance of Laboratory Trickle-Flow Reactors. Can. Eng. Sci. 35 (1980) 59-66.
86. Weekman, V. W., Jr. and J. E. Meyers. Fluid Flow Characteristics of Cocurrent Gas-Liquid Flow in Packed Beds. AICHE J. 10 (1964) 951-957.
87. Wijffels, J., B. Verloop and F. J. Zzuiderweg. Wetting of Catalyst Particles under Trickle-Bed Conditions. Advances in Chemical Science (Third Int. Symp. Chem. React. Eng.) 133 (1974) 151-163.
88. Yoshikawa, N., K. E. Iwai, S. Goto and H. Teshima. Liquid Solid-Mass Transfer and Gas Liquid Coker Flows to Beds of Small Packings. Journal Chem. Eng. (Japan) 14 (1981) 444-450.
89. Zarzycki, R. Inst. Chem. Eng. 12 (1972) 82.
90. Zhunglu, P., F. Han-Yu and J. M. Smith. Trickle-Bed Effectiveness Factors for Liquid Phase Reactants. AICHE J. 30 (1984) 818-820.

MODELLING OF TRICKLE BED REACTORS

A. GIANETTO* and F. BERRUTI**

*Dipartimento di Scienza dei Materiali ed Ingegneria Chimica, Politecnico di Torino, Torino, 10129, Italy.
**Department of Chemical Engineering, University of Waterloo, Waterloo, Ontario, N2L 3G1, Canada.

TBRs OPERATION AND THEIR USE

One of the most common reactors employed to treat continuously three phase systems (gas-liquid-solid) is the Trickle Bed Reactor. It consists of a column with a fixed bed of catalyst particles through which liquid flows in the form of films, droplets and rivulets. Gas moves cocurrently; sometimes counter-current flows are also used. Usually one reactant is introduced in the liquid phase and the other in the gas phase. The cocurrent type of TBR is schematically shown in Fig. 1.

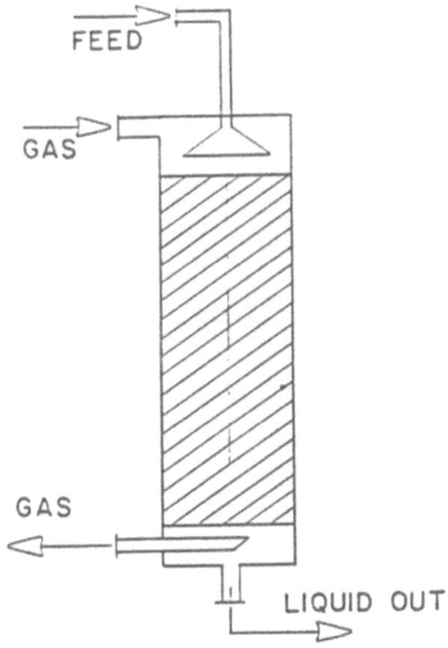

Fig. 1 - Trickle Bed Reactor

In this type of reactor, gas superficial velocity is of the order of 0.1-0.3 m/s which is low enough to avoid mechanical interactions between gas and liquid. The velocity of the liquid, in the range 1 - 8.10^{-3}m/s is still low, but sufficient to guarantee satisfactory external wetting of the catalyst particles. Table 1 shows advantages and disadvantages of Fixed Bed Multiphase Reactors. Table 2 shows the characteristic parameters of TBRs compared to the two other important multiphase reactors: Stirred Slurry and Flooded Fixed Bed Reactor.

Table 1

Advantages and Disadvantages of Fixed Bed Multiphase Reactors

Advantages	Disadvantages
1) Liquid flow approaches piston flow leading to higher conversions for most reactions.	1) Low catalyst effectiveness and selectivity due to the large particle size.
2) Low catalyst loss, permitting the use of costly catalysts.	2) Limitations on the use of viscous or foaming liquids.
3) No moving parts.	3) Risk of reactor obstruction when solid by-products are produced by side reactions.
4) Possibility to operate at high pressure and temperature because of the absence of seals.	4) Strong influence of hydrodynamics on the reactor performance.
5) Large reactor sizes.	5) Reduced range of possible gas and liquid flow rates.
6) Low liquid-solid volume ratio and hence less occurrence of homogeneous side reactions.	6) Incomplete wetting of the catalyst at low liquid flow rates and possible liquid by-pass along the reactor wall.
7) Low investment and operating costs.	7) Sensitivity to thermal effects and problems with temperature control.
8) Low pressure drop and no flooding, for cocurrent, downflow reactors.	8) Difficulties in recovery the heat of reaction.
9) Possibility of operating partially or totally in the vapor phase by varying the liquid flow rate.	

Recently there has been a trend towards higher velocities in downflow systems to increase fluid phase interactions in order to achieve better solid wetting and higher gas-liquid, liquid-solid mass transfer coefficients. Of course this is obtained at the expense of greater energy consumption, so that the choice of operating conditions must be examined for each reactor application.

Table 2

Values of the Characteristic Parameters in Three Phase Reactors

	Stirred Slurry	Trickle Bed	Flooded Fixed Bed
Catalyst loading (% vol)	~0.01	~0.5	~0.5
Liquid Hold up	0.8	0.05-0.2	0.4
Gas Hold up	0.2	0.45-0.3	0.1
Particle size (mm)	0.1	1-5	1-5
Catalyst external area (m^{-1})	500	1000	1000
Catalyst effectiveness factor	1	<1	<1
Gas-liquid interfacial area (m^{-1})	400	200-600	200
Power consumption (W/m^3)	1000	100-1000	100-1000
Maximum pressure (atm)	80	higher pressures possible	
Maximum reactor volume (m^3)	50	200	200

Recently Fixed Bed Reactors with a cocurrent liquid and gas upflow have gained increasing interest. In this case the liquid is entrained by the gas in a system with high interaction between both phases that absolutely cannot be hydrodynamically similar to the usual TBR. An analogy occasionally exists when both systems are operated with high gas and liquid flow rates. Nevertheless upward flow reactors will be considered here too. Table 3 shows criteria for the choice between Upflow and Downflow Fixed Bed Reactors.

It is well known that both Downward and Upward Fixed Bed Reactors work in different hydrodynamic regimes, depending upon the absolute and relative gas and liquid flow rates and the phase properties. For downward systems the following regimes have been observed:

a. Gas continuous flow; b. Rippled flow;	low interphase interaction
c. Pulsed flow; d. Spray flow; e. Dispersed bubble flow f. Foaming flow.	high interphase interaction

Fig. 2 shows a map of the most important hydrodynamic regimes for downward not-foaming systems. Without discussing these regimes, the conditions mostly used in industry are a., b., and occasionally c. for more inviscid systems.

For upward operation, increasing gas and liquid flow rates results in the following regimes:

a'. Submerged system with gas bubbling;
b'. Pulsed flow;
c'. Spray flow;
d'. Foaming flow.

For pulsed and spray flows suitable fluid distributors are required. The most common conditions used industrially are a'. and b'.

The influence of these hydrodynamic regimes on all the characteristic reaction parameters is highly important. As a consequence, plant modelling and design are remarkably conditioned by the regimes, because they affect pressure drop, holdup of the phases and interfacial areas, mass and heat transfer resistances, catalyst wetting, dispersion and back-mixing phenomena, residence time distribution of the phases and segregation.

Notwithstanding the above-mentioned causes of incertitude, Fixed Bed Reactors, originally employed for hydrogenation of various crude petroleum fractions and other refinery streams, have been extended to the treatment of single chemical compounds. Among these petroleum processes it is possible to include, in rough order of their importance in current refining, hydrodesulfurization, hydrofinishing, hydrocracking, hydrodenitrogenation, and hydrodemetalization.

Table 3

Criteria governing the choice between Upflow and Downflow Fixed Bed Multiphase Reactors

a) Larger pressure drops occur in upflow reactors and hence larger variations in partial pressure of the gaseous reactants.
b) Better mixing can be attained in upflow operation: heat exchange is easier, but an increase in axial dispersion is achieved.
c) At low gas and liquid rates, the upflow reactor is similar to a bubbling column, where the liquid is the continuous phase. By contrast, in the downflow reactor, the continuous phase is usually the gas phase and the liquid trickles through the bed.
d) At the same fluid flow rates, the upflow reactor gives higher volumetric gas/liquid mass transfer coefficients generally at the expenses of higher energy consumption.
e) Upflow reactors are characterized by higher liquid hold ups. Larger mass transfer resistance in the liquid and possible homogeneous side reactions must be expected.
f) Better liquid distribution on the cross-section is found for upflow reactors, as well as better heat dissipation and more uniform temperature.
g) In upflow reactors the catalyst must be held in place because at high gas velocities tends to fluidize, whereas in downflow reactors it tends to compact.
h) In upflow operation catalyst surfaces are wetted better than in downflow and catalyst effectiveness is usually increased.
i) In upflow operation slower aging of the catalyst is expected because of the more uniform catalyst wetting.

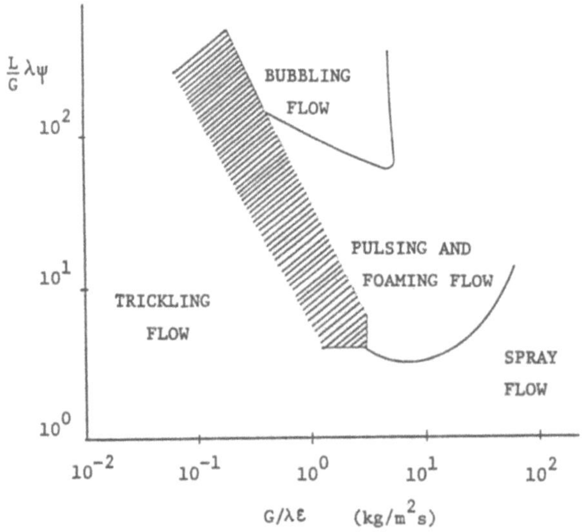

Fig.2 - Map of hydrodynamic regimes in cocurrent downflow TBRs.

The chemical industry employs hydrogenation to produce a variety of industrial chemicals such as cyclohexane, cyclohexanol, cyclohexanone, fatty alcohols, sorbitol or monomers such as hexamethyldiamine or 1,6 hexanediol. A number of new hydrogenation processes, such as coal liquefaction and synthetic crude production from bitumen are now undergoing pilot and even demonstration scale developments.

To go a little more into detail in the description of TBR operation, the first point to be considered is that the system is characterized by several mass and heat transfer steps.

Fig. 3 and 4 illustrate the mass transfer phenomena and show reactant and product concentration profiles in a Slurry and a Trickle Bed Reactor, respectively. The reaction is solid catalyzed between components A and B fed with the gas and the liquid, respectively.

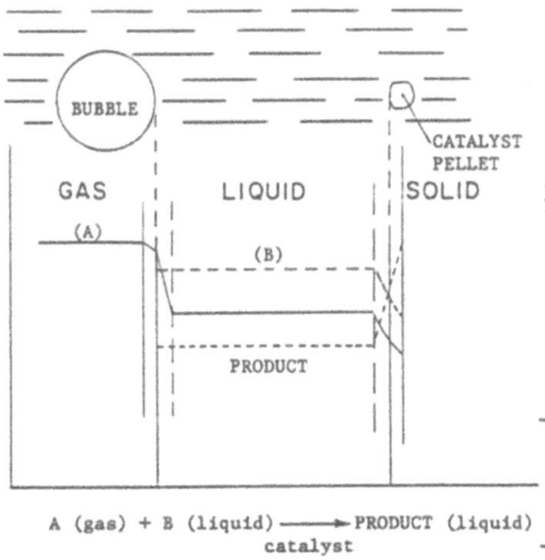

Fig. 3 - Concentration profiles in a Slurry Reactor

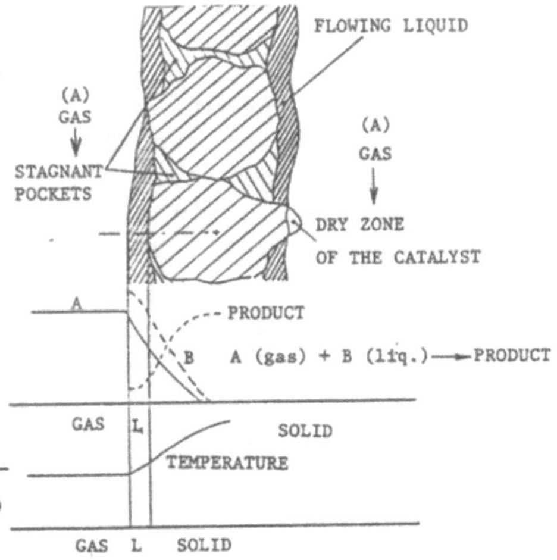

Fig. 4 - Concentration and temperature profiles in a Trickle Bed Reactor

Reactant A is first absorbed into the liquid phase and then reacts on the catalyst surface with reactant B already present in the liquid. When the catalyst is porous, both dissolved reactants diffuse into the pores, towards the center of the catalyst particle to reach the active internal sites of the solid. Reaction products diffuse in the opposite direction. In the case of TBRs, when the external wetting of the catalyst is incomplete, A can be directly absorbed in the liquid that fills the catalyst pores by capillarity. In this case less external mass transfer resistence is expected because the liquid film is absent.

Sometimes catalyst pores not covered by the liquid film are not filled with liquid due to evaporation phenomena caused by an excess of heat of reaction. If both reactants A and B have an appreciable vapor pressure at the working conditions, they can react as gaseous reactants after diffusing inside the catalyst pores. In this case inter-particle diffusion resistance is strongly reduced (at least 10 times) and the reaction rate can be very fast. These last phenomena are not significant in Slurry Reactors where the catalyst particles are always completely wetted and are much smaller - 0.1 mm (equivalent diameter) instead of 10-50 mm as in TBRs.

The mode of operation of Fixed Bed Reactors is strongly dependent upon many parameters, the main ones being flow regimes, geometry, feeding devices, reactant physical and chemical properties product and possibly solvent properties. These incertitudes affect reactor modelling and consequently design procedures.

HYDRODYNAMIC MODELS

Many models have been proposed to simulate the above-mentioned behaviour, beginning with hydrodynamics. Without going into too much detail on hydrodynamics, we will try to give more space to modelling phenomena involving the chemical reaction. A general discussion on gas and liquid flow rates and their consequences seems all that is necessary.

The first hydrodynamic problem concerns the pressure drop which affects both energy dissipation and mass transfer. Fig. 5 shows the influence of gas mass velocity on pressure drop. Empirical approaches have been used to correlate two-phase pressure drop with gas and liquid flow rates.

In the first approach, called the Lockart-Martinelli method (Sato et al. [65]), the two-phase pressure drop δ_{LG} is correlated in terms of a pressure drop enhancement with respect to single phase systems δ_L or δ_G. The relationship is:

$$\delta_{LG} = \Phi_L^2 \, \delta_L \quad (1)$$

where:

$$\Phi_L = f(\chi) \quad (2)$$

and:

$$\chi^2 = \left(\frac{\delta_L}{\delta_G}\right) \quad (3)$$

Fig. 6 shows $\Phi_L = f(\chi)$

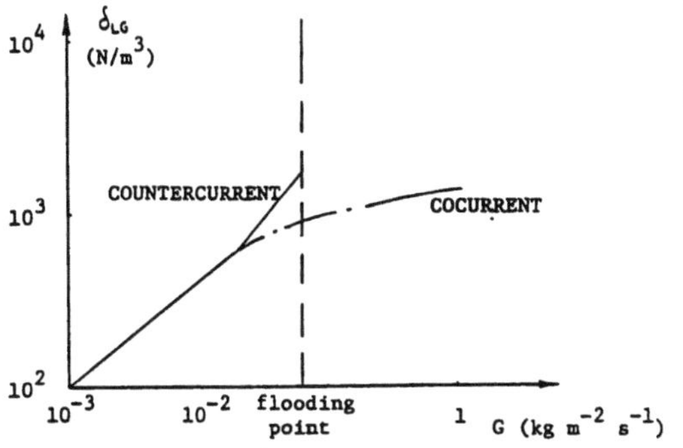

Fig. 5 - Influence of the gas flowrate on two-phase pressure drop in cocurrent and countercurrent devices

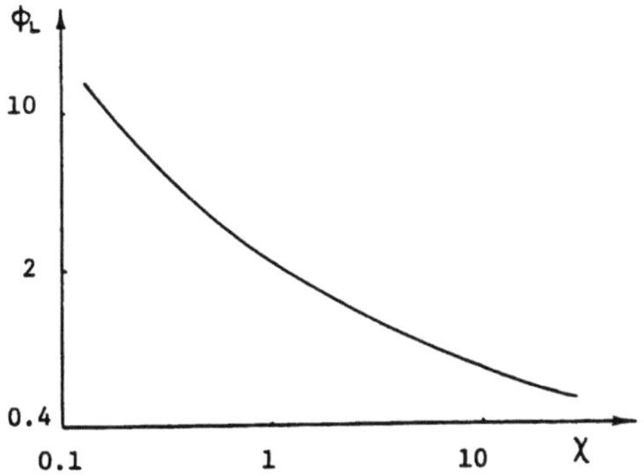

Fig. 6 - Two-phase pressure drop

An alternative approach based on gas phase flow resistence (Specchia and Baldi [78]) is valid for low interaction regimes between gas and liquid. A two phase friction factor which allows for an increase in liquid hold up with liquid rate, is defined according to an Ergun-type equation:

$$\delta_{LG} = k_1 \frac{\left(1-\varepsilon(1-\beta_S-\beta_D)\right)^2}{\varepsilon^3 \, (1-\beta_S-\beta_D)^3} \mu_G \, v_{OG} + k_2 \frac{1-\varepsilon(1-\beta_S-\beta_D)}{\varepsilon^3 (1-\beta_S-\beta_D)^3} \rho_G \, v_{OG}^2 \quad (4)$$

Constants k_1 and k_2 are dependent on packing size and shape and can be obtained by measuring δ_G for wetted packing (with static hold up) but without liquid flow. The equation is not valid for high intraction regimes.

According to Turpin and Huntington [91], the friction factor f_{LG} is an empirically determined, logarithmic function of the factor Z. The relationships are:

$$\delta_{LG} = f_{LG} \frac{2\rho_G v_{OG}^2}{d_\rho} \quad (5)$$

where: $\ln f_{LG} = 8.0 - 1.12\,(\ln Z) - 0.0739\,(\ln Z)^2 + 0.0152(\ln Z)^3 \quad 0.3 \leq Z \leq 500$ (6)

and
$$Z = \left(\frac{Re_G^{1.167}}{Re_L^{0.767}}\right)\left(\frac{\mu_{H_2O}}{\mu_L}\right)^{0.9} \qquad (7)$$

Another alternative approach correlates δ_{LG} with the energy dissipation (valid only for high interaction regimes):

$$\delta_{LG} = \frac{\varepsilon}{v_{OL} + v_{OG}}\left[E + A(\beta_G, L, v_{OG})\right]\left[1 + \beta_G\left(\frac{\rho_G}{\rho_L}\left(\frac{v_{OG}}{v_{OL}}\right)^3 - 1\right)\right] \qquad (8)$$

where: $E = \left[\frac{v_{OL}}{(1-\beta_G)}\right]\delta_L$ = rate of energy dissipation per unit of free volume for single phase liquid flow (9)

The above models are limited for particles with a diameter less than 3 mm; foaming, viscous or non-coalescing liquids behave differently.

Other correlations have been proposed by Midoux et al. [50].

Liquid holdup and pressure drop are strictly connected. The pressure drop, in fact, depends upon the size and the tortuosity of the bed channels in which the gas phase is flowing, and liquid holdup influences the size of these channels and their tortuosity.

For a bed of porous particles, the total liquid holdup can be represented as the sum of partial holdups:

$$\varepsilon_t = \varepsilon_{ext} + \varepsilon_{in} = \beta_t\ \varepsilon = \varepsilon(\beta_{in} + \beta_{ext}) \qquad (10)$$

where ε_t is measured by weighing or by dynamic techniques (tracer injections), ε is the void volume of the column, β_{in} is the internal liquid saturation (inside the particles), which is, in the absence of thermal effects, approximately equal to unity due to capillarity.

β_{ext} is the external liquid saturation (outside the particles) and can be represented as

$$\beta_{ext} = \beta_r + \beta_{fd} \qquad (11)$$

where β_r is the residual liquid saturation, or the liquid fraction remaining in the bed after its complete draining, and β_{fd} represents the free-draining or operating liquid saturation, being the liquid fraction that can be collected by draining the reactor after a quick shutting down of the liquid feed. Fig. 7 shows the behavior of β_{ext} as a function of the gas superficial velocity.

For the residual liquid saturation β_r a graphic correlation proposed by Charpentier [11] is available as a function of the Eötvos number

$$E\ddot{o} = \frac{\rho_L g d_p^2}{\sigma_L}$$

as shown in Fig. 8.

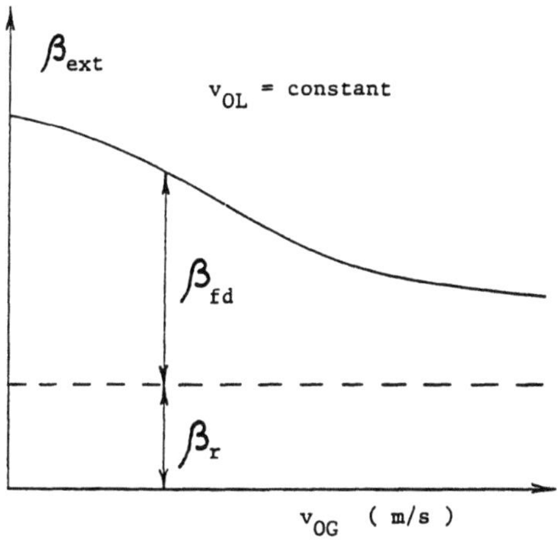

Fig. 7 - Influence of the gas superficial velocity on the external liquid saturation

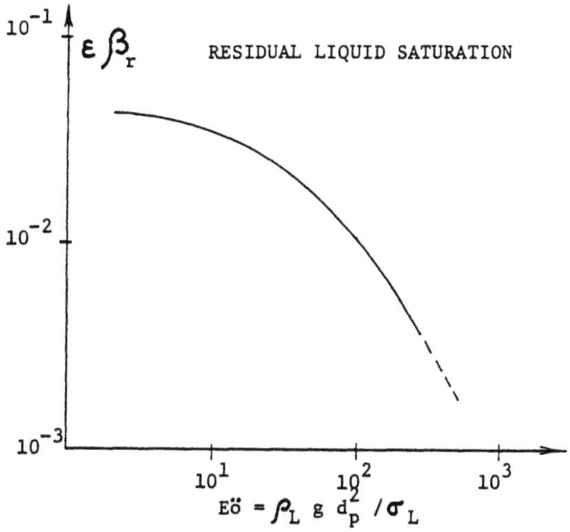

Fig. 8 - Correlation for the residual liquid saturation (Charpentier et al. (11))

Correlations for β_{fd} are available in terms of dimensionless numbers:

Downflow (low interaction regimes):

$$\beta_{fd} = 3.86 \, Re_L^{0.545} \, Ga^{*-0.42} \left(\frac{a_L d_p}{\varepsilon}\right)^{0.65} \tag{12}$$

where Ga* is a modified Galileo number, because the gravity term is replaced by gravity plus pressure drop along the reactor, responsible for liquid drainage. $(a_L d_p/\varepsilon)$ is a dimensionless group which takes into consideration particle shape and size.

Downflow (high interaction regimes):

$$\beta_{fd} = 0.125 \, (Z/\psi^{1.1})^{-0.312} \, (a_L \, d_p/\varepsilon)^{0.65} \tag{13}$$

where

$$\psi = \frac{\sigma_{H_2O}}{\sigma_L} \left[\frac{\mu_L}{\mu_{H_2O}} \left(\frac{\rho_{H_2O}}{\rho_L}\right)^2\right]^{0.33}$$

The external liquid saturation β_{ext} can be represented, for an upflow system, as:

$$\beta_{ext} = 0.6\ \chi^{0.16}\ a_S^{0.33} \qquad (14)$$

where χ is the Lockart-Martinelli parameter.

Many other correlations have been proposed attaining more or less their purpose, also for foaming and viscous liquids, but further research in this field might be very useful. In some of the correlations another geometric factor (d_p/D_c) - ratio of particle to column diameter - is considered because it affects the changing limits of the hydrodynamic regimes.

A general presentation dealing with modelling of pressure drop and holdup in different regimes is also available (Specchia [84]).

Dispersion is another important parameter that can influence the reaction conversion. It is expressed as the Bodenstein number (measured via tracer tests interpreted with hydrodynamic models: PD or PDE) which is correlated as a function of the operating variables of a TBR. However, the relation depends upon the model used. Fig. 9 shows liquid Bo for trickle operation according to two different models. Generally Bo decreases with increasing gas velocity and increases with the liquid velocity.

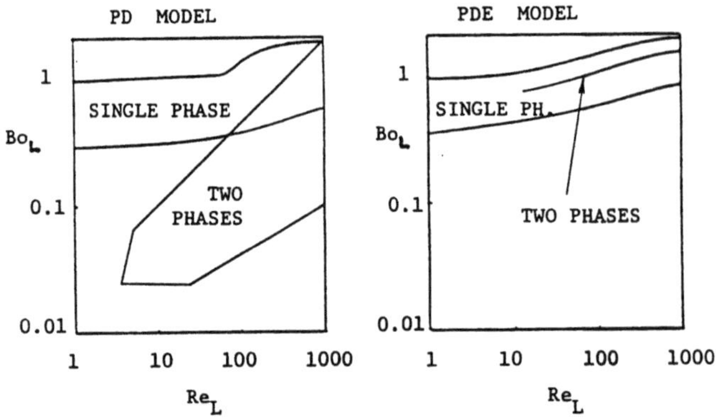

Fig. 9 - Bo_L number for Trickle Bed operation versus Re_L number according to PD and PDE models [9]

The correlations according to the PD model are:

for low interaction regimes:

$$Bo_L = 0.45\ Fr_L^{0.27} \quad \text{(downflow)} \tag{15}$$

$$Bo_L = 2.13\ 10^{-3}\ \varepsilon_{Lt}\left(\frac{Re_L}{\varepsilon_{Gt}\ d_p^{3.3}}\right)^{0.736} \quad \text{(upflow)} \tag{16}$$

for high interaction regimes (upflow and downflow):

$$Bo_L = 0.93\ \exp\left(\frac{-237}{Re_L}\right) \tag{17}$$

A review of hydrodynamic models is reported in a paper by Gianetto et al. [16], and in a lecture of H. Hofmann [34].

The models considered can be classified as follows:

A - Continuous models: each phase is regarded as a continuum. Partial differential equations provide balances for each phase: a total balance, a balance for each component, a balance for heat and possibly for mechanical energy.

B - Stagewise models: packing with its voids is regarded as a series of cells. Ordinary differential equations are used for mass and heat balances.

C - Distribution models: flow in the packing trickles according to a probabilistic law. Probability distributions are involved in the balance equations.

Some examples:

A - <u>Continuous model</u>: Piston, Dispersion, Exchange Model (PDE). Fig. 10 represents schematically this model.

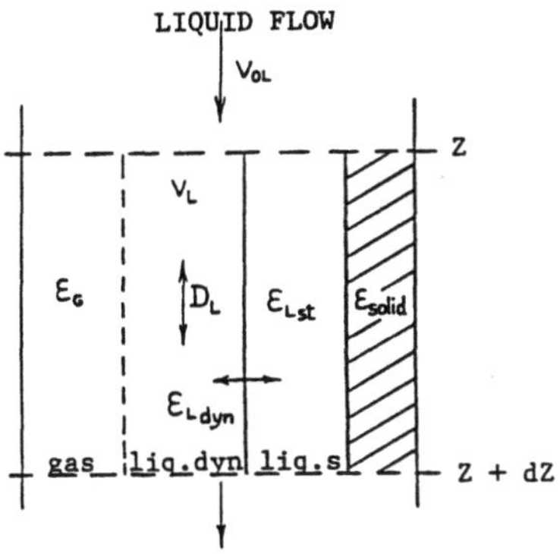

Fig. 10 - Piston, Dispersion, Exchange (PDE) model [34]

Three mechanisms are considered and three parameter values are chosen to characterize the system:

a) Dispersion coefficient D_L

b) Mean residence time τ_L

c) Percentage of stagnant region ϕ

Characteristic numbers are:

$$Bo = \frac{D_L \varepsilon_{Ldyn}}{v_{OL} d_p} \tag{18}$$

$$\theta = \frac{t}{\tau_L} \tag{19}$$

$$\phi = \frac{\varepsilon_{Lst}}{\varepsilon_{Ldyn} + \varepsilon_{Lst}} \tag{20}$$

The characterizing equations are:

$$\left[(1-\phi)D_L\frac{\partial^2 C_{dyn}}{\partial z^2} - (1-\phi)v_{OL}\frac{\partial C_{dyn}}{\partial z} - \phi\frac{\partial C_{st}}{\partial t} - (1-\phi)\frac{\partial C_{dyn}}{\partial t}\right] = 0 \quad \text{dynamic)} \tag{21}$$

$$\phi\frac{\partial C_{st}}{\partial t} = \phi\, k\, a_S\, (C_{dyn} - C_{st}) \qquad \text{(static)} \tag{22}$$

with the boundary values for pulse injection of tracer:

$$C_{dyn} = 0\,;\, C_{st} = 0 \text{ for } t<0 \text{ and all } z \tag{23}$$

$$C_{dyn} = \delta(t)\,;\, C_{st} = 0 \text{ for } t=0 \text{ and } z=0 \tag{24}$$

$$C_{dyn} = 0\,;\, C_{st} = 0 \text{ for all } t \text{ and } z=\infty \tag{25}$$

A solution can be obtained using Laplace transformation. However this model can be simplified under the following conditions:

* Stagnant holdup ε_{Lst} may be ignored in the case of high interaction regimes (i.e. PDE $\rightarrow$ PD).
* Axial dispersion is negligible for reactions where:

$$\frac{Z}{d_p} > \frac{20\, n}{Bo} \ln \frac{C_o}{C} \tag{26}$$

with n = order of reaction.
* Axial dispersion may be ignored in the presence of high phases interaction (PD $\rightarrow$ P).
* Wetting, film and pore diffusion should be included by introduction of, for instance, an efficiency factor η .
* The system can be considered as homogeneous.

B - Stagewise Model:

If there is no backmixing, solution stage by stage seems to be the most convenient. If backmixing occurs, dynamic and stagnant contributions of a phase have to be considered, the total number of influencing parameters increases and the stagewise model is less suitable but remains more flexible. The difficulty of finding the parameter values also remains (H. Hofmann [34]). Fig. 11 shows an example of a three parameter cell model for the liquid phase.

The following characteristic numbers are normally involved:

$$E = f(m, \delta_m = \frac{v_m}{t_m l_{m-1}}) \quad \text{= backmixing coefficient} \qquad (27)$$

$$\theta_m = \frac{t_m}{V_T \tau_{tot}} \quad \text{= Dimensionless time} \qquad (28)$$

$$\phi = \frac{V_T}{V_C} \quad \text{= dead volume/ideal mixed volume} \qquad (29)$$

Other coefficients can be considered, i.e. external backmixing, bypass coefficient, etc.

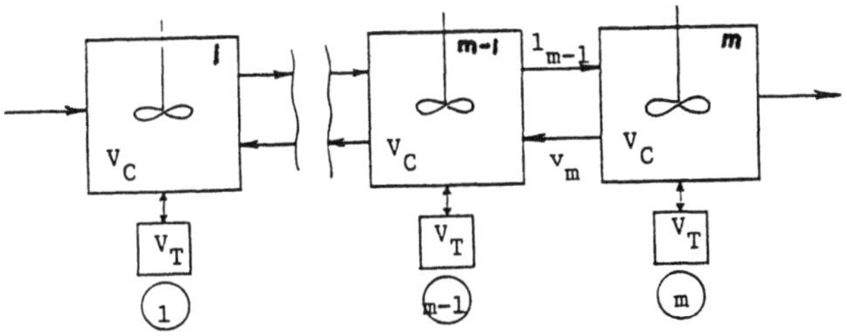

Fig. 11 - Stagewise model [34]

The model equation is:

$$\frac{\partial \overline{C}}{\partial \theta} + \overline{A}\ \overline{C} = \overline{b} \qquad (30)$$

where:

$$\left[\overline{C} = C_1, C_2, C_3, \ldots. C_m\right]^T \qquad (31)$$

$$\overline{b} = \left[m, C_0, \phi_1, \phi_2, \ldots. \phi_m\right]^t \qquad (32)$$

and $\overline{A}$ = matrix containing backmixing coefficients.

T = constant.

$\overline{C}$ = average concentration level.

The solution is available via Laplace transformation (J. Kardos and A. Pulz [39], [40], [41]).

C - Distribution Models

They are useful for trickle flow operation.

a) Time delay model (Buffham and Gibilaro [6]) (Fig. 12).

The model is characterized by three parameters:

τ_m = mean residence time
τ_i = break-through time of tracer
τ_D = mean delay time

Three characteristic numbers are used as well:

$$\frac{\tau_m - \tau_i}{\tau_D} \tag{33}$$

$$\frac{(\tau_m - \tau_i)(t - \tau_i)}{\tau_D^2} \tag{34}$$

$$\frac{\tau_m - t}{\tau_D} \tag{35}$$

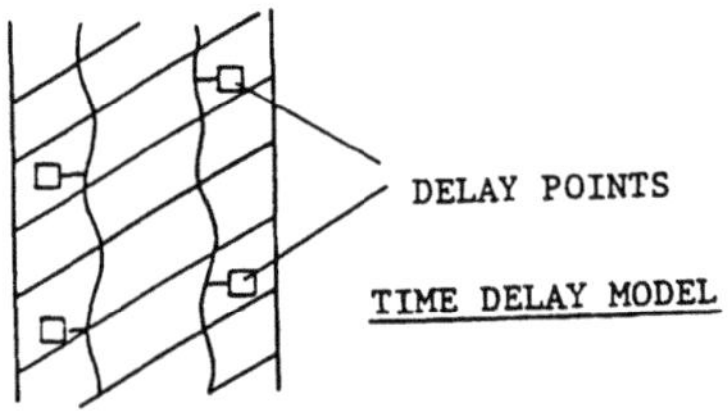

Fig. 12 - Time Delay Model [6]

b) Probabilistic model (R.W. Michell and I.A. Furzer [53]) (Fig. 13)

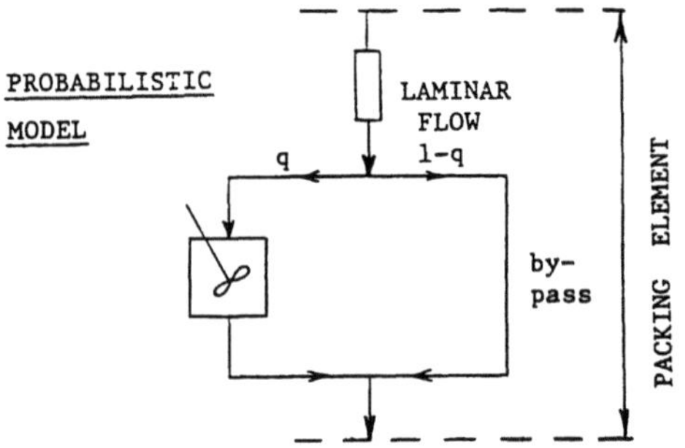

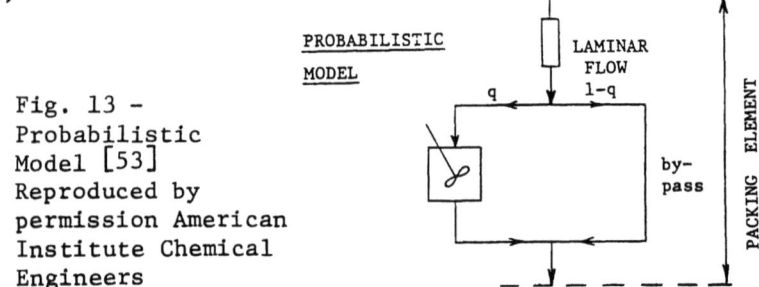

Fig. 13 - Probabilistic Model [53] Reproduced by permission American Institute Chemical Engineers

The probability of a fluid element passing through x = 1, 2, ... n contact cells (perfect mixers) is given by:

$$p(x) = \frac{n!}{x!(n-x)!} q^x (1-q)^{n-x} \tag{36}$$

where: $$q = 0.0187 \, Re_L^{0.32} \tag{37}$$

c) Stochastic models (D.K. Schmalzer and H.F. Hoelschen [71]).

The mixing process is modelled as a walk in velocity states. The basic process of the model is the transition of fluid "packets" among allowed velocity states, characterized by a set of transition probabilities. The observable phenomenon is the spatial distribution of packets, as a projection of the velocity state process.

The model is represented by:

$$U = P^T \underline{U} \tag{38}$$

where

$\underline{U}$ = vector distribution of packets among velocity states
P = matrix of velocity transition probabilities
T = constant.

CONVERSION MODELS

The first general approach to a realistic operative model of TBR consists in a combination of mass balances for all the independent components, one balance for each component for each phase present. The number of balances needed is reduced, of course, in the case of irreversible reactions. Mass balances must take into account methods of feeding the reactor, interphase mass transfer, chemical kinetics, catalyst wetting and effectiveness, longitudinal and radial dispersion, etc.; sometimes part of the data is obtained from the boundary or the initial conditions. When necessary, the model should include a heat balance and also a momentum balance. Balances can be written for the reaction:

$$\nu_A A + \nu_B B \rightarrow \text{PRODUCTS} \tag{39}$$

Assuming A and B are volatile reactants, let us define η_w as the fraction of the catalyst surface wetted by a moving liquid film, and $(1-\eta_w)$ the fraction non-wetted by a liquid layer, η_d , as mentioned before, the fraction non-wetted with dry pores, and finally

$$K_1 = \frac{1}{(\frac{1}{k_1}) + (\frac{1}{k_g})} \quad \text{= overall gas/liquid mass transfer coefficient} \tag{40}$$

Material, heat and momentum balances can be written as follows:

Mass Balance:

Gas Phase:

$$-D_{Ag}\frac{d^2 C_{Ag}}{dz^2} + v_{OG}\frac{dC_{Ag}}{dz} + (k_1 a_1)_A (H_A C_{Ag} - C_{A1}) = 0 \tag{41}$$

$$-D_{Bg}\frac{d^2 C_{Bg}}{dz^2} + v_{OG}\frac{dC_{Bg}}{dz} + (k_1 a_1)_B (H_B C_{Bg} - C_{Bg}) = 0 \tag{42}$$

Liquid Phase:

$$-D_{A1}\frac{d^2 C_{A1}}{dz^2} + v_{OL}\frac{dC_{A1}}{dz} + (k_s a_s)_A (C_{A1} - C_{As}) - (k_1 a_1)_A (H_A C_{Ag} - C_{A1}) = 0 \tag{43}$$

$$-D_{Bl}\frac{d^2C_{Bl}}{dz^2}+v_{OL}\frac{dD_{Bl}}{dz}+(k_sa_s)_B(C_{Bl}-C_{Bs})-(k_la_l)_B(H_BC_{Bg}-C_{Bl})=0 \qquad (44)$$

Solid Phase:

$$\frac{(k_sa_s)_A(C_{Al}-C_{As})}{\nu_A}=R_t=\frac{(k_sa_s)_B(C_{Bl}-C_{Bs})}{\nu_B} \qquad (45)$$

Overall Reaction Rate:

$$R_t=(1-\varepsilon)k_\nu(\eta\eta_wC^m_{As}C^n_{Bs}+\eta(1-\eta_w-\eta_d)C^m_{Al}C^n_{Al}+\eta_dC^m_{Ag}H^m_AC^n_{Bg}H^n_B) \qquad (46)$$

where $(1-\eta_w\ \eta_d)$ is the fraction of the catalyst surface without any liquid film but with pores filled with liquid, in equilibrium with the gas phase.

Heat Balance (Disregarding evaporation or condensation):

$$-\lambda_{eff}\frac{d^2T}{dz^2}+(v_{OL}\rho_LC_{pL}+v_{OG}\rho_GC_{pG})\frac{dT}{dz}=R_t(-\Delta H_R) \qquad (47)$$

This balance will be discussed later.

Momentum Balance:

$$\frac{-dP}{dz}=f(\varepsilon,\psi,d_p,D_C,\rho_L,v_{OL},\rho_G,v_{OG},\mu_L,\sigma_L,\ldots) \qquad (48)$$

Several variables occur in more than one balance equation and generate interactions. For instance, some of the rate parameters such as k_ν are strongly temperature dependent. Consequently, the equations must be solved simultaneously. This is a difficult task because the differential equations are not linear. Fortunately the system can often be simplified or substituted either by applying similarity principles by considering dimensionless groups of parameters as constants during scale up, or by using mock-ups (study of fundamental hydrodynamic and mass transfer phenomena with relatively simple but realistic experimental systems, on a 1:1 geometrical scale). Concerning other simplifications of the system, liquid and gas dispersion, D_ℓ and D_g, and axial effective thermal conductivity λ_{eff} may be neglected in large and deep Trickle Bed Reactors. If one reactant is in large excess, the mass balance equation for that component may be removed from the system, because its concentration will be practically constant throughout the bed and not kinetically controlling the mass transfer and chemical rate phenomena. Good catalyst wetting enables the contacting effectiveness η_w to be taken as unity or close to 1, and to assume η_d equal to zero.

The terms in the balance corresponding to catalyst not irrogated generally vanish. If the gaseous reactant is very soluble, pressure drop through the reactor bed will have little influence on conversion, allowing the momentum balance to be disregarded. For a low intrinsic chemical reaction rate, inter and intraparticle mass transfer phenomena do not affect the apparent reaction rate which is the intrinsic transformation rate in the reactor. Generally speaking, estimation of the apparent reaction rate and identification of permissible simplifications are unambiguous only when just one step, in the sequence of mechanisms governing the reaction, is rate limiting for the range of possible operating conditions.

An analytical solution of the balance equations is possible under severely limiting conditions. Notwithstanding the usefulness of solution simplicity for scale-up purposes, the simplifications needed to obtain an analytical solution are not always fully justified. Errors involved in the development of an analytical solution can nevertheless be compensated for by modifying the same simplified model, or by adding a safety factor.

Homogeneous Model

A rather common analytical solution results from the following simplifying assumptions:
1) liquid phase moving as piston flow; 2) absence of micro and macro thermal gradients; 3) absence of external mass transfer limitations; 4) liquid phase saturated by A (where A is the reactant stoichiometrically in excess); 5) pseudo first order kinetics with respect to B; 6) absence of stagnant liquid pockets (liquid retained by capillarity forces and not moving throughout the bed); 7) complete catalyst wetting; 8) A very soluble.
These are simplifications proposed by Sylvester [86], [87], [88], [89].

The resulting differential equation is as a consequence a simplified balance of B limited to the liquid phase. Integration of this equation with the initial condition:

$$C_B = C_{Bo} \quad \text{at } z = 0 \tag{49a}$$

leads to the so-called "pseudo homogeneous model":

$$-\ln \frac{C_{Be}}{C_{Bo}} = \frac{Z k_\nu \eta (1-\varepsilon)}{v_{OL}} \tag{49b}$$

where Z is the bed depth, k_ν is the intrinsic rate constant, η is the catalyst effectiveness factor and V_{OL} is the liquid superficial velocity. If the above hypotheses are not completely fulfilled, an apparent rate constant measured under conditions of similitude may be used. In this case the model is the "non ideal pseudo homogeneous model":

$$-\ln \frac{C_{Be}}{C_{Bo}} = \frac{Z k_{app} \eta (1-\varepsilon)}{v_{OL}} \tag{50}$$

The ratio k_{app}/k_ν is defined by Satterfield [67] as the contacting effectiveness. η_c deviates from unity when one of the above assumptions becomes unrealistic. Circumstances for which this occurs are rather difficult to predict. Satterfield has estimated from data taken on commercial hydrotreating units as a function of superficial liquid velocity V_{OL}. η_c was found to depend on the physical and chemical properties of the reactants, internal and external catalyst wetting, hydrodynamic regime, and geometrical conditions. Nevertheless η_c was found to be around 1 when $V_{OL}>$ 4-5 mm/s. For this condition the pseudo homogeneous model seems to be applicable. By taking into account that:

$$\frac{v_{OL}}{Z} = \text{(S.V.) space velocity)} \tag{51}$$

The above equation (50) becomes:

$$-\ln \frac{C_{Be}}{C_{Bo}} = \frac{k_{app}\eta(1-\varepsilon)}{(S.V.)} \tag{52}$$

suitable for a non ideal situation where k_{app} cannot be considered as a true constant. The above equation can be successfully used as a model for scale-up of a variety of hydrotreating systems such as hydrodesulfurization, hydrodenitrogenation, and metals removal. In the model C_B is usually the mole or weight fraction of the compound containing sulfur or nitrogen or metals in the feed. In the scale-up, at constant (S.V.), the same C_{Be}/C_{Bo} ratio will be obtained in the prototype reactor provided $K_{app}\cdot\eta\cdot(1-\varepsilon)$ is kept constant. Except when the chemical rate is so slow that mass transfer is not limiting, $k_{app}\cdot\eta\cdot(1-\varepsilon)$ can be maintained constant only if the hydrodynamics do not vary, that is if $V_{OL}, V_{OL}/V_{OG}, d_p, D_c$, gas and liquid properties, height of the calming zone, all remain the same. It must be concluded that both a constant (S.V.) and the same hydrodynamics must be achieved in scale-up to obtain the same degree of conversion $(C_{Bo}-C_{Be}/C_{Bo})$. Unfortunately this is virtually impossible because the height of the catalyst bed in the pilot and industrial units would have to be about the same and this is not a simple condition, because an industrial reactor consists of a series of beds with a total length of 20-30 m. Therefore the designer is faced with a choice between two options in designing an experimental system to obtain data for modelling: operating at a constant C_{Be}/C_{Bo} with low liquid and gas flow rates (and thus different hydrodynamics), mainly to explore the reaction system when it is kinetically complex, or operating with the same hydrodynamics but reaching very low conversion, to determine where the mass transfer resistances are rate controlling. This choice is a crucial problem in scaling up a reactor system. This can be solved only after determining the chemical kinetics and the rate controlling step for the particular reaction.

Widespread use of pseudo homogeneous models makes it important to examine the sources of errors involved in the use of these assumptions. This detailed discussion is also necessary when the model is used in a less simplified form, because it gives the degree of confidence of the reactor operation in a quantitative manner. The phenomena causing departure from the pseudo homogeneous model are: 1) end effects; 2) axial dispersion; 3) presence of homogeneous reactions; 4) interphase mass transfer limitations; 5) incomplete catalyst wetting; 6) micro and macro thermal gradients (inside or outside catalyst particles).

End Effects and Axial Dispersion

End effects and axial dispersion become unimportant when the catalyst bed is adequately long. For end effects, a discriminating empirical criterion, obtained by combining the relationships of Hoftyzer [35], Onda et al. [58] with the data of Herskovitz [27], Stanek and Kolev [85] and Zarzycki [95], gives:

$$Z > \frac{1}{4} \frac{D_l^2}{d_p^{0.5}} \text{ (cm)} \tag{53}$$

and for axial dispersion the criterion of Mears [47] states that when:

$$\frac{Z}{d_p} > \left[\frac{20mD_l}{v_{OL} d_p} \ln \frac{C_{Be}}{C_{Bo}} \right] \tag{54}$$

axial dispersion can be neglected and the error in the calculation of the reactor conversion is less than 5%. Some data on axial dispersion are shown in Fig. 14 by Charpentier et al. [9] expressed as Bodenstein numbers versus Reynolds numbers. Axial dispersion seems to increase when liquid velocity decreases.

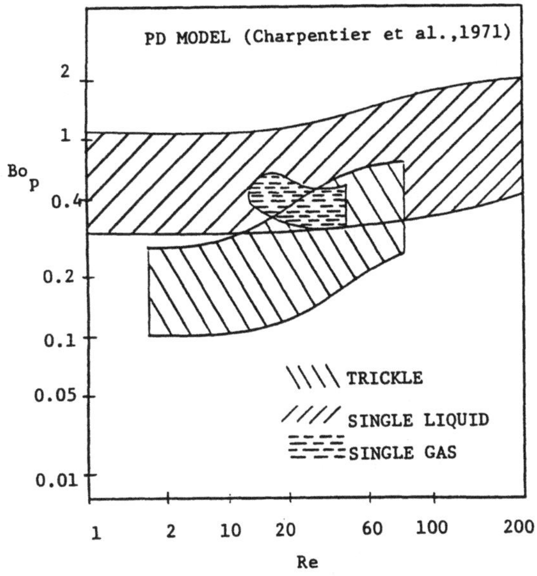

Fig. 14 - Bo versus Re_L for Trickle Beds and single gas or liquid phase reactors (Charpentier et al. [9])

Homogeneous Reactions

Homogeneous reactions can occur in the liquid phase. As an example thermal cracking may accompany certain high temperature catalytic hydrogenation reactions. In this case liquid holdup plays an important role. A quantitative measure of this effect by varying liquid flow rate must be considered with particular caution, because flow rate affects many other phenomena, namely residence time, dispersion, mass and heat transfer, etc. The influence of homogeneous reactions can be tested with inert zones localized or spread out inside the bed, obtained by using inert support. When homogeneous side reactions are negligible, the same selectivity can be maintained in the scale up, by keeping the same gas and liquid velocities, packing size, temperature and especially the same liquid hold up. Otherwise tests in smaller units must be carried out with catalyst diluted with inerts, to compensate for the lower liquid velocity and holdup.

Interphase Mass Transfer

Interphase mass transfer limitation for the apparent kinetic rate is conditioned by hydrodynamics that affects catalyst wetting, interphase turbulence and interphase areas.

Four distinct situations can occur:

1 Limiting reactant A is fed to the reactor in the gas stream.
a) For a wetted catalyst surface assuming no internal gas phase resistance (rather realistic hypothesis):

$$\frac{1}{K_s a_s} = \frac{1}{k_1 a_1} + \frac{1}{k_s a_s} \tag{55}$$

b) For a non wetted catalyst surface, the gas-solid resistance is negligible:

$$\frac{1}{K_s a_s} = \frac{1}{k_g a_s} = 0 \tag{56}$$

Catalyst pores are usually filled with liquid due to capillary forces. If this is not fulfilled due to evaporation caused by the heat of reaction, the mass transfer resistance within the pores dramatically decreases because the effective internal diffusivity is $D_g \cong 100\ D_\ell$ (η increases).

2 Limiting reactant B is fed to the reactor in the liquid stream.
a) For effectively wetted surfaces (with liquid continuously moving), liquid-solid mass transfer is the controlling mechanism:

$$\frac{1}{K_s a_s} = \frac{1}{k_s a_s} \tag{57}$$

b) For a nonwetted catalyst surface, if B is not volatile, there is no mass transfer and no reaction. If B does not have a negligible vapor pressure, the mass transfer resistance is for gas-solid transport but often this resistance is still negligible.

Correlations are available in the literature for areas and mass transfer coefficients (gas to liquid mass transfer correlations have been reviewed by Charpentier [10], Shah [73], Hirose et al. [31], Gianetto et al. [16], [17], [18] and by Specchia et al. [79], and liquid to solid mass transfer correlations by Hirose et al. [32] and Specchia et al. [80].

Unfortunately, the available correlations are not always reliable because data have often been obtained from laboratory units or small pilot plants or from measurements carried out only with water and air. Besides, the operating conditions in experimental devices are not the same as in commercial units and usually are not uniform throughout the apparatus so that the data are averages only. Usually for interacting regimes the volumetric gas-solid mass transfer coefficient is $k_\ell a_\ell > k_s a_s$. Conversely, for the continuous phase regime the above coefficients have the same order of magnitude. If the solubility of A is high, $k_\ell a_\ell$ is not important for modelling the reactor. Criteria are available to evaluate the influence of extraparticle and intraparticle mass transfer on the actual reaction rate. Nevertheless, they are also useful for modelling because they give an immediate idea of the resistance kinetically affecting the system. For instance, as demonstrated by Koros [42], [44], if reactant A is limiting, mass transfer resistance can be neglected when

$$C_{Ag} \cong 0.95\ H\ C_{Ag} = 0.95\ C_{A1} \tag{58}$$

For catalyst supported on spherical pellets and a first order reaction $R_A = k_\nu C_{As}$, the following relation is obtained:

$$\frac{K_s}{d_p k_\nu} > 3 \tag{59}$$

This relation states that external mass transfer is not the limiting step. A similar criterion was suggested by Satterfield [68].

Koros [44] obtained a criterion for neglecting external mass transfer by comparing the Weisz-Prater criterion for the absence of a diffusional limitation within the catalyst as given by Froment and Bishoff [14]:

$$\Phi^2 = \frac{\tau d_p^2 R_A}{D C_1 \varepsilon_i} < 1 \tag{60}$$

and the correlation of Hirose et al. [32], for liquid-solid mass transfer, whose lower conservative limiting value is:

$$\varepsilon\ Sh\ (Sc)^{-1/3} = 1 \tag{61}$$

Hence, the resulting relationship is:

$$(\Phi')^2 = \frac{d_p^2 R_A}{DC_1 \; Sc^{1/3}} \tag{62}$$

which gives an actual measure for neglecting external mass transfer.

Since

$$\frac{\tau}{\varepsilon_i} >> \frac{1}{Sc^{1/3}}, \tag{63}$$

$$(\Phi')^2 < \Phi^2 \tag{64}$$

If $\phi^2 < 1$, in the absence of intraparticle limitations $(\phi^1)^2 < 1$ always i.e., mass transfer resistance is neglected. This is not negligible when $\phi > 10$ (E.F. = 0.1).

Many other semiempirical methods can be reported, like the procedure suggested by Koros and Novak [42] to test the importance of intrapellet or extrapellet mass transfer resistance by varying the active metal content in the catalyst using the same hydrodynamic conditions.

Incomplete Wetting of the Catalyst

Incomplete wetting of the catalyst causes a non-uniform reactant concentration on its surface. Sometimes the pellet may not be only externally but also partially wetted internally as a result of unsatisfactory liquid distribution and of liquid evaporation due to heat of reaction.

When there is no external liquid film, internal zones below the dry surface of the catalyst can be fed with reactant B by liquid diffusion through the internal pore network and with reactant A directly from the gas phase. B can also reach the dry pores via the gas phase if it has a significant vapor pressure under the operating conditions. Many authors have discussed these phenomena and tried to evaluate their effects: Dudukovic [13], Tan and Smith [90], Mills and Dudukovic [51], Goto and Smith [21], Capra et al. [8]. The wetted surface can also be in contact with moving liquid at different velocities giving rise to different local reactant availability. All these phenomena lead to different apparent reaction rates and the effects are in practice included in the earlier discussed ratio k_{app} / k_v, if the pseudo homogeneous model is being used.

Herskovitz et al. [30], Sedriks et al. [72], Satterfield et al. [70] have proposed diagnostic procedures to determine whether or not liquid phase contacting and catalyst wetting may affect the apparent reaction rate. The first procedure, suitable for low conversions of the gas phase reactant, compares the apparent reaction rate for downflow operation, characterized by the trickling flow of the liquid, to that for upflow operation with a continuous liquid phase through a fixed catalyst bed, the gas being preabsorbed in the liquid. The second procedure observes that during start-up without prewetting the solid bed, wetting increases gradually so that it is possible to observe a corresponding variation in the conversion rate. The third diagnostic technique involves determination of the concentration of products in both the gas and the liquid phases. If the product concentration in the gas is larger than that based on equilibrium with the liquid, this indicates that direct transfer to the gas or a gas phase reaction is taking place.

In order to estimate the role of wetting on the apparent rate it is necessary to know the fraction of the total catalyst external surface wetted by the moving liquid. Onda et al. [59] have proposed:

$$f = 1 - \exp\left[-1.36\ Ga_1^{0.05}\ We_1^{0.2}\ \left(\frac{\sigma_c}{\sigma_1}\right)^{0.75}\right] \qquad (65)$$

where:

$$Ga_1 = \frac{d_p^3 g}{\nu_1^2} \qquad We_1 = \frac{v_{OL}^2 \rho_1 d_p}{\sigma_1}$$

and σ_ℓ is the liquid surface tension, while σ_c is the value of the surface tension of the liquid that gives a zero contact angle on the packing. Mills and Dudukovic [52] suggested a similar correlation for f, namely,

$$f = 1 - \exp\left[-2.6\ Re_1^{0.39}\ Fr_1^{-0.317}\ We_1^{-0.137}\left(\frac{a_v d_p}{\varepsilon}\right)^{0.11}\right] \qquad (66)$$

where $Fr_1 = v_{OL}^2/gd_p$ and a_v is the surface area per unit volume of dry packing. Both correlations can be summarized in a useful form for scale up:

$$f = 1 - \exp\left[-\xi v_{OL}^{s}\right] \qquad (67)$$

in which the liquid velocity V_{OL} is coupled with a lumped constant, ξ, replacing the other parameters and S is an empirical exponent. This equation shows that the wetted fraction f approaches 1 as V_{OL} increases. For liquid hydrocarbons Satterfield [67] has shown that $f = 1$ for $v_{OL} > 4.5\ kg/m^2s$.
By assuming that the contact efficiency $\eta_c = k_{app}/k_\nu$ coincides with f and combining an expression for f with Eq. (52). Mears [48] suggested a semiempirical "effective wetting" model to be used for scale up:

$$-\ln\frac{C_{Be}}{C_{Bo}} = (1-\varepsilon)\eta k_\nu (S.V.)^{-0.68} L^{0.32} d_p^{0.16} v_{OL}^{-0.05} \left(\frac{\sigma_c}{\sigma_1}\right)^{0.21} \qquad (68)$$

Other workers have interpreted η_c as a function of other parameters. Henry and Gilbert [26] considered η_c to depend on liquid holdup and proposed an external holdup model:

$$- \ln \frac{C_{Be}}{C_{Bo}} = (1-\varepsilon)\eta k_\nu (S.V.)^{-0.66} L^{0.33} d_p^{-0.66} v_{OL}^{0.33} \qquad (69)$$

Wehner and Wilhelm [94] took into consideration axial dispersion and proposed an axial dispersion model:

$$- \ln \frac{C_{Be}}{C_{Bo}} = (1-\varepsilon)\eta k_\nu (S.V.)^{-1} - Bo^{1} (1-\varepsilon)^2 \eta^2 k_\nu^2 (S.V.)^{-2} \qquad (70)$$

where the Bodenstein number for the liquid phase is $Bo = \frac{D_L}{v_{OL} z}$

Paraskos et al. [61], Montagna and Shah [54] and Montagna et al. [55] used a pilot plant for hydrotreating gas oil to test the various correlations. Log-plots of C_{Be}/C_{Bo} versus $1/(S.V.)$ or versus $\dot{L}$ gave straight lines for demetalization and denitrogenation reactions. On this basis, a generalized equation:

$$- \ln \frac{C_{Be}}{C_{Bo}} = \frac{L^w}{(S.V.)^{1-w}} \qquad (71)$$

where w is a fitted exponent, seems to be acceptable for scale up purposes. Unfortunately, the slopes of the straight lines and thus the exponents depend upon the feed composition, temperature, and catalyst size. Therefore, the equation has only limited use.

Physically, the effective wetting model of Mears seems to be the most realistic. It gives a correlation for the hydrodesulfurization of vacuum and atmospheric residuum and works better than the holdup model proposed by Montagna et al. [55]. These correlations do not take into account axial dispersion that probably is significant in small Trickle Bed Reactors with large catalyst pellets and low liquid flow rates. Montagna and Shah [54] observed that the bed length effect found in desulfurization can be explained by both the axial dispersion and effective wetting models. This suggests that the lumped description of hydrodynamics in Trickle Bed Reactors is inadequate and the complex phenomena taking place cannot be considered with only one parameter in any single model.

Bondi [5] offers another version of lumping intrinsic chemical rates and hydrodynamics:

$$\frac{1}{k_{app}} - \frac{1}{k_\nu} = \frac{A}{v_{OL}^b} \qquad (72)$$

A and b being fitted constants. b is about 2/3 but it varies with the reaction. The uncertainties in the fitted constants for all correlations indicate that a less empirical approach would be fruitful.

The trend is now towards a direct evaluation of contacting efficiency η_c by examining the actual physical phenomena occurring on the particles in the three-phase system, and by taking into account the mechanism by which A and B are transported to the catalyst active sites. The fundamental transport equations shown earlier are the right analytical instrument but a cumbersome computing problem can be created.

One form of lumping, the introduction of an effectiveness factor η, is widely used. The estimation of η can be difficult, especially when wetting is incomplete. To simplify the problem, it is necessarty to introduce adequate models. Two models considered with our co-workers in Torino are useful examples.

1. In the first, Sicardi et al. [75] and Capra et al. [8] assumed that reactants entering the pellet through a fraction f of the external surface, characterized by a certain reactant concentration and wetting, react in the region of the pellet immediately adjacent to that surface fraction, which occupies the same fraction f' of the volume of the pellet. Thus, for a first order reaction, the Thiele moduli for the wetted fractions f', f'' of the external surface area and the corresponding volumes will be:

$$\Phi_{f'} = 3\,\frac{f'V_p}{f'A_p}\sqrt{\frac{k_\nu}{D_{eff}}} = \Phi \quad = \text{conventional Thiele modulus} \tag{73}$$

$$\Phi_{f'} = 3\,\frac{f''V_p}{f''A_p}\sqrt{\frac{k_\nu}{D_{eff}}} = \Phi \tag{74}$$

where f', f'', ... are different wetted fractions of the pellet.

Intraparticle effectiveness factors for either f', f'' have the same functional form, because of the assumption of isotropy. The effectiveness factor for spherical pellets is:

$$\eta = \frac{3}{\Phi}\left[\frac{1}{\tanh\,\Phi} - \frac{1}{\Phi}\right] \tag{75}$$

When reactant B, fed with the liquid stream, is limiting, the boundary conditions for B on the effectively wetted (f) and non wetted (1-f) surface fractions are respectively:

$$C_{Bs} = \text{const} \neq 0 \qquad \text{for wetted surface} \tag{76a}$$

$$C_{Bs} = 0 \quad \frac{dC_{Bs}}{dr} = 0 \qquad \text{for non-wetted surface} \tag{76b}$$

The rate of reaction, for a first order reaction in B, is:

$$R_{pB} = k_{\nu B}\eta C_{Bs}\frac{fV_p}{V_p} = f\,\eta\,k_{\nu B}C_{Bs} \tag{77}$$

When the reactant A fed with the gas is limiting, the values of C_{As} at f', f'', ... vary as functions of the liquid-solid mass transfer resistances. If there is just one wetted and one non-wetted region, then:

$$R_{pA} = k_{\nu A}\eta\left[(C_{As})_f\,f + (C_{As})_{1-f}(1 - f)\right] = R_{pB} \tag{78}$$

The concentrations C_{Bs}, C_{As} in the above equations can be evaluated by equating the rates of reaction and of external mass transport. For instance:

$$K_s A_p f(C_{Bl} - C_{Bs}) = R_{pB} V_p = f \eta k_{vB} C_{Bs} V_p \tag{79}$$

C_{Bs} is then a function of $C_{B\ell}$, K_s, k_{vB}, and η.
An overall catalyst effectiveness factor η_o can be derived to take into account the effective surface wetting and the external reactant supply. This factor is defined as the ratio of the actual conversion rate and that obtained without diffusion resistance. This overall effectiveness factor may be approximated by the weighted average value for the differently wetted fractions of the pellet surface (Capra et al. [8]). If the wetting situation can be simplified into one wetted and one non-wetted surface fraction, with pellet volume fractions equivalent to surface fractions, then:

$$\eta_o = f \eta_f + (1 - f)\eta_{(1-f)} \tag{80}$$

where η_f and $\eta_{(1-f)}$ are the effectiveness factors for wetted and non-wetted surface fractions, each characterized by different external mass transfer resistances.

If reactant B is limiting, solving the intraparticle diffusion model equation with boundary conditions (76) and using the definition of effectiveness factor, results in:

$$\eta_f = \frac{2\,(Sh)_s I_1(\Phi)}{\Phi\left[\Phi I_1(\Phi) + (Sh)_s I_0(\Phi)\right]} \tag{81}$$

and

$$\eta_{(1-f)} = 0 \tag{82}$$

where I_o and I_1 are the modified Bessel functions of order zero and first, respectively, and

$$(Sh)_s = \frac{K_s d_p \varepsilon_{in}}{2D_{eff}} \tag{83}$$

If reactant A is limiting, assuming for non wetted areas a very large gas-solid mass transfer coefficient, it can be shown that

$$\eta_f = \frac{2\,(Sh)_s I_1(\Phi)}{\Phi\left[\Phi I_1(\Phi) + (Sh)_s I_0(\Phi)\right]} \tag{84}$$

and

$$\eta_{(1-f)} = \frac{2I_1(\Phi)}{\Phi I_0(\Phi)} \tag{85}$$

Figures 15 and 16 show values of η_o calculated by numerical methods (continuous lines) as functions of the fraction of the wetted area for limiting reactant B and A, respectively. On each figure the dashed lines give the effectiveness factor η_o evaluated from the approximate relationships (80) to (85).

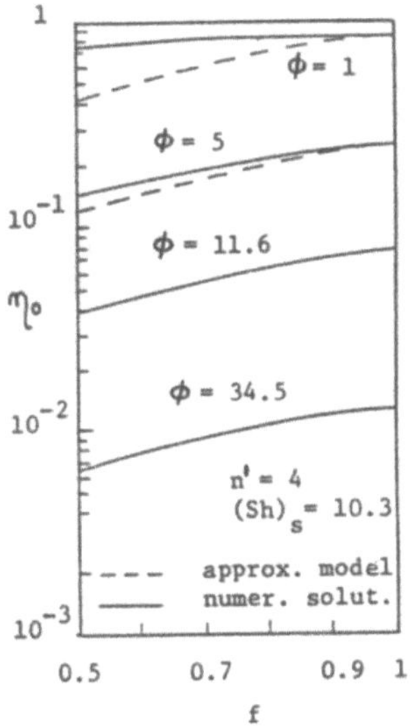

Fig. 15 - Overall effectiveness factor versus the wetted fraction of the pellet surface (B limiting)

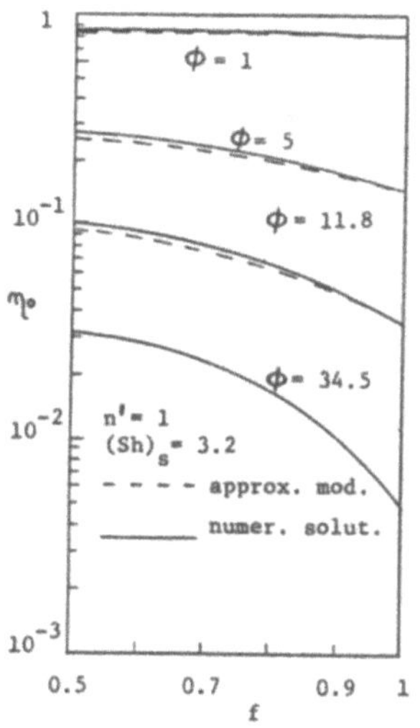

Fig. 16 - η_o versus f (A limiting)

When reactant B is limiting, the agreement is good for high values of ϕ . For low ϕ the disagreement between the exact curve and the approximations for η_o increases at low wetting (low f). For A limiting the approximate method always seems to be reliable. It is interesting to observe that when B is the limiting reactant, η_o increases with f. Conversely, for A limiting, η_o diminishes with increasing f because A is more easily supplied through dry zones of the catalyst surface. A similar approach, assuming adequate ϕ and $(Sh)_s$ values, may be used when the reaction partially occurs in the vapor phase inside dry pores.

2. A second model for identifying inadequate wetting when the limiting reactant B is not volatile is by estimation of the diffusivity ratio D_{app}/D_{eff} (Colombo et al. [12], Sicardi et al. [76], Specchia et al. [77], Sicardi et al. [75]).
D_{app} is the apparent intrapellet diffusivity in a Trickle Bed Reactor measured by a tracer response method, and D_{eff} is the effective diffusivity of the tracer inside the pellets measured when they are completely wetted by the liquid. In this latter case the diffusion path is a function of the pellet radius. For partial wetting, in a trickling operation, this path must be substantially larger because the tracer must be supplied to the poorly

irrigated zones of the pellets from the adjacent volume through the internal pore network. As a consequence, measurement of intrapellet diffusivity in partially wetted Trickle Bed packings results in a D_{app}/D_{eff} less than one. If the packing is well-wetted, the ratio approaches unity.

Figure 17, on the basis of tests carried out by dynamic methods with different tracers at the Politecnico of Torino, shows that D_{app}/D_{eff} increases with the liquid velocity V_{OL} and tends to 1. This is in quantitative agreement with a similar plot of k_{app}/k_{ν} proposed by Satterfield [67] based on data obtained by measuring catalyst effectiveness in commercial units for oil treatment.

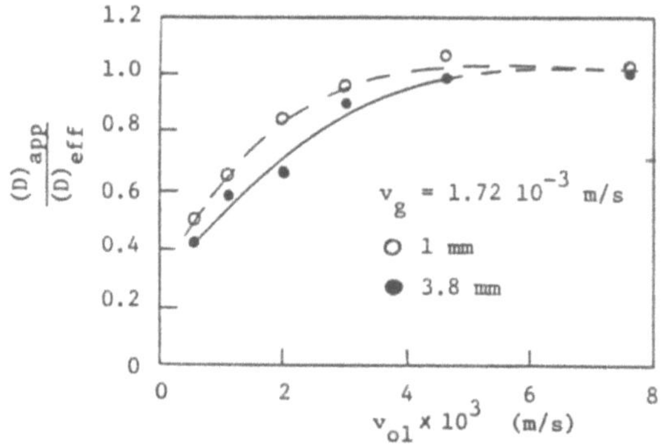

Fig. 17 - Ratio of apparent and effective diffusivities as a function of liquid superficial velocity (75). Reprinted with permission from Chem. Engng. Science 35. Copyright (1980) Pergamon Press.

This agreement is not surprising because D_{app}/D_{eff} is related to k_{app}/k_{ν}. In fact the generalized Thiele modulus for a first order reaction with a partially wetted catalyst particle, when the limiting reactant B is not volatile, can be written as:

$$\Phi^* = 3 \frac{V_p}{fA_p} \sqrt{\frac{k_\nu}{D_{eff}}} = \frac{\Phi}{f} \tag{86}$$

If, on the other hand, D_{app} is used to allow for partial wetting,

$$\Phi^* = 3 \frac{V_p}{A_p} \sqrt{\frac{k_\nu}{D_{app}}} = \Phi \sqrt{\frac{D_{eff}}{D_{app}}} \tag{87}$$

As a consequence:

$$f = \sqrt{\frac{D_{app}}{D_{eff}}} \tag{88}$$

The effective wetting model relates f to k_{app}/k_{ν}, so that

$$\frac{k_{app}}{k_\nu} = f = \sqrt{\frac{D_{app}}{D_{eff}}} \tag{89}$$

Figure 18 shows that the values of $(D_{app}/D_{eff})^{1/2}$ obtained with different packings and tracers are in good agreement. These results support the use of Eq. (66) for scale up.

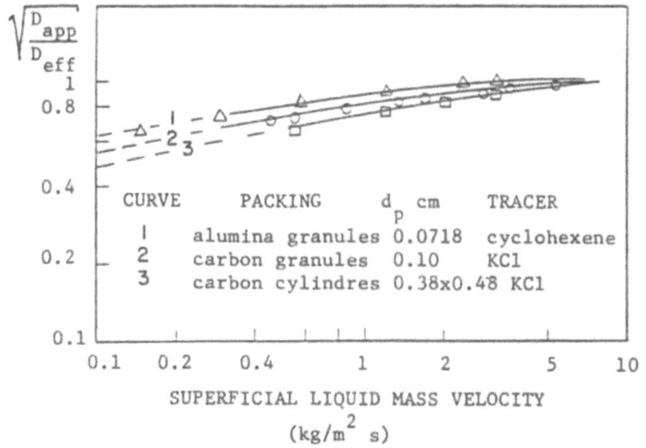

Fig. 18 - Ratio of apparent and effective **diffusivity** as a function of the superficial liquid mass velocity

THERMAL GRADIENTS

Micro and macro thermal gradients are generated within or outside the pellets by the heat of reaction. This suggests that scaling up of Trickle Bed Reactors for exothermic reactions must be done with caution to avoid temperature excursions which could cause excessive vaporization of the liquid outside and inside the pellets and give rise to local hot spots, an increase in the heat release rate and a decrease of external heat transfer coefficients.
Hanika et al. [22], Satterfield and Ozel [73], and Germain et al. [15] observed hot spots attributed to vaporization for the hydrogenation of cyclohexene to cyclohexane, benzene to cyclohexane and α-methylstyrene to cumene, respectively. Hot spots were observed to move axially through the catalyst bed. Sintering of the catalyst in industrial units has been reported by Weekmann [93].
The possibility of strong temperature gradients inside the pellets can be checked using the Prater criterion (Prater [63]). This criterion is a general one, independent of reaction order, particle size and shape.

The basis for this evaluation is the differential heat balance for a pellet:

$$D_{eff}\frac{dC}{dr}(-\Delta H_R) = \lambda_{eff}\frac{dT}{dr} \tag{90}$$

By integrating:

$$\Delta T = T - T_s = \frac{(-\Delta H_R)D_{eff}}{\lambda_{eff}}(C_s - C) \tag{91}$$

An estimate of the maximum increase in temperature within a pellet, ΔT_{max}, can be obtained by assuming that C = 0 at the pellet core. λ_{eff}, the effective thermal conductivity of the pellet, varies according to whether the pores are filled with liquid or gas since $\lambda_{eff_\ell}/\lambda_{eff_g}$ = 10-100. The range reflects the dependence of λ_{eff} on the porosity, skeletal thermal conductivity of the solid, fluid composition and total pressure. In other words, the value of λ_{eff_ℓ} for liquid filled pores ranges from 10^{-5} to 10^{-4} kcal/m·s·K. By contrast, D_{eff_ℓ}/D_{eff_g} = 10^{-2} to 10^{-3} . By introducing these ratios into Eq. (91), $(\Delta T_{max})_\ell/(\Delta T_{max})_g$ is found to range from 10^{-3} to 10^{-5} . Consequently, ΔT_{max} is negligible when the catalyst is completely wetted by liquid, but a value as high as 100 K is possible if a gas phase reaction occurs.
Mears [47] has suggested:

$$\left[\gamma\ \beta\right] < 0.05\ m \tag{92}$$

as a criterion for a uniform pellet temperature, where m is the order of the reaction, γ is E/RT . It represents the sensitivity of the reaction to temperature variations (whose value is normally bewteen 10 and 30), and β is $C_s(-\Delta H_R)D_{eff}/\lambda_{eff}\,T_s$, the Prater number divided by the maximum temperature difference between the center and the external surface of the pellet. The Prater number is > 0 for exothermic and < 0 for endothermic reactions.
The non-uniform temperature inside the pellet affects the catalyst effectiveness. Fig. 19 shows this effect as a function of Thiele modulus and Prater number.

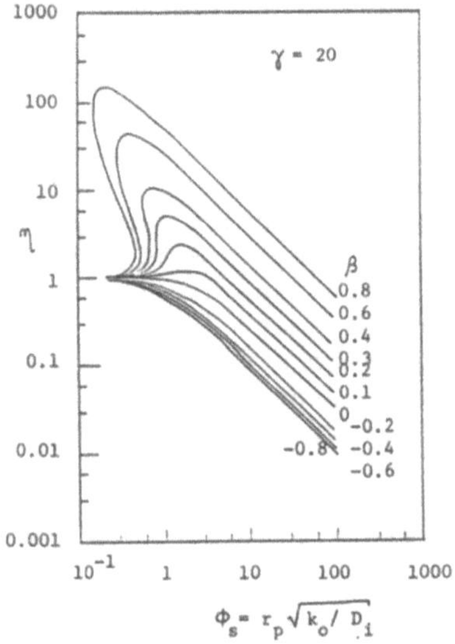

Fig. 19 - Effectiveness factor for a first order reaction in a spherical catalyst particle (Weisz and Hicks (96)). Reprinted with permission Chem.Engng.Science 17. Copyright (1962) Pergamon Press.

The fluid-solid heat transfer according to Mears [47] will control heat movement in the bed and lead to measureable temperature differences between the fluid and the catalyst pellets when the Biot number, $h_s d_p/\lambda_{eff}$, is less than 10. In the Biot number, h_s is the fluid-solid or extrapellet heat transfer coefficient. Biot numbers smaller than 10 occur in commercial scale units notwithstanding their larger fluid velocities. Catalyst pellets in Trickle Bed Reactors can be in contact with liquid or gas. An order of magnitude estimate of the expected fluid-solid temperature differences in either case can be obtained by using the analogy between mass and heat transfer (Calderbank, [7]):

$$k_s (Sc)^{2/3} = \frac{h_s}{C_p \rho} (Pr)^{2/3} \tag{93}$$

The heat balance at the boundary of the catalyst particle is:

$$k_s A_p (C_1 - C_s)(-\Delta H_R) = h_s A_p (T_s - T_1) \tag{94}$$

Combining the above two equations,

$$T_s - T_1 = \frac{(-\Delta H_R)}{C_p \rho} \left[\frac{Pr}{Sc}\right]^{2/3} (C_1 - C_s) \tag{95}$$

Assuming $(-\Delta H_R) = 3.10^{-4}$ kcal/kmole, the particle is immersed in the liquid, $C_{p\ell} \cdot \rho_L = 10^3$ kcal/m^3.K, and $(Pr/Sc)_\ell = 0.01$, eq. (95) yields

$$T_s - T_1 = 6.6 \ (C_1 - C_s) \tag{96}$$

On the other hand, if the particle is not wetted, heat transport is between solid and gas, $C_{pg} \cdot \rho_G = 1$ Kcal/m^3.K, and $(Pr/Sc)_g = 1$. On substitution,

$$T_s - T_g = 3 \ 10^4 (HC_g - C_s) \tag{97}$$

Comparison of Eqs. (96) and (97) shows that the temperature difference across the film surrounding the catalyst pellet must be very low for a fully wetted particle, but could be important for a non-wetted particle. The design engineer must ensure that scale-up of reactor diameter for highly exothermic reactions does not diminish heat transfer from the reactor, or increase evaporation of liquid and generation of hot spots. To test for these effects, a pilot plant should be operated so that evaporation can occur leading to the development of dry zones. When this condition is found detailed axial temperature measurements should be taken. Extra precautions must also be taken to avoid hot spots due to flow maldistribution. A study by Jaffe [37] showed that a shadowing effect in large reactors can cause low flow zones and very high local temperatures.

Heat released by the reaction must be carried from the reactor by the flow of both the fluids, thermal conduction through the solid particles axially and radially and exchange through the vessel walls. Models for axial and radial heat transport to ensure that the axial and radial temperature gradients will remain within specified limits are necessary for reactor operation and stability. Temperature control becomes more difficult as the diameter of the catalyst bed becomes larger. Furthermore, diameter can influence the hydrodynamic regimes, and thus the radial and axial heat transport. Limits of axial and radial temperature variations are set by process considerations. Generally speaking, these must not exceed a few tens of degrees Celsius.

The radial temperature profile has been found to depend functionally on macro Biot number, $h_w D_c/k_{eff}$, where h_w is the heat transfer coefficient at the wall, D_c is the reactor diameter, and k_{eff} is the effective thermal conductivity in the bed. Correlations for h_w and k_{eff} (Specchia and Baldi [78], Satori and Nishizaki [66]; Kau and Greenfield [38] for low interaction regimes indicate:

$$Bi \propto D_c v_{OL}^{0.25} d_p^{0.3} \tag{98}$$

Therefore the Biot number depends strongly on the reactor diameter and slightly on the liquid velocity and on the catalyst size. The longitudinal temperature profile is given by the model already presented in the general introduction concerning the axial heat balance along the reactor.

When volatile liquid is present it is convenient to use a fictitious specific heat for the gas phase in Eq. (47), defined as the variation of enthalpy of the gas saturated by the vaporized liquid with the temperature:

$$C^*_{pg} = \frac{dH_G}{dT} \tag{99}$$

This takes into account the increase of the gas enthalpy due to the latent heat of the vaporized liquid.
The equation for the longitudinal heat transport becomes:

$$-\lambda_{eff} \frac{d^2T}{dz^2} + (v_{OL}\rho_L C_{pL} + v_{OG}\rho_G C^*_{pG}) \frac{dT}{dz} = R_t(-\Delta H_R) \tag{100}$$

Radial heat transport may usually be represented by a pseudo-homogeneous model with two parameters (Specchia and Baldi [81], Specchia et al. [82], Specchia and Sicardi [83]). The catalytic bed is assumed to be a pseudo-homogeneous system characterized by an effective thermal conductivity k_r and by a heat transfer resistance located at the wall of the reactor. The corresponding coefficient is h_w. In any point in the reactor the three phases (gas, liquid, solid) are supposed to be at the same temperature. By assuming a cylindrical geometry for the TBR, when there is heat dispersion through the wall, the heat balance can be written as:

$$-\lambda_{eff} \frac{\partial^2 T}{\partial z^2} + \frac{\partial T}{\partial z} (v_{OL}\rho_L C_{pL} + v_{OG}\rho_G C^*_{pG}) = k_r(\frac{1}{r}\frac{\partial T}{\partial r}) + \frac{\partial^2 T}{\partial r^2} + R_t(-\Delta H_R) \tag{101}$$

with the boundary condition:

$$h_w(T_w - T_{R,w}) = k_r \left[\frac{\partial T}{\partial r}\right]_{r=r_w} \quad (102)$$

If the longitudinal dispersion is neglected ($\lambda_{eff} = 0$) the heat transfer through the wall can be expressed by the above pseudo-homogeneous, two parameter model (k_r, h_w). In Fig. 20 axial temperature profiles are shown (Hanika et al. [22], [23]). In the beginning, an increase of the gas rate diminishes the axial temperature gradient, then, for a certain gas rate, some catalyst pellets become dry and the rate of reaction increases giving rise to higher heat generation, higher temperature gradients and thermal instabilities.

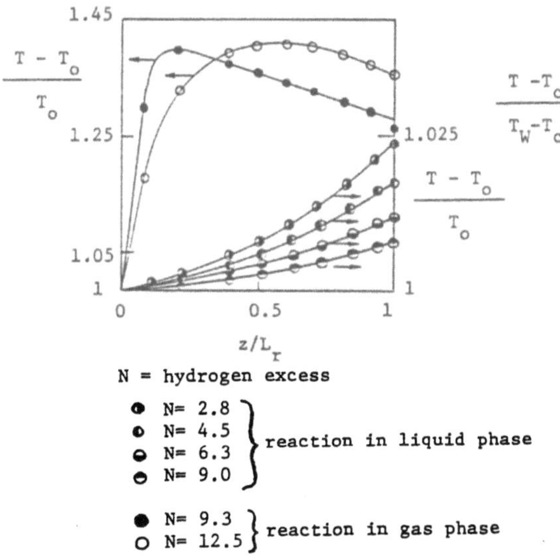

Fig. 20 - Effect of the gas flowrate on the axial temperature profile.
T_0= feed temperature (Hanika et al. (23))

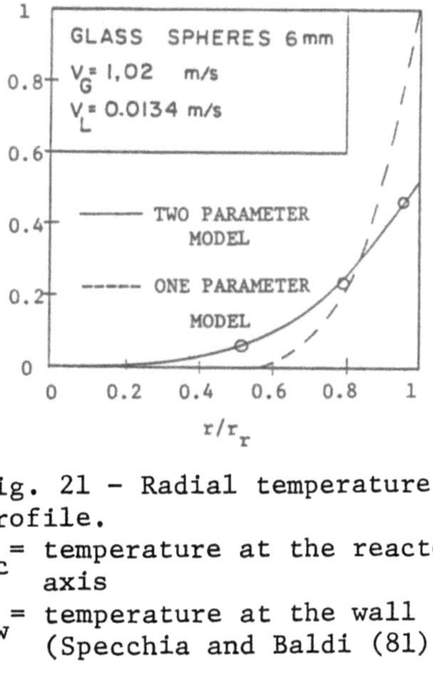

Fig. 21 - Radial temperature profile.
T_c= temperature at the reactor axis
T_w= temperature at the wall (Specchia and Baldi (81))

Fig. 21 shows the radial temperature profile found experimentally in a Trickle Bed Reactor compared to the calculated one using a pseudo-homogeneous two parameter model. The observed temperature step at the wall is due to the heat resistance at the wall. The dashed line represents the profile calculated with only one parameter (k_r).

Numerical data on k_r and h_w in three phase reactors are rather scarce in the literature (Hashimoto et al. [25], Specchia and Baldi [81]. k_r is a complex function of the gas and liquid flow rates, of the pellet size and of their bulk material properties. It is possible to write:

$$k_r = (k_r)_o + (K_r)_g + (k_r)_1 \quad (103)$$

$(k_r)_o$ is the stagnant contribution of heat conduction through the pellets and the stagnant liquid pockets near the contact points of the packing (independent of hydrodynamic conditions), whereas the convective contribution given by $(k_r)_g$ and $(k_r)_\ell$ arises from mixing in the radial direction and depends on the reactor hydrodynamics. In industrial units, due to the higher heat capacity of the liquid, $(k_r)_\ell$ is the most important parameter and $k_r \simeq (k_r)_\ell$. This is less evident in laboratory devices.

$(k_r)_o$ and $(k_r)_g$ can be evaluated, at least for low interaction regimes, with equations proposed for gas flow through a packed bed (Specchia and Sicardi [82], [83]):

$$\frac{(k_r)_o}{\lambda_g} = \frac{\varepsilon}{1.5} + \frac{(1-\varepsilon)}{0.13\varepsilon^{1.44} + \frac{2}{3}\left(\frac{\lambda_g}{\lambda_g}\right)} \quad (104)$$

and

$$\left[\frac{(k_r)_g}{v_{OG}\rho_G C_{pG} D_e}\right]^{-1} = 8.65\left[1 + 19.4\left(\frac{D_e}{D_c}\right)^2\right] \quad (105)$$

where λ_g and λ_s are the thermal conductivities of gas and solid, respectively, D_e and D_c the diameter of a sphere having the same external surface as the catalyst pellet and the diameter of the reactor, respectively. $(k_r)_1$ turns out to be the most important parameter for the radial heat dispersion. For $(k_r)_1$ Specchia and Baldi [81] have proposed two correlations according to the hydrodynamic conditions, limited to the system water-air.

$$\frac{v_{OL}\rho_L C_{pL} d_p}{(k_r)_1} = 0.041\left(\frac{v_{OL}\rho_L d_p}{\beta_T \mu_L}\right)^{0.87} \text{ l.r.} \quad (106)$$

and for high interaction regime:

$$\frac{v_{OL}\rho_L C_{pL} d_p}{(k_r)_1} = 338\left(\frac{v_{OL}\rho_L d_p}{\mu_L}\right)^{0.675} (\beta_T)^{0.29}\left(\frac{a_v d_p}{\varepsilon}\right)^{-2.7} \quad (107)$$

The gas flow rate is taken into account because it affects the external total liquid saturation β_t.

The h_w parameter depends strongly on hydrodynamics; for low interaction regimes it increases with both liquid and gas flow rates whereas for high interaction regimes it tends to an assymptotic value (Muroyama et al. [57], Specchia and Baldi [81]). Its values are higher than in the case of gas alone (about one order of magnitude) due to the strong influence of the liquid film at the reactor wall. It depends on the velocity of the film and on the fraction of the area covered by the liquid phase.

Table 4 shows the correlations proposed by Hashimoto [25], Muroyama [57] and Specchia et al.[79]. The agreement among the various correlations is not very satisfactory and more research is needed.

The radial profile in TBR can be approximated as:

$$\frac{T_i - T_w}{T_m - T_w} = 1 + \left[1 + 8\,\frac{k_r}{D_c h_w}\right]^{-1} = 1 + \left[1 + \frac{8}{Bi}\right]^{-1} \tag{108}$$

where T_i, T_w and T_m are the axial, wall and average temperatures in a given section of the reactor.
From the various correlations proposed in the literature, we find:

$$k_r \cong (k_r)_1 \propto v_{OL}^{0.1}\, d_p^{0.1}\, \beta_T^{0.9} \tag{109}$$

and

$$h_w \propto v_{OL}^{0.9}\, d_p^{-0.1}\, \beta_T^{-0.9} \tag{110}$$

Assuming as suggested by Baldi and Specchia [1], for ($d_p < 0.6$ mm), Satori and Nashizaki [66], Kau and Greenfield [38]:

$$\beta_T \propto v_{OL}^{0.3}\, d_p^{-0.3} \tag{111}$$

then the following expression for the Biot number results:

$$Bi = \frac{h_w D_c}{k_r} \propto D_c\, v_{OL}^{0.25}\, d_p^{0.34} \tag{112}$$

Table 4

Correlations for Radial Heat Transfer in Trickle-Bed Reactors

Authors	Effective thermal conductivity	Wall heat transfer coefficient
Hashimoto et al. (1976)	$\frac{k_r}{\lambda_L} = \frac{1.1\ 10^{-4}}{\lambda_L} + 0.095(Re)_G(Pr)_G \frac{C^*_{pG}}{C_{pG}} \frac{\lambda_G}{\lambda_L} + (\alpha\beta)_L(Re)_L(Pr)_L$ $(\alpha\beta)_L = \left\{\frac{1}{1.9+0.0264(d_e G_L/h_e \mu_L)(\mu_L/\mu_o)} + 0.197\right\}\frac{d_e}{d_p}$ $d_e = \frac{2}{3}\frac{\varepsilon}{1-\varepsilon} d_p$	
Muroyama et al. (1977)		$\frac{h_w d_p}{\lambda_L} = 0.012(Re)_L^{1.7}(Pr)_L^{1/3}$ (low int regime) $\frac{h_w d_p}{\lambda_L} = 0.092(Re)_L^{0.8}(Pr)^{1/3} \cdot \left[h_e\left(1+\frac{4 d_p}{6 d_r(1-\varepsilon)}\right)\right]^{-0.8}$ (high int. regime)
Specchia and Baldi (1979)	$\frac{v_L \rho_L C_{pL} d_p}{(k_r)_L} = 0.041 \frac{(Re)_L^{0.87}}{\beta_T^{0.87}}$ (low int. regime) $\frac{v_L \rho_L C_{pL} d_p}{(k_r)_L} = 338\ (Re)_L^{0.675} \beta_T^{0.29} \left(\frac{a_V d_p}{\varepsilon}\right)^{-2.7}$ (high int. regime)	$\frac{h_w D_C}{\lambda_L} = 0.0835\ Re^{0.92} Pr^{1/3}$ ($10 \leq Re \leq 1200$) $\frac{h_w D_C}{\lambda_L} = 1.23\ Re^{0.53}\ Pr^{1/3}$ ($1200 \leq Re \leq 10000$)

If the radial temperature profile depends on Bi, it has been shown that it is mainly affected by the reactor diameter and only slightly by the liquid velocity and by the size of the catalyst pellets. It is then easy to make a rough calculation of the temperature variation along a radius of the reactor. Generally speaking, a TBR is not suitable for reactions that are highly exothermic because this apparatus does not give a sufficient heat transfer due to the liquid phase. A general model for the evaluation of superheating in the reactor has not been suggested yet.

To solve problems arising in Trickle Bed Reactors caused by exothermic reactions, the engineer must first identify the critical, limiting steps for the heat transfer by using "ad hoc" criteria (discussed above), experimental tests and realistic models. Secondly, he must take steps to keep the micro and macro temperature gradients within acceptable limits for reactor productivity and stability. He must always consider the economics of the reactor, in terms of both initial capital investments and operating costs.

Mathematical models which are most suitable for dealing with three phase reactors, need to fulfill the following requirements:

1. A model ceases to be useful if it is too complex.

2. A model should have as few parameters as possible because parameters can usually be difficult and costly to evaluate.

3. The data used for design and scale up must be conservative (this is the case in using k_{app} obtained from laboratory scale equipment because of poorer hydrodynamics and wetting found in such units in comparison with industrial units). Furthermore, it is necessary to consider the source of the data and model, paying special attention to situations where interaction between physical and chemical phenomena can exist.

4. Functional stability and safety must be a first priority even if this leads to a costly conservative design. (This is important especially with respect to thermal gradients).

To simplify and, even more, to obtain greater confidence in the design procedure, further progress in the understanding of three phase reaction systems is necessary. A critical area is the development of criteria for identification of rate limiting phenomena. Papers on this subject have been presented by Satterfield [67], Hofmann [33], Gianetto et al. [16], Koros [44], and Shah [74].

The complexity of the models presented up to now, where the various phenomena have been considered almost individually, will result increased by interconnections between different simultaneous events.

An Example

An example would be useful to illustrate the difficulties met in the interpretation of results obtained using a laboratory TBR. A paper by Tukac et al. [92], stressing these problems has been submitted for publication.

The reaction considered was the catalytic oxidation of ethyl alcohol, dissolved in an aqueous alkaline solution, to acetic acid. As oxidizing agent gaseous oxygen was used. The catalyst employed was 0.5% Pd supported on alumina. Table 5 gives the properties of the catalyst. A previous kinetic study was carried out to measure the intrinsic chemical kinetic parameters of this reaction. For this purpose a tubular, packed, flooded reactor was used (oxygen was supplied after its absorption in the solution before entering the reactor). Papers on this subject have already been published: Baldi et al. [2], [3].

Table 5

Properties of the Catalyst (0.5% Pd on alumina) (Tuka et al. (1985))

Shape		cylindrical
Size: diameter	(m)	$3.24\cdot10^{-3}$
height	(m)	$3.60\cdot10^{-3}$
B.E.T. area	(m^2/ g)	100-110
External area	(m^2/ g)	1.083
Internal porosity*		0.585
Apparent particle density	(kg/m^3)	1.56
Thickness of the cat. layer*	(m)	$2.18\cdot10^{-4}$

* given by the manufacturer

As previously explained, problems arise from an uneven distribution of the liquid around the catalyst pellets in TBRs. In fact only a part of the external particle surface is effectively wetted by the flowing liquid; the remaining zones are partly in contact with semistagnant liquid pockets and partly covered by an almost motionless thin film. The chosen reaction is not significantly exothermic and the catalyst pores can therefore be assumed completely filled by liquid. No particular problem exists if the controlling reactant is fed with the liquid because the most active zones of the catalyst are very likely those characterized by a flowing liquid and the conversion rate increases as the liquid flow rate is increased (Satterfield [67], Montagna and Shah [54], Koros [43], Iannibello et al. [36]). On the other hand, when the key reactant is in the gas phase, the gaseous reactant may also be supplied to the catalyst through the motionless thin liquid film by molecular diffusion, and this mechanism may be faster than that occurring through the zones covered by a flowing liquid. Several papers have considered gas reactant controlled reactions: catalytic oxidation of sulfur dioxide on carbon pellets (Hartmann and Coughlin [24], Pavko and Levec [62], Mata and Smith [45], Berruti et al. [4]), and hydrogenation of α - methylstyrene on Pd-Al_2O_3 catalyst (Morita and Smith [56], Herskowitz et al. [30], Turek and Lange [93]). Some of these workers found that sometimes an increase of the liquid flow rate can lead to a decrease of the rate of conversion. The explanation of this phenomenon lies in the generation of more active dry zones because of reduced mass transfer resistance outside and inside the particles. Turek and Lange [93] postulated that the chemical reaction may increase gas-liquid-solid mass transfer because liquid zones, almost ineffective since they are saturated by the transferring component, become effective if the component is transformed by the reaction.

In the reactor models, only the reactant supply to the catalyst is considered; the problem of taking away the products can be, as a first approximation, disregarded. Product accumulation could reduce the rate of conversion but would make a realistic model of the reactor much more complex.

In tests carried out in Torino, oxygen was the gaseous rate controlling reactant and systematic studies of the hydrodynamics, concentrations and temperatures were performed.

The rate expression, evaluated in a previous study of the kinetics (Baldi et al. [2]), was expressed as:

$$R_s = - \eta k_r C_1 \tag{113}$$

where η = catalyst effectiveness factor, and

$$k_r = \frac{k_o C_2}{1 + k_2 C_2} \tag{114}$$

where C_1 and C_2 are the concentrations of oxygen and ethyl alcohol, respectively. The constants, expressed according to an Arrhenius-type equation were:

$$k_o = 9.564\ 10^8 \exp\left(\frac{-6030}{T}\right); \quad k_2 = 2.126\ 10^5 \exp\left(\frac{-2980}{T}\right) \tag{115}$$

with the units (m^6/sm^3_{cat}kmoles) and (m^3/kmoles),), respectively.

Fig. 22 shows the experimental Trickle Bed Reactor. The key item is the reactor which consisted of several glass jacketed sec-

tions 0.3 m long, 0.03 m I.D., each provided with a teflon redistributor for the liquid. The catalytic bed was 1.2 m long and was maintained at isothermal conditions by a thermostatized liquid circulating in the jacket. The temperature of the phases entering and leaving the catalytic bed was measured by thermocouples; the maximum difference in temperature between the two points was 0.7°C. The catalytic section was followed by a cooling section and by a gas-liquid separator. The temperature in the separator was less than 25°C at the least favourable conditions. A calming section, 0.6 m long, was provided before the catalytic section, whose purpose was to distribute the liquid over the entire section of the reactor and also to allow the phases to attain equilibrium conditions. The liquid fed to the reactor was an aqueous solution of ethanol and K_2CO_3, mixed, in certain runs, with a recirculating flow coming from the gas-liquid separator. Liquid recirculation was adopted at the higher liquid superficial velocities, in order to obtain a reliable measure of the ethanol conversion.
A wide range of ethanol concentrations in the feed tank was tested. These concentrations were adjusted as the operating conditions were changed to maintain the ethanol concentration in the bed within a certain range. Pure oxygen was used as the gas phase. It was previously saturated with water at the test temperature before entering the calming section. The range of the operating conditions is listed in Table 6.

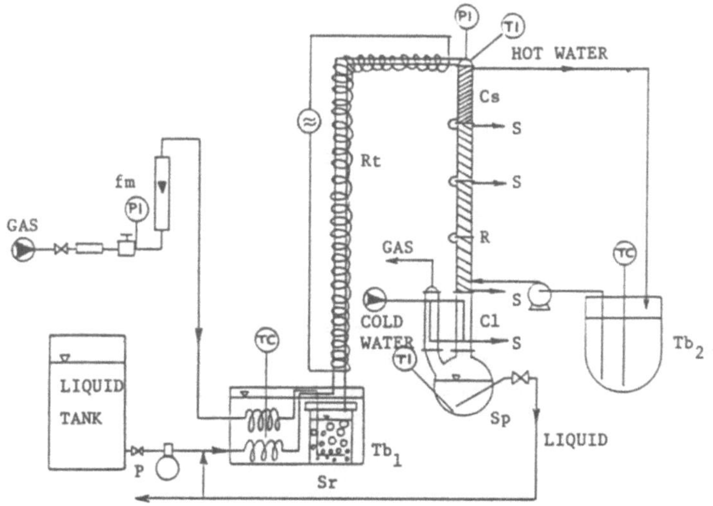

Fig. 22 - Experimental Trickle Bed Reactor.

Cl: cooling section; Cs: calming section; fm: flowmeter; P: metering pump; R: reactor; Rt: electrical resistance; S: sample outlets; Sp: phase separator; Sr: gas saturator; Tb_1 and Tb_2: thermostatic baths.

Table 6

Range of the operating variables

Temperature	45°- 80°C
Total pressure	10^5 Pa
Liquid superficial velocity	$1.5\cdot10^{-4}$ - $4\cdot10^{-3}$ m/s
O_2 flowrates	$1.5\cdot10^{-5}$ - $6\cdot10^{-5}$ m^3/s
K_2CO_3 concentration	$1.45\cdot10^{-2}$ $kmol/m^3$
Ethanol concentration at the reactor inlet	$1.1\cdot10^{-2}$ - $5.5\cdot10^{-2}$ $kmol/m^3$
O_2 partial pressure	$5\cdot10^4$ - $8.8\cdot10^4$ Pa

Gas and liquid entering and leaving the catalytic bed were assumed to be in equilibrium. The conversion was determined by analyzing samples taken at the inlet, along the lenght, and at the outlet of the reactor. Mass balances were employed to check the results. The average ethanol conversion rate $\bar{R}_r$ was calculated with the following equation:

$$-\bar{R}_t\, V = \frac{L}{\rho_L}\,(C_{2\,o} - C_{2\,e}) - N_{O_2} y_2 \qquad (116)$$

where $C_{2o} - C_{2e}$ is the difference in ethanol concentration entering and leaving the catalyst bed, and y_2 is the fraction of ethanol in the oxygen stream leaving the reactor. These average values were significant since the reaction rate along the reactor is constant because the rate is practically controlled by the concentration of absorbed oxygen in the liquid (almost constant along the reactor).

To illustrate the influence of the various parameters investigated, $\bar{R}_r$ was defined as the "kinetic" conversion rate R_k calculated at the average concentration in the bed, neglecting external mass transfer effects and assuming total catalyst utilization:

$$R_k = -(1-\varepsilon)\eta\bar{k}_r C_t P \frac{y_1}{H_1} \qquad (117)$$

with $\eta = 1$ and $\bar{k}_r$ calculated at the average value of C_2 along the reactor bed. R_k is the maximum conversion rate that could be obtained in the reactor.

Fig. 23 shows $\bar{R}_r/R_k$ at $v_{OL} = 4\cdot10^{-3}$ m/s as a function of C_2 and temperature.

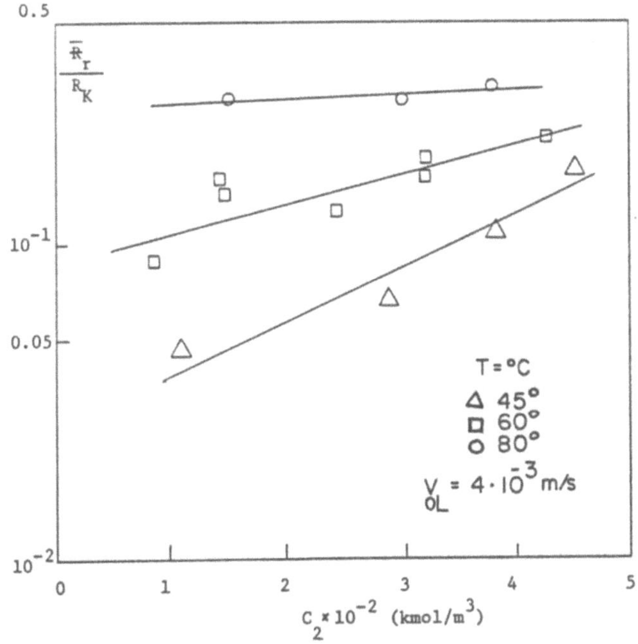

Fig. 23 - $\bar{R}_r/R_k$ as a function of C_2 and temperature (92)

The trend is the same at other velocities. The low values of this ratio show the influence of physical resistances and the effect of the hydrodynamics. There is also a marked influence of temperature and ethanol concentration, the latter especially at 45 and 60° C. $\bar{R}_r/R_k$ increases as C_2 increases tending to the values obtained at 80°C. Fig. 24 shows that the liquid velocity does not have a significant effect on the conversion rate, in contrast to the importance of the hydrodynamics.

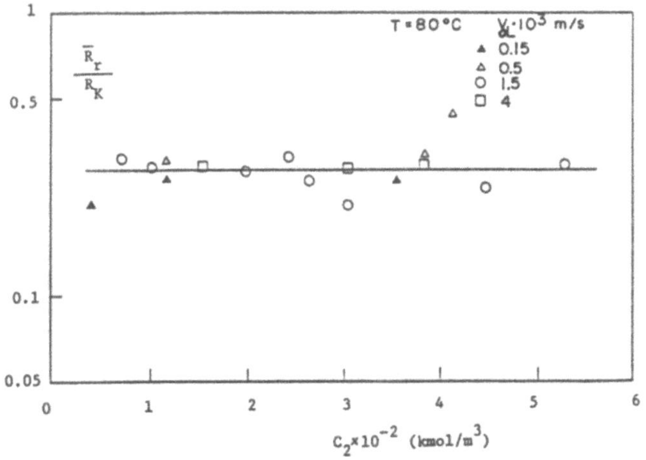

Fig. 24 - $\bar{R}_r/R_k$ versus C_2 at various v_{OL} at 80 ° C

The values plotted on Fig. 24 are for T = 80°C, but the results are similar for the other temperatures tested.

The local rate of conversion R_r can be calculated with the following differential balance equation:

$$R_r = \left(\frac{\pi_2 \bar{\gamma}}{RT} v_{OG} + C_t v_{OL}\right) \frac{dx_2}{dz} \qquad (118)$$

If two resistances in series are considered (i.e., mass transfer through the liquid covering irregularly the catalyst pellets and the actual reaction resistance including mass transfer inside the particle), and assuming oxygen is the controlling reagent due to its low solubility in the liquid, a first approximation to a kinetic equation can be written with respect to oxygen:

$$R_r = - \frac{Py_1}{H_1} \Big/ \frac{1}{(1-\varepsilon)\eta k_r C_t} + \frac{1}{\left((ka)_{ls}\right)_c C_t} \qquad (119)$$

where $[(ka)_{\ell s}]$ is the volumetric mass transfer coefficient in the presence of chemical reaction. That can be calculated by a trial and error procedure from the above equations, by numerical integration of the differential balance (118). This mass transfer coefficient was compared with that calculated from values of $(ka)_l$ and $(ka)_s$ obtained from correlations proposed by Specchia et al. [80] and Goto and Smith [20], respectively. The coefficient $[(ka)_{ls}]_p$ is:

$$\frac{1}{\left((ka)_{ls}\right)_p} = \frac{1}{(ka)_l} + \frac{1}{(ka)_s} \qquad (120)$$

The ratio $[(ka)_{ls}]_c/[(ka)_{ls}]_p$ is plotted versus C_2 in Fig. 25. The results were obtained from tests performed at 80 °C, at different liquid superficial velocities. At 45°C and 60°C the values of this ratio are smaller than those at 80° C, but have the same trend (Tukac et al. [92]).

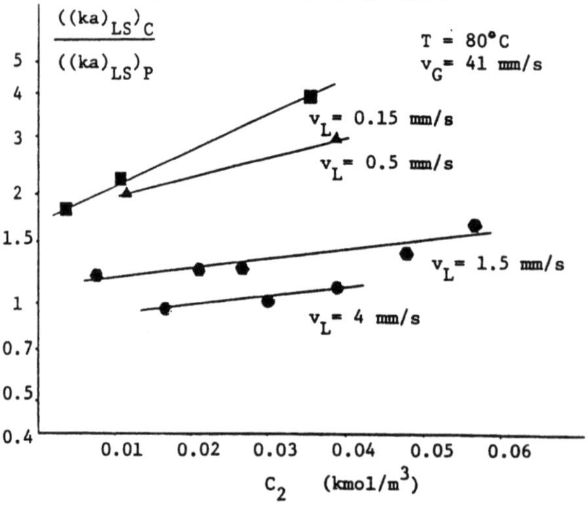

Fig. 25 - Ratio of "chemical" and "physical" mass transfer coefficients as functions of C_2 at 80 C.

The ratio $[(ka)_{ls}]_c/[(ka)_{ls}]_p$ increases as the liquid superficial velocity decreases. This is in agreement with a smaller mass transfer resistance for smaller liquid flow rates, because the surface of the catalyst is covered more and more by a thin film and not by a layer of moving liquid. At 80°C the observed values of the ratio are equal to or larger than 1. At the lower temperatures the ratio diminishes, but it is increased by increasing the ethanol concentration expecially at lower temperatures and lower liquid flow rates. In this model, all the pellets are supposed to be utilized with a uniform surface concentration of oxygen and ethanol. Actually, the local surface concentrations of both reactants may be quite different over the catalyst external surface depending not only on the local hydrodynamics but also on the intrinsic kinetic rate. By changing the temperature and the ethanol concentration, and hence the intrinsic kinetics, the distribution of concentrations may change and the catalyst utilization too. This is especially true for those zones of the catalyst covered by liquid with a very slow renewal. Ethanol is expected to be supplied to these zones essentially from the gas phase, by a series of processes involving evaporation from the dynamic zones and condensation. These processes are possibly rather fast, but at low concentration of ethanol and high concentration of oxygen may give rise to a supplementary mass transfer resistance especially for the zones characterized by fast oxygen supply. In effect, the assumption of mass transfer as rate controlling may not be true when the ethanol supply becomes small due to a decreasing in temperature and ethanol concentration in the liquid. As a consequence, an important factor may be the ratio between the partial pressures of ethanol and oxygen (i.e. p_2/p_1), that controls the concentration of the reactants in these zones.

In Fig. 26 the values of $\bar{R}_r/R_k$ are plotted as functions of p_2/p_1 at different temperatures for $v_{OL} = 4 \cdot 10^{-3}$ m/s. For the other liquid velocities the observed behaviour does not change significantly with respect to the one reported.

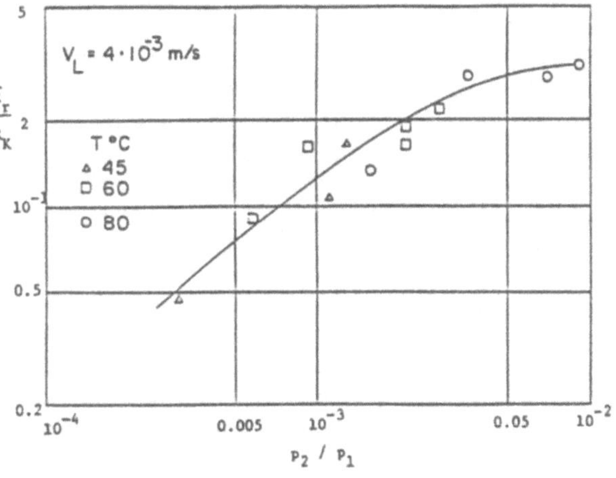

Fig. 26 - Influence of p_2/p_1 on $\bar{R}_r/R_k$ at various temperatures.

As can be seen, the proposed parameter (p_2/p_1) is able to correlate the data at different temperatures and ethanol concentrations. By increasing p_2/p_1, $\bar{R}_r/R_k$ seems to tend to a constant value. This is reasonable since at high values of the abscissa the distribution of ethanol becomes uniform and the conversion rate depends only on the supply of oxygen. In Fig. 27, data obtained at different gas flow rates are also plotted. In the range tested, the influence of v_{OG} is negligible.

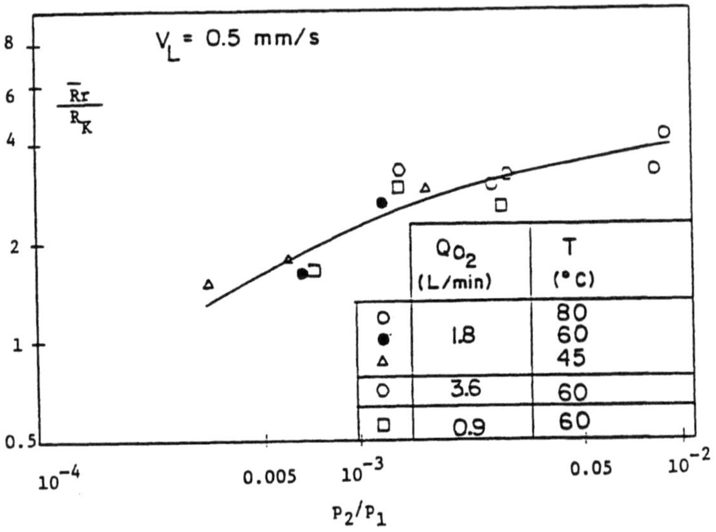

Fig. 27 - Influence of p_2/p_1 on $\bar{R}_r/R_k$ at different gas flow rates and temperatures

The results of the work described have confirmed that when the rate of conversion in a Trickle Bed Reactor is controlled by a gaseous reactant, mass transfer from the gas to the catalyst surface can take place also through zones of the liquid which are not normally effective during the process without chemical reaction. These results have also shown an influence of temperature and concentration of the liquid volatile reactant never pointed out before. This effect is difficult to interpret with a model, but may be explained by assuming that the liquid volatile reactant is supplied to the zones in contact with stagnant liquid by a mechanism of evaporation from the dynamic zones and subsequent condensation. For this process a certain mass transfer resistance exists. At low values of the ratio between the partial pressure of ethanol and oxygen, utilization of the catalyst diminishes because of a possible lower concentration of ethanol in these zones with respect

to the average one. The parameter p_2/p_1 has been shown to take into account this depletion.
The interpretation of the above results is difficult because of interconnections among the various mass transfer conditions occurring simultaneously. The proposed correlations and relationships can give an idea of the phenomena, but, except in simple cases, they are often not sufficient to allow quantitative evaluations and to supply an objective interpretation of the actual reactor operation. Therefore the results of calculations must always be taken as approximate. The case reported shows that very important parameters can arise, such as p_2/p_1. As a consequence, in modelling multiphase reactors excessive details in the measure of the conventional parameters are not useful and a model should have as few parameters as possible because their evaluation can become very difficult and subject to errors. The data used for design and scale up must always be conservative either for functional stability or for safe operation.

Notation List

A	constant in Eq. (72) (m^b / s^{b+1})
A	function Eq. (8)
A_p	external surface area of the pellet (m^2)
a_L, a_S, a_V	surface per unit volume of gas-solid, liquid-solid or dry packing (m^{-1})
Bi	Biot number $= h_w D_C/k_r$ (macro) $= h_s d_p/\lambda_{eff}$ (micro)
Bo	Bodenstein number $= D_1 \varepsilon_{Ldyn}/v_{OL} d_p$
b	constant in Eq. (72)
C	concentration $(kmole/m^3)$
$\bar{C}$	average concentration $(kmole/m^3)$
C_p	specific heat $(kcal/m^3 K)$
C^*_{pg}	defined as dH_g/dT $(kcal/m^3 K)$
D	diffusivity (m^2/s)
D_{eff}	effective diffusivity $(m^2/s) = D\varepsilon/\tau$
D_{app}	apparent diffusivity (m^2/s)
D_C	reactor diameter (m)
D_e	diameter of a sphere having the same external surface of the pellet (m)

D_1, D_g or D_L, D_G	axial dispersion for liquid and gas respectively (m^2/s)
d_p	pellet diameter (m)
E	backmixing coefficient Eq. (27)
Eö	Eötvos number = $\rho_1 \, g \, d_p^2 / \sigma_1$
f, f', f''	wetted fractions of the pellet surface
f_{LG}	friction factor
Fr	Froude number = $v_{OL}^2 / g \, d_p$
g	gravitational constant (m^2/s)
G	gas mass velocity (kg/m^2s)
Ga	Galileo number = $d_p^3 \, g / \nu_L^2$
Ga*	modified Galileo number
H_G	enthalpy of the gas phase (kcal/kg)
H	gas solubility or Henry's constant
$-\Delta H_R$	heat of reaction (kcal/kmole)
h_s	fluid-particle heat transfer coeff. ($kcal/m^2sK$)
h_w	wall heat transfer coeff. ($kcal/m^2s\,K$)
$(h_w)_o$	stagnant contribution to wall h.t.c. ($kcal/m^2s\,K$)
$(h_w)_f$	mixing contribution to wall h.t.c. ($kcal/m^2s\,K$)
k_1, k_2	constants Eq. (4)
k_0, k_2	kinetic constants Eq. (114)
k_ν, k_{app}	intrinsic and apparent rate constants for first or pseudo-first order reactions per unit of catalyst volume (s^{-1})
k_r, $(k_r)_o$, $(k_r)_1$, $(k_r)_g$	overall thermal conductivity in the particle bed, contribution through pellets and liquid pockets, convective contribution of the liquid and gas respectively (kcal/m s K)
k_g, k_1, k_s	mass transfer coefficients in the gas, liquid and liq-solid interface (m/s)
$(ka)_{ls}$	volumetric mass transfer coeff. liq-sol (s^{-1})
K_1, K_s	overall gas-liquid, fluid-solid mass transfer coeff. (m/s)
L	liquid mass velocity (kg/m^2s)
m	order of reaction
N_{O_2}	molar flow rate of oxygen (kmole/s)

n	order of reaction or total number of contact cells
n'	number of rivulets on outer surface of catalyst pellets
P, p_1, p_2	total pressure, partial pressures (kg/m s^2)
$p(x)$	probability Eq. (36)
Pr	Prandtl number
q	fraction of liquid flow rate not by-passed
R, R_s, R_t	reaction rate per unit of catalyst volume (kmole/m^3s)
R_k	kinetic conversion rate Eq. (117)
$\overline{R}_r$	average ethanol conversion rate Eq. (116)
Re_L	Reynolds number = $(\rho_L v_{OL} d_p/\mu_L)$
r_p, r	radius of the pellet and of the core of the pellet (m)
S	constant in Eq. (67)
Sc	Schmidt number $(\mu/\rho\ D)$
Sh	Sherwood number = $(k_s d_p/D)$
S.V.	space velocity (s^{-1})
T	temperature (K)
T	constant in Eq. (32)
t	time (s)
v_l, v_g, v_{OL}, v_{OG}	liquid and gas absolute and superficial velocities (m/s)
V	reactor volume (m^3)
V_p	volume of a pellet (m^3)
We_l	Weber number $=(v_{OL}^2\ \rho_L d_p/\sigma_L)$
w	constant in Eq. (71)
x, y	molar fractions in the liquid and gas
z, Z	axial coordinate and total length of the catalyst bed (m)
Z	parameter in Eq. (7)

Greek Symbols

β	Prater number = $((-\Delta H_R) C_s D_{eff}/\lambda_{eff} T_s)$
β_r, β_{fd}, β_t, β_G	residual, operating and total liquid saturation called improperly holdup, and gas saturation

β_{ext}	external liquid saturation
β_{in}	internal liquid saturation
γ	sensitivity of the reaction to temperature = E/RT_s
γ	activity coefficient
δ	pressure drop (kg/s^2m^2)
ε	bed porosity
ε_i	intraparticle holdup
ε_{st}	static holdup
ε_{dyn}	dynamic holdup
ε_t	total holdup
ε_r	residual holdup
η, η_o	catalyst effectiveness, overall catalyst effectiveness
η_w, η_d	fraction of the surface of the catalyst wetted by moving liquid and dry, respectively
η_c	contacting efficiency = k_{app}/k_ν
θ	dimensionless time = t/τ
λ, λ_{eff}	thermal conductivity, effective thermal conductivity in the pellet or in the bed (kcal/m s K)
λ	parameter in Fig. (2) = $\left(\frac{\rho_G}{\rho_{AIR}} \frac{\rho_L}{\rho_{WATER}}\right)^{1/2}$
μ_L	liquid viscosity (kg/m s)
ν	stoichiometric coefficient
ν_L	kinematic viscosity of the liquid (m^2/s)
ξ	constant in Eq. (67)
π	vapour pressure ($kg/m\ s^2$)
ρ	density (kg/m^3)
σ_1, σ_c	liquid surface tension and liquid surface tension for a packing giving zero contact angle (kg/s^2)
τ	tortuosity
τ_L or τ_m	mean residence time (s)
τ_D	mean delay time (s)
τ_i	break-through time of tracer (s)
ϕ	volume ratio in Eq. (30)
ϕ	parameter = $\varepsilon_{Lst}/(\varepsilon_{Ldyn}+\varepsilon_{Lst})$
Φ	Thiele modulus = $3\frac{V_p}{A_p}\sqrt{\frac{k_\nu}{D_{eff}}}$

Φ	Weisz-Prater criterion $=\left(\frac{\tau\, d_p^2\, R_A}{D\, C_1 \varepsilon_i}\right)^{1/2}$
Φ	external mass transfer criterion $=\left(\frac{d_p^2\, R_A}{D\, Sc^{1/3} C_1}\right)^{1/2}$
ϕ_L	parameter in Eq. (2)
ψ	parameter $= \frac{\sigma_{WATER}}{\sigma_L}\left[\frac{\mu_L}{\mu_{WATER}} \cdot \left(\frac{\rho_{WATER}}{\rho_L}\right)^2\right]^{1/3}$
χ	$(\delta_L/\delta_G)^{1/2}$

Subscripts

A, B	reactants A and B
c	chemical reaction
D, d, dyn	dynamic
g, G and l, L	gas and liquid
o, e	inlet, outlet
p	pellet or "physical", i.e. without chemical reaction
s, S and st	surface and static
t	total
ext, in	external, internal
1, 2	oxygen, ethanol

REFERENCES

1. Baldi, G. and V. Specchia, "Distribution and Radial Spread of Liquid in Packed Towers with Two-Phase Concurrent Flow: Effect of Packing Shape and Size". Ing. Chim. Ital. 12, (1976) 107-112.
2. Baldi, G., S. Sicardi, I. Mazzarino, A. Gianetto and V. Specchia, "Performance of a Laboratory Trickle Bed Reactor: Catalytic Oxidation of Ethyl Alcohol". From: Frontiers in Chem. React. Eng., edited by: L.K. Daraiswamy and R.A. Mashelkar, 1984, vol. I, p.375-95;Wiley Eastern Ltd., New Delhi.
3. Baldi, G., A. Gianetto, S. Sicardi, V. Specchia, I. Mazzarino "Oxidation of Ethyl Alcohol in Trickle Bed Reactors: Analysis of the Conversion Rate"; 32nd Chem. Eng./Conf. of the Can. Soc. for Chem. Eng., 2-5 Oct., Toronto, Canada (1983).
4. Berruti, F., R.R. Hudgins, E. Rhodes, S. Sicardi, "Oxidation of Sulfur Dioxide in a Trickle Bed Reactor: A Study of

Reactor Modelling", Can. Jour. of Chem. Eng., vol. 62, (1984) 644-650.
5. Bondi, A., "Handling Kinetics from Trickle-Phase Reactors", Chem. Techn. 1, (1971) 185-188.
6. Buffham, B.A. and L.G. Gibilaro, "A Unified Time Delay Model for Dispersion in Flowing Media", Chem.Eng.J.1,(1970) 31-36.
7. Calderbank, A.H., in Uhl and Gray (Editors), "Mass Transfer in Mixing: Theory and Practice", (1967) Academic Press, N.Y.
8. Capra, V., S. Sicardi, A. Ginetto and J.M. Smith, "Effect of Liquid Wetting on Catalyst Effectiveness in Trickle Bed Reactors", Proc. 2nd World Congress of Chem. Eng., (1981) 146-155, Montreal, Canada.
9. Charpentier, J.C., M. Bakos and P. Le Goff, "Hydrodynamics of Two-Phase Concurrent Down Flow Regimes. Liquid Axial Dispersion and Dead Zones", Proc. 2nd Conference on Appl. Physical Chem., Vol. 2, (1971) 31-47, Veszprem, Hungary.
10. Charpentier, J.C., "Recent Progress in Two Phase Gas-Liquid Mass Transfer in Packed Columns", Chem. Eng. J. 11, (1976) 161-181.
11. Charpentier, J.C., C. Prost, W.P.M. Van Swaaij and P. Le Goff, "Etude de la Retention de Liquide dans une Colonne a garnissage Arrose a contre-courant et a co-courant de Gaz-Liquide", Chim. Ind. Gen. Chem. 99, (1968) 803-812.
12. Colombo, A.J., G. Baldi and S. Sicardi, "Solid-Liquid Contacting Effectiveness in Trickle Bed Reactors", Chem. Eng. Sci. 31, (1976) 1101-1108.
13. Dudukovic, M.P., "Catalyst Effectiveness Factor and Contacting Effectiveness in Trickle Bed Reactors", AIChE J. 23, (1977) 940-944.
14. Froment, G.F. and K.B. Bischoff, "Chemical Reactor Analysis and Design", John Wiley and Sons, N.Y. (1979).
15. Germain, A.H., A.G. Lefebre and G.A. L'Homme, "Experimental Study of Catalytic Trickle Bed Reactor", Chem. React. Eng. II, ACS Monograph Series No. 133, Am. Chem. Soc. (1974) 164-174 (Washington).
16. Gianetto, A., G. Baldi, S. Sicardi and V. Specchia, "Hydrodynamics and Solid-Liquid Contacting Effectiveness in Trickle Bed Reactors", AIChE J. 24, (1978) 1087-1104.
17. Gianetto, A., G. Baldi and V. Specchia, "Absorption in Packed Towers with Concurrent High Velocity Flows. I - Interfacial Areas", Ind. Chim. Ital, 6, (1970) 125-133.
18. Gianetto, A., V. Specchia and G. Baldi, "Absorption in Packed Towers with Concurrent High Velocity Flows. II - Mass Transfer", AIChE J., 19, (1973) 916-922.
19. Goto, S., A. Lakota and J. Levec, "Effectiveness Factors of n-th order Kinetics in Trickle Bed Reactors", Chem. Eng. Sci. 36, (1981) 157-162.
20. Goto, S. and J.M. Smith, "Trickle-Bed Reactor Performance", AIChE J., 21, (1975) 706-719.

21. Goto, S., J.M. Smith, "Wetting Efficiency in a Trickle Bed Reactor", Chem. Eng. Sci. 36, (1981) 280-286.
22. Hanika, J., K. Sporka, V. Ruzicka and J. Hrstka, "Measurement of Axial Temperature Profiles in an Adiabatic Trickle Bed Reactor", Chem. Eng. J. 12, (1976) 193-197.
23. Hanika, J., V. Vosecky and V. Ruzicka, "Dynamic Behaviour of the Laboratory Trickle Bed Reactor", Chem. Eng. J. 21, (1981) 109-114.
24. Hartman, M. and R.W. Coughlin, "Oxidation of Sulfur Dioxide in a Trickle Bed Reactor Packed with Carbon", Chem. Eng. Sci., 27, (1972) 867-880.
25. Hashimoto, K., K. Muroyama, F. Fuhiyoshi and S. Nagata, "Effective Radial Thermal Conductivity in Cocurrent Flow of a Gas and Liquid Through a Packed Bed", Int. Chem. Eng., 16, (1976) 760-766.
26. Henry, H.C. and J.B. Gilbert, "Scale-up of Pilot Plant Data for Catalytic Hydroprocessing", I/EC Proc. Des. Dev. 12, (1973) 328-331.
27. Herskowitz, M., "Liquid Distribution in Trickle Bed Reactors", Ph.D. Thesis, Univ. California, (1978), Davis, California.
28. Herskowitz, M. and J.M. Smith, "Liquid Distribution in Trickle Bed Reactors. Part I: Flow Measurements", AIChE J. 24, (1978) 439-449.
29. Herskowitz, M. and J.M. Smith, "Liquid Distribution in Trickle Bed Reactors. Part II: Tracer Studies", AIChE J. 24, (1978) 450-456.
30. Herskowitz, M., R.G. Carbonell and J.M. Smith, "Effectiveness Factor and Mass Transfer in Trickle Bed Reactors", AIChE J. 25, (1979) 272-278.
31. Hirose, T.M., M. Toda and Y. Sato, "Liquid Phase Mass Transfer in Packed Bed Reactor with Cocurrent Gas-Liquid Downflow", J. Chem. Eng. of Japan, 7, (1974) 187-192.
32. Hirose, T.M., Y. Mori and Y. Sato, "Liquid-to-Particle Mass Transfer in Fixed Bed Reactor with Cocurrent Gas-Liquid Downflow", J. of Chem. Eng. Japan 9, (1976) 220-224.
33. Hofmann, H., "Multiphase Catalytic Packed Bed Reactors", Conf. Chem. Reaction Eng. 6, (1979) Ghent, Belgium.
34. Hofmann, H., "Hydrodynamics and Mass Transfer in Bubble Columns", Seminar on Multiphase Reactors - (1982) Kitchener, Ont. Canada.
35. Hoftyzer, P.J., "Liquid Distribution in a Column with Dumped Packing", Trans. Inst. Chem. Engrs. (London) 42, (1964) T-109-T117.
36. Iannibello, A., S. Marengo, G. Burgio, G. Baldi, S. Sicardi and V. Specchia, "Influence of the Liquid Flow Rate on the Conversion in a Trickle Bed Reactor", in press on I/EC Process Des. Dev.
37. Jaffe, S.B., "Hot Spot Simulation in Commercial Hydrogenation Processes", I/EC Proc. Des. Dev. 15, (1976) 410-413.

38. Kan, K.M. and P.F. Greenfield, "Pressure Drop and Holdup in Two-Phase Cocurrent Trickle Flows through Beds of Small Particles", I/EC Proc. Des. Dev. 18, (1979) 740-744.
39. Kardos, J. and A. Pulz, "Darstellung und Auswertung von Verweilzeitverteilungen nach dem Zellenmodell. I und II", Chem. Techn. 21, (1969) 216-222 and 275-279.
40. Kardos, J. and A. Pulz, "Darstellung und Auswertung von Verweilzeitverteilungen nach dem Zellenmodell, III", Chem. Techn. 28, (1976) 329-333.
41. Kardos, J. and A. Pulz, "Darstellung und Auswertung von Verweilzeitverteilungen nach dem Zellenmodell. IV", Chem. Techn. 30, (1978) 447-451.
42. Koros, R.M. and E.J. Nowak, "A Diagnostic Test of the Kinetic Regime in a Packed Bed Reactor", Chem. Eng. Sci. 22, (1966) 470.
43. Koros, R.M., "Catalyst Utilization in Mixed-Phase Fixed Bed Reactors", Proc. 4th ISCRE, Heidelberg, Apr. 6-8, (1976) 372-378.
44. Koros, R.M., "Multiphase Chemical Reactors - Design Methods", NATO Advanced Study Institute on Multi-phase Chem. React., (1980) Vineiro, Portugal.
45. Mata, A.R. and J.M. Smith, "Oxidation of Sulfur Dioxide in a Trickle Bed Reactor", Chem. Eng. J., (1981), 22, 229-235.
46. Matsuura, A., Y. Hitaka, T. Akehata and T. Shirai, "Effect of Latent Heat on the Temperature Profile for Concurrent Air-Water Downflow in Packed Beds", Kagaku Kogaku Ronbunshu, 5, (1979) 263-266.
47. Mears, D.E., "The Role of Axial Dispersion in Trickle-Flow Laboratory Reactors", Chem. Eng. Sci. 26, (1971) 361-366.
48. Mears, D.E., "The Role of Liquid Holdup and Effective Wetting in the Performance of Tricle Bed Reactors", Chem. React. Eng. II - Adv. in Chem. Ser. 133, Am. Chem. Soc. (1974) 268-275 (Washington).
49. Mears, D.E., "Tests for Transport Limitations in Experimental Catalytic Reactors", I/EC Proc. Des. Dev. 10, (1971) 541-547.
50. Midoux, N., N. Favier and J.C. Charpentier, "Flow Pattern, Pressure Loss and Liquid Holdup Data in Gas-Liquid Downflow Packed Beds with Foaming and Non-foaming Hydrocarbons", J. Chem. Eng. Japan, 9, (1976) 350-355.
51. Mills, P.L. and M.P. Dudukovic, "A Dual-Series Solution for the Effectiveness Factor of Partially Wetted Catalyst in Trickle Bed Reactors", I/EC Fundam., 18, (1979) 139-147.
52. Mills, P.L. and M.P. Dudukovic, "Tracer Methods for Evaluation of Liquid-Solid Contacting in Trickle Bed Reactors", Proc. 2nd World Congress of Chem. Eng., (1981) Montreal, Canada.
53. Mitchell, R.W. and I.A. Furger, "Trickle Flow in Packed Beds", Trans. Inst. Chem. Eng. 50, (1972) 334-341.
54. Montagna, A.A. and Y.T. Shah, "The Role of Liquid Holdup, Effective Catalyst Wetting and Backmixing on the Performance

of a Trickle Bed Reactor for Residue Hydrodesulfurization", I/EC Proc. Des. Dev. 14, (1975) 479-483.

55. Montagna, A.A., Y.T. Shah and J.A. Paraskos, "Effect of Catalyst Particle Size on the Performance of a Trickle Bed Reactor", I/EC Proc. Des. Dev. 16, (1977) 152-155.
56. Morita, S. and J.M. Smith, "Mass Transfer and Contacting Efficiency in a Trickle Bed Reactor", Ind. Eng. Chem. Fund., 17 (1978) 113-120.
57. Muroyama, K., K. Hashimoto and T. Tomita, "Pulsing Flow Characteristics of a Gas-Liquid Cocurrent Flow through Packed Beds", Karagu Kokagu Ronbunshu, 3, (1977) 612-615.
58. Onda, K., H. Takeuchi, Y. Maeda and N. Takeuchi, "Liquid Distribution in Packed Columns", Chem. Eng. Sci., 28, (1973) 1677-1683.
59. Onda, K., H. Takeuchi and H. Koyana, "Study of Mass Transfer between Phases by a Diaphram Cell", Kagaku Kogaku 32, (1976) 121-126.
60. Oshima, S., T. Takemasu, Y. Kuriki, K. Shimada, M. Suzuki, and J. Kato, "Liquid Phase Mass Transfer Coefficient and Gas Holdup in a Packed Bed Cocurrent Upflow Column", Kagaku Kogaku 3, (1977) 400-404.
61. Paraskos, J.A., J.A. Frayer and Y.T. Shah, "Effect of Holdup on Incomplete Catalyst Wetting and Backmixing During Hydroprocessing in Trickle Bed Reactors", I/EC Proc. Des. Dev. 14, (1975) 315-319.
62. Pavko, A. and J. Levex, "Wetting Efficiency in Trickle Bed Reactors", Proc 2nd World Cong. on Chem. Eng., Vol 3, 156-159, Oct. 4-9, (1981) Montreal, Canada.
63. Prater, C.D., "The Temperature Produced by Heat of Reaction in the Interior of Porous Particles", Chem. Eng. Sci. 8, (1958) 284-286.
64. Puranik, S.S. and A. Vogelpohl, "Effective Interfacial Areas in Irrigated Columns", Chem. Eng. Sci. 29, (1974) 501-506.
65. Sato, Y., T. Hirose, T. Ida, "Flow Pattern and Pulsation Properties of Cocurrent Gas-Liquid Downflow in Packed Beds", Kagaku Kogaku 38, (1974) 534-537.
66. Satori, H. and S. Nishizaki, "Studies on the Hydrodesulfurization of Residual Oil using a Test Apparatus of Larger Size. II. On the Theoretical Analysis of the Fixed Bed Reactor", Int. Chem. Eng. 11, (1971) 339-344.
67. Satterfield, C.N., "Trickle Bed Reactors", AIChE J 21, (1975) 209-214.
68. Satterfield, C.N., "Mass Transfer in Heterogeneous Catalysis", MIT Press, (1970) Cambridge, Massachusetts.
69. Satterfield, C.N., A.A. Pelossof and T.K. Sherwood, "Mass Transfer Limitations in a Trickle Bed Reactor", AIChE J., 15, (1969) 226-231.
70. Satterfield, C.N. and F. Ozel, "Direct Solid Catalyzed Reaction of a Vapor in an Apparently Completely Wetted Trickle Bed Reactor", AIChE J. 19, (1973) 1259-62.

71. Schmalzer, D.K. and H.F. Hoelscher, "A Stochastic Model of Packed Bed Mixing and Mass Transfer", AIChE J. 17, (1971) 104-110.
72. Sedriks, W. and N.C. Kenney, "Partial Wetting in Trickle Bed Reactors: the Reduction of Crotonaldehyde over a Palladium Catalyst", Chem. Eng. Sci. 28, (1973) 559-561.
73. Shah, Y.T., "Gas-Liquid-Solid Reactor Design", McGraw-Hill, (1979) N.Y.
74. Shah, Y.T., "Trickle Bed Reactors", NATO Advanced Study Institute on Multi-phase Chem. React., (1980) Vineiro, Portugal.
75. Sicardi, S., G. Baldi, C. Specchia and A. Gianetto, "Catalyst Areas Wetted by Flowing and Semistagnant Liquid in Trickle Bed Reactors", Chem. Eng. Sci. 35, (1980) 67-73.
76. Sicardi, S., G. Baldi, V. Specchia, I. Mazzarino and A. Gianetto, "Packing Wetting in Trickle Bed Reactors: Influence of the Gas Flow Rate", Chem. Eng. Sci. 36, (1981) 226-228.
77. Specchia, V., G. Baldi and A. Gianetto, "Solid-Liquid Mass Transfer in Trickle Bed Reactors", Proc. 4th ISCRE, (1976) 656-662 Heidelberg, Germany.
78. Specchia, V. and G. Baldi, "Pressure Drop and Liquid Holdup for Two-Phase Concurrent Flow in Packed Beds", Chem. Eng. Sci. 32, (1977) 515-519.
79. Specchia, V., S. Sicardi and A. Gianetto, "Absorption in Packed Towers with Concurrent Upward Flow", AIChE J, 20 (1974) 646-653.
80. Specchia, V., G. Baldi and A. Gianetto, "Solid-Liquid Mass Transfer in Concurrent Two-Phase Flow Through Packed Beds", I/EC Proc. Des. Dev. 17, (1978) 362-369.
81. Specchia, V. and G. Baldi, "Heat Transfer in Trickle Bed Reactors", Chem. Eng. Comm. 3, (1979) 483-488.
82. Specchia, V. and S. Sicardi, "Modified Correlation for the Conductive Contribution of Thermal Conductivity in Packed Bed Reactors", Chem. Eng. Comm. 6, (1980) 131-36.
83. Specchia, V. and S. Sicardi, "Trasferimento di Calore in Reattori a Sgocciolamento", Atti Acc. Sci., Torino, 114, (1980) 71-79.
84. Specchia, V., "Fluodinamica in Reattori a Sgocciolamento", Proceed. of Course on Reattori Chimici Trifasici, Soc. Chimica Italiana, 29 Nov.-3 Dec. 1982, San Donato Milanese.
85. Stanek, V. and N. Kolev, "A Study of the Dependence of Radial Spread of Liquid in Random Beds on Local Conditions of Irrigation", Chem. Eng. Sci. 33, (1978) 1049-1053.
86. Sylvester, N.D. and P. Pitayagulsarn, "Effect of Transport Processes on Conversion in a Trickle Bed Reactor", AIChE J, 19, (1973) 640-644.
87. Sylvester, N.D. and P. Pitayagulsarn, "Effect of Catalyst Wetting on Conversion in a Trickle Bed Reactor", Can. J. Chem. Eng. 52, (1974) 539-540.

88. Sylvester, N.D. and P. Pitayagulsarn, "Radial Liquid Distribution in Concurrent Two-Phase Downflow in Packed Beds", Can. J. Chem. Eng. 53, (1975) 599-606.
89. Sylvester, N.D., "Slurry and Trickle Bed Reactor Effectiveness", Water Res. 9, (1975) 447-453.
90. Tan, C.S. and J.M. Smith, "Catalyst Particle Effectiveness with Unsymmetrical Boundary Conditions", Chem. Eng. Sci. 35, (1980) 1601-1609.
91. Turpin, J.L. and R.L. Huntington, "Prediction of Pressure Drop for Two-Phase, Two-Component Concurrent Flow in Packed Beds", AIChE J. 13, (1967) 1196-99.
92. Tukac, V., I. Mazzarino, G. Baldi, A. Gianetto, S. Sicardi, V. Specchia, "Interpretation of Results obtained in a Laboratory Trickle Bed Reactor", (1984) presented for publication to Chem. Eng. Science.
93. Turek, F., and R. Lange, "Mass Transfer in Trickle Bed Reactors at Low Reynolds Number", Chem. Eng. Sci., 36, (1981) 569-579.
94. Weekman, K.W. Jr., "Hydroprocessing Reaction Engineering", Proc. 4th ISCRE, (1976) 615-647, Heidelberg, Germany.
95. Wehner, J.E. and R.W. Wilhelm, "Boundary Conditions of Flow Reactor", Chem. Eng. Sci. 6,, (1957) 89-92.
96. Weisz, P.B., and J.S. Hicks, "The Behaviour of Porous Catalyst Particles in View of Internal Mass and Heat Diffusion Effects", Chem. Eng. Sci. 17, (1962) 265-275.
97. Zarzycki, R., "The Distribution of Liquid in a Packed Column. II. The Effect of Gas Flow Velocity on the Distribution", Int. Chem. Eng. 12, (1972) 88-92.

HEAT TRANSFER IN PACKED BED REACTORS

D.L. CRESSWELL

I.C.I. NEW SCIENCE GROUP, THE HEATH, RUNCORN, CHESHIRE, ENGLAND

1 INTRODUCTION

There are several commercial gas-solid catalysed reactions in which heat transfer plays a significant, if not dominant, role in limiting the reactor productivity, lowering the process selectivity and reducing the life of the catalyst. Among these include the oxidation of ethylene, benzene, C_4 hydrocarbons and methanol, the ammoxidation of propylene, methanol synthesis (Lurgi), the hydrochlorination of methanol and steam reforming of natural gas and naphtha.

These reactions are carried out within a parallel bundle of reaction tubes, comprising either a few hundred suspended within a furnace, as in steam reforming, to many thousands contained within a reactor shell through which heat transfer oil, water or molten salts are circulated. The various tubes range from 2 cm to 10 cm in diameter, depending upon the degree of exothermicity (endothermicity) of the reactions and from a few metres to as much as 15 metres in length. Mass velocities may vary widely, too, within the range 1-20 kg per square metre of tube cross section per second, depending upon the pressure. Catalyst pellets come in all shapes and sizes, ranging from 3 mm irregular granules to near-perfectly formed rings of almost 2 cm external diameter. In spite of all these variations one factor is common; the ratio of tube internal diameter to pellet diameter tends to lie within a narrow range, say between 5 and 10.

The reasons for this are not accidental but rather have evolved, partly through experience, and partly through pilot plant studies, modelling exercises and so on, as the region within which the many trade-offs between capital, raw-material and energy costs become active. Modelling of heat transfer in packed beds has a vital role to play in enabling these optima to be better defined and more quickly defined so that development work can proceed efficiently. However, it is only one aspect of several which are of importance to the overall picture, and this should not be lost sight of in assessing the sophistication of the treatment of heat transfer necessary.

This review, therefore, will not attempt to cover the gamut of heat transfer models proposed in the literature and reviewed elsewhere [1]. Rather it will concentrate on those few models most widely used in industrial applications and attempt to lay emphasis on their reliability, on parameter values to employ and mechanistic, rather than empirical, methods of prediction, wherever possible. Some closing remarks on future trends in heat transfer research straight out of the crystal ball will be tentatively suggested.

2. GENERAL OBSERVATIONS

This section sets the scene for subsequent modelling exercises. It is intended to be mainly qualitative, although inevitably quantitative data are presented which are of direct use to the designer. It deals with the accuracy of temperature measurement itself and with the effects of tube diameter, particle diameter, particle shape, particle conductivity and absolute temperature level, as well as mass velocty, on overall heat transfer rates.

2.1 The radial temperature profile

Figure 1a displays a fairly typical set of radial temperature measurements taken in the air stream immediately leaving a packed tube, heated at the wall by condensing steam. At any given radial position, the various symbols denote point temperatures measured at different angular positions.

The angular variation of temperature, which appears random in nature, is certainly significant relative to the total radial variation, and appears to increase in the neighbourhood of the wall. Yet, no model exists which even attempts to recognise this behaviour! Bearing the size of these variations in mind one has to have some doubt on the legitimacy of sophisticated treatments which claim to be able to distinguish between solid and fluid temperatures.

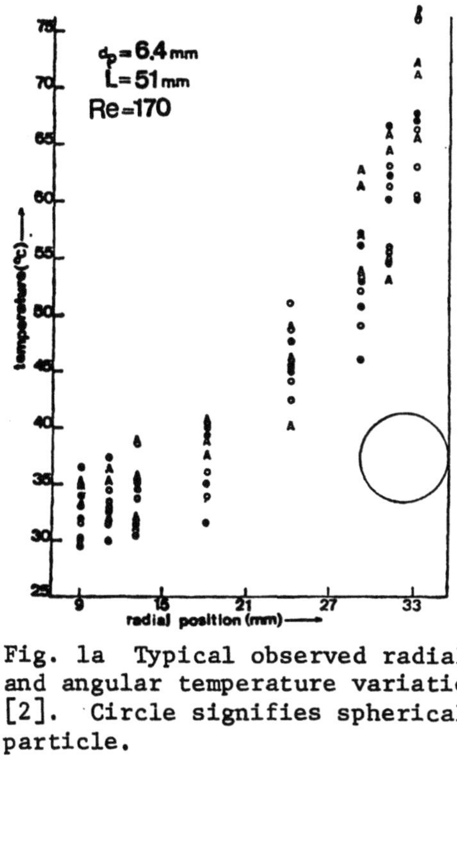

Fig. 1a Typical observed radial and angular temperature variations [2]. Circle signifies spherical particle.

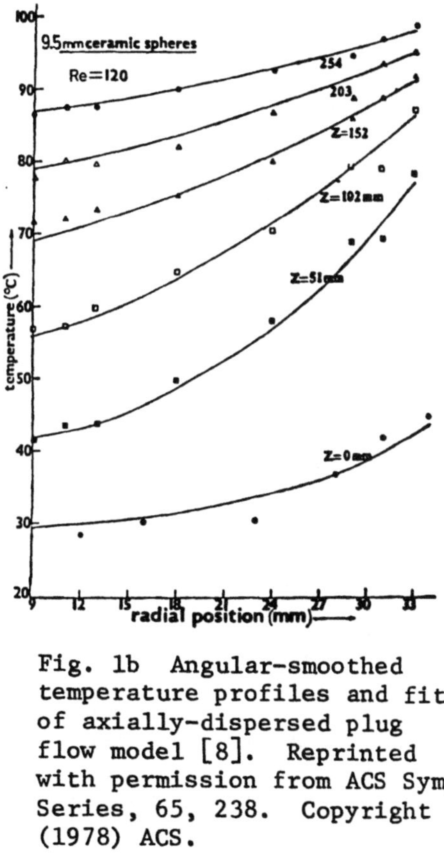

Fig. 1b Angular-smoothed temperature profiles and fit of axially-dispersed plug flow model [8]. Reprinted with permission from ACS Symp. Series, 65, 238. Copyright (1978) ACS.

If we are prepared to deal only with angular-averaged temperatures, as it seems we must, the prospect is not nearly so bleak. Remarkably, perhaps, the averaging produces smooth radial temperature profiles, even when the ratio of tube to particle diameter is fairly small. This leads us naturally to the happy prospect of applying a continuum-type model to describe radial heat transfer, characterised by an effective radial conductivity $k_{r,eff}$. Such smoothed profiles are shown in Figure 1b. These profiles do not extrapolate to the wall temperature, however, signifying a further resistance at the wall, which might be accounted for by introducing a second parameter $h_{w,eff}$, the effective wall heat transfer coefficient.

If we throw into the pot the further assumption of plug-flow, which seems a reasonable ideal in long and narrow tubes, and stir, what matures is a two-dimensional homogeneous continuum model, containing two parameters which we would prefer to determine by experiment.

It sometimes happens, however, that the tube diameter necessary is too small to make reliable radial temperature measurements. Pragmatists would then probably be content to measure an overall heat transfer coefficient U between the bed centre and the surroundings and use a one-dimensional model, rather than rely on extrapolation of heat transfer parameters.

In fact, the two models can be matched asymptotically with increasing bed depth in terms of thermal flux to give

$$U/h_{w,eff} = J_o(\alpha_1) \tag{1}$$

where α_1 is the smallest positive root of $\alpha J_1(\alpha) = Bi.J_o(\alpha)$ and $Bi = h_{w,eff}\, d_t/2.k_{r,eff}$. If, further, the radial temperature profile is taken to be parabolic (not a bad assumption, even under reaction conditions), (1) may be replaced by the simpler equation

$$1/U = 1/h_{w,eff} + d_t/4\, k_{r,eff} \tag{2}$$

2.2 The tube diameter

The potential for transferring heat radially to or from the bed is proportional to $4\,U/d_t$. Equation (2) tells us, therefore, that this potential is inversely proportional to d_t^n, where n lies between 1 and 2, depending upon the controlling heat transfer process. Heat transfer rates must be matched to heat generation/abstraction rates, as far as possible, in order to minimise temperature gradients along the bed, which are generally undesirable. At economically acceptable levels of tube productivity this may demand tubes as small as 2 cm in diameter. In general, tube diameters lie in the narrow range 2-5 cm, save for steam reforming operations where larger diameters can be used.

2.3 The particle diameter

Particle diameter has a most interesting effect on the overall heat transfer coefficient U, for a maximum in the relationship appears to exist, as shown in Figure 2. This observation is, in fact, not new. It was first spotted nearly 40 years ago by Leva and co-workers [3] and quantified by them. The maximum was predicted to always occur at the fixed ratio $d_t/d_p = 6.5$,

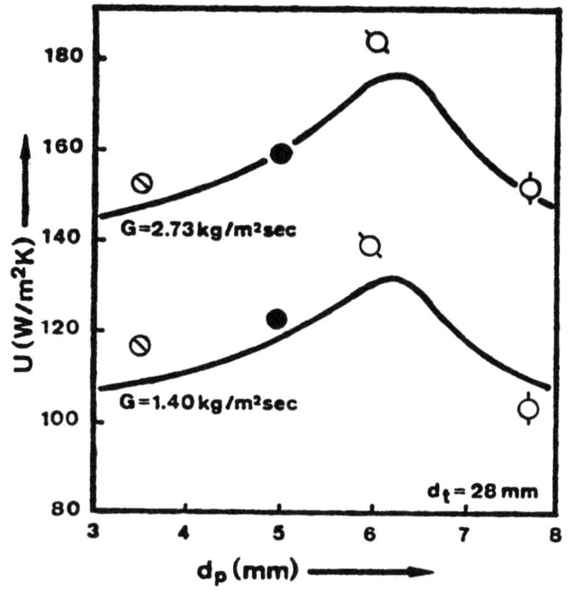

Fig. 2 The dependence of the overall heat transfer coefficient on particle diameter [4]. Reprinted with permission ACS Symp. Series 196, 527. Copyright (1982) ACS.

irrespective of d_t or G, and our own data bears this out reasonably well. However, no explanation was offered. The mechanistic model presented in Section 4 predicts the continuous curves, which follow the data fairly well. It would suggest that the major thermal resistance shifts from effective conduction through the bed to, ultimately, heat transfer at the wall as the particle size is increased. The radial effective conductivity increases with particle diameter, whereas the wall heat transfer coefficient decreases, hence the existence of an optimum in the overall coefficient, U.

2.4 The particle shape

There are more surprises in store when it comes to the matter of particle shape, although this time they cannot be fully explained. There seems to be no single shape of particle which gives the highest heat transfer coefficient at all mass velocities. At low mass velocities, where effective radial conduction is dominant, open-shaped packings, such as thin walled rings, give the highest heat transfer coefficients, probably because they encourage good radial mixing. However, at high mass velocities, where heat transfer at the wall is limiting, irregular shapes, such as Berl saddles or half-rings are the most effective, probably because they pack more densely in the neighbourhood of the wall, and therefore lead to higher interstitial velocities in the wall region. These findings are summarised in Figure 3 with air as the fluid. When heat transfer coefficient is plotted versus the bed pressure drop per unit external surface area of packing, the same trend is observed.

2.5 Absolute temperature

A considerable extrapolation in the absolute temperature at which the bulk of heat transfer measurements are reported (< 100°C) and reaction temperatures (200-800°C) can render the former data of limited value, particularly at the higher temperatures, where radiation is involved.

Some of our own data are shown in Figure 4. They indicate that the overall heat transfer coefficient increases linearly with absolute temperature in the range 220-420°C, at a rate that seems to increase with mass velocity. Only a small part of this increase can be explained by variation in fluid physical properties (dotted curve), as determined by the correlations in Section 4. The majority of the increase is ascribed to radiation and can be explained quantitatively (solid curves), as in reference 4.

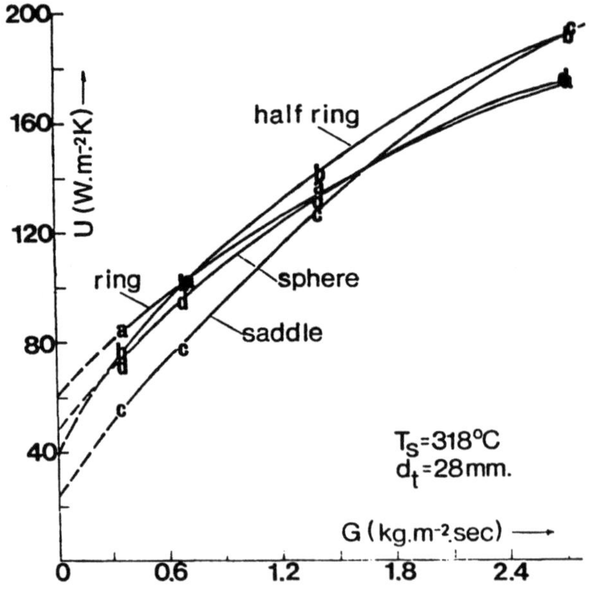

Fig. 3 The dependence of the overall heat transfer coefficient on mass velocity for various particle shapes [4]. Reprinted with permission ACS Symp.Series 196,527.Copyright (1982)ACS.

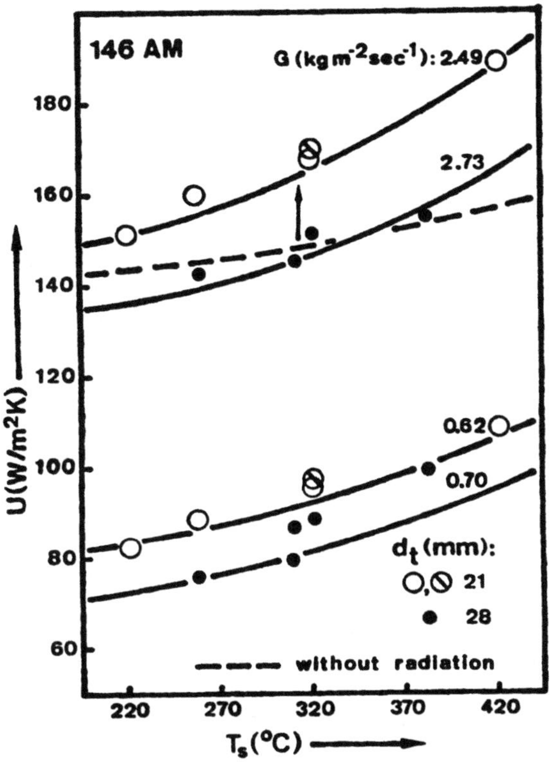

Fig 4. The effect of salt bath temperature on the overall heat transfer coefficient [4]. Reprinted with permission ACS Symp. Series 196, 527. Copyright (1982) ACS.

2.6 Mass velocity

The apparent dependence of the overall heat transfer coefficient on mass velocity is highly complex. This is because overall heat transfer is a lumping together of several underlying phenomenological heat transfer processes, each of which may have a different mass velocity dependence, and whose contributions differ proportionately as conditions are altered.

Thus, the exponent on the mass velocity dependence of U has been measured to lie within the range 0.3 - 0.6, the precise value depending upon the particle size, shape, tube diameter [4]. Mechanistic considerations would predict a value anywhere in the range 0-0.75!

2.7 Particle conductivity

Most catalyst carriers have conductivities within the range 0.3 - 7 w/m.K. Since conduction through the solid is limited by the high resistance offered by gas fillets in the neighbourhood of contact points, the particle conductivity is generally of minor consideration.

3 EXPERIMENTAL DETERMINATION OF EFFECTIVE HEAT TRANSFER PARAMETERS

3.1 Design considerations

Test beds tend to be relatively short since the bulk of the heat transfer may be accomplished in less than a metre of packing. Reactor tubes, on the other hand, are usually considerably longer and generally narrower. This has two important consequences. Firstly, the design of the apparatus must be such that entry length effects are minimised. Secondly, it may be necessary to incorporate axial, as well as radial conduction, when analysing data, although in the modelling of reactor tubes this is often unnecessary.

Entry length effects are reduced by the use of a calming section [5]. Both calming and test sections are jacketed and bolted together at the plane Z=0 in Figure 5, though thermally insulated from each other. Both sections are packed with similar solid particles to provide a continuous length of packing.

The wall temperatures in the two sections are held at uniform, but different, values, preferably by circulation of water or molten salts or condensation of steam. Electric heating is best avoided, since it is very difficult to achieve uniform wall temperatures.

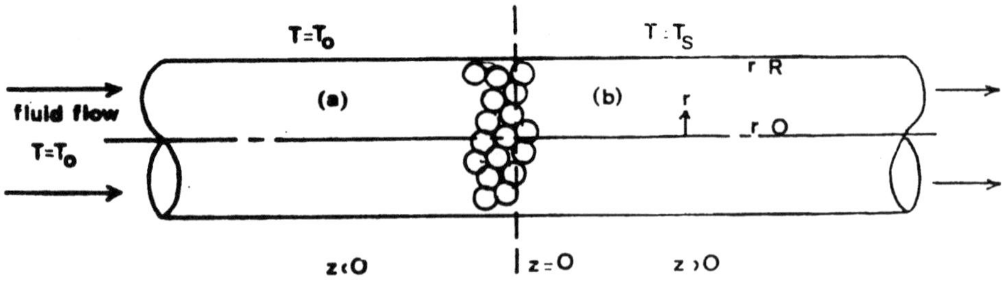

Fig. 5 Integral test bed and calming section [5]. Reprinted with permission Chem.Engng.Sci.30,261. Copyright (1975)Pergamon Press.

Radial temperature measurements are best made in the fluid immediately leaving the test bed, thus enabling angular-averaged measurements to be made without disturbing the bed [6].

With narrow tubes, this arrangement is no longer possible. However, the axial temperature profile may still be measured, preferably by means of a thermocouple sliding within a narrow capillary tube placed along the centre of the bed [4].

The effects of thermal conduction along the calming section wall, owing to heat leakage across the plane Z = 0, can have a significant influence on radial heat transfer parameters [20]. It is recommended, therefore, that this be built of glass or similar low conductivity material, rather than metal. Similarly, the cross supporting the thermocouples above the bed is preferably made out of nylon, rather than metal.

3.2 Mathematical Models

(a) One dimensional model

A heat balance on a volume element of the bed, assuming plug flow, leads to the following differential equation:

$$G\,Cp\,\frac{dT}{dZ} = \frac{4U}{d_t}(T_s - T) \qquad (3)$$

with $T = T_o$ at $Z = 0$

Integration of Eqn. (3) suggests that ln θ should depend linearly on bed depth Z, with slope $-4U/d_tGCp$, where $\theta = T_s - T$.

Wellauer's data [4] plotted in Figure 6 confirms this characteristic, but it does not extend to Z=0. Only after full development of radial temperature and velocity profiles does the one-dimensional treatment become valid. This entry depth is between 20 and 30 cms, depending upon the mass velocity.

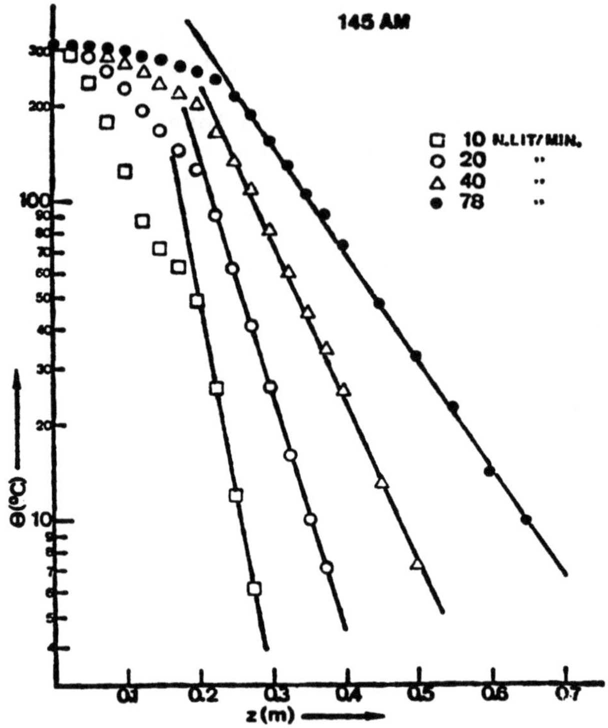

Fig. 6 Plot of ln θ vs Z showing linear dependence [4]. Reprinted with permission ACS Symp.Series 196,527.Copyright (1982)ACS.

(b) Two dimensional model

In this case a heat balance on a volume element of either the calming section or the bed leads to the equation

$$GCp.\frac{\partial T_b}{\partial Z} = k_{r,eff}.\left(\frac{\partial^2 T_b}{\partial r^2} + \frac{1}{r}\frac{\partial T_b}{\partial r}\right) + k_{a,eff}.\frac{\partial^2 T_b}{\partial Z^2} \qquad (4)$$

where it is assumed the same parameter values apply in the two sections (ie T_o and T_s are not too dissimilar).

Boundary conditions are :

$$\frac{\partial T_b}{\partial r} = 0 \qquad \text{at} \quad r = 0 \qquad (5)$$

$$- k_{r,eff}.\frac{\partial T_b}{\partial r} = h_{w,eff}.(T_b - T_s) \text{ at } r = R,\ Z>0$$

$$(6)$$

$$- k_{r,eff}.\frac{\partial T_b}{\partial r} = h_{w,eff}.(T_b - T_o) \text{ at } r = R,\ Z<0$$

$$T_b \rightarrow T_o \quad \text{as} \quad Z \rightarrow -\infty \tag{7}$$

$$T_b \rightarrow T_s \quad \text{as} \quad Z \rightarrow +\infty \tag{8}$$

$$T_b(0^+) = T_b(0^-) \tag{9}$$

$$\frac{\partial T_b}{\partial Z}(0^+) = \frac{\partial T_b}{\partial Z}(0^-) \tag{10}$$

Integration of Eqns. (4) - 10) leads to

$$\frac{T_s-T_b}{T_s-T_o} = \sum_{n=1}^{\infty} \frac{Bi.(1+A_n) \cdot J_o(\alpha_n \cdot r/R)}{A_n.(Bi^2 + \alpha_n^2).J_o(\alpha_n)} . \exp\left\{\frac{-Pe_a(A_n-1)Z}{2d_p}\right\} \tag{11}$$

where $Pe_a = GC_p d_p/k_{a,eff}$

$Pe_r = GC_p d_p/k_{r,eff}$

$Bi = h_{w,eff} \cdot R/k_{r,eff}$

and $A_n = \sqrt{1 + 16\ \alpha_n^2/Pe_a.Pe_r(d_t/d_p)^2}$

$$\alpha_n\ J_1(\alpha_n) = Bi.J_o(\alpha_n)\ ,\ n = 1,2,\ \ldots\infty \tag{12}$$

A special case of the axially-dispersed plug flow model (ADPF), given by Eqn. (11), arises when $Pe_a \rightarrow \infty$. This leads to the frequently used plug flow model solution

$$\text{(PF)}: \quad \frac{T_s-T_b}{T_s-T_o} = 2 \sum_{n=1}^{\infty} \frac{Bi.J_o(\alpha_n.r/R)}{(Bi^2+\alpha_n^2).J_o(\alpha_n)} . \exp\left\{\frac{-4\alpha_n^2 Z}{\beta^2 d_p.Pe_r}\right\} \tag{13}$$

where $\beta = d_t/d_p$.

3.3 Estimation of parameters

The dimensionless heat transfer parameters Pe_a, Pe_r and Bi are jointly estimated by fitting Eqn. (11) to measured radial temperature profiles at several bed depths by minimisation of the sum of squares function

$$S(\underline{x}) = \sum_{i=i}^{m} (T_{bi}(\underline{x}) - T_{obs,i})^2 \qquad (14)$$

where $\underline{x}^T = Pe_a, Pe_r, Bi$, and m is the total number of temperature readings.

To aid the iterative minimisation of $S(\underline{x})$ a preliminary examination of the sum of square surface is advised, using, say, only two terms in the analytic series to keep the computation practical. Some typical sections through the three dimensional surface are shown in Figure 7. They reveal a ridge running parallel to the Pe_a axis, indicating poor estimation of this parameter. Sections in the Bi, Pe_r space, on the other hand, are banana-shaped and become more steeply oriented to the Pe_r axis at low values of the mass velocity. From these contours it appears that Pe_r is best determined at lower Re and Bi at higher Re.

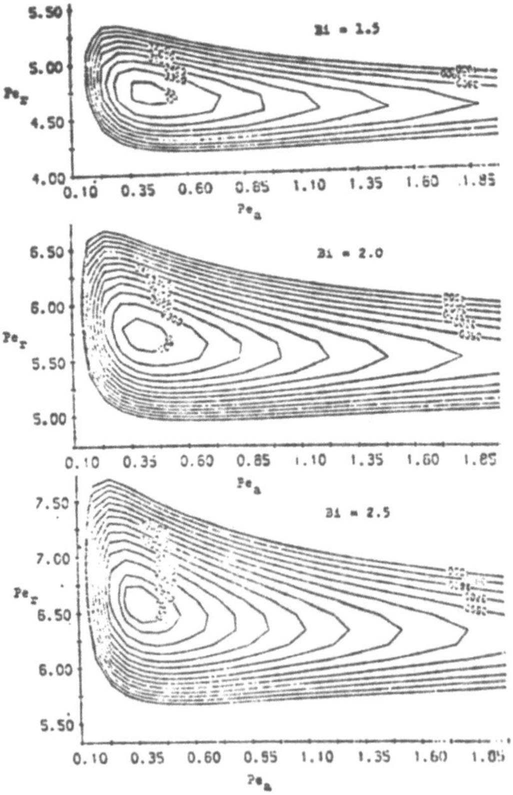

Fig. 7 Sum of squares contours in the Pe_r, Pe_a space at various values of Bi [2].

As a reasonable compromise between speed and accuracy 15 terms of the infinite series solution in Eqn (11) and 8 terms in Eqn (13) generally guarantee an accuracy of 0.1°C on temperature, which is well within the accuracy of measurement.

Any of a number of non-linear optimisation algorithms may be used to minimise $S(\underline{x})$ and convergence is usually guaranteed from good starting estimates. A modified Marquardt algorithm [7] is to be particularly recommended.

3.4 Evaluation of Models

Dixon and co-workers [8] carried out a careful statistical analysis of the ADPF and PF models. This revealed some interesting findings.

Neither model showed significant lack of fit to radial temperature profiles when fitted on a depth by depth basis. However, the plug flow model parameters were found to decrease systematically with increasing bed depth. Figure 8 shows this most clearly in the case of the effective radial conductivity. No such effect is observed with the axial conduction model, as is apparent from Figure 9.

If the two figures are superimposed, the estimates of $k_{r,eff}$ obtained from the ADPF model tend to coincide with those obtained on the longest bed using the PF model. It would seem, therefore, that axial conduction can significantly affect radial heat transfer in the short beds typically used in heat transfer studies.

The axial conduction model showed no significant lack of fit when simultaneously fitted to temperature measurements at several bed depths, covering a wide range of mass velocity, particle size and conductivity. Figure 1(b) shows a typical fit to angular-averaged measurements.

3.5 Parameter confidence intervals

Linearised 95% marginal confidence intervals on the key parameters Pe_r and Bi tend to differ greatly (Figs 10 and 11). The radial Peclet number is generally well-determined, whereas the Biot number shows considerable scatter. This is perhaps not surprising, since estimation of the latter necessitates extrapolation of uncertain temperature measurements made in the vicinity of the wall (Fig 1(a)). In spite of angular-averaging and fitting over several bed depths, much uncertainty remains.

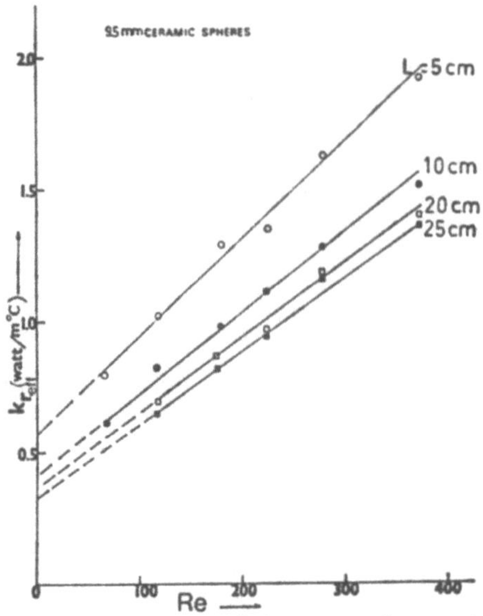

Fig. 8 - Depth dependence of $k_{r,eff}$ from the plug flow model [8]. Reprinted with permission from ACS Symp.Series, 65,238. Copyright (1978) ACS.

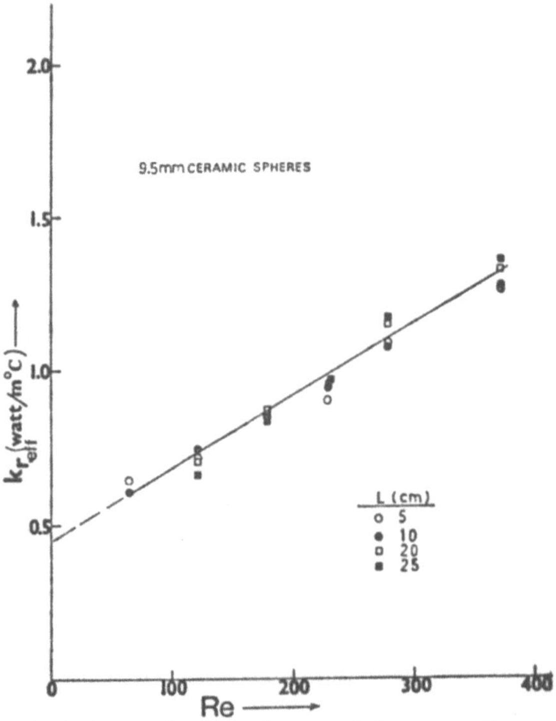

Fig. 9 - Depth independent values of $k_{r,eff}$ from the ADPF model [8]. Reprinted with permission from ACS Symp.Series,65,238. Copyright (1978) ACS.

The two parameters, too, tend to be highly correlated. While $k_{r,eff}$ is not a conductivity in the true sense, it nevertheless has a sound theoretical basis and can be predicted from underlying mechanisms quite accurately. The wall heat transfer coefficient $h_{w,eff}$, on the other hand, has continued to defy either direct measurement or prediction. Perhaps it is no more than an empirical parameter needed in the model to account for a decreasing $k_{r,eff}$ near the wall.

Possibly it was this realisation that led to Gunn and Ahmed [9] proposing recently a new model incorporating a finite wall region.

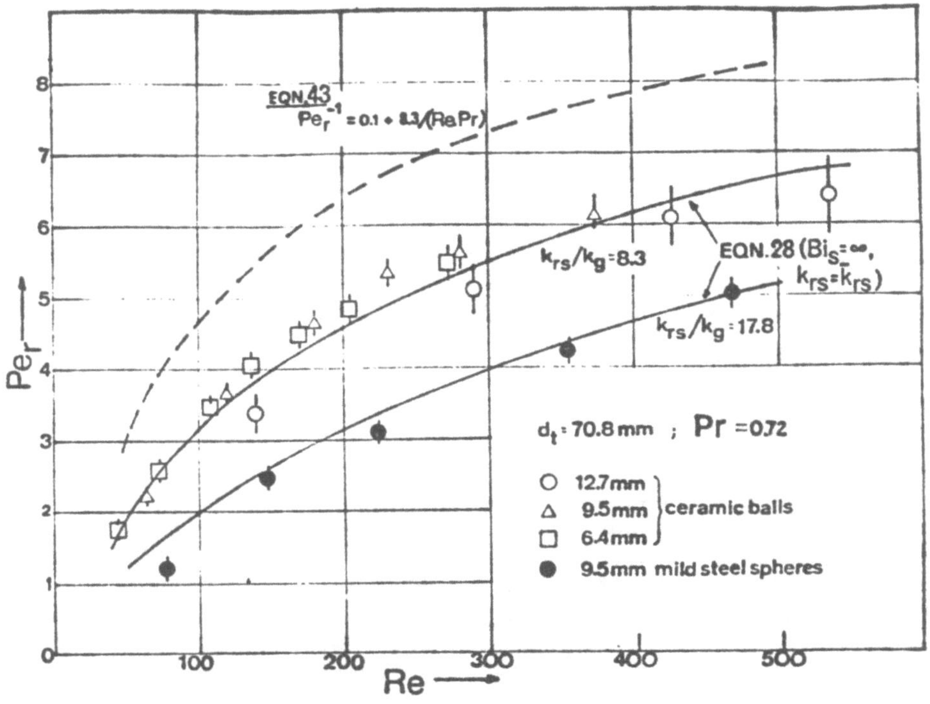

Fig. 10 - Various radial Peclet data showing 95% marginal confidence intervals (tie-lines); [8]. Reprinted with permission from ACS Symp. Series, 65, 238. Copyright (1978) ACS.

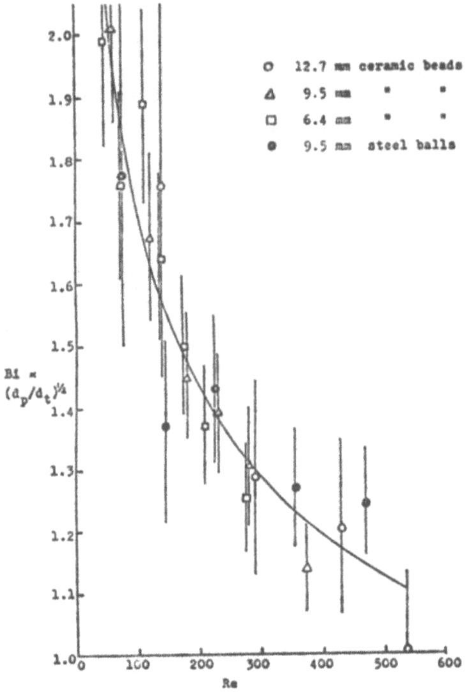

Fig.11 - Various Biot number data showing 95% marginal confidence intervals [8]. Reprinted with permission ACS Symp. Series, 65, 238. Copyright (1978) ACS.

4 MECHANISTIC MODELS OF EFFECTIVE HEAT TRANSFER

4.1 Preliminary observations

Figure (10) neatly shows up the importance of the solid for providing a direct pathway for radial heat transfer. Radial mixing experiments in beds of uniform spheres would indicate a turbulent conduction limit to Pe_r of ~10. Yet, measured Peclet numbers are consistently lower than this, and clearly depend upon the thermal conductivity of the packing. However, the effect of pellet conductivity appears to be logarithmic, rather than linear.

The equivalence of Biot numbers for metallic and ceramic packings in Figure (11) clearly indicates, too, a significant pathway for heat transfer between the solid and the wall.

It is interesting also to note from Figure (9) that the effective radial conductivity varies linearly with mass velocity and extrapolates to a finite value as Re → 0, once more indicating a direct solid contribution.

This leads us to speculate that perhaps the bed may be divided into two phases:-

(a) a fluid phase

(b) a solid phase

which exist as inter-twined continua and admit of different temperature profiles [10].

Each phase is characterised by effective radial and axial conductivities, and heat transfer coefficients are needed to account for heat exchange between the phases and the wall and between the phases themselves, as represented in Figure 12 for spherical packing.

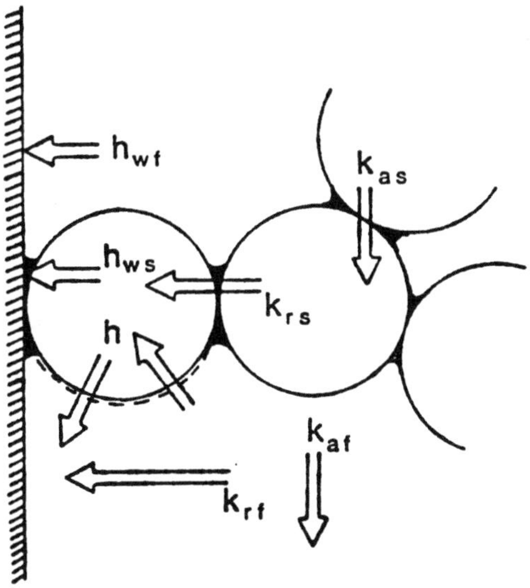

Fig. 12 - Fundamental heat transfer processes occurring within a packed bed [13]

4.2 Two dimensional heterogeneous model

Following on this suggestion the next step would be to obtain solutions for the temperature profiles in the moving and in the stationary phases and then to evaluate effective transport parameters by interpreting these results in terms of a one-phase model of the bed. This would lead to relations between effective transport parameters and what are believed to be the major underlying and independently measurable heat transfer steps.

In an actual packed bed, temperature gradients are distributed on two scales - large scale gradients in the axial and radial directions and smaller scale gradients within individual particles. It is very difficult to account fully for these two effects. Thus, the details of heat transfer between point or line-contacting pellets are difficult to describe analytically. There are two ways around this problem. Firstly, the fine structure is sacrificed by smoothing temperature gradients within pellets, leading to a continuum model capable only of describing the large scale. Or, alternatively, one may neglect point contacts between particles and treat temperature distributions within separate particles, each one influenced by the temperatures in neighbouring particles <u>only</u> via the temperature of the fluid surrounding it. Large scale gradients are conveniently modelled by treating the bed as a two-dimensional array of inter-connected mixing cells, each cell containing a catalyst particle. This picture of the packed bed was inspired by Deans and Lapidus [11], and has much to recommend it at the limit of turbulent transport.

However, in the region of Reynolds numbers of interest, heat transfer through the solid packing is significant, as already shown, and therefore we prefer a continuum description, in spite of reservations on its suitability to beds of low tube to particle diameter ratio.

A heat balance on a volume element of bed in Figure 5 leads to the following differential equations:-

Fluid phase

$$k_{rf} \cdot \left(\frac{\partial^2 T}{\partial r^2} + \frac{1}{r}\frac{\partial T}{\partial r} \right) + k_{af} \cdot \frac{\partial^2 T}{\partial Z^2} - ah(T-t) = GCp\frac{\partial T}{\partial Z} \qquad (15)$$

Solid phase

$$k_{rs} \cdot \left(\frac{\partial^2 t}{\partial r^2} + \frac{1}{r}\frac{\partial t}{\partial r} \right) + k_{as} \cdot \frac{\partial^2 t}{\partial Z^2} + ah\,(T-t) = 0 \qquad (16)$$

with boundary conditions

$$\frac{\partial T}{\partial r} = \frac{\partial t}{\partial r} = 0 \quad \text{at } r = 0 \qquad (17)$$

$$- k_{rf} \cdot \frac{\partial T}{\partial r} = h_{wf} \cdot \left(T-T_S \right) \quad \text{at } r = R,\ Z>0 \qquad (18)$$

$$- k_{rf} \cdot \frac{\partial T}{\partial r} = h_{wf} \cdot (T - T_0) \quad \text{at } r = R,\ Z<0 \tag{19}$$

$$- k_{rs} \cdot \frac{\partial t}{\partial r} = h_{ws} \cdot (t - T_S) \quad \text{at } r = R,\ Z>0 \tag{20}$$

$$- k_{rs} \cdot \frac{\partial t}{\partial r} = h_{ws} \cdot (t - T_0) \quad \text{at } r = R,\ Z<0 \tag{21}$$

$$t,\ T \rightarrow T_0 \quad \text{as } Z \rightarrow -\infty \tag{22}$$

$$t,\ T \rightarrow T_S \quad \text{as } Z \rightarrow +\infty \tag{23}$$

$$T(0^+) = T(0^-)\ ;\ t(0^+) = t(0^-) \tag{24}$$

$$\frac{\partial T(0^+)}{\partial Z} = \frac{\partial T(0^-)}{\partial Z}\ ;\ \frac{\partial t(0^+)}{\partial Z} = \frac{\partial t(0^-)}{\partial Z} \tag{25}$$

Dixon and Cresswell [12] gave a semi-analytical solution to these equations. If we confine ourselves to radial heat transfer in this discussion (i.e. set $k_{af} = k_{as} = 0$), the fluid phase temperature profile in the test section (Z>0) is given by:

$$\frac{T - T_0}{T_S - T_0} = 1 + 2\left\{\frac{Bi_f((^r/R)^2 - 1) - 2}{Bi_f + 4}\right\} \cdot \exp\left[-\left\{\gamma + \frac{\alpha N_F}{Pe'_{RF}(\frac{N_S}{8} + \alpha)}\right\}\frac{Z}{R}\right] \tag{26}$$

where $Bi_f = h_{wf} . R/k_{rf}$

$$\gamma = \frac{8 . Bi_f}{Pe'_{RF}(Bi_f + 4)}$$

$$Pe'_{RF} = GCpR/k_{rf}$$

$$\alpha = Bi_S/(Bi_S + 4)$$

$$Bi_S = h_{ws}R/k_{rs}$$

$$N_F = aR^2h/k_{rf}$$

$$N_S = aR^2h/k_{rs}$$

Applying the same collocation technique to the one-phase model equations, assuming $k_{a,eff} = 0$, gives

$$\frac{T_b - T_0}{T_S - T_0} = 1 + 2\left\{\frac{Bi((^r/R)^2 - 1)-2}{Bi + 4}\right\}. \exp\left\{\frac{-8\ Bi(^Z/R\ )}{(Bi + 4).Pe_R'}\right\} \quad (27)$$

where $Pe_R' = GCpR/k_{r,eff}$

4.3 Comparison of one-phase and two-phase models

Dixon and Cresswell preferred to match Eqn (27) directly to the fluid phase temperature profile in Eqn (26), giving eventually the pair of relations

$$\frac{1}{Pe_r} = \frac{1}{Pe_{rf}} + \frac{(k_{rs}/k_g).(Bi_f + 4)}{Re.\ Pr \qquad Bi_f} \cdot \left[\frac{8}{N_S} + \frac{Bi_S + 4}{Bi_S}\right]^{-1} \quad (28)$$

$$Bi = Bi_f \quad (29)$$

Other model matching criteria may also be employed, according to need. For instance, Eqn (27) may be matched to the solid phase temperature profile to give a relation for $k_{r,eff}$ that is particularly suitable at very low Re, where solid phase conduction tends to predominate. Alternatively, wall fluxes may be equated, but this tends to give a cumbersome set of relations. Eqns (28) and (29) are more appropriate for intermediate to high Re flow, where the fluid mixing outweighs solid conduction, as in reactor applications.

The "fundamental" parameters k_{rf}, k_{rs}, h_{wf}, h_{ws} and h can all be determined independently and thus the relations (28) and (29) provide fully predictive equations for $k_{r,eff}$ and $h_{w,eff}$.

A review of the available literature for estimating the "fundamental" parameters, complete up to 1979, is given in [12]. However, there have been interesting developments since then throwing some light on the less well-determined steps and these are reviewed briefly here.

4.4 Solid conduction

The recent experimental study by Melanson and Dixon [13] provides estimates of k_{rs} and the solid phase Biot number Bi_S for spheres, cylinders and rings in beds of low tube to particle diameter ratio. It also throws up important questions about the mechanisms of solid conduction and wall heat transfer in such beds which seem to defy current theories.

These authors employed an identical technique to that used previously by Yagi and Kunii [14]. Steady state heat flow and radial temperature profiles were measured in an annular bed, the walls of which were maintained at different known temperatures, the fluid (air) being stagnant. This technique permits the simultaneous estimation of wall heat transfer coefficients and the effective conductivity in the bed centre ($k^o_{r,eff}$), whereas most data report bed-average values $\bar{k}^o_{r,eff}$, which lumps in the two wall resistances.

(a) Bed-average values ($\bar{k}^o_{r,eff}$)

Melanson and Dixon [13] found that the bed-average stagnant conductivity, defined by

$$Q = -2\pi . \bar{k}^o_{r,eff} . \frac{(b - a)}{\ln (b/a)} . (T_b - T_a) \tag{30}$$

could be predicted to ±20% by the Kunii-Smith [15] formula over a wide range of conductivities and pellet shapes. This model was definitely preferred to that of Bauer and Schlünder [16].

Omitting the radiation terms for this comparison, the Kunii-Smith equation may be stated

$$\frac{\bar{k}^o_{r,eff}}{k_g} = \varepsilon + \frac{1 - \varepsilon}{\left(\ell_v/d_{pv} + \frac{2}{3\kappa} \right)} \tag{31}$$

where

$$\phi = \frac{\ell_v}{d_{pv}} = \begin{cases} \phi_2 \text{ for } \varepsilon<0.26 \\ \phi_2 + (\phi_1 - \phi_2).\dfrac{(\varepsilon - 0.26)}{(0.476-0.26)} & \text{for } 0.26<\varepsilon<0.476 \\ \phi_1 \text{ for } \varepsilon>0.476 \end{cases}$$

$$\text{and} \quad \emptyset_1 = \frac{0.333 \cdot (1 - \frac{1}{\kappa})^2}{\ln\{\kappa - 0.577(\kappa-1)\} - 0.423 \cdot (1-\frac{1}{\kappa})} - \frac{2}{3\kappa}$$

$$\emptyset_2 = \frac{0.072 \cdot (1 - \frac{1}{\kappa})^2}{\ln\{\kappa - 0.925 \cdot (\kappa-1)\} - 0.075 \cdot (1-\frac{1}{\kappa})} - \frac{2}{3\kappa}$$

$$\kappa = \frac{k_p}{k_g}$$

For non-spherical particles d_{pv} is replaced by the diameter of an equal-volume sphere (neglecting hollow spaces, as in rings for example).

(b) Bed-centre values ($k^o_{r,eff}$)

The ratio of bed-centre to bed-average conductivities $k^o_{r,eff}/\bar{k}^o_{r,eff}$ is plotted vs. the ratio of the annulus diameter to the particle diameter d_a/d_{pv} in Figure 13. Quite a good correlation is found in spite of the uncertainty in defining a proper d_{pv} for rings, especially, and bearing in mind the wide range of particle thermal conductivities. The two parameters are effectively equal for $d_a/d_{pv} > 10$, reflecting the dominance of bed conduction to wall heat transfer resistances. However, within the range of commercial interest, say $5 \leq d_a/d_{pv} \leq 10$, $k^o_{r,eff}/\bar{k}^o_{r,eff}$ may exceed 2!

There is simply no way to explain this difference by making minor modifications to ε or $\emptyset$ in the Kunii-Smith model. Natural convection cannot explain the result, since our own measurements indicate $k^o_{r,eff}$ to be largely independent of T_b-T_a, within reason.

It is possible that $k^o_{r,eff}$, unlike $\bar{k}^o_{r,eff}$, is not a bed property, which would call into question the very continuum nature of the packed bed for small d_t/d_p ratios. Rao and Toor [17] report further experimental data which seem to throw doubt on the continuum description. Here then is an area ripe for further study.

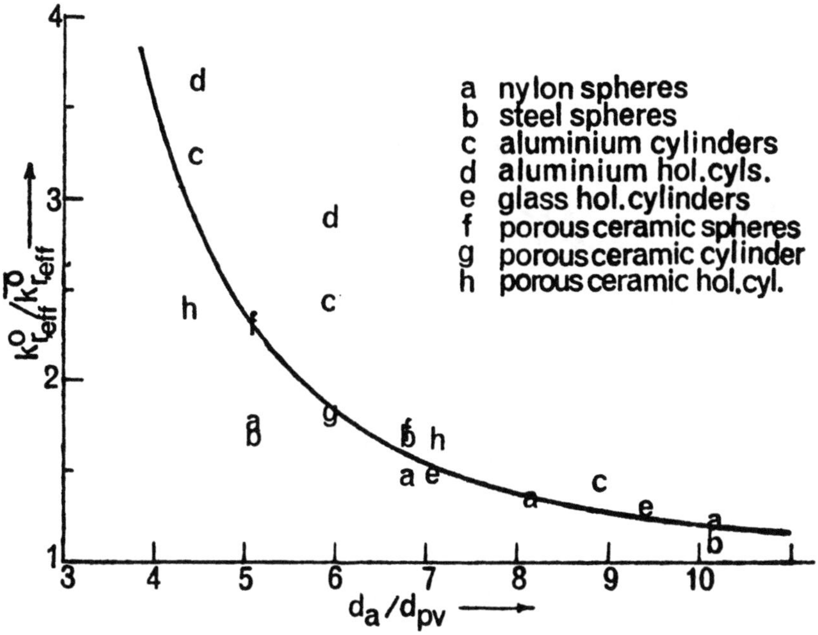

Fig. 13 - Ratio of bed centre to bed-averaged stagnant bed conductivities [13]

(c) Wall Biot numbers (Bi^o)

To lend a degree of smoothing to this parameter, which is often difficult to estimate reliably, the inner and outer wall Biot numbers (Bi_a and Bi_b) of the annular bed are averaged to give

$$Bi^o = \frac{Bi_b + (^b/a).Bi_a}{2} \tag{32}$$

and plotted vs. the ratio d_a/d_{pv} in Figure 14. The solid curve is merely an attempt to smooth the data by eye, since predictive formulas are rather inaccurate. The various symbols are depicted in Figure 13.

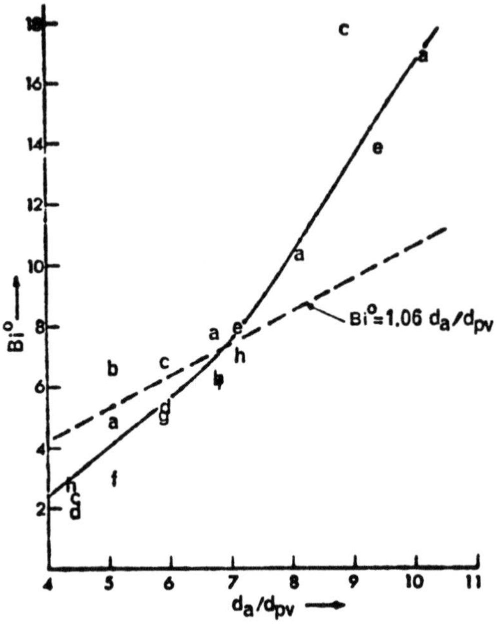

Fig. 14 - Stagnant bed Biot numbers [13]
Symbols as in Fig. 13

One such predictive formula is shown by the dotted line, and was derived for a close-packed array of spheres in contact with a flat wall [10]. It gives a rather poor approximation to observed values.

4.5 Heat transfer between the fluid and the wall

A recent experimental study by Colledge and Paterson [18] presents an interesting attack on the difficult problem of determining the rate of heat transfer between the fluid and the wall. These authors measured the rate of evaporation of an organic liquid swelling agent from a thin (ca. 0.1 mm) polymer coating on the inside wall of a packed bed to provide average mass tansfer coefficients over the whole surface. They analysed their data using a two-dimensional homogeneous continuum model with assumed values of the radial dispersion coefficient D_r taken from the literature [19].

The relationship between j_D and the Reynolds number was shown to be independent of the particular swelling agent used, and was correlated by

$$j_D = \frac{Sh_{wf}}{Re.Sc^{1/3}} = 0.523.(1- d_p/d_t) . Re^{-0.262} \qquad (33)$$

for low Sc gaseous systems. Eqn (33) was derived for spheres within the range $1.88 < d_p < 10.37$ mms and cylinders $4.58 < d_p < 9.16$ mms, packed in a tube of 29.6 mm I.D., and covering the Reynolds number range $100 < Re < 500$.

By heat and mass transfer analogy, $j_D = j_H$, where

$$j_H = \frac{Nu_{wf}}{Re.Pr^{1/3}} \qquad (34)$$

and, since $Bi_f = Nu_{wf} \cdot \frac{R}{d_p} \cdot \frac{Pe_{rf}}{Re.Pr}$ (35)

it follows from (33)-(35) that

$$Bi_f = 0.262 . ((d_t /d_p) -1).Pe_{rf}.Pr^{-2/3}. Re^{-0.262} \qquad (36)$$

Dixon and co-workers [19] place Pe_{rf} between 8 and 11 for turbulent radial mass transfer, irrespective of the tube to particle diameter ratio. Substitution into (36) gives lower and upper bounds on Bi_f given by

$$Bi_f = (2.1 - 2.9) . ((d_t /d_p)-1).Pr^{-2/3}. Re^{-0.262} \qquad (37)$$

This work is important in that it now enables the relationship

$$Bi = Bi_f \qquad (29)$$

between effective heat transfer parameters in the one-phase model, and fluid phase parameters in the two-phase model, to be directly tested without extrapolating the experimental conditions.

5 TESTING OF MECHANISTIC MODELS

We now turn to the central question concerning the evaluation of mechanistic heat transfer models. It will be prudent to begin with the small tube-to-particle diameter case because of the central role it plays in practice. Here, we shall be concerned primarily with testing the two-phase model predictions given by Eqns (28) and (29). Finally, we shall examine the adequacy of the model for predicting overall U values for narrow diameter tubes and compare it with alternatives claimed to be wide-ranging.

5.1 Effective wall heat transfer (Bi)

The relation Bi = Bi_f, given by Eqn (29), can now be directly tested on a gaseous fluid system (air) by comparing the Bi_f data of Colledge and Paterson [18], obtained from mass transfer experiments, with measured Bi data for heat transfer from two sources [8, 20].

Taking Pr = 0.72 for air, Eqn (37) becomes

$$Bi_f \cdot \left(\frac{d_p}{R}\right) = (5.23 - 7.22).(1 - (^{d}p/d_t))\ .\ Re^{-0.262} \tag{38}$$

For $^{d}t/d_p$ within the range 5 to ∞, the term 1 - $^{d}p/d_t$ varies only between 0.8 and 1. Therefore, choose a mid-range value of 0.9 and simplify Eqn (38) to

$$Bi_f \cdot (d_p/_R) = (4.71 - 6.5).Re^{-0.262} \tag{39}$$

The lower and upper bounds of Eqn (39) are shown by the solid curves in Figure 15 and compare favourably with measured Bi data. The data for large steel balls lie systematically above the rest, but are still on the extremities of Eqn (39).

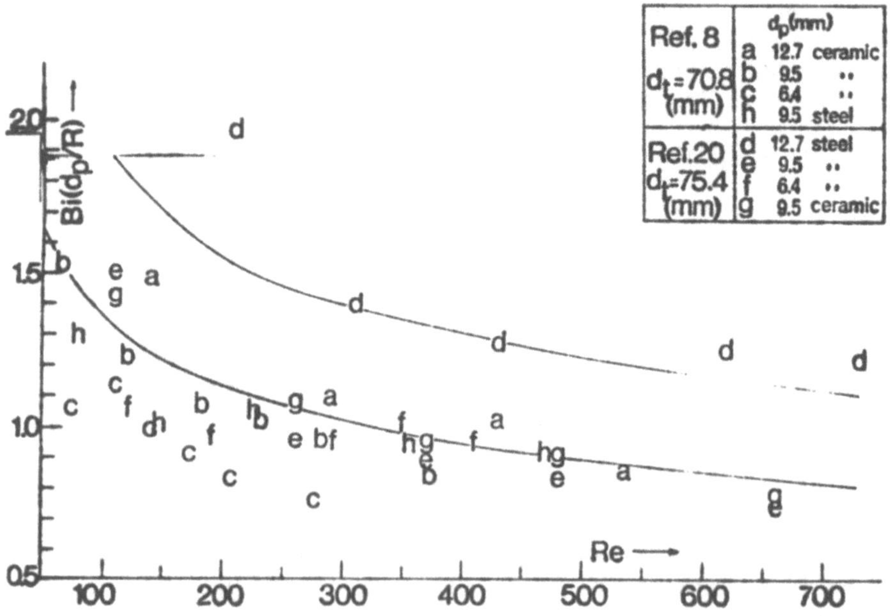

Fig. 15 Test of Eqn (29) relating one phase and two-phase model wall heat transfer coefficients for various spherical packings.

Thus, Eqn (29) appears to be broadly satisfied over a relatively wide range of particle size, conductivity and Reynolds number. It offers a potentially powerful means of predicting $h_{w,eff}$ from radial and wall mass transfer data, together with the effective radial thermal conductivity, which can usually be estimated quite accurately.

5.2 Effective radial conductivity (Pe_r)

A most telling examination of radial heat transfer can be distilled from a recent series of papers by Dixon and co-workers [13, 19, 20, 21], who systematically studied the various underlying transport processes - radial mass transfer and mass transfer at the wall (from which heat transfer parameters may be estimated by heat and mass transfer analogies), together with effective radial solid conduction and solid-to-wall heat transfer. The relevant parameter values to be inserted into Eqns. (28) and (29) are summarised in Table 1 below. All the experiments were conducted on a tube of 75.4 mm I.D.

d_p (mm)	k_{rs}/k_g	Bi_s	Pe_{rf}
Steel Spheres			
6.4	20.75	17.3	$\simeq$10
9.5	30	6.2	10
12.7	27.1	6.7	10
Ceramic spheres			
9.5	11.6	6.1	$\simeq$10

Table 1. Heat transfer parameters for evaluation of the effective radial thermal conductivity.

Dixon's Biot number data were determined by measuring the rate of dissolution of a coating of benzoic acid into a flowing water stream. Taking $Pe_{rf} = 10$, which was found to be fairly constant, and setting $Pr = 0.72$ for air, his equation reduces to

$$Bi_f = 6.22 \cdot (d_t/d_p - 1) \cdot Re^{-0.39} \qquad (40)$$

which agrees fairly well with Colledge and Paterson's form (Eqn. 38), in view of the large extrapolation in physical properties between water and air.

The only process not investigated was heat transfer between the solid and fluid phases. The parameter N_S was therefore estimated from literature data given previously [12] :-

$$N_S = \frac{1.5(1-\varepsilon)\,(d_t/d_p)^2}{(k_{rs}/k_g)\left\{\frac{1}{Nu_{fs}} + \frac{0.1}{(k_p/k_g)}\right\}} \qquad (41)$$

where $$Nu_{fs} = \frac{0.255}{\varepsilon} \cdot Pr^{1/3} \cdot Re^{0.665} \qquad (42)$$

A sample calculation of predicted values of Bi and Pe_r from Eqns. (28) and (29) is presented below in Table 2.

Re	Nu_{fs}	N_s	A	Bi_f	Bi (obsvd)	Pe_r	Pe_r (obsvd)
110	13.0	24.6	0.80	6.9	5.95	2.47	2.52
260	23.1	43.6	0.99	4.9	3.16	3.86	4.00
370	29.1	55.1	1.08	4.3	3.51	4.51	4.89
480	34.7	65.5	1.15	3.9	3.29	5.0	5.23
660	42.8	80.9	1.19	3.4	2.90	5.72	5.61

Table 2. Sample calculation for 9.5 mm steel spheres

The comparison between predicted and observed parameters reveals several interesting features. Firstly, the Peclet number Pe_r is well predicted (to within 10%). This is generally the case in our experience. The Biot number, on the other hand, is overestimated by 20% in this example. In general, an overestimate of between 15 and 30% is typical, reflecting the difficulty in both measurement (Fig. 11) and prediction of this parameter.

The factor

$$A = \frac{1 + 4/Bi_f}{1 + 4/Bi_s + 8/N_s}$$

which pre-multiplies the effective solid conductivity increases with the Reynolds number (Table 2), but appears to lie close to unity within the Re range in which solid conduction is significant. This suggests that the simpler formula,

$$1/Pe_r = 1/Pe_{rf} + (k_{rs}/k_g)/Re.Pr \qquad (43)$$

favoured by many earlier workers, offers a good approximation to Eqn. (28), providing k_{rs} is properly interpreted as the 'central core', rather than the 'bed average' effective solid conductivity.

It is clear that for metallic packing, effective solid conduction is still contributing 50% to radial heat transfer at Re $\simeq$ 500. For low conductivity catalyst carriers, on the other hand, this figure is nearer 25%.

The term $8/N_S$ is generally much smaller than $4/Bi_S$ in the solid conduction term, so that inter-phase heat transfer is of marginal importance.

Some earlier comparisons of predicted and observed Peclet numbers [12], before Bi_S data became available, are shown in Figure 10 The assumption $Bi_S = \infty$, which does not now seem justified, is finely balanced by the use of "bed-averaged" solid conductivity figures, lower than 'central-core' values reported in Table 1, to give a fortuitously good prediction.

5.3 Overall heat transfer coefficient (U)

An exhaustive examination of overall heat transfer coefficients, defined by Eqn. (1), covering 15 combinations of packings and tube diameter, was carried out by Wellauer and co-workers [4]. The two-phase model relations were able to predict U values to within ± 10%, as shown in Figure (16).

The averaged normalised standard error

$$\left[\sum_{i=1}^{60}\left(\frac{U_{obs,i} - U_{pred,i}}{U_{obs,i}}\right)^2 \Big/ 60\right]^{1/2}$$

and the maximum error over 60 runs are given in Table 3.

Also given are similar statistics for other models which have appeared in the recent heat transfer literature, and which are claimed to be wide-ranging. The one-phase model predictions are obtained by replacing Eqn. (28) with Eqn. (43).

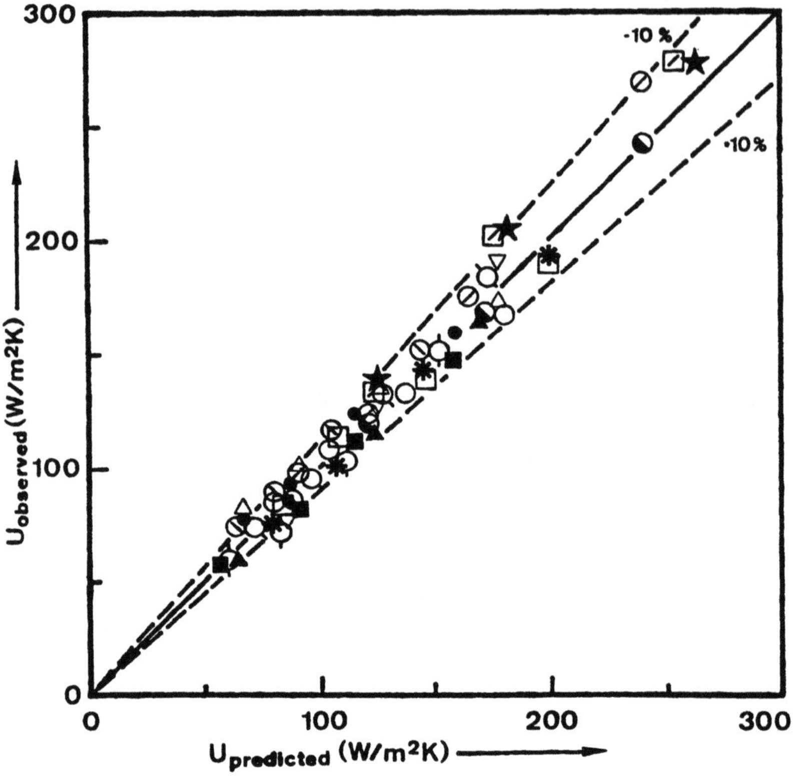

Fig. 16. Comparison of observed and predicted U values [4] Reprinted with permission ACS Symp. Series 196, 527. Copyright (1982) ACS.

Model	Mean error (%)	Max. error (%)
Two-phase	6.8	-14.0
One-phase	18.0	-42.0
Specchia and co-workers [22]	51.5	-
Kulkarni and Doraiswamy [23]	51.0	-
Schluender and co-workers [24,25]	41.0	-

Table 3 Comparison of model predictions

Previous one-phase continuum models [22], [23], [24,25] and [16], which are all based on "large diameter tube" heat transfer do not extrapolate to narrow diameter tubes. These equations systematically underpredict the overall heat transfer coefficient by 40-50%, on average. On the other hand, a one-phase model, employing the proper 'central core' apparent solid conductivity [13], and utilising wall heat transfer data measured on beds of low tube-to-particle diameter ratio [8], shows a mean error of only 18%.

In a recent reactor study, Arntz and co-workers [26] speculated that the overall heat transfer coefficient under reaction conditions was some 50% larger than that measured under non-reactive conditions, in order that the model match observations. In view of the findings given above, this is probably not due to some unexplained effect of the reaction on heat transfer rates, but simply the result of an inappropriate choice of heat transfer correlation.

6 MODELLING OF CHEMICAL REACTORS

Having traced through the heat transfer models and given some consideration to the experimental evaluation of heat transfer parameters the time is now opportune to consider the chemical reactor and, in particular, heat transfer therein.

To illustrate the importance of heat transfer in chemical reactors an actual industrial study is exemplified.

6.1 Yield considerations

We are usually concerned with the question of optimising the yield of a particular desired product in a complex reaction pathway, for example, maleic anhydride in the partial air oxidation of n-butane - a system of considerable current interest.

The series of reactions involved, while undoubtedly very complex, can be boiled down to the classical consecutive-parallel form

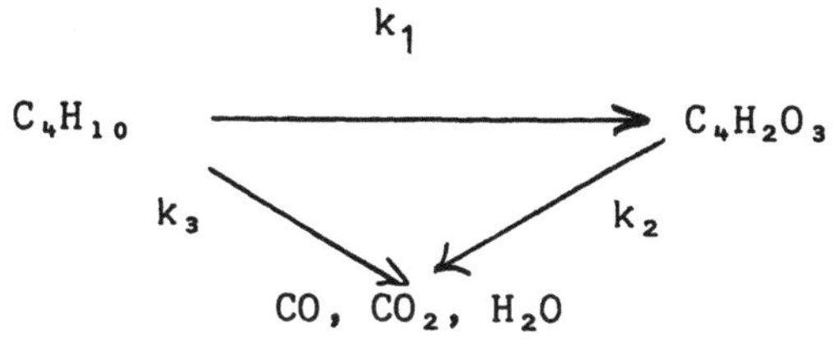

with $E_1 \simeq E_3$, but $E_2 \gg E_1$

Immediately, therefore, proper control of reaction temperature is seen as vital to maximise the yield of maleic anhydride in this series of highly exothermic reactions. At the same time, a high conversion of n-butane is desirable, since its recovery from the product stream and recyle would be prohibitively expensive for atmospheric air oxidation. To further complicate the design problem catalyst deactivation, leading to both activity and selectivity losses, has to be recognised.

6.2 Modelling Issues

Two recent papers deal with the question of model parsimony in relation to yield matters involving highly exothermic reactions [27,28]. Figure 17 shows one such product and temperature distribution observed along a single full-scale pilot plant tube. Also shown by the continuous curves are the predictions of a heterogeneous reactor model, accounting for both intra-particle and interphase concentration and temperature gradients, but lumping radial heat transfer into an overall heat transfer coefficient (U). Several different types of study were needed to provide independent estimates of the intrinsic reaction kinetics, pore diffusion and heat transfer within the bed.

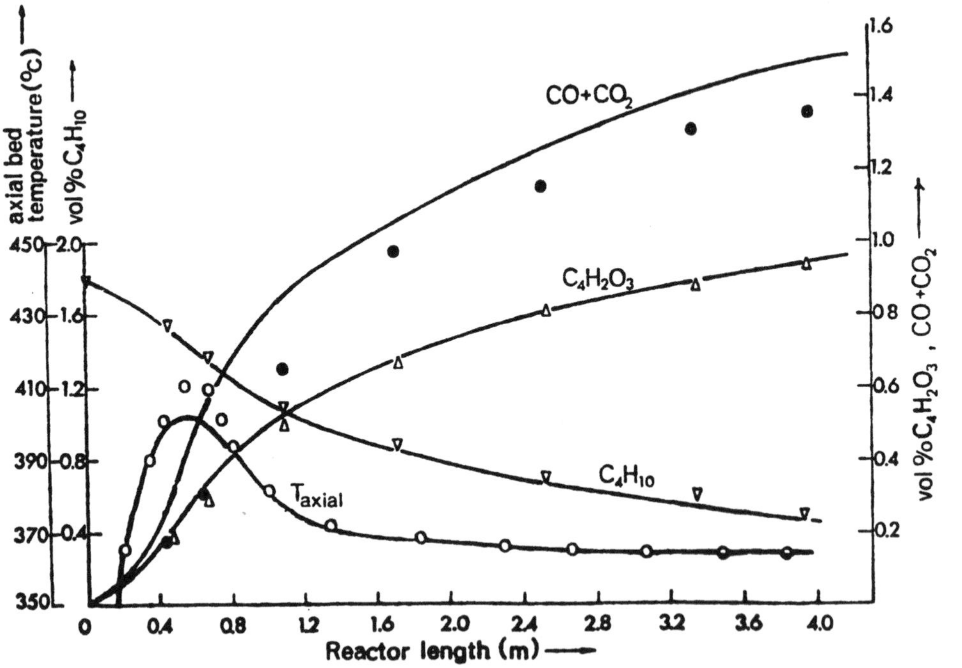

FIG 17. Observed and predicted reactant, product and temperature distributions in a full-scale reactor tube [28]

Clearly, in spite of the simplified heat transfer treatment, good agreement is found. More detailed observations of the model, shown in Table 4, highlight the balance of macroscopic and microscopic gradients.

BED DEPTH (cms)	TEMPERATURE GRADIENTS (°C)			EFFECTIVENESS FACTORS			SELECTIVITY (MOLE %)
	gas/wall	gas/solid	intra-particle	η_1	η_2	η_3	
20	11.5	1.8	0.4	0.62	2.80	0.98	77.1
40	36.5	2.6	0.6	0.66	1.72	0.95	71.9
60	38.2	2.4	0.5	0.72	1.38	0.94	68.8
100	23.9	1.4	0.3	0.82	1.15	0.95	65.8
200	12.4	0.7	0.1	0.89	1.06	0.95	62.4
400	6.1	0.4	-	0.90	1.02	0.92	58.8

Table 4. Heterogeneity of the reactor model. Tube rate 1687 N.lit./hr; salt bath temp = 400°C; inlet butane concentration = 1.82%.

It would appear that neither intra-particle nor inter-phase temperature gradients develop to any significant extent, even for this set of highly exothermic reactions. The balance of the radial temperature gradient appears to lie at the wall, thus leading to the conclusion that the simplest one-dimenional homogeneous continuum model is adequate to describe heat transfer within the bed. This is perhaps surprising when one considers the degree of detail in many literature studies of heat transfer in packed beds, including our own!

Significant intra-particle concentration gradients can develop, however, which lead to a lowering of catalyst effectiveness factors for reactant degradation reactions 1 and 3, but a raising of effectiveness for the product destruction reaction 2. The effectiveness factors also vary significantly along the bed, which probably necessitates a detailed treatment of the intra-particle reaction-diffusion problem, since this bears not only on activity but also selectivity.

Selectivity to maleic anhydride inevitably falls along the bed, but its rate of decline is exacerbated by both pore diffusion and radial heat transfer. Clearly, there is much still to go at as far as selectivity is concerned.

6.3 Exploiting the Model

Attempts to optimise reactor yield tend to push the reactor to the limits of operability (ie high inlet hydrocarbon concentration coupled with high salt bath temperatures). The catalyst deactivation is greatly accelerated, so that such potential gains may be short-lived, even if they are realiseable in the first place. Case 1 in Figure 18 indicates a typical result.

However, once an understanding of the heat transfer and pressure drop characteristics of different sizes and shapes of catalyst pellets is at hand, a simple practical solution is offered. This involves the development of a dual catalyst support system, comprising, for example, thin-walled rings at the front of the reactor, offering high heat transfer to pressure drop, and small solid cylinders in the tail, providing improved global activity. The result is a considerably flattened temperature distribution, providing enhanced yield and lower hot-spot temperatures. A lowering of hot-spot temperature of 20°C is estimated to double the catalyst life.

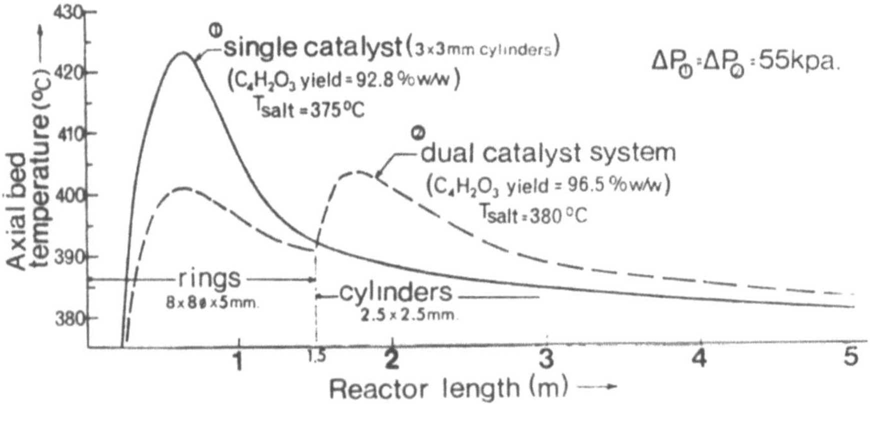

Fig 18. Comparison of single and dual catalyst systems.

7 FUTURE DEVELOPMENTS

The subject has received nearly 40 years of steady development and is now in a mature stage in many respects. However, opportunities remain for further scientific underpinning, and in the applications area, in particular. The topics listed below are just some of those that I have run up against in industrial practice where data are lacking. It is by no means an exhaustive list.

7.1 Scientific Underpinning

a) Measurements under extreme conditions.

There are few data reported where both the temperature and the throughput (mass velocity) are high, as for instance in steam reformers (T = 800°C, G = 20 kg/m².sec). Thus, large extrapolations of existing data are necesasary to model one of the most important and interesting practical examples of heat transfer in packed beds.

b) The importance of bed voidage

The vast majority of heat transfer studies have been on beds of spheres or cylinders, for which the bed voidage is approximately 40%. The measurements do not extrapolate satisfactorily to beds made up of thin-walled Raschig rings, Lessing rings, Pall rings, wagon wheels and micro-monoliths, which may lead to voidages as high as 95%. These shapes offer considerably reduced pressure drops, but their heat transfer properties are poorly understood. Bed voidage and particle shape are variables that enter into realistic optimisations of reactor performance.

c) Transient heat transfer

Far too few experimental data are to hand on transient heat transfer in non-adiabatic packed beds, which tend to limit model development for reactor start-up and control studies. One important exception is the work of Levec and Carbonell [29], which challenges some of our pre-conceptions about dynamic models.

d) Fluid flow in packed beds of low tube-to-particle diameter ratio

Several workers have inferred the existence of a radial velocity profile in a packed bed of low tube-to-particle diameter ratio from measurements of the fluid velocity at the bed exit [30]. However, their results are in considerable disagreement. A semi-theoretical study, using a modified Brinkmann model [31], indicates the existence of a steep maximum in the velocity next to the wall, but this remains unsubstantiated. Non-intrusive measurements of gas velocity within the packed bed are needed before a proper evaluation of the interactions of radial velocity, radial heat transfer, conversion and reaction selectivity are forthcoming.

7.2 Applications

a) Boiling water reactors/thermosiphons

Heat transfer test rigs are usually designed with a constant wall temperature to simplify analysis of the data. The emphasis is invariably placed on minimising heat transfer resistances on the coolant side. Yet, there are important industrial examples which employ boiling fluids to remove the reaction exotherm, and achieve circulation of the coolant by thermo-siphoning. Heat transfer regimes on the coolant side may vary from liquid phase convection, nucleate boiling to even film boiling along different lengths of the reactor tubes. Modelling of such reactors requires a more integrated approach than that normally adopted. Perhaps we may borrow ideas from the nuclear industry [32]?

b) Radial flow reactors

Commercial radial flow reactors are finding increasing use in ammonia and methanol syntheses - classical exothermic reversible reactions, because of their increased capacities, low pressure drops, good control of reaction temperature and ease of scale-up in the axial direction. In current designs it is usual to separate the adiabatic beds by inter-stage coolers or quenches. However, in a new type of methanol converter [33], shown in Figure 19, boiler tubes are arranged in a number of concentric rows within the catalyst bed, producing an arrangement of adiabatic reaction zones and "heat-removed" reaction zones. This design permits a very close approach of the temperature distribution to the theoretical ideal (Figure 20).

It would be relatively straightforward to extend this concept to complex exothermic and "selectivity-led" reactions, with large potential benefits.

At some stage, however, it would be necessary to conduct heat transfer measurements, since it is doubtful if heat transfer correlations for axial flow reactors can be extrapolated to this case. Here then is an ideal opportunity for innovative measurement and modelling, which could help to lead us into new technology.

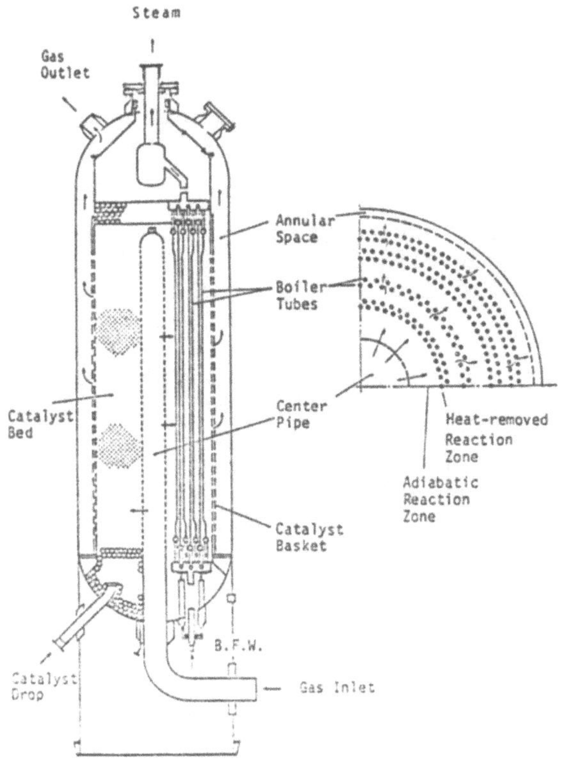

Fig 19. Arrangement of the radial flow reactor [33].

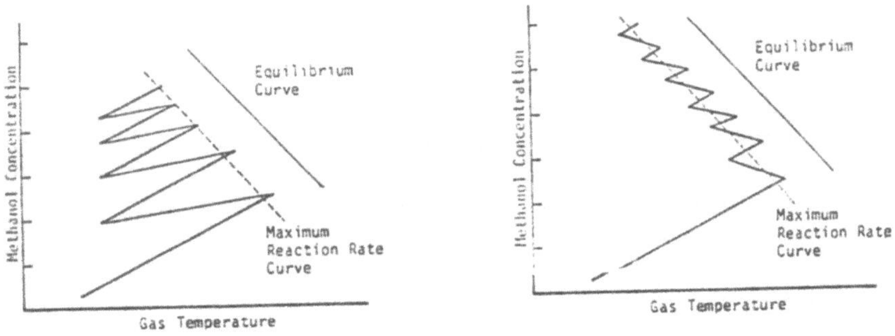

(a) Quench type reactor (b) Radial flow reactor [33]

FIG 20. Reaction temperature profiles

ACKNOWLEDGEMENT

I would like to acknowledge here the invaluable contributions made by colleagues to references bearing my name and theirs, in particular to A G Dixon, W R Paterson, T Wellauer, R K Sharma and E J Newson. I am also indebted to them for allowing me to see and review the fruits of their latest researches ahead of publication.

Acknowledgements are also due to the American Chemical Society for permission to publish Figures 1b, 2, 4-6, 8-11, 16 and to Ohsaki and co-workers for Figures 19 and 20.

NOTATION

All heat transfer parameters are defined in terms of total area (void and non-void) normal to the direction of heat transfer.

a	specific interfacial surface area.
a	outer radius of inner concentric cylinder.
b	inner radius of outer concentric cylinder, stagnant bed experiments reported in reference 13.
Bi	apparent Biot number, $h_{w,eff}\, d_t/2\, k_{r,eff}$
Bi_a, Bi_b	stagnant bed Biot numbers at the walls of the annular bed, defined in reference 13.
Bi^o	weighted average of Bi_a and Bi_b, defined in Eqn 32.
Bi_f	fluid to wall Biot number, $h_{wf} d_t/2\, k_{rf}$ (two-phase model).
Bi_s	packing to wall Biot number, $h_{ws} d_t/2k_{rs}$(two-phase model).
C_p	fluid specific heat.
d_a	diameter of annular bed, 2.(b-a).
d_p	diameter of spherical packing.
d_{pv}	diameter of equal volume sphere, neglecting hollow spaces in the packing (eg the central hole in a Raschig ring!)
d_t	tube inside diameter.
D	molecular diffusivity of swelling agent employed in wall mass transfer experiments.
E	intrinsic reaction activation energy.
G	superficial mass velocity.
h	effective fluid to packing heat transfer coefficient.
h_{fs}	fluid to solid heat transfer coefficient.
$h_{w,eff}$	effective wall heat transfer coefficient (one-phase model).
h_{wf}	fluid to wall heat transfer coefficient (two-phase model).
h_{ws}	packing to wall heat transfer coefficient (two-phase model).
j_D	j-factor for wall to fluid mass transfer, defined in Eqn 33.
j_H	j-factor for wall to fluid heat transfer, defined in Eqn 34.

k	specific reaction rate constant.
k_g	fluid molecular conductivity.
k_p	solid particle conductivity.
$k_{a,eff}$	effective axial conductivity (one-phase model).
k_{af}	fluid phase axial conductivity (two-phase model).
k_{as}	solid phase axial conductivity (two-phase model).
$k_{r,eff}$	effective radial conductivity (one-phase model).
k_{rf}	fluid phase radial conductivity (two-phase model).
k_{rs}	solid phase radial conductivity (two-phase model).
$k^o_{r,eff}$	stagnant bed effective radial conductivity (one-phase model).
$\bar{k}^o_{r,eff}$	mean stagnant bed effective radial conductivity (including wall region thermal resistances).
k_{wf}	fluid to wall mass transfer coefficient.
L	length of packed test section.
l_v	effective fluid fillet thickness (Kunii/Smith model [15])
N_F	interphase heat transfer group, a R^2h/k_{rf}.
N_s	interphase heat transfer group, a R^2h/k_{rs}.
Nu_{fs}	fluid to solid Nusselt number, $h_{fs}.d_p/k_g$.
Nu_{wf}	fluid to wall Nusselt number, $h_{wf}.d_p/k_g$.
Pe_a	effective axial Peclet number, $GC_pd_p/k_{a,eff}$ (one-phase model).
Pe_r	effective radial Peclet nunber, $GC_pd_p/k_{r,eff}$ (one-phase model).
Pe_{rf}	radial fluid Peclet number, GC_pd_p/k_{rf} (two-phase model).
Pr	Prandtl number, $\mu C_p/k_g$.
Q	quantity of heat transferred between annular cylinders, defined in Eqn (30).
r	bed radial co-ordinate.
R	tube inside radius.
Re	Reynolds number, Gd_p/μ.
Sc	Schmidt number, $\mu/_\rho D$.
Sh_{wf}	Sherwood number for wall mass transfer, $k_{wf}.d_p/D$.
t	solid phase temperature (two-phase model).
T	fluid phase temperature (two-phase model).
T_b	temperature (one-phase model).
T_0	calming section wall temperature.
T_s	test section (or salt bath) wall temperature.
U	overall heat transfer coefficient (one-phase model).
Z	bed length co-ordinate.
β	d_t/d_p.
ε	bed voidage
κ	k_p/k_g.
η	effectiveness factor:ratio of reaction rate integrated over pellet to rate evaluated at local bulk conditions.
μ	fluid viscosity.
ρ	fluid density.
θ	$T_s - T$.

REFERENCES

1 Froment G F., "Progress in fundamental design of fixed bed reactors", International Chemical Reaction Engineering Conference, Pune (India) 9-11th January (1984), Vol 1, p 12.

2 Dixon A G., "Heat transfer in packed beds of low tube/particle diameter ratio". Ph.D. Thesis. University of Edinburgh (1978).

3 Leva M, Weintraub M, Grummer M and Clark E L., "Cooling of gases through packed tubes", IEC. 40, 747 (1948).

4 Wellauer T, Cresswell D L, Newson E J., "Heat transfer in packed reactor tubes suitable for selective oxidation", ACS Symp. Series, 196, 527 (1982).

5 Gunn D J, Khalid M., "Thermal dispersion and wall heat transfer in packed beds", C.E.S. 30, 261 (1975).

6 De Wasch, A P., Froment G F., "Heat transfer in packed beds", C.E.S. 27, 567 (1972).

7 Subroutine E04 FBF, NAG Library, NAG Ltd. Oxford.

8 Dixon A G., Cresswell D L, Paterson W R., "Heat transfer in packed beds of low tube/particle diameter ratio", ACS Symp. Series 65, 238 (1978).

9 Gunn D J, Ahmad M M., "The Characterisation of radial heat transfer in fixed beds", First UK National Heat Transfer Conference, Leeds. 3-5 July (1984).

10 Olbrich W E., "A two-phase diffusional model to describe heat transfer processes in a non-adiabatic packed tubular bed", Chemeca 70, Proceedings of a Cenference, Melbourne and Sydney, Aug 19-26, (1970) p101, Butterworth, London, England (1971).

11 Deans H A, Lapidus L., "A computational model for predicting and correlating the behaviour of fixed-bed reactors", A.I.Ch.E.J. 6, 656 (1960).

12 Dixon A G, Cresswell D L., "Theoretical prediction of effective heat transfer parameters in packed beds", A.I.Ch.E.J. 25, 663 (1979).

13 Melanson M M, Dixon A G., "Solid conduction in low DT/DP beds of spheres, pellets and rings", Int.Jl. Heat and Mass Transfer, to be published, (1985).

14 Yagi S, Kunii D., "Studies on heat transfer near wall surface in packed beds", A.I.Ch.E.J. 6, 97 (1960).

15 Kunii D, Smith J M., "Heat transfer characteristics of porous rocks", A.I.Ch.E.J. 6, 71 (1960).

16 Bauer R, Schlünder E U., "Effective radial thermal conductivity of packings in gas flow. Part II. Thermal conductivity of the packing fraction without gas flow", Int Chem Eng 18 189 (1978).

17 Rao S M, Toor H L., "Heat transfer between particles in packed beds", IEC Fund. 23 294 (1984).

18 Colledge R A., Paterson W R., "Heat transfer at the wall of a packed bed : a j-factor analogy established", p 103, 11th Annual Research Meeting on Heat Transfer and Catalysis and Catalytic Reactors, The Institution of Chemical Engineers, University of Bath, 9-10th April (1984).

19 Dixon A G, Di Costanzo M A., Soucy B A., "Fluid-phase radial transport in packed beds of low tube-to-particle diameter ratio" Int.Jl.Heat and Mass Transfer 27 1701 (1984).

20 Dixon A G., "The length effect on packed bed effective heat transfer parameters", submitted to the Chemical Engineering Journal.

21 Dixon A G, Labua L A., "Wall-to-fluid coefficients for fixed bed heat and mass transfer", submitted to the International Journal of Heat and Mass Transfer.

22 Specchia V, Baldi G, Sicardi S., "Heat transfer in packed bed reactors with one-phase flow", Chem.Eng.Commun. 4 361 (1980).

23 Kulkarni B.D, Doraiswamy L K., "Estimation of effective transport properties of packed bed reactors", Catal.Rev.Sci Eng., 22 431 (1980).

24 Schluender E U, Hennecke F W., "Waermeuebergang in beheitzen oder gekuhlten Rohren mit Schuettungen aus Kugeln, Zylindern und Raschig-Ringen", C.I.T., 45 277 (1973).

25 Bauer R, Schluender E U., "Effective radial thermal conductivity of packings in gas flow. Part I : Convective transport coefficient. Int.Chem.Eng., 18 181 (1978).

26 Arntz D, Knapp K, Prescher G, Hoffman H., "Catalytic air oxidation of propylene to acrolein : Modelling based on data from an industrial fixed-bed reactor", ACS Symp.Ser. 196 3 (1982).

27 Sharma R K, Cresswell D L, Newson E J., "Selective oxidation of benzene to maleic anhydride at commercially relevant conditions", I.S.C.R.E. 8 p353, Edinburgh, Sept (1984); Inst.Chem.Eng.Symp.Series No 87.

28 Sharma R K, Cresswell D L, Newson E J., "Kinetics and reactor modelling in full scale pilot plant production of maleic anhydride by oxidation of η-butane", A.I.Ch.E. 1984 Annual Meeting, San Francisco, paper No 96b.

29 Levec J, Carbonell R G., "Longitudinal and Lateral thermal Dispersion in packed beds : Part I Theory; Part II Comparison between Theory and Experiment", A.I.Ch.E.J. (in press).

30 Lerou J J, Froment G F., "Velocity, temperature and conversion profiles in fixed bed catalytic reactors", Chem.Eng.Sci 32 853 (1977).

31 Vortmeyer D, Schuster J., "Evaluation of steady flow profiles in rectangular and circular packed beds by a variational method", Chem.Eng.Sci. 38 1691 (1983).

32 Gay R R , Gitnick B J, Ansari F, "Modelling of steady state flow in heated (BWR) parallel channels", A.I.Ch.E.Symp.Series - Heat Transfer, Orlando (1980) p139.

33 Ohsaki K, Nishimura Y, Miki H, Eto M., "Development of a new methanol reactor", A.I.Ch.E. 1984 Spring National Meeting Anaheim, California 20-24 May.

THE MEASUREMENT OF VOID FRACTION PROFILES IN PACKED BEDS

J. J. Lerou* and G. F. Froment

Laboratorium voor Petrochemische Techniek
Rijksuniversiteit Gent, Belgium
(*) E. I. Du Pont de Nemours & Co., Inc.,
Central Research & Development Department,
Experimental Station, Wilmington, Delaware 19898

INTRODUCTION

Although the nonuniform structure of packed beds was known since a long time, it was never fully investigated nor applied before. In recent years however, there is a trend towards a more accurate description of the bed's structure to explain and to simulate the behavior of fixed beds. Lerou and Froment [1] showed how predicted temperature and concentration profiles in reactors may differ depending upon whether nonuniform voidage profiles were accounted for or not. Carbonell [2] and Chang [3] used the structure of the bed to explain dispersion phenomena. Vortmeyer et al. introduced the structural characteristics of the bed to improve their models for creeping flow in reactors [4] and for the theoretical prediction of fixed bed chemical reactor performance [5,6]. Cohen and Metzner [7] accounted for nonuniformities of the bed in the pressure drop relationship for both linear and non-Newtonian fluids. Beasly and Clark [8] found that spatial variations in void fraction significantly influenced on the dynamic response of both fluid and solid temperature in the simulation of a packed bed for thermal energy storage. Snaddon and Dietz [9] took the structure of the bed into account when they measured the interstitial flow intensification within packed granular bed filters.

Lerou and Froment [1] demonstrated that the structure of the packed bed can be defined by its radial void fraction profile. In order to have adequate and reliable data on void fraction profiles an experimental measuring system was constructed, which is presented here together with suggestions for subsequent analysis.

EXPERIMENTAL EQUIPMENT

To date many methods have been used to measure porosity profiles. The most common technique consisted of pouring wax or an epoxy resin into the bed to maintain the particles into position for subsequent machining of radial increments. The void fraction of each section is calculated from a simple material balance [10-13]. Thadani and Peebles [14] used the different adsorption of X-rays in the sphere material and the matrix material to measure the local void fraction. The solidified column was segmented instead of machined. Each segment was radiographed and the point void fractions were determined by photometric reading of the radiograph emulsion point densities. A different approach was taken by Ridgway and Tarbuck, [15-17] who developed a technique based on the rapid rotation of a cylindrical bed, in which a measured amount of liquid was introduced. The local void fraction is calculated from the thickness of the liquid layer which is formed against the wall and which is measured by means of a cathetometer. Schuster and Vortmeyer [18] inserted in the cylindrical container a piston whose bottom plane was covered with an adhesive tape. The piston was lowered to touch the top layer of the bed and then removed. The packing material, which stuck to the piston was photographed. This procedure was repeated several times, until a statistical analysis was possible. An other method by Buchlin et al., [19] is based on the fluorescence of a slightly impure organic liquid and on the refractive index matching of the packing particles. The method requires the use of particles made of a transparent material. The liquid has to be sensitive for fluorescent reemission by light. When observing the illuminated part of the bed within a spectrum limited to the fluorescence only, the regions filled with fluorescent liquid remain visible whereas enlightened glass particles remain dark. Hence the voidage distribution within the illuminated part of the bed becomes apparent. By using a laser beam which is transformed into a laser sheet of 1.1 mm in thickness the bed can be scanned. A camera is used to record the image of each slice of the bed.

The packing, investigated in the present study, was catalyst material for which also the heat transfer characteristics had to be measured. Machining was not suited because of the brittleness of the particles. Using a piston with an adhesive plane was not feasible since the particles were too large. The fluorescence method could not be used since the particles were not transparent. The technique used in this study is based on a further development of the method used by Ridgway and Tarbuck [15].

The equipment consists of a chromium plated steel tube, internal diameter of 10 cm and 1.25 cm wall thickness. The tube rotates inside two SKF ball bearings. This unit is mounted on a plate formed by two U-profiles, 10 mm thickness, which is fixed on two circular profiles. This enables the whole equipment to be placed in a vertical position for loading and unloading of the packing. The tube is rotated by means of two asynchronous electrical motors with a total installed power of 1.5 hp. The speed of rotation is controlled by a transformer: any RPM between 0 and 1950 can be obtained. A digital RPM-meter (IES-Moviport C117) measures the number of revolutions.

A known quantity of water is injected in the tube and is centrifuged to the wall, where it forms an annular ring. From the thickness of the layer and the known geometry of the container the void fraction calculation is straightforward. The thickness of the layer is measured by means of a pressure transducer which is mounted on the wall of the tube. One chamber of the transducer is connected to the inside of the tube, the other one to the atmosphere. The liquid inside the rotating tube exerts a static pressure on the wall. The signal of the transducer is transmitted from a rotating device to a fixed one by means of a Vibrometer. This instrument, which operates with mercury contacts, transmit the signals without adding noise. The pressure transducer is calibrated by rotating the empty tube at a fixed RPM with known amounts of water. From the geometry of the tube the film thickness is easily calculated and a calibration line established.

The experiment is carried out as follows. The tube is loaded upright using any packing procedure: dumping the particles, dumping in water, vibrating the tube etc. The tube is closed by means of a piston, which enables the packing to be slightly compressed and to fix the particles in the tube. The tube is placed horizontally and a known amount of water is added. The tube is rotated at the desired RPM, and the output voltage of the pressure transducer is recorded. Then a new amount of water is added and the procedure is repeated to extend the layer over which the voidage is measured toward the axis. The maximum layer thickness at 1700 RPM is about 35 mm.

RESULTS AND DISCUSSION

Three different packings were investigated. Details are given in Table I. Figures 1 to 3 show typical examples of the measured void fraction profiles.

TABLE I. Packing Characteristics

Glass Spheres	d_p = 6.1 mm	ε = 0.366	d_t/d_p = 16.2
Ceramic Spheres	d_p = 18.7 mm	ε = 0.461	d_t/d_p= 5.35
Raschig Rings	d_{ext} = 24.4 mm d_{int} = 10.5 mm h = 23.0 mm	ε = 0.577	

The void fraction profile of a packing consisting of spheres is shown in Figure 1. The void fraction is 1 at the wall and sharply decreases with increasing distance from the wall. A first minimum is found at approximately half a particle diameter from the wall. Further away from the wall the profile shows a damped sinusoidal behavior. Against the confining wall a rather well structured layer of spheres touching the wall is formed. The minimum of the void fraction is encountered at approximately half a particle diameter from the wall since the spheres belonging to the first layer touch each other at about this distance from the wall. Further away from the wall the void fraction increases. It cannot increase until unity, however, since the second row of spheres rests in the cups formed by the spheres in the first row. Proceeding in from the wall this pattern is repeated and, since each row is more random than the one which precedes, the voidage oscillates around the mean value with a decreasing amplitude, which is damped out at about 4 to 5 particle diameters from the wall. Figure 2 shows a similar behavior for larger spheres. The location of the first minimum can be predicted by means of a model in which the tube to particle diameter ratio, d_t/d_p, plays an important role:

1. the smaller the value of d_t/d_p, the larger the distance between the wall and the first minimum.

2. the d_t/d_p ratio determines the number of spheres which can be accommodated in a cross section of the bed to touch the wall.

   for $d_t/d_p = 7.391$ $n = 19$ $\varepsilon_{min} = 0.123$

   for $d_t/d_p = 7.393$ $n = 20$ $\varepsilon_{min} = 0.0935$

   The last value is also the minimum value of the void fraction for a close hexagonal packing against a flat wall.

The above observations hold for a ideal packing of identical spheres. A real packing is more open because of sterical hindering in the build up of the packing. This can be seen from the values of the experimentally found first minima:

for $d_t/d_p = 16.4$ $\varepsilon_{min} = 0.22$

for $d_t/d_p = 5.35$ $\varepsilon_{min} = 0.18$

Raschig rings were also investigated. A typical result is shown in Figure 3. There is no oscillating behavior in this case. There is a first minimum at a distance of about 1 cm and a second one at about 2.25 cm. A similar result was found by Roblee et al. [11]. This result can be explained by considering all possible configurations of the rings at the wall as was done originally by Kondelik et al. [13] to explain the void fraction profile of a packing of equilateral cylinders.

Since there is at the wall a mixture of all possible configurations, it is obvious that the first minimum and the first maximum are less prounounced than for a packing of spherical particles.

APPLICATION

The information concerning the void fraction profiles can be used in different ways. Cohen and Metzner [7] divided the cross section of the tube in three regions: a wall, a transition and a bulk region. For each region a mean value of the void fraction was calculated and a hydraulic radius was defined which was used in a pressure drop correlation. Martin [20] divided the bed into two regions: a wall and a bulk region. He calculated for both different flow rates and a different rate of heat transfer. Carbonell [2] also used a two zone model for his analysis of the dispersion phenomena. In more recent work Vortmeyer et al. [5,6] tried to use the complete radial void fraction profile, and so did Chang [3]. They followed the same itinerary outlined by Lerou and Froment [1] and Marivoet et al. [21]. Starting from the void fraction profile the radial velocity profile is calculated. With both profiles the effective thermal conductivity is established and the temperature and concentration profiles can be calculated by means of a two dimensional pseudo homogeneous model for the reactor.

In a first step the void fraction profile has to be correlated in terms of the aspect ratio. Govindarao and Froment [22] approached the problem by expressing the void fraction in a layer in terms of the contributions to the solid volume in a layer by spheres with centers lying in feasible neighbouring layers. Singleton [23] made a similar, more statistical, approach: he assessed the location of all solid material in terms of distribution functions.

In a second step the expression for the void fraction has to be incorporated into fluid flow equations of the Ergun or the Brinkman type to obtain velocity profiles which enable more rigorous simulations of the heat and mass transfer processes in packed beds.

NOTATION LIST

d_t	tube diameter	mm
d_p	particle diameter	mm
d_{ext}	outside diameter of Raschig ring	mm
d_{int}	inside diameter of Raschig ring	mm
h	height of Raschig ring	mm
n	number of spheres touching the wall	
ε	void fraction	

REFERENCES

1. Lerou, J. J. and G. F. Froment. Velocity, Temperature and Conversion Profiles in Fixed Bed Catalytic Reactors. Chem. Eng. Sci., 32 (1977) 853-861.
2. Carbonell, R. A. Flow Nonuniformities in Packed Beds: Effect on Dispersion. Chem. Eng. Sci. 35 (1980) 1347-1356.
3. Chang, H. C. A non-Fickous Model of Packed Bed Reactors. AIChE J. 28 (1982) 208-214.
4. Kalthoff, O. and D. Vortmeyer. Ignition/Extinction Phenomena in a Wall-Cooled Fixed Bed Reactor. Chem. Eng. Sci. 35 (1980) 1637-1643.
5. Vortmeyer, D. and R. P. Winter. Impact of Porosity and Velocity Distribution on the Theoretical Prediction of Fixed-Bed Chemical Reactor Performance. ACS Symp. Ser. 196 (1982) 49-61.
6. Vortmeyer, D. and R. P. Winter. Die Bedeutung der Strohmungsverteilung fur die Modellierung von chemischen Festbettreactoren bei hoheren Reynoldszahlen. Chem.-Ing.-Techn. 55 (1983) 950-951.
7. Cohen, Y. and A. B. Metzner. Wall Effects for Laminar Flow of Fluids through Packed Beds. AIChE J. 27 (1981) 705-715.
8. Beasly, D. E. and J. A. Clark. Transient Response of a Packed Bed for Thermal Storage. Int. J. Heat Mass Transfer 27 (1984) 1659-1669.
9. Snaddon, R. W. L. and P. W. Dietz. Interstitial Flow Intensification within Packed Granular Bed Filters: Experiments and Theory. Ind. Eng. Chem. Fundam. 23 (1984) 147-153.
10. Kimura, M. K. Nomo and T. Kanera. Chem. Eng. Japan 10 (1953) 397.
11. Roblee, L. H. S., R. M. Baird and J. W. Tierney. Radial Porosity Variations in Packed Beds. AIChE J. 4 (1958) 460-466.
12. Benenati, R. F. and C. B. Brosilow. Void Fraction Distribution in Beds of Spheres, AIChE J. 8 (1962) 359-361.
13. Kondelik, P. J., J. Horak and J. Tesarova. Heat and Mass Transfer in Heterogeneous Catalysis. Variations of Local Void Fraction in Randomly Packed Beds of Equilateral Cylinders. Inc. Eng. Chem. Proc. Des. & Devel. 7 (1968) 250-252.

14. Thadani, M. C. and F. N. Peebles. Variation of Local Void Fraction in Randomly Packed Beds of Equal Spheres. Ind. Eng. Chem. Proc. Des. & Devel. 5 (1966) 265-268.
15. Ridgway, K. and K. J. Tarbuck. Randomly Packed Beds of Spheres of Different Sizes. J. Pharm. Pharmac. 18 (1866) 1586-1595.
16. Ridgway, K. and K. J. Tarbuck. Brit. Chem. Eng. 12 (1967) 584-589.
17. Ridgway, K. and K. J. Tarbuck. Voidage Fluctuations in Randomly Packed Beds of Spheres Adjacent to a Containing Wall. Chem. Eng. Sci. 23 (1968) 1147-1155.
18. Schuster, J. and D. Vortmeyer. Ein Einfaches Verfahren zur Naherungsweisen Bestimmung der Porositat in Schuttungen als Funktion des Wandabstandes. Chem.-Ing.-Techn. 52 (1980) 848-849.
19. Buchlin, J. M., M. Riethmuller and J. J. Ginoux. A Fluorescence Method for the Measurement of the Local Voidage in Random Packed Beds. Chem. Eng. Sci. 32 (1977) 1116-1119.
20. Martin, H. Low Peclet Number Particle to Fluid Mass and Heat Transfer in Packed Beds. Chem. Eng. Sci 33 (1978) 913-919.
21. Marivoet, J., P. Teodoroiu and S. J. Wajc. Porosity, Velocity and Temperature Profiles in Cylindrical Packed Beds. Chem. Eng. Sci. 29 (1974) 1836-1840.
22. Govindarao, V. M. H. and G. F. Froment. Voidage Profiles in Packed Beds of Spheres. (to be published - Chem. Eng. Sci. (1985)).
23. Singleton, F. D. Ph.D. Thesis, University of Rochester (N.Y.) (1971).

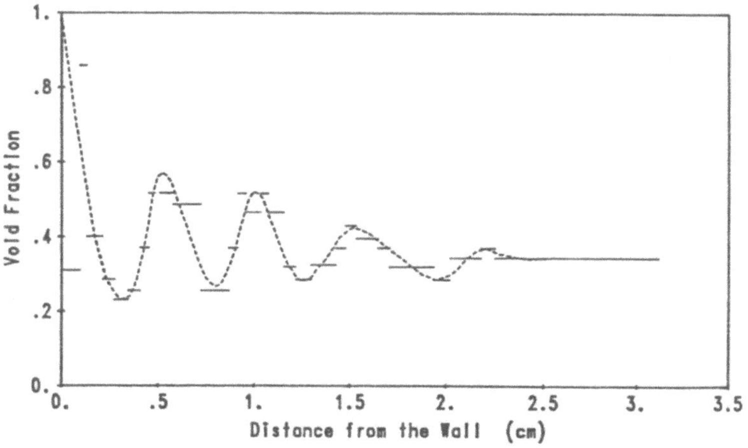

Figure 1. Measured radial void fraction profile of a bed packed with small spheres.

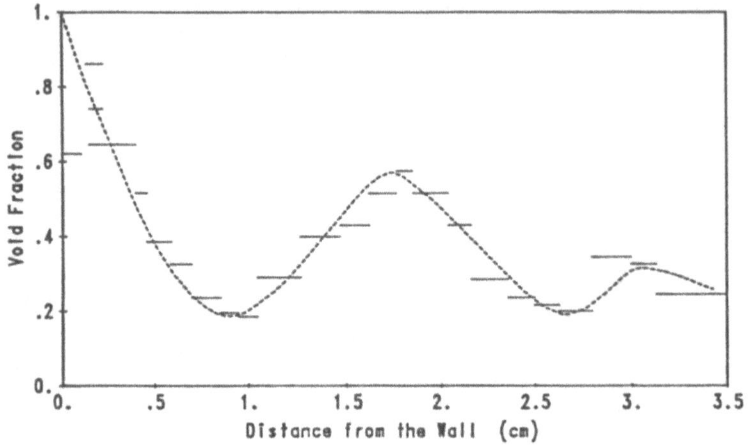

Figure 2. Measured radial void fraction profile of a bed packed with large spheres.

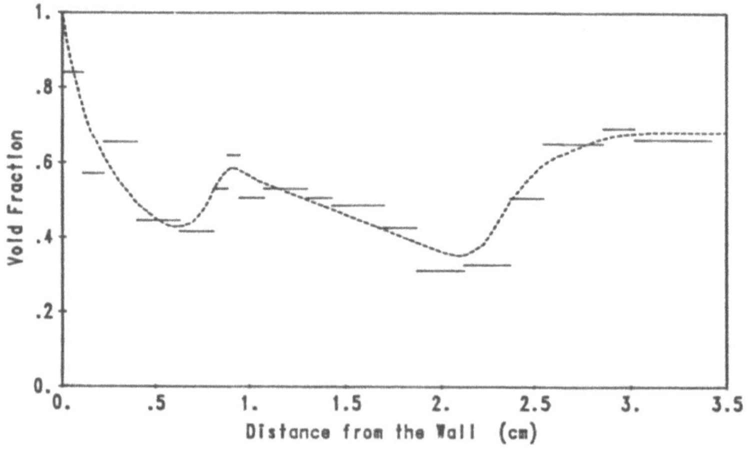

Figure 3. Measured radial fraction profile of a bed packed with Raschig rings.

CONVERTING METHANOL INTO GASOLINE IN A NOVEL PSEUDOADIABATIC CATALYTIC FIXED BED REACTOR

A. Ravella*, H. de Lasa*, E. Rost**

*Faculty of Engrg. Sci., The University of Western Ontario
London, Canada
**Facultad de Ingenieria, Universidad de La Patagonia
Argentina

Introduction

Studies have indicated that the influence of the coolant flow should be considered in detail for the design and operation of multitubular fixed bed catalytic reactors [1], [2], [3], [4], [5], [6]. These contributions describe the axial temperature profile as functions of the coolant flow rate, and give a detailed description of a characteristic operating regime named Pseudoadiabatic Operation (PO). The PO of a catalytic fixed bed reactor for exothermic reactions is a regime where the axial temperature increases steadily with the bed length so that the highest temperature in the unit is reached at the reactor outlet. The PO takes place when the non-boiling cooling fluid (molten salts, molten alloys or organic heat-transfer fluids) is cocurrently circulated with respect to the reactants and when at the same time the operating parameters are such that the heat generated is always greater than the heat removed by the coolant [1].

The PO concept should substantially modify the design and the operation of exothermic multitubular reactors [3], [4], [5], [6]. The simplicity of the PO contrasts with the much more complex instrumentation and control strategies required to sense and control conventional "hot spots" (temperature maxima in the axial direction) that develop under non-PO. In fact, the prediction of the magnitude and of the exact position of the hot spot is quite uncertain, making the design of highly exothermic fixed bed reactors susceptible to important errors. These important errors can influence both the selectivity prediction and the assessment of reactor runaway conditions. The problem of sensing "hot spots" becomes a critical one in multitubular reactors when a non-boiling coolant is circulated under cross flow condition [2]. In this case each tube row of the bundle has a "hot spot" located at a different axial position. One row of the bundle has the highest "hot spot", making this row the critical one of the reactor [2]. All these problems are eliminated under the PO regime because all the tubes in the reactor have the same temperature profile thermal symmetry and all "hot spots" are located at the reactor outlet.

Recently, a pseudoadiabatic catalytic multitubular reactor which converted methanol into gasoline was modeled [6] using kinetic data reported by Liederman et al [7]. It was shown that the average temperature predicted with a unidimensional pseudohomogeneous model was in good agreement with the T_{av} resulting from a bidimensional pseudohomogeneous representation. At the same time it was observed that $T_{r=0}$, temperature along the reactor centerline, was given with good accuracy by Pereira Duarte and Lemcoff's equation [8].

$$T_{r=0} = T_{av} + \frac{(T_{av} - T_c)\, 0.25\, Bi}{1 + 0.25\, Bi} \qquad (1)$$

The objective of the present study is to show that the agreement between one and two dimensional pseudohomogeneous models is satisfactory in the PO regime, and that this agreement is still valid when the intraparticle diffusional effects are considered. With these results it is further confirmed that high methanol conversion and easily controllable hot spots can be advantageously combined in the design of a multitubular PO catalytic unit.

Pseudohomogeneous Models

The example considered in this contribution is a multitubular methanol conversion reactor having the following characteristics (Table 1): particle diameter 2mm, total number of tubes 3000, internal tube diameter 2.09 cm, external tube diameter 2.66 cm, distribution of tubes: square pitch with an equivalent diameter of 0.726 cm, coolant: Na-K molten eutectic (Na:22%, K:78%). Unidimensional and bidimensional pseudohomogeneous models can be considered for the simulation of each one of the different tubes of a multitubular catalytic reactor unit. The characteristic and basic assumptions involved in both the unidimensional and bidimensional pseudohomogeneous models were described in a previous contribution [6]. In the same contribution the numerical technique used for solving the set of partial differential equations was described [6].

de Lasa et al. [6] showed, using the bidimensional model, that the temperature profiles in a fixed bed catalytic reactor where reactants and coolant circulated cocurrently could be classified as follows (Figure 1):

a) Conditions where both average temperatures (full lines) and maximum temperatures (broken lines) are always increasing with the axial reactor position. For instance, this is the case at p^o = 1 atm. This regime of full PO is named POT (pseudoadiabatic operation for any temperature) because all the temperatures inside the unit, including the different values of the maxima curve, fulfil the condition $T < T_\infty$.

b) Conditions where only the average temperatures satisfy the requirement of being smaller than the T_∞ value. Examples of this PO for the average temperature (POAT) is the curve of Fig. 1 corresponding to an initial methanol partial pressure of 2 atm. In fact, in this case, the maxima curves show hot spots higher than T_∞ while T_{av} is always smaller than T_∞.

c) Non-Pseudoadiabatic operating conditions where both the average temperature and the maxima curve do not satisfy the condition $T < T_\infty$. This type of temperature profile, characteristic of hot spot operation, was named by Soria Lopez et al.[1] and de Lasa [3] MFARP (T curves with a maximum or hot spot in a finite axial reactor position).

It is possible to observe in the same Figure 1 that significant differences between the predictions of the unidimensional and bidimensional models can be found in the MFARP regime. Under these conditions the unidimensional model provides only a preliminary assessment of pressure and temperature profiles. Figure 1 shows that the temperature calculated with the bidimensional model in the MFARP regime at a given p exceeds the ones predicted by the one dimensional representation. This certainly confirms our previous observations [6]. However, if the temperature along the reactor length is evaluated (T - z plane), the T assessed at a given z is higher for the one dimensional representation than for the two dimensional representation.

In the POT and POAT, the agreement is within 1.5°C (Figure 1) for W_C = 10kg/s. It can then be stated considering the results of the present study and the results of a previous contribution for W_C = 15 kg/s [6] that unidimensional and bidimensional pseudohomogeneous models provide for a multitubular methanol conversion reactor in the 10-15 kg/s coolant flow range essentially the same temperature predictions when either POT or POAT is selected. The unidimensional model could then be used in combination with eq. (1) to estimate T_{av} and $T_{r=0}$ under Pseudoadiatabic Operation.

Heterogeneous Models

The mathematical models required for the simulation of packed bed catalytic reactors have to include the possible reactant partial pressure and temperature gradients inside the catalyst pellets.

In doing so, a distinction between the conditions in the bulk of the gas phase and the catalyst has to be established. Models of such kind are named heterogeneous models because they specifically consider the contribution to the reactor behaviour given by the solid phase (catalyst) and by the gas phase, respectively. To be able to take advantage of the heterogeneous descriptions the mass and heat balances in a differential thickness of a pellet and in a differential bed thickness have to be solved simultaneously. However, the need for solving simultaneously the heat and mass balance inside the pellet can be avoided under some conditions. These special conditions are identified using the Prater number, Pr, or the maximum temperature rise inside the pellet.

$$Pr = \frac{C_s(-\Delta H)\ D_{eff}}{k_{eff}T_s} \tag{2}$$

Calculation of the Prater number requires knowledge of the effective diffusivity of the reactant in the catalyst, D_{eff}, and the thermal conductivity of the pellet, k_{eff}. The D_{eff} parameter has to be evaluated from the properties of the reacting gas stream combined with the physical description of the pellet structure.

In order to justify some of the assumptions made in the present study it is important to visualize that each one of the pellets was prepared as follows: 1μm particles of active ZSM-5 catalyst were mixed with inert particles and/or binder of the same size (1μm). This assumption allows an estimation of a macropore size of about 0.44μm.

In the pseudohomogeneous models described in the first section of this paper and by de Lasa et al. [6] the diffusional limitations in the macropore structure of the pellet was neglected. In fact, the rate expression proposed by Liederman et al. [7] was obtained in a fluid bed where the dimensions of the zeolite particles were in the range of 50 μm.

$$r_m = \frac{k_m p}{1 + K\ p^{\circ}} = k_m^{*}\ p \tag{3}$$

In the present work, it has been conservatively assumed that the change in particle size from 50 μm to 1 μm does not modify the overall kinetic rate expression given by eq. (3). As a consequence, the effectiveness factor for the pellet is determined by the diffusional resistance in the macropores formed by the spacing between the particles.

An analysis of the controlling regime for the diffusional processes in the catalyst pellets showed that molecular diffusion was the dominant mode of transport in the 0.44 μm macropores. The molecular diffusivity to estimate D_{eff} was calculated with the following relationship [9].

$$D_{1,2} = \frac{0.001858\ T^{3/2}\ [(M_1 + M_2)/(M_1 M_2)]^{1/2}}{P\ \delta^2_{1,2}\ \Omega_D} \tag{4}$$

In the present study it is considered that methanol (component 1) is diffusing countercurrently to n-hexane, a typical compound characterizing the products of methanol conversion (component 2). Two important parameters involved in the denominator of eq. (4), $\delta_{1,2}$ and Ω_D had to be evaluated. The $\delta_{1,2}$ term is the force constant in the Lennard-Jones potential function while Ω_D is the collision integral. $\delta_{1,2}$ is independent of temperature and pressure, Ω_D only changes 10% in the temperature range considered for the simulations and can be evaluated with an average value. Under these conditions eq. (4) became:

$$D_{1,2} = 1.41 \times 10^{-5}\ T^{3/2}/P \quad (cm^2/s) \tag{5}$$

and the D_{eff} diffusivity parameter is given by the following expression ($\beta = 0.4$, $\tau = 4$ [9]):

$$D_{eff} = D_{1,2}\ \frac{\beta}{\tau} = 5.08 \times 10^{-7}\ T^{3/2}/P \quad (m^2/h) \tag{6}$$

A second parameter that had to be estimated was k_{eff}, the effective thermal conductivity of the particle. In the present study a k_{eff} = 0.18 kcal/h m°C value was assumed [9]. It is pointed out, however, that errors in k_{eff} were not too significant. This is because the estimated Pr number was always very small, $Pr \leq 0.02$ or, in other words, the maximum temperature change inside the pellet was in the most unfavourable condition 2% of the bulk gas temperature.

On these grounds, the influence of diffusional effects inside the pellets, assuming an isothermal effectiveness factor, was considered. This gave for the rate of reaction inside each catalyst pellet the following expression:

$$r_{eff} = r_m\ \eta \tag{7}$$

where $\eta = \frac{3}{\phi_s}\ [\frac{1}{\tanh \phi_s} - \frac{1}{\phi_s}]$, $\phi_s = R_p \sqrt{\frac{k_v}{D_{eff}}}$, $k_v = k^*_m \rho_B \frac{RT}{1-\varepsilon}$

In the heterogeneous model k_v and D_{eff} were evaluated at each reactor position as functions of temperature and pressure. Later, ϕ_s and η were calculated and finally r_{eff} was assessed with eq. (7).

Figure 2 shows a comparison of the results obtained with the one and two dimensional models when diffusional limitations in the solid pellets (heterogeneous models) were considered. These curves correspond to the operating condition of W_c = 10 kg/s and for three different inlet methanol partial pressures - 1 atm, 3 atm and 5 atm. Continuous lines represent the average temperatures predicted with the two-dimensional model. Dotted lines are the average temperatures predicted with one-dimensional model. Broken and broken-dotted lines are the temperatures at the centerline calculated with the two and the one-dimensional models, respectively. The agreement between the models is very good when the reactor operates under PO conditions both for POT and POAT. It is, however, possible to observe on the same Figure 2 that the two models show important differences when the MFARP regime is approached (p° = 5 atm).

In the case of heterogeneous models, the PO limiting conditions separating the PO and MFARP curves cannot be directly determined by simple algebraic equations because of the more complex relationship between the reaction rate and the temperature and pressure. However, consideration of the diffusional processes inside the pellets transforms some MFARP curves into PO conditions. A typical example is the operating condition p° = 3 atm, W_c = 10 kg/s reported in Figures 1 and 2. The average temperature shows a "hot spot" using a pseudohomogeneous model while the same condition becomes Pseudoadiabatic employing the heterogeneous representations. In other words, consideration of the heterogeneous models for methanol conversion into gasoline shows that the PO extends well beyond the predictions resulting from the pseudohomogenous models. Then the "a priori" equations derived by de Lasa et al., [6] in a previous contribution, giving the limiting conditions p_ℓ° and T_ℓ, provide a conservative estimate of the operating conditions leading to the PO regime.

$$p_\ell^\circ = \frac{A}{B}\left(1 + \frac{C_2}{D}\right)(T_\ell - T_o) \tag{8}$$

$$T_\ell = \frac{a}{b - \ell n(1 + Kp_\ell^\circ) + \ell n(A/D)} \tag{9}$$

Figure 3 indicates how the consideration of diffusional transfer limitations within the solids modifies the p-T trajectories. The results shown in Figure 3 were obtained for one dimensional models using the pseudohomogeneous and heterogeneous approaches. Here again we present two examples (p° = 4 atm and 5 atm) where MFARP conditions with a pseudohomogeneous model become PO with the heterogeneous representation. Moreover, Table 2 compares the partial pressure of methanol and temperature predicted for various reactor axial and radial positions. At the same time the values of η and its changes with z are reported. Comparing the partial pressure of methanol at z = 3m for W_C = 15 kg/s it is possible to visualize how the conversion is affected by the intraparticle diffusional phenomena.

p° = 1 atm, z = 3m, Δx = 0.0105

p° = 2 atm, z = 3m, Δx = 0.0269

p° = 3 atm, z = 3m, Δx = 0.0468

p° = 4 atm, z = 3m, Δx = 0.0662

Consequently, if the initial partial pressure of methanol is below 3 atm, the effect of the intraparticle diffusional phenomena is quite limited. For p° = 3 atm errors in the predictions of methanol conversion at the reactor outlet were about 4.68%.

Conclusions

(i) The conversion of methanol into gasoline in a multitubular catalytic reactor was appropriately simulated using a one-dimensional pseudohomogeneous model for PO conditions. This was shown by comparing the model predictions with the ones of a two-dimensional pseudohomogeneous representation.

(ii) The temperature at the reactor centerline was effectively predicted employing eq. (1) developed by Pereira Duarte and Lemcoff [8].

(iii) For the more critical MFARP conditions, the one dimensional and bidimensional pseudohomogeneous models gave small differences of the predicted temperature for $p^\circ > p^\circ_\ell$ and significant differences for $p^\circ \gg p^\circ_\ell$.

(iv) Thermal effects inside the catalyst pellets can be neglected (Prater dimensionless number very small).

(v) Mass transfer internal diffusional limitations may become significant in 2 mm particles. Consideration of mass transfer internal diffusional limitations showed that the PO extends well beyond the limiting condition given by p°_ℓ and T_ℓ (eqs. (8) and (9)).

(vi) Both heterogeneous and homogeneous models confirmed that high methanol conversion and easily controllable hot spots are compatible conditions in a multitubular reactor designed using the Pseudoadiabatic Reactor Concept.

Acknowledgement

Support from the Natural Sciences and Engineering Research Council of Canada (NSERC), Consejo Nacional de Investigaciones Cientificas y Tecnicas (CONICET)-SUBCYT, Argentina, is gratefully acknowledged.

Notation

A $\rho_B M\,P/(u_s\rho_G)$ (atm.kg$_p$.h/kmol.m)
a kinetic parameter, E/R (K)
B $(-\Delta H)\rho_B/(u_s\rho_g C_{pg})$ (Kg$_p$ h K/kmol m)
b natural logarithm of the pre-exponential factor
Bi Biot number = $\alpha_w R/k_{er}$
C_p heat capacity (kcal/kg K)
C_s reactant concentration at the pellet surface (kmol/m^3)
C_2 $2U/u_s\,\rho_g C_{pg} R$ (1/m)
D $2\pi R\,t_n U/(W_c\,C_{pc})$ (1/m)
$D_{1,2}$ diffusion coefficient of a mixture of two gases (m^2/h)
D_e $2\,R_1(4/\pi-1)$, shell side equivalent diameter (m)
D_{eff} effective diffusivity = $D_{1,2}\,\beta/\tau$ (m^2/h)
dp particle or pellet diameter (m)
E energy of activation (kcal/kmol)
K kinetic parameter (^{1}/atm)
k thermal conductivity (kcal/h m K)
k_{eff} effective thermal conductivity of the pellet (kcal/h m K)
k_{er} effective radial thermal conductivity of the bed (kcal/h m K)
k_m kinetic constant (kmol/kg$_p$ h atm)
k^*_m $k_m/(1+kp^\circ)$ (kmol/kg$_p$ h atm)
k_v volumetric kinetic constant (m^{3}gas/m$_p^3$ h)
L total reactor length (m)
M molecular weight (kg/kmol)
P inlet total pressure (atm)
p methanol partial pressure (atm)
p° inlet methanol partial pressure (atm)
p°_ℓ methanol inlet partial pressure giving the PO limiting conditions (atm).
Pr Prater dimensionless number
r_{eff} effective rate of methanol conversion = $\eta\,r_m$ (kmol/kg$_p$h)
rm rate of methanol conversion (kmol/kg$_p$h)
R internal tube radius (m)
R gas constant $\left(\frac{m^3\ atm}{kmol\ K}\ \text{or}\ \frac{kcal}{kmol\ K}\right)$
R_p pellet radius (m)
R_1 external tube radius (m)
t_n total number of tubes
T temperature inside the reactor tubes (K)
T_{av} average temperature in the cross-section (K)
T_c coolant temperature (shell side) (K)
T_ℓ PO limiting temperature (K)
$T_{r=0}$ temperature at tube centerline (K)
T_s temperature at the pellet surface (K)
U overall heat transfer coefficient (Kcal/h.m^2.K)
$(1/h_{out} + 1/\alpha_w + \Delta r/k_{er})^{-1}$, bidimensional model
$(1/h_{out} + 1/\alpha_w + R/4\,k_{er})^{-1}$, unidimensional model
u_s superficial gas velocity (m/h) defined at T=298K, at the inlet total pressure

W mass flow (kg/s)
x conversion (-), $(p_{out} - p^{\circ})/p^{\circ}$
z axial coordinate (m)

Greek Letters

α_w heat transfer coefficient near the wall (catalytic bed side) ($kcal/h\ m^2K$)
β Pellet voidage (-)
ΔH Enthalpy of reaction (kcal/kmol)
η Effectiveness factor (-)
ε bed voidage (-)
τ tortuosity factor (-)
$\delta_{1,2}$ force constant in Lennard-Jones potential equation
Ω_D collision integral
ϕ_s Thiele modulus for spheres = $R_p(k_v/D_{eff})^{1/2}$.
ρ density (kg/m^3)
μ viscosity (kg/m h)

Subscripts & Superscripts

av average
B bed
c coolant
g gases
ℓ limiting condition for PO regime
m methanol
o inlet conditions
∞ at infinite reactor length
p Pellet
s at pellet surface
1 component representing reactants
2 component representing products

References

1. Soria Lopez, A., H. de Lasa and J.A. Porras, "Parametric Sensitivity of a Fixed Bed Catalytic Reactor. Cooling Fluid Flow Influence". Chem. Eng. Sci., 36, 285-291 (1981).
2. de Lasa, H.I., L.K. Mok and A. Soria Lopez, "Oxidation of Ortho-xylene in a Catalytic Packed Bed Reactor. The Cross Flow Operation. The Critical Row of Tubes". Proceedings Second World Chem. Engng. Conference, 31st CSChE Meeting, Montreal, Vol. III 297-300 (1981).
3. de Lasa, H., "The Pseudoadiabatic Operation. A Useful Tool for Eliminating Hot Spots of Catalytic Fixed Bed Reactors". Proceedings 32nd CSChE Meeting, Vancouver 954-964, Vol. 2 (1982).
4. de Lasa, H., "Application of the Pseudoadiabatic Operation to the Catalytic Fixed Bed Reactors. Case of the Orthoxylene Oxidation". Can. J. Chem. Engng. 61, 710-718 (1983).
5. de Lasa, H., "Pseudoadiabatic Reactor for Exothermal Catalytic Conversions". Canadian Patent Application 4478807 (1984).

6. de Lasa, H., A. Ravella and E. Rost, "Pseudoadiabatic Operation of a Fixed Bed Catalytic Reactor for the Conversion of Methanol into Gasoline". Proceedings XVI ICHMT Symposium "Heat and Mass Transfer in Fluidized Beds". Yugoslavia. To be published by Hemisphere Publishing Corporation (1984).
7. Liederman, D., S.M. Jacob, S.E. Voltz and J.J. Wise. "Process Variable Effects in the Conversion of Methanol to Gasoline in a Fluid Bed". Ind. Eng. Chem. Process Des. Dev., 17, 3, 340-356 (1978).
8. Pereira Duarte, S.I. and N.O. Lemcoff. "Analysis of Fixed Bed Catalytic Reactor Models". Private Communication (1982).
9. Satterfield, C., "Mass Transfer in Heterogeneous Catalysis". M.I.T. Press, Cambridge and London England (1970).

Table 1

Parameters and Reactor Characteristics for the Pseudohomogeneous Unidimensional Catalytic Reactor Model

a	= 10555 K	L	= 3m
b	= 13.87	$2R$	= 0.0209 m
C_{pc}	= 0.234 kcal/Kg K	$2R_1$	= 0.0266 m
C_{pg}	= 0.520 kcal/Kg K	r_m	= $k_m p(1 + Kp^\circ)$
De	= 0.00727 m	t_n	= 3000
d_p	= 0.002 m	u_s	= 1365 m/h (defined at 298K)
k_c	= 22.93 kcal/m h K	μ_c	= 0.637 kg/m h
k_g	= 0.0576 kcal/h m K	μ_g	= 0.072 kg/m h
K	= 1.5 atm^{-1}	ΔH	= - 10690 kcal/Kmol
ρ_b	= 1300 kg/m^3	ρ_c	= 847 kg/m^3

Table 2

Predictions of Temperature and Methanol Partial Pressure Unidimensional Homogeneous and Heterogeneous Models, Wc = 15 kg/s.

p° atm.	z m	P_{av} HOM (atm)	P_{av} HET (atm)	T_{av} HOM (K)	T_{av} HET (K)	$T_{r=0}$ HOM (K)	$T_{r=0}$ HET (K)	η
1	1	0.563	0.575	643.1	642.3	648.5	647.5	0.972
	2	0.239	0.255	653.5	652.8	656.5	655.8	0.965
	2.5	0.146	0.160	655.7	655.2	657.7	657.3	0.963
	3	0.087	0.098	657.0	656.7	658.2	658.0	0.962
2	1	1.363	1.392	652.7	650.6	660.3	657.8	0.959
	2	0.530	0.620	687.4	683.2	694.7	690.2	0.918
	2.5	0.258	0.337	694.1	691.4	698.4	696.0	0.904
	3	0.117	0.170	696.6	695.2	698.7	697.7	0.898

Table 2 (continuation)

p^{o} atm.	z m	P_{av} HOM (atm)	P_{av} HET (atm)	T_{av} HOM (K)	T_{av} HET (K)	$T_{r=0}$ HOM (K)	$T_{r=0}$ HET (K)	η
3	1	2.301	2.334	651.8	649.7	659.0	656.3	0.956
	2	0.929	1.167	715.3	702.4	728.2	713.1	0.875
	2.5	0.317	0.567	730.6	722.4	737.6	730.5	0.828
	3	0.094	0.230	732.4	729.5	734.6	733.4	0.810
4	1	3.288	3.321	648.4	646.6	654.4	652.2	0.957
	2	1.598	1.957	727.2	706.9	744.0	718.8	0.860
	2.5	0.372	0.956	765.4	745.0	776.6	757.3	0.759
	3	0.063	0.327	763.9	759.4	765.8	765.2	0.715
5	1	4.291	4.322	645.0	643.5	650.0	648.2	0.959
	2	2.573	2.927	720.0	701.3	735.3	712.1	0.870
	2.5	0.521	1.623	799.2	753.8	819.2	769.1	0.727
	3	0.043	0.524	791.6	784.1	793.4	793.0	0.633

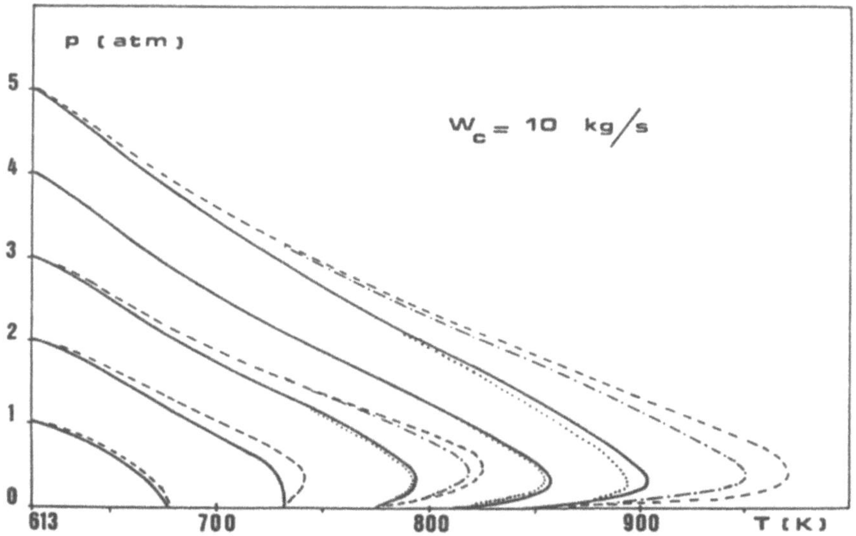

Figure 1. Comparison of the Bidimensional and Unidimensional Pseudohomogeneous Model Predictions. The full lines and dotted lines represent the average temperature for the Bidimensional and Unidimensional models, respectively. The broken and broken-dotted lines represent the $T_{r=0}$ for the bidimensional and unidimensional models. $T_{r=0}$ for the unidimensional model was assessed with eq. (1).

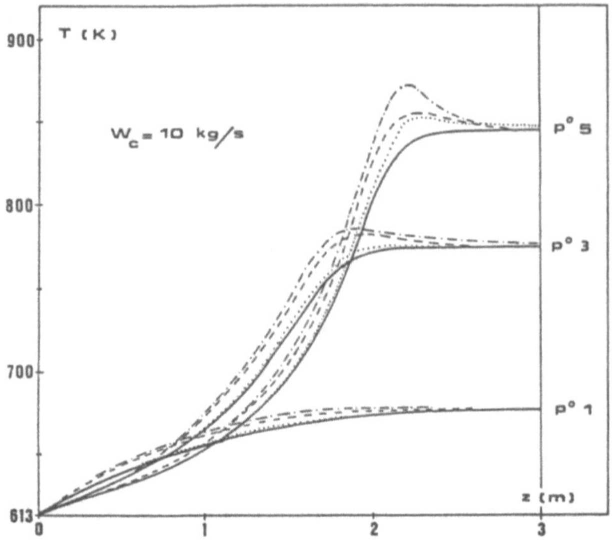

Figure 2. T(z) functions in a methanol conversion catalytic reactor using Bidimensional and Unidimensional Heterogeneous Models.
Curve references in the caption of Figure 1.

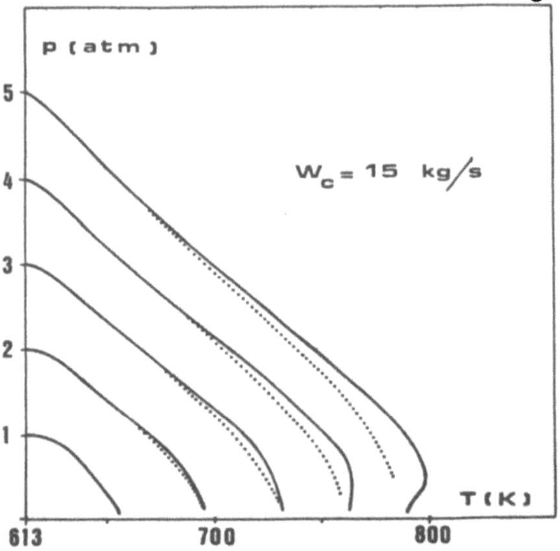

Figure 3. Comparison of the Unidimensional Pseudohomogeneous Model and Unidimensional Heterogeneous Model Predictions. The full lines and dotted lines represent the average temperature for the Pseudohomogeneous and Heterogeneous Models, respectively.

MALDISTRIBUTION IN THE RADIAL-FLOW FIXED BED REACTOR

C-.S. YOO AND A.G. DIXON

WORCESTER POLYTECHNIC INSTITUTE, WORCESTER, MA 01609

1. INTRODUCTION AND BACKGROUND

Radial-flow reactors provide high capacity without increased pressure drop or greatly increased vessel dimensions. This is done by holding the catalyst in a basket forming an annular bed, and causing the gas to flow between the central tube and the outer annulus (separating the catalyst bed and the external pressure shell). This presents the gases with a high flow cross-sectional area per bed volume, so that high throughput can be achieved without increased gas velocity, the catalyst bed can be relatively thin and thus give a low pressure drop. This low pressure drop means that smaller pellets can be used with higher effectiveness factors, and the reactor can be run at a lower total pressure, allowing the use of a centrifugal compressor and thus lower power consumption.

The gas flow in radial beds can be either from the central tube outwards (centrifugal or CF-flow) or from the outer annulus inwards (centripetal or CP-flow). In addition, the flow in the central tube can either be in the same direction as that in the outer annulus (Z-flow) or counter to it (π-flow). There are thus four operating configurations for a single-bed radial-flow converter.

The distribution of gas to the catalyst bed, influenced by the change in direction of gas flow from axial to radial, can have an important effect on reactor performance. In the event of gas maldistribution there will be a spread of residence times in the bed, which can adversely affect both conversion and selectivity [1][2], and may cause local hot-spots. It is possible to force the gas flow to be uniform, for example by using nozzles to feed the gas to the bed, however, it is of interest to see whether the unit can be designed and operated so as to minimize or eliminate maldistribution, without introducing special distributors.

From the results of several studies [2-5] some design rules can be presented to achieve a uniform flow distribution. Broadly, the designer should try to increase flow resistance through the bed, while decreasing flow resistance in the channels so that the pressures in the center tube and outer annulus change in the same way, maintaining a constant pressure drop across the bed [2,4]. Some specific rules are

i) favor π-flow over Z-flow as it always gives a better distribution

ii) if the channel flow resistance << bed flow resistance, then for π-flow the distribution can be made uniform by setting

$$\gamma = \frac{\text{central tube cross-sectional area}}{\text{outer annulus cross-sectional area}} = 1$$

iii) if channel flow resistance is not negligible, then there is an optimal value of γ, as discussed by Chang et al [6], which will minimize the maldistribution, again for π-flow.

Clearly, increasing bed resistance, and hence pressure drop, will alleviate the maldistribution problem, however this may compromise the rationale for choosing radial-flow in the first place. In addition, there may be several design variables that may be manipulated to achieve uniform flow, and it is not clear which are feasible or effective over a realistic range of values.

The object of the present work is to determine which operating and design variables can improve flow distribution without greatly increasing bed pressure drop.

2. HYDRODYNAMIC MODELS

The earliest published results on flow in radial beds were by Russian authors, recognizing the increasing use of such reactors in their chemical industry. Sycheva et al [7] conducted an experimental study of the radial pressure drop in the bed, and found an empirical modification of the Ergun equation to be satisfactory for CP flow in beds of spheres, with the friction factor f dependent upon Reynolds number Re (evaluated at bed inlet, since the gas velocity is a function of radial position, from the continuity equation), particle size and the inner and outer bed radii, R_1 and R_2. Subsequent studies have assumed the validity of an unmodified Ergun equation for f when needed [4,5]. To investigate the effects of reactor geometry, gas flow rate and operating mode on gas distribution, equations are needed for flow in the central tube, through the packed bed, and in the outer annulus. The approach to obtaining such equations has been via macroscopic balances on a differential axial segment of the flow channel, as the difficulty of formulating a correct boundary condition for a perforated wall has precluded use of microscopic balances.

Dil'man et al [8] developed a model of fluid movement in a channel with permeable walls from the macroscopic mechanical energy balance. This was later extended to apply to a radial-flow reactor by Genkin et al [3] for the CF-flow case. The equations were valid for isothermal incompressible flow in a horizontal cylindrical channel; they were extended by Kaye [5] to eliminate the restriction to a horizontal tube and to cover all four operating modes. Chang and Calo [2] have also presented an extension to include both axial and radial flow in the catalyst bed, although they concluded that axial flow was small.

Dil'man et al [8] found good agreement between theory and experiment for the laminar flow of water in a tube fitted with capillaries. Genkin et al. [3], using a full radial bed, and Kaye [5], using a 120° wedge, verified that the equations predicted the axial center-tube and outer annulus velocities well for gas flow. Kaye also showed that the bed pressure drop was accurately predicted.

The use of the macroscopic mechanical energy balance for radial reactor flow modelling has been criticized by Li [4], partly due to some assumptions made in deriving the equations, but mainly because of the assumptions regarding the friction factor. Li suggested that the use of a macroscopic momentum balance was more appropriate, involving a wall force friction factor $\lambda^{(m)}$, rather than the energy dissipation friction factor $\lambda^{(k)}$. Predictions of both models agree well with experimental data, however, and it is more convenient to use the energy balance, as correlations for $\lambda^{(k)}$ are readily available, which is not the case for $\lambda^{(m)}$.

The mechanical energy balances for the center-tube and outer annulus are linked, through the pressure terms, by the equation for pressure drop across the bed. This may be integrated over the bed width and the equations combined to give a single equation for the dimensionless axial velocity in the center-tube, v_z, which may be written in the general form [5]

$$v_z'v_z'' + A_1v_z'v_z + A_2(1-v_z)v_z' + A_3v_z^2 + A_4(1-v_z)^2 + A_5v_z'' = 0 \qquad (1)$$

subject to $v_z(0)=1$ and $v_z(1)=0$, with a suitable definition of the dimensionless axial c-ordinate z. The coefficients A_i are complicated functions of flow rate, bed geometry, inlet and outlet tube geometries etc., and depend on the flow configuration; they have been tabulated by Kaye [5]. The axial center-tube velocity is related to the radial velocity by the continuity equation, which gives

$$v_r \propto v'_z \qquad (2)$$

so that for constant v_r (uniform gas flow) v_z must be a linear function of z, and the boundary conditions dictate that

$$v_z = 1 - z \qquad (3)$$

in this case.

If Eqn. (3) is substituted into Eqn. (1), a condition on A_1, A_2, A_3 and A_4 for uniform gas flow can be found, which involves the axial coordinate z. This can only be satisfied by $A_1=A_2=A_3=A_4=0$. Now each A_i is inversely proportional to bed resistance, so increasing this to large values will give uniformity. Alternatively, if bed resistance is not large enough, some other conditions must be met. For π-flow $A_2=A_4=0$, which favors this flow configuration. If bed resistance is much larger than open tube resistance, $A_3 \simeq 0$, and then A_1 becomes zero if we set $\gamma=1$. Thus the design criteria given previously can be derived directly from equation (1).

3. SIMULATIONS

Equation (1), with the associated boundary conditions, is a non-linear second-order boundary-value ODE. This was solved by the method of collocation with piecewise cubic Hermite polynomial basis functions for spatial discretization, while simple successive substitution was adequate for the solution of the resulting nonlinear algebraic equations. The method has been extensively described before [9], and no problems were found in this application.

Kaye [5] found that the model parameters A_1-A_5 depended mainly on three dimensionless groups: Re, γ and β. The Reynolds number is based on center-pipe diameter and reactor inlet velocity, γ is the area ratio defined earlier and β is a measure of the flow resistance of the catalyst bed and baskets. The results of changing these three parameters are shown in Figures 1-3, and are similar to those found by Kaye [5], thus validating the solution obtained. Only CF-flow is shown, as there is little difference between CF- and CP-flows [2]. Both π- and Z-flows are shown, and as expected π-flow always gives less maldistribution, measured by the deviation of v_z from the 45° line (Eqn. (3)). It is interesting to note that π-flow nearly always gives lower bed reduced flow, while Z-flow gives upper bed reduced flow. The ammonia synthesis catalyst used in radial-flow reactors is subject to settling, which could induce lower bed by-passing, which would be exacerbated by π-flow. Thus Z-flow may still be worth considering, if settling is a major problem.

In Figure 1 it is shown that the maldistribution is insensitive to Re for Re<2000 and Re<200,000, but between these limits is a region of worsening distribution as Re increases. In Figure 2 the profile for π-flow becomes nearly ideal if $\gamma=1$, but $\gamma>1$ leads to lower bed by-passing, while $\gamma<1$ gives upper bed by-passing. For Z-flow ideality can only be approached as $\gamma \to 0$, which is clearly impractical. Figure 3 simply confirms that increasing bed resistance improves flow uniformity.

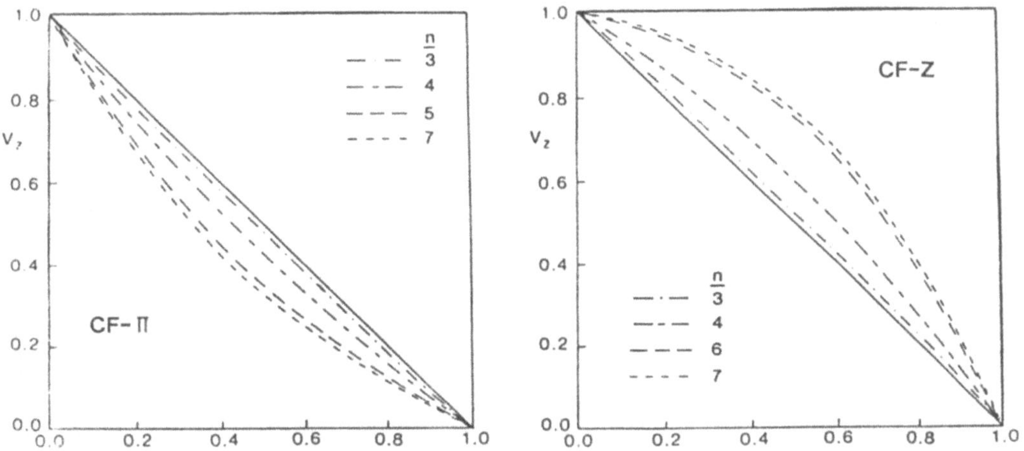

Figure 1. Effect of Reynolds Number on Maldistribution
(β = 0.94, γ = 1.5, Re = 2 x 10^n)

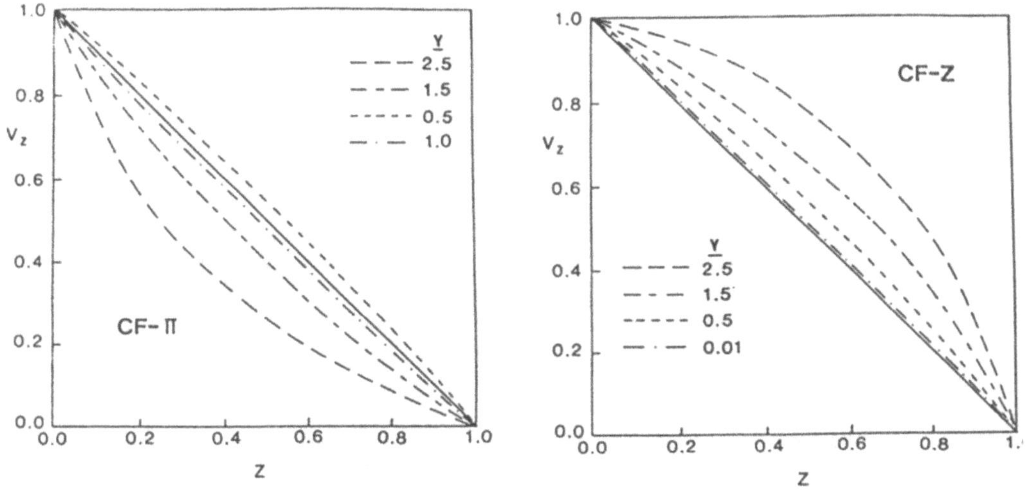

Figure 2. Effect of Area Ratio on Maldistribution
(β = 0.94, Re = 2 x 10^4)

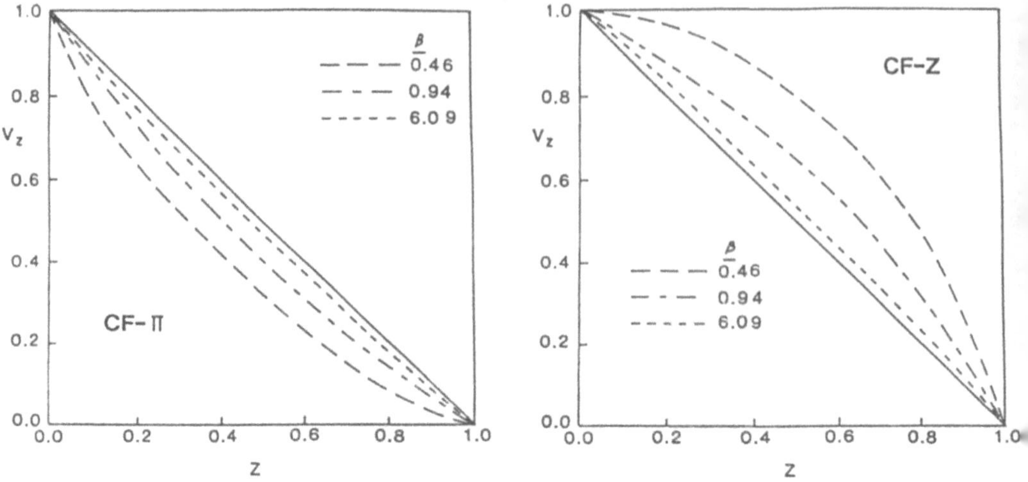

Figure 3. Effect of Bed Resistance on Maldistribution
(γ = 1.5, Re = 2 x 10^4)

The simulations are thus seen to be in agreement with the earlier design rules. The next section will address the question of which variables can and should be changed to prevent maldistribution, and what will be the cost of so doing.

4. ANALYSIS OF DESIGN VARIABLES

The two-level factorial design method is very convenient in screening for the important variables affecting a measured quantity, and assessing their relative importance [10]. In this method a number of variables or factors are chosen, and a range of values selected for each. Only the extreme limits of the ranges are used (denoted as + and - values). The experiments or simulations are run for all possible combinations of factor values (2^n experiments for n factors) and the values of quantities of interest, the responses, calculated. A computational procedure then determines which factors significantly influence each response. Interactions between factors may be estimated; if higher-order interactions are not large then some subset of the experiments may be sufficient to give the main effects, known as a fractional factorial design.

In this work a 2^{6-1} fractional factorial design was run on the factors shown in Table 1, requiring 32 model simulations. The ranges of the variables were chosen to cover conditions likely in industrial practice. The factors γ and π/Z were included to assess the magnitude of their effects; ε, dp and $\emptyset$ would all be expected to influence bed resistance. The bed width was kept constant, so, for a given γ, specifying R fixed radial geometry. The variables ε and dp were taken as independent, to reflect use of different particle shapes and packing methods. The responses were the pressure drop, ΔP, across the bed and baskets, and a measure of maldistribution, given by

$$M = \int_0^1 \left| \left(1 - \left| \frac{v_r}{v_{r,max}} \right| \right) \right| dz \qquad (4)$$

	Design variables	Variable levels -	Variable levels +	Units
1.	Basket open area fraction, ϕ	0.05	0.5	--
2.	Pellet diameter, dp	4.0	10.0	mm
3.	Bed void fraction, ε	0.32	0.42	--
4.	Center tube/annulus area ratio, γ	1.0	1.5	--
5.	Outer shell radius, R	0.6	0.85	m
6.	Flow direction, π/Z	Z	π	

Table 1. Factorial design variables and ranges of values covered.

The results are shown in Figures 4 and 5, in which the main factors and interactions which do <u>not</u> fall onto the straight lines on the Normal probability plot are those which <u>can</u> be distinguished from random error and are therefore significant. The main factors correspond to the design variables of table 1 and are denoted by single-digit numbers. The interactions are denoted by two-digit numbers.

5. SUMMARY OF DESIGN GUIDELINES

The useful insights obtained from the fractional factorial design study are as follows:

i) the basket fractional open area has no detectable effect on the pressure drop or maldistribution (note that this is a uniform basket).

ii) moderate improvements in flow distribution can be obtained by choosing π-flow and by setting $\gamma=1$, with no penalty in increased pressure drop.

iii) improvements in flow distribution can be obtained by increasing R and decreasing ε and dp, but these also lead to pressure drop penalties. It is more effective to change R than ε and dp.

In general the magnitude of the pressure drop is not high, and in view of the other advantages of small catalyst pellets it may be an attractive option to simply increase bed resistance by decreasing dp, and accept the ΔP penalty.

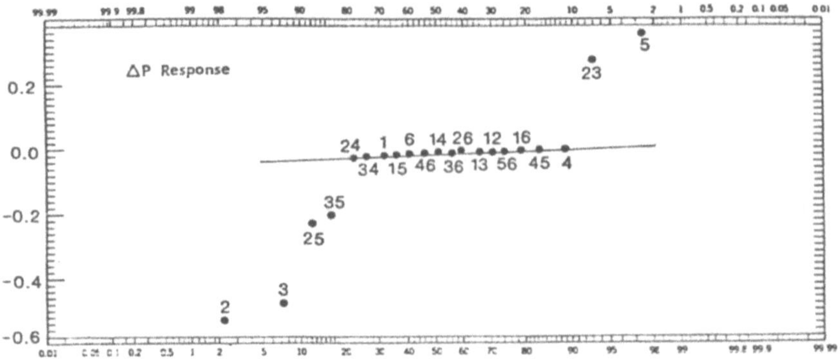

Figure 4. Significance of design variables and their interaction for maldistribution
(see Table 1 for key to numbering)

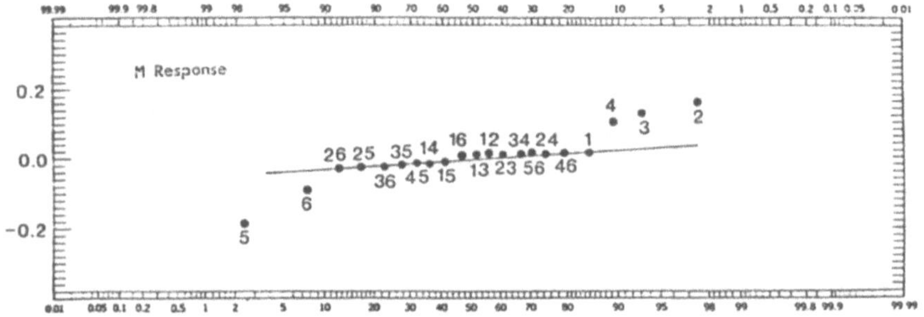

Figure 5. Significance of design variables and their interaction for maldistribution
(see Table 1 for key to numbering)

NOTATION

A_1, A_2, A_3, A_4, A_5	constants in equation (1)
CF	centrifugal (outwards) flow
CP	centripetal (inwards) flow
d_p	pellet diameter (mm)
f	friction factor
M	measure of maldistribution
ΔP	dimensionless bed pressure drop
R	outer shell radius (m)
R_1	inner bed radius (m)
R_2	outer bed radius (m)
Re	Reynolds number
v_z, v_z', v_z''	dimensionless axial velocity (primes denote differentiation with respect to z)

v_r dimensionless radial velocity
z dimensionless axial coordinate
Z co-current flow

β related to catalyst bed flow resistance
ε bed void fraction
$\lambda^{(k)}, \lambda^{(m)}$ energy dissipation, wall force friction factors
γ center tube/annulus cross-sectional area
π countercurrent flow
ϕ basket open area fraction

REFERENCES

1. Ponzi, P.R. and L.A. Kaye. Effect of Flow Maldistribution on Conversion and Selectivity in Radial Flow Fixed-Bed Reactors, AIChE J. 25 (1979) 100-108.
2. Chang, H.-C. and J.M. Calo. An Analysis of Radial Flow Packed Bed Reactors. ACS Symp. Ser. 168 (1981) 305-329.
3. Genkin, V.S., V.V. Dil'man and S.P. Sergeev. The Distribution of a Gas Stream over the Height of a Catalyst Bed in a Radial Contact Apparatus. Int. Chem. Eng. 13 (1979) 24-28.
4. Li, C.-H. Flow Distribution in Radial Flow Packed Bed Reactors. (Ph.D. Thesis, Univ. of Washington, Seattle, 1978).
5. Kaye, L.A. Fluid Distribution in Radial Flow, Vapor Phase, Fixed Bed Reactors. Paper 12E AIChE Annual Meeting, Miami Beach, FL (1978).
6. Chang, H.-C.,M. Saucier and J.M. Calo. Design Criterion for Radial Flow Fixed-Bed Reactors. AIChE J. 29 (1983) 1039-1041.
7. Sycheva, A.M., B.Z. Abrosimov, S.M. Mel'nikov and I.G. Fadeev. The Resistance of a Bed of Packing during Radially Directed Flow. Int. Chem. Eng. 10 (1970) 66-70.
8. Dil'man, V.V., S.P. Sergeev and V.S. Genkin. Description of Stream Movement in a Channel with Permeable Walls on the Basis of an Energy Equation. State Scientific-Research and Planning Institute of the Nitrogen Industry Translation from Teor. Osn. Khim. Tekhn. 5 (1971) 564-571.
9. Dixon, A.G. Solution of Packed-Bed Heat-Exchanger Models by Orthogonal Collocation using Piecewise Cubic Hermite Functions (MRC Technical Summary Report #2116, Mathematics Research Center, Univ. of Wisconsin-Madison, 1980).
10. Box, G.E.P., W.G. Hunter and J.S. Hunter. Statistics for Experiments. (New York, Wiley (1978)).

MATHEMATICAL MODELLING AND SENSITIVITY ANALYSIS OF HIGH PRESSURE POLYETHYLENE REACTORS

Costas Kiparissides and Harilaos Mavridis

Department of Chemical Engineering
Aristotle University of Thessaloniki
Thessaloniki, Greece

1. INTRODUCTION

The high-pressure polymerization of ethylene in tubular reactors is a process of considerable economic significance. A low density polyethylene (LDPE) reactor is designed to produce a variety of polymers each of which has distinct properties and end-use applications. One of the most important problems encountered in operating an LDPE reactor is to select the optimum operating conditions that maximize the reactor productivity at the desired product quality. To accomplish optimal operation of an LDPE tubular reactor or designing a new one, it is necessary to know the sensitivity of system responses relative to the variations of the controlling process variables.

In the present study, a comprehensive model for low density polyethylene tubular reactors is developed. The model includes physical property variations along the reaction path and can predict the monomer conversion, initiator conversion, number and weight average molecular weights as well as the temperature and pressure profiles along the reactor length. To analyse the effects of design, operating and kinetic parameters on the reactor performance, a systematic procedure is developed based on the sensitivity differential equations derived for the system state variables. It is shown that sensitivity analysis can lead to a systematic search for selecting the optimal operating conditions that maximize the reactor productivity.

2. POLYMERIZATION TUBULAR REACTOR MODEL

High-pressure polyethylene tubular reactors are traditionally jacketed tubes, with a large length-to-diameter ratio which ranges from 1000:1 to as high as 10000:1. Reactor temperature varies somewhere between 100 and 300°C and the reactor operates at pressures of 2000-3000 atm. Ethylene polymerization is a highly exothermic process. Approximately one-half of the heat of reaction is removed through the reactor wall by a heat transfer fluid which circulates through the reactor jacket. This results in a nonisothermal reactor operation.

Free-radical high-pressure LDPE tubular reactors can be modelled in terms of a system of non-linear differential equations. Over the past twenty years, several modelling studies have been reported in the literature (Thies and Schoenemann [1], Agrawal and Han [2], Chen et al. [3], Thies [4], Lee and Marano [5], Donati et al. [6], Goto et al. [7], Yoon and Rhee [8], Gupta et al. [9]). In the present work, a detailed mathematical model is developed for an LDPE tubular reactor. The variation of the physical properties of the reaction mixture with position is accounted for. The elements of the model are the reaction mechanism, the mass, energy and momentum balances, and the moments of live and dead polymer distributions. Polyethylene is produced via the following set of elementary reactions:

Initiation: $I \xrightarrow{k_d} 2R(o)$

Propagation: $R(x)+M \xrightarrow{k_p} R(x+1)$

Termination by Combination: $R(x)+R(y) \xrightarrow{k_{tc}} P(x+y)$

Disproportionation: $R(x)+R(y) \xrightarrow{k_{td}} P(x)+P(y)$

Transfer to Monomer: $R(x)+M \xrightarrow{k_{trm}} P(x)+R(o)$

Transfer to Polymer: $R(x)+P(y) \xrightarrow{k_{trp}} P(x)+R(y)$

Transfer to Solvent: $R(x)+S \xrightarrow{k_{trs}} P(x)+R(o)$

β-Scission: $R(x) \xrightarrow{k_\beta} P(x)+R(o)$

Based on the above kinetic mechanism, the steady-state mass, energy and momentum balances for an ideal plug-flow LDPE reactor are written:

Mass balance for initiator:

$$u \frac{dC_I}{dx} + k_d C_I = 0 \tag{1}$$

Mass balance for monomer:

$$u \frac{dC_M}{dx} + k_p C_M \lambda_o = 0 \tag{2}$$

Mass balance for solvent:

$$u \frac{dC_S}{dx} + k_{trs} C_S \lambda_o = 0 \tag{3}$$

Energy balance:

$$\rho C_p u \frac{dT}{dx} - (-\Delta H_P) k_P C_M \lambda_o - \frac{4U}{D_i}(T_w - T) = 0 \tag{4}$$

Moments of the live polymer distribution:

$$u \frac{d\lambda_o}{dx} - \{2k_d f C_I - (k_{tc}+k_{td})\cdot\lambda_o^2\} = 0 \tag{5}$$

$$u \frac{d\lambda_1}{dx} - \{k_P C_M \lambda_o - (k_{tc}+k_{td})\cdot\lambda_o\lambda_1 - \lambda_1(k_{trs}C_S + k_{trm}C_M + k_\beta) - $$
$$- k_{trp}(\lambda_1\mu_1 - \lambda_o\mu_2)\} = 0 \tag{6}$$

$$u \frac{d\lambda_2}{dx} - \{2k_P C_M \lambda_1 - (k_{tc}+k_{td})\cdot\lambda_o\lambda_2 - \lambda_2(k_{trs}C_S + k_{trm}C_M + k_\beta) - $$
$$- k_{trP}(\lambda_2\mu_1 - \lambda_o\mu_3)\} = 0 \tag{7}$$

Moments of the dead polymer distribution:

$$u \frac{d\mu_o}{dx} - \{k_{td}\lambda_o^2 + \tfrac{1}{2}k_{tc}\lambda_o^2 + \lambda_o(k_{trs}C_S + k_{trm}C_M + k_\beta)\} = 0 \tag{8}$$

$$u \frac{d\mu_1}{dx} - \{(k_{tc}+k_{td})\lambda_o\lambda_1 + \lambda_1(k_{trs}C_S + k_{trm}C_M + k_\beta) + $$
$$+ k_{trp}(\lambda_1\mu_1 - \lambda_o\mu_2)\} = 0 \tag{9}$$

$$u \frac{d\mu_2}{dx} - \{k_{td}\lambda_o\lambda_2 + k_{tc}(\lambda_o\lambda_2 + \lambda_1^2) + \lambda_2(k_{trs}C_S + k_{trm}C_M + k_\beta) + $$
$$+ k_{trp}(\lambda_2\mu_1 - \lambda\ \mu_3)\} = 0 \tag{10}$$

Momentum balance:

$$\frac{dP}{dx} + 4f\frac{\rho u^2}{2D_i} = 0 \tag{11}$$

The moment equations (7) and (10) depend on a higher order moment, μ_3, which requires a closure method in order to solve the moment differential equations. The closure method used is that of Hulburt and Katz [10] according to which μ_3 is expressed as

$$\mu_3 = \frac{\mu_2}{\mu_0\mu_1}(2\mu_0\mu_2-\mu_1^2) \qquad (12)$$

The molecular weight averages $\overline{M}_n$ and $\overline{M}_w$ and the polydispersity index PD are calculated by the following equation:

$$\overline{M}_n = \frac{\mu_1+\lambda_1}{\mu_0+\lambda_0}\,MW \quad ; \quad \overline{M}_w = \frac{\mu_2+\lambda_2}{\mu_1+\lambda_1}\,MW \quad ; \quad PD = \frac{\overline{M}_w}{\overline{M}_n} \qquad (13)$$

Numerical values of the kinetic, design and operating parameters of the reactor model are reported in reference 12. It should be noted here that the inclusion of the differential equations (5)-(7) for the live polymer species increases the stiffness of the system equations (1)-(11). This was proved by calculating the ratio $(|\lambda_i|_{max}/|\lambda_i|_{min})$ of the smallest eigenvalue over the largest eigenvalue of the Jacobian matrix of the system equations. For the selected parameters, this ratio was found to be of the order of 10^4. It should be noted here that for the chosen kinetic parameters, no significant difference in the calculated results was observed when the moment differential equations (5)-(7) were replaced by the corresponding algebraic equations (QSSA) (14)-(16):

$$\lambda_0 = \left\{\frac{2k_d f C_I}{k_{tc}+k_{td}}\right\}^{\frac{1}{2}} \qquad (14)$$

$$\lambda_1 = \frac{k_p C_M \lambda_0 + k_{trp}\lambda_0\mu_2}{k_{trs}C_S+k_{trm}C_M+k_\beta+k_{trp}\mu_1+(k_{tc}+k_{td})\lambda_0} \qquad (15)$$

$$\lambda_2 = \frac{2k_p C_M \lambda_1 + k_{trp}\lambda_0\mu_3}{k_{trs}C_S+k_{trm}C_M+k_\beta+k_{trp}\mu_1+(k_{tc}+k_{td})\lambda_0} \qquad (16)$$

3. PARAMETRIC ANALYSIS OF REACTOR PERFORMANCE

The reactor model equations (1)-(13) permit a realistic calculation of monomer and initiator concentrations, temperature and pressure profiles, number average, $\overline{M}_n$, and weight average molecular weight, $\overline{M}_w$, under typical operating conditions. A modified fourth order Runge-Kunta routine was employed for the numerical integration of equations (1)-(11). This routine (Villadsen and Michelsen [11]) estimates the local error and adjusts the integration step according to the value of the error. Figures 1-8 show some representative model results calculated for a set of operating conditions applicable to industrial LDPE reactors. From Figures 1 and 2, it can be seen that the polymerization is incomplete when the jacket temperature is below 115°C, which is too low to preheat the reaction mixture. An increase in the jacket temperature from 115°C to 120°C yields a substantial increase in the monomer conversion. However, further increase in T_w causes a decrease in monomer conversion due to the early consumption of initiator. The problem of selecting the optimal wall temperature to maximize monomer conversion has been

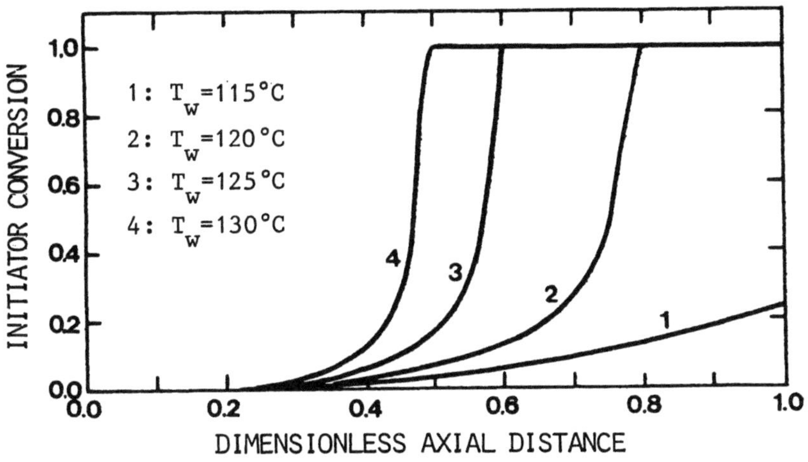

FIG. 1: Effect of wall temperature on initiator conversion ($C_{Io}=4x10^{-4}$ gmol/L)

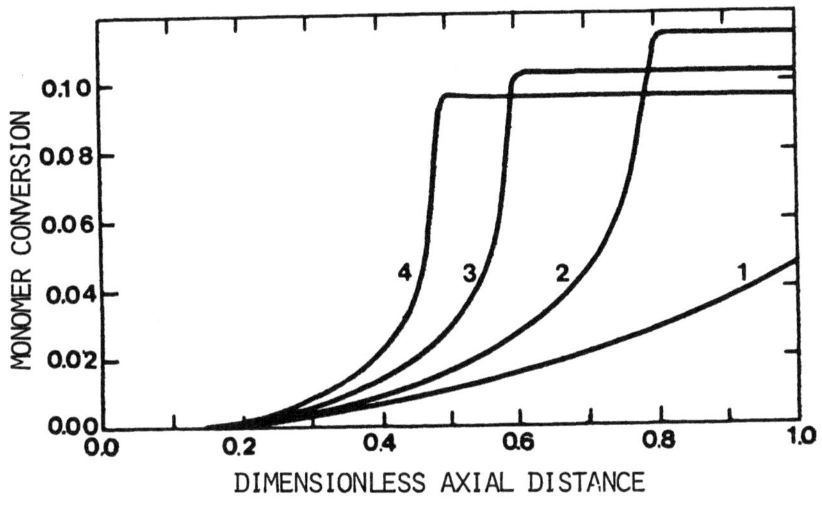

FIG. 2: Effect of wall temperature on monomer conversion profile

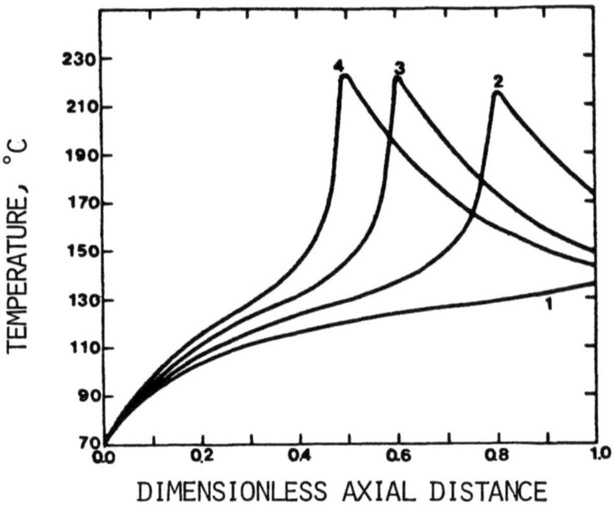

FIG. 3: Effect of wall temperature on reactor temperature profile

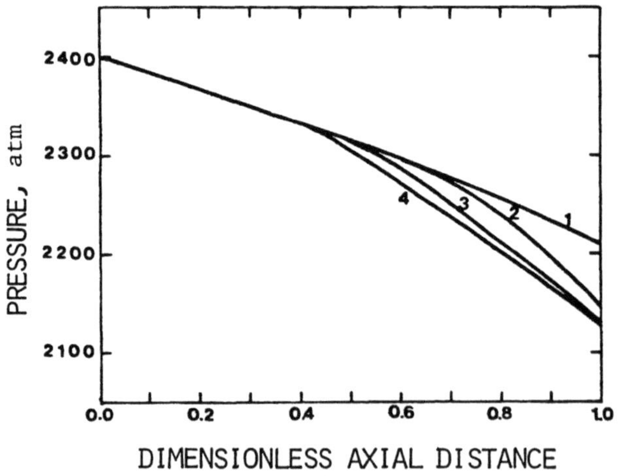

FIG. 4: Effect of wall temperature on reactor pressure profile

solved by Mavridis and Kiparissides [12]. They showed that the optimal wall temperature results in a dead-end polymerization near the reactor exit. In dead-end polymerization, the temperature profile reaches its maximum at the reactor exit, Figure 3.

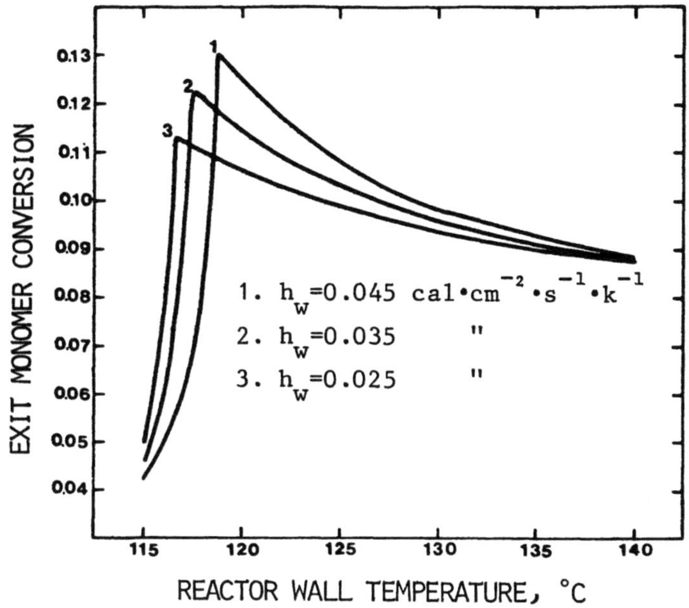

FIG. 5: Effect of wall temperature on monomer conversion at the reactor exit

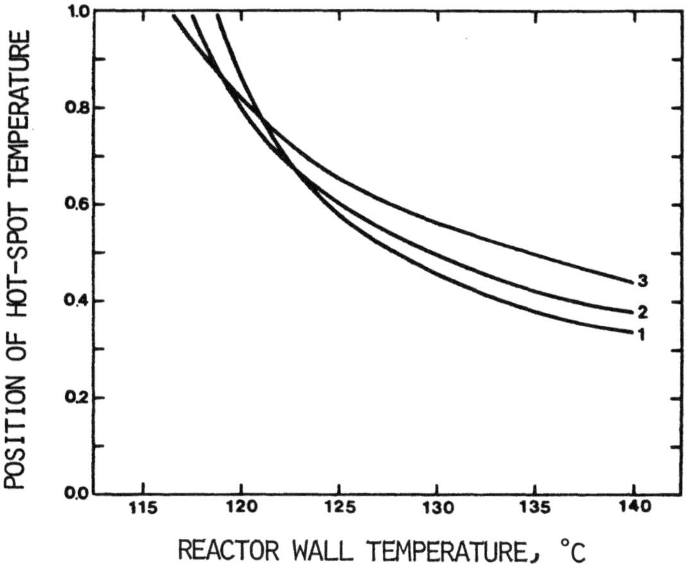

FIG. 6: Effect of wall temperature on the position of hot-spot temperature for various values of h_w

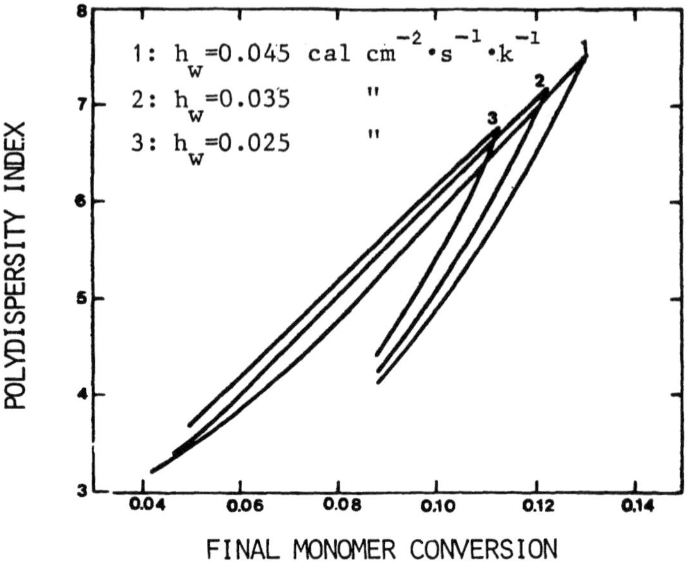

FIG. 7: Final monomer conversion-polydispersity index relation. Each point in curves 1, 2 and 3 corresponds to a different wall temperature.

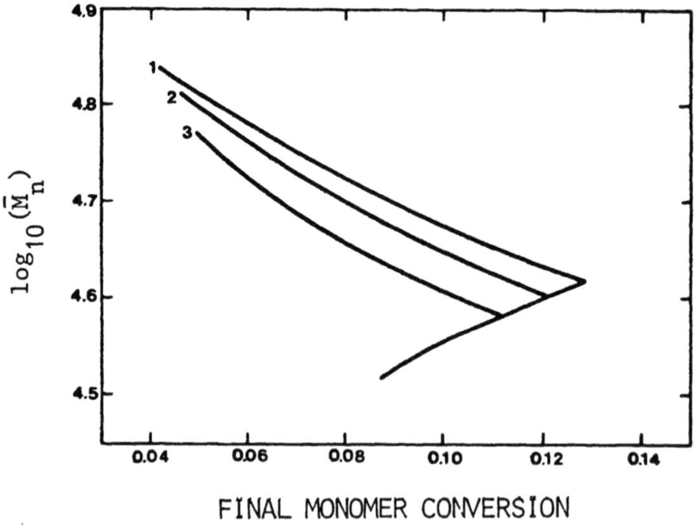

FIG. 8: Final monomer conversion-log($\bar{M}_n$) relation. Each point in curves 1, 2 and 3 corresponds to a different wall temperature.

Figure 4 shows the effect of T_w on the pressure profile along the reactor length. Figures 5 and 6 show the effect of T_w on final values of monomer conversion and position of hot-spot temperature for various values of wall heat transfer coefficient, h_w. Finally, Figures 7 and 8 show the relationships between final monomer conversion-polydispersity index and final monomer conversion-log(M_n) for various values of h_w.

4. SENSITIVITY ANALYSIS

To accomplish optimal operation of an existing LDPE tubular reactor or to design a new reactor system, it is necessary to know the sensitivity of system responses caused by deviations of system parameters from their nominal values. It is well known that polymerization reactors are very sensitive to changes in operating and kinetic parameters. Therefore, it is very important to know the underlying relationships between the system outputs and the model parameters.

The polymerization of ethylene in a tubular reactor can be described by a system of ten ordinary differential equations (1)-(10). Because the state variables ($C_I \sim 10^{-4}$, $T \sim 10^2$, $\mu_0 \sim 10^{-3}$, $\lambda_0 \sim 10^{-8}$ etc) differ considerably in magnitude, new dimensionless variables are introduced to ease the sensitivity analysis of the LDPE reactor. In terms of the new variables, equations (1)-(10) are written:

$$\frac{d\underset{\sim}{y}}{dz} = \underset{\sim}{f}(\underset{\sim}{y}, \underset{\sim}{p}) \quad ; \quad \underset{\sim}{y}(0) = \underset{\sim}{y}_0 \tag{17}$$

where $\underset{\sim}{y}$ is the vector of the dimensionless state variables

$$\underset{\sim}{y} = (y_1, y_2, \ldots, y_{10})^T = (y_I, y_M, y_T, \mu_0, \mu_1, \mu_2, \lambda_0, \lambda_1, \lambda_2, y_S)^T \tag{18}$$

The vector p contains the design, operating and kinetic parameters of the reactor model and is defined as

$$\underset{\sim}{p} = (U, T_w, C_{Io}, T_o, C_{so}, D_i, k_{do}, E_d, k_{po}, E_p, k_{tco}, E_{tc}, k_{tdo}, E_{td}, k_{trmo}, E_{trm}, k_{trpo}, E_{trp}, k_{trso}, E_{trs}, k_{\beta o}, E_\beta)^T \tag{19}$$

A means of assessing the effects of parameter uncertainties upon the model output variables is the sensitivity coefficient.

The sensitivity coefficient, φ_{ij}, is defined as the rate of change of the value of the ouput variable y_i with respect to perturbations in the parameter p_j.

$$\varphi_{ij} = \frac{\partial y_i}{\partial p_j}, \quad i=1, 2 \ldots 10 \quad ; \quad j=1, 2 \ldots 22 \tag{20}$$

The sensitivity coefficient, equation (20), indicates the magnitude and direction of change of the output varible y_i caused by deviations of the parameter p_i from its nominal value. For a vector of output variables $\tilde{y}$, the matrix of the derivatives of the system responses with respect to the system parameters $\tilde{p}$ is called the sensitivity matrix.

$$\Phi = \left(\frac{\partial y_i}{\partial p_j}\right) = \left(\frac{\partial \tilde{y}}{\partial \tilde{p}}\right) \tag{21}$$

The sensitivity coefficient will depend on the axial distance z. However, since equation (17) cannot be integrated analytically the variation of the sensitivity coefficients will be determined by integration of the sensitivity differential equations which are derived next.

Let us consider the system of equations (17). For small perturbations in p, a first-order approximation for the corresponding change in the state variables $\tilde{y}$ is

$$\Delta \tilde{y} = \sum_{j=1}^{r} \tilde{\varphi}_j \Delta p_j + \ldots \tag{22}$$

where $\tilde{\varphi}_j$ is the sensitivity vector

$$\tilde{\varphi}_j = \left(\frac{\partial y}{\partial p_j}, \frac{\partial y}{\partial p_j}, \ldots \frac{\partial y_n}{\partial p_j}\right)^T \tag{23}$$

By differentiation of equation (20) we obtain the corresponding sensitivity differential equations in terms of the model nonlinear functions f_j.

$$\frac{d\varphi_{kj}}{dz} = \sum_{i=1}^{n} \frac{\partial f_k}{\partial y_i} \cdot \varphi_{ij} + \frac{\partial f_k}{\partial p_j} \tag{24}$$

where k, i=1, 2, ... 10 and j=1, 2, ... 22.

The initial conditions of these equations (24) are:

$$\varphi_{ij}(o) = \begin{cases} o \text{ if } p_j \text{ is not an initial condition} \\ \delta_{ij} \text{ if } p_j \text{ is an initial condition} \end{cases} \tag{25}$$

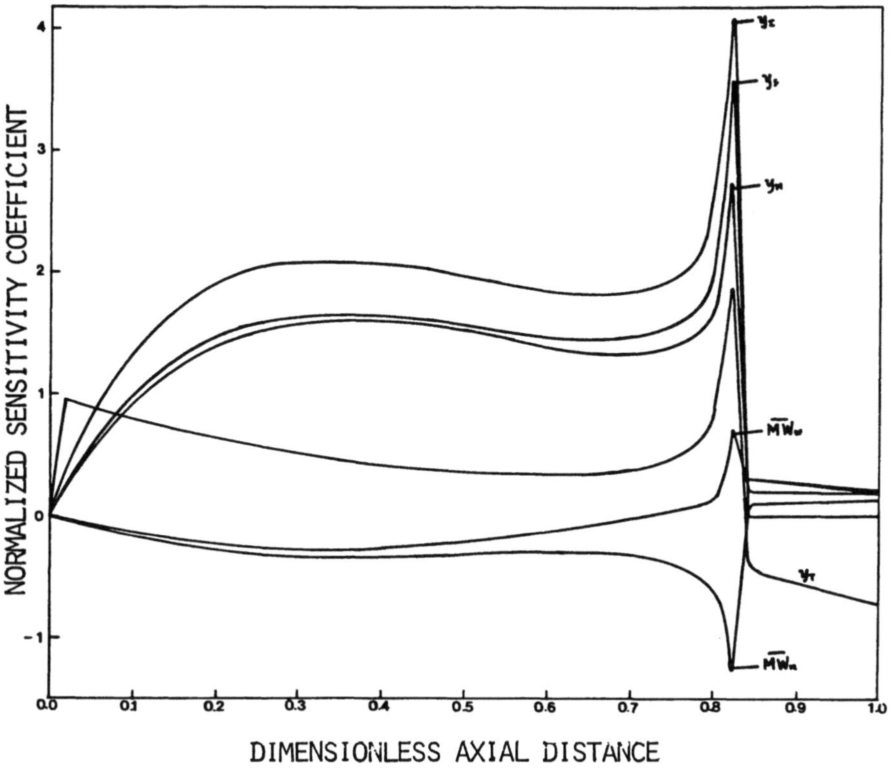

FIG. 9: Effect of perturbations in the heat-transfer coefficient on system responses

The total number of sensitivity equations is equal to the number of state variables times the number of model parameters (i.e. 10x22). These equations are derived in reference [13]. The variation of the sensitivity coefficients along the reactor is determined from the numerical integration of the sensitivity differential equations (24) and model equations (1)-(10). It should be noted that the sensitivity equations (24) are extremely stiff. Thus, extra care must be taken in integrating these equations. Accordingly, a multi-step predictor corrector method suitable for stiff differential equations was used.

5. RESULTS OF THE SENSITIVITY ANALYSIS

Results of the sensitivity analysis of model responses y with respect to the design and operating parameters (U, T_w, C_{Io}, $\tilde{T}_o$, C_{So}, D_i) are summarized in Table I and Figures 9 to 12. In these Figures,

PARAMETER (P_j)	$\rho_{y_M,j}$	$\rho_{\bar{M}_n,j}$	$\rho_{\bar{M}_w,j}$	$\rho_{PD,j}$	$\rho_{y_I,j}$	$\rho_{y_S,j}$	$\rho_{y_T,j}$
U	0.230	0.119	0.198	0.079	$-7.18.10^{-13}$	0.196	-0.684
T_w	-7.708	-4.489	-12.730	-8.241	$-4.15.10^{-11}$	-5.948	-34.6
$(C_I)_o$	0.033	-0.412	- 0.350	0.061	$-2.71.10^{-12}$	0.131	-0.745
T_o	-7.906	-4.622	-13.0	-8.381	$-1.29.10^{-11}$	-6.087	-41.75
$(C_S)_o$	0.	-0.196	- 0.218	-0.414	0.	0.	0.
D_i	-0.231	-0.119	- 0.198	-0.0793	$1.32.10^{-11}$	-0.197	0.684

TABLE I: Normalized sensitivity coefficients at the reactor exit

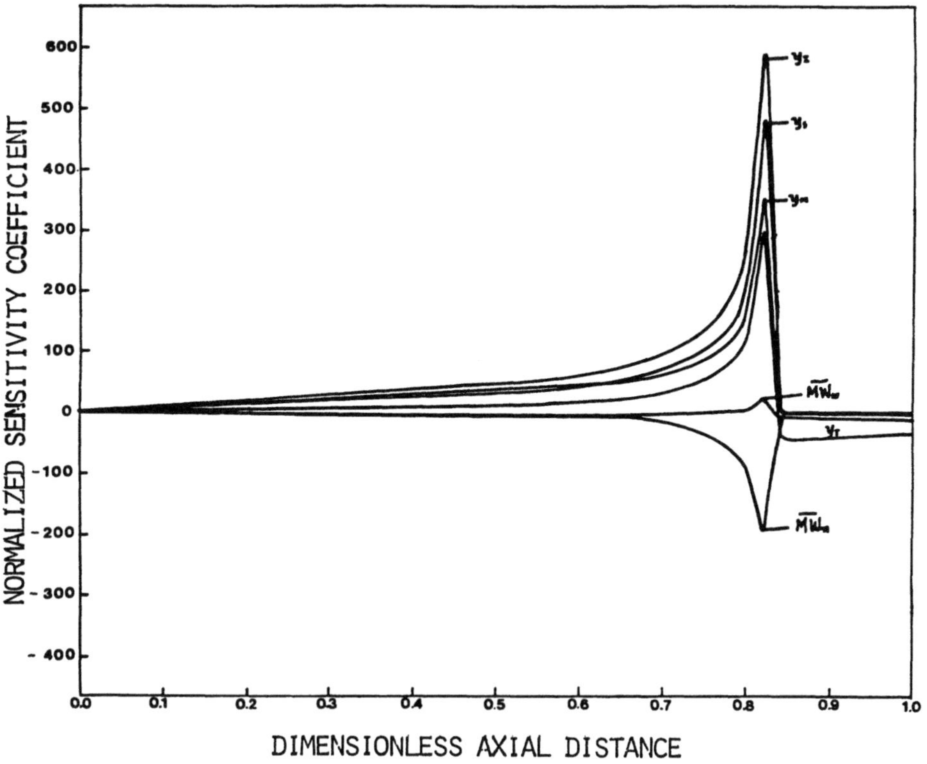

FIG. 10: Effect of perturbations in the wall temperature on system responses.

the normalized sensitivity coefficients

$$\rho_{ij} = \frac{\partial y_i / y_i}{\partial p_j / p_j} = \varphi_{ij}\left(\frac{p_j}{y_i}\right) \tag{26}$$

are plotted as a function of reactor length. The reason for plotting the normalized sensitivity coefficient ρ_{ij} instead of φ_{ij}, is to show the relative influence of parameter perturbartions on the state variables. The numerical values of the sensitivity coefficients φ_{ij} are calculated from equation (26), using the results of Figures 9-12 and the nominal values of the state variables shown in Figure 13. Note that, for a positive sensitivity coefficient, a positive change of the parameter from its nominal value will result in an increase of the output variable. On the other hand, a negative sensitivity coefficient indicates that a positive change in the parameter will cause a decrease in the output variable.

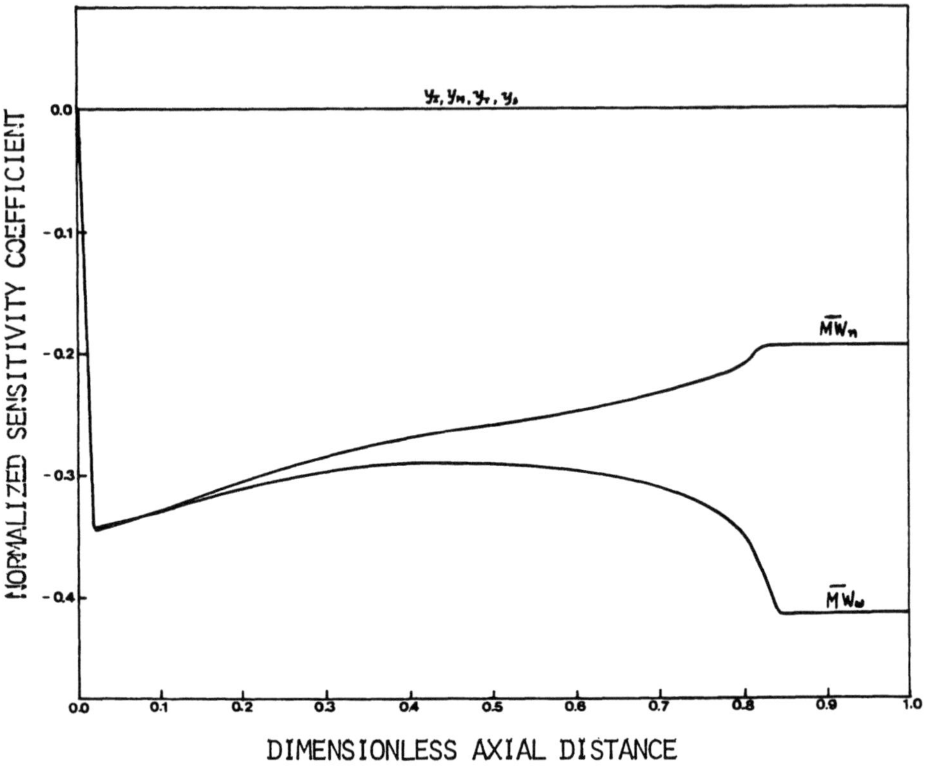

FIG. 11: Effect of perturbations in the initial solvent concentration, C_{so}, on system responses.

The effects of perturbations in the overall heat transfer coefficient U on the model responses are shown in Figure 9. It can be seen that the system responses are very sensitive near the hot-spot temperature. It is interesting to note that the sensitivity coefficients for $\bar{M}_n$, $\bar{M}_w$ and T change sign along the reactor length. This means that, depending on the location in the reactor, a positive variation in U might cause either a decrease or an increase in these variables. The values of the normalized sensitivity coefficients at the reactor exit are listed in Table I. The results of Table I show that as the heat transfer coefficient increases, the exit monomer conversion, $\bar{M}_n$, $\bar{M}_w$ and PD increase while the temperature of the reaction fluid at the reactor exit decreases. This implies that conversion improvements due to heat transfer can be made. However, the degree of improvement of monomer conversion due to heat transfer will depend on the initiator system that is used for the polymerization. The results of the sensitivity analysis are in full agreement with earlier reported results by Lee and Marano [5], Chen et al. [3], Gupta et al. [9].

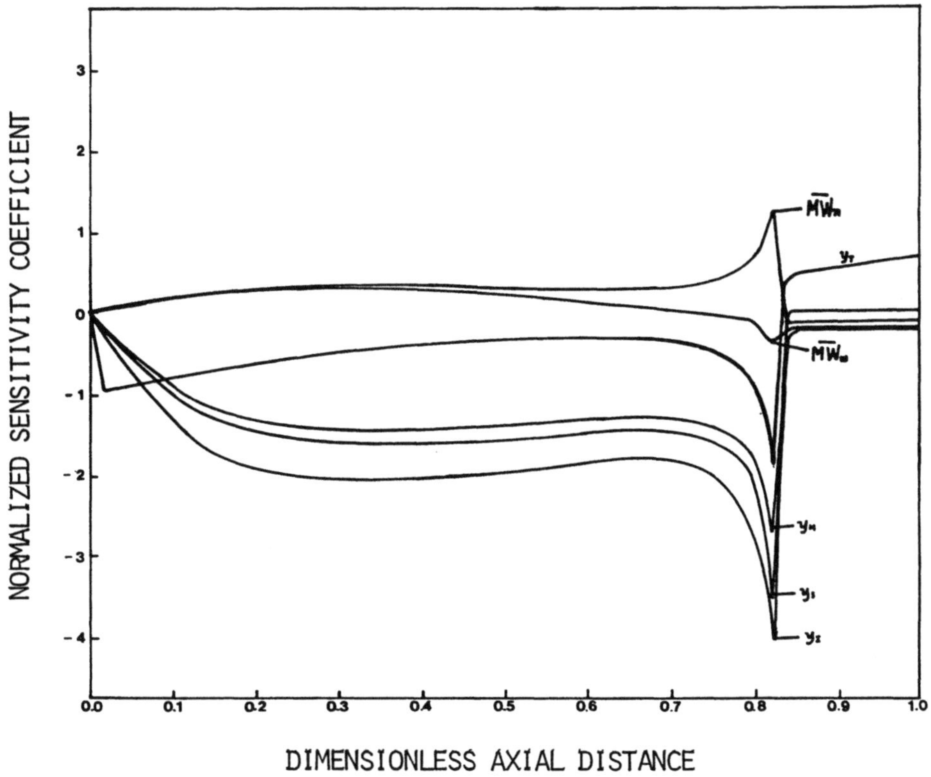

FIG. 12: Effect of perturbations in the tube diameter, D_i, on system responses.

The effect of reactor wall temperature perturbations on the system responses are shown in Figure 10. From the results in Table I, it can be seen that the output variables decrease as the reactor wall temperature increases. It is of interest to note that the last result is based on the assumption that reaction conditions and the initial initiator concentration are chosen so that a peak temperature occurs before the reactor exit. The final values of normalized sensitivity coefficients of the system variables with respect to the initial initiator concentration are shown in Table I. From the listed final values of ρ_{ij} it can be seen that an increase in $(C_I)_o$ brings about an increase in monomer conversion and a decrease in $\bar{M}_n$, $\bar{M}_w$ and reaction temperature.

The effect of solvent perturbations on the system variables is shown in Figure 11 and Table I. It is clear from these results that solvent perturbations have no effect on monomer conversion, initiator conversion or reaction temperature. On the other hand, solvent

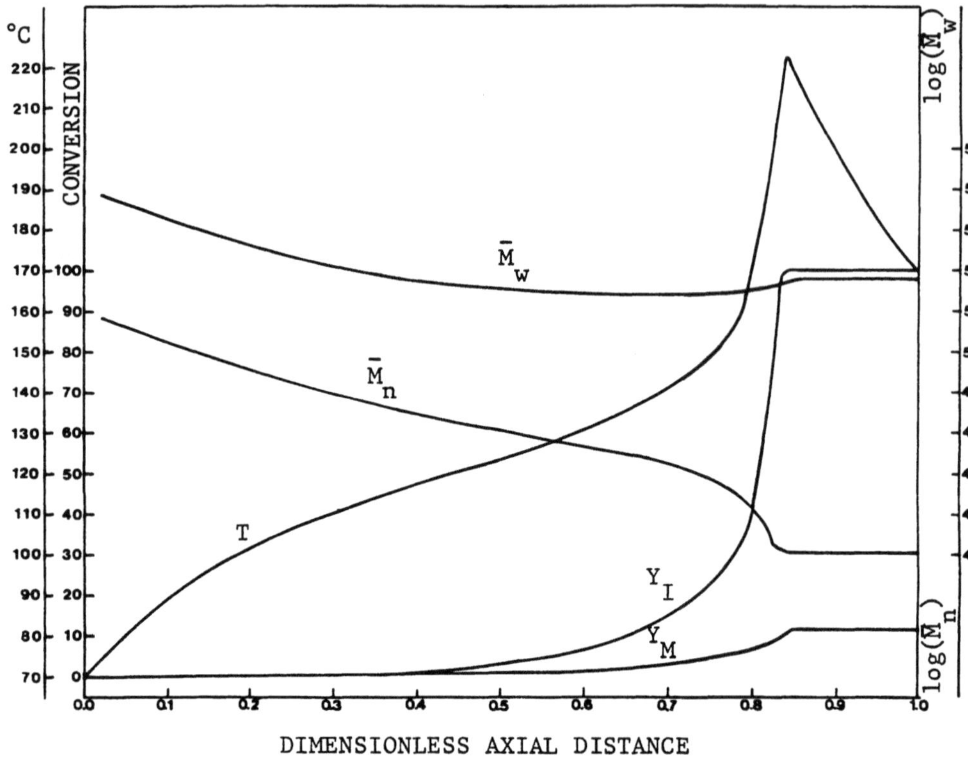

FIG. 13: Nominal reactor temperature, monomer conversion, initiator conversion, number-anerage molecular weight and weight-average molecular weight profiles.

variations do affect the number average and weight average molecular weights. In fact a positive variation in (C_{so}) will cause a decrease in the output variables $\bar{M}_n$ and $\bar{M}_w$.

Finally, the effect of perturbation in the design parameter D_i is shown in Figure 12. It can be seen that perturbations in D_i have exacty the opposite effect on the output variables to that obtained by deviations in the overall heat transfer coefficient U. This is explained by the fact that both parameters affect the rate of heat transfer. Thus, an increase in D_i will bring about a decrease in the exit monomer conversion, $\bar{M}_n$ and $\bar{M}_w$, and will cause an increase in the exit temperature of the reaction fluid. The results in Figure 12 indicate that a small reactor diameter will improve the heat transfer characteristics of the process and increase the final monomer conversion. In addition to the above qualitative treatment of the normalized sensitivity coefficients one can make a quantitative interpretation of the results of Figures 9-13 and Table I. As an example, suppose the effect of perturbations in T_w on the output variables y_M, $\bar{M}_n$ and PD. For a small perturbation in T_w (i.e. 1%), the sensitivity coefficient $(\partial y_i/\partial p_j)$ can be approximated by $(\Delta y_i/\Delta p_j)$. Therefore, equation (26) becomes:

$$(\text{\% change in } y_i) = (\rho_{ij})\text{x}(\text{\% change in } p_j) \tag{27}$$

Suppose an 1% positive change of the reactor wall temperature from its nominal value of 393°K. Using the values of the normalized sensitivity coefficients of Table I and equation (27), we obtain the following quantitative changes for y_M, $\bar{M}_n$ and PD:

$$\begin{aligned}(\text{\% change in } y_M) &= -7.708\% \\ (\text{\% change in } \bar{M}_n) &= -4.489\% \\ (\text{\% change in PD}) &= -8.241\%\end{aligned} \tag{28}$$

These results show that an 1% change in T_w (from 393 K to 396.93 K) will bring about a decrease in monomer conversion from 0.1138 to 0.105, a decrease in $\bar{M}_n$ from 32024 to 30586, and a decrease in the final polydispersity index PD from 4.655 to 4.271 provided that a peak temperature occurs before the reactor exit.

Results of the sensitivity analysis of the system responses with respect to the model kinetic parameters (k_{do}, E_d, k_{po}, E_p, k_{tco}, E_{tc}, k_{tdo}, E_{td}, k_{tmo}, E_{tm}, k_{tpo}, E_{tp}, k_{tso}, E_{ts}, $k_{\beta o}$, E_{β}) are reported in Mavridis and Valtaras [13].

NOMENCLATURE

C_I	:	concentration of initiator in the reactor, gmol/L
C_M	:	concentration of monomer, gmol/L
C_p	:	specific heat of the reaction mixture, cal/(g·°C)
C_s	:	concentration of solvent, gmol/L
D_i	:	inside tube diameter, cm
$(-\Delta H_p)$	:	heat of polymerization, cal/gmol
f	:	Fanning friction factor, in equation (11)
f	:	efficiency factor for initiator
I	:	initiator species
k_d	:	rate constant for initiator decomposition, s^{-1}
k_p	:	rate constant for chain propagation, L/(gmol·s)

k_{tc} : rate constant for termination by combination, L/(gmol•s)

k_{td} : rate constant for termination by disproportionation, L/(gmol•s)

k_{trm} : rate constant for transfer to monomer, L/(gmol•s)

k_{trp} : rate constant for transfer to polymer, L/(gmol•s)

k_{trs} : rate constant for transfer to solvent, L/(gmol•s)

k_{β} : rate constant for β-scission reaction, s^{-1}

M : monomer species

$\bar{M}_n$: number average molecular weight

$\bar{M}_w$: weight average molecular weight

MW : molecular weight of monomer

P : pressure in equation (11), atm

PD : polydispersity index

P(x) : dead polymer chains of length x

R(x) : live polymer chains of length x

S : solvent species

T : reactor temperature

T_w : reactor wall temperature

u : linear velocity of the fluid in the reactor, cm/s

U : overall heat-transfer coefficient, $cal/(cm^2 \cdot s \cdot {}^\circ C)$

x : axial distance

y_I : fractional initiator conversion

y_M : fractional monomer conversion

λ_κ : kth moment of the live polymer distribution, gmol/L

μ_κ : kth moment of the dead polymer distribution, gmol/L

ρ : density of fluid, g/L

REFERENCES

1. Thies, H. and K. Schoenemann. Kinetic Description of a Complex Reaction under Locally Different Conditions by a System of Inhomogeneous Differential Equations ; Chemical Reaction Engineering Proceedings, 2nd International Symposium on Chemical Reaction Engineering, Amsterdam (1972).

2. Agrawal, S. and C.D. Han. Analysis of the High Pressure Polyethylene Tubular Reactor with Axial Mixing, AIChE J., 21 (1975), 449-465.
3. Chen, C.H., J.G. Vermeychuk, J.A. Howell and P. Ehrlich. Computer Model for Tubular High-Pressure Polyethylene Reactors, AIChE J., 22 (1976), 463-471.
4. Thies, H. Strategy for Modelling High-Pressure Polyethylene Reactors. 86th National AIChE Meeting, Houston, Texas, April (1979).
5. Lee, K.H. and J.P. Marano. Free-Radical Polymerization: Sensitivity of Conversion and Molecular Weights to Reactor Conditions ; Polymerization Reactors and Processes (ACS Symposium Series 104, J.N. Henderson and T.C. Bouton, eds.), American Chemical Society, Washington, D.C., (1979), 223-251.
6. Donati, G., L. Marini, G. Marziano, C. Mazzaferri, M. Spampinato and E. Langianni. Mathematical Model of Low Polyethylene Tubular Reactor ; Chemical Reaction Engineering (ACS Symposium Series 196, J. Wei and C. Georgakis, eds.), American Chemical Society, Washington, D.C., (1981), 579-590.
7. Goto, S., K. Yamamoto, S. Furui and M. Sugimoto. Computer Model for Commercial High-Pressure Polyethylene Reactor Based on Elementary Reaction Rates Obtained Experimentally ; Journal of Appl. Polym. Science: Appl. Polym. Symposium, 36 (1981), 21-40.
8. Yoon, B.J. and H. Rhee. A Study of the High Pressure Polyethylene Tubular Reactor, Chem. Eng. Commun., 34 (1985), 253-265.
9. Gupta, S.K., A. Kumar and M.V.G. Krishnamurthy. Simulation of Tubular Low-Density Polyethylene, Polymer Eng. and Sci., 25 (1985), 37-47.
10. Hulburt, H.M. and S. Katz. Some Problems in Particle Technology: A Statistical Mechanical Formulation, Chem. Eng. Sci., 19 (1964), 555-574.
11. Villadsen, J. and M.L. Michelsen. Solution of Differential Equation Models by Polynomial Approximation (Prentice-Hall, N.J. 1978).
12. Mavridis H. and C. Kiparissides. Optimization of a High-Pressure Polyethylene Tubular Reactor, Polymer Process Eng., 3 (1985), 263-290.
13. Mavridis H. and S. Valtaras. Mathematical Modeling and Optimization of a High-Pressure Polyethylene Tubular Reactor, Diploma Thesis, University of Thessaloniki (1984).

DYNAMIC ANALYSIS OF AN ETHANE CRACKING REACTOR

Kirkbir, F., Kisakurek, B.

Ankara University, Turkey

I. INTRODUCTION

Tubular reactors are one of the basic types of chemical reactors. Such reactors can either be packed with catalyst or be empty, depending on the reaction system considered. Some examples for catalytic fixed bed tubular reactors are ethylene, sulphur or naphthalene oxidation reactors. On the other hand, hydrocarbon thermal cracking, ethylene polymerization reactors are typical examples for empty tubular reactors.

Control of such reactors are very important from the point of view of increasing the product quality and operating at optimum conversion and safety. The first step in the control of equipment is to analyze the system dynamically and to represent this mathematical behaviour by a model. For tubular reactors this model usually consists of several partial differential equations which are mainly developed from conservation equations related to mass, energy and momentum. Since analytical solutions of these are not usually possible, they are solved by numerical techniques.

1.1. Literature Survey

Much research in this field is in the area of "learning models". This is because of the several phenomena that affect the modelling of chemical reactors such as instabilities, multiple steady states, oscillations, etc. The purpose of controlling is faced with several such as catalyst activity, may not be possible.

i. In some cases, the measurement of some of the reactor variables, such as catalyst activity, may not be available.

ii. On line measurements, such as outlet concentrations, are very difficult to get and costly.

iii. The reactor may contain several points of measurement and several input variables which may be controlled.

iv. The time delay in the control loop prevents the stabilisation of a chemical reactor. As expected, there is a lot of literature on this subject which involves several physical phenomena and mathematical solutions of developed models. A comprehensive survey on this field is given by Gilles (10), Ray (16), and Schmitz (17).

Dynamic analysis of tubular reactors(both packed or empty) has the following main characteristics:

1. Usually, the equation of motion is not considered (1),(19), (20). On the other hand, Gilles (10) comments that in a very fast and highly exothermic reaction, the internal feedback of the reactor is determined by the hydrodynamic process alone; thus, he uses the equation of motion.

2. To decrease the computational time, a one dimensional model is considered. Sundaram and Froment (23) developed a heat transfer correlation from the two dimensional model and used it in a one dimensional model to obtain more accurate results.

3. In some cases dispersion terms are also neglected (29).

4. Physical parameters are assumed to be constant (5),(1) (9). Brooks (2) criticises this assumption.

5. The accumulation of heat within the wall is neglected. Eiberger (4),(5) and Georgakis (9) report that, although this assumption is reasonable for homogenous liquid phase reactions, it is not valid for homogenous gas phase reactions (1),(19).

6. In some cases, the time dependent term of the continuity equation is neglected (4),(11).

7. Georgakis et.al. (9) claim that in many instances of multiple steady states, the state with the highest conversion may be unstable and, because of this,it is impossible to operate the reactor without control.

8. Due to the notable decrease in computation time, orthogonal collocation methods are used more frequently in recent years (1), (6),(11),(19),(21),(22).

In this work, thermal cracking reactors of ethylene plants are investigated dynamically. The main attention is given to the cracking coil. The modelling of the furnace itself is not considered.The tubular reactor is modelled by assuming the external wall temperature profile or heat flux profile of the coil. There is no report on dynamic analysis of this type of reactor in the literature except Jackman and Aris's work (12). The literature is mainly interested in the steady state modelling of these reactors and control.

These reactors can be modelled in three ways;(i) by direct experimental simulation, (ii) by equivalent reactor volume concept, and (iii) by conservation equations. Since the flow rates used in simulation experiments are much lower than those in real reactors (Re=1000), direct experimental simulation is subject to errors. The second method tries to overcome these problems arising due to non-isothermal and non-isobaric behaviour of the reactor by choosing a reference temperature and pressure which yields the same conversion as the actual reactor with the given temperature and pressure profile. In spite of this, it does not consider the effects of total pressure and dilution. Hence, it is not accurate enough. Froment and his co-workers (3),(8),(22),(28) use this concept to model thermal cracking reactions and their parameters. Shah (18) investigated the effects of temperature and pressure on product distribution by using the conservation equations and obtained excellent agreement with the industrial data.

2. DYNAMIC MODEL

The following basic assumptions are used in the model.

i. Plug flow is assumed due to the high fluid velocity. Dispersion for both mass, energy and momentum transport are also negligible because of ratio of length to diameter for the reactor.

ii. Kinetic and potential energy terms are negligible when compared to the convection terms.

iii. Ideal gas is assumed. This is reasonable since temperatures are at acceptable levels and the pressure is low.

On the basis of these assumptions, one can write the conservation equations as,

$$\frac{\partial C_J}{\partial t} + \frac{\partial}{\partial z}(C_J v) = \sum_J S_{iJ} R_i \tag{1}$$

$$\rho \hat{C}_P \frac{\partial T}{\partial t} + \rho \hat{C}_P v \frac{\partial T}{\partial z} = \frac{4U}{D}(T_w - T) - \sum_i (\Delta H_i) R_i \tag{2}$$

$$\rho \frac{\partial v}{\partial t} + \rho v \frac{\partial v}{\partial z} = -\frac{\partial P_t}{\partial z} - \frac{4}{D} f' \left(\frac{1}{2}\rho v^2\right) \tag{3}$$

The boundary conditions are:

$$\begin{aligned} &t \leq 0 \; ; \quad C_J = C_J(z), \quad T = T(z), \quad P_t = P_t(z) \\ &t > 0 \text{ at } z = 0 \; ; \quad C_J = C_{Jo}, \quad T = T_o, \quad P = P_{to} \end{aligned} \tag{4}$$

The boundary conditions can be derived (13) from steady state equations.

2.1. Method of Solution

The method of lines (14) is used as the numerical technique. In this method, by "finite differencing" the space variable (here axial length of reactor), the reactor is divided into a number of cells. Then the partial differential equations are converted into ordinary differential equations where time is the only independent variable. Each cell corresponds to a continuous stirred tank reactor.

If the continuity equations are rewritten for each cell or CSTR, the following ordinary differential equations are obtained (13).

$$\frac{dN_{J,n}}{dt} = F_{J,n-1} - F_{J,n} + R_{J,n}V_n \tag{5}$$

$$HL_n C_{P_n} \frac{dT_n}{dt} = \sum_i F_{J,n-1}(H_{J,n-1} - H_{J,n}) - V_n \sum_i (\Delta H_i R_i)_n + \left[u(T_w - T)\right]_n \pi D \Delta Z_n \tag{6}$$

$$HL_n \frac{dv_n}{dt} = \rho_{n-1} v_{n-1}(v_{n-1} - v_n) S + (P_{n-1} - P_n)\frac{S}{\alpha'} - \pi D\left[\Delta Z \frac{1}{2} \rho v^2 f\right]_n \tag{7}$$

where subscript n represents the "nth cell" and HL is holdup.

2.1.1. Algoritm of Solution. Before starting the numerical integration, the following conditions have to be specified.

1. Composition, temperature, pressure and flow rate of the inlet stream before the disturbance is applied.
2. Scale and type of disturbance.
3. External tube wall temperature or heat flux profiles.
4. Reactor dimensions given as length of straight tube and bends, bend radius, bend angle, reactor length and diameter, and wall thickness.

Once the above conditions are specified, the outlet composition, temperature, pressure, and flow rate of the tubular reactor can be estimated in the following manner.

1. Steady state model equations are integrated to obtain the initial profiles.
2. With these initial conditions the dynamic model equations are integrated to obtain the conditions in the reactor at various axial positions. This is illustrated in Fig. 1.

2.1.2. Simplifications in Integration Step. The dynamic tubular

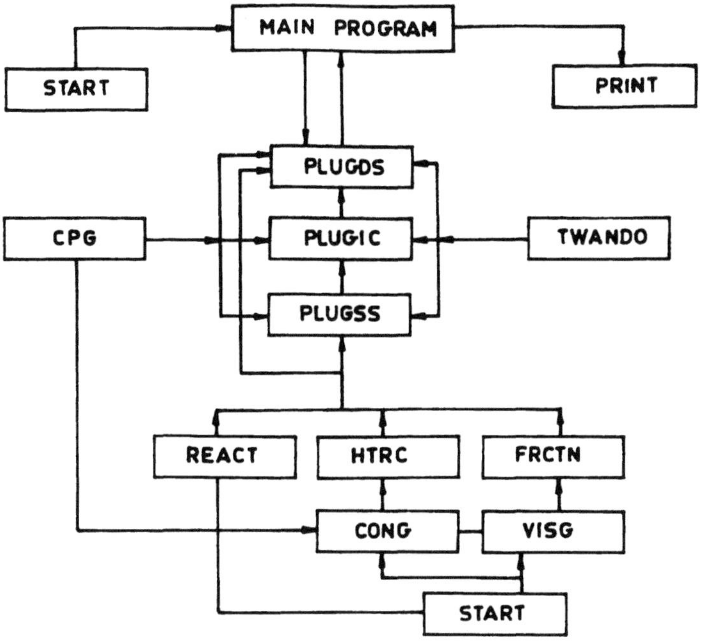

Fig. 1 - Flow Diagram for Dynamic Tubular Reactor Model.

reactor model can be simplified to decrease the computation time by considering the following points.

1. Main reactions are considered only. In ethane cracking, the model is related to five reactions and nine components. But only two reactions and six components are the main ones.

2. The equation of motion is not considered. After the development of the pressure profiles from the steady state model, either the pressure in each cell is assumed to stay constant or it can be obtained from some motion independent expressions.

3. After determining the heat transfer coefficient and friction factor from steady state model equations, either one or both of them can be assumed as constants in each cell.

4. Thermophysical properties such as viscosity, heat capacity and thermal conductivity of pure components and mixtures are estimated. Either one or two or all of these can be assumed as constants in the steady state or dynamic equations.

3. EXPERIMENTAL WORK

A laboratory scale tubular reactor was built to obtain data for the thermal cracking of hydrocarbons. However, sufficient data

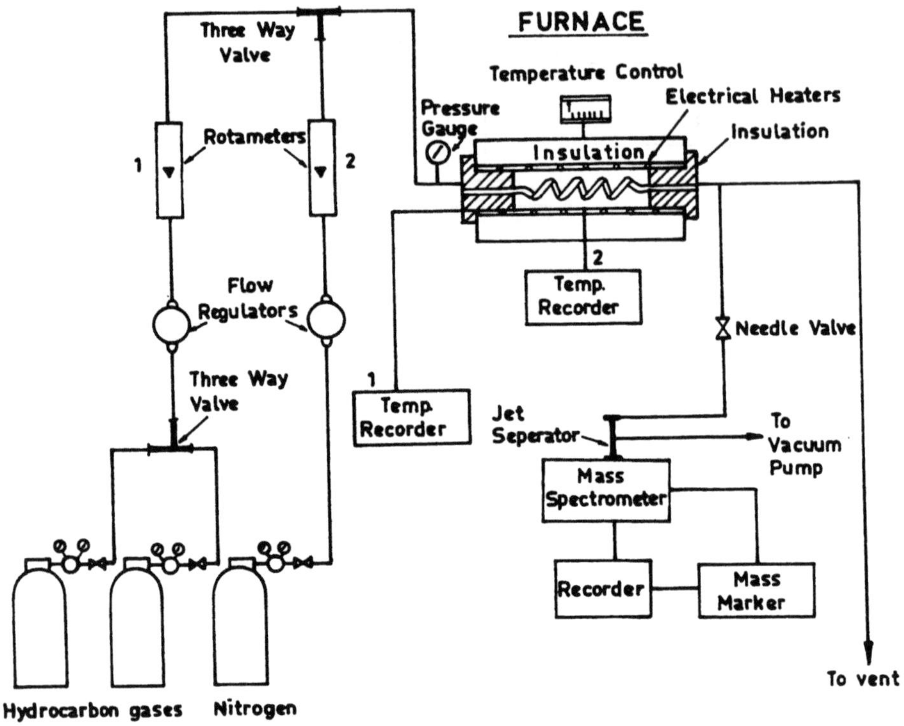

Fig. 2 - Experimental Set-up.

can not be properly obtained at present. Instead, model predictions were made for step changes in feed composition, temperature, and flow rate. The steady state data reported by Sundaram and Froment for an industrial reactor(23),(24) is used to check the consistency of the present model. The comparison of operating conditions for industrial and laboratory reactors is given in Table. 1.

The schematic diagram of the experimental setup is given in Fig. 2. Gaseous hydrocarbon mixture is passed through a three-way valve, flow regulator with needle valve and calibrated rotameters where the flow rates are measured. Mixtures containing 22% and 50% ethane in nitrogen are prepared in two gas cylinders. In this way, it is possible to give step changes to the composition by passing from one gas cylinder output to another by use of the three way valve. The pressure gauge measures the inlet pressure to the tube.

The tubular reactor is made of stainless steel,type 316. It is constructed in the form of a coil about 4 cm. in diameter. In the Hereaus tube furnace, the silica tube is heated electrically to supply heat. The inside diameter of this tube is 6.5 cm and the heated length is 55 cm. The temperature profiles are measured by

TABLE 1 - Industrial vs. Laboratory Reactor

Variables	Ind./Lab.	Variables	Ind./Lab.
Reactor Length, m	88.25/2	Inlet Gas Temp.,°C	652/18
Inside Diameter, m	10.8 /0.2	Inlet Gas Press.,atm	2.9/0.9
Wall Thickness, cm	0.8 /0.5	Press.Drop, atm	1.37/neg.
Length of S.Prt.,m	9.38 /-	Ethane Flow, gmol/s	18.48/var.
Length of a Bend,m	0.48 /-	Steam Dil.,kg/kg eth.	0.51/ -
Radius of a Bend,m	0.15 /-	Nitrogen Dil., mol%	- /
Bend Angle	180 /-	Conversion, %	59.2/57-68
Mater. of Const.	S.S /S.S	Residence Time, s	0.35-0.8/1.5-8

TABLE 2 - Steady State Simulation Runs

	1	2	3	4	5	6	7
Outlet Temp., °C	830.0	831.7	836.1	831.1	823.4	813.4	817.0
Conversion, %	59.2	57.8	59.9	57.0	52.5	47.0	48.9
Yield, wt %							
Ethylene	48.20	46.89	48.46	48.37	43.15	39.06	40.48
Ethane	40.80	42.24	40.09	43.02	47.48	52.98	51.10
Hydrogen	3.42	3.63	3.75	3.64	3.31	2.97	3.09
Methane	3.17	2.65	2.81	1.36	2.21	1.82	1.96
Computation Time,s	-	9.5	37.8	6.9	9.7	9.5	9.8

TABLE 3 - Unsteady State Simulation Run No. 1

Step Disturbance = (first row)-(0.54 mole fr.)

First S.S Ethane Concentration	Ind.	0.0	0.1	0.3	0.5	0.7
Outlet Temp. °C	830	814	814	819	826	831
Outlet Press. atm	1.51	2.68	2.65	2.62	2.60	2.59
Conversion, %	59.2	64.6	60.0	57.9	58.9	62.1
Yield, wt %						
Ethylene	48.2	34.9	37.7	43.0	46.9	48.4
Ethane	40.8	35.4	40.0	42.1	41.1	37.9
Hydrogen	3.42	2.68	2.88	3.29	3.60	3.73
Methane	3.17	1.47	1.51	1.76	2.01	2.18
Computation Time,s	-	300	275	271	270	273

thermocouples and the furnace is controlled by a temperature controller. Due to high temperature operation, even with proper furnace insulation, there is approximately a 100 °C difference between the center and ends of the furnace. The outlet stream from the reactor is continuously passed to a magnetic sector mass spectrometer; Du-Pont Instruments 25-490B Mass Spectrometer/Gas Chromatograph Setup. Details of this system can be found elsewhere (13).

The ethane is diluted with nitrogen. Two mixtures, 22% and 50% in ethane, are prepared at total pressure of 10 atm. Nitrogen is also used for purging purposes and to prevent oxidation of the tube. The pressure of the mass spectrometer is kept at desired levels for analysis during the experiments by a fine needle valve. The rest is purged to the atmosphere.

Cracked gases are passed to the mass spectrometer/gas chromatography system. The residence time between mass spectrometer and furnace is measured by injecting acetone as tracer to nitrogen at the sample port placed at the furnace outlet. In this way,the residence time is decreased to 4-5 seconds. The analysis can be carried out by taking a spectrum of the cracked gas. The peaks are interpreted by measuring their heights. For dynamic experiments only the peak corresponding to the mass number (mass to charge ratio) of 30 is sufficient. This peak is the third highest peak and is unique according to the components that are present in the ethane cracking system. During the experiments with 22% ethane mixture, coke formation was observed at center locations at temperatures of 800-850 °C. Possible causes are low flow rates corresponding to long residence time and possible local increase of the reactor temperature due to the contact of the cracking tube with the inside furnace wall.

4. RESULTS AND DISCUSSION

The aim of this study is to develop a predictive model that simulates for control of a tubular ethane cracking reactor. The model can be used to estimate the dynamic behaviour of the system when desired or undesired disturbances are applied to one of the process variables. If this behaviour is known mathematically for sucn cases, then control of the reactor becomes relatively easy.

Experimental data for an industrial thermal cracking reactor were obtained from the steady state data of Sundaram and Froment (24). The dynamic behaviour of this reactor is examined by giving step disturbances to the inlet composition, temperature and flow rates at the first fictituous steady state and then observe how it passes to the second steady state. For instance, the first steady state with inlet feed temperature of 450 °C is chosen. Then at time zero, a step change is applied to temperature by increasing it to

652 ^{o}C, at which the conversion data is reported. The response to this input is obtained in this manner from the dynamic model. The final composition values at the second steady state are calculated when time goes to infinity. Then the results are compared with the literature data.

The discussions are mainly in terms of the conversion of the feed since this has a very important effect on the product distribution which can be obtained from yield versus conversion diagram directly. This is due to the fact that these data are nearly independent of temperature.

4.1. Steady State Simulations

In this section, the effect of several estimation methods and reaction models used in the modelling on the product distribution are outlined and the results are compared with the available industrial data at steady state, which is given in the first column of Table 2 (23). Other columns gives the results for different simulation runs. Predictions are in good agreement with the industrial data.

4.1.1.Thermophysical Property Estimation. For pure components there is not a large difference in thermophysical properties estimated by different methods except for thermal conductivity of water. The main problem is the thermal conductivity estimation of mixtures.The method of Friend and Adler (15) is suitable due to its simplicity. However, the method of Mason and Saxena as given by Touloukiau (26) estimates higher thermal conductivities and requires larger computation time than the previous method but seems to be more accurate.

4.1.2. Reaction Model. The molecular reaction scheme is taken from Sundaram and Froment (22). The model is composed of 5 reactions related to the main products,namely, ethylene,hydrogen, methane and propylene, and is summarized below

$$
\begin{array}{llll}
1. & C_2H_6 \underset{}{\overset{k_1}{\rightleftharpoons}} C_2H_4 + H_2 & 4. & C_3H_6 \overset{k_4}{\rightleftharpoons} C_2H_2 + CH_4 \\
2. & C_2H_4+C_2H_6 \xrightarrow{k_2} C_3H_6+CH_4 & 5. & C_2H_2 + C_2H_4 \xrightarrow{k_5} C_4H_6 \\
3. & 2C_2H_6 \xrightarrow{k_3} C_3H_8+CH_4 & &
\end{array}
\tag{8}
$$

with the following rate expressions,

$$R_1 = k_1\left[\frac{Y_3P_t}{RT} - \frac{Y_2Y_8}{K_1}\left(\frac{P_t}{RT}\right)^2\right] \tag{9a}$$

$$R_2 = k_2\left[Y_2Y_3\left(\frac{P_t}{RT}\right)^2\right] \tag{9b}$$

$$R_3 = k_3\left[Y_3\left(\frac{P_t}{RT}\right)\right] \tag{9c}$$

$$R_4 = k_4\left[Y_4\left(\frac{P_t}{RT}\right) - \frac{Y_1Y_5}{k_2}\left(\frac{P_t}{RT}\right)^2\right] \tag{9d}$$

$$R_5 = k_5\left[Y_2Y_5\left(\frac{P_t}{RT}\right)^2\right] \tag{9e}$$

The comparisons given in Table 2 are obtained from simulation runs that use Sundaram and Froment's heat transfer correlation(24), five reaction models(22),and Friend and Adler's method for mixture thermal conductivity estimation (15). The first column represents experimental steady state data from industry (23).The second column represents two dimensional simulation with Sundaram and Froment's heat transfer correlation

$$\frac{hD}{k'} = 0.023\ Re_b^{0.8}Pr_b^{0.4} + 429.2\ Z^{-0.0706} - 544.3(x/100.0)^{-0.0437} \tag{10}$$

The third column uses Mason and Saxena method(26) for mixture thermal conductivity estimations. The fourth column gives the results for the simplified reaction model and thermal physical properties of the mixture. In this run, the first two reactions are considered in the five reaction model of Sundaram and Froment (22). Thus, the following two reaction model, given below, give reasonably accurate results.

$$C_2H_6 \rightleftharpoons C_2H_4 + H_2 \tag{11}$$

$$C_2H_4 + C_2H_6 \longrightarrow C_3H_6 + CH_4 \tag{12}$$

Columns 5, 6, and 7 represent Sieder-Tate correlation, the latter modified by Whitaker (27), and simulation with Dittus-Boelter heat transfer correlation (15), respectively. The coke formation reactions are not considered in the model since there is very little information on the kinetics of coke formation mechanism. A similar comparison was also made by Sundaram and Froment with main emphasis being given to two dimensional models (24).

4.1.3. Heat Transfer to Coils. If the wall temperature profile is used in the simulations, the heat transfer from the wall to the reacting gas has to be evaluated from correlations which are generally derived from pure heat transfer measurements. The reaction will distort the temperature profile and cause deviations in estimates (22). For this reason, Sundaram and Froment (24) developed a new heat transfer correlation for ethane cracking. It is noted that this new correlation gives better results than the others with the given external wall temperature profile. However, when the external tube wall temperature profile is increased by 25 ^{o}C throughout the reactor, the same exit conversion can be obtained using Dittus-Boelter correlation.

4.1.4.Numerical Solution of Steady State Equations.The simple Euler method is used in numerical integration. Reasonable results are obtained for a minimum step size of 50 cm. Step sizes over 250 cm. causes instability.

4.2. Dynamic Simulations

Step disturbances are given to the feed composition, temperature and flow rate. The results are outlined elsewhere (13). To decrease the comp tational time, model equations were further simplified as follows:

1. First, the heat transfer coefficient and friction factor were estimated for steady state simulations, and then they were assumed as constants in the transient state for each cell.
2. The continuity equation gives,
   i. Total Mass Balance:

$$\frac{dHL_n}{dt} = F_{n-1} - F_n + V_n \sum_i R_{i,n} \tag{13}$$

   ii. Component Mass Balance:

$$\frac{dY_{i,n}}{dt} = \frac{F_{n-1}(Y_{i,n-1} - Y_{i,n}) + R_{i,n}V_n - Y_{i,n}\sum R_{i,n}V_n}{HL_n} \tag{14}$$

The integration of this equation gives the change in holdup of the nth cell with respect to time. However, it does not give any information about how the outlet molar flow rate, F , changes with time. Rearranging it, one may obtain,

$$F_n = F_{n-1} + V_n \sum_i R_{i,n} - \frac{dHL_n}{dt} \quad \text{with} \quad R_i = \sum_J S_{Ji} R_J \tag{15}$$

so that the last term in the righthand side is negligible when compared to the others.This considerably reduces the computation time.

3. The first two terms on the right-hand side of the equation of motion give numerically unstable results at simulations which require very small time increments corresponding to excessive computation times (0.00125 s). To overcome this difficulty, quasi stationary velocity profiles are assumed. The simulated results for pressure profiles deviates from the industrial data, nevertheless these deviations do not effect the results significantly (13).

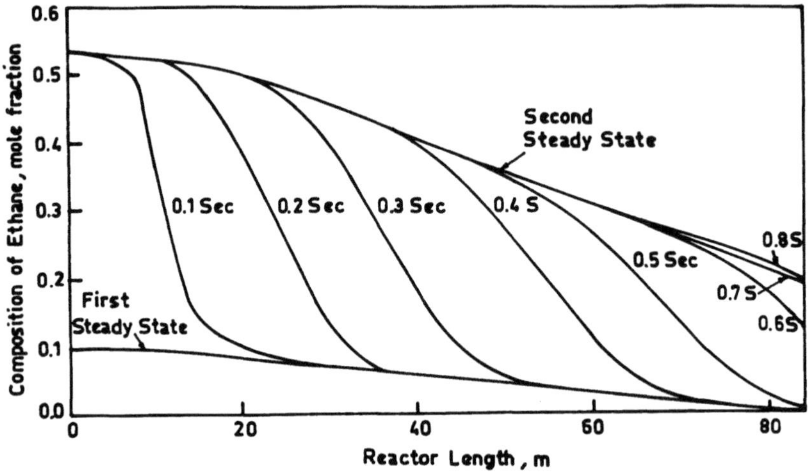

Fig. 3 - Predicted Concentration Profiles at Transient-state for Step-disturbance to Feed Composition as 0.1 - 9.539 mole fraction of Ethane for Industrial Scale Reactor. Second Steady-state is reached in 0.8 seconds.

4.2.1. Disturbance to Feed Composition. Fig. 3 illustrates the results of step disturbances given to the feed composition. First steady state is taken at several inlet concentrations of 0.1 mole fraction of ethane. Step disturbances are given as 0.5391 which is data reported from an industrial scale unit (24). Since the inlet compositions at first steady state approaches the second steady state value of 0.5391, as time increases to infinity (the point at which the unsteady state profile becomes the steady state profile), it is concluded that the model is in good agreement with the reported data. However,if the second steady state value taken is very different from the first steady state value (large magnitude in step input), then the deviations between the model predictions and the data of Sundaram and Froment start to occur. There are two reasons for this behaviour;(i)the misleading estimation of pressure and/or, (ii) the misleading estimation of heat transfer coefficients. As indicated above, the estimation of pressure is obtained from quasi-stationary assumption and the constant friction factor assumption for the transient state. If the inlet concentration of ethane at the first and second steady states differ from each other considerably then the assumption of constant overall heat transfer coefficient for each cell fails. Step disturbance affects the outlet response in approximately 0.4-0.5 seconds which corresponds to residence times of about the same order of magnitude (7). Table 3 gives the comparison of product distributions and outlet conditions obtained from dynamic simulations as time goes to infinity. The step disturbance is the first row - 0.54 mole fraction.

4.2.2. Disturbance to Feed Temperature. There is no significant effect (13) of the feed temperature on the dynamic behaviour of the reactor. Since high heat transfer rates are considered in such reactors, the gas mixture increases its heat capacity immediately no matter what the feed temperature is.For feed temperatures higher than the temperature of the second steady state, the same argument can be made indicating that the effect of the zone where the gaseous mixture changes its heat capacity is negligible.

4.2.3. Disturbance to Feed Flow Rate. Since the flow rate of the feed at the first steady state approaches the flow rate of the second steady state, the results from dynamic simulations approach the industrial data. The reason to these deviations are, again, the misleading assumption of constant overall heat transfer coefficient during the transient state. For example, for a feed flow rate of 19 gmole/s, the estimated overall heat transfer coefficient is 0.012 $cal/s.cm^2.{}^oC$ at steady state. However, for feed flow rate of 34.3 gmole/s (industrial data), it is around 0.015 $cal/s.cm^2.{}^oC$.

4.2.4. Numerical Solution of Dynamic Model Equations. The dynamic model equations are highly non-linear partial differential equations.They are converted into a set of ordinary differential equations by the method of lines. There are two parameters in the solution procedure. The first one is the length increment size selected for finite differencing and the other is the time increment size selected for the solution of ordinary differential equations.

According to predictions, the system passes from one steady state to another in about one second. Therefore, to avoid numerical instability,the time increment step of 0.005 second is selected for 50 cell divisions. The length increment size does not effect the results significantly. For example,for a feed concentration disturbance from 0.1 to 0.5391 mole fraction ethane,the modelling with 50 cells give 59.2% conversion which corresponds to; 813 ^oC outlet temperature and 37.3, 40.8, 1.48, and 2.85 yields as wt % for ethylene, ethane, methane,and hydrogen, respectively.The modelling with 100 cells gives 60.0% conversion with 814 ^oC as the exit temperature. The corresponding composition figures are 37.7, 40.0, 1.51, and 2.88, respectively.

5. CONCLUSIONS

The dynamic model developed in this study is in agreement with the industrial data for the concentration disturbances between 0.3-0.7 mole fraction of ethane in feed. It is also in agreement with temperature disturbances. However, deviations occur when step disturbances are given to the feed flow rate. The main reason for these deviations are believed to be due to the overall heat transfer correlation. Due to the lack of complete and accurate data, the latter is assumed to be constant during the transient state.

The model predicts the dynamic behaviour in 90 seconds of computer time calculations using a Burroughs mainframe computer. On the other hand, the disturbance is restricted to a duration of about 1 second. For this reason, the model developed here should better be considered as a "learning model". It can be used to predict the performance of cracking in fixed-bed reactors at which the transient periods are larger, as in fixed-bed reactors. The model shows that there are two important parameters in the control strategy of thermal cracking reactors: (i) feed flow rate, and (ii) feed composition. To use this model for fixed bed reactors, the heat transfer coefficients at the wall have to be changed and the void fraction parameter and the conservation equations related to mass and energy transfer within the catalyst particles should be included in the analysis.

NOTATION LIST

C	Molar concentration, gmole/cm^3
C_p	Heat capacity of mixture per unit mass, cal/g.^{o}K
D	Inside diameter of pipe, cm
f', f	Friction factor, for straight portions of reactor tube
F	Molar flow rate, gmole/s
G	Mass flow rate, g/cm^2.s
h	Heat transfer coefficient, cal/cm^2.s.^{o}K
H	Enthalpy, cal/gmole
HL	Holdup, gmole
ΔH_i	Heat of reaction, cal/gmole
k'	Thermal conductivity, cal/cm.s.^{o}K
k_i	Rate constant, 1/s or cm^3/gmole.s
n	Cell number
N_J	Number of moles of component J in each cell
P	Pressure, atm
Pr	Prandtl number, $C_p\mu/k'$
R	Ideal gas constant
Re	Reynolds number, $Dv\rho/\mu$
R_i	Reaction rate, gmole/cm^3.s

S Cross sectional area, cm^2
S_{iJ} Stoichiometric coefficient of component J in reaction i
t Time, s
T Temperature, ^oK
U Overall heat transfer coefficient, $cal/cm^2.s.^oK$
v Fluid velocity, cm/s
V Volume, cm^3
x Percent conversion, %
Y Mole fraction
Z Axial coordinate, cm
ΔZ Length of the cell

Greek Symbols
α' Conversion factor, $9.87 \times 10^{-7} cm.s^2.atm/g$
μ Viscosity, g/cm.s
ρ Fluid density, g/cm^3

Subscripts
i Reaction
J Component
m Mixture
o Reactor outlet
t Total
w Wall

LIST OF REFERENCES

1. Bonvin,D., R.G.Rinker and D.A.Mellichamp. "Dynamic Analysis and Control of a Tubular Reactor at an Unstable State". Chem.Eng. Sci. v35 (1980) 603-612.
2. Brooks,B.W. "Start-up and Dynamic Behaviour of a Chemical Reactor". Chem. Eng. Sci. v34 (1979) 1417-1419.
3. Buekens,A.G. and G.F.Froment. "Thermal Cracking of Isobutane". IEC.Proc.Des.Dev. v10 (1971) 309-315.
4. Eigenberger, G. "Zur Modellbildung und Dynamik des Homogenen Rohrreaktors". Chemie Ing. Tech. v46 (1974) 11-19.
5. Eigenberger, G. Adv.Chem.Series, v133 (1974) 477-488.
6. Finlayson,B.A."The Method of Weighted Residuals and Variational-Principles". Academic Press, New York (1972).
7. Froment,G.F."Thermal Cracking for Olefins Production. Fundamentals and Their Application to Industrial Problems". Chem. Eng. Sci. v36 (1981) 1271-1282.
8. Froment,G.F.,B.O.Van de Steene, P.J.V.Berghe and A.G.Goossens. "Thermal Cracking of Light Hydrocarbons and Their Mixtures", AIChE.J. v23 (1977) 93-106.
9. Georgakis,C., R.Aris, and N.R.Amundson, "Studies in the Control

of Tubular Reactors". Chem.Eng.Sci. v32 (1977) 1359-1369.
10. Gilles,E.D. "Reactor Models".4th Int./6th Euro.Symp.on Chemical Reaction Engineering. (1976).
11. Hansen,K.W. and S.B.Jorgensen."Dynamic Modelling of a Gas Phase Catalytic Fixed-bed Reactor-I,II". Chem.Eng.Sci. v31 (1976) 579-586; 587-598.
12. Jackman,A.P. and R.Aris."Optimal Control for Pyrolytic Reactors 4th Euro.Symp.on Chemical Reaction Engineering. (1968).
13. Kırkbir,F. "Dynamic Analysis of a Tubular Reactor". M.Sc.Thesis Middle East Technical University, Turkey. (1984).
14. Kısakürek,B."A Predictive Model for Dynamic Distillation".Chem. Eng.Comm. v20 (1983) 157-
15. Perry,R.H. and C.H.Chilton. "Chemical Engineers Handbook". 5th ed. McGraw-Hill, Tokyo (1973).
16. Ray,W.H. "Fixed-bed Reactors.Dynamics and Control".2nd Int/5 th Euro.Symp.on Chemical Reaction Engineering. Amsterdam (1972).
17. Schmitz,R.A. "Multiplicity, Stability,and Sensitivity of States in Chemically Reacting Systems.A Review".Adv.Chem. Am.Chem.Soc. Washington (1975) 156-212.
18. Shah,J.M. "Ethylene Production". IEC. v59 (1967) 70-85.
19. Sharma,C.S. and R.Hughes. "The Behaviour of an Adiabatic Fixed-Bed Reactor for the Oxidation of Carbon Monoxide-I,II". Chem. Eng.Sci. v34 (1979) 613-624; 625-634.
20. Sorensen,J.P."Experimental Investigation of The Optimal Control of a Fixed-bed Reactor". v32 (1977) 763-774.
21. Stewart,W.E. and J.P.Sorensen. "Transient Reactor Analysis by Orthogonal Collocation". 2nd. Int./5th Euro. Symp. on Cnemical Reaction Engineering". Amsterdam (1972).
22. Sundaram,K.M. and G.F.Froment. "Modelling of Thermal Cracking Kinetics.Thermal Cracking of Ethane,Propane and Their Mixtures" Chem.Eng.Sci. v32 (1977) 601-609.
23. Sundaram,K.M. and G.F.Froment. "A Comparison of Simulation Models for Empty Tubular Reactors".Chem.Eng.Sci.v34(1979) 117-124.
24. Sundaram,K.M. and G.F.Froment. "Two-dimensional Model for the Simulation of Tubular Reactor for Thermal Cracking". Chem. Eng. Sci. v35 (1980) 364-371.
25. Thoenes,D.,"Current Problems in the Modelling of Chemical Reactors". Chem.Eng.Sci. v35 (1980) 1840-1853.
26. Touloukiau,Y.S.,P.E.Liley and S.C.Saxena (Eds.)"Thermo-physical Properties of Matter". The TPRC Data Series. v3 IFI Plenium.NY. (1970).
27. Whitaker,S. "Forced Convection Heat Transfer Correlations for Flow in Pipes,Past Flat Plates,Single Cylinders,Single Spheres, and Flow in Packed Beds and Tube Bundles". AIChE.J. v18 (1972) 361-371.
28. Van Damme,P.S., S.Narayanan and G.F.Froment. "Thermal Cracking of Propane and Propane-Propylene Mixtures. Pilot Plant versus Industrial Data". AIChE.J. v21 (1975) 1065-1073.
29. Van Doesburg,H. and W.De Jong. "Transient Behaviour of an Adiabatic Fixed-bed Methanator-I". Chem.Eng.Sci. v31 (1976) 45-51.

DESIGN OF AMMONIA AND METHANOL SYNTHESIS REACTORS

IB DYBKJÆR

HALDOR TOPSØE A/S

1. INTRODUCTION

Ideally, the design of a chemical reactor is a strictly logical process, in which the optimum solution to the problem is determined from experimental and other data by well-defined procedures. The real world does, however, not work in this way. This is clearly shown, if not by other evidence, then by the multitude of different solutions which have been found in reactor design for the same problem. All the solutions are normally technically and economically very reasonable - and none of them can claim to be clearly superior in all cases.

In the following, an attempt is made to illustrate on the basis of methods applied at Haldor Topsøe A/S some of the methods used and the difficulties encountered in reactor design. This is done by considering two examples, which in many ways are quite similar, namely the synthesis of ammonia and methanol. Both of these examples are from important industries, where the competition is fierce, and where hundreds of millions of dollars would be the reward for the development of a clearly superior process and reactor design.

2. THE FRAMEWORK

Let us first, in order to place the problem of reactor design in perspective, consider a few facts about the ammonia and methanol industries.

The total installed capacity for production of ammonia worldwide is about 135 million tons per year, corresponding to about 370 plants each with a capacity of 1000 tons per day (the plants actually existing have, of course, a range of capacities from less than 100 TPD to above 1500 TPD). The ammonia is mainly used for production of fertilizers, and the annual growth rate is in the next decade expected to be 3.2% per year. It has been estimated that there will (including replacement of outdated capacity) be a demand for construction of about 80 new plants (1000 MTPD) before 1995 [1].

The capacity for methanol production is more modest, about 17 million tons per year corresponding to about 50 plants each with a capacity of 1000 tons per day [2]. In the methanol industry, existing plants have a range of capacities from very small installations to 2500 TPD).

The methanol industry is today characterized by a significant overcapacity, and this situation will continue at least until the end of the decade. If an expected widespread use of methanol as a blending agent in gasoline does in fact develop, there will, however, be a demand for a very significant addition to the production capacity in the next decade. In some forecasts, a demand of above 50 million tons per year has been predicted in the year 2000 [3].

The total cost of establishing a grass root ammonia production facility including all offsites and all services could typically be about 150 million US dollars for a capacity of 1500 MTPD and a location in an industrialized country (based on data from [4]. The figure would be approximately the same for a 2000 MTPD MeOH plant. Out of this, the cost of the process equipment for the whole unit (fob) could be 25-35%, or about 40-50 million dollers. The cost of the synthesis reactor(s) could be 1-3 million dollars depending on the design. This is, of course, not an insignificant amount of money, but it is clear from the above figures that the cost of the synthesis converter is of minor importance for the overall economics of the projects. If a relatively expensive reactor can allow savings in the cost of other equipment or reduction in operating cost, then it can very often be economically justified. There has in fact in recent years been a clear tendency to install larger and larger and therefore more and more expensive synthesis reactors, especially in ammonia plants.

In many cases, the size of the reactor has not been determined by its cost, but simply by transport limitations. Economic analyses have shown that the largest reactor, which could be brought to the site, was the best choice.

The above does, of course, not mean that it is not of prime importance to be able to predict as accurately as possible how the performance of an ammonia or methanol synthesis reactor will be at given process conditions. The synthesis reactor is the heart of the plant, and if it does not perform as expected then the plant will be unable to reach capacity.

3. FACTORS TO CONSIDER IN REACTOR DESIGN

In text books about chemical reaction engineering and similar subjects, the discipline "reactor design" is often interpreted as meaning "calculation of necessary reactor (catalyst) volume". There are, however, as rightly pointed out by e.g. Rase [5] several other factors to consider in reactor design - at least when this is interpreted as all information required as an instruction to a workshop for manufacture of a reactor for a specific practical application. Some of the more important points are listed in Table 1.

TABLE 1

FACTORS TO CONSIDER IN REACTOR DESIGN

Chemistry and thermodynamics
- the reaction, possible side reactions
- thermodynamics (equilibrium, heat of reaction, properties of reactants and products)

Reaction kinetics
- intrinsic + diffusion parameters
- pellet kinetics

Properties of catalyst
- thermal stability
- particle size and shape
- pore system
- mechanical strength
- change in properties during activation and operation

Process optimization
- energy recovery
- pressure drop
- limitations in volumetric flow

Process control
- temperature control
- flow distribution
- safety aspects, SU/SD

Mechanical design
- materials of construction
- workshop manufacture
- sea and land transport
- erection

The basic data - the chemistry and the thermodynamics for the process being considered - will define limits of a broad range of conditions at which the reaction is possible at all.

The kinetics of the reaction and the properties of the catalyst, especially the thermal stability, will further narrow the range of possible reaction conditions and define a "window" of possible operating parameters. Process optimization, energy efficiency, and safety aspects will then determine at what conditions within the "window" the reactor should operate to give the optimum result. And then mathematical models are used to determine how big the reactor must be to obtain the performance (conversion and pressure drop) determined by the process optimization. Instrumentation is then considered, proper materials of construction are selected, catalyst loading and unloading is considered, possible transport limitations are determined, workshop manufacture is considered, and at last the design of the reactor is completed. The procedure is, of course, iterative since the reactor cost is one of the parameters in the economical optimization, but, as mentioned above, often a factor of minor importance for the overall result.

It must again be emphasized that this only means that the cost of the reactor may be of minor importance. It is extremely important to be able to predict in a reliable way how the reactor will perform since failure to perform as desired may completely jeopardize the economics of the production.

Let us - after these rather extensive introductory remarks - turn to the specific problems encountered in design of ammonia and methanol synthesis reactors. We shall not endeavour to treat all of the above mentioned aspects, but mainly concentrate on the initial steps and on the basis of this illustrate how the various principal types of reactors can be applied in these syntheses. We shall discuss the reaction kinetics for the reactions and the calculation of reactor performance and some of the problems encountered in the calculation of reactor performance. The mathematical procedure used for the computer calculations is discussed by Christiansen and Jarvan [6] in a separate presentation in this volume.

4. SELECTION OF OPERATING PARAMETERS

The ammonia and methanol syntheses are in many ways quite similar reactions, and they are as a consequence carried out in the same type of equipment. Both reactions are exothermic, and high equilibrium conversion is favoured by low temperature and high pressure.

In both cases, only partial conversion can be obtained at realistic conditions, and in order to approach full conversion a recycle system is used. The detailed lay-out is, of course, different in different cases. As illustrations, the lay-out of a Haldor Topsøe Low Energy Ammonia Synthesis Loop is shown schematically in Fig. 1 and of a Haldor Topsøe Low Energy Methanol Synthesis Loop in Fig. 2.

Topsøe Low Energy Ammonia Process
Layout of Synthesis Loop

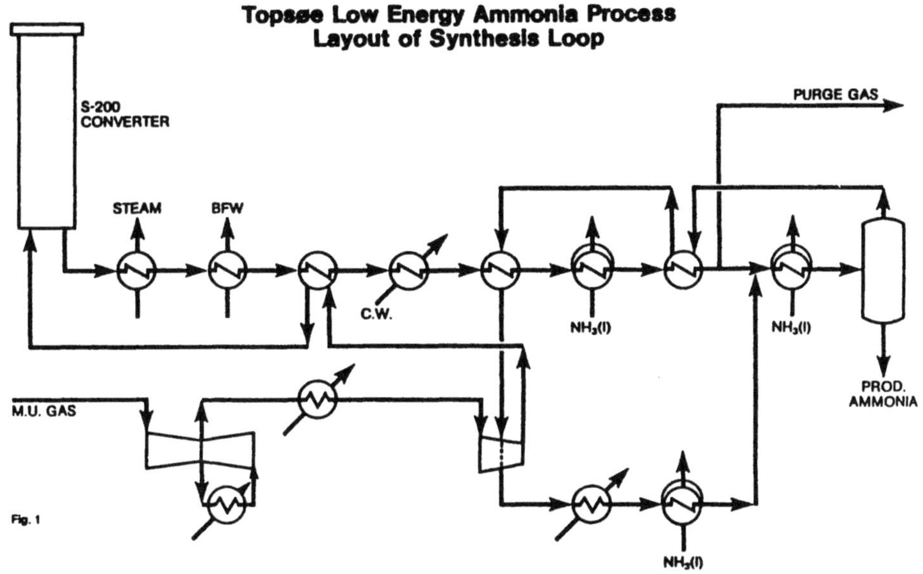

Fig. 1

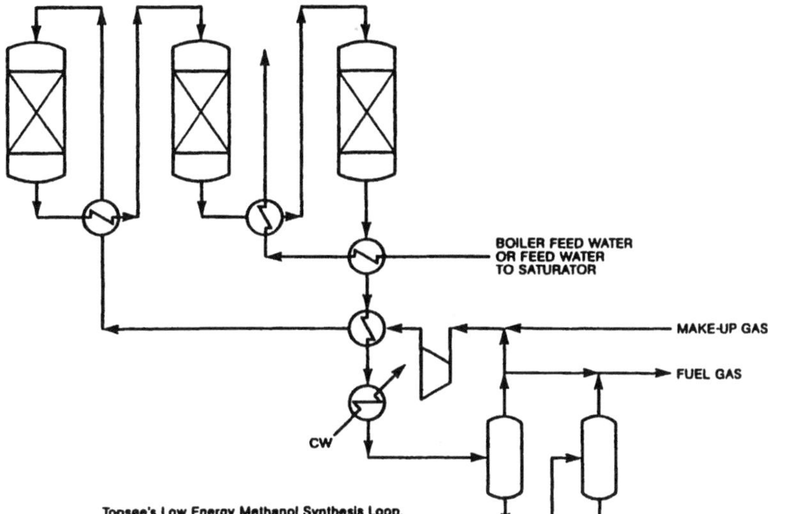

Fig. 2. Topsøe's Low Energy Methanol Synthesis Loop.
(Other arrangements of heat recovery are possible).

In both cases, synthesis gas is added to a recycle loop where reactor feed gas is passed through a reactor and heat recovery system. Product is separated and the unconverted gas is recycled to the reactor system. At some point a purge gas is taken in order to prevent accumulation of inerts, excess reactants and/or byproducts in the recycle system. Detailed descriptions of the process layout and discussions concerning energy consumption, etc. are given elsewhere [7] and [8] and will not be repeated here.

4.1 Ammonia Synthesis

The ammonia synthesis catalyst has remained essentially unchanged for many years, and there has been no significant change in the permissible range of operating temperature, which is about 350-550 ^{o}C. A wide range of operating pressure has been used in practice, from less than 100 bar to 1000 bar. The trend in modern plants has been to select operating pressure in the lower range; typical operating parameters for modern synthesis loops at two different pressures are given in Table 2:

TABLE 2

OPERATING CONDITIONS, NH_3 LOOPS AT 220 BAR AND 140 BAR (1000 MTPD)

Inlet pressure, bar	220	140
Inlet flow, Nm^3/h	407,000	500,000
Inlet mole % NH_3	3.8	4.1
Outlet mole % NH_3	19.9	17.1
Inert level, inlet, mole %	12	8
Separator temp., ^{o}C	-5	-5
Effective cat. vol., m^3	60	100

The operating pressure is, of course, partly imposed by considerations concerning catalyst stability and activity. But the combined effects of the unfavourable equilibrium conversion at low pressure and the low temperature required to condense the ammonia product may be more important. Table 3 shows this by illustrating two effects of decreased pressure:

- at low operating pressure, very low operating temperature is required to obtain a resonably high concentration at reactor outlet. Note that at the relevant temperature level, a decrease in temperature of 100 ^{o}C may cause a reduction in reaction rate of about a factor of 10 (at same ammonia concentration).

-at low operating pressure, a very low temperature is required to obtain reasonable recovery of ammonia from the recycling gas. Refrigeration for product condensation is normally achieved by cooling with ammonia, and this limits applicable separation temperature to about -25°C (ammonia boils at atmospheric pressure at -33°C).

TABLE 3

Some Effects of Pressure in Ammonia Synthesis

Pressure of the Reaction Mixture	bar	10	50	100	200	500
Temperature at which equilibrium conc. of NH3 is 20 vol% (starting composition: $H_2/N_2 = 3/1$, no inerts)	°C	283	374	425	488	600
Temperature at which gas phase conc. of NH3 over condensed NH3 is 3 vol%	°C	-54	-28	-16	-4	+6
Concentration of NH3 in vapour phase over condensed NH3 at -10°C	mol%	*)	6.7	3.9	2.4	1.7

*) No condensation

It might be argued that one could allow a higher concentration of ammonia in the recycling gas and by increasing the concentration in the converted gas still obtain a reasonable conversion. This would, however, increase the required catalyst activity (or volume for constant activity) very significantly since the ammonia synthesis reaction is on existing types of catalyst strongly inhibited by the product ammonia.

A further point which favours elevated operating temperature and thereby elevated operating pressure is the risk of poisoning. Ammonia synthesis catalyst is extremely sensitive to poisoning by oxygen-containing compounds such as H_2O and CO. The poisoning is caused by a reversible adsorption of oxygen species on the active sites, and the equilibrium is such that at temperatures below about 350°C, almost complete coverage - and therefore almost complete deactivation of the catalysts - is obtained even at concentrations of oxygen-containing compounds about or below 1 ppm. [7] . Such low concentrations are very difficult (and expensive) to achieve, and as a consequence the risk of poisoning causes a practical lower limit to the temperature and thereby to the pressure.

The conclusion of the above points is that even a quite significant improvement in the catalyst activity would not cause any significant change in operating conditions in an ammonia synthesis loop. This was discussed further elsewhere [9] .

4.2 Methanol Synthesis

In methanol synthesis, a very significant change in the typical operating conditions was seen in the late sixties, when Cu-based synthesis catalysts were introduced. It was at that time not unknown that Cu-based catalysts have very high activity for methanol synthesis; such catalysts were described by Natta [10] , and they had in fact been used industrially, e.g. in Japan, since the forties. The Cu-based catalysts are, however, very sensitive to poisoning, especially by S and Cl, and this prevented widespread use until efficient feedstock purification methods were developed mainly as part of the technology for steam reforming of naphtha.

In the traditional high pressure -or rather high temperature-methanol synthesis processes Zn-Cr-catalysts were used. These catalysts require temperatures in the range 330-410°C, and at these temperatures pressures above 300 bar are required to obtain reasonable equilibrium conversion even in stoichiometric CO/H_2 mixtures with low CO_2 concentration.

The Cu-containing catalysts have somewhat lower thermal stability and operating temperatures above 300°C are hardly used. The operating pressure depends on the gas composition as discussed below, but is generally in the range 40-100 bar. The pressure is determined solely from considerations about equilibrium conversion and reaction rate. The conditions in the separator are not determining since it is always possible with water cooling to obtain reasonably low methanol concentration in the gas leaving the separator.

The synthesis gas composition depends on the raw material and on the process used in the synthesis gas preparation. Three examples are shown in Table 4. In the table is also given the conditions including gas composotion at reactor inlet and the methanol concentration at reactor outlet. These date were determined un such a way that the adiabatic temperature increase by reaction of the inlet gas to a close approach to equilibrium from a reasonable inlet temperature does not bring the catalyst to unacceptable temperatures.

TABLE 4

		Case 1	Case 2	Case 3
Synthesis gas: vol %	H_2	67.77	73.77	69.67
	CO	28.75	15.48	17.84
	CO_2	2.88	6.61	10.76
	Inerts	0.60	4.14	1.74
Reactor feed: vol %	H_2	71.81	82.11	61.18
	CO	10.01	3.16	4.43
	CO_2	3.07	1.79	5.72
	CH_3OH	1.15	0.36	0.33
	H_2O	0.03	0.13	0.07
	Inerts	14.03	12.46	28.28
Pressure, Bar		45	25	80
Temperature, ^oC		220	220	220
Reactor effluent: vol %	CH_3OH	6.06	4.58	5.78
Temperature, ^oC		254	234	242

Case 1: Synthesis gas produced from coal or heavy oil by gasification, CO-conversion and acid gas removal.

Case 2: Synthesis gas produced from natural gas by steam reforming.

Case 3: Synthesis gas produced from natural gas by steam reforming followed by catalytic partial oxidation.

The difference in the outlet concentration of methanol is obtained because of differences in the heat of reaction depending on the CO/CO_2 ratio and of differences in the heat capacity of the reaction mixture. Conditions were chosen such that the same adiabatic temperature increase (60^oC) was obtained in the first reactor in all three cases.

Fig. 3 shows the equilibrium lines and the inlet and outlet conditions for the three cases.

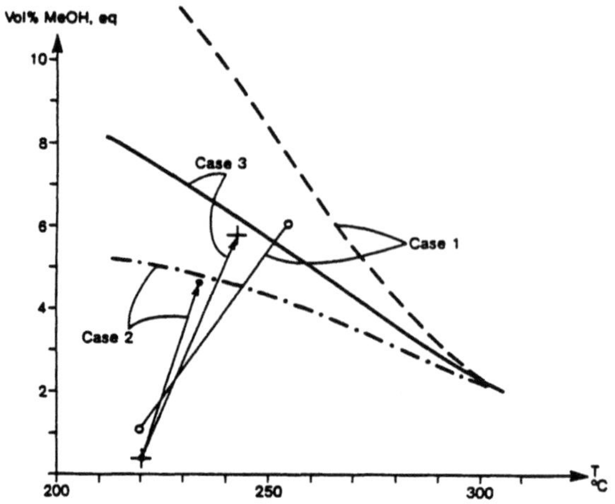

Fig. 3. Methanol Synthesis. Equilibrium Lines and Reactor Inlet/Outlet Conditions. 3 Cases, cfr. Table 4.

The figure illustrates that in spite of the differences in gas composition and pressure, the "task" of the reactor system is very similar in the three cases. This is of course because of the limitations imposed by the properties of the synthesis catalyst which is the same in all cases.

5. PRINCIPAL REACTOR TYPES

In both ammonia and methanol syntheses, the desirable conditions around the reactor system as described above are such that it is not possible in one adiabatic step to go from the inlet to the outlet conditions in the reactor. Some type of cooling is required. The cooling can in principle be applied in three different ways:

- internal cooling with cooling tubes in the catalyst bed or with catalyst in tubes surrounded by a cooling medium.
- quench cooling by injection of cold gas between adiabatic beds (or into a catalyst bed at different locations).
- external cooling by heat exchange between adiabatic beds.

All three types of reactors have been used in many different versions in large industrial units, both in ammonia and methanol synthesis. Some of the reactor types, which have been used, are described in [11] and [12] .

The principal difference between the three reactor types may be discussed with reference to plots such as Fig. 4.

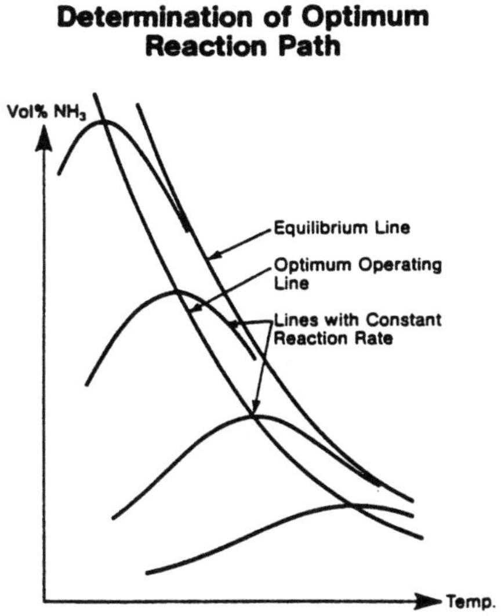

Fig. 4

This plot shows the equilibrium line for ammonia synthesis, adiabatic reaction lines, lines with the same reaction rate (there is a difference of one order of magnitude between the rate lines) and the optimum reaction path (i.e. the temperature at which the reaction rate is maximum at each ammonia concentration). It is clear that if this optimum reaction path were followed from inlet to outlet concentration in the reactor system, then the minimum catalyst volume would be required.

In Fig. 5, A, B and C typical operating lines (lines showing corresponding temperature and ammonia concentration in the reactor) are shown for three examples as indicated below the figures.

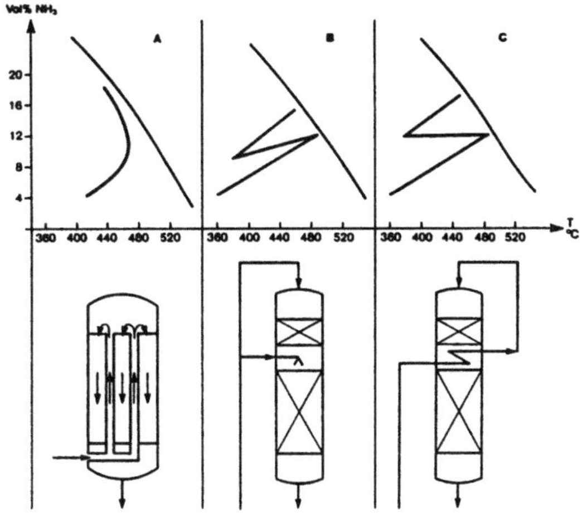

Fig. 5. Ammonia Synthesis. Form of Temperature Profile with 3 Types of Reactors.

It is seen that the operating lines for the internally cooled reactor is closest to the optimum reaction path and that the operating line for the quench reactor is the poorest approach to the optimum line. As a consequence, the required catalyst volume is normally smallest for the internally cooled reactor and largest for the quench cooled reactor.

Fig. 6A, B and C show operating lines for the three types of reactors in methanol synthesis (confer Table 4, case 2). The situation is the same here. The internally cooled reactor gives the best approach to the optimum operating line and may as a consequence be designed for the smallest catalyst volume, whereas the quench cooled reactor requires the largest volume. It is not, however, possible to base a choice between the reactor types solely on the required catalyst volume. As indicated in Table 1, a number of other considerations must be taken into account.

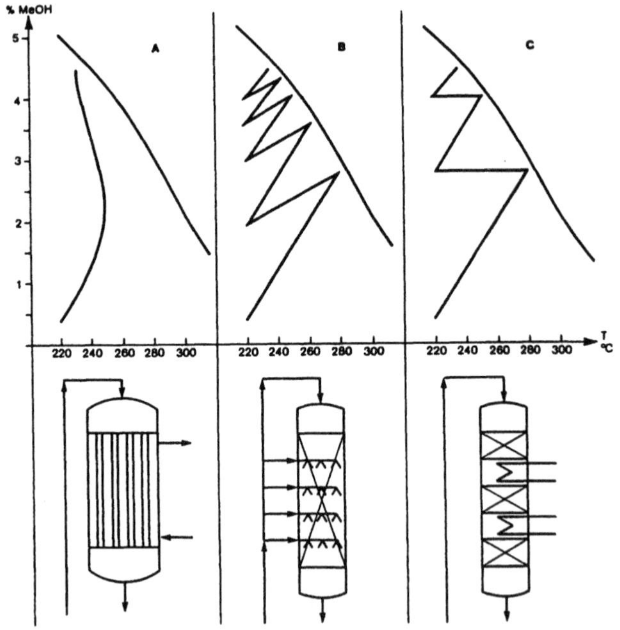

Fig. 6. Methanol Synthesis. Form of Temperature Profile with 3 Types of Reactors.

In Haldor Topsøe's ammonia and methanol synthesis processes a series of adiabatic beds with indirect cooling between the beds is normally used, at least in large plants. In smaller plants internally cooled reactors are considered. In ammonia synthesis, the Topsøe solution is today the so-called S-200 converter (Fig. 7) and [6] . This converter type, which is a further development of the S-100 quench-type converter, was developed in the mid seventies; the first industrial unit was started up in 1978, and today about 20 are in operation or on order. Both the S-100 and the S-200 reactors are radial flow reactors. The radial flow principle offers some very specific advantages compared to the more normal axial flow. It does, however, also require special catalyst properties. The advantages of the radial flow principle and the special requirements to the catalyst are summarized in Table 5.

S-200 Radial Flow
Ammonia Synthesis Converter
without Lower Heat Exchanger

Fig. 7.

TABLE 5

Radial Flow Reactor

Advantages:

- Is designed for low pressure drop
- Permits use of small catalyst particle size with high efficiency.
- Can be scaled up to large capacity without excessive reactor diameter.

Requirements:

- Small catalyst particle size should be advantageous (low efficiency on large particles).
- Production of small catalyst particles must be technically and economically feasible.
- The catalyst must have suitable mechanical properties.

The ammonia synthesis catalyst and the ammonia synthesis process are well suited for the application of the radial flow principle. The process lay-out is such that pressure drop is critical, the effectiveness factor on a large catalyst particle is significantly below unity, the catalyst is mechanically strong and does not shrink or settle significantly during reduction and/or operation, and the catalyst can be made available with a very small particle size.

In methanol synthesis, the case for radial flow converters is less obvious. Topsøe has earlier proposed [8], the use of one-bed radial flow converters in large methanol plants. Later analyses, partly based on a change of catalyst type, have, however, led to the conclusion that axial flow should be preferred even in very large methanol synthesis converters. The reasons for this difference in reactor concept between ammonia synthesis and methanol synthesis are in the differences between the properties of the catalysts. As mentioned above, the ammonia synthesis catalyst is ideally suited for the radial flow principle. This is not true for the methanol synthesis catalyst. The reasons for not using the radial flow principle in methanol synthesis are the following:

- It would be very difficult to produce the methanol synthesis catalyst with significantly smaller particle size than what is used industrially (because it is a tabletted catalyst).

- The methanol synthesis catalyst is mechanically much weaker than the ammonia synthesis catalyst, and it shrinks significantly during the activation, which must necessarily take place in the synthesis reactor.

It has been found that the potential advantages by applying the radial flow principle in methanol syntheses cannot justify the potential risks imposed by especially the last point mentioned above, and it has as a consequence been decided to design the methanol synthesis reactor system as a series of normally three axial beds with cooling between the beds. In large plants, the reactors will be separate vessels with boilers or heat exchangers between the reactors. In smaller plants, it may be possible to have more than one catalyst bed inside the same pressure shell. An alternative solution which is valid for smaller plants is to use a boiler water reactor with catalyst in tubes.

6. CALCULATION OF REACTOR VOLUME

In the preceeding paragraphs, it has been discussed what kind of considerations must be taken into account before the operating conditions around the reactor system are determined, and the mechanical type of reactor is chosen. There are, of course, many other factors as described in Table 1 which have not been discussed in any detail, but let us now assume that we have determined a set of operating conditions for a reactor, and that we have chosen a mechanical construction principle for the reactor. It then remains to calculate the catalyst volume which will at the specified operating conditions and in the selected reactor configuration perform as required by the process. In order to do this a reliable kinetic expression is required for the reaction on the specific catalyst, reliable information about the influence of ageing and poisoning on the reaction rate, and of course, a reliable calculation procedure.

6.1 Kinetics

The kinetics of the ammonia synthesis have been studied extensively by Haldor Topsøe and others. The kinetic expression used by Topsøe in design calculations has been published by Nielsen, Kjær and Hansen [13] . The reaction rate depends on the gas composition in such a way that it is maximum at a H_2/N_2 ratio slightly below the stoichiometric ratio. The reaction is inhibited by product ammonia and decreases significantly with decreasing temperature.

Chemically, the ammonia synthesis reaction is very simple. There is only one reaction, with no possibility for side reactions. No selectivity problem exist. The situation in methanol synthesis is different. The reaction mechanism and the reaction kinetics for methanol synthesis on copper catalyst has been studied and some results were presented by Bøgild Hansen et al [14] . Since this presentation was not printed, a brief summary including the reaction rate equations for the methanol synthesis reaction, the shift reaction and the reverse shift reaction is quoted below.

> "Studies of chemisorption of hydrogen, water, carbon monoxide, and carbon dioxide alone and in sequence on a Cu-Cr-Zn low temperature methanol synthesis catalyst show that the catalyst surface contains two different types of active sites. Hydrogen and water are chemisorbed in competition on one type, carbon monoxide and carbon dioxide on the other. The results for Cu-Zn-Al catalysts follow the same pattern.
>
> Rates of formation of methanol and water at 230°C depend strongly on hydrogen pressure. Rate of methanol synthesis is almost independent of partial pressure of carbon monoxide and carbon dioxide at low conversions. At high carbon dioxide content in the feed gas the rate depends strongly on conversion; at low carbon dioxide content it is almost independent of conversion. This indicates that the methanol formation is inhibited by water but not by methanol. Initial rates of formation of methanol and water are equal indicating that methanol is formed by hydrogenation of carbon dioxide. Rate of reverse shift reaction is low compared to rate of methanol synthesis. Rate of shift reaction is higher when favoured by equilibrium.
>
> A reaction mechanism is suggested which involves dissociative chemisorption of hydrogen and water in competition on one type of active sites and chemisorption of carbon dioxide on the other type. Chemisorption of carbon dioxide is so strong that it prevents chemisorption of carbon monoxide. Chemisorbed carbon dioxide and hydrogen are in equilibrium on the surface. Reverse shift takes place by dissociation of the reaction product into carbon monoxide and a chemisorbed hydroxyl-species. The shift reaction is taking place by reaction between carbon monoxide from the gas phase and hydroxyl-species on the surface. Methanol is formed by step-wise hydrogenation of chemisorbed carbon dioxide.

If it is determined that the rate determining step for methanol formation in this step-wise hydrogenation is

$$CO_2H\cdot S(1) + H\cdot S(2) \longrightarrow COH\cdot S(2) + OH\cdot S(2)$$

with S(1) and S(2) denoting active sites of the two types, then the following rate equations can be derived assuming equilibrium chemisorption of carbon dioxide, hydrogen and water:

Methanol formation:

$$r_M = k_1 \cdot \frac{K_{CO_2}\cdot a_{CO_2}}{(1+K_{CO_2}\cdot a_{CO_2})} \cdot \frac{(K_{H_2}\cdot a_{H_2}) \cdot (1-\beta_1)}{\left[1+(K_{H_2}\cdot a_{H_2})^{0.5} + K_{H_2O}\cdot a_{H_2O}/(K_{H_2}\cdot a_{H_2})^{0.5}\right]} \quad (1)$$

Reverse shift:

$$r_{RS} = k_2 \cdot \frac{K_{CO_2}\cdot a_{CO_2}}{(1+K_{CO_2}\cdot a_{CO_2})} \cdot \frac{(K_{H_2}\cdot a_{H_2})^{0.5}\,(1-\beta_2)}{1+(K_{H_2}\cdot a_{H_2})^{0.5} + K_{H_2O}\cdot a_{H_2O}/(K_{H_2}\cdot a_{H_2})^{0.5}} \quad (2)$$

Shift:

$$r_S = k_3 \frac{K_{H_2O}\cdot a_{H_2O}\cdot a_{CO}\cdot(1-\beta_3)}{(K_{H_2}\cdot a_{H_2})^{0.5}(1+K_{CO_2}\cdot a_{CO_2})\cdot\left[1+(K_{H_2}\cdot a_{H_2})^{0.5} + K_{H_2O}\cdot a_{H_2O}/(K_{H_2}\cdot a_{H_2})^{0.5}\right]} \quad (3)$$

where a denotes activities and β_1, β_2, β_3 are the equilibrium terms. If $K_{CO_2} \cdot a_{CO_2} \gg 1$, the equations can be simplified to:

$$r_M = k_1' \frac{a_{H_2}\,(1-\beta_1)}{1+K_1\cdot a_{H_2}^{0.5} + K_2\cdot a_{H_2O}/a_{H_2}^{0.5}} \quad (4)$$

$$r_{RS} = k_2' \frac{a_{H_2}^{0.5}(1-\beta_2)}{1+K_1\cdot a_{H_2}^{0.5} + K_2\cdot a_{H_2O}/a_{H_2}^{0.5}} \quad (5)$$

$$r_S = k_3' \frac{a_{CO}\cdot a_{H_2O}}{a_{H_2}^{0.5}} \cdot \frac{(1-\beta_3)}{1+K_1\cdot a_{H_2}^{0.5} + K_2\cdot a_{H_2O}/a_{H_2}^{0.5}} \quad (6)$$

The proposed mechanism is, of course, rather speculative, and the corresponding rate equations may be unduly complicated. But they can, with proper values of the constants, explain all findings reported above rather well".

6.2 Catalyst ageing and poisoning

The influence of catalyst ageing and poisoning is normally taken into account by multiplying correction factors to the pre-exponential factor in the rate constant. The correction factors used in ammonia and methanol synthesis design calculations are proprietary and connot be disclosed. It can be mentioned, however, that under normal operating conditions the ageing of an ammonia synthesis catalyst is rather slow. The catalyst is also well protected against poisoning, and very long lifetimes with almost constant performance can be achieved, and the useful catalyst life in excess of ten years is not uncommon. In methanol synthesis, the situation is different. Even in poison-free atmosphere, a significant decrease in catalyst activity with time is seen, and it is necessary to design the synthesis loop in such a way that the loss of catalyst activity can be compensated for by increased operating temperature and pressure. In addition to this, the risk for poisoning is greater in methanol synthesis than in ammonia synthesis, and as a consequence shorter lifetimes are relevant for this catalyst, typically in the order of three years.

The most important poisons in ammonia synthesis are oxygen-containing compounds (confer paragraph 4.1) and -especially in older plants- sulphur compounds. In methanol synthesis the sulphur and chlorine compounds are most often responsible for poisoning in industrial plants. A discussion of ageing and poisoning mechanisms in ammonia and methanol synthesis is given in [15] .

6.3 Calculation Procedure

In calculations of catalyst volumes, optimization of reactor performance and analysis of industrial data, Topsøe uses a general purpose program called REACTOR. The program has a modular structure and communicates with other programs and databases developed by Topsøe and contained in a general data processing package called GIPS (General Integrated Programming System).

The structure of the program REACTOR appears from Fig. 8. Input data are given in the form of data groups which define the problem. A list of types of data groups are shown in Fig. 9.

Fixed-bed Catalytic Reactor Program
Structure of Program

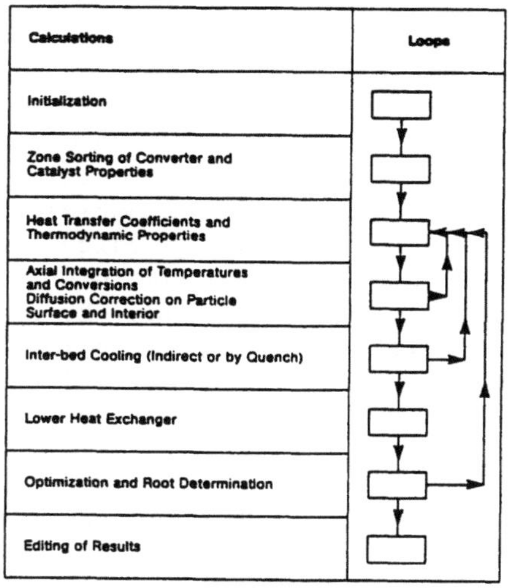

Fig. 8.

Fixed-bed Catalytic Reactor Program
Input Data Groups

	Data Group
Mandatory	Main Data Group
	Reaction Rate Expression
	Converter Mechanical Data
	Catalyst Data
	Feed Gas Stream
	Operation Data
Optional	Catalyst Zone Data
	Catalyst Property Catalog
	Lower Exchanger Data
	By-product Formation
	Inter-bed Cooling (Indirect or by Quench)
	Diffusion Control Data
	Intermediate Profile Points
	Fixed Heat Transfer Coefficients
	Fixed Temperature Profile
	Adjustment Data: Rate Constant Fitting Temp. or Conversion Stop
	Input Printing Control
	Output Printing Control

Fig. 9.

In the reaction rate data group, either a general kinetic expression can be used, or built in expressions for e.g. ammonia or methanol synthesis can be specified.

In a similar way, the mechnical reactor configuration can be specified, or built in specifications for various types of reactors can be sued. With the sued of optional data groups it is possible to specify interbed cooling, to include calculation of lower (feed effluent) heat exchanger, to carry out an optimization of inlet temperature(s) to the catalyst bed(s), to calculate required catalyst volume to obtain a specified conversion, etc. As a further important feature it is possible to include -by specifying an optional data group- calculation of diffusion restrictions in the catalyst particles and to determine the effectiveness factor at each point in the catalyst bed.

After the input has been read and sorted, heat transfer coefficients and other thermodunamic data are calculated at the beginning of each catalyst zone. Temperature and conversion profile in the catalyst bed is then calculated by an axial integration. The mathematical model used in the integrations is described in [6] . This model allows in principle the determination of diffusion restrictions and calculation of effectiveness factors for each reaction in cases where several reactions take place simultaneously. In such cases the concept of effectiveness factor may become rather dubious as shown below for the methanol synthesis, and this may be reflected in difficulties in the calculations.

After integration of the catalyst bed(s) - if so desired after interbed cooling indirectly or by quench and calculation of lower heat exchanger - the sequence may be repeated in order to determine optimum operating temperatures. When the desired calculation has been performed, the results are edited and printed in a report.

6.4 Ammonia Synthesis

The calculation procedure described above and in [6] gives as a result a complete description of the conditions in the reactor including temperature and concentration profiles, pressure drop, reaction rates, gas enthalpies, equilibrium temperatures, effectiveness factors, etc. Furthermore, radial temperature and concentration profiles in catalyst particles and across the gas film surrounding the particles may printed for selected levels in the catalyst bed. Fig. 10 and 11 show some results obtained by simulating the performance of an adiabatic catalyst bed for the same inlet and outlet conditions (cfr. Table 2, first example) specifying two different catalyst particle sizes.

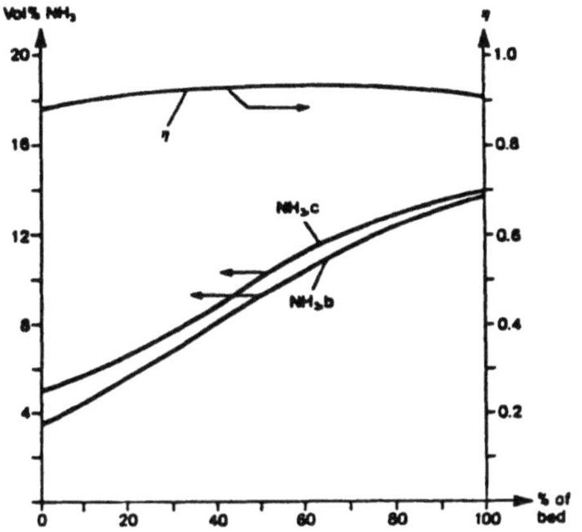

Fig.10. Ammonia Synthesis. Effectiveness Factor (η) and NH_3 Concentration in the Bulk (NH_3,b) and in Centre of Catalyst Pellet (NH_3,c) in First Catalyst Bed with 2.5-3mm Particles (cfr, Table 2, Case 1).

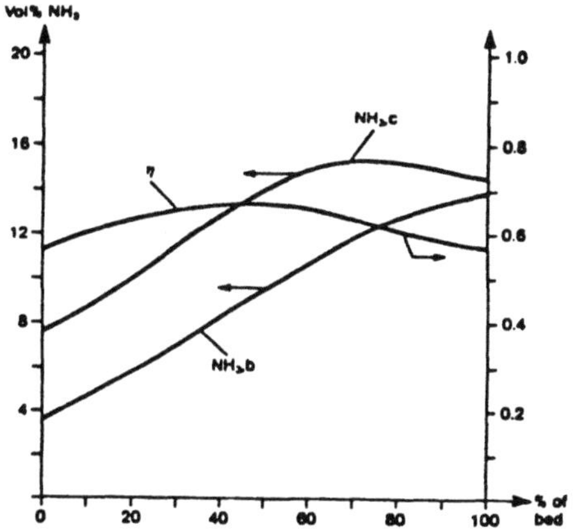

Fig.11. Ammonia Synthesis. Effectiveness Factor (η) and NH_3 Concentration in Bulk (NH_3,b) and in Centre of Catalyst Pellet (NH_3,c) in First Catalyst Bed with 6-10 mm Particles (cfr. Table 2, Case 1).

It is seen that with small catalyst particles (Fig. 10) the effectiveness factor is almost unity, and the concentration inside the particles is consequently close to the bulk concentration. With larger particles (Fig. 11), the effectiveness factor drops to about 60%, and the difference between the concentration inside particles and in the bulk phase becomes larger. The concentration in the bulk phase increases smoothly from the inlet to the outlet, whereas the concentration inside the particles runs through a maximum indicating that the reaction runs close to equilibrium in the center of the particles in the last part of the catalyst bed.

The calculated catalyst volume for 6-10 mm particles is approx. 50% larger than for 1.5-3 mm particles corresponding to the difference in effectiveness factor. The pressure drop in the catalyst bed is, on the other hand, about 4 times larger with small particles than with the larger particles. This is, however, of no consequence when radial flow reactor is used since the pressure drop in the catalyst bed is with this reactor type insignificant even with the small particles.

6.5 Methanol Synthesis

When the calculation procedure and the mathematical model described above and in the appendix is applied for calculation of the performance of a methanol synthesis reactor, interesting results are obtained because of the special reaction system. In the mathematical model, the following 2 reactions are specified, each with its own kinetic expression:

$$CO_2 + 3H_2 \rightleftharpoons CH_3OH + H_2O \quad (7)$$

$$CO_2 + H_2 \rightleftharpoons CO + H_2O \quad (8)$$

The two reactions are both limited by equilibrium, and they are coupled in such a way that the position of the equilibrium for one depends on the conversion for the other reaction.

Some consequences are shown in Fig. 12.

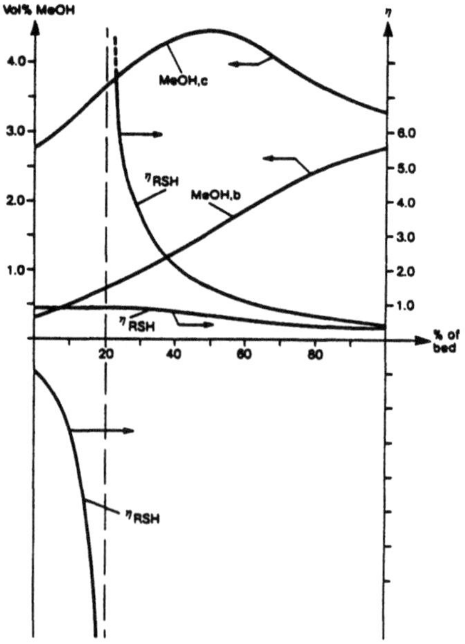

Fig. 12. Effectiveness Factor for Methanol Synthesis (η MeOH) and for Reverse Shift (ηRSH) and Methanol Concentration in Bulk (MeOH,b) and in Centre of Catalyst Pellet (MeOH,c) in First Reactor (cfr. Table 4, Case 2).

It is seen that the effectiveness factor for the methanol synthesis decreases smoothly from about 80% at inlet to about 40% at outlet. The bulk concentration of methanol increases through the reactor, whereas the concentration inside the particles runs through a maximum about 50% down in the catalyst bed. This is all quite similar to the picture obtained for ammonia synthesis.
But the shift reaction behaves in a peculiar way. The calculated effectiveness factor drops in the upper part towards very large negative values. It approaches an asymptote, changes to large possitive values and drops to a reasonable level at reactor outlet. The reason for this behaviour is evident from Fig. 13.
In the inlet layer the gas composition in the bulk phase is such that the equilibrium temperature for the shift reaction is below the actual temperature.
There is a potential for the forward shift reaction. This means that the shift reaction would run in one direction at bulk conditions but in the other direction inside the particles. Negative effectivences factor is therefore calculated. In the lower part of the catalyst bed the equilibrium temperature is in all points above the actual temperature. This means that the direction of the shift reaction is the same in bulk and inside particles, and therefore positive effectiveness factors are calculated. It is clear that in cases as this the concept of "effectiveness factor" has become rather meaningless due to the coupling between the reactions. Mathematically, the asympotic singular point is difficult to handle, and in many cases problems are encountered in the calculations. Therefore, it may be more convenient in cases with strong coupling between reactions as described above to develop kinetic equations valid for whole particles with possible diffusion effects built into the equations.

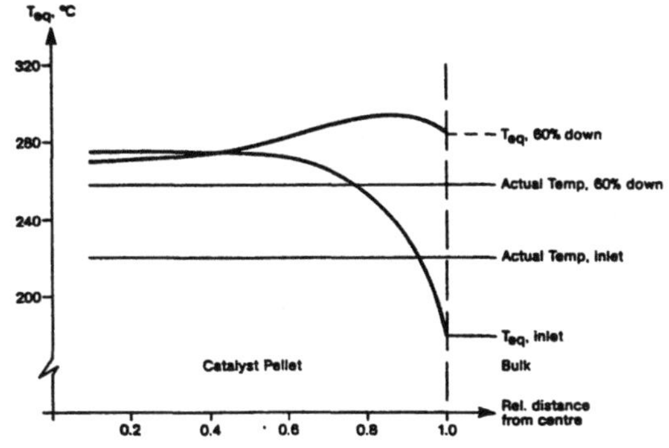

Fig. 13. Methanol Synthesis: Shift Equilibrium Temperature as Function of Local Gas Composition in Interior of Catalyst Pellet at Inlet First Catalyst Bed and 60% Down in the Bed.

7. Summary and Conclusion

Reactor design is a complex science which involves many diciplines ranging from mathematical modelling to materials science, from "flow sheeting" to thermodynamics. In the above an attempt has been made to describe how the desired conditions in the reactor system are determined by optimization of the process lay-out, how a mechanical type of reactor may be selected, and how the performance of the selected reactor type may be simulated using a suitable mathematical model.

Many points important in reactor design have not been discussed. Nothing has been said about selection of materials of construction, mechanical construction, instrumentation and process control, integration of heat recovery into the overall plant energy system, etc. But it is hoped that the paper will anyhow give an idea about the complexity of the problems encountered in the reactor design and about the solutions which are applied in order to obtain satisfactory results in industrial practice.

References

1. Sheldrick, W.F.: "World Fertilizer Review and the Changing Structure of the International Fertilizer Industry", Australian Fertilizer Manufacturer's Conference, Perth, Nov. 1984.

2. "Emergency Energy and Chemical Applications of Methanol: Opportunities for Development Countries". The World Bank, Washington, 1982.

3. Marshall, E.F,: Future Changes in Methanol Use and Methanol Synthesis. Presented at the American Chemical Society Symposium on Synthesis Gas as an Industrial Feedstock (August 25, 1981).

4. Stratton, A. and M. Teper: "The Economics of Producing Ammonia and Hydrogen", IEA Coal Research, Dec. 1984.

5. Rase, H.F.: "Chemical Reactor Design for Process Plants". John Wiley & Sons, New York, 1977.

6. Lars J. Christensen and Jørgen E. Jarvan: "Transport Restrictions in Catalyst Particles with several Chemical Reactions.

7. Dybkjær, I. and E.A. Gam: "Energy Saving in Ammonoa Synthesis - Design of Converters and High Activity Catalysts", CEER 13 (9) 29-35, 1984.

8. Dybkjær, I.: "Topsøe Methanol Technology", CEER 13 (6) 17-25, 1981.

9. Dybkjær, I., and E.A. Gam: "Benefits of Highly Active Ammonia Synthesis Catalyst", Paper read at AIChE Ammonia Symposium, San Francisco, NOvember, 2629, 1984.

10. Natta, G. "Synthesis of Methanol" in: Catalysis, vol. 3 (Emmet, P.H. ed.), Reinhold, New York 1955.

11. "The Design of Ammonia Converters", Nitrogen 140, 30, Nov,Dec 1982.

12. Zardi, U.: "Review these Developments in Ammonia and Methanol Reactors", Hydrocar. Proc. 61, 8, 129, Aug. 1982.

13. Nielsen, A., J. Kjær and B. Hansen' "Rate Equation and Mechanism of Ammonia Synthesis at Industrial Conditions", J. Cat. 3, 1, 68-79, (1964). With erratum, J. Cat. 7, 2, 208, (1967).

14. Hansen, J.B., P.E. Højlund Nielsen, and I. Dybkjær: "Synthesis of Methanol on Cu-based Catalysts". Paper presented at AIChE 1981 Annual Meeting, New Orleans, Nov. 1981.

15. Rostrup-Nielsen, J.R., and P.E. Højlund Nielsen: "Catalyst Deactivation in Synthesis Gas Preparation and important Syntheses", in "Deactivation and Poisoning of Catalysts (Budar, J. and H. Wise, eds.) Marcel Dekker, in press.

LECTURERS

- A.A.C.M. Beenackers, Department of Chemical Engineering, University of Groningen, The Netherlands and W. Van Swaaij,Twente University of Technology, The Netherlands.
- M.A. Bergougnou, C. Briens, Faculty of Engineering Science, The University of Western Ontario, London, Ontario, Canada.
- D. Cresswell, New Science Group, ICI, England.
- M. Crine, Laboratoire de Genie Chimique, Université de Liège, Belgium.
- W. Deckwer, Fachbereich Chemie, Universitat Oldenburg, Oldenburg, Germany.
- H. de Lasa, S.L.P. Lee, Faculty of Engineering Science, University of Western Ontario, London, Ontario, Canada.
- M. Dudukovic, Department of Chemical Engineering, Washington University, St. Louis, U.S.A.
- I. Dybkjaer, Haldor-Topsoe, Lyngby, Denmark
- J.R. Grace, Department of Chemical Engineering, University of British Columbia, Vancouver, Canada.
- H. Hofmann, Institute of Technical Chemistry, University of Erlangen - Nuremberg, Erlangen, West Germany.
- R. Koros, Exxon Research and Engineering, Florham Park, New Jersey, U.S.A.
- A. Rodrigues, Department of Chemical Engineering, University of Porto, Porto, Portugal.
- V. Specchia, A. Gianetto, Politechnico di Torino, Italy and F. Berrutti, University of Waterloo, Canada.
- M. Ternan, CANMET/ERL, Ottawa, Ontario, Canada.
- J. Villermaux, Laboratoire des Sciences du Genie Chimique, CNRS-ENSIC, Nancy, France

PARTICIPANTS

- A. Akgerman, Department of Chemical Engineering, Texas A&M University, College Station, Houston, Texas, U.S.A.
- A. Akyurtlu, Department of Chemical Engineering, University of Kentucky, Lexington, KY 40506-0046.
- J. Akyurtlu, Department of Chemical Engineering, University of Kentucky, Lexington, KY 40506-0046.
- A. Da Graca Alexandre Leitao, Department of Chemical Engineering, University of Porto, 4099 Porto Codex, Portugal.
- J. Asua, Departamento de Quimica Tecnica, Universidad del Pais Vasco, Alza, 20017 San Sebastian, Spain.
- S. Feyo de Azevedo, Department of Chemical Engineering. University of Porto, rua dos Bragas, 4099 Porto Codex, Portugal.
- N. Bakhshi, Department of Chemical Engineering, University of Saskatchewan, Saskatoon, Saskatchewan, S7N 0W0, Canada.
- M. Bellut, Department of Chemical Engineering, University of Waterloo, Waterloo, Ontario, Canada N2L 3G1

- F. Berrutti, Department of Chemical Engineering, University of Waterloo, Waterloo, Ontario, Canada N2L 3G1.
- C. Briens, Faculty of Engineering Science, University of Western Ontario, London, Canada N6A 5B9
- T. de Bruijn, CANMET, 555 Booth Street, Ottawa, Ontario, Canada K1A 0G1
- A. Catros, Centre de Recherche ELF, BP22, 69360 St. Symphorien d'Ozon, France
- J. Chaouki, Ecole Polytechnique, Department of Chemical Engineering, P.O. Box 6079, Station A, Montreal, Quebec, Canada H3C 3A7.
- C. Costa, Department of Chemical Engineering, University of Porto, 4099 Porto Codex, Portugal.
- A. Dixon, Department of Chemical Engineering, Worcester Poly-technic Institute, Worcester, MA 01605,U.S.A.
- L. Ferreira, Faculdade de Engenharia, CP 1756, Luanda, Rep. Popular de Angola.
- J. Feimer, Esso Petroleum Canada, P.O. Box 3022, Sarnia, Ontario, Canada N7T 7M1.
- R. Fox, Department of Chemical Engineering, Durland Hall-KSU, Manhattan, KS 66506 U.S.A.
- P. Galtier, Institut Francais du Petrole, CEDI BP No. 3, 69390 Vernaison, France.
- R. Gierman, Koninklyke/Shell-Laboratorium, Postbus 3003, Amsterdam, The Netherlands.
- J.F.P. Gomes, L.N.E.T.I. (Laboratorio Nacional de Engenharia e Tecnologia Industrial), Combustion Group, rue Alves Redol, 1000 Lisboa, Portugal.
- J.C. Gonzalez, Intevep SA, Departamento Desarrollo de Procesos, Apartado 76343, Caracas 1070A, Venezuela.
- W.R.A. Gossens, Chemical Engineering, SCK-CEN Boeretang 200, B-2400, Mol, Belgium.
- K. Grosser, Department of Chemical Engineering, North Carolina State University, Raleigh, NC 27645-7905, U.S.A.
- J.E. Gwyn, Shell Development Col, Westhollow Research Center, M-2400, Houston, Texas 77450, U.S.A.
- J. Hazlett, Chemical Engineering Section, Division of Chemistry, National Research Council of Canada, M-12, Montreal Road, Ottawa, Ontario, Canada K1A 0R9
- K. Kinnari, Research and Development, STATOIL, Forus, P.O. Box 300, N-4001 Stavanger, Norway.
- C. Kiparissides, Department of Chemical Engineering, Aristotle University of Thessaloniki, Thessaloniki, Greece.
- B. Kisakurek, Fen Fakultesi, Kimya Muhendisligi Bolumu, Ankara University, Tandogan, Ankara, Turkey.
- K. Kucukada, Chemical Engineering Department, Middle East Technical University, Ankara, Turkey.
- P. Kytonen, Neste Engineering, Neste Oy, SF-06850 Kulloo, Finland.

- T. Lee, Coal Research Department, Alberta Research Council, P.O. Box 429, Nisku, Alberta, Canada. TOG 2GO
- J.J. Lerou, Central Research Development Department, DuPont de Nemous, Experimental Station E262/222, Wilmington, DE 19898, U.S.A.
- D. Lynch, Department of Chemical Engineering, University of Alberta, Edmonton, Alberta, Canada. T6G 2G6.
- R. Marzin, INTEVEP S.A., Departamento Desarrollo de Procesos, Apartado 76343, Caracas 1070A, Venezuela.
- A. Nunes dos Santos, Chemical Engineering Department, Faculdade de Ciencias e Tecnologia, Universidade Nova de Lisboa, 2825 Monte de Caparica, Portugal.
- R. Oliveira Quinta Ferreira, Departamento de Engenharia Quimica, Universidade do Porto, 4099 Porto Codex, Portugal.
- B. Ozum, Coal Research Department, Alberta Research Council, Devon, Alberta, Canada TOC 1EO.
- M. Perrier, Ecole Polytechnique, Department of Chemical Engineering, P.O. Box 6079, Section A, Montreal, Quebec, Canada H3C 3A7
- M.C. Phillips, Shell Laboratorium, Postbus 3003, Amsterdam, The Netherlands.
- S. Pratsinis, Department of Chemical Engineering, UCLA, 553 Boelter Hall, Los Angeles, California 90024.
- R. Ramachandran, Montsanto, Box 1311, Texas City, Texas 77590, U.S.A.
- G. Rios, Laboratoire de Genie Chimique, Université des Sciences et Techniques du Languedoc, Place Bataillon, 34060 Montpellier Cedex, France.
- A. Sapre, Mobil Research and Development, Billingsport Road, Paulsboro, NJ 08066, U.S.A.
- H. Sastre, Departamento de Quimica, Universidad de Oviedo, 33007, Oviedo, Spain.
- C.A. Sereno, Faculdade de Engenharia, C.P. 1756, Luanda, Angola.
- D.A. Shaw, Monsanto, CS6F, 800 N. Lindbergh, St. Louis, MO. 63167, U.S.A.
- H. Skoczylas, Dow Chemical (Canada) P.O. Box 3030, Sarnia, Ontario, Canada N7T 7M1
- D. Soveran, Saskatchewan Research Council, 515 Henderson Drive, Regina, Saskatchewan, Canada S4N 5X1
- B. Uysal, Chemical Engineering Department, Middle East Technical University, Ankara, Turkey.
- W. Van Swaaij, Twente University of Technology, 7500 AE Enschede, Pos. Bus. 217, The Netherlands.
- R. Willson, Esso Petroleum Canada, 55 St. Clair Ave. West, Toronto, Ontario, Canada M4V 2Y7
- A. Zoulalian, Laboratoire du Genie des Procédés, Université de Nancy I, B.P. 239, Vandoeuvre-les-Nancy Cedex, France.

LOCAL COMMITTEE (Faculty of Engineering Science, University of Western Ontario, London, Canada)

- T. Baron
- D. Berg
- R. Collyer
- B. Freel
- N. Freychet
- C. Huynh
- T. Ityokumbul
- D. Kraemer
- P. Lee
- A. Mahay
- A. Prakash
- A. Ravella
- M. Roach
- J. Stephenson
- R. Sumner
- S. Urquhart

INDEX

MIX
Papier aus verantwortungsvollen Quellen
Paper from responsible sources
FSC® C105338

If you have any concerns about our products,
you can contact us on
ProductSafety@springernature.com

In case Publisher is established outside the EU,
the EU authorized representative is:
Springer Nature Customer Service Center GmbH
Europaplatz 3, 69115 Heidelberg, Germany

Printed by Libri Plureos GmbH
in Hamburg, Germany